LOW-RANK COAL TECHNOLOGY

LOW-RANK COAL TECHNOLOGY

Lignite and Subbituminous

G.H. Gronhovd
E.A. Sondreal

Grand Forks Energy Technology Center

John Kotowski
George Wiltsee

Energy Resources Company, Inc.

NOYES DATA CORPORATION
Park Ridge, New Jersey, U.S.A.
1982

Library of Congress Catalog Card Number: 82-3433
ISBN: 0-8155-0896-4
ISSN: 0270-9155
Printed in the United States

Published in the United States of America by
Noyes Data Corporation
Mill Road, Park Ridge, New Jersey 07656

10 9 8 7 6 5 4 3 2 1

Library of Congress Cataloging in Publication Data
Main entry under title:

Low-rank coal technology.

(Energy technology review, ISSN 0270-9155 ;
no. 79)
Bibliography: p.
Includes index.
1. Lignite. 2. Coal. I. Gronhovd, G. H.
II. Series.
TP329.L68 662.6'222 82-3433
ISBN 0-8155-0896-4 AACR2

Foreword

This book contains a comprehensive review of low-rank coal technology and, in addition, in condensed form, an overview of findings and/or recommendations on the occurrence and properties of low-rank coals (subbituminous coal and lignite), existing and potential markets and economics, regulatory and environmental issues, and suggested RD&D (research, development, and demonstration) projects. The book is based on selected portions of a study prepared by Energy Resources Company, Inc. and Grand Forks Energy Technology Center. The study was initiated to contribute to the development of subbituminous coal and lignite.

The U.S. energy economy must undergo fundamental changes if it is to provide a renewed foundation for national security and economic progress. In the immediate future, coal will play a much greater role than it does today. Development of low-rank coal resources in the West and Gulf Coast regions of the United States will help in achieving this goal.

The production and use of low-rank coals for electric power generation has increased tenfold in the last decade, and now totals over 150 million tons per year. Successful establishment of a synthetic fuels industry along with continued growth in power production could push the use of low-rank coal to 1 billion tons per year by the year 2000. It is essential that this growth occur in an environmentally acceptable manner.

Low-rank coal deposits are distinguished from bituminous coal by lower heating value, higher moisture content, physical properties, generally lower sulfur content and a predominance of alkaline rather than acidic ash components. These and other properties of low-rank coal affect the technologies for their extraction, preparation, direct use, and conversion, and justify a separate focus on low-rank coals in national R&D efforts.

Part I of the book contains a brief resource characterization plus summaries of regulatory constraints, environmental factors and considerations, market analyses, and research and development evaluations for low-rank coal. Part II, the main part of the book, is a thorough technology evaluation covering extraction; transportation; preparation, handling, and storage; and processing and utilization.

The information in the book is from the following documents:

> *Low-Rank Coal Study—National Needs for Resource Development. Volume 1—Executive Summary* prepared by John Kotowski and George Wiltsee of Energy Resources Company, Inc., Walnut Creek, California, and G.H. Gronhovd and E.A. Sondreal of Grand Forks Energy Technology Center, Grand Forks, North Dakota, for the U.S. Department of Energy, November 1980.
>
> *Low-Rank Coal Study—National Needs for Resource Development. Volume 3—Technology Evaluation* prepared by John Kotowski and George Wiltsee of Energy Resources Company, Inc., Walnut Creek, California, and G.H. Gronhovd and E.A. Sondreal of Grand Forks Energy Technology Center, Grand Forks, North Dakota, for the U.S. Department of Energy, November 1980.

The expanded table of contents is organized in such a way as to serve as a subject index and provides easy access to the information contained in the book.

> Advanced composition and production methods developed by Noyes Data are employed to bring this durably bound book to you in a minimum of time. Special techniques are used to close the gap between "manuscript" and "completed book." In order to keep the price of this book to a reasonable level, it has been partially reproduced by photo-offset directly from the original reports and the cost savings passed on to the reader. Due to this method of publishing, certain portions of the book may be less legible than desired.

Contents and Subject Index

PART I: OVERVIEW

PART II: TECHNOLOGY EVALUATION

Part I

Overview

The information in Part I is from *Low-Rank Coal Study—National Needs for Resource Development. Volume 1—Executive Summary* [DOE Report FC/10066-T1 (V.1)], prepared by John Kotowski and George Wiltsee of Energy Resources Company, Inc., Walnut Creek, California, and G.H. Gronhovd and E.A. Sondreal of Grand Forks Energy Technology Center, Grand Forks, North Dakota, for the U.S. Department of Energy, November 1980.

Introduction

Low-rank coals -- lignite and subbituminous -- are those which have been subjected to the least amount of metamorphic change during the coal-forming process. As such, they retain greater fractions of moisture and volatile matter from the original peat material, and contain less fixed carbon, than the high-rank coals -- bituminous and anthracite. The primary measure used to classify the lower ranks of coal is heating value. Lignite is defined (in this country, by the American Society for Testing and Materials) as coal with a heating value less than 8,300 Btu/lb, on a moist, mineral-matter-free basis (m,mmf). Subbituminous coal ranges in heating value from 8,300 to 11,500 Btu/lb (m,mmf). By comparison, bituminous coal and anthracite range from 10,500 to over 15,000 Btu/lb (m,mmf). Other important characteristics distinguish the low-rank coals from high-rank coals, as will be discussed in following sections.

Low-rank coals represent a major, and largely untapped, energy resource for this country. Very extensive deposits of lignite and subbituminous coal exist in the western states, the Gulf coast, and Alaska, as shown on Figure 1. Major deposits of low-rank coal are also found in many other countries, most notably the USSR, Australia, Canada, and the central and eastern European nations. Worldwide coal statistics indicate that low-rank coals account for roughly one-third of the total resource and current production tonnages.

Figure 1

Low-Rank Coal Regions of the United States

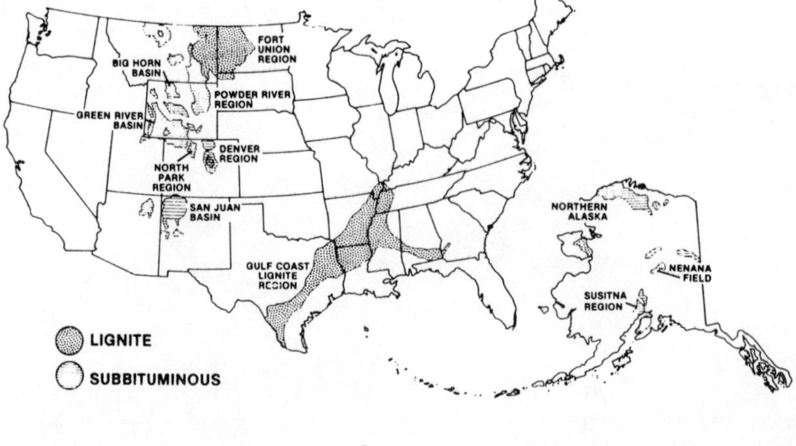

This report recommends a comprehensive national research, development, and demonstration (RD&D) program to enhance the development of low-rank coals. The major conclusion of this study is that the unique properties of these coals affect the technologies for their extraction, preparation, direct use, and conversion and justify a separate focus on low-rank coals in the national RD&D efforts. Why are the recommendations produced from this study important now? The timing appears critical for several reasons:

1. Large quantities of economically recoverable reserves make coal one of the few indigenous energy sources capable of relieving our overdependence on imported oil during the next 20 years.
2. Expanding the use of coal rapidly enough to make a difference (e.g., tripling by the year 2000) will necessarily require that much of the new production come from western low-rank coal areas.
3. Accomplishing this kind of growth in coal production and use without unacceptable delays and damage to the environment will severely challenge the technical and financial capabilities of the industries and governments involved.
4. Very little attention has been paid to the special problems of low-rank coals in previous and existing national coal RD&D programs.

The lack of attention to low-rank coal is not surprising when U.S. historical coal production rates are examined, as in Figure 2. Prior to 1970, low-rank coals accounted for 1-2 percent of the total annual U.S. coal production . Utilization of these resources was important only in local areas and was at small scale. Production of low-rank coals mushroomed during the 1970's, accounting for nearly all of the growth in coal production achieved during that decade. In 1980, low-rank coals account for 24 percent of the total U.S. coal production, and are poised for further growth. By 1990, these coals may account for 40 to 50 percent of a greatly expanded total.

Figure 2

Historical U.S. Coal Production

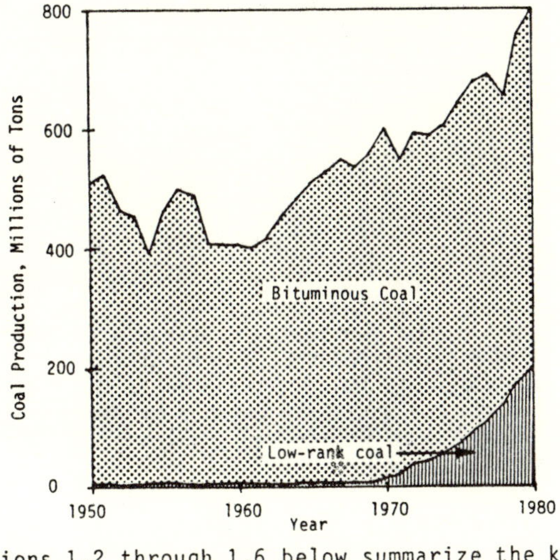

Sections 1.2 through 1.6 below summarize the key findings and recommendations of the low-rank coal development analysis.

Occurrence and Properties of Low-Rank Coals

The quantities of lignite and subbituminous coal in the U.S. are vast, as summarized in Table 1. Over 1 trillion tons of identified resources have been located and inferred by geologists. Over 100 billion tons occur close enough to the surface in favorable mining areas to be classified as the economically recoverable strippable reserve base. Another 108 billion tons of subbituminous coal are classified as the economically recoverable reserve base by underground mining. (It should be noted that these quantities are roughly comparable to the bituminous coal resources in the U.S.: 0.75 trillion tons of identified resources, 47 billion tons of strippable reserve base, and 182 billion tons of underground reserve base.)

Table 1
United States Low-Rank Coal:
Identified Resource and Reserve Base Estimates
(Billion Short Tons)

LIGNITE

Region	Identified Resources[a]	Strippable Reserve Base[b]
Fort Union	465.3	26.3
Gulf	68.3	11.6
Denver	10.0	2.9
Others	0.2	n.a.[c]
TOTAL	543.8	40.8

SUBBITUMINOUS COAL

Region	Identified Resources[a]	Strippable Reserve Base[b]	Underground Reserve Base[b]
Powder River	238.1	57.5	97.2
San Juan	50.6	5.5	0.9
Alaska	110.2	0.5	4.8
Others	147.2	4.4	4.8
TOTAL	546.1	67.9	107.7

Notes: [a]Identified resources include Demonstrated (measured and indicated) and inferred resources 2-1/2 feet or more thick to an overburden depth of 3000 feet.

[b]The Reserve Base includes coal that is technically and economically minable at this time. Criteria for minimum seam thickness and maximum overburden thickness vary from state to state.

[c]Estimate not available.

When these reserve base quantities are multiplied by appropriate recovery factors (i.e., about 50 percent for underground mining and 90 percent for surface mining) and by average heating values, a comparison of recoverable reserves on a Btu basis can be made (see Figure 3). As indicated, the recoverable reserves of low-rank coals and peat together are greater than the reserves of high-rank coal; these in turn are vastly larger than our remaining reserves of oil and gas.

Figures 4 and 5 present a summary classification of the strippable reserve base quantities of lignite and subbituminous coal according to three important characteristics: thickness of the coal seams, sulfur content, and ash content of the coal. These generalized classifications serve to illustrate some of the favorable attributes of the U.S. low-rank coals.

Figure 3

Recoverable Fossil Fuel Reserves in the United States

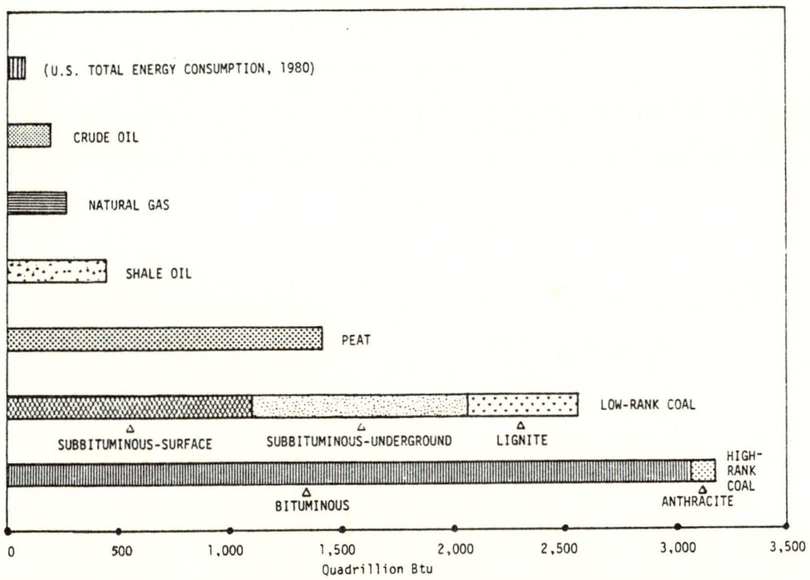

Figure 4

Classification of U.S. Lignite Strippable Reserve Base

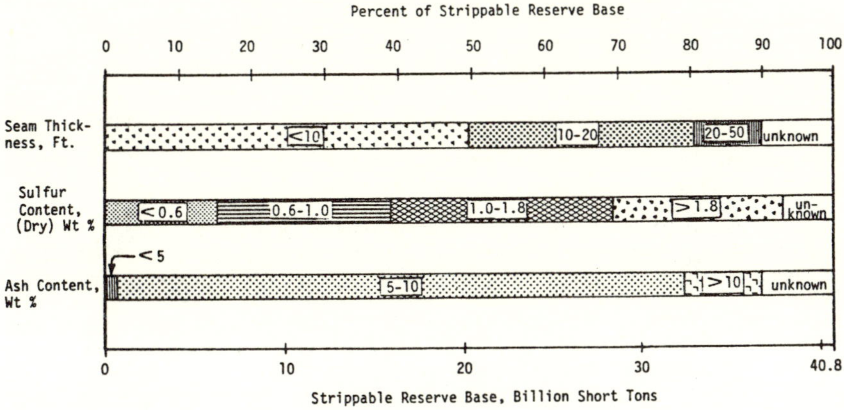

Figure 5

Classification of U.S. Subbituminous Coal Strippable Reserve Base

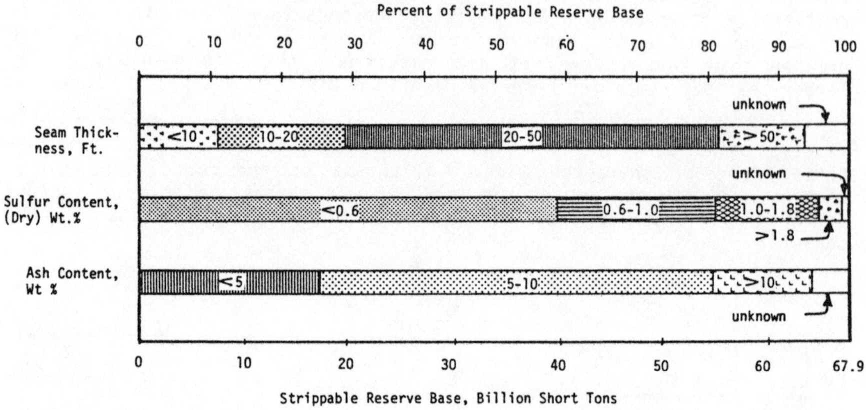

Many of the low-rank coal seams (particularly subbituminous) are thick in comparison to eastern bituminous coal seams. The low-rank coals are generally low in sulfur content compared to eastern bituminous coal (less than 1 percent sulfur compared to 2-5 percent on a weight basis; on a Btu basis, the difference is smaller because of the lower heating value of low-rank coal). Ash content is more difficult to generalize: Fort Union lignite and Powder River Basin subbituminous coal tend to be low-to-moderate in ash; Gulf lignite and San Juan Basin subbituminous coals tend to be moderate-to-high in ash content.

One universal property of coal (all ranks) is variability. Not only are there significant variations in the physical and chemical properties of different coal deposits; there are usually wide variations within a given coal seam as well. This fact should be recognized when interpreting any summarized data about the characteristics of coals, such as the data in this section on U.S. low-rank coals.

Examining the composition of coal more specifically, differences in both the organic matter and the inorganic matter distinguish U.S. low-rank coals from bituminous coal. Table 2 shows properties of vitrinites (the predominant form of pure organic matter in coal) as a function of rank. The effects of metamorphic change on heating value, and moisture content, as well as on the carbon/oxygen ratio, are evident.

The low heating values of low-rank coals affect the costs (per Btu) of extraction, transportation, combustion, and conversion processes. The high oxygen content tends to be present in functional forms such as hydroxyl and carboxyl groups; these are generally much lower or absent in bituminous coal. Hydroxyl groups are believed to be a major cause of the very high reactivity of low-rank coals. The carboxyl groups tend to occur in salt form; protons have been replaced by metal cations such as calcium and sodium, leading to an important form of "inherent mineral matter" in low-rank coals.

The moisture content in low-rank coals is partly trapped in pores and capillaries, and partly bound in the molecular structure. Application of heat is the only practical means for removing this inherent moisture. Depending on the drying method employed, the resulting changes in the coal structure and physical/chemical properties can be very significant. When exposed to the air, the surface of low-rank coal dries and slacks, producing dust which can cause fugitive emissions problems.

Table 2

Selected Properties of Vitrinites (Pure Organic Matter) In U.S. Coals of Different Ranks

	Lignite	Subbituminous	Bituminous
Moist, Mineral-Matter-Free Basis:			
Heating Value, Btu/lb	6,700	9,000	12,500
Moisture, Wt.%	35	25	<10
Dry, Mineral-Matter-Free Basis:			
Carbon, Wt.%	69	74.6	83
Hydrogen, Wt.%	5.0	5.1	5.5
Oxygen, Wt.%	24	18.5	10

The properties of inorganic matter (ash-forming minerals) in U.S. low-rank coals are in many respects unique, and are at least as important as the organic matter in affecting combustion and conversion technologies. Table 3 presents average chemical compositions of ash samples from lignite, subbituminous, and bituminous coals.

Table 3

Average Ash Compositions From U.S. Coals of Different Ranks (Weight Percent of SO_3 - Free Ash)

		Lignite	Subbituminous	Bituminous
Acidic Components	SiO_2	24.9	39.4	48.1
	Al_2O_3	14.1	21.1	24.9
	Fe_2O_3	11.5	10.1	14.9
	TiO_2	0.5	0.8	1.1
	P_2O_5	0.4	0.4	--
Alkali Components	CaO	31.2	20.1	6.6
	MgO	8.7	5.6	1.7
	Na_2O	8.2	2.1	1.2
	K_2O	0.5	0.3	1.5

One important difference is the much higher proportion of alkali elements in the ash from low-rank coals. These components (particularly sodium) alter the melting behavior of ash and lead to unique boiler fouling and slagging problems. Alkaline ash reacts with sulfur compounds, a property that can be used to reduce SO_2 control costs in wet scrubbing of flue gas, as well as in fluidized bed combustion. Fly ash from low-rank coals often has high electrical resistivity, making it difficult to collect in electrostatic precipitators.

The alkali elements in low-rank coals are believed to have catalytic properties, speeding up gasification and possibly liquefaction reactions. The alkaline mineral matter in low-rank coals tends to be largely inseparable from the organic matter by standard coal washing (float/sink) techniques.

These comments, and Table 3, are generally more applicable to the northern low-rank coals (Fort Union lignite and Powder River Region subbituminous coal) than to those farther south (such as San Juan Region subbituminous coal and Gulf lignite). Occurrences of high-silica ash and largely extraneous (separable) mineral matter are relatively common in the latter regions.

Fort Union Region Lignite

The Fort Union Region has been referred to as the largest coal basin on earth. Very large strippable lignite deposits are found in western North Dakota and eastern Montana. The seams are moderately thick, and stripping ratios are very favorable in areas currently being mined. The lignite is primarily used as boiler fuel by minemouth utility plants, and by industrial and institutional steam plants. Many of these boilers experience severe ash fouling on the convection section tubes, due to the prevalence of high sodium contents in much of the Fort Union Region lignite ash.

Table 4 presents average analyses of lignite samples from mines in the region. As an indication of the variability of these sample analyses, moisture content ranged from 32.5 to 43.6 percent; ash content ranged from 3.2 to 12.1 percent; and the lignite heating value ranged from 5880 to 7580 Btu/lb.

Table 4

Average Analyses of Fort Union Region Lignite Mine Samples (212 Sample Locations)

Proximate Analysis, Wt.% (As Received)		Ultimate Analysis, Wt.% (Dry, Mineral-Matter-Free)	
Moisture	37.2	Hydrogen	4.9
Volatile Matter	26.3	Carbon	71.9
Fixed Carbon	30.3	Nitrogen	1.1
Ash	6.2	Oxygen	21.0
		Sulfur	1.1

Heating Value, Btu/lb 6,820

Gulf Lignite

Lignite deposits in the Gulf Region occur primarily in Texas, Louisiana, Arkansas, Mississippi, and Alabama. The largest present commercial deposits are found in sediments of the Wilcox Group with thin overburden. Seams are quite irregular in thickness, and generally less than 10 feet thick. Some of the Texas deposits have very favorable mining and reclamation conditions, and are being rapidly exploited by utility companies to feed minemouth electric power plants. Most of these plants have not experienced the severe ash fouling problems associated with high-sodium Fort Union lignite. However, the ash content of the Gulf lignite is generally higher than the Fort Union lignite, and in some plants the ash is quite abrasive due to high silica levels.

Table 5 presents average analyses of Gulf lignite samples, representing a composite from all of the states mentioned above. The average properties shown represent somewhat poorer quality lignite than is currently being mined by utilities in Texas. As a general rule, the quality of Gulf lignite improves in moving from east to west. For example, lignite samples from Alabama average about 4,600 Btu/lb; from Arkansas, about 5,000 Btu/lb; and from Texas' Wilcox Formation, over 7,000 Btu/lb.

Table 5

Average Analyses of Gulf Lignite Samples
(495 samples)

Proximate Analysis, Wt.% (As Received)		Ultimate Analysis, Wt.% (Dry, Mineral-Matter-Free)	
Moisture	30.8	Hydrogen	6.1
Volatile Matter	30.8	Carbon	62.1
Fixed Carbon	24.9	Nitrogen	1.1
Ash	13.5	Oxygen	28.8
		Sulfur	1.9
Heating Value, Btu/lb	5,803		

Powder River Region Subbituminous Coal

The Powder River Region lies just to the southwest of the Fort Union Region, in the states of Montana and Wyoming. The Wyoming portion of the region contains the thickest coal seams in the U.S., with some seams averaging 70 feet in thickness. The largest-producing coal mines in the country are also found in this region.

The subbituminous coal in the Powder River Region is generally of high quality, and is particularly valued for its low sulfur content. A large portion of the coal produced in this region is transported to midwestern and south central utilities for use as "compliance" coal in meeting SO_2 emissions standards.

Table 6 presents average analyses of subbituminous coal mine samples from the Powder River Region. Variability of the coal properties is wide, as in other regions: moisture content of the samples ranged from 20 to 31 percent; ash content, from 3 to 16 percent; heating value, from 7,360 to 9,610 Btu/lb; and sulfur content ranged from 0.1 to 3.5 percent.

Table 6

Average Analyses of Powder River Region
Subbituminous Coal Mine Samples
(79 sample locations)

Proximate Analysis, Wt.% (As Received)		Ultimate Analysis, Wt.% (Dry, Mineral-Matter-Free)	
Moisture	25.4	Hydrogen	5.1
Volatile Matter	29.6	Carbon	75.6
Fixed Carbon	38.7	Nitrogen	1.2
Ash	6.3	Oxygen	17.3
		Sulfur	0.8
Heating Value, Btu/lb	8,820		

San Juan Basin Subbituminous Coal

The San Juan Basin in New Mexico and Colorado contains both subbituminous and bituminous coal. The low-rank coal is near the top of the heating value range for subbituminous coal. Samples range in heating value from 8,900 to 11,900 Btu/lb, and in moisture content from 10 to 20 percent. Ash content of the coal is generally greater than 10 percent; the ash tends to be relatively low in alkali elements compared to other low-rank coals. Table 7 presents average analyses of subbituminous coal samples from the San Juan Basin.

Table 7

Average Analyses of San Juan Basin Subbituminous Coal Samples (82 samples)

Proximate Analysis, Wt.% As Received	
Moisture	12.8
Volatile Matter	33.1
Fixed Carbon	40.6
Ash	13.5
Heating Value, Btu/lb	10,020

Ultimate Analysis, Wt.% (Dry, Mineral-Matter-Free)	
Hydrogen	5.5
Carbon	77.6
Nitrogen	1.4
Oxygen	14.3
Sulfur	1.2

Markets and Economics

During the past 60 years, the markets for coal in the United States have shifted substantially, while the total consumption has remained generally between 400 and 600 million tons/year. Railroad and residential/commercial direct use of coal -- two very large markets at one time -- have all but disappeared. Industry use has been relatively constant for the last 30 years, with about half being coking coal for metallurgical production -- a market reserved for high quality bituminous coal.

Electric Utilities

Electric utility use of coal has grown steadily for the last 30-40 years. Low-rank coals played a minor role in that market until the late 1960's, primarily because the low-rank coal resources are remote from industrial centers, and cheap oil and gas were abundantly available. Beginning in the late 1960's, electric utility use of low-rank coals began to grow very rapidly, as shown on Figure 6.

Why did this happen? Several technological and regulatory developments changed the relative economics of low-rank coal use by the electric utility industry:

1. Very large pulverized coal boilers, as well as extremely large strip mining machines, were adapted for low-rank coals. These allowed the industry to realize the economies of scale required to make low-rank coal use competitive.

2. Electric power transmission systems and regional reliability councils developed to the point that vast regions of the U.S. became interconnected and interdependent in terms of power supply. Thus, minemouth power plants in the low-rank coal regions could penetrate large markets for the first time. (See Figure 7).

3. Limits were placed on emissions of SO_2 from coal-fired power plants in the early 1970's. This created a huge demand in the midwest for low-sulfur western coal, despite the high transport costs.

4. The price and long-term supply uncertainty of OPEC oil and domestic gas rose rapidly in the early 1970's. This created a strong incentive to switch from these fuels to coal for power generation, particularly in Texas.

Figure 6

Low-Rank Coal Production by Region, 1968-1978

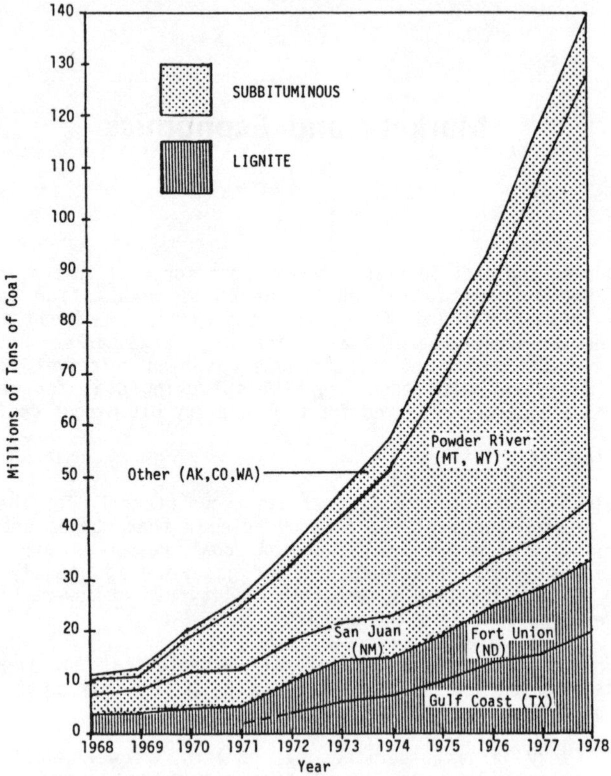

Figure 7

Major Electric Interties in the U.S. (mid-1970's)

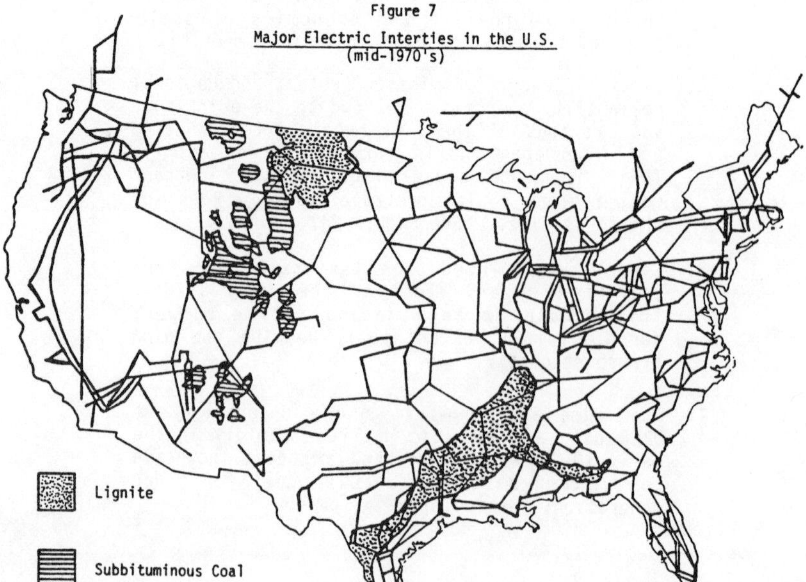

In this dramatic growth process (generally 15-20 percent per year for low-rank coal overall), lignite has remained a regional fuel, used almost exclusively within the borders of the Fort Union and Gulf lignite regions. In contrast, subbituminous coal has penetrated markets far from the coal mines. About 18 percent of the Powder River Region subbituminous coal produced in 1978 was used in power plants in Montana and Wyoming. The remainder (65 million tons of "compliance" coal) was shipped by unit train to power plants in twenty other states.(see Figure 8). During the same year, about 92 percent of the San Juan Region subbituminous coal production was burned by utilities in New Mexico and Arizona; small amounts were shipped to Missouri and Texas.

Figure 8

Shipments of Powder River Region Subbituminous Coal to Electric Utilities, 1978

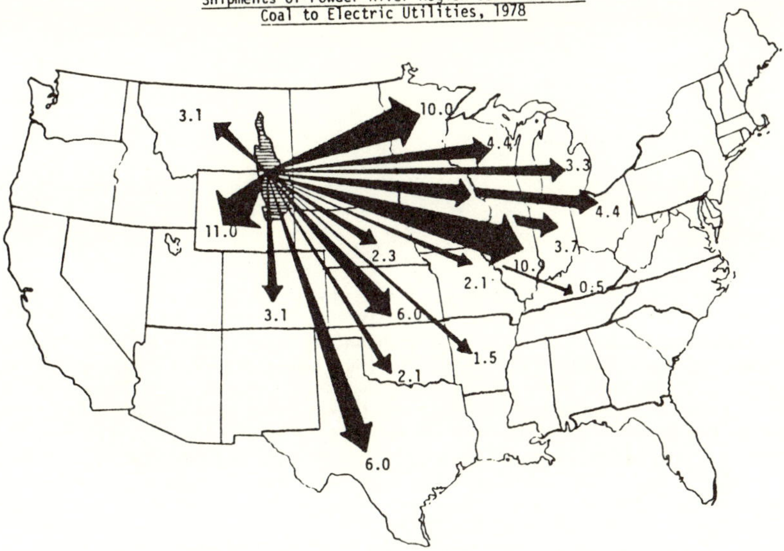

Table 8 summarizes fossil fuel deliveries to electric utilities in 1978. Two of the major reasons for rapid penetration of the utility market by low-rank coals are evident on the table: low delivered fuel cost compared to all alternatives, and low sulfur content compared to bituminous coal. The figures shown reflect national averages; substantial regional variations occur. For example, the cost (per Btu) of subbituminous coal at the minemouth is comparable to the value shown for lignite; depending on the distance shipped, transport costs can double or triple the delivered cost of subbituminous coal.

Table 8

Fossil Fuels Delivered to Electric Utilities, 1978

	Fuel Delivered			Average	Average
	10^{15} Btu	(1000 tons)	Average Btu Per Pound	Weight Percent Sulfur	Price ¢/10^6Btu
Oil	3.74		18,000	1.0	215.6
Gas	3.12		21,800	--	143.8
Coal:					
Bituminous	8.05	354,019	11,368	2.2	121.8
Subbituminous	1.66	90,519	9,169	0.57	79.1
Lignite	0.40	30,611	6,605	0.68	45.0

Table 9 shows in round numbers the cost incentive for midwestern utilities to use western low-sulfur coal rather than local high-sulfur coal plus scrubbers. (Tables 8 and 9 do not account for differences in the boiler plant costs caused by differences in coal properties. For a given steam output, boilers and pulverizers are larger and more expensive when low-rank coal is used.)

Table 9

Cost of Low-Sulfur Coal Compared to Illinois Coal Plus Flue Gas Desulfurization (FGD), 1978
(¢/10^6Btu) (hypothetical plant, Illinois site)

	Low-Sulfur Western Coal	Illinois Bituminous Coal
Coal Cost, f.o.b. mine	40	80
Transport Cost (unit train, Decker, MT to Chicago, IL)	90	
FGD Levelized 30-yr cost	--	80-130
TOTAL	130	160-210

Without question, the electric utilities will continue to be the most important market for low-rank coals. Coal and nuclear power represent the only proven and domestically abundant options that can be relied upon to meet growing electricity demand during the next several decades.

The challenges will be: 1) to improve the reliability and environmental performance of conventional pulverized coal-fired plants, and 2) to introduce new technologies (e.g., gasification/combined cycle, fluidized bed combustion) that substantially improve overall conversion efficiency, environmental control capability, water consumption, and other such factors.

A key question affecting future utility demand for low-rank coal is whether the 1979 revised New Source Performance Standards for SO_2 emissions from new coal-fired power plants will negate the advantage that western low-sulfur coal has had in midwestern markets. The new standard requires scrubbers (or other sulfur removal systems) on all new plants, with a sliding scale of 70 to 90 percent removal (see Figure 9).

Low-sulfur coals will have a cost advantage in the FGD portion of new power plants due to the lower sulfur removal requirement. However, this advantage will be much less than it is for existing plants, and may or may not outweigh the long-distance delivery costs of western coal, plus the incremental costs of handling, storing, pulverizing, and burning low-rank coal.

Another critical factor affecting the future use of western low-sulfur coal in distant markets will be transportation costs. Deregulation of railroad ratemaking is proceeding. This is likely to result in even larger increases in coal transportation rates than have occurred recently. If coal slurry pipelines are built, this would tend to decrease rates.

In addition to the New Source Performance Standards, ambient air quality standards will constrain new power plant construction in areas where concentrations of pollutants are close to the established limits. Western North Dakota already has such a problem with respect to SO_2 concentrations.

Figure 9

New Source Performance Standards for SO_2 Emissions from Electric Utility Steam Generators

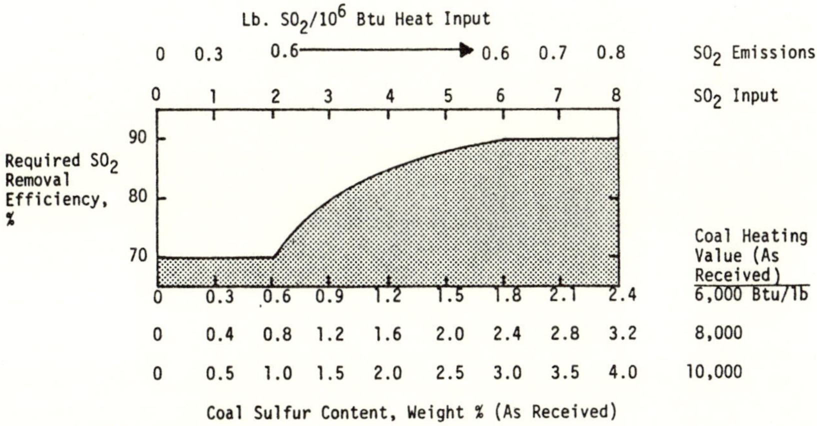

Industrial Markets

Industrial markets for low-rank coal are currently very small, amounting to only about one percent of total production, or 1-2 million tons/year. This utilization is primarily in industrial or institutional boilers near the coal deposits. There is some use of subbituminous coal in the midwest and west by industrial companies, including some non-boiler applications such as cement plants.

The basic problem for low-rank coal in penetrating industrial markets is the inability to compete with oil and gas in most regions, even though the cost of the fuel (per Btu) is lower. An industrial-size coal-fired boiler (either stoker or pc) is about 3 times higher in capital cost than an oil- or gas-fired boiler.

Even in cases where the total calculated costs to produce a pound of steam from coal are comparable or slightly better than from oil or gas, many firms reject the coal option. Reasons most frequently cited are the uncertainties and difficulties in meeting environmental regulations, and the inconvenience of handling solid fuel. The cost incentive has to be strong before low-rank coal can penetrate this market.

For industrial areas distant from the low-rank coal deposits, a major impediment is the inability of relatively small coal consumers to achieve the economies of scale that utilities achieve. These economies apply to the whole extraction and delivery, utilization, and cleanup system. For example, single-car rail rates per ton of coal delivered are double (or more) the unit train rates negotiated by utilities.

Solving this problem would require the development of a "whole-saling" infrastructure for low-rank coal -- large, efficient delivery systems to industrial distribution centers. Options to consider would be: 1) central combustion facilities feeding a steam and power distribution network (a natural extension of electric utilities' current operations); and 2) central gasification facilities feeding syngas to individual boilers. One advantage of these options compared to individual direct use would be the convenience and economy of centralizing all important environmental control and disposal operations. Alternatively, minemouth coal conversion operations, combined with long-distance product transportation, might achieve similar economies of scale. In any of these cases, a prerequisite to raising the capital required for such projects would be the existence of a firm market for the products. Ensuring this steady, long-term demand in compact areas may require the development of new cooperative investment and financing approaches.

Calculations show that the one major industrial area where coal is currently cost-competitive with oil and gas for industrial-scale steam generation is the industrial midwest (i.e., Illinois, Michigan, Ohio). However, in the absence of SO_2 emission regulations for such plants, local bituminous coal has a cost advantage over western subbituminous coal in this market.

During the 1980's (and beyond), oil and gas prices will continue to rise faster than coal prices. As this occurs, the cost of steam from coal will eventually become competitive with steam from oil and gas in areas other than the midwest.

One high-potential growth area for industrial use of low-rank coal is the Gulf coast. By the mid-1980's it is projected that Gulf lignite will be competitive with oil and gas as conventional boiler fuel. And, if fluidized bed combustion technology proves out, the cost incentive to use Gulf lignite (as well as "imported" western coal) could be substantial. The huge petrochemical industry in the Gulf region also represents a very good potential market for low-rank coal (via synthesis gas) in the next 10-20 years.

Synthetic Fuel Markets

Coal-based synthetic fuels are expected to begin contributing to our energy supply in significant amounts starting in the 1990's. The electric utility industry may be a major customer, using coal-derived gas in highly efficient combined cycle plants, and coal liquids in existing oil-fired boilers and new peaking units. Once synthetic fuels pass the cost "crossover" point with respect to petroleum and natural gas, penetration of synthetic fuels into many of the markets now served by oil and gas could occur rapidly. The transportation and distribution infrastructures for oil and gas products from coal are in place.

Several characteristics of low-rank coals (e.g., high reactivity, low sulfur content, non-caking properties) make them generally favorable feedstocks for many of the coal conversion processes. This fact, coupled with the low extraction cost per Btu, indicates that low-rank coal could account for a very large portion of the eventual synfuels market.

The first commercial SNG-from-coal plant in the U.S. is expected to be the Great Plains Gasification Associates project, which will use North Dakota lignite. Almost all of the SNG projects that have been proposed during recent years (most of which have been cancelled or deferred due to high costs and the regulatory climate) intended to utilize western low-rank coals.

The prospects for commercial production of synthetic liquids from coal appear to be farther in the future than for SNG. In addition, the advantages of low-rank coals as feedstocks to direct liquefaction processes are not as obvious as the advantages in first-generation gasification processes. Preliminary economic tradeoff studies indicate that low-rank coals could be at least as attractive as bituminous coal for liquefaction, but these calculations are based on inadequate process data. The future market for low-rank coals as liquefaction feedstocks must be considered highly uncertain until more RD&D has been performed.

Summary

One scenario for the future use of coal -- derived from Air Electric Power Research Institute long-range planning study -- is shown on Figure 10. Other forecasts (by DOE, the National Coal Association, and others) are similar. (The low-rank coal share shown on the graph is an estimate produced in this study. It is not intended as a forecast, but rather as an indication of the potential for growth.)

Figure 10

A Future Scenario for the U.S. Coal Industry

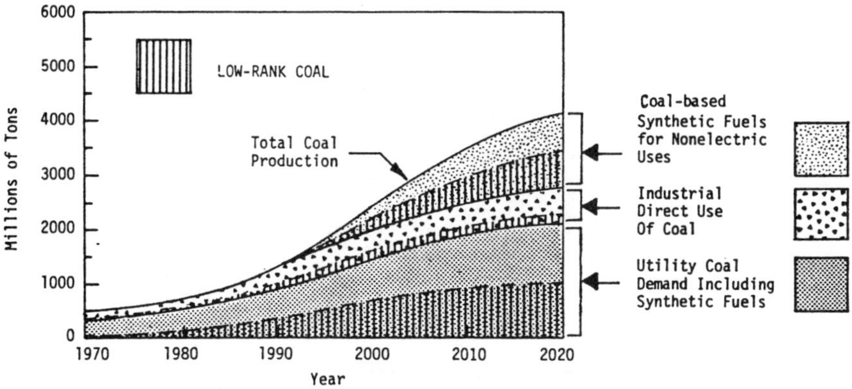

Under this scenario, total U.S. coal production would triple by the year 2000 to about 2 billion tons/year. This is an average annual growth rate of 4.7 percent -- much higher than the historical rate.

The low-rank coal portion of the total could grow to the order of 50 percent, or 1 billion tons/year, by 2000. To achieve this production rate, low-rank coal use would have to continue growing much faster than total coal use, at a rate of approximately 8-10 percent per year for the next 20 years.

Major conclusions of the market analysis for low-rank coals are as follows:

1. Low-rank coal growth is mainly constrained by demand. Competing fuels have the edge in many markets. Low-rank coal mining capacity currently exceeds production by a substantial amount, and is projected to do so at least through the mid-1980's.

2. The primary demand constraints are caused by: a) the location of low-rank coal reserves far from major markets; b) the high cost per Btu of transporting low-rank coals over long distances; c) the costs and uncertainties of using coal due to environmental standards; and d) the lack of a coal distribution infrastructure for industry; and e) the high cost of converting oil-or-gas-fired equipment to coal.

3. In the recent past, demand for low-rank coal in the electric utility market has been created by a combination of regulatory and technological developments. These encouraged the utilities to exploit the basic advantages of low-rank coals: low extraction costs and low sulfur content compared to bituminous coal.

4. The first-generation technology for low-rank coal development -- large-scale surface mining, unit train transport, large pc-fired boilers, and major regional power grids -- is likely to continue as the major low-rank coal utilization system for some time. However, the continued rapid growth of this existing market and technology is being challenged by new, more stringent environmental control requirements. Cost-effective solutions to these problems require technology development.

5. Similar considerations will apply in the longer-range future. Even though the nation needs rapid coal development to reduce its dependence on imported oil, demand for coal-derived substitute fuels in major oil markets will be constrained. Product quality and cost, and environmental, health, and safety issues associated with synfuels could limit their marketability. Government policy initiatives and new waves of technology will be the keys to removing these constraints.

6. As coal use expands, supply-side constraints will also emerge. Impacts of greatly expanded surface mining and coal use in the arid and semi-arid west, water availability, and inadequate coal transportation capabilities are the major factors that could hold back continued growth. Cost-effective solutions to these problems also require technology development.

Regulatory and Environmental Issues

All energy production, delivery, and utilization systems in the U.S. are being increasingly shaped by the actions of the federal government and state regulatory agencies. For example, implementation of the Fuel Use Act by the Department of Energy will have wide-ranging effects on the selection of fuels and boilers by utilities and industry. Currently the federal government is establishing the Synthetic Fuels Corporation to initiate the large-scale production of oil and gas substitutes.

Coal is inherently a dirty fuel. Each step in the sequence from extraction through utilization involves some environmental risk. These risks have come under increasing regulatory scrutiny -- a trend that can be expected to continue as coal use expands. Technology development is being forced; costs are being driven up.

In both of these areas -- government policies regulating production or use of fuels, and environmental standards -- the requirements represent moving targets to those who must comply. The uncertainty created in industry by constantly changing public policies tends to retard decision-making, and affects research and development needs.

Low-rank coals share many generic environmental/regulatory issues with all coal. There are also some unique considerations due to the location and properties of the coal. Table 10 summarizes the major issues of concern.

Much of the western low-rank coal is mined in areas that possess natural beauty but lack rainfall. In some areas of the west, water resource allocation decisions that will significantly affect coal development will be made only with great difficulty due to the many competing uses.

Reclamation is difficult in many of the arid low-rank coal-bearing regions, and its ultimate success may require many years to ascertain. At the other extreme, such as the current lignite mining area in Texas, revegetation occurs quickly with abundant rainfall. In this region, there is no need for segregation of topsoil during mining (as there is in most other areas) to aid in re-establishing ground cover.

Transportation of western subbituminous coal has been a regulatory issue since the mid-1970's due to the monopolistic position of some railroads, their pricing policies and merger applications, and their opposition to proposed slurry pipeline projects. Deregulation of rail rates is being considered in Congress. Higher rates would enable the railroads to improve and maintain unit train trackage, but would tend to depress demand for low-rank coal in the midwest and south. Conversely, lower transportation rates and additional competing capacity (i.e., slurry pipelines) would increase both demand and supply.

Table 10

Environmental Issues in Low-Rank Coal Development

Air Quality

1. SO_x, NO_x, and particulate emissions from power plants and conversion plants.
2. Odors, hazardous trace organic emissions from conversion plants.
3. Fugitive dust from mining, preparation, storage piles, rail cars.
4. Visibility; ambient pollutant concentrations; acid rain; CO_2 buildup.

Water Quality

1. Leaching to aquifers from mined or gasified areas.
2. Liquid effluent disposal from conversion processes.
3. Water from slurry pipelines and coal preparation plants.

Water Quantity

1. Cooling water required for combustion and conversion plants.
2. Process water required for gasification plants.
3. Water required for slurry pipelines.

Land Productivity

1. Productivity of reclaimed mined land in arid and semi-arid regions of the West.

Solid Waste Disposal

1. Ash disposal from combustion and conversion processes.
2. Sludge disposal from wet and dry SO_2 scrubbers, and and coal preparation plants.
3. Mine spoil piles.

Land Use

1. Land out of agricultural production or other use during mining or underground coal gasification.
2. Plant site area required for combustion and conversion.

Noise

1. Combustion, conversion, mining, and transportation facilities.

Public Safety

1. Potential exposure to hazardous material in conversion plants and UCG.
2. Grade crossing hazards and derailments in rail transport.
3. Subsidence following underground mining and UCG.

Federal and state air quality regulations are major technology-forcing issues for coal-fired power plants. The tightening of air emission standards for new boilers in the 1979 NSPS has sent low-rank coal-fired power plant designers back to the drawing boards to improve the control of SO_2, NO_x, and particulate emissions (see Table 11). Continued tightening of some of these standards (i.e., NO_x) as well as development of new standards for hydrocarbons, trace elements and solid waste can be expected to force additional changes -- not only in add-on cleanup modules, but in the combustion units themselves. Ambient air quality standards and visibility rules may also add to technology requirements.

Table 11

New Stationary Source Performance Standards
For Electric Utility Steam Generating Units, December 1971 and June 1979

Pollutant		December 1971 Standards	June 1979 Standards
SO_2:	Emission limit	1.2 lb/10^6 BTU	1.2 lb/10^6 BTU (based on a 30-day rolling average)
	Reduction in potential emissions		90% when emissions are 0.6 lb/10^6 BTU or greater
			70% when emissions are less than 0.6 lb/10^6 BTU
Particulate Matter:	Emission limit	0.1 lb/10^6 BTU	0.03 lb/10^6 BTU
	Opacity of emission	20%	20% (based on a 6-minute average)
			Based on a 30-day rolling average:
NO_x:	Emission limits	(i) 0.70 lb/10^6 from the combustion of coals except lignite.	(i) 0.50 lb/10^6 BTU from the combustion of subbituminous coal, shale oil, or any solid, liquid, or gaseous fuel derived from coal.
		(ii) 0.80 lb/10^6 BTU from the combustion in a cyclone furnace of any fuel containing more than 25 percent, by weight, lignite which has been mined in North Dakota, or Montana.	(ii) 0.80 lb/10^6 BTU from combustion in a slag tap furnace of any fuel containing more than 25 percent, by weight, lignite which has been mined in N. Dakota, S. Dakota, or Montana.
		(iii) Combustion of a fuel containing more than 25 percent, by weight, coal refuse is exempt from the NO_x standards and monitoring requirements.	(iii) Combustion of a fuel containing more than 25 percent, by weight, coal refuse is exempt from the NO_x standards & monitoring requirements.
		(iv) 0.60 lb/10^6 BTU from the combustion of lignite except as stipulated in (ii) above.	(iv) 0.60 lb/10^6 BTU from the combustion of any solid fuel not specified in (i), (ii), or (iii) above.

Notes: 1. These standards apply to electric utility steam generating units capable of combusting more than 73 megawatts heat input (250 million BTU/hour) of fossil fuel.

2. A major difference between the December 1971 and June 1979 NSPS is in compliance testing. The June 1979 NSPS require continuous stack monitoring and a 30-day rolling average for the SO_2 and NO_x emissions. The December 1971 required that emission monitoring only be performed at the beginning of plant operation and thereafter when the EPA deemed it necessary.

A major thrust in the recommended RD&D program for low-rank coal is to develop improved combustion environmental control technology (ECT). On one hand, it should be possible to take advantage of some of the unique properties of low-rank coal in meeting the standards (i.e., SO_2 removal by alkaline ash). On the other hand, certain aspects of ECT for low-rank coal (i.e., particulate capture and disposal) may present more difficult technical problems compared to bituminous coals.

Similar needs for focused research apply to the ECT requirements for low-rank coal gasification, liquefaction, and pyrolysis processes. The unique physical and chemical properties of the coal, as well as western location factors, create special problems and opportunities in controlling air, water, and solid emissions.

Nearly all recent studies of our national energy future and journal articles on Federal coal-related issues cite the need for expeditious new leasing programs for western coal. This is basically a "get on with it" versus "ban development and pollution" political issue. Similar (but global rather than national) concerns surround the acid rain and upper atmospheric CO_2 phenomena. Acid rain and CO_2 buildup are not unique to low-rank coals, nor to this country's policy agenda.

Key Low-Rank Coal Technology Issues

If low-rank coal use is to continue its rapid expansion, improvements are needed in existing technology to meet tightening standards and to enhance efficiency and reliability. There is also a need for new waves of technology which will allow coal to play the transition role expected of it in the next 40-50 years. These needs are driven by the market forces discussed earlier, as well as the environmental and regulatory requirements and constraints.

A thorough analysis of the technology currently applied (or potentially applicable) to low-rank coal development has been performed. Emphasis was placed on the unique physical and chemical properties of low-rank coals, and the effects or requirements imposed on technologies by these properties. The conclusion is that the key technical issues for low-rank coals are different from those for bituminous coal, and require a separate focus.

Figure 11 illustrates in shorthand form some of the important effects of low-rank coal properties on various technologies. When examined in conjunction with the specific needs for improvement of the technologies themselves, these effects generate a long list of key issues, which are summarized below. These issues, in turn, dictate the major elements of the recommended RD&D program for low-rank coals; these are presented in the next section.

Conventional combustion and associated environmental control technology is the area of most urgency. Pulverized coal-fired boiler technology is well-developed and has for the most part been adapted successfully to low-rank coals. Utilities in general need to improve the reliability and availability of boilers due to rising cost pressures.

One coal-specific issue affecting boiler reliability is ash fouling of boiler tubes. This problem is most severe with high-sodium lignite from the Fort Union Region. The costs of downtime in these boilers are being determined. A key question now is whether any of the proposed solutions -- such as revised boiler design, ion exchange removal of sodium, or injection of additives to the boiler -- are cost-effective.

In the area of sulfur emission control from power plant stacks, the combination of low sulfur content and high ash alkalinity frequently found in low-rank coals presents interesting opportunities. As a general strategy, it would be desirable to get away from conventional wet scrubbing with its many operating problems. Dry scrubbing offers lower capital costs and, potentially, better reliability. The higher reagent costs compared to wet scrubbing can be tolerated if the sulfur content of coal is sufficiently low . Field tests of the first commerical spray dryers in power plants, plus bench-scale research to develop improved sorbents or systems, should advance this technology.

Figure 11

Effects of Low-Rank Coal Properties on Technology As Compared to Bituminous Coal

	EXTRACTION				TRANSPORT		PREP, HANDLING & STORAGE			DIRECT UTILIZATION AND CONVERSION						
LOW-RANK COAL PROPERTIES	SURFACE MINING	RECLAMATION	UNDERGROUND MINING	UCG	RAIL AND BARGE	SLURRY PIPELINES	DRYING	CRUSHING AND PULVERIZATION	HANDLING AND STORAGE	CONVENTIONAL COMBUSTION	COMBUSTION ECT	FBC	GASIFICATION	LIQUEFACTION	PYROLYSIS	MHD
OCCURRENCE	+	−	−	+	−	−					±					
PHYSICAL PROPERTIES					−	−	−	−	−				−			
HIGH MOISTURE CONTENT				?	−	−	−	−	−	−	+	−	±	−	−	−
LOW SULFUR CONTENT	+	+	+	+						+	±	+	+	±	+	+
HIGH ALKALI CONTENT				?		?				±	±	±	±	?		
HIGH REACTIVITY	−		−	+	−		−		−	+		+	+	±	+	
NON-CAKING PROPERTIES				+						+		+	+		+	

LEGEND: (+) Generally positive effect on that technology; (−) Generally negative effect on that technology; (±) Either positive or negative effect, depending on circumstances; (?) Suspected to have an effect but not well characterized

Meeting New Source Performance Standards for SO_2, NO_x, and particulate emissions are not the only questions utilities will face in building new plants. Ambient air quality standards could force utilities to consider retrofitting existing plants to obtain SO_2 emission offsets. Different control options would apply to retrofit situations as compared to new plant designs; these need to be developed, taking advantage of the favorable low-rank coal properties.

The new particulate emissions standard of 0.03 lb/10^6 Btu is expected to present a major problem to utilities burning low-rank coals. Electrostatic precipitators have a difficult time collecting high-resistivity, fly ash from low-rank coals. Various techniques and devices have been proposed to improve ESP performance, but it is not clear yet if they can meet this standard. Alternatives to ESP's, such as fabric filters and various novel devices, have little or no experience on low-rank coal fly ash; basic data are needed. Special attention should be given to sampling, analyzing, and designing collection systems for very fine particulate matter.

Several aspects of NO_x emissions control from low-rank coal combustion may be unique: 1) conventional NO_x control techniques such as overfire air and low excess air could make ash fouling problems worse; and 2) dry sorbents used for SO_2 control may absorb some NO_x.

Disposal of solid wastes from power plants is an area where standards are currently evolving and data are needed. Specific questions about low-rank coal include: 1) leachability of high-sodium fly ash; 2) ash-alkali scrubber sludge fixation requirements; and 3) disposal methods for soluble (sodium-based) spent sorbents from spray dryers.

Trace elements and organic compounds can be emitted from coal-fired power plants. The nature of these emissions, and their fate and effects in the environment are not known. The unique compositions and forms of minerals in low-rank coals suggest that these compounds might be distributed differently in the effluents from low-rank coal plants. A basic problem in this area is the lack of adequate sampling and analytical procedures.

Technologies that are expected to be important to low-rank coal utilization in the future include fluidized bed combustion, gasification, liquefaction, and pyrolysis. In each of these technologies, there are specific concerns relating to the properties of low-rank coals.

Fluidized bed combustion offers a number of potential advantages over conventional pulverized coal combustion. These include flexibility to handle widely varying fuels, capture of sulfur by limestone in the fluidized bed, high combustion efficiency, compact size, and lower NO_x emissions due to reduced combustion temperatures. Low-rank coals with high alkali-to-sulfur ratios offer a significant additional advantage: the ability to absorb sulfur on the alkaline ash and meet emissions standards with little or no added limestone (see Figure 12).

Optimization of FBC technology for low-rank coals involves several key issues: selection of the best overall design configuration for a highly reactive, sulfur-retaining fuel; understanding the properties of an ash-rich fluidized bed with no limestone; preventing agglomeration of solids in the bed due to high-sodium content; and understanding sulfur retention as a function of coal and ash properties and operating conditions. For pressurized FBC, there is a question concerning the effects of alkaline ash (particularly, high-sodium ash) on hot gas cleanup systems.

Coal gasification technology involves many types of processes. Low-rank coals are preferred feedstocks to the first-generation, fixed-bed gasification processes such as Lurgi. This is due primarily to the noncaking behavior of low-rank coals; high reactivity and low-sulfur content are also advantageous. On the negative side, very high-moisture coals are not desirable for fixed-bed gasification. Because the first commerical gasification plants are likely to use low-rank coals, any of the residual problems with the technology (such as waste water treating) are important for low-rank coals.

Figure 12

Percent Sulfur Retention as a Function of Total Alkali-to-Sulfur Ratio in AFBC

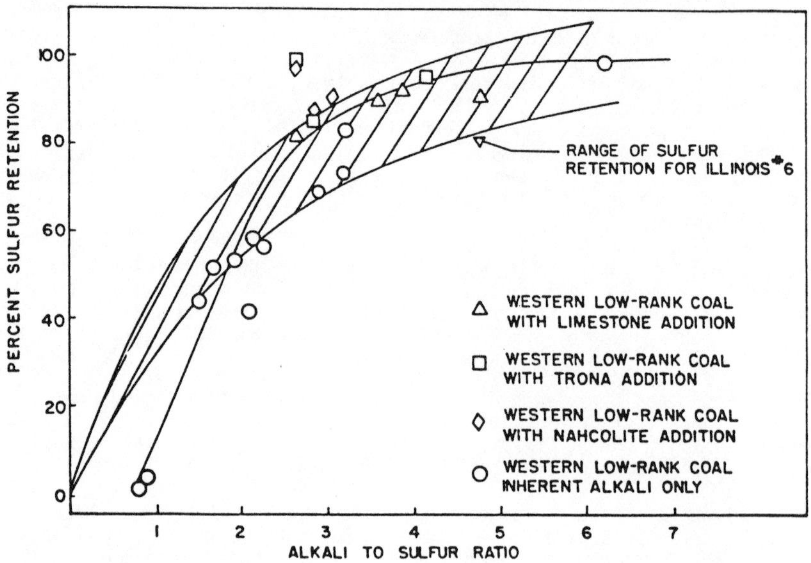

The major driving force behind the development of advanced fluidized bed and entrained flow gasification processes has been the desire to use caking bituminous coals. Thus, data on how to select and apply new technologies for low-rank coal tend to be unavailable. For these emerging technologies, the low-rank coal issues are process-specific and detailed, rather than large central issues. For example: slurry feeding a high-moisture coal to a Texaco gasifier could be a problem; or the alkali recovery unit in the Exxon catalytic gasification process could be adversely affected by the inherent alkali elements in low-rank coals.

The behavior of low-rank coal (especially lignite) has been shown to be different from bituminous coal in all of the developing direct liquefaction processes. High moisture content increases the reactor pressure requirement. High oxygen content leads to high CO_2 production, which also increases pressure requirements. Low-rank coal reacts readily with carbon monoxide, possibly favoring the use of syngas rather than pure hydrogen. The high calcium content in low-rank coal ash causes deposition of calcium carbonate solids in liquefaction reactors. Quality of the liquid products appears to differ for low-rank coals, particularly the high viscosity of the bottoms fraction (see Figure 13).

The implication is that low-rank coal liquefaction, once optimized, will involve different process design parameters and operating procedures from those for bituminous coal. There is need for additional RD&D work on low-rank coal liquefaction, including basic research into process mechanisms and product properties, engineering/economic optimization studies, and testing of low-rank coals in the large existing pilot plants.

Coal pyrolysis is a technology that can be used if the objective is to make a combination of products: fuel gas, liquid fuels or specialty chemicals, and char for direct use or briquetting. Data from early studies on low-rank coal pyrolysis indicate problems with both the yields and quality of liquid products. A complete review of existing data and an evaluation of the feasibility of low-rank coal pyrolysis would be worthwhile.

Figure 13

Effects of Coal Type and Process Conditions on Viscosity of EDS Residual Liquefaction Bottoms

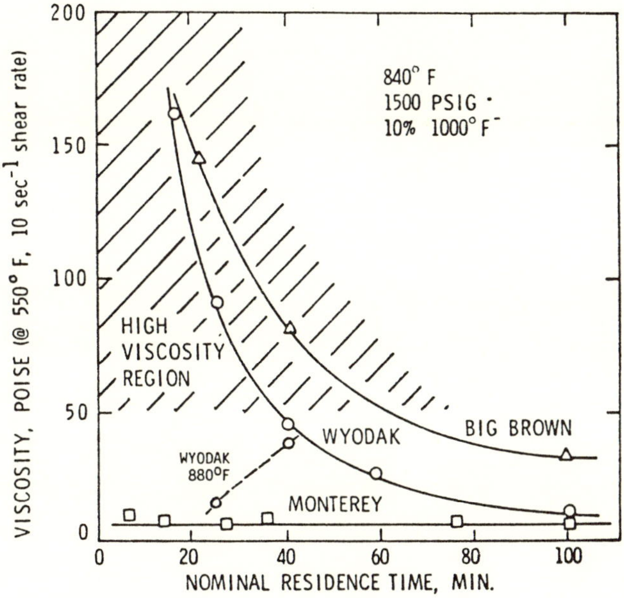

In the areas of preparation, handling, and storage, a number of problems and opportunities exist due to the generally unfavorable handling characteristics of low-rank coal.

The high moisture content causes problems in many areas, and suggests the need for coal drying processes to meet specific needs. These include: 1) drying to improve transport economics, where overall system cost tradeoffs and dried coal stability are the key questions; 2) drying techniques that limit moisture reabsorption to prepare coal for slurrying into pipelines or entrained flow gasifiers; 3) drying techniques that preserve the coal reactivity for liquefaction instead of collapsing active surface area; and 4) drying to very low levels of moisture as required for feed to magnetohydrodynamic reactors.

Incentives also exist to develop low-rank coal cleaning processes for sodium, sulfur, or ash removal. Conventional gravity separation techniques can be usefully adapted to some low-rank coals which have appreciable extraneous ash contents. In general, the potential for sulfur removal from low-rank coals by washing is low. Ion exchange has been shown to be feasible for sodium reduction, and may find applications in the Fort Union Region to alleviate boiler fouling problems.

Low-rank coal extraction and transportation technologies are generally well-developed. Most of the technical issues involve optimization of systems and cost reduction. Reclamation of surface-mined land in arid regions continues to provide research challenges. Dewatering of coal slurries and treatment, reuse, or disposal of water are areas of concern for proposed coal slurry pipelines.

RD&D Program Recommendations

Based on a detailed review of the key technical issues, market needs, and various constraints to development, a national RD&D program for low-rank coal is recommended. The approach taken was to identify major research topics, state their objectives, identify supporting data and/or ongoing related research, and indicate how the results of the research would be integrated with other work to advance low-rank coal development. Within each research topic the approach to the work has been stated only in general terms. Definition of the step-by-step details of the work has been left to the organizations that will perform the RD&D.

To indicate the general level of importance or urgency associated with each recommended project, "Priority I" and "Priority II" designations have been assigned. Taken as a whole, the Priority I program represents RD&D that is considered essential to the advancement of low-rank coal development. It is in essence a zero-base program recommendation. The Priority II program represents RD&D that is considered very important, and could greatly enhance the longer-term development of low-rank coal. Other worthwhile ideas and projects were considered, but were relegated to a lower priority level and excluded from the recommended program.

The overall emphasis of the program is on exploratory, technology, and engineering development efforts which make systematic use of the knowledge and understanding gained from basic and applied research. This emphasis is illustrated on Figure 14, which classifies the recommended RD&D in each technology area into various phases, defined briefly as follows:

1. Basic Research - efforts to increase knowledge and quantitative understanding of natural phenomena. An example is the recommendation for improved petrographic characterization of low-rank coals.

2. Applied Research - systematic study to devise processes or systems of possible (but uncertain) practical utility. An example is the recommended development of standard analytical methods to characterize coal-derived liquids.

3. Exploratory Development - efforts to explore possible innovation in a particular technology area, and to assess whether further development is warranted. An example would be the study of coal mineral catalysis in recycle liquefaction processes.

4. Technology Development - development of proces-

Figure 14

Classification of Recommended RD&D Program Into Phases of RD&D

TECHNOLOGY AREA / PHASE OF RD&D	BASIC RESEARCH	APPLIED RESEARCH	EXPLORATORY DEVELOPMENT	TECHNOLOGY DEVELOPMENT	ENGINEERING DEVELOPMENT	DEMONSTRATION	COMMERCIALIZATION PRODUCT AND OPERATION
EXTRACTION AND TRANSPORTATION		II	I II	I II	I	I	
PREPARATION, HANDLING & STORAGE		II	I II	I II	I II	II	
CONVENTIONAL COMBUSTION		I	I II	I II	I II	I II	I
COMBUSTION ENVIRONMENTAL CONTROL TECHNOLOGY			I II	I II	I II	I II	I
FLUIDIZED BED COMBUSTION			I	I II	I II	I	
GASIFICATION				I II	I	I	I
LIQUEFACTION & PYROLYSIS			I II	I II	I II		
BASIC COAL SCIENCE	I II	I II	I II				

Legend: I - Priority I
II - Priority II

ses or subsystems at laboratory scale, as well as preliminary studies encompassing systems analysis, trade-offs, cost-benefits, and environmental analyses. An example is the recommended small-scale testing to select additives for ash fouling control in boilers.

5. Engineering Development - detailed design, construction, and test for performance, producibility, and reliability of system prototypes and pilot plants. An example is the recommended acquisition and testing of novel devices for fine particulate collection.

6. Demonstration - verification of economic and environmental viability for commercial application, through design, construction, test and evaluation of large-scale systems. An example would be the recommended demonstration of a direct ignition system for pulverized low-rank coal fired boilers.

7. Commercialization, Production, and Operation - self explanatory. An example would be the recommended systematic evaluation of the first utility spray dryer FGD units.

Brief summaries of the recommended RD&D in each major technology area are presented below.

Extraction

As shown on Table 12, the Priority I recommendations for low-rank coal extraction involve reclamation techniques for surface mined land, new and improved surface mining techniques, and underground coal gasification.

Table 12

Recommended RD&D For Low-Rank Coal Extraction

Priority I	Priority II
1. Surface Mined Land Reclamation	4. Dewatering of Mine Area and Groundwater Control
2. Surface Mining a. Techniques for multiple thin seams, thick seams and deeper overburden b. Optimization of equipment specifications c. Cost reduction through operations research and systems engineering	5. Underground Mining a. Mining thick seams b. Mining under unconsolidated overburden c. Dewatering and groundwater control
3. Underground Coal Gasification a. Aquifer disruption and groundwater contamination b. Subsidence and gas leakage c. Linking techniques d. Coal seam characterization and process monitoring	

Reclamation studies should involve a combination of field work and monitoring, laboratory or controlled growth experiments, and modeling.

Improvement of surface mining techniques requires the application of open pit or other methods to difficult mining situations such as multiple seams, thick or pitching seams, and deeper seams. European techniques such as bucketwheel excavators combined with conveyor belt systems are difficult to apply to the overburden and climate conditions in the Northern Great Plains, for example. Design and operation of earth-moving machinery, as well as overall mine planning, could be enhanced by the development of computerized design, monitoring, and control systems.

Underground coal gasification has future potential as a method for extracting coal energy from deeper seams that would be uneconomical to mine. It also promises to provide a relatively inexpensive source of coal-derived syngas. Several major areas of research need to pursued, as noted on the table, before this technology can be considered sufficiently reliable or environmentally sound. DOE has an ongoing program addressing these research needs, and several industry projects are underway as well.

The Priority II recommendations in Extraction include: 1) development of better mine dewatering methods and groundwater control systems, especially for deeper coal seams; and 2) development of appropriate underground mining technology for thick western coal seams, which often have unconsolidated overburden and act as major aquifers.

Transportation

The recommended RD&D for low-rank coal transportation is limited to one transport mode -- coal slurry pipelines (see Table 13). Rail, barge, truck, and conveyor systems will continue to evolve and improve, but do not generate any major research needs (with the exception of coal dust, reactivity and freezing problems, which are addressed under Coal Preparation).

Table 13

Recommended RD&D For Low-Rank Coal Transportation

Priority I	Priority II
1. Slurry Dewatering a. Separation of coal fines b. Treatment of separated water c. Utilization of treated slurry water 2. Slurry Pipeline Water Requirements	3. Slurry Pipeline Reliability a. Restarting slurry flow b. Freeze protection c. Ruptured slurry pipelines d. Distances over which coal suspension can be maintained.

In the Priority I area, slurry dewatering at the receiving end of a coal pipeline is a problem that could benefit from scientific investigation. Laboratory tests indicate that during transport some low-rank coals (especially lignite) become very finely divided, making separation by standard gravity techniques ineffective. Investigation of the surface properties of coal fines and of chemical means to enhance separation processes would be worthwhile. Site-specific studies of the treatment and utilization alternatives for the separated water should also be made.

A major issue with respect to slurry pipelines in arid regions of the west is the utilization of large amounts of water to transport the coal. Studies to minimize water requirements should include investigations of slurry media other than water -- oil, methanol, or CO_2, for example.

In the Priority II area for low-rank coal transporation, the recommendations include systems studies, simulations, and some laboratory research to address questions about slurry pipeline reliability.

Preparation, Handling, and Storage

The Priority I recommendations for low-rank coal preparation research include two projects on coal drying, two on coal cleaning or beneficiation, and an investigation of fines generation during crushing and handling (see Table 14).

Table 14

Recommended RD&D for Low-Rank Coal Preparation, Handling, and Storage

Priority I	Priority II
1. Coal Drying For Conversion Processess	6. Briquetting or Pelletizing
2. Chemical Cleaning Processes for Low-Rank Coals	7. Preparation and Beneficiation Techniques Applied to Slurry Pipeline Systems
3. Physical Cleaning Processes for Low-Rank Coals	8. Waste Disposal from Coal Beneficiation
4. Coal Drying to Improve Transport Economics	9. Control of Dust, Oxidation and Spontaneous Combustion
5. Optimized Crushing & Handling Equipment to Minimize Fines Generation	10. Freeze Control
	11. Comminution Techniques

Drying of low-rank coals for use in conversion processes requires that different techniques be used to meet different process requirements. Techniques that limit reabsorption would be useful for processes using slurry feed systems: drying to very low moisture levels is required by the MHD process; drying the coal without collapsing its pore structure is required for liquefaction. Techniques exist or have been proposed for all of these purposes; they need to be verified.

Chemical cleaning processes for low-rank coals would include: 1) ion exchange for sodium removal; 2) removal of potentially hazardous elements; and 3) sulfur and ash removal. Processes have been partially developed for several of these purposes, but need further testing and economic tradeoff studies to verify their usefulness.

Physical coal cleaning processes would be applicable to low-rank coals containing significant amounts of extraneous mineral matter. Multi-solvent approaches have been suggested, and should be investigated. Techniques applicable to separation of fine particles, as well as magnetic separation techniques, should be tested for applicability to low-rank coals. To determine the overall potential or need for this technology, an assessment of the trend towards extraction of lower-grade low-rank coal (as better deposits are mined out) should be done.

Drying of low-rank coal to improve long-distance transport economics has two major problems: high costs, and difficulty in handling and storage of dried coal. A state-of-the-art review of various thermal, gas and liquid phase drying techniques is needed. Effects of dried coal on boiler design need to be determined. Cleanup and utilization of the separated water should be investigated.

The generation of coal fines during low-rank coal handling and crushing needs to be studied to: 1) determine the specific procedures that generate fines, and 2) develop modifications to reduce production of, and/or utilize fines.

A large number of Priority II projects are recommended in Preparation, Handling, and Storage, reflecting the need for improvement in the physical and handling properties of low-rank coals. Briquetting technology has been developed and applied to many types of coal, and could be applied once again in this country if markets for the product existed.

The study of slurry preparation techniques should incorporate the results of coal drying studies to limit moisture reabsorption, as well as other means to improve the properties of low-rank coal slurries. The concept of incorporating thermal, chemical, and mechanical beneficiation techniques into a slurry pipeline system should be investigated.

Concentrated wastes from coal beneficiation processes should be characterized to determine RCRA requirements applying to their disposal.

Techniques to control dust, oxidation, and spontaneous combustion problems have been developed, but these problems still occur. Additional work on new cost-effective solutions would be worthwhile.

Control of freezing problems in coal transport and handling systems is a persistent problem that could be amenable to solution by novel techniques.

Finally, improvement of low-rank coal comminution (crushing and grinding) technology is a continuing need.

Conventional Combustion

Table 15 lists the recommended RD&D for improvement of conventional low-rank coal combustion technology. The Priority I projects address the problems of ash fouling and slagging, and the opportunity to decrease our oil consumption by developing a direct coal ignition system.

Table 15

Recommended RD&D For Low-Rank Coal Conventional Combustion

Priority I	Priority II
1. Ash Fouling and Slagging Mechanisms	4. Improved Boiler Cleaning Procedures
2. Control of Fouling and Slagging With Additives	5. Temperature Limitation Vs. Boiler Corrosion
3. Direct Ignition of Pulverized Coal Without Oil	6. Improved Stoker Furnace for Small Applications

The study of ash fouling and slagging mechanisms builds on the knowledge gained from years of research at the Grand Forks Energy Technology Center. Recommended research elements include developing analytical techniques to predict the fouling and slagging potential of coal samples, and the effects of boiler design or operating procedure modifications.

One potential method to ease removal of ash fouling deposits would be the injection of additives to the boiler. This requires a sequence of: 1) basic studies to determine mechanisms of additive reactions with ash; 2) small-scale testing to screen and select promising additives; and 3) demonstration of the best additives in operating boilers.

A direct ignition method to replace the current oil ignition systems in pulverized-coal-fired boilers would be designed to take advantage of the high reactivity of low-rank coal.

The Priority II recommendations for Conventional Combustion include the development of improved boiler cleaning procedures, an investigation of extending tube metal temperature limits, and development of an improved stoker furnace for small institutional uses.

Improvement of boiler cleaning procedures (on line) requires developing criteria for predicting the ease of deposit removal as well as methods for more accurately determining sootblower requirements. Improved methods for deposit removal during boiler outages would be useful.

Research to determine the temperature limitations of alloys in relation to external corrosion in the convection pass of low-rank coal-fired boilers might lead to the ability to operate at higher steam temperatures, thus increasing the steam cycle efficiency. Probes could be utilized to measure corrosion rates in operating boilers. Experiments should be conducted with additives that might extend the upper temperature limitation of alloys.

Studies to develop an environmentally, economically, and technically acceptable stoker furnace for institutional use of low-rank coals should be initated. These studies would include an evaluation of currently available stokers to identify needed improvements; development of improved furnace designs, as well as improved low-rank coal fuels; and demonstration of the operability of prototype systems.

Combustion Environmental Control Technology (ECT)

As shown on Table 16, the Priority I recommendations for low-rank coal combustion ECT involve the key areas of stack gas cleaning for SO_2, NO_x, and particulate emissions control, and solid waste disposal. A systems analysis project is recommended to evaluate integrated environmental control systems for low-rank coals. The tendency has been to add control systems to power plants in series without much regard for their interactions. By considering the interactions among alternative control systems, it might be possible to devise more efficient, reliable, and cost-effective methods to achieve the multiple cleanup objectives.

Table 16

Recommended RD&D For Low-Rank Coal Combustion Environmental Control Technology

Priority 1	Priority II
1. Integrated Environmental Control Systems	9. Solid Waste Utilization
2. Improved Spray Dryer and Dry Sorbent Systems	10. Retrofit SO_2 Reduction Techniques Such as Lime/Limestone Injection
3. Improved Particulate Control Methods	
4. Fine Particulate Control Technology	
5. Solid Waste Disposal Procedures	
6. Improved Reliability of Ash Alkali Wet Scrubbing	
7. Trace Elements and Organic Compounds in Flue Gas	
8. Improved Procedures for NO_x Control	

Spray dryer and dry sorbent systems are believed to be capable of meeting SO_2 removal standards for many low-rank coals, and to be preferable to wet scrubbers in terms of cost and reliability. A research project should be conducted to: 1) evaluate the performance of the first commercial units; and 2) develop improved systems -- both hardware and sorbents -- through a combination of laboratory, pilot, and field tests.

Research into improved particulate control methods for low-rank coal involves a number of critical elements: 1) continued investigation of means to improve ESP performance, such as advanced charging techniques, fly ash conditioning, and improved fly ash removal methods; 2) testing and improvement of of fabric filter technology, including improved bag cleaning methods, fabrics and fabric finishes, and reduced baghouse size and pressure drop; and 3) developing and proving new concepts.

The study of fine particulate control technology is given separate emphasis because regulations are still evolving and sampling, analytical, and control methods for very fine particles are not well developed.

Solid waste disposal procedures are being scrutinized in general, due to the new RCRA regulations and concern about hazardous wastes. The solids from low-rank coal-fired power plants -- bottom ash, fly ash, and wet or dry scrubber sludges -- tend to have unique characteristics and require focused research.

The ash alkali wet SO_2 scrubbing process has significant cost advantages over conventional lime/limestone scrubbing for low-rank coals with high alkali-to-sulfur ratios. However, continuing work to improve the reliability and efficiency of the process is needed.

An area of growing concern, but lacking data, is the emission of trace elements and organic compounds in the flue gas of coal-fired power plants. The first step is to develop sampling and analytical procedures that can measure these emissions from low-rank coal-fired plants.

Another area of growing concern for fossil fuel combustion plants is NO_x emissions. Increasingly stringent standards for this difficult-to-control combustion product are anticipated. Possible emission control techniques include both combustion system modifications and add-on flue gas treating devices. In both cases, the unique burning properties and emissions associated with low-rank coals could impact the NO_x control system design.

The Priority II recommendations for combustion ECT include: 1) investigation of possible ways to utilize, rather than dispose of, solid wastes from low-rank coal combustion; and 2) development of effective retrofit SO_2 reduction techniques that could be applied to existing power plants.

Fluidized Bed Combustion

The Priority I recommendations for RD&D on fluidized bed combustion of low-rank coals deal primarily with the special problems associated with alkaline ash (see Table 17).

Table 17

Recommended RD&D For Low-Rank Coal Fluidized Bed Combustion (FBC)

Priority I	Priority II
1. Agglomeration of Solids in Fluidized Bed	7. Coal and Sorbent Feeding and Distribution
2. Sulfur Retention by Inherent Alkali in Low-Rank Coals	8. Staged Combustion for NO_x Control
3. Design Configuration of FBC Optimized for Low-Rank Coal	9. Temperature, Gas, and Solids Distribution in Low-Rank Coal FBC
4. Properties of Limestone-Deficient, Ash-Rich Fluidized Bed	
5. Hot-Gas Cleanup and Turbine Reliability for Pressurized FBC	
6. Materials Problems and Selection for Low-Rank Coal FBC	

Tests on some low-rank coals with high sodium content have shown a tendency for the bed solids to agglomerate into large chunks. Further studies are needed to determine the conditions and mechanisms that cause or prevent agglomeration, including the selection of bed materials and sorbents.

Additional information is needed on sulfur retention in low-rank coal FBC, as a function of coal and ash properties, pressure, other operating conditions, bed materials, and sorbents.

Studies to determine the best design configurations for different coals and for different applications (i.e., large utility versus small industrial boilers) are needed. These include evaluation of different overall systems (i.e., fluidized bed versus circulating bed) as well as identification of optimal design parameters for low-rank coals (e.g., bed depth, velocity, and heat transfer surface).

The ability of many low-rank coals to absorb SO_2 on the alkaline ash creates a need for studies on the properties of an ash-rich bed with little or no added sorbent. Data needed include the long-term operability of an ash-rich bed, physical and chemical properties, heat transfer coefficients and thermal diffusivity.

A critical R&D area for pressurized FBC is hot gas cleanup, which is required to protect turbine blades from corrosive/erosive attack. It is expected that low-rank coals, with highly alkaline ash, present unique hot gas cleanup problems.

Because of the unique chemistry of low-rank coal FBC, studies to identify if problems exist with materials of construction in these systems are needed. This would include in-bed and above-bed heat transfer surfaces, as well as air distributors and cyclones.

In the Priority II area for FBC, the recommended RD&D includes: 1) optimizing coal and sorbent feeding and distribution systems; 2) evaluating staged combustion systems for NO_x control; and 3) basic studies of temperature, gas, and solids distributions in fluidized beds. In all of these areas, the high reactivity and alkalinity characteristic of low-rank coals are expected to have significant effects.

Gasification

As shown in Table 18, the Priority I recommendations for RD&D on low-rank coal gasification cover a number of topics. The first project is an engineering/economic analysis of both existing and developing gasification processes to select and optimize process(es) for low-rank coals.

Table 18

Recommended RD&D For Low-Rank Coal Gasification

Priority I	Priority II
1. Process Adaptation for Low-Rank Coals	7. Effects of Pressure, Temperature, and Atmosphere on Evolution and Destruction of Volatile Matter
2. Wastewater Treatment for Process Effluent	8. Distribution Coefficients of Soluble Organics in Wastewater
3. Slag Behavior	9. Slurry Feeding of Low-Rank Coal to High-Pressure Gasifiers
4. Catalytic Effects in Low-Rank Coal Gasification	
5. Slag, Ash, and Residue Leaching Characteristics and Immobilization	
6. Minimizing Health Effects of Coal Liquids	

Wastewater treatment studies are needed to determine whether differences in coal properties or gasification processes affect the quality of the wastewater and the required treatment. This would include development of laboratory methods for screening novel treatment techniques, as well as demonstration of new methods at pilot scale.

Slag behavior is important in a number of high-temperature gasification processes. Critical areas include the effects of slag on refractory materials, as well as flow properties (viscosity) as a function of temperature. The alkali-rich mineral matter in low-rank coals causes different slag behavior compared to the iron- and silica-rich slags from eastern coals.

Studies of catalytic effects in low-rank coal gasification could lead to improved processes. This would include both the catalytic activity of inherent mineral matter in the coal, plus possible interactions between the inherent mineral matter and added catalysts.

Proper disposal of solid wastes from gasification plants requires that data be obtained on the leaching characteristics of slags, ashes, and wastewater treating residues. As required, techniques to immobilize leachable components should be developed.

Some coal liquids produced as gasification by-products are known to have carcinogenic or toxic effects. Characterization of these liquids from low-rank coal gasification should be pursued, as well as methods to insure that humans are not exposed.

In the Priority II area for gasification, some basic studies are recommended that could result in significant improvements in the technology. These include: 1) studies of the effects of pressure, temperature, and atmosphere on evolution and destruction of volatile matter from low-rank coals; 2) determination of distribution coefficients of soluble organics in gasifier wastewater streams; and 3) methods to increase the solids content of high-moisture coals in slurry feeding systems to high-pressure gasifiers.

Liquefaction

A large number of RD&D efforts are recommended in the area of low-rank coal liquefaction, reflecting both the many unresolved issues for liquefaction technology in general, and the unique problems associated with the properties of low-rank coal organic and mineral matter (see Table 19). In the Priority I group, the first project is an engineering/economic analysis of the best process configuration and operating conditions for low-rank coals. This would utilize results from many of the other recommended projects.

Table 19

Recommended RD&D For Low-Rank Coal Liquefaction

Priority I	Priority II
1. Process Adaptation for Low-Rank Coal	9. Bottoms Viscosity Studies
2. Recycle Solvent Studies	10. Effects of Staged or Continuous Removal of Gas
3. Syngas and Hydrogen Effectiveness Studies	11. Fate of Nitrogen
4. Calcium Carbonate Formation in Reactors	12. Effects of Staged Temperature Exposure
5. Coal Moisture Content and Drying	13. Disposable Catalyst Approaches
6. Coal Mineral Catalysis in Recycle	14. Corrosion of Stainless Steel by Coal Liquids
7. Bottoms Recycle	15. Erosion in Liquefaction Systems
8. Minimize Health Effects of Coal Liquids	16. Mathematical Reactor Model to Account for Mixing and Turbulence Effects

Recycle solvent studies are needed to determine the effects of solvent composition on the rate of hydrogenation of the coal, and on the yields and quality of products.

Building on earlier work using syngas ($CO + H_2$) for low-rank coal liquefaction, more detailed studies comparing the pros and cons of syngas versus hydrogen should be conducted.

Additional efforts are needed to determine the conditions under which calcium carbonate reactor solids are formed (i.e., effects of pressure) in liquefaction reactors.

More data are needed to understand the effects of coal moisture content (as fed to the reactor), and the effects of various methods of coal drying, on liquefaction yields and products.

An evaluation of catalytic effects of coal mineral matter in recycle liquefaction needs to be made with regard to the unique low-rank coal mineral forms.

Recycle of heavy bottoms material has been shown in limited testing to significantly improve yields of distillate products from lignite. Additional work on this concept should be done.

Health effects of coal liquids must be minimized in coal liquefaction processes, as in other conversion systems.

The Priority II RD&D recommendations for liquefaction include: 1) liquefaction bottoms viscosity studies (effects of coal and recycle solvent, process conditions, nitrogen, oxygen, and hydrogen content); 2) reactor configuration studies, such as staged removal of gas, staged temperatures, and effects of mixing and turbulence; 3) nitrogen fate determinations in both hydrogen and syngas liquefaction; 4) evaluations of disposable catalyst approaches for low-rank coals; and 5) corrosion and erosion in liquefaction systems.

Pyrolysis

In the Priority I area for coal pyrolysis, an evaluation of the feasibility of low-rank coal pyrolysis is recommended (see Table 20). This would include a review of data from earlier studies, selection of process configurations for analysis, and engineering/economic tradeoff studies of potential processes.

Studies to minimize health effects of coal liquids are required for coal pyrolysis processes as well.

Table 20

Recommended RD&D for Low-Rank Coal Pyrolysis

Priority I	Priority II
1. Feasibility of Pyrolysis for Selected Low-Rank Coals	3. Improved Analytical Methods for LRC Pyrolysis Studies
2. Minimizing Health Effects of Coal Liquids	4. Improved Pyrolysis Product Properties and Yields

In the Priority II area, the recommendations include: 1) development of improved analytical methods for pyrolysis studies, such as tests for char reactivity and pyrolytic oil characterization; and 2) investigation of methods to improve the chemical and physical properties of low-rank coal pyrolysis products as well as product yields, including maximization of BTX chemical feedstocks by flash hydropyrolysis.

Basic Research

As indicated on Table 21, there are many basic research studies that could enhance low-rank coal and peat technology significantly in the long run. In general, there is a basic need in all coal regions for detailed resource and coal seam characterization efforts. These would delineate the properties of specific reserves and resources in terms of geology, organic and mineral matter composition, variability, washability, and so on. There is also a need for establishing a suite of standard low-rank coal samples for reference and comparison purposes among coal research laboratories.

Table 21

Recommended Basic Research for Low-Rank Coals and Peat

Priority I	Priority II
1. Resource and Coal Seam Characterization	11. Surface Characteristics of Low-Rank Coal and Peat Fines
2. Standard Low-Rank Coal Samples	12. Reactivity of Low-Rank Coals at 1200-1800°F in FBC
3. Petrographic Characterization	13. Impacts of Drying Methods on Rheological Properties of Low-Rank Coal-Water Mixtures
4. Reactions Between Alkali Materials and Sulfur	14. Kinetics and Reaction Mechanisms of Low-Rank Coals and their chars with H_2O, H_2, CO and CO_2
5. Surface Tension of Coal Slags	15. Fate of Oxygen and Nitrogen Components in Coal and Peat Conversion Systems
6. Composition and Characteristics of Ashes and Slag from Low-Rank Coals and Peat	16. Reactivity of Peat in Various Atmospheres
7. Analytical Characterization of Liquefaction Solvents	
8. Coal Liquefaction Catalysis	
9. Oxidative Depolymerization of Low-Rank Coal	
10. Toxicity of Coal Liquids	

Petrographic characterization of low-rank coals must start with development of techniques and classification systems that apply to these coals. Then, systematic efforts can begin to relate petrographic data on low-rank coals to reactivity and yields in conversion systems.

Basic studies on the reactions between alkali elements and sulfur that occur in low-rank coals could have applicability in a number of areas, such as conventional combustion, fluidized bed combustion, wet and dry flue gas scrubbing.

Similarly, exploratory studies on surface tension of low-rank coal slags could lead to chemical or physical improvements in wet-bottom combustion or gasification systems.

In a similar vein, experimental and theoretical studies on the composition and characteristics of ashes and slags from low-rank coals and peat will improve our understanding of why they behave as they do and how their properties can be altered or used to advantage.

Standard analytical methods, which are consistent among laboratories, for characterizing solvents derived from low-rank coals would be very helpful to all of the researchers in coal liquefaction. A reproducible measure of solvent quality (hydrogen donor ability) would allow for better comparison among alternative processes.

A state-of-the-art review of low-rank coal liquefaction catalysis is needed to understand possible catalytic mechanisms and to formulate a catalyst development program.

Oxidative depolymerization processes should be explored as a concept for converting low-rank coals to liquid fuels or chemicals.

Toxicity studies on coal liquids are needed to provide basic data to help in devising means for minimizing adverse health effects.

In the Priority II area, a number of additional basic research projects are recommended for low-rank coals and peat. These include: 1) studying the surface characteristics of low-rank coal and peat fines with regard to the applicability of froth flotation beneficiation processes; 2) evaluating the reactivity of low-rank coals at 1200 - 1800°F in fluidized bed combustion; 3) studying the effects of various drying methods on the rheological properties of low-rank coal-water mixtures for slurry pipeline and high-pressure feeding application; 4) conducting comprehensive basic studies of the kinetics and mechanisms of reactions of low-rank coals and their chars with H_2O, H_2, CO and CO_2, including catalyst effects; 5) determining the fate of oxygen and nitrogen components in coal conversion systems (and their roles in product quality); and 6) studying the reactivity of peat in various gaseous and liquid atmospheres to support process development efforts.

Part II

Technology Evaluation

The information in Part II is from *Low-Rank Coal Study–National Needs for Resource Development. Volume 3–Technology Evaluation* [DOE Report FC/10066-T1 (V. 3)], prepared by John Kotowski and George Wiltsee of Energy Resources Company, Inc., Walnut Creek, California, and G.H. Gronhovd and E.A. Sondreal of Grand Forks Energy Technology Center, Grand Forks, North Dakota, for the U.S. Department of Energy, November 1980.

Introduction and Summary

The utilization of all forms of coal takes place only as a result of several processing steps. First, the raw material must be extracted from the earth, which is almost always done by strip mining for low-rank coals. The next step may be either transportation to its point of use, or beneficiation (cleaning) at the mine mouth followed by transportation. Coal beneficiation may be oriented at separating mine dilution (essentially dirt, rocks, etc.) from the coal, or removing sulfur, moisture or mineral matter from the coal itself.

At this point, the coal is ready for use. The utilization method currently accounting for the largest use of coal is direct combustion in pulverized coal and stoker type furnaces. Fluidized bed combustion is an emerging technology which appears to offer a number of advantages over the conventional coal burning techniques, while still being considered a type of direct combustion. The combined result of rising crude oil prices and research in coal gasification, liquefaction, and pyrolysis indicate that these technologies will become increasingly important as means for supplying the country's energy needs. In reviewing America's coal resources as a source of supply for this synthetic fuels industry, it is clear that low-rank coals can and should play an important part in filling this future need.

Other potential uses for low-rank coals such as magnetohydrodynamics, and production of carbon products (carbon black, carbon fibers, adsorbents, electrodes, etc.) are interesting possibilities but are not sufficiently close to commercialization or do not represent a large enough market to warrant detailed consideration in this report.

During and following utilization of the coal, several forms of environmental control are generally necessary, although thorough coal cleaning before use could potentially eliminate most of these requirements. If the coal is burned, some form of sulfur control is usually required which may or may not be integral with particulate control devices. Solid wastes, including ash and sulfur scrubber sludge, may also require treatment before disposal.

Each step from extraction through utilization and environmental control represents a potential bottleneck to increased use of low-rank coals. Investigation and research represent one way of understanding the issues and resolving the problems which limit low-rank coal utilization.

The key issues for each technology step are summarized below. Complete discussions of these key issues are presented in the introduction to each sub-chapter for the technology of interest.

Extraction

As the first link in a chain of coal utilization steps, coal mining has the potential to determine the overall rate at which coal is used, regardless of the application. The current market for low-rank coals is not constrained by mine capacity; mines on the average are operating at approximately ten percent below maximum output. Long term forecasts of coal demand indicate that the current sluggish rate of new mine openings may limit coal supply capability in the next several decades.

Aside from these economic and political problems, the mining industry faces problems of a technological nature. However, these technology related issues are considered to be small in comparison to those faced by utilization systems (combustion, gasification, liquefaction and pyrolysis). Instead of requiring major efforts in basic and process-related research, extraction technology issues will most likely be resolved through applied research and engineering design efforts addressing the specific problems of low-rank coal extraction.

Low-rank coals occur in a wide variety of deposits, many of which can be surface mined. Stringent reclamation requirements are now in effect for surface-mined land. These require the development and proving of techniques which precede, coincide with, and follow the mining operation to restore the land surface to desired long term conditions. In addition, better techniques are needed for mining multiple thin seams, thick seams, and seams with deep overburden. Current practice in these cases does not produce an optimum mining operation in terms of cost, environmental impact, and land reclamation.

Surface mining operations also have room for improvement in the techniques currently used to specify equipment based upon core samples and other data. The result will be a better match between equipment and task, resulting in a more efficient and cost effective operation. Further cost reductions are expected from operations research and systems engineering studies of mine operations.

Underground coal gasification (UCG) is being developed to exploit deep coal seams for which conventional surface or underground mining is infeasible. For technical reasons, UCG appears to be most applicable to thick continuous coal seams of high gaseous permeability and reactivity. Such formations are abundant in western low-rank coal deposits. Several key technical issues are already being addressed, including: limitation of aquifer disruption and groundwater contamination; limitation of subsidence and gas leakage; reliable, cost effective techniques for linking feed and product boreholes; and techniques for improved coal seam characterization and process monitoring.

Groundwater control and mine dewatering are other areas of concern in western coal extraction. High concentrations of alkali and alkaline earth salts (sodium, calcium, and magnesium, in particular) create a problem of alkaline mine drainage.

As is true for surface mining, the properties of low-rank coals do not greatly influence the choice of underground mining technology. (This is determined more by geology, manner of deposition and groundwater characteristics.) This technique is not expected to substantially contribute to the extraction of low-rank coals within the next twenty years due to significantly poorer worker productivity (less than one-third that of surface mining). The degree to which underground mining will contribute to future production will depend on the resolution of issues relating to thick seam mining, mining under unconsolidated overburden, and mine dewatering.

Transportation

Coal transportation is another step to coal utilization in which delays or undercapacity can potentially impact all users regardless of technology employed. Western coal is currently transported via rail, waterway, and on-site truck and conveyor systems. The absence of a waterway system in the West requires that low-rank coals first be transhipped (by rail). One coal slurry pipeline is operating and numerous other coal slurry pipelines have been proposed for subbituminous coal, but permitting delays have postponed construction starts.

Because there is very little experience with coal slurry pipelines, several technical issues may need to be resolved if this technology is to become a major mode of low-rank coal transportation. Problems of concern include: the separation, treatment, and proper disposal of slurry water from the finely divided coal at the receiving end of a pipeline; reduction of the total water consumption in slurry pipelines located in arid regions of the west; and various questions relating to the reliability of pipeline flow.

In general, no major technical hurdles stand in the way of increased low-rank coal transportation. However, recent steep increases in rail rates illustrate the need for improvements in transportation economics. Through the application of technology, two basic approaches to this problem can be taken. First, bulk handling systems and/or specialized transport systems can be applied to obtain economies of scale or economies of optimum design. This approach is exemplified by slurry pipelines and the unit train concept. The second approach to improved transport economics is to upgrade the quality of the coal prior to transport. An example of this approach is minemouth processing to reduce moisture content and thereby increase heating value. Technology of this nature is addressed in the section on coal preparation, handling, and storage.

Coal Preparation, Handling, and Storage

This category includes a wide range of coal treatment methods applied after mining and before utilization to achieve a variety of physical and/or chemical changes in the coal.

Coal preparation includes comminution (all size reduction techniques), mineral matter control (for removal of sulfur, ash, mine dilution, or specific mineral components such as sodium), moisture reduction (for slurry dewatering, and removal of both surface and bound moisture). Briquetting and pelletizing may also be included under preparation, since these processes produce a fuel product with superior handling and storage characteristics. Blending is a preparation and handling technique which consists of combining coals of different physical and chemical properties. The problems faced in the storage of mined coal over long periods of time include strategies for inventory control, techniques to control windage loss and freezing, and methods for preventing spontaneous combustion and for retaining desirable coal properties (such as reactivity).

In contrast to the majority of bituminous coal mined in the United States, low-rank coals are generally subjected to little or no preparation. The low extraneous mineral matter content typical of low-rank coal is one reason for this; low inherent sulfur and low heating value are also reasons. The net effect of these characteristics is to yield a coal which does not respond in a cost-effective manner to the beneficiation processes which have been developed for bituminous coals.

The economic attractiveness of using these processes on low-rank coals, and for developing new techniques which are specifically designed for low-rank coals, will increase with rising energy costs and tightening environmental regulations. Some of the potential opportunities for the application of beneficiation technology not previously applied to low-rank coals include: 1) reduction of moisture content to obtain improved transportation and utilization economics; 2) reduction of sodium content in high fouling low-rank coals by ion exchange to reduce boiler operating and maintenance costs; and 3) selective or general reduction of mineral matter content by physical or chemical means, either for unusually contaminated coals or in preparation for certain conversion processes. In anticipation of these changes, research efforts should focus on moisture reduction techniques for low-rank coals and slurries, applications of ion exchange (and other chemical cleaning processes) to low-rank coals (especially coal fines), and the use of gravity separation techniques on low-rank coals. Development of these technologies also implies a concern for proper waste disposal techniques in each circumstance.

Briquetting and pelletizing techniques are potentially valuable in expanding the utilization of low-rank coals because of their superior handling, storage, and utilization characteristics. A better understanding of briquette or pellet formation and subsequent physical and chemical changes will aid in further enhancing the attractiveness of this technology. Briquetting and pelletizing may be one attractive method for dealing with fines generated during handling and comminution, and for reducing the handling problems associated with dried lignite.

Utilization Technologies

The end objective of extraction, transportation and preparation efforts is the utilization of the coal in one of several modes. Conventional direct combustion accounts for the overwhelming use of current coal production, although fluidized bed combustion, gasification and liquefaction will likely become important consumers of low-rank coal in future decades. Low-rank coal properties will uniquely affect the design and application of these technologies to western coals.

Conventional Direct Combustion

Conventional combustion includes coal combustion in cyclone, stoker, and pulverized coal type furnaces. Pulverized coal furnaces are in widest use among utilities. Cyclone furnaces were introduced because of their ability to burn coals with low ash fusion temperatures and to recover a high percentage of the coal ash as bottom slag. However, it was found that at the high temperatures required to maintain slag flow, a significant amount of the sodium present in the ash was volatilized and contributed to ash fouling problems. In addition, high operating temperatures favor the formation of nitrogen oxides. For these reasons, cyclone burners have been all but abandoned with the exception of several units which are still operating on North Dakota lignite. Stoker firing is generally limited to smaller applications of less than 100,000 lb/hr steam.

The special properties of low-rank coals influence virtually all aspects of direct combustion. Of primary importance are high moisture levels (low heating values) and properties of low-rank coal mineral matter.

Low heating values increase the tonnage of coal required for a given power output in comparison to higher rank coals. This imposes a requirement for greater capacities in coal transport equipment, handling equipment, pulverizers, stack gas cleaning devices and disposal efforts. Retrofit of bituminous coal boilers to burn low-rank coals also means a significant derating in output.

High sodium levels found in the mineral matter of some low-rank coals (particularly North Dakota lignites) promote slag formation by lowering ash fusion temperatures, and aggravate boiler tube fouling. However, when high calcium and/or magnesium levels also occur, these effects of sodium are somewhat negated. Continued research into the mechanisms of ash fouling and slagging, and into the control of these problems through the use of additives or other means, would be worthwhile.

The high reactivity and nonagglomerating characteristics of low-rank coals allow for a larger particle size during combustion while still assuring complete burnout. In addition, these properties provide an opportunity for development of direct coal ignition systems for low-rank coal utility boilers, which would replace the oil-ignition systems currently utilized.

Combustion Environmental Control Technology (ECT)

Federal and state standards for both plant emissions and ambient concentrations of air and water pollutants have become major determining factors in the design and siting of coal-fired power plants. Environmental control requirements for plants constructed in the 1980's will account for more than half of the plant costs. Low-rank coals have several unique physical and chemical properties that affect the selection, design, and operation of appropriate control systems. The tendency in the utility industry to date has been to add control systems to power plants in series without much regard for their interactions. Thus, one potentially fruitful area of investigation is to take a systems approach to the overall problem of coal selection and preparation, boiler design, control device design, and disposal or utilization of wastes or by-products. Proper integration of the solutions to the many interrelated problems involved in generating electric power from low-rank coal while meeting all environmental requirements could save the industry billions of dollars over the next several decades.

Low sulfur levels are an important advantage of many western low-rank coals, reducing the extent of flue gas desulfurization necessary to meet environmental standards. Some of these coals offer the possibility of using their alkaline ash as a reagent for flue gas desulfurization, thus eliminating the need for limestone or other sorbents. Spray dryer and dry sorbent SO_2 scrubbing systems have high potential applicability to low-rank coals because of their chemistry. Improvement of existing ash-alkali wet scrubbing technology would also be desirable.

The current trend in utilities toward baghouses instead of electrostatic precipitators reflects the difficulty of meeting stricter particulate emission standards. Low-rank coal fly ash often has a high electrical resistivity (although the effect is not as severe with high sodium coals), an undesirable characteristic when operating electrostatic precipitators. A number of important research topics should be pursued in the area of improved particulate control methods for low-rank coals. The study of fine particulate control technology should be given special emphasis because regulations for respirable particles are still evolving, because the chemistry of low-rank coal fine particulate matter may be unique, and because sampling, analytical, and control methods for very fine particles are not well developed.

Three other key ECT areas exist for low-rank coal combustion in which regulations, measurement techniques, relevant data, and control technology are just beginning to evolve. These are: 1) the reduction of nitrogen oxide emissions; 2) disposal and/or utilization of solid wastes and sludges from power plants; and 3) control of emissions of trace elements and organic compounds in the flue gas from low-rank coal combustors.

Fluidized Bed Combustion

Atmospheric fluidized bed combustion (AFBC) is a potentially attractive alternative to the conventional pulverized coal boiler with its tail-end scrubbing systems. In general, fluidized bed coal combustion offers lower NO_x emissions, in situ sulfur recovery, increased combustion efficiency, reduced tube fouling, lower excess air requirements, smaller combustor size, improved fuel versatility, modular construction and easily handled by-product material. These advantages over conventional combustion techniques are due primarily to the intense turbulence in the bed, the relatively low and uniform combustion temperature, and the comparatively long solid residence times in the bed without a long linear flow path requirement. The use of low-rank coals in fluid bed combustors also offers the potential for sulfur capture without added sorbents. In addition, an FBC can tolerate a wide variation in coal properties as is typical of low-rank coals.

Several problems exist with regard to low-rank coal use in fluid bed combustors. Most are related to unique western coal composition and properties, but others arise out of the tendency for waste materials to be leached with groundwater. For example, the very characteristics of western coals which permits substantial self-control of sulfur, high alkali content, also contributes to waste disposal problems, high temperature corrosion, and gas stream components which must be removed to a very high degree to prevent turbine blade damage in PFBC applications. Another problem presented by low-rank coals in some cases is that limestone addition is required to prevent bed agglomeration even when it is not needed for sulfur control.

Gasification

Low-rank coals are suitable feedstocks for all three basic gasifier types: fixed bed, fluidized bed and entrained flow reactors. In general, low-rank coals are superior gasification feedstocks which display non-caking characteristics, high reactivity, and lower sulfur contents in comparison to bituminous coals. Their natural variation in ash composition also creates a different and highly variable viscosity-temperature relationship and behavior toward refractory materials which must be accounted for in gasifier design and specification of operating parameters. Differences in the organic structure of the coal also dictate different wastewater treating requirements than would be indicated for bituminous coals. The higher moisture contents and fines fractions often found in low-rank coals indicate fixed bed applications less strongly because of the large volumes of wastewater produced and the difficulty in feeding fines to these gasifiers. However, other advantages of fixed bed devices, especially when operated in the slagging mode, may outweigh these drawbacks.

Liquefaction

Coal liquefaction includes the manufacture of liquid products by two basic process types: direct and indirect liquefaction. Direct liquefaction processes react coal with hydrogen or synthesis gas in a process derived liquid solvent medium. Suitably designed catalysts can be used to influence rate and selectivity in both primary liquefaction and secondary treatment for product upgrading.

Indirect liquefaction is in actuality a gasification process as far as the coal is concerned. Coal is first gasified (with oxygen) to produce a synthesis gas which is then catalytically reformed to produce a spectrum of liquid fuels and oxygenated chemical products.

At present, the unique properties of low-rank coals are not sufficiently well-defined to allow optimal application of direct liquefaction processes developed for bituminous coals. It is known for example that low-rank coals (particularly lignite) react very readily with carbon monoxide, but the conditions under which this characteristic is best utilized are not currently known. Similarly, high moisture and oxygen contents and low sulfur contents affect optimal process conditions (higher pressures due to excess water vapor and CO_2), liquefaction chemistry and product distribution, but the most cost-effective way of dealing with the variables is ill-defined.

The alkali and alkaline earth components of low-rank coal mineral matter may catalyze liquefaction reactions, a fact of particular importance in processes which do not employ added catalysts. On the other hand, ash agglomerates which cause operational problems have been discovered in liquefaction reactors by several researchers.

Low-rank coals have been observed to produce heavy bottoms products of very high viscosity, creating operational difficulties and limiting the scope of use for this product. Several feasible solutions to the problem exist, but additional experimentation and operating experience are needed.

Low-rank coals show promise as liquefaction feedstocks because of their low cost, high reactivity, natural catalytic activity and rapid reaction with CO. Research is required to determine the best possible ways of exploiting these advantages, while dealing with possibly greater requirements for reducing gas and higher process pressures. Regardless of the final determination of optimal process conditions for a given coal, it is clear that liquefaction processes designed for bituminous coals will not function optimally when processing low-rank coals. Therefore, the potential for low-rank coals as liquefaction feedstocks should be assessed only under circumstances which fully demonstrate their potential.

Pyrolysis

Pyrolysis is defined as the transformation of a substance into one or more substances by the application of heat alone. When considering low-rank coal pyrolysis, the primary issues are product yields and quality. Due to the large quantities of inherent moisture and oxygen within low-rank coals, a considerable quantity of gas is produced by pyrolysis, and comparatively lesser quantities of valuable liquids and char. However, studies on lignite pyrolysis have shown an upgrading in heating value from the raw coal to the solid product, yielding a char with nearly the same Btu content as bituminous chars. Furthermore, the chars from low-rank coals are more reactive (in some cases pyrophoric) and may therefore be uniquely suited to certain applications. The low sulfur advantage held by many low-rank coals is also retained in the char.

As is the case in liquefaction, the pyrolysis of low-rank coals requires different process conditions to achieve optimal results than those used for higher rank coals.

Because of the importance of the trade-off between low raw material cost and low pyrolysis yields, an engineering/economic study of this issue should be made before undertaking any significant research programs in low-rank coal pyrolysis.

Summary

It is clear that low-rank coals have tremendous potential for filling future energy needs in the United States. Their unique properties indicate that they are preferred feedstocks in some applications, and require that process designs developed for bituminous coals be re-evaluated to achieve optimal results from low-rank coals. A vigorous program of research and development will be required to exploit the full potential of this valuable national resource.

Extraction

3.2.1 Introduction and Summary

The total identified resource of U.S. low-rank coal is over 1 trillion tons. Approximately 10 percent of this resource is classified as the strippable reserve base. This is the portion of the resource that is of greatest immediate interest, because virtually all low-rank coal extraction is currently accomplished by surface mining.[a] As shown in Table 3.2.1.1, the 1977 production of low-rank coals was 111 million tons, or roughly 0.1 percent of the strippable reserve base. In the future, the technologies of underground mining and underground coal gasification (UCG) will be applied to the deeper low-rank coal resources if market conditions permit. Therefore, this study addresses all three of these coal extraction techniques.

Surface mining technology is not greatly affected by the characteristics of the coal itself, although the intrinsically lower heating value of low-rank coal compared to bituminous coal restricts extraction operations to those with very low cost per ton[b] (typified by very large, mechanized, efficient strip mines with low ratios of overburden-to-coal). Much more important to surface mine design are geographic factors such as climate, rainfall, terrain, reclamation potential, and surface water characteristics; and geologic factors such as types of overburden, coal seam depth, dip, thickness, and continuity, and groundwater systems. These latter (geologic) factors are also the most important determinants affecting the underground extraction techniques. In addition, underground mining is affected by some physical properties of the coal and overburden, primarily those which determine the competence of the mine roof. Underground coal gasification technology is greatly affected by the geologic and hydrologic factors, and by essentially all of the physical/chemical properties of the coal and overburden (such as permeability, reactivity, swelling or shrinking behavior, etc.). UCG appears to be most applicable to deep, thick, continuous low-rank coal seams of high permeability and reactivity which abound in the western United States.

A number of key technical issues relating to the improvement of these low-rank coal extraction technologies are currently of importance.

---

[a]This will apparently continue to be the case for at least the next decade. For example, of the "Planned New Coal Mine Development and Expansion 1979-1988" in Montana, North Dakota, New Mexico, Texas, and Wyoming, 98.1 percent of the expected coal production will be surface mined.[2]

[b]Another important factor constraining the competitive cost of low-rank coal extraction is the large distance that often separates the resource from its markets, which results in significant contributions to the price of delivered coal by transportation costs. (This contribution is magnified by the low heating value of the fuel.)

Table 3.2.1.1

Total U.S. Low-Rank Coal Resources, Strippable Reserve Base, and Production Estimates

(billion short tons)

	Identified Resources[a]	Strippable Reserve Base[a]	1977 Production[b]
Lignite	543.8	40.8	0.028
Subbituminous	546.1	67.9	0.083
Total	1,089.9	108.7	0.111

[a]Source: *Low-Rank Coal Study—National Needs for Resource Development. Volume 2—Resource Characterization* [DOE/FC/10066-T1(V.2)].

[b]Source: Reference 1.

The surface mining issues relate to: (1) increasingly stringent surface mined land reclamation requirements; (2) the continuing pressure on mine operators to increase productivity and yields and decrease unit production costs; and (3) the extension of surface mining technology to more difficult geographic and geologic circumstances. The underground mining key issues recognize that the depositional characteristics of western low-rank coals differ from the underground-mined eastern bituminous coals, and that mere extension of the existing technology to the low-rank coal resources will not be adequate in many cases. The technology development needs of underground coal gasification are more basic, since large-scale, long-term technical feasibility of a UCG process has yet to be demonstrated outside of the Soviet Union. The key issues which have been identified are listed below and discussed in the paragraphs that follow:

1. Surface Mining:

    a. Techniques for revegetation and soil rejuventation

    b. Techniques for multiple thin seams, thick seams, and deeper overburden

    c. Techniques for optimizing equipment specifications based on laboratory core analysis

    d. Cost reduction through Operations Research and Systems Engineering

    e. Dewatering of mine area and groundwater control

2. Underground Mining:

    a. Techniques for mining thick seams

    b. Techniques for mining under unconsolidated overburden

    c. Dewatering of mine area and groundwater control

3. Underground Coal Gasification:

   a. Control of aquifer disruption and groundwater contamination

   b. Control of subsidence and gas leakage

   c. Reliable, cost-effective linking techniques

   d. Techniques for improved coal seam characterization and process monitoring

1. Surface Mining

1a. Techniques for Revegetation and Soil Rejuvenation

The Surface Mining Control and Reclamation Act of 1977 requires reclamation of all surface mined lands and lands disturbed at the surface by underground mining, to a usefulness as high or higher than previously. Prior to obtaining an operating permit, reclamation plans must be submitted which include stockpiling of topsoil, regrading, isolation of toxic spoils or horizons, water management, spoil and disturbed land stabilization, and revegetation.

Reclamation is most effective when planned in advance of and implemented in conjunction with mine development. For example, regrading practices can affect the success of revegetation. Burial of toxic spoils and recovering of the surface with topsoil is critical to provision of the necessary supply of organic material for plant growth. Small terraces and slight depressions reduce erosion and increase infiltration, which facilitates plant growth. Mulches such as straw and wood chips or chemical binders may also be used to control erosion until the vegetation is established. Additives like lime, fertilizer, treated fly ash, sewage sludge, or compost may be used to condition acidic or nutrient-poor soils. A vegetative cover should be established as soon as possible after grading to minimize erosional losses.

Revegetation is difficult in the arid areas of the west where only half of a nine- or ten-inch rainfall may be available for plant utilization. The planting of seedlings, sod, and transplants has been more successful than has seeding in these areas. An estimated 5 years are required to return a Wyoming strip mine to productive cropland, and ten years are required to return it to productive range land.[7] Substantial amounts of water could be required for irrigation in some areas, causing potential water usage conflicts.

Reclamation of lands under which western low-rank coals are deposited is amenable to solution, at a cost. The problem is to reduce the cost. To some degree, better equipment to reduce spoils handling time and help contour the surface more efficiently would reduce costs. Specialized genetically manipulated and bred seedlings and new planting techniques would also reduce costs by accelerating revegetation. New data being generated from research on revegetation in these regions will be very helpful. However, for the immediate future, aggregation and dissemination of already existing data and studies in revegetation, planting, soil treatment, plant strains, etc., is needed.

1b. Techniques for Multiple Thin Seams, Thick Seams, and Deeper Overburden

Thick seams, multiple thin seams (both dipping and horizontal), and deeper overburden present significant problems which require new equipment, more complex development sequences, and probably new technologies in order to be more cost competitive with the more "conventional" strippable low-rank coal resources. All of these characteristics are found to some degree in major western low-rank coal resources and necessitate one common feature in the successful extraction of the coal - some form of extended pit. In conventional area mining, the rate of overburden removal and extraction of the exposed coal seam occur at comparable rates so that the "pit" nature of the mine can be minimized and rapid and concurrent reclamation can be accomplished. This concurrent exposing of seams and covering of mined-out areas is not easy for the above-mentioned seam characteristics using conventional equipment.

Limitations on drag line boom length and angle of soil repose determine the maximum depth that can be mined by simple overcasting. Extended pit overburden removal requires extensive spoil rehandling and/or the use of frontloaders, shovels, strippers and hauling trucks in various combinations. However, the use of bridge conveyors capable of cross-pit handling of spoils has been developed by Europeans and is currently employed in Southwest Africa. The conveyor bridge eliminates the need for spoils rehandling. The Department of Energy is currently planning to sponsor design and construction of such a conveyor system.[4]

1c. Techniques for Optimizing Equipment Specifications Based on Laboratory Core Analysis

Surface mines utilize many different types of equipment in essentially every conceivable combination. This large earth-moving machinery is so expensive, and its efficient operation is so important to the cost of the mined coal, that proper equipment design and selection is of paramount importance. This involves selecting the proper types and combinations of equipment, as well as optimizing the design features (size, capacity, cutting edges, etc.) of each piece of equipment, to fit the local conditions. Mining and equipment engineers are developing and using increasingly sophisticated computer programs to aid in this process.

During the premining planning phase, a three-dimensional "picture" of the resources is obtained through the use of mapping, coring, laboratory analysis, and geophysical logging techniques, combined with computerized interpretation of the data. Factors such as seam quality, thickness, depth, continuity, and characteristics of overburden and partings are determined in detail.

Among other uses, these data serve as input to computer models which can simulate the operating characteristics of most commercially available surface mining equipment. Given the overburden characteristics, etc., the models can compare alternatives and solve for optimal mine and equipment designs and costs.

These models can be extended in two basic directions: (1) equipment design, and (2) overall mine optimization or more innovative mine development schemes. An example of the first type would be the selection of optimum boom length, bucket capacity, etc., for a drag line, given extensive data on the coal seam and overburden characteristics.

An example of the second type of optimization problem might involve the selection of the best combination of equipment types for the extended bench method of stripping overburden. One option would be to use two drag lines, with the first one removing a portion of the overburden. The second drag line would operate on a bench created by leveling the spoil from the first machine, removing the remaining overburden and rehandling a portion of the initial spoil. However, the first drag line is not being employed very economically in this scheme, and a truck-shovel combination in place of this drag line might be more efficient. In this case the stripper shovel and trucks working ahead of the drag line would create the bench and pave the way over the rolling terrain.

1d. Cost Reduction Through Operations Research and Systems Engineering

Operations Research involves the application of numerical methodologies (usually via computer) to simulate a process or operation. The use of computer models to help design equipment and choose the optimum equipment mix to develop a mine (as described in the preceding subsection) is an example of operations research. These methodologies can also be used to develop operating schedules and strategies that optimize management's "objective function" in response to changing conditions such as prices, regulations, labor costs, etc. Examples of different management objectives include: (1) maximum internal rate of return; (2) minimum labor costs; or (3) minimum downtime. Depending on the choice of objective function, different mining schedules or strategies might be appropriate under a given set of conditions.

These types of computer models can be used in conjunction with industrial engineering methods (such as time and motion studies) to improve operating efficiencies as well. Analysis of observed mining operations and comparison with idealized "theoretical" operations can locate sources of variance which might be amenable to improvement. This type of analysis can be valuable in any phase of the operation, from equipment assembly through revegetation. For example: (1) The time required to assemble drag lines is considerable (months); the computer model can investigate the use of modular subassemblies which are bolted together in the field rather than welded, cutting assembly time to one-tenth the former time. (2) In deeper surface mines, it has been found that actual overburden removal rates are less than predicted by bucket capacity of the drag lines due to the need for rehandling. In these situations, drag lines with longer booms and smaller buckets can be more productive due to the absence of rehandling. (3) To aid operators of drag lines during operation, as well as to obtain performance data, small onboard computers are starting to be employed. Their potential future impact on productivity is expected to be very great (in one study as much as 9 percent).[3]

As these examples indicate, the combination of computer-aided design and computer-aided operations (being introduced in many other industries as well) will make it progressively less expensive for coal mining companies to practice ongoing innovation. When properly integrated, the digital output from the "design" programs can serve as direct input to the "operations" programs, streamlining the process of improving mine productivity.

1e. Dewatering of the Mine Area and Groundwater Control

According to the Surface Mining Control and Reclamation Act of 1977, hydrologic effects of surface and underground mining are to be controlled. Features of the hydrologic system to be protected include depth to groundwater, location of surface water drainage channels, flow regimes and groundwater recharge capacity.

Acid mine drainage, which is commonly associated with surface mining of eastern bituminous coal, is not a major problem in most areas of the west because of low precipitation rates, low sulfur content of the coal, and remoteness of streams. However, there are exceptions to these regional generalizations; isolated cases of acid mine drainage have been reported in Colorado and Montana.

More commonly, alkaline drainage is a problem in the western coal fields. There, overburden and coal deposits are characterized by high concentrations of sodium, calcium, magnesium, CO_3, HCO_3, SO_4 (from overburden sulfate rather than from pyrite oxidation as in acid drainage), and chlorine. The dilution potential of many western streams is marginal, because of ephemeral or reduced summer and winter flows. Associated with alkaline mine drainage are elevated levels of dissolved solids and such constituents as sulfate.

The principal sources of water in the mine area are surface water runoff, rain-related runoff, and dewatering of shallow groundwater aquifers. The first factor can be minimized by diverting the surface water source. The second problem is variable, but often small because much of the western low-rank coal is in semi-arid regions where rainfall is very light. Dewatering or disrupting groundwater aquifers can be a very significant problem, especially if the coal seam itself is an aquifer (which is quite common in the west). Coal-bed groundwater, even if not of drinking-water quality, is frequently used for livestock watering. In the northern Great Plains, groundwater is used heavily for both domestic and livestock water supplies. Subirrigation of alluvial valley floors is vital to agriculture in some western states, and the disturbance of groundwater systems in these areas is of particular concern.

Generally when mine area water-related problems occur, a sump is formed in the mine floor, and the water is pumped to the surface and treated according to the appropriate Federal effluent guidelines. One possible area for research is to study the feasibility of sealing aquifer zones in coal strip-mining areas to prevent or reduce discharge of groundwater to excavations and halt dewatering of aquifers.

Post-mining restoration of coal-seam aquifers is a difficult problem which (like rehabilitation of the surface) must be addressed in the mine plan. Normal procedures involve deposition of plant-toxic material and permeable material at the bottom of the pit (where the coal was). If the overburden contains clays or shales, these could serve to dam the aquifers when deposited in the pit, and subsequently cause rises in the water tables up-gradient of the mined areas and decreases in water tables down-gradient of the mined areas. It might be possible to segregate and deposit material of similar permeability to the original coal, and thus construct artificial aquifers in places where coal seams served this purpose before mining. However, the use of such methods to patch the gaps in coal-seam aquifers is not well developed at the present time.[5,6]

2. Underground Mining

2a. Techniques for Mining Thick Seams

Conventional underground mining technology has a practical upper limit to the thickness of seam which can be mined - approximately 7 feet. Western low-rank coal seams are commonly greater than 12 feet thick. In order to efficiently extract coal from thick western seams, new or modified technologies will be required.

Three methods are currently receiving attention as well as Department of Energy funding: (1) caving by pillar extraction, (2) multislice longwall mining, and (3) longwall caving. Caving by pillar extraction involves the controlled caving of the coal seam roof, by undercutting the seam, using room and pillar techniques. Longwall caving employs the controlled collapse of the coal seam roof, which is undermined by longwall techniques. In multislice longwall mining the coal seam is extracted from the top down in a series of longwall slices or passes, using conventional longwall equipment. If underground mining of western low-rank coals is to be expanded, these or similar techniques will have to be refined, and the conditions of their optimum use and selection determined.

2b. Techniques for Mining Under Unconsolidated Overburden

Roof support in underground mining operations is without doubt the most important single concern for mine management.[8] The presence of unconsolidated overburden presents significant problems to underground mining. In all conventional underground mining techniques, the roof must be supported before the face can be advanced; some mine roofs are self-supporting while others are maintained with the help of artificial supports - unconsolidated overburden would necessitate the latter.

The central requirement in determining support requirements is the geological section related to the immediate roof strata (about 10-15 feet). From the geological section, the problem is to determine the anchor horizon for the rock bolt and the density of roof supports. The closest competent bed determines the anchor horizon, and, consequently, the length of bolt required. The search for the competent bed depends on the height of the roadway and the length of bolt that can be installed. Where a competent bed is absent, provisions must be made for additional conventional support.

Thus, either the coal seam itself acts as the anchor horizon (implying thick seams of structurally sound coal) or additional supports are required for unconsolidated overburden. The former case is wasteful of coal, and the latter is time-consuming and expensive. New techniques would probably be required to mine these deep seams - and/or some new roof-supporting systems developed.

2c. Dewatering of Mine Area and Groundwater Control

Coal mine drainage is practically nonexistent at some mines and causes severe problems at others. Mine drainage can be either from surface runoff or infusion from underground aquifers. Precautions such as avoiding the siting of shafts and boreholes in low spots on the surface will prevent surface runoff water from entering the mine. Water influx from aquifers is harder to avoid, particularly when the coal seam itself is a major aquifer.

Water can be removed and/or controlled by diversion tunnels, gravity, or pumps. Use of water diversion tunnels appears attractive from a conceptual standpoint, but invariably proves economically infeasible.[9] Water removal from existing underground mines is almost always achieved through the use of pumps and gravity. The degree of contamination of the water depends on the duration of its exposure to the coal. As a result, plastic pipes (for runoff) and rapid removal of the water from sumps or other traps is the current industrial practice. Upon abandonment, mines are "sealed," preventing further air and/or water infusion.

In general, the control of water during mining is a centuries-old problem; the technologies have been "fine-tuned" through experience so that little basic research would be necessary. The treatment of mine drainage effluents is a problem, but one of cost reduction. There does appear to be some research needed in the area of mine sealing upon abandonment, but more in the application and perfecting of existing methodologies.

3. Underground Coal Gasification

3a. Control of Aquifer Disruption and Groundwater Contamination

Aquifer disruption problems anticipated in a large UCG plant are similar to those associated with underground mining. Removal of the coal and subsidence of overlying strata will disturb any groundwater systems in those rock strata. In general, the thicker the coal seam removed, the more movement likely to be experienced by the aquifer. Because there is little that can be done to prevent or control the natural fracturing and settling process, it will be important to obtain good hydrologic data throughout the area to be affected by a proposed UCG plant, and to predict the impacts of the operation on the flow of groundwater. If these consequences are not acceptable due to existing or planned uses of the water, another site will have to be selected.

In addition to disruption of groundwater flows, UCG plants could create unique water contamination problems. Gasification residues and byproducts could dissolve in the groundwater flowing through a previously gasified zone. Both soluble organic compounds (e.g., phenols) and inorganic compounds (e.g., sulfates) would be involved. Limited amounts of field data on the extent of this potential problem have been obtained to date. There is some evidence from both the field and the laboratory to indicate that as the contaminated water flows through surrounding unreacted coal, the organic species are absorbed by the coal within a relatively short distance and time.[9] The fate of the inorganic constituents is less clear.[10]

More data and better quantitative models are needed to reliably predict the impacts of UCG operations on groundwater systems. This requires: (1) a conscientious effort to monitor hydrological conditions and groundwater quality before, during, and after UCG field tests; (2) incorporation of the data into predictive models; and (3) supporting laboratory research.

3b. Control of Subsidence and Gas Leakage

The design of a large-scale underground gasification system must take subsidence into account; the considerations are similar to the development of an underground mine. The layout of injection and production wells can either leave coal pillars between parallel gasification channels to support the overburden and isolate each gasifier from the other (analogous to room-and-pillar mining, with similarly low resource recovery); or allow a number of interconnected gasifiers to advance in a broad front with subsidence occurring immediately behind (analogous to longwall mining). The Soviets appear to have worked with both systems. Operations in this country have not reached large enough scale to test either type of operation yet.

In the case of steeply dipping beds (SDB), very severe but localized subsidence may occur due to the large vertical component of the coal being removed. The Soviets observed some massive subsidence in their SDB plants, but this did not disrupt the operations when the air injection was through the underburden.[11] However, as might be expected when large ground movements occur, some gas leakage was observed through cracks to the surface.

The problems with subsidence may be classified as (1) operational, and (2) environmental. Operational problems would include damage to borehole casings, wellheads, and piping systems within the subsiding zone. Gas leakage would reduce the thermal efficiency of the process and thus cause lost revenues. Environmental problems would include surface depressions, and (in some areas depending on the nature of the overburden) unpredictable subsidence events over a period of years following the gasification operation. Leakage of syngas would have a minor effect on local air quality. It might also present a safety hazard, if ignition of the hot, hydrogen-containing gas occurred in an area containing plant personnel. To resolve this issue it will be important to collect field data, develop operating experience, and improve predictive models of subsidence in various geological settings. The usefulness of field data from small-scale field tests is limited, however. Subsidence is a site-specific phenomenon, as well as being a function of plant scale.

3c. Reliable, Cost-Effective Linking Techniques

A critical part of UCG technology is the creation of permeable pathways, or links, in the coal seam between the injection and production boreholes. These links can be created by reverse combustion, directional drilling, and other techniques. To ensure an efficient gasification process, it is important that these links be formed reliably and in the bottom portion of the coal seam. Reverse combustion, the cheapest and most direct linking method, is unfortunately sometimes unreliable. Multiple linkage paths have been observed, as well as paths at the top of the coal seam.[12] This latter problem leads to an inefficient operation in which the gasification zone "overrides" the coal seam, recovering only a portion of the coal and producing low-quality gas. Directional drilling is a more positive and reliable technique, but it is also more expensive. In addition, a successfully completed directionally drilled well did not result in efficient gasification in a recent test at Hoe Creek, Wyoming.

Further field development efforts are required before linking techniques can be considered sufficiently reliable for commercial UCG operations. In addition to improving the two techniques mentioned, researchers are working on other ideas such as water jet drilling and explosive shaped charges.

3d. Techniques for Improved Coal Seam Characterization and Process Monitoring

Because the underground coal seam is the gasification reactor, obtaining complete geologic and hydrologic information on the site is critical to the success of the subsequent operations. Initially, it is necessary to confirm that the coal seam meets basic criteria such as continuity. This can be done through a combination of remote and downhole surveying techniques (e.g., seismic surveying). Drilling, coring, and logging is performed to develop the necessary geophysical and hydrological data. In some cases, the conduct of air injection/production tests, and possibly reverse burn linking tests, would be considered part of the site characterization and evaluation phase as well.

One critical problem is to keep the cost of the site characterization effort within reason, while still obtaining enough detailed data to design the process appropriately for the specific site. This requires the development of improved characterization, as well as interpreting and modeling test data obtained from well-characterized sites.

To meet this objective, the use of extensive arrays of surface and downhole diagnostic instrumentation to observe the process during small-scale field tests is warranted. The most useful of these systems to date have been downhole thermocouple arrays, used in combination with a variety of other in situ and remote (seismic, electromagnetic) measurements. Post-test coring or excavation has also proved informative. From these efforts, a reproducible model of the process is beginning to emerge.[12]

In future commercial plants, cost considerations will dictate that downhole instrumentation be minimized. Some remote sensing techniques may be developed to the point where they can provide useful diagnostic information at low cost.

References - Section 3.2.1

1. Lin, K., Coal Data 1978, National Coal Association, Washington, D.C., 1980.

2. Coal Age, February 1980, p. 101.

3. Coal Age, January 1980, pp. 126-136.

4. Coal Age, December 1979, pp. 74-75.

5. Van Der Leeden, F., L. Cerrilo, and D. Miller, Groundwater Pollution Problems in the Northwestern United States, Environmental Protection Agency, EPA 660/3-75-018, May 1975.

6. Grim, E.C., and R.D. Hill, Environmental Protection in Surface Mining of Coal, EPA-670/2-74-093, October 1974.

7. Doyle, W.S., Strip Mining of Coal - Environmental Solutions, Noyes Data Corporation, Park Ridge, N.J., 1976.

8. Menchen, W.R., et al., Underground Coal Mining - An Assessment of Technology, EPRI AF-219, July 1976.

9. Mead, S.W., H.C. Ganow, and F.T. Wang, "Groundwater and Subsidence Investigations of the LLL In Situ Coal Gasification Experiments," in Proceedings of the 4th Underground Coal Conversion Symposium, Steamboat Springs, Colorado, July 17-20, 1978, SAND 78-0941, U.S. Department of Energy, printed June 1978.

10. Mattox, C.F., and M.J. Humenick, "Organic Groundwater Contaminants from UCG," in Proceedings of the 5th Underground Coal Conversion Symposium, Alexandria, Virginia, June 18-21, 1979, Conf. No: 790630, U.S. Department of Energy, printed May 1979.

11. Gregg, D.W., Ground Subsidence Resulting from Underground Gasification of Coal, Lawrence Livermore Laboratory, UCRL-52255, March, 1977.

12. Brandenburg, C.F., "Field Implementation of UCC Research," in Proceedings of the 5th Underground Coal Conversion Symposium, Alexandria, Virginia, June 18-21, 1979, Conf. No: 790630, U.S. Department of Energy, printed May 1979.

3.2.2 Surface Mining

3.2.2.1 Technology Description

Surface mining involves: (1) removal of the topsoil and overburden; (2) extraction of the coal seam; (3) replacement of the overburden and topsoil; and (4) reclamation of the land. Each of these steps, as well as some important preliminary operations, will be described briefly below.

Exploration and Evaluation of the "Prospect"[1,2]

The steps involved in a typical sequence for the exploration and evaluation of a coal resource in the western states are shown in Table 3.2.2.1. The objective is to obtain an accurate three-dimensional map of the prospective mine, reflecting seam depth, thickness, continuity and pitch (if possible), and the nature of the overburden, as well as potential surface and groundwater problems.

Table 3.2.2.1

Typical Search for a New Coal Deposit in Western U.S.A.

Regional Appraisal (Stage #1)	Detailed Reconnaissance (Stage #2)	Detailed Surface Investigation of Target Area (Stage #3)	Detailed Three-Dimensional Physical Sampling of Target Area (Stage #4)
O-Geologic compilation for "marketing" area*	F-Field check of sections containing coal seams*	F-Detailed mapping of outcrops	F-Drilling*-Logging*
	F-Reconnaissance drilling for stratigraphy and coal thickness	F-Detailed stream sediment and/or geochemical survey	L-Mineralogical, Chemical analyses and physical tests on samples, cores and cuttings*
	F-Chemical and calorific check of outcrops or drill samples (not badly burned)		F-Down-hole geophysical surveys
			L-Amenability tests on coal samples for intended use
			O-Reserves computations*
			O-Preliminary Valuation*
			F-Investigation of water problems and water availability for plants*
			F-Investigation of suitability of ground for plant, tailings, dump and town sites
			F-Test pit, shaft sinking or tunneling to obtain bulk samples

Legend: O = office study; F = Field investigation; L = Laboratory tests
* = activity or method which is indispensable
Source: Reference 1

From an operator's perspective, the area for improvement is in the obtaining of more and better data at less cost. The infusion of better coring and logging equipment and techniques from the oil industry has reduced costs and increased efficiency. Also, the application of coal geophysics and computers is making better interpretation and use of the data possible.

Included in the evaluation is an analysis of market needs and location, transportation availability and cost, and product quality versus utility specifications. Because of the huge investment in machinery involved in large western strip mines, the "market" ofter consists of a minemouth power plant or long-term contract(s) to ship the coal to a remote location.

Evaluation of Possible Environmental Impacts and Controls

Inherent in the process of determining if a prospect is "economic" is the evaluation of barriers and costs involved in environmental permitting and compliance. This has been especially true since the Strip Mining Act of 1977 was passed, in which operators' responsibilities and requirements were specified in considerable detail.

The prudent economic evaluation of a mine must include careful consideration of the potential costs of reclamation. Aspects such as climate, rainfall, topsoil, natural vegetation, current and projected land use, and terrain must all be determined and evaluated for their impacts on reclamation efforts.

Surface Mine Design

Following the preliminary economic and environmental evaluations that result in a decision to develop a mine, mining engineers and reclamation specialists work together with management to design a detailed mining and reclamation plan before mining is begun. Two basic surface techniques are most commonly used to extract western low-rank coal: (1) area mining, and (2) open pit mining.[a] Area mining involves removing the topsoil and overburden in strips (box-cuts) typically 1 mile long and 100 feet wide. In the simplest cases (shallow seams on flat terrain), the spoil is handled only once by simple overcasting, as illustrated in Figure 3.2.2.1. Extraction of deeper, thicker, dipping, or multiple seams requires use of the more complex open pit mining techniques,[b] in which larger strips (e.g., 1,000 feet wide) are excavated and overburden is shifted from benches within the pit to uncover the coal.

In essentially all low-rank coal surface mines, topsoil and overburden are kept separate. (Exceptions occur in Texas where soil and rainfall conditions allow revegetation to proceed without the need for separate handling of topsoil.) Blasting is used as required to loosen both the overburden and the coal, which is then loaded on trucks. Reclamation is accomplished by replacing and grading the overburden, replacing the topsoil, and revegetating.

Although the initial capital costs for equipment can be very high (so as to exclude many small operators), the cost to extract a ton of coal by surface mining is low, and the productivity per man-day (25-30 tons) is correspondingly high. The cost of reclamation can also be kept within

---

[a]Contour mining (and one other technique - auger mining) are used mainly in the Appalachian region to extract relatively thin coal seams from rolling to very steep terrain.

[b]Among the world's most impressive open pit operations are the brown-coal mines in the Rhineland area near Cologne and Aachen, West Germany. Seams 65-320 feet thick are being extracted to depths that will exceed 1,650 feet. Extraction and transportation systems for overburden and coal are highly sophisticated, as are the reclamation methods.[4,5]

reasonable limits if properly managed in the predevelopment stages. These are significant economies of scale as indicated by the trend to larger, faster equipment and the fact that 22 of the 25 largest coal mines in the United States (1978) are western low-rank coal surface mines.[6]

Figure 3.2.2.1

Area Strip Mining with Concurrent Reclamation

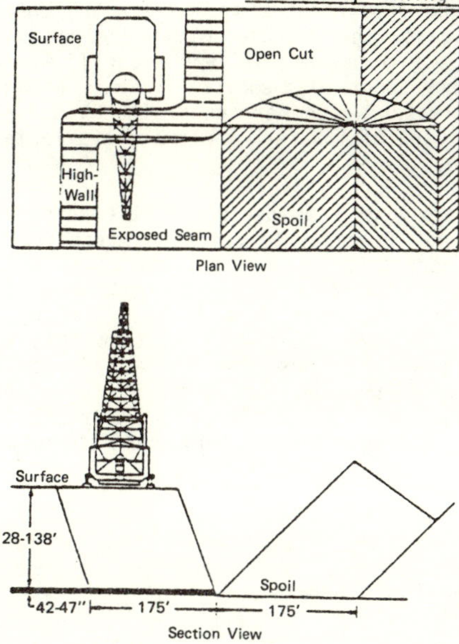

Stripping with a Dragline.

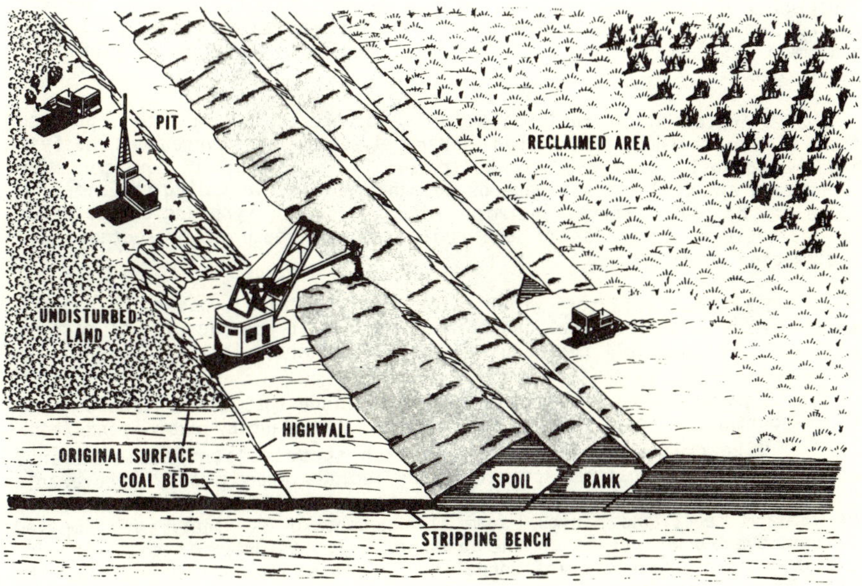

Source: Reference 3

There is considerable variability from mine to mine in terms of seam thickness, depth, rate of extraction, etc., and also in terms of the current economic limits on these important variables. Nevertheless, examination of a few statistical averages is instructive. The most important single measure of a surface mine's economic attractiveness is the overburden ratio, usually expressed as cubic yards of overburden removed per ton of coal extracted.[c] The average overburden ratio for all low-rank coal surface mines which reported overburden removed in 1977 was about 4.4. Averages for individual states ranged from about 2 for Alaska and Montana up to about 8 for Texas. In contrast, the average overburden ratio reported for all bituminous coal surface mined in 1977 was about 17.4, with individual state averages ranging from about 10 (Georgia) to 28 (Alabama).[6] This difference reflects both the greater value per ton of the higher-rank coals, and the fact that huge quantities of low-rank coals are available in thick, shallow seams.

The current limits on low-rank coal seam depth and thickness being extracted in the United States do not necessarily represent technical or equipment limitations, but rather economic or market limits. This is particularly evident in the context of the very deep and thick brown coal deposits being surface mined in other countries such as West Germany and Australia. The abundance of U.S. low-rank coals ensures that this will continue to be the case for many years.[d] Thus, the primary forces responsible for continuing advancement of western coal surface mining technology are the need for cost reduction (productivity improvement) and the need to meet increasingly stringent environmental standards.

Equipment Selection

Proper selection and design of the giant earth-moving equipment used in surface mines is an integral part of the mine planning process. The number of equipment options and their possible combinations has been increasing steadily. Computerization of both equipment design and operation is being practiced more and more widely, and is indicative of the increasing complexity and sophistication of surface mining equipment. Improved reclamation practices and the increased extraction of deeper, thicker, dipping, or multiple seams are forcing an increase in mobile equipment usage, notably in situations requiring selective handling of overburden.

If required for reclamation purposes, the topsoil must be separated from the overburden and stored separately from the overburden wastes. Usually topsoil is removed by bulldozers. The choice of equipment for overburden and coal removal depends on many factors, including the size and shape of the coal reserve; the mine production rate; type and character of the overburden; thickness of the coal seam and depth of overburden; and reclamation requirements.[2]

Some of the factors which favor the respective types of excavators are listed below. The final choice must be made to use the type of machine which can best meet all the requirements of any particular mine.

---

[c]The densities of these materials are such that the thickness ratio of overburden to coal closely approximates the quantity cubic yards/ton.

[d]The criteria used to define each state's "strippable reserve base," which are designed primarily to include economically recoverable coal, generally place maximum overburden thickness limits on low-rank coals of 125-250 feet, and minimum seam thicknesses of 3-5 feet, with implied maximum overburden ratios on the order of 5:1. As noted in the introduction to this chapter, current annual production of low-rank coal represents considerably less than 1 percent of the strippable reserve base. Also, Averitt has suggested that within the lift and swing limits of existing (1974) machinery a 30:1 overburden ratio is technically feasible as a maximum for present and near-future strip mining.[7]

Draglines:

- Have great flexibility and range in handling overburden and in movement about the area.
- Can handle much deeper overburden than a shovel of comparable size.
- Are capable of digging a much deeper box cut than a shovel of comparable size.
- Normally operate on the high ground ahead of the mine cut and not in the pit, which allows more flexibility for pit operations.
- Weigh less than shovels of comparable bucket capacity and have much better flotation characteristics.
- Are capable of handling material with poor stacking stability.
- Have lower initial cost per cubic yard of bucket capacity, especially in larger machines.
- Have better operating time.
- Require less maintenance.

Shovels:

- Operate on the coal, reducing roadway costs.
- Require less surface and highwall preparation and less drilling and blasting.
- Have lower power cost per cubic yard.
- Are capable of moving larger pieces of material.
- Have lower rope cost per cubic yard of bucket capacity than draglines.

Excavating Wheels (Bucket Wheel Excavators):

- Provide continuous excavation of highwall material.
- Can move material much greater distance from highwall to spoil than either dragline or shovel.
- Must have unconsolidated highwall material to be used effectively.
- Reduce reclamation cost.

Crawler Tractors and Rubber-Tired End Loaders:

- Are highly mobile.
- Are diesel powered, requiring no power distribution system.

- Are capable of moving overburden any distance required.
- Are capable of doing reclamation work simultaneous with stripping.

In present mining situations, the application of crawler tractors and rubber-tired end-loaders has been primarily in outcrop or contour mining operations with comparatively small annual production. Most western surface mining is conducted with shovel or dragline or a combination of the two types of machines, depending on the requirements of each operation. There are a few areas where the wheel excavator has proven successful due to the character of the overburden (e.g., at the Centralia mine in Washington state). However, in most instances, the wheel excavator is used in conjunction with either a shovel or a dragline to move the unconsolidated material which is normally encountered in the upper portion of the overburden. The trend has been to go to progressively larger machines. Draglines now exist with 220-cu-yd buckets, and shovel buckets range up to 180 cu yd.

The development of better structural and electrical materials, need for higher hourly productivity, and the necessity of mining to greater depths of overburden have been the prime reasons for the development of larger equipment sizes.

One problem which has occurred with the advent of these larger machines has been the decrease in operating time. This has been caused, primarily, by the increased size and weight of the components and repair parts. Parts which could be handled by hand on the smaller machines must now be handled by machine. Any routine repair or maintenance work on these large machines therefore takes more time out of the operating schedule.

As in the case of the excavators, the equipment used to load the coal depends on surface mine design features; in addition, the haulage or transportation system is an important determinant. Typically, if trucks are used for haulage, shovels or front loaders are employed.

The coal is transported from the pit by trucks or conveyor systems. Trucks are most flexible but consume large amounts of energy. Conveyors are less flexible and require high production rates to be justified economically, but are less costly when optimally employed.

Surface mining involves the use of huge and costly earth-moving equipment which must be operated by well-trained and highly-skilled operators. To a large degree the work force is not "in place" where most low-rank coal is mined and not experienced in the new and often unique equipment which is to be employed. As a result training is necessary so that the equipment is operated properly, efficiently and safely. Training is one area of improvement which is expected to pay off in the short term (as well as help eliminate shortages of this quality of worker).

Reclamation

The correct planning of a surface mine includes a detailed reclamation plan in the predevelopment stage. The choice of equipment and mine design are determined to some degree by reclamation requirements. Each surface mine has its own peculiarities and the reclamation plan needs to be tailored to fit the area involved. The ultimate objective of the reclamation process is to re-establish surface conditions that (1) at least approach the productive potential for which the reclaimed area is to be used; (2) do not disrupt major drainage patterns; and (3) blend with the surrounding landscape.

Important steps in all surface mine reclamation efforts are:[8]

1. Knowing the characteristics of the material to be revegetated.

2. Proper shaping and topsoiling of the spoil material.

3. Selecting species or varieties adapted to the soil, climate, and planned use.

4. Seeding or planting at the proper time, depth, and rate.

5. Proper fertilization as determined by soil tests.

6. Modification of the micro-climate by appropriate procedures such as pitting, furrowing, mulching, and snowfencing, etc.

7. Management after establishment to maintain productivity.

Reclamation can be made more efficient and less costly by the better use of data on forestry, biology, and agriculture. For example, cloning and tissue culture techniques are being employed to develop thousands of plants, indigenous to a specific mine site, from a cutting of one of the plants. The use of pooled information on climate, vegetation, toxicity of overburden, etc., will accelerate reclamation efforts. Also, graders and mulchers specifically designed for surface mine reclamation can reduce costs and increase efficiency.

3.2.2.2 Environmental Control Technology

The potential environmental impacts of a surface mining operation are site-specific, but can be categorized in general terms as follows:

1. Land impacts and modification of drainage patterns, requiring sound reclamation practices.

2. Potential slope instability associated with spoil and refuse disposal.

3. Contamination from mining waste (e.g., leaching of soluble acids, gases, and minerals that may be toxic) affecting water quality and aquatic ecosystems.

4. Modification or disruption of aquifers as a result of mining through them (causing either local or regional groundwater impacts).

5. Reduced availability of water in the area due to the water consumed in mining and reclamation operations (as well as possible supply disruptions due to aquifer modification).

6. Air contamination (e.g., dust, other emissions from machinery).

Land impacts are perhaps the most important of the factors listed. Major attention is being given to these problems, and PL 95-87 establishes a program to resolve them as western coal mining expands. The current legislative trend is to require restoration of the disturbed area to its approximate original contour with all spoil ridges and highwalls eliminated and no depressions left to accumulate water. Contour grading does not mean that all areas must be leveled, but rather that the profile of the land must be put back to approximately the way it was before the strip mining began. To accomplish contour grading, the spoil from the first cut is graded so as to blend into the contour of the adjoining land. Successive spoil piles are then graded with all material pushed toward the last cut, where it is deposited in the final pit. Long slopes on the graded spoil must be interrupted by terraces and/or diversion ditches. All of the diversions and terraces must be constructed according to sound engineering principles and must end in suitable outlets.

Removal and placement of the overburden are critical in environmental control. The nontoxic, nonacid, and fertile material should be stockpiled for later spreading or placed on top of the less desirable spoils already mined. The placement of the spoil should assure that long, steep slopes are avoided, that the material is not subject to slippage, and that it does not produce high peaks difficult to regrade.

Several states require the operator to separate topsoil from the subsoil and to stockpile the two types separately so they will not be mixed during the excavation process. When mining is completed, the materials can then be put back in their original sequence and the surface revegetated to prevent erosion. Some operations remove the topsoil and immediately spread it on areas recently graded, thus handling the material only once. This provision ensures that the best soil for plant growth is on top and not indiscriminately mixed with subsoils, which often contain toxic materials.

Some form of tillage of the site before planting is usually necessary. Any tillage measures must follow the contour of the slope and run parallel to the divisions or terraces. Chemical improvement of the soil in the form of liming and fertilizers is often needed for rapid establishment of vegetation.

Highwalls can cause environmental problems. An unstable highwall that sloughs off can ruin the natural drainage in a strip area. Material falling off the highwall can dam up channels and thereby prolong the contact between water and toxic material, or even force the water to seep through toxic spoil piles. Sloughing highwalls can open up new toxic materials to weathering. Problems such as these can often be overcome by grading the spoil back against the highwall and "knocking off" the top of the highwall.

3.2.2.3 Effects of Low-Rank Coal Properties

As indicated earlier, the physical and chemical properties of the coal itself have relatively little impact on the technology requirements for surface mining. In North Dakota, some selective mining is practiced to extract lignite from portions of a seam with relatively low sodium content. Thus, coal properties can affect the mining plan in some cases. Much more important are the geographical and geological factors associated with the occurrence of low-rank coals (e.g., stripping ratio, overburden characteristics, seam thickness and continuity).

Low-rank coals occur in vast areas of the United States in a wide variety of depositional and geographic environments (refer to Volume 2 - Resource Characterization). Only a few selected examples of challenging surface mining problems associated with low-rank coal reserves are listed below. In the following section on current mining operations in various low-rank coal regions, some additional illustrative examples of techniques used to surface mine various low-rank coal deposits are presented.

Thick Seams

Large quantities of low-rank coal (particularly the subbituminous coal in the Powder River Basin, for example) exist in very thick seams ranging up to 100 feet or more. These coal beds pose difficult problems for conventional U.S. surface mining techniques such as area mining by dragline. Extraction of such thick seams requires use of open pit or terrace-pit techniques.

Multiple Seams

The existence of multiple seams of coal separated by relatively thin interburden layers is also common in the western low-rank coal regions. These situations also require the use of open pit techniques.

Steeply Dipping Seams

Coal seams that dip at angles greater than 30° (in some cases, vertical or overturned) are found in abundance, generally at the margins of major basins. Typically these seams are also subject to severe folding and faulting due to the mountain-building activity associated with them. A variety of open pit techniques, some of them quite innovative, are being applied to these difficult-to-mine coal seams.

3.2.2.4 Current Status

Recent production statistics for surface-mined coal in the low-rank coal regions are shown in Table 3.2.2.2. In 1978 the total increased to 154 million tons, which represents about 24 percent of the total U.S. coal produced. The five states currently producing the most low-rank coal - Wyoming, Montana, Texas, North Dakota, and New Mexico - are also the states in which the most rapid growth in capacity is occurring (with the exception of Montana).

Table 3.2.2.2

Surface Mine Production Statistics for Low-Rank Coal-Producing States, 1977

State	Surface Mine Count	Surface Production, 1,000 Short Tons				Overburden Ratio, Cubic Yards/Ton[b]	Acres Mined[b]	Acres Reclaimed[b]
		Lignite	Subbituminous	Bituminous	Total			
Alaska	1	--	705	--	705	2.02	93	180
Arizona	3	--	11,059	--	11,059	3.84	459	483
Colorado	20	--	1,751	5,953	7,704	6.52	778	1,085
Montana	8	328	26,898	--	27,226	2.38	549	586
New Mexico	5	--	9,198	1,145	10,343	6.17	512	1,017
North Dakota	10	12,028	--	--	12,028	3.97	665	1,774
Texas	6	15,865	--	--	15,865	8.00	1,593	1,784
Washington	2	--	5,040	17	5,057	6.66	93	203
Wyoming	16	--	28,654	16,724[a]	45,378	3.30	1,653	1,483
Total	71	28,221	83,305	23,839	135,365	4.42[c]	6,395	8,595

[a]Much of the Wyoming coal is in the "overlap" rank cantegory (10,500-11,500 Btu/lb moist, mineral matter-free) in which the subbituminous A and high-volatile C bituminous coals fall. Although this production is classified as bituminous coal in reference 6, other sources such as reference 18 classify these coals seams as subbituminous.

[b]Not all mines report these data; however, for the states listed, the quantities shown represent between 80 and 100 percent of the total production.

[c]Quantity shown is the average.

Source: Reference 6.

Table 3.2.2.3 lists the currently operating surface mines that are producing subbituminous coal and lignite in the United States. As shown, the typical size of these operations is several million tons/year, with the Amax Belle Ayr mine in Wyoming leading the list at 18.1 million tons/year. The following sections summarize the operations, practices, and problems associated with surface mining operations in the highest producing states.

Table 3.2.2.3

Major Lignite and Subbituminous Mines in the United States

Company	Mine Name	Rank	Production (1978) (Short tons)
Wyoming			
Amax	Belle Ayr	Sub	18,065,664
Arch Mineral	Seminoe 1	Sub[b]	2,500,000
Arch Mineral	Seminoe 2	Sub[b]	2,800,000
Arch Mineral	Medicine Bow	Sub[b]	3,100,000
Big Horn	Big Horn 1	Sub	2,838,862
Bridger	Jim Bridger	Sub[b]	5,175,540
Carter Mining	Rawhide	Sub	2,620,000
Carter Mining	Caballo	Sub	144,510
Glenrock Coal	Dave Johnston	Sub	3,358,899
Resource Exploration	Hanna Basin	Sub[b]	900,000
Rosebud Coal Sales	Rosebud	Sub[b]	2,868,048
Sunoco	Cordero	Sub	3,800,000
Thunder Basin	Black Thunder	Sub	1,983,334
Wyodak Resources	Wyodak	Sub	850,190[a]
Kemmerer Coal Company	Elkol-Sorenson	Sub	4,061,794
FMC Corporation	Skull Point	Sub[b]	893,000
Montana			
Decker Coal Company	Decker (W and E)	Sub	9,073,592
Western Energy Company	Rosebud	Sub	10,576,000
Knife River Coal Mining Company	Savage	Lignite	300,001
Westmoreland Resources	Absaloka	Sub	2,554,201
Peabody Coal Company	Big Sky	Sub	2,064,886
Texas			
Alcoa	Sandow	Lignite	2,008,198
I.C.I.	Darco	Lignite	271,381
Texas Utilities	Big Brown	Lignite	5,298,285
Texas Utilities	Monticello	Lignite	6,966,468
Texas Utilities	Martin Lake	Lignite	5,971,750
North Dakota			
Baukol Noonan Inc.	Center	Lignite	3,400,000
Baukol Noonan Inc.	Noonan	Lignite	540,000
Consolidated Coal Co.	Velva	Lignite	289,678
Consolidated Coal Co.	Glenharold	Lignite	3,686,094
Falkirk Mining Co.	Falkirk	Lignite	142,308
Husky Industries	Mine No. 2	Lignite	135,000
Knife River Mining Co.	Gascoyne	Lignite	2,871,839
Knife River Mining Co.	Beulah	Lignite	1,887,267
North American Coal Corp.	Indian Head	Lignite	911,743

[a]1977 production.

[b]Heating value of this coal is between 10,500 and 11,500 Btu/lb (moist, mineral-matter-free) - see note a on Table 3.2.2.3.

(cont'd)

Table 3.2.2.3 (cont'd)

Major Lignite and Subbituminous Mines in the United States

Company	Mine Name	Rank	Production (1978) (Short tons)
New Mexico			
Pittsburgh & Midway Coal Mining Company	McKinley	Sub	2,992,958
Utah International Inc.	Navajo	Sub	8,000,000
Western Coal Company	San Juan	Sub	2,613,030
Arizona			
Peabody Coal Company	Black Mesa	Sub[b]	2,515,820
Peabody Coal Company	Kayenta	Sub[b]	6,771,768
Colorado			
Sigma Mining Company	Canadian Strip	Sub	148,560[a]
Kerr Coal Company	Marr Strip No. 1	Sub	500,000
Colowyo Coal Company	Colowyo	Sub	1,000,000
Empire Energy Company	Williams Fork Strip No. 2	Sub	242,096
Utah International Company	Trapper	Sub	345,948[a]
Washington			
Washington Irrigation & Development Company	Centralia	Sub	4,700,000

[a]1977 production.

[b]Coal is classified by some as bituminous.

Source: References 6,9-14.

Wyoming[14]

With vast and diverse resources of subbituminous coal, Wyoming has a large and growing number of surface mines which utilize all major mining methods and types of equipment. In the Powder River Basin, practically all surface mining is performed with truck-shovel combinations. In the Hanna coal field, draglines are used. In the Rock Springs area and other western Wyoming surface mining operations, various combinations of draglines, truck-shovels, and scrapers are in use.

The most commonly used surface mining method in Wyoming is the open pit method, although the area and contour mining techniques are also employed. Active surface mines have highwalls up to 150 feet, with the average between 40 and 60 feet. The open pit mines generally use scrapers and shovels for overburden removal, whereas the contour and area stripping operations most frequently use draglines.

Amax Coal Company's Belle Ayr mine near Gillette began operating in 1973, producing coal from the 70-foot thick Roland/Smith seam. The mine produced over 18 million tons in 1978, and about 15 million tons in 1979. Overburden averaged 118 feet thick during 1979. The open pit is nearly 2 miles long and 1 mile wide. Overburden is stripped by five 18- to 23-cu-yd-capacity shovels, loaded into 120-ton-capacity trucks, and backfilled into areas where mining has been completed. Coal is removed in two separate benches by 40-cu-yd-capacity shovels, and hauled by trucks to the crushing plant. Similar equipment and techniques are used at Amax Coal Company's nearby Eagle Butte mine, where overburden averages 80 feet thick, with combined coal seams of 131 feet. The open pit measures nearly 1.5

miles long, 1 mile wide and 200 feet deep. Reclamation activities in the semi-arid climate at these Powder River region mines involve seeding with more than 20 types of grasses, shrubs, forbs, and trees.

Arco Coal Co.'s Black Thunder mine, also near Gillette, is extracting a subbituminous coal seam averaging 68 feet thick. Overburden varies from 20 to 220 feet. Shovels are used to dig both the overburden and the coal. Overburden is continuously backfilled into previously mined areas and recontoured.

Kemmerer Coal Co. mines 12 major seams near Kemmerer, Wyoming with an aggregate thickness of 300 feet; the coal seams dip at angles of 17 to 22 degrees. Three basic equipment types are used in different parts of the mining operations (an unusually large fleet of equipment): truck-shovel combinations, draglines, and scraper-dozer combinations. Draglines are used in pits mining up to three seams. Where coal seams are thick and topography is favorable, the dragline makes two passes, placing overburden on both the highwall and outcrop side of the cut. Scraper-dozer teams are used to remove interburden between seams and to deepen and widen portions of the pit beyond the reach of the dragline. In pits where more than three seams are mined, truck-shovel combinations are used because the pit gets too large and unmanageable for the dragline and scraper-dozer.

The Black Butte Coal Co. mine near Rock Springs, Wyoming was the first western surface mine to start operating (August 1979) under the new Office of Surface Mining (OSM) regulations. More than 40 federal and state permits were required to start the operation. During the mine's life, 11 pits will be developed, with three or four active at one time. Thirteen coal seams between 3 and 35 feet thick will be mined. The use of draglines assisted by truck-shovel teams at Black Butte is probably a first in the coal industry. The main reason is that under OSM regulations, the dragline operation must be maintained on the down-dip side of the outcrops to protect the nesting areas of wildlife. Spoil cannot be cast over the outcrops containing these nesting areas.

The truck-shovel team is used to remove the first bench of the overburden, which is initially hauled to out-of-pit disposal areas. The dragline then moves in to open up the box cut with ample room to place spoil. Once the dragline begins spoiling, the trucks dump their loads between the peaks of the dragline spoil, finishing the contour. After the dragline completes the first cut, it continues to take a series of parallel, successively deeper cuts until the cutoff depth of 130 to 150 feet is reached. Beyond this depth, rehandling spoil becomes a problem. Pit lengths range from 1 to 4 miles.

Overland conveyors are being installed at the pits to reduce the coal hauling distance. A 19,800-foot-long overland conveyor will move coal from the western area of the property to the unit-train loading facility.

Montana[14]

Four large subbituminous coal surface mining operations, and one small lignite mine, account for essentially all of Montana's current coal production. Since the passage of the 30 percent severance tax in 1975, no new mines have been installed. Standard mining practice and equipment selection have not been fully established. The size of overburden removal equipment had not been of the magnitude used in the Midwest until Westmoreland Resources assembled a 110-cu-yd dragline at the Absaloka mine in 1979. The two draglines at that mine remove overburden from a suite of four subbituminous coal seams which total 58 feet thick. Front-end loaders dig the coal and load it into 100-ton-capacity haulage trucks.

At Western Energy Co.'s Rosebud mine, a 23-foot subbituminous coal seam is exposed by two walking draglines of 60-cu-yd capacity. Overburden is 100 feet thick. Three shovels load the coal into haul trucks.

Texas[9]

All lignite mines in Texas (except one) are supporting minemouth electric power generating stations and hence require large area mines to meet fuel needs. The land topology is generally flat and unconsolidated, lending itself well to area mining practices. Neither the overburden nor the lignite require blasting for removal. Consequently, the surface mining operation consists of overburden removal, lignite loading and mine haulage.

The mines are prestripped using dozers and/or scrapers. The overburden is then handled by sidecasting with a dragline. Stripping is performed using the "box cut" method. Pit widths are generally 120 feet and lengths range from 0.6 to 2.0 miles. Maximum depth is typically 150 feet but may go as deep as 250 feet in a multiseam mine. Operation usually runs on a three-shift-per-day, seven-day-per-week basis. When more than one seam is encountered, more complex open pit methods are applied.

Lignite is commonly excavated using 12-to-20-cu-yd electric mining shovels to load trucks. Rubber-tired front-end loaders and hydraulic shovels are also used. Their use may increase in the future because the higher mobility allows them to adapt to changing conditions rapidly. Bucket wheel excavator systems are also being considered for lignite loading. In this case, lignite would probably be loaded into a hopper feeding a conveyor belt for haulage.

Mine haulage is generally by trucks in the 85- to 120-ton class. Tractors with bottom-dump trailers are common although some end-dump trucks have been planned. Haul distances at the Big Brown mine are 5 to 10 miles one way and maintenance problems have developed on the haul road. Other transport systems include a 2-mile conveyor belt at the Sandow mine and combination truck-unit train haulage at Monticello and Martin Lake.

In general, the overburden is mixed and it is not necessary to handle topsoil separately to revegetate mined lands. Rainfall is adequate for revegetation and no problems with acid drainage are reported. After leveling is complete, spoil areas are fertilized and sprigged with Coastal Bermuda-grass or seeded with Crimson clover. Although it is early to reach conclusions (none of the reclaimed areas have been approved for release), reclamation efforts appear to have been especially successful in Texas due to the plentiful rainfall and possibly the properties of the spoil material.

North Dakota[15]

In North Dakota, locating the first strip pit on a lignite bed so that only commercial quality lignite is obtained can prove difficult. Often the lignite along the outcrop of the bed is of inferior quality as it contains high moisture and high ash, and has a low heating value and a low ash fusion temperature. Such low-quality lignite can be found for as much as half a mile from the outcrop towards the center of the deposit. It is necessary to delineate the line on the lignite bed where the first pit should be located so that only high-quality lignite is exposed when the overburden is removed. Ordinarily, this is accomplished by first drilling test holes and determining the actual limits of the lignite bed. Additional test holes are then drilled from the outcrop towards the center of the bed to a point where analyses of the drill samples indicate the lignite is of a commercial quality. The point where high-quality lignite is found is designated as the commercial lignite boundary.

Another problem that can prove to be very expensive and sometimes difficult to solve is that of mine drainage. Most stripping in North Dakota follows the contour of the land; the overburden is relatively shallow at the outcrop with the elevation of the surface usually increasing toward the center of the bed. Pits are normally located on the outer perimeter of the lignite bed, and drainage from the higher interior elevations has a tendency to collect in the open cut. Water accumulation in the pit can consist of both surface runoff and groundwater. Every effort is made to channel all surface runoff around the stripping operation.

Past analyses have indicated that the ash content in the top 6 inches of a lignite seam has a tendency to be high. Usual practice is to remove at least part of this top layer by cleaning the bed with bulldozers and scrapers immediately after exposure by stripping. All material loosened by these operations is removed. In addition, the surface of the portion of the seam set aside for each day's loading is again cleaned through use of front-end loaders and self-loading scrapers. A few years ago, a practice was inaugurated of sweeping the surface with mechanical brooms just before drilling, blasting, and loading.

One severe problem faced by the lignite mining industry in North Dakota is the fact that peak production occurs during the coldest winter months when temperatures can plunge to -40° F. Ordinary metals exhibit peculiar characteristics at these temperatures and have a tendency to become brittle. During extremely cold temperatures, large loading booms would sometimes snap; iron and cast steel gears in a tipple would shatter and fly through the air much like shrapnel; and huge walking cam guides on a large walking dragline would snap and break even though these castings weighed several tons each. With the advent of new specification steels, and the practice of heating critical components during very cold weather, this condition has been alleviated to some extent.

Keeping haul roads free from snow has also caused problems. Elevated roadways and snow fences have been used to attempt to minimize this problem. Also, the bottom clay underlying lignite beds in some areas of North Dakota is not capable of supporting the ordinary loading shovel, as the bearing pressure of the tracks exceeds the bearing strength of the clay. At the South Beulah mine, it is necessary to use large flotation tracks on the loading shovel in order to prevent the shovel from sinking into the bottom.

Material shortages and long delivery times often plague expansion efforts in the industry. Also a shortage of skilled labor in the area creates problems with new developments.[16]

Reclamation problems are receiving considerable attention at this point.[17] One of the most severe concerns currently facing reclamation deals with differential subsidence and piping erosion. In general, this is caused by the slumping and settling of reshaped spoils, and erosion along subsurface cracks which result in cavities beneath the spoil surface. These problems seem amplified when spoils are sodic and, thus, structurally unstable. Many questions concerning these phenomena and their prevention remain.

Another area of concern is erosion control on newly spread surface soils. Because of the lack of vegetation to protect the surface, the reclaimed land is susceptible to wind and water erosion. Technology is available whereby this erosion can be reduced. Practices like mulching, shaping to less steep slopes, and prompt establishment of vegetation can greatly reduce erosion losses.

Even though the replacement of suitable plant growth material over sodic and saline spoils has greatly reduced the problems associated with reclamation of these spoils, concern still exists for controlling salts.

High adsorbed sodium results in slow water infiltration and high salinity results in unfavorable water availability for plant use. Research is under way to better understand the movement of salt and water at the interface between saline or sodic spoils and replaced surface soil. These problems are especially important in areas where very limited amounts of suitable soil materials are available for stockpiling and replacement.

Another major concern of reclamation is long-term sustained production of annual and perennial crops. Observations available at present are based on only a few growing seasons. The time factor in evaluating the success of reclamation has not been adequately determined. Research currently in progress is designed to monitor production of reclaimed land for several years over a variety of weather cycles. Only by such evaluation can long-term productivity of reclaimed land be estimated. Closely allied with sustained production is the determination of the best management practices for reclaimed land. This includes factors like soil fertility, cultural practices, livestock stocking rates and grazing period, natural succession of species, and species adaptability to reclaimed lands.

New Mexico[13]

Strip mining of coal in New Mexico is carried out on terrain that ranges from relatively flat surfaces to rolling hills, and in many places amid "badlands" formed by steep-walled gullies separated by ragged rock ridges.

Natural vegetation varies from sparse grass and desert shrubs to small junipers and pinons in some of the higher areas on the borders of the San Juan Basin. All the strip mines are in arid to semiarid areas that receive about ten inches of precipitation a year. Thus for the most part, only the native semiarid vegetation types grow on the original soil or on reclaimed spoil banks.

Most strip mining there is performed using the "box cut" method. After the mining is completed, the terrain looks like knobby ridges and valleys. Unless grading or leveling is done, erosion during rains tends to wash ridge material into the valleys and, hence, partially fill them. As with most western low-rank coals, most New Mexico coals are low sulfur; thus no problem with acid water arises.

The overburden in many parts of New Mexico includes hard sandstone that requires blasting for removal. Walking draglines are used for extraction of this material. Then shovels and front-end loaders holding 10 to 16 tons load the coal into trucks for transport.

Revegetation of spoil piles by reseeding has not always been successful in northwestern New Mexico. At best, artificial reseeding provides a temporary cover until natural vegetation re-establishes itself. After about 20 years, it is hoped that natural vegetation will support as many cattle and sheep as before the strip mining operation.

References - Section 3.2.2

1. Pfleider, E.P. (Editor), Surface Mining, AIME Mudd Series, New York, 1972.

2. Cassidy, S.M. (Editor), Elements of Practical Coal Mining, AIME Mudd Series, New York, 1973.

3. Grim, E.C., and R.D. Hill, Environmental Protection in Surface Mining of Coal, EPA-670/2-74-093, October 1974.

4. Nephew, E.A., Surface Mining and Land Reclamation in Germany, ORNL-NSF-EP-16, May 1972.

5. Teuscherer, H.J., "New Developments in Opencast Mining Technology at Rheinische Braunkohlenwerke AG, Koeln," in Technology and Use of Lignite, Proceedings: Bureau of Mines - University of North Dakota Symposium, Grand Forks, North Dakota, May 9-10, 1973, USBM I.C. 8650, 1974, pp. 20-33.

6. Lin, K., Coal Data 1978, National Coal Association, Washington, D.C., 1980.

7. Averitt, P., Coal Resources of the United States, January 1, 1974, USGS Bulletin 1412, 1975.

8. Lang, R.L., "Strip Mine Rehabilitation Problems and Research in Wyoming," in Practices and Problems of Land Reclamation in Western North America, edited by M.H. Wali, University of North Dakota Press, Grand Forks, 1975.

9. Russell, J.E., "Surface Mining Technology, Practices, and Plans for Gulf Coast Lignite," in Technology and Use of Lignite, Proceedings of the Tenth Biennial Lignite Symposium, Grand Forks, North Dakota, May 30-31, 1979, GFETC/IC-79/1, pp. 388-425.

10. Jones, D.C., J.E. Schultz, and D.K. Murray, Coal Resources and Development Map of Colorado, Department of Natural Resources, State of Colorado, 1978.

11. Miller, N., The Geological Survey of Wyoming, Wyoming Coal Fields, 1978.

12. 1979 Keystone Coal Industry Manual, McGraw-Hill Publ., New York, 1979.

13. Schilling, J.H., et al., Survey of Surface Mining in New Mexico, State Bureau of Mines and Mineral Resources, Circular 44, Socorro, New Mexico, 1971.

14. Coal Age, May 1980, pp. 93-137.

15. Kane, A.S., "Operational Aspects of Lignite Mining," in Technology and Use of Lignite, Proceedings: Bureau of Mines - University of North Dakota Symposium, Grand Forks, North Dakota, April 27-28, 1967, Bureau of Mines Information Circular 8376, May 1968.

16. Murray, R.E., "Large-Scale Surface Mining on the Northern Great Plains," in Technology and Use of Lignite, Proceedings of a Symposium, Grand Forks, North Dakota, May 14-15, 1975, GFERC/IC-75/2, 1975, pp. 286-293.

17. Ries, R.E., F.M. Sandoval, and J.F. Power, "Reclamation of Disturbed Lands in the Lignite Area of the Northern Plains," in Technology and Use of Lignite, Proceedings of a Symposium, Grand Forks, North Dakota, May 18-19, 1977, GFERC/IC-77/1, 1977, pp. 309-327.

18. Glass, G.B. Wyoming Coal Fields, 1978, Geological Survey of Wyoming, Public Info. Circular No. 9, 1978.

3.2.3 Underground Mining

3.2.3.1 Technology Description

Virtually no low-rank coal is extracted by underground mining in the United States today. Vast quantities of western low-rank coal resources lie at depths beyond the current economic limits for surface mining. Some of these resources may be extracted by underground mining in the future; thus a brief synopsis of the relevant technology is presented here. Application of underground mining techniques to U.S. low-rank coals will require new or adapted techniques to deal with the thick seams and unconsolidated overburden encountered in many areas.

There are three different types of underground mines, depending on the manner of making the opening from the surface to the coal seam: drift mines, shaft mines, and slope mines. These three methods of accessing the coal seam are illustrated in Figure 3.2.3.1.

Figure 3.2.3.1
Types of Underground Coal Mines

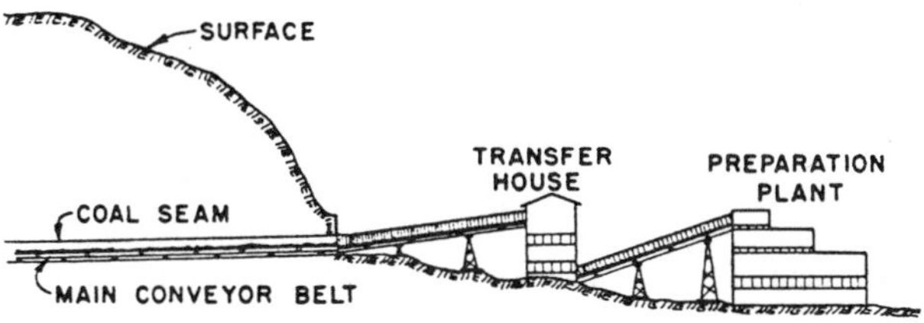

Drift Mine

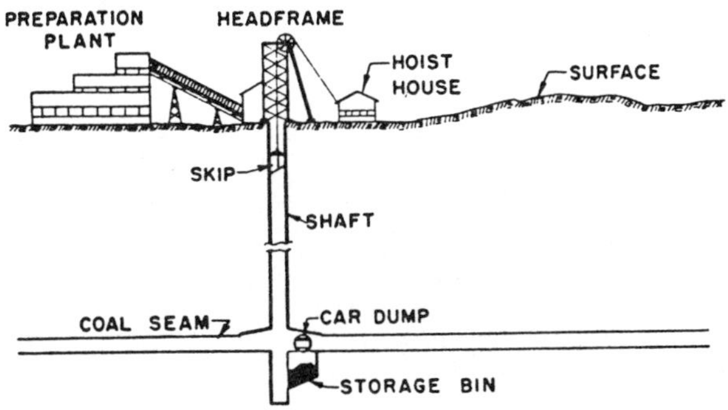

Shaft Mine

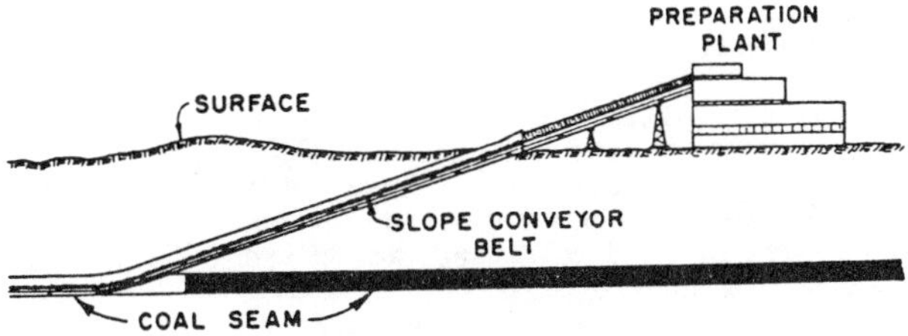

Slope Mine

A drift mine can be used in those situations where a nearly horizontal coal seam outcrops on the side of a hill. The portal drift (mine entry) is merely an extension of the underground entry system. This type of mine is generally the easiest and cheapest to open because no excavation through rock is required. Transportation of coal to the outside may be by track haulage, belt conveyor, or by battery-powered rubber-tired equipment.

A shaft mine is constructed when the surface terrain is fairly level and the coal deposit is quite deep. Two or more vertical shafts are sunk to the depth of the deposit: one, to house the elevator that will haul coal to the surface and transport miners and equipment in and out of the mine; the other, to provide ventilation. A large fan at the top of the shaft controls the circulation of air, drawing away both stale air and gases.

In a slope mine the access adit is driven down to the coal on a gradual incline, which allows machinery to run, either under its own power or with assistance from a hoisting cable, from the surface to the coal face. Miners usually enter and leave the mine in rail cars, and coal is carried to the surface either in cars or on conveyor belts. Slope mines are most effective when a coal seam lies close to the surface but too deep to be surface mined, especially in hilly areas. The workings of these mines often extend for miles underground.

Sometimes an individual mine will have all three types of openings, i.e., drift, slope, and shaft. For instance, the coal haulage might come to the outside through a drift opening. As the mine develops under heavier cover, additional openings become necessary at intervals for ventilation, and for portals to shorten the traveling time for men and supplies. A slope might be used for the portal where the cover is not too great and a shaft, or shafts, for air. Where the cover becomes very thick a point is reached where it is more economical to use shafts for men, supplies, and equipment.

Room and Pillar Mining[1]

The room and pillar system of underground coal mining leaves undisturbed blocks of coal in a regular grid to prevent collapse of overhead rock strata. Approximately one-half of the coal is normally left behind in pillars to prevent subsidence. (In some cases, room and pillar mining may be followed by pillar robbing on the retreat, allowing the roof to collapse. This technique leads to subsidence and cannot be practiced in areas where the resulting surface damage cannot be tolerated.) Room and pillar mining can be further classified into the conventional mining method, and continuous mining.

In conventional room and pillar mining the coal is extracted in a sequence of steps, each of which involves specialized equipment: (1) Undercutting; (2) drilling; (3) blasting; (4) loading; (5) transporting and (6) roof support.

In undercutting (center cut and top cut can also be employed), a horizontal slice of about 7 inches is cut out from the bottom seam to allow loose coal to fall downward and not into the work area when the face is blasted. The block of coal outlined is then drilled using mobile powered drills or hand-held electric or hydraulic drills. The blasting is done either with compressed air or explosives. The broken coal is gathered by a loading machine, or in some cases both gathered and transported by specially designed equipment. Normally a shuttle car is used for transport to a belt conveyor or mine car loading point. Roof support, the last step in the conventional mining process, is the most critical for the safety of the operation. Wooden timbers, steel crossbars on posts, or most commonly, roof belts, are installed. A roof bolting machine drills holes in the roof and installs expansion bolts to bind the layers of overhead strata together. Ventilation is then extended and the coal face is ready for the next cycle.

The conventional mining system is the most labor intensive, and requires relatively good roof conditions in order to withstand the stresses of blasting. Blasting and undercutting produce large amounts of dust as well.

In the continuous mining system, a single machine called a "continuous miner" breaks the coal mechanically and loads it for transport. Roof support is then installed, ventilation is advanced, and the coal face is ready for the next cycle. Generally the advance in a single "cut," usually about 18 feet, is limited by the length of the machine, so as to keep the operator under supported roof at all times. Efforts have been made, with varying success, to mount roof bolting equipment on continuous miners. Where this has been successful, the advance per cut is limited only by ventilation requirements.

Continuous mining systems can be employed where fragile roof conditions would prevent blasting. Presently, approximately 60 percent of the coal from underground mines (bituminous) is taken by continuous mining machines.

Upon retreat from the face, the pillars can be partially or totally extracted depending on surface subsidence problems. Total extraction results in roof collapse and major subsidence.

Partial extraction can be accomplished in a number of ways. Diagonal cuts can be made in the pillar; an auger drill can be used to extract the coal; or sophisticated equipment, specifically designed for pillar robbing, such as the "miniwall" system from Ingersoll-Rand[2] can be employed to increase the yield of coal extracted from the mine.

Sublevel Caving by Pillar Extraction[3,4]

Western coal seams are commonly greater than 12 feet thick. Conventional or continuous mining systems are not commonly applied to seams this thick (7 feet is considered by some to be a reasonable upper limit). Sublevel caving by pillar extraction is a proposed modification of the normal room and pillar method to include recovery of the top coal (above the 8-foot openings) in a specific sequence. Top coal would be recovered on retreat by drilling and shooting its entire thickness in increments or "falls" and loading the fall with a narrow drum-type continuous miner. The minimum seam thickness for efficient use of the method is estimated to be 12 feet; it may be applicable to seams up to 40 feet in thickness.

Longwall Mining[1,5]

Longwall mining utilizes conventional room and pillar continuous mining equipment to develop panels of coal which may be a mile long, 600 feet wide and 4 or more feet high. A high-horsepower mining machine removes the coal by shearing or plowing a slice of coal up to 30 inches thick from the face of the panel. The broken coal is moved to the headpiece side of the panel by the face conveyor and out of the mine by the main haulage belt conveyor. Self-advancing hydraulic roof supports, lined up parallel to the face, move up one by one as the shear or plow travels across the face. The overburden (gob) collapses behind the roof supports, thus relieving the pressure on the coal face.

The efficiency of the longwall technique has progressively improved through introduction of self-advancing supports and the flexible-armored face conveyors. This fully mechanized version, using imported equipment, was introduced to the United States in 1960. During the past several years, equipment manufacturers, singly and in conjunction with European manufacturers, have made several advancements toward improved productivity and safety.

The increasing interest in longwall mining is attributable to the following factors:

- Improved economics (lower cost per ton)
- Health and safety considerations (compliance with laws and regulations; roof support, ventilation and dust requirements)
- Increased recovery of coal (present overall recovery closely approximates the room and pillar rate with a greater potential as new techniques and equipment are developed)
- Predictable subsidence
- Adaptability to thick and multiple seams
- Capability to mine at greater depths

There are several constraints which continue to limit application of this technology:

- The non-fracturing and non-caving characteristics of rock strata above the abundantly available shallow depth seams do not permit mining by the longwall method.
- Reluctance of mine operators to commit the large initial capital investment, which can run as high as 10 million dollars to initiate operations. Mine size must be larger than 1 million tons per year to justify the capital investment.
- Surface subsidence, although more predictable and uniform, can only be tolerated in certain unpopulated areas.

Several DOE-sponsored studies are currently under way to identify and eliminate problems which limit productivity and to increase the economic attractiveness of longwall mining. One such area for improvement is decreasing the setup time for the equipment on each new face. Studies and research projects are also directed toward automation techniques and toward field trials and demonstrations of improved longwall equipment and techniques.

More work is also necessary on the mining of the thick western low-rank coal seams. The practical limit on the longwall equipment as applied in Europe is about seven feet. Techniques such as multislice longwall mining and longwall caving mining need to be further developed to see if and how they can be applied to the thicker western seams.[5]

Shortwall Mining

Shortwall mining is similar to longwall mining and has been extensively employed in Australia. The difference is the face length (typically 150 feet) and the face equipment (usually continuous miners and shuttle cars). This method is somewhat less capital intensive than longwall mining because the same coal extraction and conveying equipment can be used in mine development and production mining operations. Coal recovery is comparable to the longwall method. Shortwall mining, like longwall mining, can be used only in areas in which subsidence is acceptable.

Underground Hydraulic Mining

Underground hydraulic mining of coal involves the use of high-pressure (2000 psi) water jets to break the coal loose and convey it from the face as a slurry, either in open flumes, steel pipes, or in flexible hose pipelines. Hydraulic mining is especially applicable to steeply pitched or unusually thick seams of coal because of the following:

- Conventional coal mining machinery cannot handle seams thicker than about 10-12 feet (except through techniques such as sublevel caving).
- Hydraulic mining moves the operator safely away from rockfalls in the vicinity of the working face.
- Respirable dust generation levels are lower.

At the present time, hydraulic coal mining is not practiced in the United States. In Canada, Kaiser Resources is presently operating a 4000-ton-per-day hydraulic mine in British Columbia; in the Soviet Union, ten hydraulic mines are in operation in the Donets and Kuznetsk coal basins. Hydraulic mining is also practiced in the Federal Republic of Germany. Demonstrations of hydraulic mining are being planned in the United States. However, one potential problem is the availability of sufficient water in the western locations.

3.2.3.2 Environmental Control Technology

The three principal environmental concerns associated with underground mining are mine drainage, solid waste disposal, and subsidence. In addition, underground mining is an especially hazardous occupation.

Mine drainage results from the infiltration of surface water through the surrounding strata to the mine. Mine drainage must be pumped back to the surface, where it is potentially a pollutant, to avoid mine flooding. The composition of mine drainage varies with the specific geologic area but is often acidic (although it has been reported basic in some western operations) with high sulfate content, and high levels of aluminum, calcium, magnesium, and manganese. Treatment of mine drainage is site-specific and can be expensive. Treatment may require one or more of the following:

Neutralization

Aeration

Chemical treatment to precipitate dissolved solids

Removal of suspended solids

The requirement for treatment can be reduced in some cases by preventing infiltration through water diversion, mine sealing, and surface restoration to promote runoff rather than penetration.

Problems associated with solid mine waste disposal include discharge of particulates, noxious gases, acid mine runoff, erosion and landslides. The accepted practice is to lay down refuse in compacted layers and to cover the pile with earth. This technique protects the refuse from leaching by rainwater and allows vegetation cover to be established. In a small number of cases, profitable use of the solid waste has been achieved through backfilling of abandoned mines, the manufacture of bricks and cinder blocks, the construction of secondary roads, and combustion as a low-grade fuel for generation of electricity.

Subsidence is controlled by leaving pillars of coal in mined areas to support the overburden. Pillars cannot be removed in areas where the resulting surface subsidence will damage buildings, farmland, groundwater and lakes or streams. Subsidence is of less concern in rugged areas with few buildings or little farming. It is in such areas that fuller coal recovery can be accomplished.

The safety of underground coal mining is likely to be enhanced by two developments. The first of these is the fielding of the "miner-bolter," a continuous mining machine with an integrated roof bolter. This machine will aid the control of the roof in the immediate area of the working face, the most hazardous area. The second development is the implementation of coalbed methane degasification practices. Pilot projects to drain methane from coal beds in advance of mining operations are now under way. Should such practices become widespread, they will reduce the probability of catastrophic mine explosions.

3.2.3.3 Effects of Low-Rank Coal Properties

As in the case of surface mining, the properties of the western low-rank coal do not per se influence the type of underground mining to a significant degree. However, thick seams, which are characteristic of large portions of the western coal resources, do greatly influence the method of mining. Other important variables include:

- Amount of cover over the coal seam.
- Characteristics of the coal seam including: (a) the thickness of the coal seam; (b) the nature and location of any impurities or irregularities; (c) the dip (or pitch) of the coal seam; (d) the nature and direction of any definite cleavage in the coal seam; (e) the hardness of the coal; (f) the structural strength of the coal; and (g) the presence and amount of explosive gas (methane).
- Quantity of water likely to be encountered in the mining operation.
- Nature and strength of the roof and floor rock strata surrounding the coal seam; in many western locations, incompetent, unconsolidated rock formations prevail.
- Any previous or concurrent mining in seams above or below the coal seam to be worked.

For example, a recent study which examined seam height, methane problems and roof condition found that, contrary to what was becoming industry practice, conventional room and pillar mining and blasting and loading could be more productive than continuous mining in seams thicker than 6 feet.[6] The study found that in terms of cost and productivity, conventional mining is much more sensitive to changes in seam height than the continuous system.

3.2.3.4 Current Status

The Department of Energy is sponsoring several R&D projects to develop and prove underground mining techniques for thick low-rank coal seams.[7] These include:

1. Longwall multilift methods - mine design planning will soon be completed for the first U.S. application tests.

2. Longwall in steeply pitching seams - field tests in Colorado are under way.

3. Longwall in sublevel caving - development is continuing for construction and field testing of a prototype machine for this method.

4. Sublevel caving with pillar extraction - field trials will be initiated soon.

5. Advanced mining systems, including high-volume hydraulic jet mining; a variable wall miner for panel mining; and a borehole mining system using high-pressure water and a slurry pump.

References - Section 3.2.3

1. Cassidy, S.M. (Editor), Elements of Practical Coal Mining, AIME Mudd Series, New York, 1973.

2. Coal Age, February 1979, p. 74.

3. Coal Mining and Processing, February 1979, p. 70.

4. Coal Mining and Processing, February 1979, p. 61.

5. Otto, R.H., Three Potential Longwall Mining Methods for Thick Coal Seams in the Western United States, Bureau of Mines IC 8792, 1979.

6. Menchen, W.R., et al., Underground Coal Mining - An Assessment of Technology, EPRI AF-219, July 1976.

7. Fossil Energy Program Summary Document, DOE/ET-0087, March 1979.

3.2.4 Underground Coal Gasification

3.2.4.1 Technology Description[1]

Underground coal gasification (UCG) technology development has been conducted in various countries for over 50 years. Although it has been shown to be technically sucessful in the Soviet Union on a commercial scale, its economic feasibility in a free market under environmental and societal restraints has yet to be proven. Prior to 1972, free-world experimentation in UCG was technically and economically unsuccessful due to the choice of what now are known to be unfavorable geologic settings (i.e., thin bituminous coal seams).

UCG can potentially use coal which is economically or technically infeasible to mine because the coal is too thick, deep, dirty (with non-coal occlusions), high-ash or wet; because it has an excessive angle of dip; or because it has unpredictable and poor overburden characteristics that make mining unsafe. Conditions that favor UCG include: (1) thick seams (5 feet minimum, over 10 feet preferred); (2) non-swelling, reactive, relatively permeable coal (i.e., low-rank coal); (3) sufficiently thick overburden to cap the underground reactor (about 200 feet minimum), preferably with an impermeable clay layer immediately above the coal seam; and (4) good seam continuity (not highly faulted or fractured).

Coal is gasified underground by drilling boreholes into the seam and injecting air (or oxygen and steam) into the underground reaction zone. The coal is partially oxidized, producing low- or medium-Btu gas. The hot gas is forced through the seam to the exit boreholes and is carried to the surface where it is cleaned and upgraded for use.

The natural permeability of a coal seam is too low to sustain the high gas flow rates required for gasification. Thus, a critical part of the technology is the creation of permeable pathways, or links, in the coal seam between the injection and production boreholes. These links can be created by reverse combustion, directional drilling, or other techniques.[a] For low-rank coals, which shrink and disaggregate upon heating, it is particularly important and effective to produce the links near the bottom of the coal seam. Coal immediately above the gasification zone breaks away and, in effect, creates an underground packed bed reactor. Gasification then proceeds from the bottom to the top of the seam, utilizing nearly all the coal.

If a linkage channel is formed at the top of the seam, a much smaller portion of the coal is gasified; in addition, the injected air tends to bypass the gasification zone, causing production of very low quality gas. Thus, the existence of a linkage channel low in the seam ensures that the coal and air are effectively reacted, and that the product gas has a high heating value that remains fairly uniform with time.

These principles are illustrated in Figure 3.2.4.1 which shows a side view of a gasification zone in a low-rank coal seam. The linkage path has been formed at the bottom of the coal seam. As gasification proceeds from right to left, coal falls into the gasified cavity and creates a highly reactive rubble zone. As the system expands, the gasification zone gradually encompasses the full thickness of the coal seam and moves in a broad front towards the outlet well.

Figure 3.2.4.2 shows a plan view of a field development as applied by the Soviets to a brown coal deposit near Moscow, which allows controlled extraction of the entire coal resource. The dotted lines show the location of the underground linkage channels formed in the coal by reverse combustion in preparation for gasification. The production phase of gasification is carried out by forward gasification in the channels. A combination of convective and dispersive effects in a thermally evolving system permits lateral movement of oxygen a considerable distance from the injection wells, and accounts for the experimentally observed wide sweep widths in low-rank coal. Commercially attractive well spacings are believed to be longer than the 75 feet shown in this figure; spacings of 100 to 150 feet would be more appropriate in the United States.

[a]Other linking techniques used or attempted in the past include electrolinking, hydrofracturing, explosive fracturing, and pneumatic pressure fracturing. Explosive (shaped charge) or hydraulic drilling techniques have been proposed and partially developed.[2]

Figure 3.2.4.1

Conceptual View of a Channel During Gasification

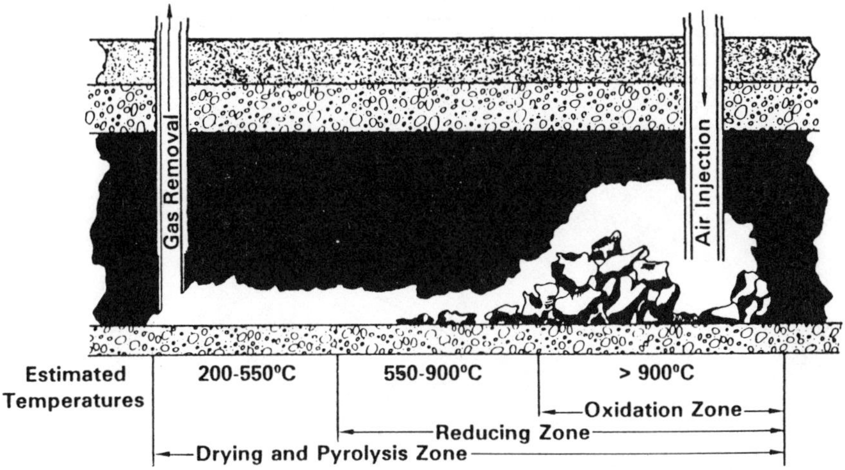

Source: Reference 1

Figure 3.2.4.2

Plan of a Shaftless Generator, Moscow Region Plant

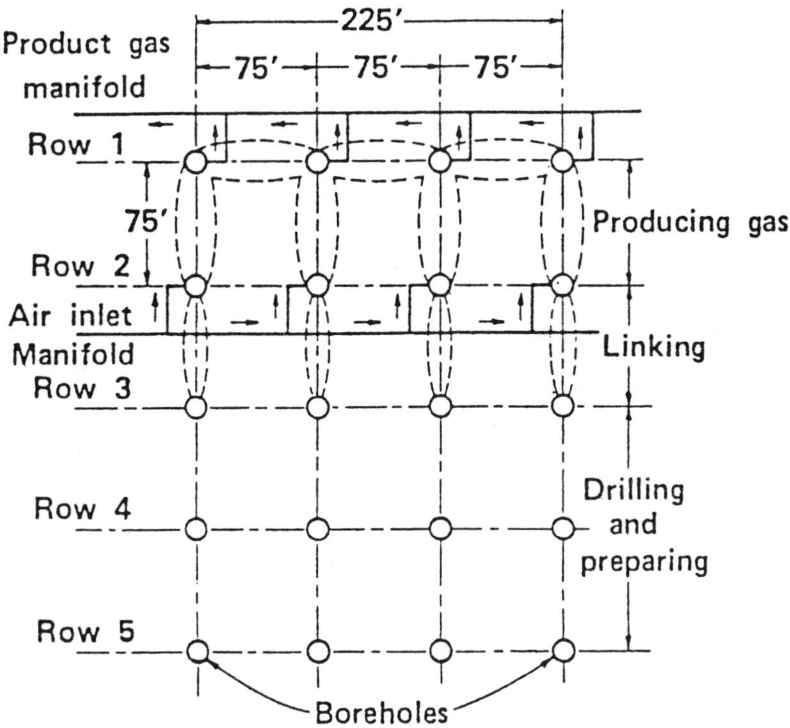

Source: Reference 1

In steeply dipping coal seams, the same basic principles are applied in a different geometric setting (see Figure 3.2.4.3). Gas production boreholes are drilled at a slant from the outcrop to stay in the coal seam. Air is injected through boreholes drilled to the seam through the underburden or overburden (the former is shown in the figure, and has the advantage of keeping the injection wells out of the subsidence zone). The gasification zone advances updip, and coal rubble falls into the channel; product gas flows to the surface through the slant-drilled holes in the coal seam.

Figure 3.2.4.3

UCG in a Steeply Dipping Bed

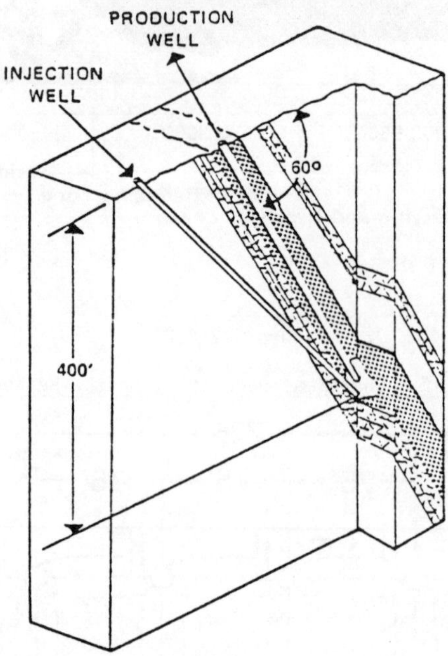

Source: Reference 1

In many years of field work (starting in the early 1930's and culminating in commercial scale operations in the 1950's), the Soviets proved that underground coal gasification is technically feasible.[3,4] Large-scale, continuous operations were successfully conducted with both the flat-lying and steeply dipping configurations illustrated above. The resource recovery efficiency (energy in clean gas to energy in coal) was typically 40 to 50 percent, which was the product of a 70 to 80 percent sweep efficiency and a 60 to 70 percent thermal efficiency. The Soviets were able to maintain reasonably stable gas heating values and production rates over periods of several years. Small-scale field tests in the United States in the past 7 years have also shown the basic technical feasibility of the process in low-rank coals; these field projects are summarized in Section 3.2.4.4.

The key process operations in underground coal gasification are briefly discussed below. In Section 3.2.4.3, the effects of coal properties on these operations are amplified.

Site Characterization and Evaluation

Because the underground coal seam is the gasification reactor, obtaining complete geologic and hydrologic information on the site is critical to the success of the subsequent operations. Initially, it is necessary to confirm that the coal seam meets the basic criteria such as continuity. This can be done through a combination of remote and downhole surveying techniques (e.g., seismic surveying). Drilling, coring, and logging is performed to develop the necessary geophysical and hydrological data. In some cases, the conduct of air injection/production tests, and possible reverse burn linking tests, would be considered part of the site characterization and evaluation phase as well.

Linking

As described previously, it is necessary to provide links (air flowpaths) between the injection and production wells; for efficient gasification, these links must be located near the bottom of the coal seam. The cheapest and most direct method known at present is reverse combustion. The wells are cased into the bottom portion of the coal seam, and air flow is established between them. The coal is ignited at the production well, and the flame front propagates towards the injection well (thus the name reverse combustion) at a rate of 6-10 feet per day. A channel with an effective diameter on the order of 3 feet is formed between the wells. This method is sometimes unreliable, because multiple linkage paths as well as paths at the coal seam/overburden interface are sometimes formed.[5]

A more positive and reliable technique (but more expensive) is the use of directional drilling, as illustrated in Figure 3.2.4.4. Once the deviated borehole is completed, the vertical holes are connected to it, and a very quick reverse burn step is used to enlarge the hole prior to gasification. Improvements in directional drilling technology, particularly in guidance and monitoring systems, will make this a more cost-effective technique.

Figure 3.2.4.4

Elevation View of Directional Hole

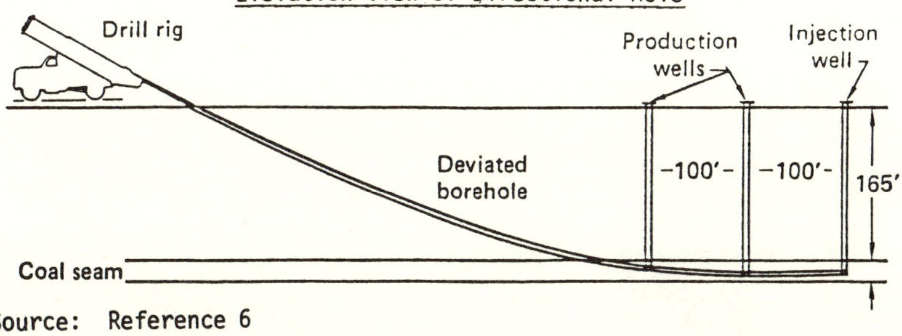

Source: Reference 6

Both the Soviets and U.S. experimenters have successfully utilized the two linking techniques described above. Because of the importance of the linking step to the cost and reliability of a UCG plant, additional techniques are being studied as well. These include the use of water jet drilling and explosive "shaped charges."

Gasification

When the links are complete between the injection and production wells, the air (or steam and oxygen) injection rate is increased and the forward burn (gasification) phase is initiated (refer again to Figure 3.2.4.1). The primary considerations during gasification are to maintain good control over the raw gas production rate and composition (heating value). This is accomplished (assuming a well-placed link path) by maintaining the oxidant injection rate at the proper level to control the rate of water influx into the gasification zone.

The design of a large-scale gasification system must take subsidence into account. In essence, the layout of injection and production wells can be established such that: 1) coal pillars are left between parallel channels to support the overburden and isolate each gasifier from the other (analogous to room and pillar mining, with similarly low resource recovery); or 2) no coal is left between the channels, and the interconnected gasifiers advance in a broad front with subsidence occurring immediately behind (analogous to longwall mining). The Soviets appear to have worked with both systems. Operations in this country have not reached large enough scale to test either type of operation yet.

In the case of steeply dipping beds, very severe but localized subsidence will occur due to the large vertical component of the coal being removed. The Soviets observed some massive subsidence when gasifying steeply dipping coal seams, but this did not disrupt the operations when the air injection was through the underburden (see Figure 3.2.4.5).[7]

Figure 3.2.4.5

Gasification of a Thick Sloping Bed
(Yuzhno-Abinskaya)
Behavior of Roof

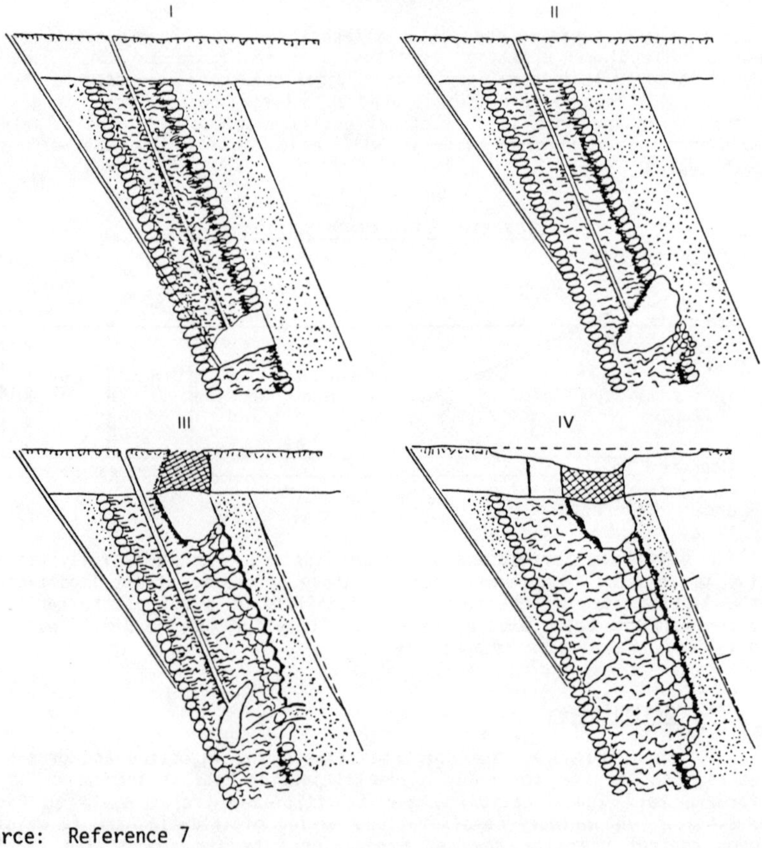

Source: Reference 7

Product Gas Cleanup and Utilization

The operation of an underground coal gasifier is closely analogous to the fixed bed gasifiers described in Section 3.5.3. The primary differences in the raw gas produced are: (1) typically, UCG product gas has somewhat lower heating value because of system inefficiencies; (2) the tar/water fraction, while similar to that obtained from a fixed bed gasifier, tends to be less in quantity and lighter in quality (due to apparent distillation effects in the product removal system); and (3) particulate loadings, including some occasional large chunks, are greater in UCG.

Product treatment and upgrading systems that can be utilized in UCG are essentially the same as those in surface-based gasifiers. Refer to Section 3.5.3 for a discussion of the various options for producing SNG, chemical feedstocks, or clean fuel gas.

Field Instrumentation and Monitoring

In a process development such as those currently being conducted in the United States, the use of extensive arrays of surface and downhole diagnostic instrumentation to observe the process is warranted. The most useful of these systems to date have been downhole thermocouple arrays, used in combination with a variety of other in situ and remote (seismic, electromagnetic) measurements. From these efforts, a reproducible model of the process is beginning to emerge.[5]

In future commercial plants, cost considerations will dictate that downhole instrumentation be utilized only in wells drilled for other necessary purposes, if at all. Some remote sensing techniques may be developed to the point where they can provide useful diagnostic information at a low cost. Otherwise, it will be necessary to rely primarily on a good basic process understanding developed through on-site experience and experiments.

Post-Burn Monitoring

Again, in a process development effort, extensive post-burn coring or excavation is justified in some cases to determine the exact extent of the reaction zone, etc. In addition, water quality and subsidence monitoring should be continued for a significant period of time. In a commercial operation, site monitoring will be generally restricted to the latter type as required for environmental compliance.

3.2.4.2 Environmental Control Technology

The major environmental impacts associated with a large scale UCG operation are:

1. Subsidence (in severe cases, accompanied by leakage of process gas to the atmosphere).

2. Groundwater impacts - both disruption of aquifer flow through subsidence, and contamination with by-products of gasification (e.g., phenols).

3. Land use effects (drilling, wellheads, piping, etc.)

4. Air quality effects associated with the surface-based process operations.

The first two items above are of great enough concern that they may ultimately limit the large-scale development of UCG.

As mentioned earlier, the system design can determine, to a large extent, the nature and timing of the subsidence pattern. The seam thickness, depth, and dip, as well as the overburden characteristics, are also critical variables. To date, no direct observation of surface subsidence during UCG has been made in this country due to the small scale of operations. (However, the surface has subsided shortly after completion of UCG experiments at Hoe Creek, near Gillette, Wyoming.) Interpretation of Soviet data, plus the analogous experience gained from underground mining, indicates that good engineering principles can be applied to control the subsidence pattern as desired.

Aquifer disruption problems are essentially the same as those associated with surface or underground mining. Removal of the coal and subsidence of overlying strata will, in most cases, disturb groundwater systems, whether the coal is removed by mining or by gasification. The solution to this generic problem lies mainly in proper site characterization and selection.

The complementary problem of groundwater contamination through contact with gasification products is, however, unique to UCG. Field data on this problem are very limited to date. The data that do exist (supported by laboratory and modeling efforts) indicate several interesting facts:[8,9] (1) the surrounding unreacted coal acts as an effective sorbent for most of the bothersome organic species such as phenols (concentrations fall off very rapidly both with distance and with time); (2) the inorganic species (sulfates, etc.) that are leached from the ash left in the reaction zone when groundwater flow is restored may be of greater concern.

Both the land use and air quality effects associated with UCG are essentially similar to those of surface-based gasification plants, and will not be discussed in detail here.

3.2.4.3 Effects of Low-Rank Coal Properties

Since UCG is a combination extraction/conversion process, the properties of coal that affect this technology include both the geologic/geographic factors that affect coal extraction, and the chemical and physical properties that govern the behavior of coal during combustion and gasification. As a general conclusion, the combined effects of the factors listed below cause low-rank coals to be the preferred resource for UCG.

Seam Thickness

Seam thickness is one of the most important parameters affecting UCG. As far as is known today, process performance and economics improve as seam thickness increases. For seams less than 10 feet thick, the effect is very pronounced; according to Soviet data, gasification of seams less than about 5 feet thick produces gas of unacceptably low heating value (see Figure 3.2.4.6).[4] This is attributed to the excessive loss of heat to the overburden and underburden, which reduces gasification efficiency dramatically. This effect becomes relatively unimportant above seam thicknesses of about 15 feet; however, very thick seams still have the advantage that more coal is accessed per borehole, which has a direct impact on gas cost. To date, most of the low-rank coal UCG experiments in the United States have been conducted in seams 25 and 30 feet thick, with generally good results. One test in a 100-foot seam obtained excellent results (see Section 3.2.4.4).

Figure 3.2.4.6

<u>Effect of Coal Seam Thickness and Water Influx Rate on Gas Heating Value</u>

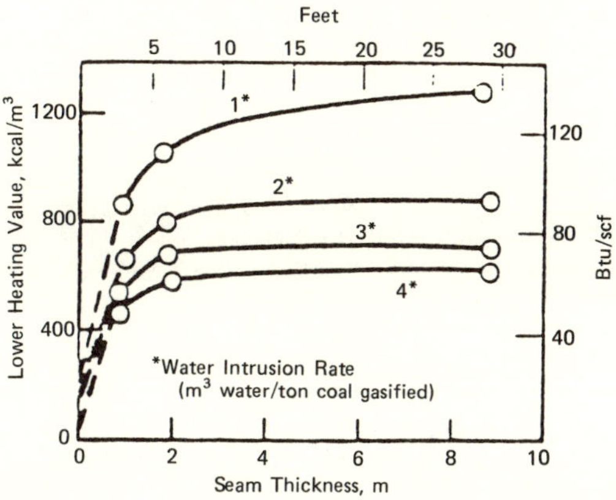

Source: Reference 4

Hydrology

The rate of water influx to the gasification zone is another basic process variable. Figure 3.2.4.6 also indicates this effect, based on Soviet data. Over the range of water intrusion rates shown there, gas quality declines as the water influx rate increases. Again, this is attributed to excessive heat loss in the system (heat required to boil the excess water). In theory there is an optimum water rate associated with gasification, based on the chemistry of the steam/char reaction and others (see section 3.5.3). In practice, coal seams which are dry enough to allow operation near the optimum may be rare.

Positive control over the water influx rate is difficult to achieve. The Soviets utilized dewatering pumps in some of their very wet locations. Given the right site conditions, it should be possible to maintain some control over the water influx rate by adjustments in air injection rate. In large-scale systems experiencing subsidence, large quantities of water could enter the gasification zone from overlying aquifers.

Seam Continuity

It is important that the coal seam be continuous over the area to be gasified. Large faults, fractures, washouts, and other discontinuities can interrupt the process advance. Inclusions in the coal seam, such as shale stringers and partings, can also affect linking and gasification operations.

Depth

The effects of depth on the process have not been completely characterized. It is necessary to have a positive seal between the surface and the gasification zone, which translates to approximately 200 feet of overburden. Beyond that, one known effect of depth is to increase the drilling cost. In addition, the hydrostatic head increases roughly at the rate of 1 psi/foot of depth, so that higher pressure operation is allowed in deeper coal seams.

Dip

As described earlier, the process used for steeply dipping beds (greater than about 30° angle of dip) is somewhat different than for flat-lying coal seams. The SDB process configuration has some inherent advantages, such as minimum drilling cost per ton of coal assessed, and the ability to locate all boreholes out of the subsidence zone.

Permeability

The native permeability of the coal seam to gases and water is of obvious importance. Low-rank coal seams have permeabilities that are orders of magnitude higher than bituminous coal seams. It is therefore significantly easier and less expensive to conduct UCG operations in low-rank coal seams.

Caking/Swelling Characteristics

As bituminous coals are heated, they pass through a temperature range in which they become plastic, swell to several times their initial volume, and agglomerate to form a sticky mass. Low-rank coals do not have this property; on the contrary, when air-dried and heated, low-rank coals have a tendency to shrink and crumble. This is a very desirable property for UCG, allowing the formation of a "packed bed" underground reactor with excellent gas-solid contact and high coal surface area exposed for reaction.

Reactivity

Low-rank coals are highly reactive (both to oxidation and gasification reactions) compared to higher-rank coals. This is another factor strongly favoring low-rank coals for UCG. Ignition of the coal seam, which can sometimes be a troublesome operation, is significantly easier in low-rank coals. In addition, the char formed from pyrolysis of low-rank coal is much more reactive than the char from high-rank coals (see section 3.5.4). This means that utilization of the injected air or oxygen should be significantly better in a low-rank coal seam, since it will tend to react immediately upon contact with fresh coal or char, rather than bypass the gasification zone to react with the product gas.

Mineral Matter

The alkaline mineral matter in the U.S. low-rank coals is hypothesized to have catalytic effects on gasification reactions. If true, this would enhance the reactivity of the coal and favor more efficient use of reactants to produce high-quality product gas.

Site Location

This factor, while not a property of the coal, can be treated as such in considering UCG technology because the process is by definition located where the resource is located. As in surface gasification systems, the product options are: (1) low-Btu fuel gas produced by air injection, which must be used on-site due to high transport costs; (2) medium-Btu fuel or synthesis gas produced by steam-oxygen injection, which has a somewhat wider range of transportability (several miles at least); (3) SNG produced by upgrading item 2 to pipeline gas quality, for long-distance transmission and substitution in conventional natural gas utilization systems; and (4) liquid fuels or chemicals produced from synthesis gas (i.e., indirect liquefaction).

Except in the case where SNG, liquid fuels, or chemicals are to be produced on-site, it is very desirable that the coal to be gasified is located close to the user of the product gas. In the U.S. low-rank coal regions, there are relatively few locations where major energy markets coincide with the coal (one such location is Texas; another may be Washington state). Thus it may be expected that UCG plants might first be located in these areas. Alternatively, plants located in remote areas (e.g., Wyoming) would feed their gas to "minemouth" stations for production of electricity, SNG, liquid fuels, or chemicals, which would then be transported to the markets.

3.2.4.4 Current Projects

Most of the UCG field experiments conducted in the United States have been under the auspices of the DOE R&D program. However, a significant number of industry projects are also being conducted, as summarized below.

DOE Program[1]

There are four field projects in the DOE Program: (1) low-Btu gasification of low-rank coal, directed by the Laramie Energy Technology Center (LETC), (2) medium-Btu gasification of low-rank coal, directed by Lawrence Livermore Laboratory (LLL), (3) Eastern coal technology, directed by Morgantown Energy Technology Center (METC), and (4) steeply dipping beds, directed by Gulf R&D Company. Sandia Laboratories is providing diagnostic field instrumentation support to LETC and LLL. There also is a supporting laboratory program, including participation by LLL, LETC, University of Wyoming, METC, University of Washington, West Virginia University, Oak Ridge National Laboratory, Argonne National Laboratory, Los Alamos Scientific Laboratory, University of Texas, and University of Alabama.

Low-Btu Gasification of Low-Rank Coal (LETC)

Since 1972, the Laramie Energy Technology Center has been using reverse combustion as the linking technique in a series of successful field tests at Hanna, Wyoming. The subbituminous coal seam is 30 feet thick. The performance parameters of these tests are summarized in Table 3.2.4.1.

Table 3.2.4.1

Low-Btu Gasification of Low-Rank Coal Test Series at Hanna, Wyoming

Test	Gas Quality (Btu/scf)	Production Rate (scf/day)	Duration (days)	Thermal Efficiency %	Tons of Coal Gasified
Hanna 1	126	1,600,000	180	-	4,000
Hanna 2-1	152	2,700,000	35	83	1,260
Hanna 2-2	175	8,500,000	25	89	2,520
Hanna 2-3	138	12,000,000	38	76	4,200
Hanna 3	138	10,000,000	38	76	2,850
Hanna 4A	90	13,000,000	80		
Hanna 4B	------------------	not yet published	------------------		

Source: Reference 5

In the course of these field tests, 11 out of 12 reverse combustion linkages were completed successfully. The Hanna 2, Phase 2 test was in many respects the most successful UCG test ever conducted; a constant and high heating value of product gas and excellent thermal efficiency were obtained. This series of tests not only demonstrated process feasibility but identified the important design and operational factors necessary for successful gas production from a thick, flat-lying low-rank coal seam.

Medium-Btu Gasification of Low-Rank Coal (LLL)

Since 1972, the Lawrence Livermore Laboratory has been studying permeability enhancement (linking) techniques and steam-oxygen gasification. LLL has completed three underground coal gasification tests at the Hoe Creek site near Gillette, Wyoming, as shown in Table 3.2.4.2. The Felix No. 2 coal seam contains subbituminous coal and is about 25 feet thick. Explosive fracturing was used to link Hoe Creek No. 2. An override situation developed with Hoe Creek No. 2, but this was overcome by injection near the bottom of the seam, which allowed for 58 days of forward gasification. A 2-day oxygen burn was executed with no safety or operational problems and produced a more efficient gasification than with air. Successful new diagnostics used in this test included a suite of subsidence measurements and high-frequency electromagnetic detection of the burn front. Hoe Creek No. 3 was a steam/oxygen gasification test using a directionally drilled channel to link the three process wells (refer back to Figure 3.2.4.4). This test was recently completed (November 1979); data had not been published at the time of this writing.

Table 3.2.4.2

Medium-Btu Gasification of Low-Rank Coal Test Series at Hoe Creek, Wyoming

Test	Gas Quality (Btu/scf)	Production Rate (scf/day)	Duration (days)	Thermal Efficiency %	Tons of Coal Gasified
Hoe Creek No. 1	110	1,700,000	11	73	128
Hoe Creek No. 2	106	3,300,000	58	68	2,300
(oxygen test)	(263)		(2)		
Hoe Creek No. 3	Directionally drilled link, steam/oxygen; completed 11/79				

Source: Reference 5

Eastern Coal Technology

The Morgantown Energy Technology Center is attempting to develop a viable underground gasification process for the recovery of energy from swelling bituminous coal. Early work involved laboratory simulation studies, modeling support, and a low-priority field effort. In the field effort at Pricetown, West Virginia, a directionally drilled well was completed in the 6-foot thick Pittsburgh coal steam for a horizontal distance of 500 feet. METC has recently completed Pricetown 1, a minitest for investigation of reverse combustion linkage in swelling bituminous coal. Successful operation was achieved, but data had not been published at the time of this writing.

Steeply Dipping Beds (SDB)

DOE signed a contract with Gulf Research and Development Co. in September 1977 to conduct underground coal gasification in a steeply dipping coal seam. The site near Rawlins, Wyoming has a subbituminous coal seam with a dip of 64 degrees and an average thickness of 20 feet. Burn No. 1 started in October 1979 and consisted of two process wells linked by slant drilling (refer again to Figure 3.2.4.3).

Field Instrumentation Support

Diagnostic instrumentation and interpretation of the results of field tests are vital to the understanding of the UCG process. Sandia Laboratories has concentrated on the development of instrumentation and process control techniques for the program.

Sandia's efforts have been primarily associated with LETC's experimental program at Hanna, Wyoming. Major instrumentation efforts have been fielded on the Hanna 2 and Hanna 4 experiments with over 600 channels of information recorded from instrumentation wells and extensive surface arrays (see Figure 3.2.4.7).

Figure 3.2.4.7

Diagnostic Instrumentation Fielded at Hanna 2 and 4

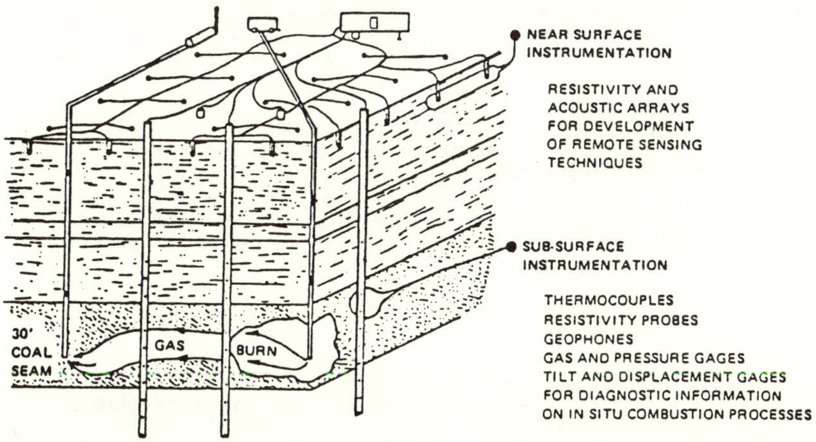

Source: Reference 1

Of all the information obtained, the thermal data have been the most informative. They have delineated the location of the reverse combustion link(s), outward and upward progress of the gasification zone, and the utilization of the coal within the experimental area.

Detailed thermal analyses have contributed to process understanding. On-site, real-time data reduction has allowed the constant monitoring of test progress so that timely decisions on process control could be made. Also, gas sampling techniques have been proven successful in obtaining in situ gas compositions and gas pressure distributions.

Results from the remote monitoring techniques were also promising. Passive acoustic monitoring has been used to locate the source of signals occurring in the strata just above the coal seam, and correlations with complete removal of coal and subsidence have been made. An induced seismic method, based upon detecting changes in signals caused by the process from borehole to borehole, detected the edges of the affected regions. The most promising remote technique is based upon the large change in electrical resistivity which occurs in the coal seam during the UCC process. Development of remote techniques is vital because commercial operations will require less costly surface diagnostics rather than downhole instrumentation.[2]

Supporting Research [1]

The DOE field projects are supported by a broadly based laboratory and modeling program to develop a better understanding of the process, as summarized briefly below:

- Argonne National Laboratory - Steam-char kinetics
- Oak Ridge National Laboratory - Block pyrolysis experiments
- LETC and University of Wyoming - Forward and reverse combustion process models; economic parameter models; laboratory block simulations of UCG
- Lawrence Livermore Laboratory - Two-dimensional coal recovery model; product gas composition model; laboratory gasifier facility; laboratory, modeling, and field studies on UCG groundwater and air pollution; and subsidence
- West Virginia University - Coal conductivity measurements; one-dimensional process model; thermo-mechanical simulation model
- University of Washington - Laboratory work on stress/mass transfer coupling; theoretical studies of reverse combustion
- University of Alabama - Physical properties of char; shrinking core model
- University of Texas - Mathematical model of override conditions; mechanical and ion exchange properties of Texas lignite overburden
- Science Applications, Inc. - Analysis and interpretation of METC's laboratory UCG simulation data; subsidence models
- SRI International - Market evaluation
- Booz, Allen & Hamilton Inc. - Preliminary UCG pilot plant design and cost estimate; coal resource summary

Industry UCG Projects[11]

In the United States, private sectors and government have a history of cooperation in underground coal gasification. The U.S. Bureau of Mines, in cooperation with the Alabama Power Company and Sinclair Coal Company, performed a number of UCG tests near Gorgas, Alabama, from 1947 to 1959. Following these tests, there was no activity in UCG until 1967, when a UCG test was carried out by Cities Services Oil Company in Oklahoma bituminous coal, and in 1969 when Gulf Research and Development Company performed a UCG test in a bituminous Kentucky coal seam.

Following the encouraging results obtained in the DOE projects, several energy companies, including the Resource Sciences Corporation, Gulf, Exxon, The Atlantic Richfield Company and Texas Utilities Company, began detailed internal studies. Gulf Research and Development Company and a consortium including the Research Sciences Corporation, Rocky Mountain Energy Company, Amoco, duPont, and Pacific Gas and Electric Company performed critical reviews of the Lawrence Livermore Laboratory's medium-Btu project at no cost to the government. Rocky Mountain Energy Company donated the site at Hanna, Wyoming where LETC has conducted successful UCG tests. Personnel from Gulf R&D Company and from Exxon cooperate with LETC on a no-funds-exchanged basis to the benefit of both government and industry.

Three organizations have recently concluded significant field tests; these are summarized below.

Basic Resources, Inc. (A Subsidiary of Texas Utilities Co.)

In March 1975, Texas Utilities Company signed a license agreement with Licensintorg of the U.S.S.R. for technical documentation and assistance in UCG. A technological test was designed and executed in 1976 to prove technical feasibility near Texas Utilities' Big Brown Steam Electric Station in Freestone County, Texas. The test was classified as a success. The experiments were then moved to the Tennessee Colony site in Anderson County, Texas. A two-phase field test, including both air and oxygen/steam injection, was carried out from August 1978 to March 1979 in a multichannel, multiwell configuration. Reverse combustion was employed to link the wells. The lignite seam is 7-1/2 feet thick at an approximate depth of 270 feet. The steam-oxygen feasibility field test was conducted in cooperation with Air Products and Chemicals, Inc., in a four-well parallel operation. Gas with a higher heating value of up to 265 Btu/scf was produced during the test and an average value of 230 Btu/scf was obtained during the main portion of the burn. Steam-oxygen volumetric ratios between 0.5 and 4.0 were used.

Atlantic Richfield Company (ARCO)

In 1978 ARCO designed, fielded and executed a successful field gasification test in a 100-foot-thick subbituminous coal seam near Reno Junction, Wyoming. The coal bed depth was about 630 feet with a 50-foot shale overburden and a hydrostatic head of 300 feet. The test included: (1) reverse combustion/forward gasification with air, (2) control of water influx, and (3) environmental monitoring. The DOE informally cooperated with ARCO on process instrumentation and data handling. The DOE, through the Lawrence Livermore Laboratory, formally supported the ARCO test with high-frequency electromagnetic detection of the propagation of the burn front.

Ignition took place in September 1978. Two 75-foot reverse combustion links were propagated, the second during forward gasification. Forward gasification lasted for a period of 2 months with air injection flows of up to 3000 scfm. The average product gas higher heating value for the test was over 200 Btu/scf, which is very high for air injection. The test was terminated on November 20, 1978.

Texas A&M University

Texas A&M University, supported by a group of companies, conducted a UCG test on University property in 1977. In cooperation with a consortium, a second test was recently carried out near Rockdale, Texas. The lignite is 14 feet thick at a depth of 235 feet. A reverse burn link over a 50-foot distance was completed and forward gasification was conducted.

In addition to these projects, the Resource Sciences Corporation's subsidiary, Williams Brothers Process Services, Inc., is actively involved in client-sponsored studies of economics, cleanup and gas utilization for UCG, including studies of air gasification of Wyoming coal and steam/oxygen gasification of Texas lignite. The Public Service Company of New Mexico, together with the University of New Mexico and the Los Alamos Scientific Laboratory, have been supported by DOE to initiate prospective UCG site characterization studies in the San Juan Basin of New Mexico.

The natural gas industry provides formal, direct support to DOE via the Gas Research Institute (GRI). GRI co-sponsors and co-funds the LLL UCG project for medium-Btu gasification.

References - Section 3.2.4

1. U.S. Department of Energy Underground Coal Conversion Program Description, DOE-ET-0100, June 1979.

2. Stephens, D.R., "Costs of Drilling, Completing and Linking Process Wells for Underground Coal Gasification as a Function of Linking Method, Coal Bed Thickness and Depth," in Proceedings of the 5th Underground Coal Conversion Symposium, Alexandria, Virginia, June 18-21, 1979, Conf. No: 790630, U.S. DOE, printed May 1979.

3. Olness, D.U. and D.W. Gregg, The Historical Development of Underground Coal Gasification, Lawrence Livermore Laboratory, Rept. UCRL-52283, 1977.

4. Gregg, D.W., R.W. Hill, and D.U. Olness, An Overview of the Soviet Effort in Underground Coal Gasification. Lawrence Livermore Laboratory, Rept. UCRL-52004, 1976.

5. Brandenburg, C.F., "Field Implementation of UCC Research," in Proceedings of the 5th Underground Coal Conversion Symposium, Alexandria, Virginia, June 18-21, 1979, Conf. No: 790630, U.S. DOE, printed May 1979.

6. Hill, R.W., et al., "LLL 1979 Field Program," in Proceedings of the 5th Underground Coal Conversion Symposium, Alexandria, Virginia, June 18-21, 1979, Conf. No: 790630, U.S. DOE, printed May 1979.

7. Gregg, D.W., Ground Subsidence Resulting from Underground Gasification of Coal, Lawrence Livermore Laboratory, UCRL-52255, March 29, 1977.

8. Mattox, C.F., and M.J. Humenick, "Organic Groundwater Contaminants from UCG," in Proceedings of the 5th Underground Coal Conversion Symposium, Alexandria, Virginia, June 18-21, 1979, Conf. No: 790630, U.S. DOE, printed May 1979.

9. Mead, S.W., H.C. Ganow, and F.T. Wang, "Groundwater and Subsidence Investigations of the LLL In Situ Coal Gasification Experiments," in Proceedings of the 4th Underground Coal Conversion Symposium, Steamboat Springs, Colorado, July 17-20, 1978, SAND 78-0941, U.S. DOE, printed June 1978.

10. Development of In Situ Coal Gasification Technology for Production of Low and/or Medium Btu Gas, Department of Energy Program Opportunity Notice 03-80ET14130, November 1979.

11. Stephens, D.R., "The Private Sector Involvement in Underground Coal Gasification," in Proceedings of the 5th Underground Coal Conversion Symposium, Alexandria, Virginia, June 18-21, 1979, Conf. No: 790630, U.S. DOE, printed May 1979.

3.2.5 Selected References - Extraction

The following publications are particularly informative on the subjects of coal extraction by surface mining, underground mining, or underground gasification; or on the effects of low-rank coal properties on those processes. Many other publications, as listed under References at the end of each section, provide data on specific subjects within these technical areas.

1. Cassidy, S.M. (Editor), Elements of Practical Coal Mining, AIME Mudd Series, New York, 1973, 614 pp.
   This book is designed as a training manual and reference for persons in or expecting to enter the coal mining industry. It is a beginner's book on the elements of practical coal mining that is suitable for vocational schools, high schools, community colleges, etc. However, it is also extremely comprehensive and informative on all aspects of surface and underground coal mining. Emphasis is placed on mining techniques and equipment, with numerous photographs, tables, and charts to illustrate and clarify the material presented. Each of the 23 chapters was written by an expert with practical (rather than academic) first-hand experience and knowledge of the subject.

2. Pfleider, E.P. (Editor), Surface Mining, AIME Mudd Series, New York, 1972, 1061 pp.
   This book is an authoritative text on surface mining, intended for use in mineral engineering schools and as a comprehensive reference book for the industry. It is not limited to coal mining (although there is a 60-page section specifically covering coal); rather, all of the techniques and equipment utilized in the extraction of minerals by surface mining are covered, including research and development efforts and organizational/management aspects of the technology. Eighty-four individual authors, each a recognized leader in his particular field of surface mining, contributed individual chapters. Extensive reference lists are provided.

3. Grim, E.C., and R.D. Hill, Environmental Protection in Surface Mining of Coal, EPA-670/2-74-093, October 1974, 276 pp.
   This in-house EPA report is the result of information obtained from a review of related literature and assembled by personal inquiry and on-site examination of both active and inactive surface mining operations. Premining planning is emphasized and particular attention is given to incorporating mined-land reclamation into the mining method before disturbance. New mining methods that will maximize aesthetics and minimize erosion, landslides, and deterioration of water quality are discussed. Blasting techniques and vibration damage controls are recommended. Methods of land reclamation including spoil segregation, placement, topsoiling, grading, burying of toxic materials, and revegetation are noted. Technology for the control of erosion and sediment in the mining area is presented in detail. Guidelines for planning, location, construction, drainage, maintenance, and abandonment of coal-haul roads are included. Costs are given for different degrees of reclamation and remedial measures for controlling pollution from surface mines. Reduction in costs through preliminary planning are cited. Water quality change is discussed in detail. Preventive and treatment measures are recommended. Research needs are listed as a separate section of the manual. Extensive reference lists are provided.

4. Menchen, W.R., et al., Underground Coal Mining - An Assessment of Technology, EPRI AF-219, July 1976, 457 pp.
This report was prepared by Hittman Associates, Inc. in association with Pennsylvania State University and George Washington University, to: provide government and industrial decision-makers with an understanding of alternative technologies available for increasing the production of coal by underground mining in the time frame 1974-2000; apply the methodology, techniques, and processes of technology assessment to domestic underground mining of coal; identify advanced technologies and systems for increased production of deep mined coal and develop alternative R&D and institutional strategies that may make such technologies viable; identify major areas of uncertainty facing the future of the underground coal mining industry, determine the sensitivity of increased production to these uncertainties and formulate strategies and timetables for the resolution of the uncertainties; identify major alternatives for filling institutional gaps, if any, between the various industries involved in the extraction and utilization of underground mined coal; and determine alternative levels of underground mine productive capacity by coal mining region in the United States.

5. U.S. Department of Energy Underground Coal Conversion Program Description, DOE-ET-0100, June 1979, 69 pp.
This report presents a summary of the DOE research program in underground coal gasification, and was prepared by DOE's program management team, including the major field and laboratory project managers. It is a useful reference in that it presents, in executive summary form (with references), the current status of the technology, the major environmental, economic, and institutional constraints and uncertainties, and descriptions of the various industrial and government projects comprising the DOE program.

6. Proceedings of the * Underground Coal Conversion Symposium:
*Fifth - Alexandria, Va., June 18-21, 1979, 449 pp.
Fourth - Steamboat Springs, Col., July 17-20, 1978, 526 pp.
Third - Fallen Leaf Lake, Calif., June 6-9, 1977, 522 pp.
Second - Morgantown, West Va., August 10-12, 1976, 586 pp.
Together, these symposia proceedings constitute a massive and detailed technical data base on all aspects of underground coal gasification technology, with emphasis on the most recent work conducted in the United States. Individual papers vary in their depth, precision, and accuracy; however, no other single source of data on this subject is comparable. In addition to the large number of papers included in the proceedings themselves, extensive reference lists are provided which cover practically the entire domestic and foreign literature on underground coal gasification.

Transportation

3.3.1 Introduction and Summary

The transportation of western low-rank coal is affected by a very high volume (or weight)-to-value ratio due to its low heating value. Because of this high ratio, transportation costs can represent a considerable fraction of the delivered cost.[a] Bulk handling systems and/or specialized transportation methods can be applied to obtain economies of scale or economies of optimum design and utilization. A prime example of bulk handling is the unit train concept, which combines the efficiency of large scale "dedicated" shipments with specialized loading and unloading facilities. Slurry pipelines are an example of a specialized transportation system. Barge and rail combinations (rail transport from mine to barge) are examples of optimum utilization of low cost barge transportation. Within the mining area, large haul trucks and conveyor networks are well suited for short haul transport, and offer extreme flexibility.

All of the above-mentioned transportation methods are discussed in the following sections, and the status of their development is summarized below.

With the possible exception of slurry pipelines, the various technical approaches to coal transportation are well established. Critical issues in transportation are legal and economic in nature -- two areas known for their relative instability. For this reason, much of the material presented under "current status" subsections is involved with significant ICC rulings, and many issues are still pending. Therefore, up-to-the-minute reviews of certain legal developments (e.g., ICC decontrol of railroads, railroad mergers) are best obtained from periodical literature.

Railroad

The unit train concept has matured over the past two decades to where, in 1977, over 178 million tons of coal were shipped by unit trains. One-fourth of this tonnage was shipped out of western low-rank coal states such as Montana, North Dakota, and Wyoming.[2] Although shipments utilizing larger capacity rail cars and longer dedicated shipments (integral trains) have been considered, the present combination of 100-ton hopper cars and 100-car unit trains will most likely remain the typical mode of long distance rail shipments of western coal.

---

[a]For example, subbituminous coal from Colstrip, Montana (8600 Btu/lb) has a contract price of \$8.50/ton FOB mines and a unit train transportation cost of \$11.12/ton delivered in Minneapolis, Minnesota, or a transportation cost of 57 percent of the delivered price. Higher priced coal (due to higher Btu content) would have proportionately lower transportation costs.[1]

Future coal movements will place a great reliance on unit trains. The railroad industry possesses an extensive in-place network of tracks reaching to most of the major energy-consuming sections of the nation. This track system has the capability and flexibility to transport large quantities of coal over long distances. Unit trains also appear to be a relatively energy-efficient mode of transport. Of importance to the western states is the relatively small amount of water required by railroads as compared to the amounts required by slurry pipelines and mine-mouth power plants. Western railroads will be a major beneficiary of increased coal consumption.

The greatest uncertainty affecting railroad movement of coal will be the ability of the railroads to control operating costs and the problems relating to vehicle traffic interference at grade crossings. There are no apparent cost-cutting technological advances in the future that would be comparable to the advent of the unit train. Large increases in rail rates will particularly affect the long-haul markets for Western coal. This may be a major driving force for the development of a Texas lignite mining industry. High rates will also encourage the development of coal slurry pipelines.

Barge

Transporting coal by barge is one of the most economical methods available, providing both the coal origin and the destination are near barge facilities. In the case of western low-rank coals, coal mines may be 500 to 800 miles from the nearest inland waterway. Rail-to-barge shipments (combining unit-train delivery of coal to barge loading facilities) are not uncommon, and at least 12 million tons of Montana subbituminous coal are delivered to barges and towed upstream to utilities near Chicago and Detroit (by way of Lake Michigan).

Barge transportation can move approximately 60,000 tons of coal in one tow. Barge groups may contain as many as 45 barges at one time and can be towed by one boat at speeds of about 5 miles per hour. Integrated tows provide an efficient method of transporting large tonnages of coal over long distances. Barge transportation becomes most efficient when few or no locks are present on the route and when annual tonnages exceed 800,000 tons per year.[3]

As with unit trains, operating costs make up a large percentage of the total costs associated with a tow. Barge lines do not face increased costs for roadway upgrading, as do the railroads. In addition, the barge industry is subject to less ICC regulation than the railroads, and has more freedom to dictate rate structures. Therefore, future barge rate increases should parallel the general rate of inflation, and should become increasingly attractive with respect to rail rates.[3]

Trucks and Conveyors

The technical state-of-the-art in truck and conveyor design has not changed dramatically over the past few decades. Trucks have increased in size and capacity; conveyors are now using more durable belt materials and are becoming somewhat more specialized in design. The significant changes occur in the application of these transportation modes within large surface-mining operations. With the aid of computerized calculations, surface mine planning has become both an art and a science, and more mine operators are turning from trucks to conveyors for more efficient movement of overburden and coal.

Although the number of different applications and combinations for conveyors (and truck/conveyor systems) is potentially unlimited, several popular applications involve: continuous removal of blasted overburden from deep surface mines; cross-pit transport (a significant improvement over truck haulage in this application); load-matching with a bucketwheel excavator; and loading and unloading operations paired with trucks, trains, shovels, or dozers.

Conveyor systems on the market today are available in self-propelled and shiftable models to facilitate rapid relocation and maximum flexibility. Almost any length of conveyor is available up to 3000 feet, but special applications utilize single units 4000 feet in length.

Trucks are a very flexible means of transportation and will continue to perform as vital links in the surface mine movement of coal. However, as transportation fuel and other operating costs increase, more use will be made of conveyor transport of materials, from simple one-section operations up to elaborate networks of conveyors transporting material over several miles.

Slurry Pipelines

The movement of coal by slurry pipeline has yet to become an established transportation option. Slurry pipelines are subject to a number of possible technical and legal problems:

1. Separation and cleanup of the slurry water at the receiving end of the pipeline has not been adequately demonstrated for all locations. At the Mohave power plant (receiving end of the only operating coal slurry pipeline in the U.S.), clarified slurry water from thickener tanks is used in the plant's cooling system. Previously, the unclarified water containing 5-20 percent residual solids had been evaporated in holding ponds on the desert. These techniques may not be acceptable at other locations.

2. Small-scale tests of slurry transport of lignite indicate it will produce excessive fines due to the deterioration of particle size during transport. Fines smaller than 40 microns are difficult to dewater at the receiving end of the pipeline, and may have to be disposed of in settling ponds as waste.

3. If flow is stopped for any reason, the coal particles will eventually settle out of suspension. Restarting a settled slurry requires extremely powerful pumps to overcome static pressures in the pipe, and the success of completely re-suspending the coal has not been established.

4. Pump wear and coal particle abrasion are items of concern.

5. Obtaining water rights for slurry preparation needs is a time consuming procedure and can possibly halt development of a pipeline proposal. On the other hand, operating pipelines would consume 15,000 to 20,000 acre-feet of water annually in western regions known for extended droughts and limited water resources.

Slurry pipelines are very capital intensive. In fact, about 70 percent of the pipeline shipping costs are related to capital expenditures. Pipelines are thus economical only when shipping large quantities and constant throughputs. When transport distances are greater than 500 miles and annual shipments approach 10 million tons per year, the slurry pipeline shows an advantage over other competing forms of coal transportation.[3]

Coal slurry pipelines, whether actually constructed or not, will present a threat to the present unit train dominance in coal transportation. In Cadiz, Ohio, an inoperative coal slurry pipeline has been an effective deterrent to rail freight rate increases. In light of recent rate hike requests by railroads serving the western coal fields, it is possible that western coal slurry pipelines may become a reality within the next decade.

Transportation Costs Comparison

On a cost per ton-mile basis, barges remain the most cost-effective mode of coal transportation, followed by unit trains and slurry pipelines. Figure 3.3.1.1 compares each transportation option as a function of distance. Slurry pipeline costs are rough estimates, since only one, the Black Mesa pipeline, is actually operating.

Figure 3.3.1.1

Cost Comparison per Ton-Mile of Various Coal Transportation Systems

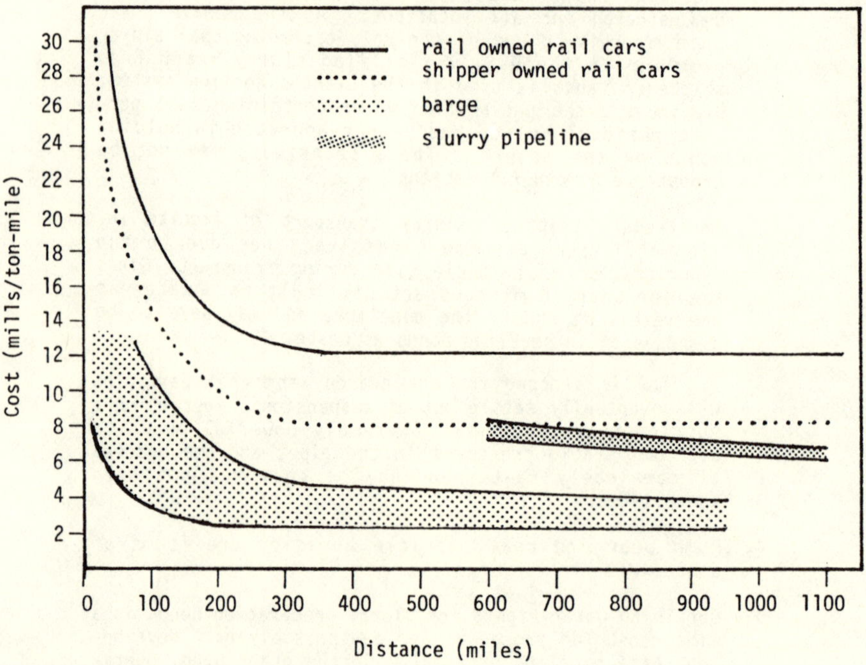

Source: Rail: Reference 3
Barge: Reference 4
Pipeline: Reference 5

It is interesting to note that the energy efficiency of all three major modes of coal transport is approximately the same. As shown in Table 3.3.1.1, railroads have a slight advantage. The steel wheel on steel rail system produces the least wasteful friction during transport, surpassing the floating barge.

Table 3.3.1.1.

Energy Efficiencies of Various Coal Transportation Systems

Transport System	Energy Consumption (BTU per ton-Mile)
Railroad	370
Slurry Pipeline	410[a]
River Barge	440
Truck	2500

[a]Excluding coal grinding

Source: Reference 3

Conversion Product Transportation

An alternative to the transportation of coal directly from the mine to the user is the conversion of the coal to a more valuable and/or more easily conveyable product. Examples of this conversion are minemouth conversion of the coal to electricity and transmission of the electricity, and the conversion of coal to a gas or a liquid and the transmission of the product via pipeline. The economics and operations of pipeline and electric transmission lines are well established. In addition, any effects of low-rank coal properties on these technologies are observed in the conversion operation rather than the product transport phase.

The decision of whether to transport the coal or the product is, of course, a partial function of coal properties (primarily heating value); many other site/route-specific variables are also important. Most energy transportation studies have concluded that it is generally cheaper to transport the coal than to transport electricity. Based simply on the existing industry practice, the economic cutoff point between transporting coal and utilizing it at the minemouth is currently somewhere between subbituminous coal and lignite. Most lignite-fired power plants in the U.S. today are at the minemouth. A large percentage of the subbituminous coal is transported to demand centers; however, a number of minemouth subbituminous coal-fired power plants which supply power to remote markets exist or are planned.

Key Issues

The key issues in western low-rank coal transport arise primarily from regulatory and environmental concerns. The technologies are well established except for coal slurry pipelines. In the latter case, the necessary hardware has been developed for other industries; operational problems that are still unresolved present the most significant technical problems among all transportation modes.

Table 3.3.1.2. presents a summary outline of identified key issues in low-rank coal transportation. Detailed discussions of these issues can be found within the following sections on each transport system.

Table 3.3.1.2

Key Issues in Low-Rank Coal Transportation

Railroad		
	Technical	- no significant issues.
	Regulatory	- current decisions by ICC and Congress regarding rate hikes and less regulatory control. The effects of these issues on future coal traffic are not clear.
	Environmental	- socio-economic effects of increased unit train traffic on small western towns.
Barge		
	Technical	- no significant issues.
	Regulatory	- barge transportation is not significantly affected by regulatory constraints. This may change in the future as barges are required to absorb a larger fraction of waterway maintenance. Price structures for coal shipments by barge will be affected.
	Environmental	- no significant issues.
Trucks and Conveyors		
	Technical	- no significant issues.
	Regulatory	- no significant issues.
	Environmental	- no significant issues.
Slurry Pipelines		
	Technical	
		- dewatering coal fines at the receiving end.
		- on-site utilization of treated slurry water.
		- minimizing pipeline water requirements.
		- restart problems if slurry flow is disrupted.
		- freeze prevention of slow moving slurry.
		- maintaining coal suspension over long transport distances.

(continued)

Table 3.3.1.2 (continued)

Regulatory	- obtaining right-of-way along pipeline route.
	- water acquisition rights.
Environmental	- treatment of slurry liquor after coal/water separation at receiving end.
	- reduce water displacement from water-scarce areas in the West.
	- minimize frequency and impact of ruptured slurry pipelines.

References - Section 3.3.1

1. Coal Week, September 10, 1979.

2. Lin, K. Coal Traffic Annual, 1978, National Coal Assoc., Washington, D.C., 1979.

3. Roe, D.E., A.W. Karr, K.R. Lemmerman, and J.E. Sinor, Solid Fuels for U.S. Industry, Vol I, Cameron Engineers, Denver, Colorado, March 1979.

4. Phillips, P.J., Coal Preparation for Combustion and Conversion, EPRI AF-791, May, 1978.

5. Rieber, M., S.L. Soo, and others, Comparative Coal Transportation Costs: An Economic and Engineering Analysis of Truck Belt, Rail, Barge, and Coal Slurry and Pneumatic Pipelines, Vol. 3, Coal Slurry Pipelines, USBM/DOI/FEA, Contract No. J0166163, August 1977.

3.3.2 Railroad Transport

3.3.2.1. Technology Description [1,2]

The unit train is the most efficient railroad transport mode being used today for movement of large quantities of coal from western surface mines to distant markets (utilities). Some coal is also moved by single car (100 tons or less), multiple car (typically 1500 tons), and by trainload (5000 tons or more) shipments. Each of these methods has distinct operational characteristics, primarily in equipment use, which are reflected in rates. By definition, a unit train consists of dedicated equipment (locomotives and cars) moving in continuously scheduled cycles between one loading point and one unloading point. A typical unit train carries 10,000 tons of coal. Automated loading and unloading facilities are integral parts of the unit train transport system as well. Western subbituminous coal mines load two to eight unit trains per day (2 to 4 hours per each 100-to-115-car train), thus accomplishing a nearly continuous flow of coal from the mine to the customer.

The high productivity of a unit train results from the combination of dedicated equipment, schedule operations, absence of switching en route, and quick turnaround. Productivity or efficiency of the transport system can be measured in a number of ways (all of which translate ultimately into cost per ton-mile or Btu-mile). One measure is the tonnage that a single car carries over the course of one year; a car in unit train service generally carries five to six times the amount it would carry in any other type of service.

A typical coal car has a volume of 3,600 to 4,000 cubic feet and is rated at a nominal 100-ton net capacity. By Association of American Railroads rules, a 100-ton car has a maximum weight, fully loaded, of 263,000 pounds on four axles. Modern cars have a dead weight (empty) of approximately 60,000 pounds, so the actual load-carrying capacity of such cars may be slightly in excess of the nominal 100 tons.[a]

The selection of a particular type of car for a unit train is made in conjuction with the selection (or availability) of the unloading facilities at the destination. Coal cars are of two basic types, gondolas and hoppers. Gondolas are flat-bottom open cars (although coal cars with hinged covers were introduced recently) and must be unloaded by means of rotary dumpers which clasp and rotate the cars. Gondolas equipped with rotary couplers at one end provide the shortest dumping cycles; otherwise the cars must be uncoupled for dumping.

Hopper cars are equipped with sloping plate bottoms and gates that open for a quick discharge. Triple-hopper cars are used most commonly, but quadruple hopper cars, with higher discharge rates, are gaining wider acceptance. For greatest rapidity in unloading, cars with longitudinal discharge gates opening a substantial fraction of the bottom are available. They offer remote and automatic operation of the gates and facilitate the simultaneous emptying of several cars over specially designed unloading trestles or hoppers.

Hopper cars are the most common type in use today, but the lighter (and somewhat cheaper) gondola cars with rotary dump systems appear to be favored for future use. Either type of car is compatible with loading facilities at the mine, which are usually set up so that coal from a continuously flowing, overtrack, pneumatically controlled surge hopper fills the cars as they pass beneath it.

Unit trains require locomotives with high traction abilities, in addition to motive power ranging from 1800 to 3600 hp per engine. Typical locomotive requirements for a 10,000-ton unit train range from 3 to 5 3,000 hp units; up to seven power units may be required on routes with steep grades.

Studies of the railroad transport requirements for western coal are in general agreement that major technological changes are unlikely to occur in the next decade.[1,2] The trends of the previous two decades - larger cars, more powerful locomotives, increased total tonnage per train, and refinements in speed and automation of loading and unloading facilities - are expected to continue to some extent, but not at the same pace. Today's 100-ton capacity cars (132 tons total weight fully loaded) have an excellent capacity-to-total-weight ratio, but impose severe stresses on existing rails and roadbeds which generally are suffering from years of deferred maintenance. Experience with 125-ton cars (157 tons fully loaded) has been very unfavorable from the standpoint of track maintenance. Unit trains significantly longer than 100 cars (10,000 tons) have been proposed and may be used in a few cases, but consideration of coupling gear strength and dynamics of long trains indicates that the 10,000-ton shipment will remain typical. Today's typical heavy-duty,

[a]The bulk density of coal ranges from 47 to 67 lb/ft^3, depending on the type of coal, lump sizes, and degree of compaction. Typical bulk densities for low-rank coals loaded into rail cars range from 50-55 lb/ft^3; the effects of moisture content (drying) and coal rank on bulk densities are so small as to be negligible.[3,4].

six-axle locomotive (rated at 3,000 hp net and weighing 200 tons) is well adapted to the loads and speeds currently prevailing, and no major changes are anticipated.

Forecasts of greatly expanded western coal production and transportation have been analyzed with respect to potential impacts or bottlenecks in the railroad system. Freight-car and locomotive builders can meet projected demands for new equipment (although temporary shortages may occur). While most railway lines in the U.S. are underutilized, several major coal-carrying routes are being operated near capacity, and may not be able to support much heavier traffic without rehabilitation, modernization of train-control systems, lengthening and relocation of passing sidings, rail renewals, and, in some cases, electrification.

The two anticipated developments in rail movement of coal are coal car leasing and pre-shipment coal drying. Of the two, the leasing of coal cars is already an established option for anxious coal companies and utilities frustrated by coal movement delays caused by shortages of rail cars. The boom in coal car leasing has stemmed from the growing time lag between increased coal development and improved rail service. Leasing, as compared to actually purchasing the cars, avoids the responsibilities of policing and maintenance of the cars by the coal company and/or the power company.

Among many advantages of using leased coal cars is a reduced freight rate charges from the rail carrier. The railroads pay a mileage compensation when the shipper provides his own cars. This allowance is paid to the leasing company and is passed on to the lessee through a reduced lease cost. The compensation is based upon the type of car and, sometimes, its age. A typical coal car might be allowed 3.75 cents per mile.[5]

Although the reduction in shipping cost is attractive, a coal company could not justify leasing coal cars through mileage compensation alone. The main objective of leasing cars is to assure a supply. ICC regulations dictate that railroad-owned cars be allocated equally. Therefore, when a shortage occurs all rail lines are shorted by the same percentage. For example, during a 50 percent shortage, the railroad would allot 25 cars to a company needing 50, and five to a company that usually receives 10.

Railroad companies have mixed reactions regarding other companies that lease rail cars. Railroads with strong capital backing say they can keep pace with rail car demand without outside intervention. They see leased cars robbing railroad companies of additional revenue. Other railroads, however, appear optimistic with the arrangement, since the financial burden and responsibility of maintenance is removed from the railroads.[5]

The actual savings realized from the use of shipper-owned (or leased) cars are represented in Figure 3.3.2.1.

Generally, it costs 25 to 35 percent more to ship coal by railroad-owned coal cars than with customer-owned (leased) cars.[6] Regardless of who owns the cars, costs per ton-mile decrease significantly with increasing distances up to about 300 miles, then level off to between 0.7 and 1.3 cents/ton-mile. It is important to realize that freight rates for unit trains are not fixed prices: the final rates are negotiated between railroad and coal company representatives, and are easily affected by the immediate availability of other competitive means of transportation.

Drying low-rank coals prior to shipment is another possible way to reduce transportation costs per unit of energy shipped. Determining the actual savings, however, encounters several interesting situations.

Figure 3.3.2.1

Western Unit Train Rates
(1979)

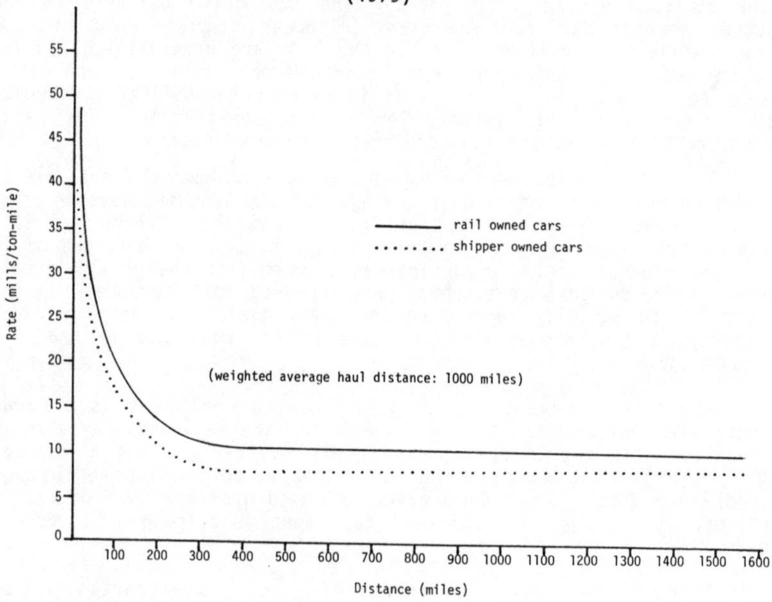

Source: Reference 6

Reducing the moisture content of lignite from 35 wt.% to 21 wt.% results in an increase in heating value (per pound) of over 20 percent. However, the difference in transportation costs may not necessarily justify this value increase. The main determinant is that coal is shipped - and rates determined - on a per-ton basis, not a Btu basis. In addition, weights (per load) are generally based on bulk densities, which, when coupled with the fact that moisture content has little impact on the bulk density of low-rank coals, could result in a negligible reduction in per-ton rates.

A further deterrent is that shipping rates are negotiable: railroads may consider dried coal a "prepared" fuel product instead of a "raw" fuel and charge a proportionally increased rate. This, however, was not done for a shipment of solvent-refined coal (heating value of about 16,000 Btu/lb) from Tacoma, Washington, to Alabama. In this instance, the SRC was shipped at a "raw coal" rate. Savings in shipping costs were thus realized by 1)avoiding a higher shipping rate, and, 2) since the product itself had a higher dollar value, shipping costs represented a smaller fraction of the delivered price of the fuel. If this particular shipment has established a precedent for future shipments of dried or cleaned coal, the savings in shipping costs may be more readily realized.

The potential economic advantages of shipping dried coal are illustrated by the following example. Transportation savings are realized on a delivered-Btu basis: if each ton of shipped coal has a higher heating value, then fewer tons are needed to satisfy the Btu requirements of the consumer. Savings are realized not as a per ton-mile rate reduction, but as a reduced shipment. For example, a 100 car unit train carrying 10,000 tons of lignite at 35 wt.% moisture contains the same Btu content as a train carrying 5 wt.% H_2O in 69 cars (6840 tons).

In different terms, a 1000 MW power plant needing 4.25 million tons of as-mined lignite per year (35 wt.% moisture) requires 425 unit train shipments annually. The same plant, if designed to accept a drier coal, would need only 290 train loads (10,000 tons each) of 5 wt.% moisture lignite. Assuming a drying cost of $2.66 per ton of as-mined lignite[a], Table 3.3.2.1 illustrates the potential savings realized by drying lignite prior to shipment. This example computes only the transportation savings, and assumes a similar negotiated rate for the dried coal as for the as-mined coal. As determined in the table, over five million dollars could be saved annually in transportation costs alone. In addition to these savings, the dried lignite will command a higher price from the consumer (utilities).

Table 3.3.2.1

Cost Analysis of As-Mined vs. Dried Lignite Transport By Unit Train

	As-Mined 35 wt.% H_2O	Dried 5 wt.% H_2O
Basis		
130×10^9 Btu/yr (For a 1000 MW electric generating station)	4.25×10^6 tpy (6300 Btu/lb)	2.91×10^6 tpy (9208 Btu/lb)
Number of annual unit train shipments (10,000 tons/train)	425	291
1000 mile transport @ $0.0125/ton-mile	$125,000	$125,000
Annual transportation costs (millions)	$53.1	$36.4
Cost of coal drying @ $2.66/ton (millions)	$0.00	$11.3
Total cost per year (millions)	$53.1	$47.7
Net annual savings from coal drying (millions)		$ 5.4

Source: Reference 7

[a]Estimated incremental capital costs for drying equipment: $21.19 million
Incremental operating costs: $7.70 million
Capital costs amortized for 15 years at 15 percent interest for a handling capacity of 4.25 million tons per year of 35 wt.% moisture lignite. Dried lignite has a moisture content of 5 wt.% moisture. Data from Reference 7.

Savings may be calculated for any freight rate and distance by the following formula:

$$\text{Savings} = 1 - \left(\frac{\text{Btu as-mined}}{\text{Btu dried}}\right) \times \left(\begin{matrix}\text{freight rate}\\ \text{(\$/ton-mile)}\end{matrix}\right) \times \left(\begin{matrix}\text{distance}\\ \text{(miles)}\end{matrix}\right)$$

3.3.2.2. Environmental Control Technology

There are few environmental control problems unique to railroad shipment of low-rank coal. The most significant problems are due to characteristic properties of low-rank coals and include problems such as fugitive dust, spontaneous combustion, and freezing. These problems are dicussed in more detail in the following section. Rail transportation, in general, creates its own environmental concerns: exhaust from the large diesel engines will emit typical combustion products (CO, CO_2, HC, NO and particulates) and locally high noise levels; frequent unit train traffic can effectively sever small western towns built around rail lines. These problems can be rectified by routine engine maintenance, pollution control equipment, and improved exhaust mufflers. Train tracks can be routed around small towns (though at considerable expense) to avoid disruption of vital municipal services.

3.3.2.3. Effects of Low-Rank Coal Properties

Low-rank coals have a higher moisture content and lower heating value than bituminous coals, hence larger quantities of lignite and (to a lesser degree) subbituminous coal must be shipped to provide an equivalent amount of delivered energy. Even though low-rank coals are less expensive than bituminous coal on a per-ton basis, the transportation costs, as a percentage of delivered energy are significantly higher.[a]

The Grand Forks Energy Technology Center (GFETC) and Commonwealth Edison of Chicago jointly conducted tests to evaluate the transportability of partially dried subbituminous and lignite coals.[3] During 1974, 400 tons of each coal type were dried in a commercial scale dryer at Pekin, Illinois, and shipped to Grand Forks, North Dakota, for stockpiling. Tests were performed to: determine dried coal density; oxygen content within the loaded cars; dust emissions during transit; moisture change during transit; and the temperature change during transit. The lignite was dried from 39 to 22 percent moisture, and the subbituminous coal from 26 to 16 percent moisture.

Results from the GFETC/Com. Ed. tests indicate that the bulk densities of dried subbituminous and lignite are comparable to as-mined bulk densities. For both coals, the average densities increased during transit due to settling: 50.6 to 56.0 lb/ft^3 for dried subbituminous coal, and 47.0 to 51.1 lb/ft^3 for dried lignite.

Dried low-rank coals are extremely reactive and can increase in temperature to the point of spontaneous combustion. Available oxygen around the coal fragments increases the chances for combustion. As an indication of the high reactivity, test results show that the oxygen concentration in the void gases between particles in the loaded rail cars decreases rapidly. Within one hour, the oxygen content of the void gas in dried subbituminous coal decreased from 21 to 4 percent, and in dried lignite, from 21 to 1 percent. If no additional oxygen (from air) reaches the coal, internal pile temperatures remain relatively stable and hence reduce the possibility of spontaneous combustion.

---

[a]Presently, about two-thirds of the cost of mining and delivering western (Wyoming) coal to an eastern (Texas) consumer is spent on transportation.[8]

Air leakage did occur in some of the cars transporting dried subbituminous coal. The bottom dump doors of some cars had become warped from use and did not seat properly. At these openings, the coal eventually ignited during shipment, although only that coal within several inches of the door closure actually ignited. Ignition also occured in the rail car of as-mined (undried) subbituminous coal around similarly damaged dump doors. No evidence of ignition was found at the tops of the cars. The shipments of lignite experienced no ignition problems because the bottom doors were repaired and sealed prior to loading.

Dust emissions from the open hopper cars were three times higher for the dried coals than for the as-mined coals. In each case, however, treatment of the dried subbituminous and lignite loads with oil coatings (ranging from 2.1 gal/ton up to 6.1 gal/ton) suppressed dust emissions to about half those from as-mined coal losses.

The dried coals did not regain appreciable amounts of moisture during shipment, even though the subbituminous-loaded rail cars were subjected to a two-inch rainstorm. Coal samples from the surface layers showed a slight increase in moisture content immediately after the rain, but the top layer returned to its previous dry level after two days of storage.

A major problem encountered in rail shipments of low-rank coals is agglomeration by freezing. Results from various experiments by GFETC personnel[3,9] indicate that both undried and dried loads of low-rank coal are susceptible to freezing. The GFETC/Com.Ed. tests point out that the elevated temperatures of dried lignite (loaded directly from the thermal dryer) caused a continued release of water vapor, which ultimately condensed on the cold car walls and adjacent particles. This problem could be minimized by cooling the coal to ambient temperatures prior to loading. In addition, oil coatings reduced the reactivity and hence reduced increasing temperatures within the load.

As-mined low-rank coal can freeze in the car because the freshly mined coal particles have a higher vapor pressure of water than in the adjacent void spaces. As a result, water vapor is released until an equilibrium is established. In freezing winter weather, this vapor condenses and initiates a frost layer which continues to grow and agglomerate the coal particles. Tests have shown that drying a portion of the lignite and selectively placing it at the car walls effectively reduces the degree of agglomeration of the raw lignite in the center of the car.[9] The dried particles apparently absorb the moisture vapor released by other particles and hence reduce frost formation.

3.3.2.4 Current Status[10,11]

Current trends in the railroad industry do not indicate massive expansion of rail lines into new areas, but rather the accelerated business mergers of formerly competing railroads. For example, the recently completed 116-mile section of rail built by Burlington Northern to serve the coal fields in Wyoming's Powder River Basin is the longest stretch of new rail built since 1931. The two dominant western rail carriers, Burlington Northern and Union Pacific, are working to further extend their control and their coal market network.[a] The proposed end-to-end mergers will reportedly make more efficient use of rail company equipment and its mileage,

[a]In 1979, 80.2 million tons of coal were shipped by Burlington Northern, most of it from the Powder River Basin. Union Pacific shipped 24 million tons in the same year, much of it from southern Wyoming bituminous surface mines. 1980 shipments by UP are expected to top 30 million tons.[10]

although the effect of such mergers on coal transportation and freight rates is not clear.

A significant merger plan recently announced is the combination of the Missouri Pacific Railroad (MOPAC) with Union Pacific (UP), which would create a 21,200 mile railroad stretching from Chicago to the Gulf of Mexico and from Seattle to St. Louis. UP also intends to acquire the Western Pacific (WP) railroad, which would give coal-hauling carrier access to San Francisco, adding to its Los Angeles and Portland outlets to the Pacific.

Although the aggressive moves by UP appear to give them a large coal-hauling advantage, much of the coal UP carries already moves on the lines of the MOPAC and WP railroads. The other two large coal-hauling lines in the West, the Burlington Northern (BN) and the Santa Fe, should now face some real competition. BN has effectively monopolized coal shipments out of the Powder River Basin; now low-rank coal mine operators and distant utilities may be offered a broader selection of shipment options.

Burlington Northern, however, is also entertaining merger proposals. BN hopes to acquire the St. Louis - San Francisco Railway (Frisco), which would enable BN to move coal from Wyoming directly to the steel mills in Alabama and the Gulf Coast.

Another rail-related issue is the liquidation of the Chicago, Milwaukee, St. Paul, and Pacific Railway, otherwise known as the Milwaukee Road. The Federal Railroad Administration (FRA) hopes to keep the rescue of the Milwaukee in the private sector, and is willing to provide federal assistance to private buyers, if necessary. The Burlington Northern and Union Pacific railroads are interested in the coal-carrying segments of the Milwaukee, as is the state of Montana, which wants to buy the lines and then lease them to a railroad to make sure the BN does not have a monopoly on coal carriage in Montana.

In the wake of recent deregulations of the airline and trucking industries, the railroads have launched a strong Congressional campaign to remove most federal controls on rail rates. The Senate easily passed a rail deregulation measure, but the issue is meeting growing opposition in the House.

The deregulation bill would let railroads set rates without Interstate Commerce Commission intervention until they reach a point where the ratio of revenue to variable costs is 200 percent. The proposed amendment by utility, consumer, and other "captive" shippers, sets the ratio of 160 percent, triggering ICC rate review above that level.[11]. At the time of this writing (August 1980), the final results of this legislation are still pending further debate. Rail deregulation, if approved, could significantly increase what many coal buyers feel are already high freight rates. Delivered coal prices would increase and could pave the way for expanded interest and development of alternative means of transport, such as coal slurry pipelines.

References - Section 3.3.2

1. Phillips, P.J. Coal Preparation for Combustion and Conversion, EPRI AF-791, May 1978.

2. Roe, D.E., A.W. Karr, K.R. Lemmerman, and J.E. Sinor, Solid Fuels for U.S. Industry, Volume II, March, 1979.

3. Paulson, L.E., S.A. Cooley, C.J. Wegert, and R.C. Ellman, Experiences in Transportation of Dried Low-Rank Western Coals, presented at SME Fall Meeting, Salt Lake City, September 1975, SME/AIME Transactions, Dec. 1976, pp. 300-305.

4. Lowry, H.H., (editor), Chemistry of Coal Utilization - Supplementary Volume, John Wiley & Sons, Inc., New York, 1963.

5. Coal Age, December 1979.

6. Roe, D.E., A.W.Karr, K.R. Lemmerman, and J.E. Sinor, Solid Fuels for U.S. Industry, Volume I, March 1979.

7. Coal Age, May 1980, pp. 149-157.

8. Draffin, C.W., "Future Coal Transportation", in Proceedings of Critical Issues in Coal Transportation Systems, NAS/NRC, 1979.

9. Ellman, R.C., J.W. Belter, and L. Dockter, Freezeproofing Lignite, USBM RI 6677, 1965.

10. Coal Age, March 1980.

11. Business Week, August 11, 1980.

3.3.3. Barge Transport

3.3.3.1 Technology Description

The inland waterway system consists of the Great Lakes and 2500 miles of navigable waterways on the Mississippi-Gulf System, including the Ohio and Illinois rivers.

Inland waterways are currently used to transport about 11 percent of the annual U.S. coal production, and 25 to 30 percent of this tonnage utilized other modes of transport (e.g., rail-to-barge shipments).[1] In 1976, domestic barges transported more than 129 million tons of coal over inland waterways, which represents about 21 percent of all commodities shipped by barge.[2]

Relatively little western low-rank coal is shipped by barge, primarily because the navigable waterways are far from western coal regions. Coal from western mines that does travel by water is transhipped by rail from the mine to the barge. An example of rail-barge transport involves the Burlington Northern, the Chicago and Illinois Midland railroads, and the Valley Line (barge) Company. Annually, about five million tons of subbituminous coal leave Decker, Montana, by rail and are brought to Havana, Illinois where barges tow the coal to Commonwealth Edison's three electric generating facilities located near Chicago on the Illinois River.[3]

River transport of coal is accomplished primarily by towboats pushing rafts of barges. The towboats range in size from less than 1000-hp up to 10,500-hp, averaging about 5000-hp. Larger towboat designs in the future are unlikely due to limited channel depths and lock sizes.

Most barges used for shipping coal are of the open hopper type. The barge generally has a double skin construction with the cargo hold being free of any obstructions to allow for easy unloading by clam shell buckets or continuous bucket unloaders. There has been little standardization in the sizes of barges, except that due to the size of the existing locks on the rivers. The size and number of barges used depends on the size of the locks and width and depth of the river. The most common barge is roughly 200 feet in length and 35 feet wide, and has a cargo capacity of 1500 tons.

Actual towing operations involve lashing the flotilla of barges together to form a single unit. Average tow speeds are on the order of 5 miles per hour. A towboat may push a single barge or as many as 45 barges at a time. When passing through lock systems, barge groups of 4-wide and 3-long or 3-wide and 4-long are used depending on the size of the locks and the size of the barge. Integrated tows provide an efficient method of transporting large tonnages of coal over long distances. As with the unit train system, the integrated barge should be used to a single destination so that the towboat and integrated barges will remain intact during the entire transport cycle. This usually means that the tow will probably be owned by the shipper or contracted out over a long period of time.[4]

Great Lakes coal is transported primarily in self-unloading dry bulk carriers ranging up to 1000 feet in length with a capacity of 60,000 tons. As in indicator of the size of these carriers, the tonnage capacity is roughly equivalent to six unit trains of coal.

3.3.3.2 Environmental Control Technology

There are no unique environmental concerns associated with barge transport of low-rank coals. Fugitive dust emissions during transport are minimal because of the slow tow speeds (5 mph). High winds may pick up quanitifiable amounts of coal dust, but the impacts would be considered insignificant. Dust emissions are likely to occur during loading and unloading operations. These releases can be controlled by the use of conventional dust suppression and collection methods.

Other possible environmental concerns from barge transport include: exhaust emissions from tugboat engines; possible release of coal quantities into waterway; and increased deterioration of river and canal banks due to waves. None of these concerns is a unique consequence of low-rank coal shipments.

3.3.3.3 Effects of Low-Rank Coal Properties

Water transport of coal is affected in virtually the same manner by the properties of low-rank coals as rail transport. Since shipping costs are on a per-ton basis, potential savings could result from drying the coal prior to shipment. Increased dust emissions, as in rail transport, then become a more significant concern.

Western low-rank coals traveling on inland waterways must first be transhipped by unit train from the mine to the loading dock. The additional handling requirements aggravate the effects of excessive fines fractions and accompanying dust emissions.

3.3.3.4 Current Status

The western coal fields, unlike those of the mid-western and Appalachian regions, do not have access to navigable waterways. However, with coal moving as much as 800 to 1200 miles by rail, increased use of rail-barge combinations is definitely feasible.

New interesting combinations are developing on reasonably large scales, with western subbituminous coal moving by rail to points on the

Great Lakes or the Mississippi and Ohio rivers, and by barge to the final destination.

Other possibilities less often considered are rail movement to northern Great Lake points from mines of the Northern Great Plains and by lake steamers to points south and east.

In 1973, Decker Coal of Montana signed a contract with Detroit Edison to deliver 180 million tons of coal over a 26-year period for use at the St. Clair-Belle River plants. After careful analysis of data, a rail-to-water route with the terminal serving as the supply buffer was chosen. Although this route was 300 miles longer than an all-rail route, studies showed that $6-$8 per ton could be saved. Congested urban rail yards along the route coupled with the fact that no one railroad had continual track rights were cited as major factors against an all-rail route. Therefore, Burlington Northern Railroad hauls the coal 800 miles from Montana to the terminal site where 1000-ft self-unloading dry bulk carriers continue the journey across Lake Superior, through the Poe Lock at Sault Ste. Marie and into Lake Michigan for delivery to Detroit Edison plants.

The Hall Street Coal Transfer Terminal in St. Louis is another of the more recently completed installations. Designed by Dravo Corp. in conjunction with ACBL Western Inc., the 10-million tpy facility is the first high capacity rail-to-barge transfer terminal to be built on the western shore of the Mississippi River.[5]

Aside from geographical constraints and freezing, the use of the waterways is subject to other problems. For example, waterways carry other commodities (such as grain) and represent a pathway for boats (such as pleasure boats) other than barges. This can lead to congestion and delays at key locks, especially true if coal transportation by waterways is increased. At the present time, eight locks on the waterway systems, including lock 26 on the Upper Mississippi, one lock on the Monongahela, and one on the Kanawha rivers, have special capacity problems. Lock 26 is of definite significance to rail-barge combinations because it has a special impact on traffic from the Upper to the Lower Mississippi. It is located in Alton, Illinois, to the north of St. Louis. Grain has been the most important commodity moving through Lock 26. Should substantial amounts of coal move downstream from points north of St. Louis, long delays would undoubtedly occur.[6]

In addition to capacity constraint problems (which cause delays), the barges are subject to disruption by strikes and disasters (such as flooding, accidents, etc.). Although barges are quite energy-efficient, the price of shipping would be subject to inflationary pressure due to some fuel and labor inputs. In short, price stability cannot be guaranteed.

Irregardless of price stability, barges remain the most cost-effective way to move coal between points on inland waterways. The barge industry for the most part is currently unregulated. (Only 15 percent of the total ton-miles of barge commerce is regulated by the ICC.)[4] This gives the inland waterways more freedom of operation and price negotiation than is available to other transportation industries.

Figure 3.3.3.1 shows the range of barge transport costs as a function of distance. Specific line-haul costs are difficult to obtain due to the unregulated nature of the industry. Rates may be higher than the range indicates due to general inflation of operating costs, and the possible future initiation of increased regulations.

Figure 3.3.3.1

Range of Barge Transportation Rates (1975)

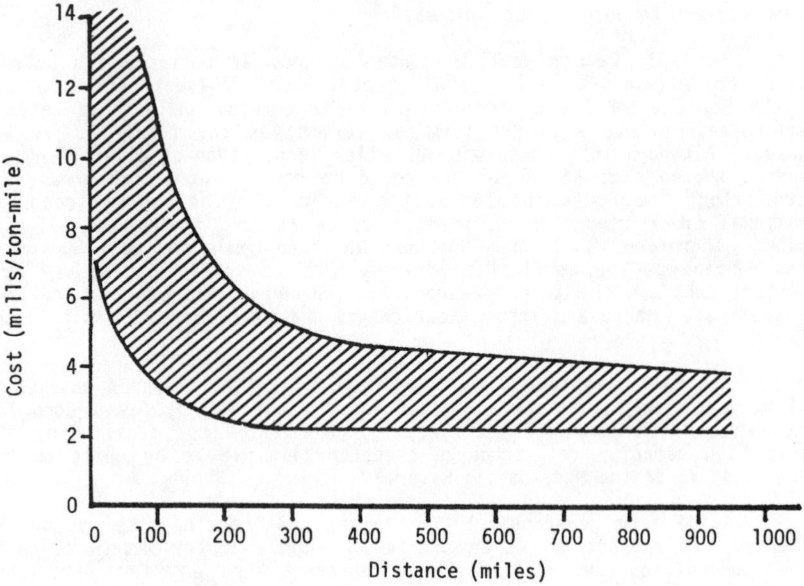

Source: Reference 7

References - Section 3.3.3

1. Report to Congress: Coal Development - Promises, Uncertainties, EMD 77-43, U.S. General Accounting Office, Wash. D.C.

2. Lin, K., Coal Traffic Annual, 1978, National Coal Assoc., Wash. D.C., 1979.

3. Desai, S.A., and J. Anderson, Rail Transportation Requirements for Coal Movement in 1980, USDOT-TSC-OST-76-32, December 1976.

4. Roe, .D.E., A.W. Karr, K.R. Lemmerman, and J.E. Sinor, Solid Fuels for U.S. Industry, Volume I, Cameron Engineers, Denver, Colorado, March 1979.

5. Coal Age, July 1979.

6. Mechanical Engineering, September,1979.

7. Phillips, P.J., Coal Preparation for Combustion and Conversion, EPRI AF-791, May 1978.

3.3.4 On-Site Trucks and Conveyor Belts

3.3.4.1 Technology Description[1,2,]

Truck and conveyor movement of surface-mined low-rank coals is a relatively small, yet vital link in the overall coal movement system.

Haul trucks have increased in capacity over the past decade in efforts to keep pace with larger dragline and shovel capacities. With the aid of computers and the use of highly sophisticated stress-and-fatigue analysis, it is possible to design surface-mining machines on an ever-increasing scale. However, manufacturers report that the mining shovel, and the haul truck, have probably approached their maximum size. The average capacity for large haul trucks seems to be about 170 tons.[1]

Trucks used in western surface-mining operations are not used outside of the mining area. Those that could fit on rural highways would not only severely damage the roadbed, but would immediately become an uneconomical method for coal movement outside of the mine area.

The current transportation breakthrough within mine sites is the efficient utilization of conveyor belt networks to move coal and overburden. Operational and maintenance costs associated with trucks have been high and continue to climb with fuel costs and inflation. Transportation costs in the mine site can be significantly reduced and production increased with the proper use of conveyors instead of trucks.

In addition to their other advantages, conveyors avoid pit congestion. Truck traffic delivers intermittent loads over 100 tons each and transport is affected by physical pit limitations (steep grades, indirect routing) and inclement weather. On the other hand, conveyor sections can be connected to create continuous, even delivery of coal or overburden over distances impassable by truck. The following examples illustrate the varied and efficient applications of conveyor systems proven in use.

- A series of 48-inch wide conveyors made an efficient system for transporting blasted overburden out of Bethlehem Mines Corporation's Panther Valley anthracite mines, Tamaqu, Pennsylvania. As the pit deepened, 310-foot sections were added to the 500-foot top section, making a line about 1500 feet long. On the pit floor, 10 shuttle conveyors, each 100 feet long, are used to reduce haul distance of the front-end loaders. At the top of the pit, a traveling 120 foot radial stacker-conveyor spreads the material for final disposal by dozer. Total system length is about 2500 feet, with a potential elevation differential of 600 feet. The linear speed of the conveyor is 485 fpm, hauling about 1000 cubic yards/hour.[2]

- As an example of cross-pit conveyor systems, the Jim Bridger mine near Rock Springs, Wyoming combines the two most efficient coal/overburden-moving systems: draglines and conveyors. The system consists of two conveyor units, a mobile hopper/feeder to receive overburden from the dragline, and a mobile bridge conveyor that spans the pit and deposits spoil at rates up to 4000 cubic yards/hour. Both units advance on self-propelled crawlers as mining progresses.

- Conveyors are well suited for use with continuous mining equipment, such as the bucketwheel excavator (BWE). A large surface coal mine in southwestern Washington uses movable conveyors to help a bucketwheel excavator move stripped overburden from the highwall to the reclamation pit, located over a mile away. The BWE discharges material directly onto a mobile transfer conveyor, known as a bandwagon. The bandwagon belt capacity is matched to the BWE output and dumps the spoil onto a 2500-foot shiftable conveyor. This in turn feeds a 4000-foot conveyor which carries the overburden away from the mining area to the fourth and fifth sections of belt. After traveling up to 9000 feet along five conveyors, the freshly stripped overburden is stacked for future use.

- Conveyor systems have also been successfully paired with shovels and dozers, and are also used for loading spoil or coal (in addition to non-coal commodities) into haul trucks or rail cars.

3.3.4.2 Environmental Control Technology

Worker safety is more the issue than severe damage to the natural environment. With both trucks and conveyors, high noise levels can be minimized by proper maintenance and ear protection for mineworkers. Falling objects from loading or conveying operations can be controlled by belt guards and careful monitoring. Hard hats must be worn at all times when around such equipment.

Dust emissions are a significant problem during loading operations, although the use of belt conveyors should reduce dust problems otherwise associated with intermittent truck loading/unloading, and with the higher speed movement of the large trucks over dusty mine roads. Conveyors do not operate at high enough speeds for dust emissions to become a serious problem.

Increased use of conveyors instead of diesel-powered trucks will reduce the exhaust concentrations within the mine area, and should cut down on traffic-related injuries and noise.

3.3.4.3 Effects of Low-Rank Coal Properties

The major effect of LRC properties on truck and conveyor transport relates to fugitive dust emissions resulting from the generally higher fines fractions found in LRC's as compared to bituminous coals. The high reactivity and potential spontaneous combustion of low-rank coals should not present any problems during truck and conveyor transportation.

3.3.4.4 Current Status

Conveyor systems are beginning to take the place of traditionally-used haul trucks in large western surface-mining operations. Modern mining plans are incorporating shiftable and self-powered conveyors designed to work with conventional stripping and digging equipment. New mining machines now utilize built-in conveying systems, such as specialized excavating machines with continuously rotating cutters or bucketwheels with their own feeding conveyors.

The evolution of truck design has produced a wide assortment of capacities, with the maximum capacity haul truck now leveling off at 170 tons. Trucks will continue to play an important role in surface mining operations, although skyrocketing fuel and maintenance costs will force mine planners to reassess their present use of trucks and develop more efficient transport operations.

References - Section 3.3.4

1. Coal Age, July 1980.

2. Coal Age, July 1979.

3.3.5 Slurry Pipelines

3.3.5.1 Technology Description

The concept of moving coal by pipeline is not new. The first U.S. coal slurry patent was granted in 1891 to Wallace C. Andrews, who also exhibited a working model of a slurry pipeline at the Columbia World's Fair in Chicago in 1890. Since then, two coal slurry pipelines have been built in the U.S.: the Ohio Coal Pipeline, which operated successfully for six years until 1963, when it was shut down after forcing a reduction in rail rates for coal deliveries in that part of the country; and the Black Mesa Pipeline in Arizona, which is operated by Southern California Edison and has been in continuous operation since 1970.[1]

The Black Mesa Pipeline carries a coal-water slurry from the Black Mesa coal fields near Kayenta, Arizona, to the Mohave Generating Station located near Laughlin, Nevada, 273 miles to the west. The (bituminous) coal is pulverized at the mine, mixed with water in a 1:1 ratio, and pumped through the 18-inch diameter pipeline at a rate of 560 tons of coal/hour, at speeds near 3.5 mph (slightly more than 5 feet per second). At this rate a ton of coal completes the trip in about three days.[2]

The effect of frictional resistance and elevation changes produces a large pressure drop that must be overcome by the slurry pumping system. In the Black Mesa Pipeline, the pressure drop can reach a maximum pressure of 1500 psig. Several pumping stations along the mainline are used to boost pipeline pressures as needed. Generally, booster stations are required every 70-100 miles.

Slurry received from the Black Mesa Pipeline is directed to a battery of storage tanks; "active" tanks are equipped with paddle-type agitators that provide continuous suspension of the slurry, while "inactive" storage tanks allow the coal solids to settle. The inactive storage provides sufficient capacity for a 12 day supply of coal.

Slurry drawn from the active tanks is dewatered by 20 Dynacone centrifuges per generating unit (for a total of 40 centrifuges). These centrifuges are oriented so that each coal pulverizer is directly fed from two centrifuges. Dewatered coal leaves the centrifuges containing a moisture content of approximately 20 percent.

The separated slurry water has the appearance of a black slimey liquor, and is often referred to as "ink", or underflow. Ink may not legally be dumped into surface water supplies nor may it be pumped into disposal wells unless the disposal site is effectively quarantined from groundwater resources. Earlier reports indicated that the underflow contained up to 20 percent solids, most of which were less then 40 microns in size. This underflow was discharged to a waste water evaporation pond.[3]

Reports now indicate that the centrifuges are more effective in dewatering the slurry solids; the separated slurry water now contains only 5 to 6 percent solids.[4] Instead of ponding this effluent, the ink is fed to thickener tanks where the solids are chemically separated from the water. The resulting sludge (underflow) is pumped to the boilers and burned as supplemental fuel, and the clarified water is used in the plant's cooling system.

Based on the limited operating experience of coal slurry pipelines and development plans for future pipeline operations, several other potential problems have been identified in addition to the water separation and cleanup problems. The major areas of concern, in roughly decreasing order of significance, are:

- water availability
- pipeline rupture
- pipeline abrasion
- slurry freezing

Water Availability

Even though sufficient water supplies could be made available in the west for use in coal slurry pipelines, existing water rights account for every available gallon in many states. Proposed slurry pipelines transporting Powder River Region coal to destinations in Texas or Arkansas may require up to 20,000 acre feet of water annually (approximately 6.5 billion gallons).[5] During normal climatic conditions, this water represents a reduction in some present or future alternative use (e.g., agriculture, commercial growth); during extended drought conditions, water tables may be depleted faster than they recharge, or, quite possibly, insufficient water would be available for sustained slurry flow.

Water consumption by slurry pipelines can be minimized at the downstream end by utilizing the separated water in other plant processes. Such is the case with the Black Mesa Pipeline. Unfortunately, this remedial approach still drains valuable water from typically water-short regions. Water consumption of slurry pipelines is often compared to the water needs of coal conversion plants and to minemouth power plants, which indicates that pipelines require only one ton of water per ton of coal, whereas combustion and conversion (gasification or liquefaction) require approximately 7 tons and 2 tons, respectively. The point missed by these comparisons is that when compared to other coal transportation options (namely railroads), slurry pipelines represent the only significant water consumer. No currently proposed pipelines include provisions for a return water system, although this has been suggested as a possible method.

Pipeline Rupture

Pipeline failure is not a unique event. For example, there were 1,373 failures of natural gas pipelines in 1975, an improvement over 1974's record of 1,477 failures.[6]. While natural gas leaks are more dangerous than coal slurry leaks, coal lines are more difficult to restart. There can be no guarantee against breakage. The Black Mesa Pipeline burst in two places on February 8, 1977.[7] Slurry escaped into the desert east and west of Kingman, Arizona. The amount of escaped slurry was not reported.

Pipeline ruptures, besides creating potentially significant environmental concerns, disrupt delivery of coal to the dependent utilities downstream. In cases where coal is delivered solely by pipeline, power outages could result within the utility service area if on-site fuel storage is consumed before the damaged pipe can be returned to service.

Slurry Freezing

Pipelines that have a slowly moving slurry may be subject to freezing in northern areas during periods of severe cold. Most of the pipeline would be underground but some portions must be exposed. Heated sections or heavy insulation provides a possible solution at an added cost. If the slurry freezes, expansion would result in a pressure increase of more than 300 psi. More importantly, a frozen section can act as a plug. Even the presence of a "slush" may reduce pipeline slurry velocity to unacceptably low speeds.

Estimation of the danger of pipeline freezing is probably similar to that of estimating the probability of drought conditions.[2] Given the historical climatic variations for any particular region, the chance of, say, a ten day period of sub-zero (or any arbitrary temperature) conditions can be statistically computed and correlated with slurry flow disruptions.

Pipeline Abrasion

Inherent in pipeline movement is friction and the resulting abrasion of both the coal particles and the pipeline wall. While coal particles become smaller along the line, increasing the slurry viscosity somewhat, the apparent lack of statistical data indicates that this may not be a significant problem. Also, pipe surface abrasion is not normally a problem when slurry moves at low velocities.[2]

3.3.5.2 Environmental Control Technology

The principal environmental control concerns associated with coal slurry pipelines include: treatment and disposal, or utilization, of the coal fines and liquor remaining after slurry separation; potential rupture of pipelines and the resulting spill; and the minimization of water requirements for a slurry preparation, particularly in water-scarce areas in the west.

Of the three concerns, operating experience with the Black Mesa Pipeline has demonstrated that slurry "ink" can be effectively treated by chemical separation of the suspended solids and water.

The chance of pipeline ruptures can be reduced with thorough monitoring and prompt maintenance. Spills will inevitably occur, though the volumes of slurry released will depend on the flow rate, pressure, pipeline burial depth, overburden density, elapsed time until detection and pipeline shutdown, and the proximity of slurry holding ponds.[5]

Water use requirements will remain high unless future experiments can successfully transport coal at water/solids ratios well below the current 1:1 limit. Techniques which might reduce water requirements include the use of additives and changing the particle size distribution. Perhaps eventually, as the economic value of water becomes a more significant variable, slurry pipelines may be required to recycle slurry water back to the pipeline origin, thus operating as "closed loop" transport media.

3.3.5.3 Effects of Low-Rank Coal Properties

The high inherent moisture content of low-rank coals constitutes an economic disincentive to use of the slurry pipeline transportation mode. In order to obtain a solid/liquid slurry ratio in the acceptable range for pumping (on the order of 1:1), the total moisture content (slurry water plus inherent coal moisture) can approach 70 percent. Such a mixture has a low Btu value per pound for transportation over long distances. Another significant low-rank coal property affecting slurry pipeline transport is the tendency for coal particles to degrade during handling. Lignite particles undergo more physical degradation than subbituminous or bituminous coals. In a study of the economic feasibility of hydraulically transporting Sandow (Texas) lignite, Lammers and co-workers simulated coal slurry transport over various distances (34 to 114 miles), dewatered, then thermally dried the coal charges.[8] Their results indicate that in the micron size range, the lignite suffered significantly more degradation than a bituminous coal that was also tested. Table 3.3.5.1 illustrates the effect initial size consist has on degradation; for each lignite type tested, the larger size particles (1/2 X 0 inch) produced a higher percentage of fines. The effects of thermal drying on dust loss and dry coal recovery are shown in Table 3.3.5.2 for eight different tests. Note that the raw lignite sample (no simulated transport), even after thermal drying, lost only one percent of its weight as dust. The bituminous sample exhibited much less dust production after thermal drying than similarly transported lignites, and almost twice as much bituminous coal was recovered after the tests than lignite.

Size degradation of low-rank coal particles fed into a coal slurry can actually improve the flow characteristics of the slurry. smaller particles (fines) become more uniformly suspended in the slurry and have a greater tendency to remain suspended over a wider range of flow rates. Conversely, such degradation severely aggravates the dewatering problem at the receiving end. Current dewatering practices, centrifugation and filtration, are strongly affected by the size distribution of coal particles, where increasing proportions of fines decreases efficiency. Data from the Black Mesa Pipeline indicates that coal fines (<40 micron size) present a difficult dewatering problem, and that for every 1 percent increase in <40 micron fines a 1 percent increase in filter cake moisture is experienced.[2] Considering the fact that the Black Mesa Pipeline carries bituminous coal, it is anticipated that low-rank coal transport by slurry pipeline will present a more serious dewatering problem due to higher levels of entrained fines.

Table 3.3.5.1

Degradation Rates of Four Texas Lignites in Simulated Hydraulic Transport

Texas Lignite Sample (County)	Size (inch)	Increase in Percentage of Minus-50 Mesh
Rusk	1/8 X 0	14.7
	1/4 X 0	27.8
	1/2 X 0	38.0
Titus	1/8 X 0	31.8
	1/4 X 0	52.8
	1/2 X 0	61.6
Henderson	1/8 X 0	17.9
	1/4 X 0	31.4
	1/2 X 0	37.8
Freestone	1/8 X 0	20.0
	1/4 X 0	36.0
	1/2 X 0	44.6

Source: Reference 8

Table 3.3.5.2
Observations During Thermal Drying of Hydraulically Transported Lignite and Bituminous Coal

Drier Feed Source	Nominal Size In.	Miles Transported	Total Moisture, % Before Dryer	Total Moisture, % After Dryer	Dust Loss, % of Surface-Dry Feed	Dry Coal Recovery, % of Surface-Dry Feed
Lignite						
Test 1	1/2 X 0	114	49.4	4.5	28.5	39.5
3	1/2 X 0	34	44.7	2.8	16.0	46.3
4	1/4 X 0	114	49.4	6.3	26.5	40.5
5	1/4 X 0	72	49.4	5.3	14.8	46.9
6	1/2 X 0	71	47.2	5.3	17.0	45.7
Bituminous Test 7	1/2 X 0	72	12.8	0.6	2.1	90.8
8	1/4 X 0	73	9.2	0.2	5.2	87.9
Raw Sandow Lignite	1/4 X 0	0	34.3	4.2	1.0	64.3

Source: Reference 8

3.3.5.4 Current Status

The increased use of slurry pipelines to transport coal has been successfully blocked by strong opposition from western water concerns and railroad companies. To date, seven coal slurry pipelines have been proposed to supply fuel for the utility market. Of the seven, three plan to carry low-rank subbituminous coal, all from the Powder River Region near Gillette, Wyoming. These three routes are shown, along with the two existing coal slurry pipelines, in Figure 3.3.5.1. One of these, a Wyoming to Louisiana pipeline built by Energy Transportation Systems, Inc.(ETSI), has reached the legal milestone allowing construction to begin.

Figure 3.3.5.1
Planned Western Low-Rank Coal Slurry Pipeline Projects And Existing Pipelines

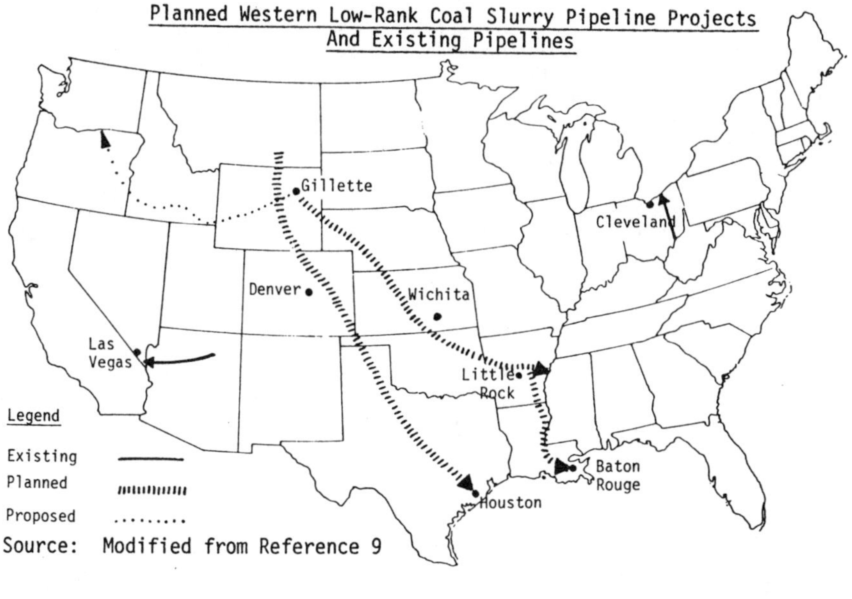

Source: Modified from Reference 9

An important item for construction of a long-distance coal slurry pipeline is obtaining the right-of-way. It is necessary to obtain permits for crossing highways, rivers, canals, railroads, and public lands. Most of these permits and easements are routinely acquired and pose few problems to the construction of a pipeline.

The major obstacle is the railroads, which historically have refused crossing permits to competitors. Since 1875, the railways have mounted large, well-funded opposition to repeated Congressional attempts at eminent domain for pipelines. In 1978, the Coal Pipeline Act (H.R. 1609) recommended the issue once again to the House of Representatives, but intensive railroad opposition successfully defeated the bill. Even so, progress was made over prior years in that the bill was at least brought to a vote. Due to increasing national concern over energy, however, proponents of the legislation feel confident that Congress will eventually pass the bill.[1]

The planned coal slurry pipeline by Energy Transportation Systems, Inc. (ETSI) is a landmark case where clever planning succeeded in obtaining the necessary rights-of-way, and discovered that in many cases the railroads are granted easements only and the subsurface rights are actually owned by others. ETSI thus obtained subsurface rights-of-way from the individual landowners and initiated litigation in federal and state courts to assure the prevention of future railroad interference.[1] Of the 65 lawsuits filed, ETSI officials report that all have been won, and construction planning can now proceed.[10]

Construction is scheduled to begin in 1982, and will reportedly begin carrying up to 37.5 million tons of subbituminous coal in 1984. A U.S. Bureau of Land Management environmental impact statement, as well as a coal evaluation program, is scheduled for completion in 1981.

Texas Eastern is proposing a pipeline to transport coal from the Powder River Basin to power plants and industries along the Texas Gulf Coast, primarily the Houston area. The 38-inch pipeline will stretch some 1300 miles and carry up to 25 million tons of coal per year. To supply the water necessary for slurry preparation, Texas Eastern proposed to the Wyoming legislature that untapped surface water be diverted from the Little Big Horn River, to be shared for state approved uses and the slurry pipeline. (The governor, Ed Hershler, vetoed the bill.) In addition to this action, Texas Eastern is obtaining key right-of-way options along the proposed routes in states that do not have eminent domain statutes. They are also maintaining contacts with present and potential users of Powder River coal.

Texas Eastern officials hope for construction to begin in late 1982 with operation scheduled for 1985.[11]

The Snake River Coal Slurry Pipeline Project was proposed in early 1974 to meet an anticipated demand for coal in Washington, Oregon, and Idaho. Subsequent demand analyses, however, have determined that the earlier expectations of coal demand were too optimistic, and thus the pipeline plan has been tabled indefinitely.

As proposed originally by Northwest Pipeline Co. of Salt Lake City and Gulf Interstate Engineering Co. of Houston, the slurry line would have utilized the right-of-way of Northwest's natural gas line for much of its length.

The proposed pipeline route starts in the coalfields near Gillette, Wyoming, and passes through other Wyoming mining areas en route to the principal delivery points at Boise, Idaho, and Boardman, Oregon. In addition to the many pick-up and delivery points, about three million tons of coal would be shipped down the Columbia River from Boardman to industrial users in Oregon and Washington.

The original economic projects for the pipeline indicated that during its first year the costs of transporting coal would be about equal to rail transportation. Thereafter, the slurry pipeline would have an economic advantage because of the escalating cost of rail.[9]

Figure 3.3.5.2 presents cost analysis results for coal shipments in slurry pipelines. Four estimates are shown according to the price of water and the possible need for a slurry water return pipe. The use of a water return system represents a major economic setback for slurry pipeline economics. Such a proposal is not anticipated, however, in the development of first-generation pipeline systems out of the West.

Figure 3.3.5.2

Coal Slurry Pipeline Transportation Costs

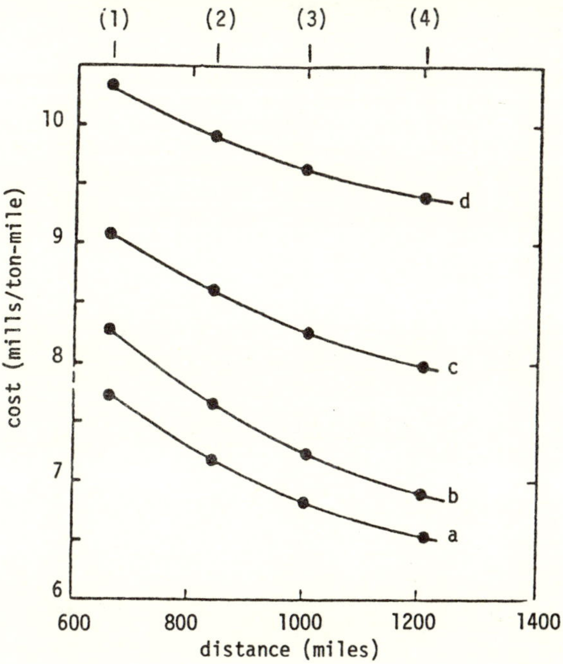

Notes

a. Assuming water costs of $1.00/1000 gallons, no return

b. Assuming water costs of $2.50/1000 gallons, no return

c. Assuming piping and pumping costs (@ $240,000/mile) for slurry water return

d. Assuming piping and pumping costs (@ $480,000/mile) for slurry water return

(1) New Mexico to Louisiana

(2) Montana to Washington

(3) Wyoming to Chicago, Illinois

(4) Wyoming to Texas

Source: Reference 2

Neglecting the obvious deleterious effects of right-of-way and water acquisitions, coal slurry pipelines appear to be competitive with rail car shipments over 800 miles. Slurry pipeline transport, when operating properly, delivers a reliable, steady supply of fuel; trains have a documented history of rail car shortages, derailments, labor disputes, and possess the somewhat monopolistic power to vary freight rates as they see fit.

The latter point may repeat an interesting situation for pipelines: the first coal slurry pipeline (in Ohio) was deemed economical (and hence constructed) based on existing rail freight rates. After construction and considerable legal action, the railroads reduced their rates and effectively drove the pipeline out of business. The same situation could occur in the West. The railroads are now losing battles over rights-of-way to pipelines, and an economic decision such as rate reduction may represent a final chance to preserve their transportation dominance.

The Black Mesa Pipeline has operated successfully and economically, but is must be noted that no railroad line exists that could have provided a similar service. The distance between the Black Mesa coal mines, located 120 miles north of the nearest railroad, and the Davis Dam, 30 miles closer, gave the pipeline a 2:3 distance advantage.

References - Section 3.3.5

1 Mechanical Engineering, December 1979.

2 Rieber, M., S.L. Soo, and others, Comparative Coal Transportation Costs: An Economic and Engineering Analyses of Truck, Belt, Rail, Barge, and Coal Slurry and Pneumatic Pipelines, Vol. 3, Coal Slurry Pipelines, USBM/DOI/FEA, Contract No. J0166163, August 1977.

3. Dina, M.L., Operative Experiences at the 1580 MW Coal Slurry Fired Mohave Generating Station, Southern California Edison Company, 1976.

4. Cobb, D.B., C.O. Giles, J.D. Hornbuckle, and F.O. Beavitt, "Coal Slurry Storage and Reclaim Facility for Mohave Generating Station", in Coal Technology '79 Conference Proceeding, Vol. I, Houston, Texas, November 6-8, 1979.

5. A Technology Assessment of Coal Slurry Pipelines, Office of Technology Assessment (OTA), PB-278-675, March 1978.

6. Oil and Gas Journal, June 14, 1976.

7. Mohave Valley News, Bullhead City, Arizona, February 23, 1977.

8. Lammers, G.C., R.R. Allen, D.J. Donaven, E.O. Wagner, and V.F. Parry, A Study of the Feasibility of Hydraulic Transport of a Texas Lignite, USBM RI5404, 1958.

9. Coal Mining and Processing, August, 1979.

10. Rocky Mountain News, Denver, Colorado, Thursday, January 24, 1980.

11. Coal Age, July, 1979.

3.3.6 - Selected References

1. Rieber, M., S.L. Soo, and others, Comparative Coal Transportation Costs: An Economic and Engineering Analysis of Truck Belt, Rail, Barge, and Coal Slurry and Pneumatic Pipeline, USBM/DOI/FEA, Contract No. J0166163, August 1977, 8 volumes.

The comparative costs of competing transport modes are dealt with in terms of engineering/economic analyses of the facilities and operating systems of each coal transport mode. The report is divided into two main sections: (1) long distance coal transport and (2) gathering and distribution systems. Within each section economic and technological comparisons are made of the available modes. These are used as the basis for comparisons among the modes. Inter-modal compatibility is studied with respect to both mixed trunk line shipment and with respect to feeder to trunk line and trunk line to distribution systems. The engineering and technological data form the bases for the costing and economic analyses.

The cost basis includes all necessary processing, loading, and unloading facilities needed for transport by each mode. Emphasis is placed on cost optimality and the ability to increase capacity.

2. Roe, D.E., A.W.Karr, K.R. Lemmerman, and J.E. Sinor, Solid Fuels for U.S. Industry, Cameron Engineers, Denver, Colorado, March 1979, 3 volumes, 1017 pp.
This three-volume document discusses the technical, economical, environmental, and legal aspects of coal utilization. Specific sections include: energy forecasts of fuel prices; coal resources; transportation modes; industrial boilers; cogeneration; coal gasification; and environmental constraints. The report is very useful for its general, overview nature of coal transportation options. However, data sources are not well documented or referenced.

3. Phillips, P.J., Coal Preparation for Combustion and Conversion, EPRI AF-791, May 1978, 364 pp.
While the major emphasis is placed on coal preparation and beneficiation, this study includes a well-organized section on rail and barge coal transportation technologies and costs. Following the examination of costs, the effects of coal beneficiation on transportation modes and costs are quantified by an example.

4. Paulson, L.E., S.A. Cooley, C.J. Wegert, and R.C. Ellman, Experiences in Transportation of Dried Low-Rank Western Coals, presented at SME Fall Meeting, Salt Lake City, September 1975, SME/AIME Transactions, Dec. 1976, 20 pp.
The Grand Forks Energy Research Center (GFERC) and Commonwealth Edison of Chicago jointly conducted tests in which 400 tons of subbituminous and lignite coals were dried in a commerical scale dryer, oil sprayed and cooled, then shipped from Pekin, Illinois, to Grand Forks, North Dakota, and stockpiled. Cars containing raw coal and dried coal, which had not been sprayed, were also transported for comparative purposes. During transit, observations were made of changes in the coals' moisture content, dust losses, and temperature.

5. Ellman, R.C., J.W. Belter, and L. Dockter, Freezeproofing Lignite, USBM RI 6677, 1965, 28 pp.
The Bureau of Mines investigated the factors that cause agglomeration of lignite by freezing during winter shipment and sought methods to avoid it. Tests showed that lignite may be freezeproofed by removal of the relatively small quantity of moisture involved in forming the frost crystals which cement particles together, or by controlling the crystal characteristics. Adding dried lignite fines proved to be an effective and superior freezeproofing method. Commerical adoption of this method has been successful.

Preparation, Handling, and Storage

3.4.1 Introduction and Summary

Coal preparation, handling, and storage encompasses a wide spectrum of processes and procedures applied to coal between the mine and its final utilization. These include numerous physical, chemical and thermal techniques which can be combined in an appropriate processing circuit to produce the desired product. The technologies involved in coal preparation can be separated into several categories as follows:

Comminution - This category encompasses all size reduction techniques including primary breaking to process run-of-mine coal to a minimum top size, secondary crushing to produce a uniform size distribution suitable for a preparation circuit or transportation, and grinding to prepare the coal for extensive cleaning processes or for use in combustion.

Mineral Matter Control - This category includes all techniques and equipment for controlling (usually reducing) various undesirable mineral constituents of the raw coal such as sulfur components, ash producing components (mine dilution products or inherent minerals) and components affecting utilization (sodium).

Moisture Reduction - This category includes the processes and equipment used to reduce the moisture content of the coal. Processes include mechanical techniques for separating solids from slurries and reducing moisture retained on coal particle surfaces, and thermal drying for removal of moisture bound within the coal structure.

Briquetting and Pelletizing - These processes are used to manufacture a solid fuel with superior handling and combustion characteristics, primarily for small industrial and commercial markets.

Storage - This category includes the techniques and strategies for maintaining mined coal in storage for long periods of time.

Blending - The process of combining coals of differing physical and chemical properties to obtain desired properties.

Examination of this list of technologies reveals two fundamental types: 1) those that are essential steps in the normal utilization of low-rank coal; and 2) those that are optional, and may be used to upgrade the coal, improving its utilization characteristics. The economics of U.S. low-rank coal extraction and utilization have dictated that no preparation techniques are practiced except for those which are absolutely essential. For example, the only preparation techniques normally applied to low-rank coals burned by utilities are primary crushing at the mine, storage, and pulverization at the power plant.

In contrast, the majority of the bituminous coal utilized in this country is beneficiated in coal preparation plants. Most commonly, these plants employ physical separation processes that remove mineral matter from the coal based on differences in specific gravity. Mechanical, and sometimes thermal, drying processes are incorporated into these plants to reduce the surface moisture of the washed coal to an acceptable level. Pyritic sulfur compounds are among the minerals partially removed in the gravity separation processes; stricter standards on SO_2 emissions have fostered numerous process development efforts to improve the desulfurization of high-sulfur eastern coals.

The primary reasons for the absence of these types of coal preparation plants for low-rank coals are: 1) the typically low extraneous mineral matter content of low-rank coal; 2) the high inherent moisture content of low-rank coals; and 3) the relatively low sulfur content of low-rank coals. In other words, the "washability" characteristics of presently mined low-rank coals are such that little or no improvement in ash or sulfur content is obtained (while surface water is added) by commonly used beneficiation processes. Whatever value is added to the coal rarely exceeds the processing cost.

As energy costs rise and environmental standards tighten, the economic attractiveness of low-rank coal upgrading options will tend to improve. Some of the potential opportunities for the application of technology not previously applied to low-rank coals include: 1) reduction of coal moisture content to obtain improved transportation and utilization economics; 2) reduction of sodium content in high-fouling low-rank coals by ion exchange to reduce boiler operation and maintenance costs; and 3) selective or general reduction of mineral matter content by physical or chemical means, either for unusually "dirty" low-rank coals or in preparation for certain conversion processes.

A number of key technical issues relating to the preparation, handling, and storage of low-rank coals are currently of importance. Most of these issues relate to potential future applications of coal upgrading technology. A few issues involve improving techniques or solving problems encountered in current practice. The key issues which have been identified are listed below and discussed in the paragraphs that follow:

1. Application of Moisture Reduction Techniques to Low-Rank Coals and Slurries

2. Applications of Ion Exchange and Chemical Cleaning Processes

3. Application of Gravity Separation Techniques to Low-Rank Coals

4. Application of Briquetting or Pelletizing Techniques to Low-Rank Coals

5. Waste Disposal from Coal Beneficiation and Slurry Dewatering Plants

6. Fine Coal Cleaning

7. Fines Generation During Handling and Comminution

8. Handling of Dried Lignite

1. Application of Moisture Reduction Techniques

Slurry Dewatering

Removal of moisture, which is used as a transport medium in slurry pipelines, is of critical importance to the utilization of low-rank coals in this transportation mode. The total coal moisture content may be considered the sum of surface and inherent moisture levels. Current mechanical dewatering techniques (including filtration and centrifugation) cannot remove inherent moisture. Surface moisture can be reduced, but only to approximately 10 weight percent by mechanical dewatering. This may be acceptable for coals having inherent moisture contents of only 10 percent, but in the case of low-rank coals, inherent moisture levels of 25 percent and greater are common. After mechanical dewatering to 10 percent surface moisture, a total moisture level of 35 percent or more could still occur with low-rank coals. This may result in serious operational problems in direct combustion or other utilization processes. Therefore, either mechanical dewatering efficiencies must be increased, or thermal means must be used to further reduce moisture levels at slurry pipeline destinations.

Currently, filtration and centrifugation are the two mechanical dewatering techniques in use. The ultimate effectiveness of mechanical dewatering is dependent upon the wettability of the coal and its surface area. Coals which are hydrophilic (very wettable) and have high surface areas, do not perform well in mechanical dewatering systems. Unfortunately, lignites fall into both of these categories. However, the theoretical limit to mechanical dewatering performace as a function of surface area, wettability and other factors is not well characterized. Such knowledge will be reqiured to determine whether research is justified to produce incremental increases in mechanical dewatering system performance, or should be directed at other dewatering techniques.

Thermal Drying

In thermal drying processes, moisture reduction rates are generally constant as a function of time at a given temperature, as long as surface moisture exists. After surface moisture has been removed, the rate of decline in moisture content drops substantially (see Figure 3.4.1.1).

Figure 3.4.1.1

Rate of Decline in Moisture Content

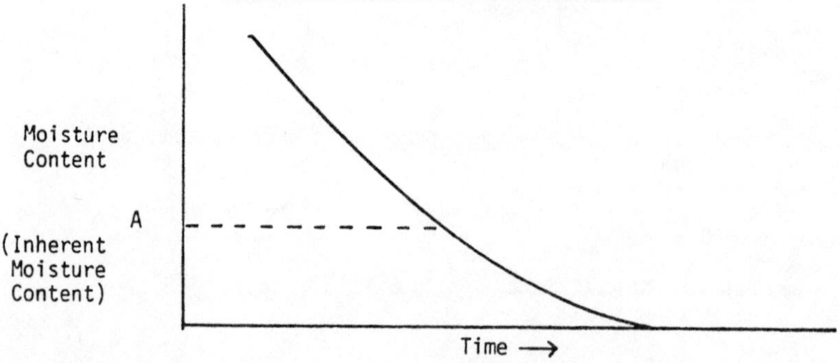

In bituminous coals, where inherent moisture levels are less than 10 percent, further drying is not necessary for conventional applications. For low-rank coals however, point A in Figure 3.4.1.1 corresponds to approximately 25-30 percent inherent moisture levels. The costs of drying substantially below this level increase inversely with the decreasing rate of moisture loss after point A.[6] For the most part, drying applications of low-rank coals are concerned with the "falling rate" region of the curve in Figure 3.4.1.1, i.e., removal of inherent moisture.

Optimal drying techniques may well depend on the end use of the coal. For example, drying for coal liquefaction will require that a minimum of deactivation (with respect to participation in liquefaction reactions) occur during the process. Drying for preparation of slurries will be aimed at maximizing the solids concentration in the slurry, and will therefore seek to limit moisture reabsorption in the coal. Similarly, drying for rail transportation will be oriented at limiting moisture reabsorption, but from vapor phase water instead of liquid water as in the case of slurry pipelines. An understanding of the mechanisms operating in each of these end uses will be important in making the proper choice of drying technique, since different options will effect different chemical and morphological changes within the coal matrix.

As an example, the Exxon Donor Solvent Coal Liquefaction Pilot Plant facility is currently testing a hot oil type dryer. The hot oil technique contacts coal feed with a process generated solvent oil at high temperature. Heat transfer from the oil vaporizes the coal moisture (which is collected by condensation and would be available for water treatment and process recycle in a full scale plant). The dried coal is thereby intimately mixed with the slurrying agent, and ready to be charged to the liquefaction reactor while suffering little or no deactivation. Another alternative considered by Exxon (and others), drying with a hot inert gas, avoids contact of the coal with atmospheric oxygen which is at least partly responsible for the deactivation effects observed. However, drying with hot inert gases was rejected by Exxon for several reasons. First, the heat capacity of gases is realatively low, which results in a loss of efficiency. In addition to greater energy requirements environmental concerns require scrubbing of the dryer offgas to prevent particulate emissions.

The primary concerns about the properties of dried low-rank coal are high reactivity and dustiness (leading to handling and storage problems), and reabsorption of moisture. In a test conducted by the Grand Forks Energy Technology Center,[8] 400 tons of Rosebud subbituminous coal and Gascoyne lignite were partially dried in a commercial scale dryer, oil sprayed, cooled and shipped by rail approximately 800 miles. After stockpiling the coal, it was concluded that storage and handling requirements do not appear to differ significantly from those required for as-mined coal (although greater compaction has been needed). Furthermore, although more than 4 inches of precipitation had fallen on the pile during the two year test period, there was no indication of moisture penetration.

A clearer understanding of the tradeoffs between thermal drying costs and reduced costs associated with transportation and utilization of dried coals is needed before the applicability of thermal drying can be considered. As part of such an investigation, several aspects of the problem would require separate study: 1) the state-of-the-art of thermal drying processes, including their effects on the physical and chemical behavior of low-rank coals; 2) effects of (various) dried coal properties on utilization process design and cost; 3) availability of freight rate structures for raw and dried coal that would allow the increased heating value material dried coal to be delivered at a lower cost per BTU; and 4) demonstration of handling, transport, and storage systems for this very dusty and reactive material.

2. Applications of Ion Exchange and Chemical Cleaning Processes

The reserves of North Dakota lignite are sufficiently large to provide a reliable source of supply for electric power generation within the state and in neighboring states. However, high sodium levels found in lignite from many North Dakota mines have been shown to cause severe ash fouling problems in utility boilers.[1] Some Texas lignites, as well as some Western subbituminous coals, also contain high enough sodium levels to cause ash fouling problems.

The occurrence of sodium in these coals exists as a uniform distribution throughout the organic matrix. For this reason, and because the sodium cations are chemically associated with the humic acids present in the coal,[2] physical separation techniques are ineffective for removing sodium. However, due to their chemical state, the sodium ions may be "exchanged" or replaced by other positively charged ions, similar to the operation of an ion exchange resin. Candidates for exchange of sodium in this application must be plentiful and cheap; among those being considered are ions of hydrogen (H^+) and calcium (Ca^{2+}).

Bench scale experiments and process design studies have been conducted at the University of North Dakota and the Grand Forks Energy Technology Center.[1] A preliminary economic analysis of a system based on sulfuric acid (H^+ cation exchange with sodium) has estimated a very rough cost (excluding profit) of $1.30 per ton of lignite for a 1.58 million ton per year facility (1979 dollars). The process design was based on a Beulah lignite feed containing 8.5 percent Na_2O in the ash, and calls for a reduction to 4 percent in the final product (coals having less than 4 percent Na_2O in the ash are not severely fouling coals). It may also be expected that reductions in other minerals will occur during the process.

The costs associated with ion exchange beneficiation must be recoverable by users of the treated lignite through savings in capital costs of facilities, reduction in boiler tube cleaning, improved heat transfer (and efficiency), lowered downtime, and reduced purchase of power to meet load requirements. The tradeoff between ion exchange costs and resultant savings at the power plant has not yet been well defined. One of the problems in this tradeoff analysis is determining the cost to a utility of downtime or other problems caused by high-sodium coal.

Due to the high reactivity of low-rank coals, investigators working in the field of ion exchange must also be sensitive to undesirable chemical and physical changes occurring within the coal as a result of interactions with the ion exchange medium. Increases in moisture content may affect transportation economics.

Chemical cleaning processes might also be applied to low-rank coals to remove the trace quantities of uranium and other potentially hazardous wastes that otherwise might be released to the atmosphere or leached from a disposal site. However, until some substantial cost is assigned to the release of these materials to the environment, or cleanup is mandated, no economic driving force will support the development of such processes.

3. Application of Gravity Separation Techniques to Low-Rank Coals

Gravity separation techniques for coal beneficiation have been in commercial use for decades and are part of a well-established technology for improving the quality of higher-rank coals. The topic encompasses jigs, tables, hydroclones, cyclones and dense media techniques. Due to their mechanism of operation these methods effect a physical separation on the basis of density differences and are therefore used for recovering coal from mine overburden, ash, and pyrites. (It should be noted that froth flotation, which effects separation of the coal from minerals by differences in surface properties rather than density, appears to be inappropriate for most low-rank coals due to their unique forms of inherent mineral matter.)

Due to the fact that western coals are inherently low in sulfur and ash, and that they generally are surface mined from uniform thick seams (resulting in inclusion of a minimum of mine overburden), these beneficiation techniques are generally not applied to low-rank coals at the present time. In addition, the wet processing involved in many gravity separation techniques would increase the already high moisture content of low-rank coals, compounding the transportation expense and utilization problems associated with high moisture coals.

Currently, only two coal preparation plants exist in the western states, and both are in operation due to unusual circumstances. Bituminous coal from an underground mine in Hanna, Wyoming, required beneficiation because of unacceptable amounts of refuse in the run-of-mine coal. The other beneficiation operation is applied to a surface mine in Centralia, Washington, where overburden and refuse are found in the subbituminous coal due to thin, erratic seams.

In the future, some western coals will probably be beneficiated for both local and eastern markets. Incentives for this practice will appear as SO_2 regulations tighten, and also as lower quality seams are mined. Beneficiation for local markets will become more common as more deep mines are opened, and as more erratic seams are mined (generally in the Gulf Coast regions).

The types of physical separation techniques that would be most appropriate for low-rank coals have not been determined. No technical reasons are known that would prevent the application of conventional wet gravity separation techniques such as jigs. However, washability data (available only for a small percentage of low-rank coals mined) indicate that relatively small improvements in mineral matter content will be accomplished, even when the cleaning incorporates the more complex fines washing techniques. Dry gravity separation techniques (such as fluidized beds) may have special appeal for western low-rank coals due to the low water availability in some regions and the undesirability of adding surface moisture to a high-moisture coal.

The extent to which gravity separation techniques come into use for low-rank coals will depend on a number of site-specific factors, including the answers to questions such as the following:

- Will lower quality low-rank coal seams be opened, and how much mine dilution will be added as a result?

- What is the economic tradeoff between minemouth sulfur removal and enhanced SO_2 scrubbing at the end use point?

- Is there an economic incentive for ash removal at the mine rather than paying increased transportation costs and disposal at the point of use.?

- Are water supplies in western coal regions adequate to support large scale coal cleaning activities?

4. Application of Briquetting or Pelletizing Techniques to Low-Rank Coals

Handling and storage problems associated with low-rank coals include spontaneous heating, oxidative degradation and excessive dusting. An opportunity to reduce or eliminate these problems as well as improve transportation and utilization economics exists through the use of briquetting or pelletizing processes. These processes are widely applied to coal of all ranks throughout the world. However, the markets served by this type of product tend to be very small in this country.

Although low-rank coal briquettes of high mechanical strength have been produced in Europe by hot and cold briquetting without the use of binders, mechanical strength can be enhanced with the use of these agents. Binder addition is generally no more than 10 percent by weight, and usually consists of a petroleum or coal derived distillate. Essentially all experimental briquetting work that has been done on U.S. low-rank coals has involved the use of binders. There is currently no significant production of briquettes from low-rank coals in the U.S.

Binding may be accomplished without the addition of external agents by subjecting some coals to pyrolysis or liquefaction conditions. Coal retorting has been shown to produce a char product which may be pressed (while hot) to strong, durable and smokeless briquettes. Because tar production from lignite pyrolysis is generally only 1 to 3 percent of the feed, additional coal distillates have been used as supplementary binders.[10]

Strength and wear resistance of finished briquettes have also been improved by the use of coatings on the finished briquettes. The use of these polymeric or resinous coatings has also been shown useful for dust prevention.

Briquette quality is strongly affected by coal moisture content, particle size and hardness. Lignite dried by the Fleissner process or in electrically heated rotary driers has optimum briquetting charactertistics at moisture contents of 10 to 16 percent. At a constant moisture content, smaller particles (0-1mm) are superior briquetting feeds than larger particles (0-2mm), as are softer coals.[9]

Due to the likely inclusion of moisture reduction as part of a pelletizing or briquetting process, it has been suggested that sodium removal from lignites by wet ion exchange techniques would be profitably integrated in a briquetting process.[11] Moisture inclusion as a result of aqueous ion exchange would be removed in the pelletizing process to produce a low-fouling, low moisture fuel with superior handling and storage properties.

5. Waste Disposal from Coal Beneficiation and Slurry Dewatering Plants

If physical or chemical coal cleaning plants and slurry pipelines are used in the future to upgrade and transport low-rank coals, one problem common to all of these plants will be the proper treatment and disposal of the concentrated waste streams. The types of materials rejected from these plants will include:

1. Coarse and fine refuse from physical coal cleaning processes, consisting primarily of wet, concentrated mineral matter separated from the coal; sludge from water clarification circuits; magnetite from dense-medium processes, and chemical reagents from froth flotation processes.

2. Spent chemicals from ion exchange or other chemical cleaning processes, contaminated with various coal-derived organic and inorganic species.

3. "Ink" separated from the coal slurry at pipeline terminals, which consists of very finely divided coal particles (20 percent smaller than 40 microns) suspended in water.

These solid or slurry waste streams will be subjected to increased scrutiny as a result of the hazardous waste provisions of RCRA. Most of the potential toxic elements contained in coal tend to be concentrated in the waste streams from beneficiation plants. Investigations into the chemistry of refuse disposal piles and the associated groundwater contamination possibilities are at a very early stage of development. A problem that has been noted in eastern bituminous coal refuse piles is high acidity (from oxidative degradation of pyrite and marcasite), with pH levels often falling below 2. At that level of acidity, many trace elements readily dissolve. Various treatments for neutralization or isolation of these acidic piles are being tested.[5] Because of the lack of experience in cleaning low-rank coals, it is not known whether the chemistry of the solid waste is similar, or whether high alkalinity might be the problem.

For waste streams such as the ink from slurry pipelines, investigations into the use of various surface-active reagents to aid in the coagulation and separation of the fine particles should be pursued. In addition to coagulants, the use of immiscible solvents or salts should be explored. Basic studies of the surface properties of low-rank coal fines, and applications of colloidal chemistry, would support the development of possible processes to minimize the disposal problems and environmental hazards.

6. Fine Coal Cleaning

The most common coal cleaning practice has been to separate the run-of-mine coal according to size, perform a cleaning operation on that fraction above a minimum size, and recombine the cleaned coal with the fine. As such, fine coal cleaning is a relatively new procedure in eastern coal cleaning plants. However, although the technology is still developmental, the trend is in this direction because the cleaning of coal fines allows a cleaner final product.

The issue of fine coal cleaning is germane to low-rank coal because of the finely dispersed nature of the mineral matter in these coals. This will require that large coal pieces be broken down below a maximum top size to insure that a sufficient amount of the total mineral matter is exposed to the beneficiation agent. However, the exact top size which will allow effective removal will depend both on the coal and the process. Quantitative details such as this are not well characterized at the present, and the general applicability of higher-rank fine coal cleaning techniques to low-rank coals is still unknown.

Extensive programs in fine coal cleaning are currently being sponsored by the Department of Energy and the Electric Power Research Institute. Technology areas include magnetic separation, froth flotation and particle agglomeration. Unfortunately, however, low-rank coals are not included in these programs at present.

Washability data indicate that mineral matter in some low-rank coals is evenly distributed among the various particle sizes. However, insufficient data of this type exist, and a basic characterization program on float-sink separations of low-rank coals would be useful. The presence of a greater fraction of the mineral matter in the fines would add additional weight to the importance of fine coal cleaning for low-rank coals, and their inclusion in the current test program.

7. Fines Generation During Handling and Comminution

The size distribution within a coal sample is important in determining its storage characteristics, transportability, and performance in direct combustion and other utilization processes. Low-rank coals exhibit a dustier nature than higher rank coals and produce fines in response to weathering and as a result of evaporation of surface moisture. However, there remains uncertainty as to the specific handling and comminution procedures that tend to generate excessive fines, or conversely, tend to minimize fines generation.

According to one industry source,[3] one cannot answer the question without knowledge of the specific coal and comminution process. Another source[4] claims that: 1) as-mined lignite and bituminous coals produce approximately the same quantity of fines; 2) crushing dried lignite (less than 30 percent moisture) creates more fines than as-mined lignite; and 3) crushing as-mined east Texas lignite will produce more fines than as-mined North Dakota lignite (testing done on Beulah lignite). The type of equipment used in the crushing operation also affects the quantity of fines produced. Three common crusher types are listed below in ascending order of fines produced:

a. Rotary breaker
b. Double roll crusher
c. Single roll crusher

Unfortunately there are no published data which provide comparisons of low-rank and bituminous coal behavior during crushing. In light of the importance of the issue, a quantitative investigation is warranted.

8. Handling of Dried Lignite

Three principal problems can occur during the handling of dried lignite - spontaneous heating, oxidation, and windage loss. All three concerns are related to the generally finer particle sizes encountered with dried lignites, and spontaneous heating and oxidation are additionally related to the high reactivity of low-rank coals. (Note that steam drying, such as the Fleissner process, tends to produce a more stable lump of lignite, for reasons that are not fully understood.)

Spontaneous heating may occur as a result of oxygen diffusion to a coal surface, particularly when available surface areas and ambient temperatures are high. Occurrence of this phenomenon is particularly hazardous in storage conditions because the formation of "hot spots" within the coal pile may lead to open burning. Even in cases where this is not an end result, oxidation of the coal results in a degradation of fuel quality and should therefore be avoided.

The chemical mechanism of spontaneous heating requires the presence of water molecules, since atmospheric oxygen does not react directly with carbon at these temperatures.[7] Combustion to CO and CO_2 proceeds by way of chain reactions where carbon reacts first with the more active -OH radical. Thus, the danger of spontaneous heating is minimized at extremely low coal moisture contents (<0.5%) as a result of this phenomenon, and also at very high moisture contents due to the quenching action of water.

High rates of windage loss (dust entrainment) can occur with dried lignite because of the breakdown of the physical structure that accompanies drying of the coal. As is the case with spontaneous combustion or oxidation, this problem can be controlled by the proper use of storage pile compaction procedures, application of coatings, size segregation, and closed storage or transport facilities. Briquetting or pelletizing might also be considered as an alternative to handling dried lignite.

The largest-scale testing of dried lignite handling done to date has been the cooperative testing of railcar shipments by GFETC and Commonwealth Edison of Chicago.[8] There is a need for commercial-scale demonstration of handling and storage of dried lignite, including all operations from the drier itself, through live and dead storage, to the boiler.

Other Issues

A number of less important issues have been identified with respect to the technologies for preparation, handling, and storage of low-rank coals. These are listed below, and in general, information relevant to these issues is incorporated in the appropriate sections of the report which follow:

1. Freeze Control
2. Control of Dust, Oxidation, and Spontaneous Combustion
3. Slurry Preparation Techniques
4. Magnetic Separation Techniques
5. Measurement Techniques/Indices for Comminution
6. Comminution Equipment and Methods
7. Wet Grinding Techniques

Basic Coal Characterization Needs

Associated with many of the key issues discussed above are needs for basic data on the properties of low-rank coals. Examples of the types of information required to support the development of coal preparation technologies applicable to low-rank coal are as follows:

1. Forms of Mineral Matter and Organic Salts in Coal
2. Characterization (Physical and Chemical) of Slurry Components
3. Washability Data
4. Surface Characteristics of Fines
5. Variability Study on Coal Seams (e.g., sodium content)
6. Petrographic Characterization of Low-Rank Coals

References - Section 3.4.1

1. Baria, D.N., W.R. Kube, and L.E. Paulson. Conceptual Design of a Commercial Plant for Sodium Removal from Lignite, January 1980.

2. Paulson, L.E. and W.W. Fowkes. Changes in Ash Composition of North Dakota Lignite Treated by Ion Exchange, U.S. Department of the Interior, Report #7176, September 1968.

3. Personal Communication with Dr. Peter Luckie, Penn State University.

4. Personal Communication with Mr. Bournville, McNally Pittsburgh Company.

5. Wangen, L.E., et al. Control Technology for Coal Cleaning Wastes, in Proceedings of the ASCE/PRC-EPRI Workshop "Solid Waste Research and Development Needs for Emerging Coal Technologies," San Diego, CA, April 23-25, 1979, p. 79.

6. Ziesling, G.R., K. Forgaard, J. LeFever, B. Rhodes, and V. Griffiths. Drying of Low-Rank Coal for MHD Application, presented at the 19th Biennial Lignite Symposium, Grand Forks, N.D., May 31, 1979.

7. Duda, Walter H. Cement Data Book, Bauverlag Gmblt, Wiesbaden, W. Germany.

8. Paulson, L.E., S.A. Cooley, W. Wegert, and R.C. Ellman. Experiences in Transportation of Dried Low-Rank Western Coals, SME Fall Meeting, Salt Lake City, September 1975, Preprint 75-B-337, SME/AIME Transactions, December 1976, 20 pp.

9. Goksel, M.A. and L. Valentyik. Production and Evaluation of Lignite Pellets, presented at the 16th Biennial Conference of the Institute of Briquetting and Agglomeration, San Diego, CA, August 6-8, 1979.

10. Oppelt, W.H., et al. Thermal Pretreatment and Pelletizing of North Dakota Lignite, Bureau of Mines R.I. 5382, January 1958.

11. Grant, Andrew J. and R.E. McKeever. Pelletizing and Drying of Lignite, presented at Coal Technology '78, Houston, Texas, October 1978.

3.4.2 Comminution

3.4.2.1 Technology Description

Comminution is the generic name for the crushing and grinding processes used to achieve controlled size reduction of coal. Crushing processes begin with relatively large coal top sizes (up to 10 inches) and produce coal with top sizes as small as 1/8 inch. Grinding (or pulverization) processes begin with feed to top size of about 1/2 to 3/4 inch and produce finely divided coal on the order of 75 microns.

The two basic reasons for reducing the particle size of coal are: 1) to facilitate handling or cleaning processes, and 2) to meet the demand for specific size distributions. For example, fixed-bed gasification processes require size distributions with a top size of about 2 inch but with minimal fines less than 1/4 inch. Pulverized coal-fired furnaces typically require that about 70 percent of the coal pass through a 200-mesh screen, which has openings of 74 microns.

Crushing may be applied to coal at the mine to reduce the as-mined top sizes to standard sizes for shipping or preparation. Crushers are also used at powerplants where delivered coal is crushed from 2 inch top size to 1/2 to 3/4 inch for feeding to the pulverization equipment. Grinding (pulverization) operations are always conducted at the point of utilization due to the problems involved in transporting and handling pulverized coal (e.g., dust emissions, explosion hazard).

Crushing and grinding technology is as old as coal utilization itself. Equipment is designed and selected to handle specific feed properties and produce the desired lump or particle sizes based largely on empirical knowledge and experience. Design correlations (such as the modified Hardgrove grindability index, discussed later) have been established over a period of years for coals with unique physical properties, such as the low-rank coals. This evolutionary process has included the "brute force" technique of building large safety margins into early designs, and using data from those operations to reduce the margins in subsequent designs.

Theory[1]

Coal breakage results from stress or strain being applied by the action of some moving part of the machine. If the induced strain exceeds the elastic limit of the material, breakage will occur and that energy is released in the form of heat. If the induced strain does not reach the elastic limit of the coal, the energy is released as energy of resilience. Although the energy input to the breaking process is relatively straightforward to measure, it is very difficult to carefully account for the disposition of that energy in the complex breaking process. Considerable efforts have been made to obtain exact measurements of energy flows, but none have been entirely successful.

Two theories of comminution which have long existed and which have been used with a varying degree of success are the Rittinger and Kick theories. The Rittinger theory asserts that the useful work accomplished in crushing and grinding is directly proportional to the new surface area produced, and to the reciprocals of the new particle diameters. In other words the useful work increases in geometrical ratio as the product particle size decreases. Gaudin has enlarged the Rittinger theory to include the concept of surface energy. Essentially, Gaudin states that the efficiency of a comminution operation is the ratio of the surface energy produced to the kinetic energy expended.

The Kick theory states that the work required is proportional to the reduction in volume of the particles concerned. Kick based this theory primarily on stress-strain relationships for cubes under compression. According to the theory, the work required for reduction to different sizes is proportional to (log F/P)/log 2, where F and P are the diameters of the feed and product particles respectively.

Both of these theories are at least partially unsatisfactory because crushing is known to be both surface and volume related. The absorption of evenly applied stresses is proportional to the volume being stressed; however, breakage is initiated at a crack tip, usually on a surface where a concentration of stresses exists. Once a crack tip is formed, all surrounding stresses are concentrated in the tip, which rapidly extends throughout the particle. This splits the particle and results in a break. The energy flow and stressed condition of the rock create additional crack tips, resulting in a breakage pattern. Very little if any additional external energy need be applied to brittle material to cause the break after the first crack tip is formed.

By using the concept of the formation of crack tips in a comminution operation, Bond developed a new "Third Theory of Comminution." This theory asserts that the total work useful in breakage which has been applied to a stated weight of homogeneous broken material is inversely proportional to the square root of the diameter of the product particles. The theory also postulates that correlation between different materials should be made by the use of a work index, Wi, which is the calculated kwh per ton applied to reducing material of infinite particle size to 80 percent passing 100 microns. Thus the work index establishes relative reduction resistance of a material in the size range tested and the relative mechanical efficiencies of different machines and processes. For any values of F and P when W is the kwh per ton required to break from feed particle diameter F to product particle diameter P, the total work input W_t is proportional to $1/\sqrt{P}$ and W is proportional to $1/\sqrt{P} - 1/\sqrt{F}$. Thus

$$W_i - W \frac{(\sqrt{F})}{(\sqrt{F}-\sqrt{P})}\sqrt{P/100} = 0 \quad \text{or}$$

$$W = W_i \frac{(\sqrt{F}-\sqrt{P})}{(\sqrt{F})}\sqrt{\frac{100}{P}}$$

Crushing

Crushing operations take place in several stages between the mine and final use of the coal. Primary breaking, which usually takes place at or near the mine, reduces the coal top size to between 4 and 8 inches. Secondary crushing reduces the coal top size to the 1-1/2 inch range, and usually takes place place at the power plant upstream of the pulverizer. For uses other than pc-fired boilers, screening crushers reduce the products of the secondary crushing process to the final commercial sizes, such as 1 x 3/8 inch stoker coal.

Each lump of coal is broken by contact with other lumps or by direct contact with the moving parts of the crusher. As the lump size decreases, more contacts are required per unit mass. Consequently, the capacity of a particular crushing machine is less for small sizes than for larger sizes, since smaller particles must remain in the crusher for longer periods of time to sustain the required number of contacts.[1]

Reduction is accomplished by machine action that involves various principles, such as impact, compression, splitting, shearing, or attrition. Impact is the sharp, instantaneous blow resulting from a moving body striking another body. Compression involves the more gradual application of pressure between two surfaces. Splitting is accomplished by penetration of the coal. Shear is usually associated with the use of combinations of several of these principles, such as impact and compression in single-roll crushers. Attrition reduces a material by subjecting it to an abrasive or rubbing action.[1,7]

There are many types of crushers available, each of which uses some combination of these operating principles. Table 3.4.2.1 lists the most common generic types of crushers utilized for coal. Coals typically have crushing and grinding characteristics in the soft to semi-hard range with respect to other rocks. Crusher designs utilized almost exclusively for hard rock applications, such as jaw or gyrating crushers, are not included in Table 3.4.2.1. Much more detailed descriptions of the different crusher types, including schematic and photographic illustrations, are available in references such as 1 and 7.

Screening

An inherent part of coal comminution technology is screening or sizing, which sorts the particles by size ranges. The principle of operation is quite simple: particles below a certain size pass through the screen deck while the remainder pass over the deck. Despite this readily observable screening action, the process contains too many variables and interactions to be amenable to theoretical or analytical treatment (see Table 3.4.2.2).

Table 3.4.2.1

Types of Coal Crushers

Crusher	Description	Principles of Operation	Comments
Single-Roll Crusher	Coal is squeezed between a revolving crushing roll and a stationary breaker plate equipped with renewable wear plates. The roll has a series of long teeth spaced at intervals, with various short teeth covering the entire crushing surface.	The long slugger teeth act as feeders and also penetrate the lumps (splitting). The smaller teeth make the proper size reduction, utilizing mainly impact, shear, and compression in their operation.	One of the oldest, and perhaps the simplest type of crusher. Capable of handling run-of-mine feed and reducing it to 1 1/2 by 0-in. without stalling. Operates at slow speeds (40-60 rpm). Handles wet, sticky, or frozen feed. Produces a minimum of fines.
Double-Roll Crusher	Coal is crushed between two revolving rolls. Teeth are designed to mesh, and the spacing between rolls is adjustable. Hawk-billed teeth are used for primary crushing; pyramidal, cone-shaped, or cross-tooth designs are used for secondary crushing; smmoth or corrugated rolls can be used for fine coal grinding or pulverization.	Meshing teeth rely mainly on impact, with a minimum of attrition or shear, to reduce the material. Compression between the rolls is undesirable and is a sign of improper maintenance or adjustment.	Ideally suited for harder coals. Provides high volume production, accurate sizing, and produces a minimum of fines. Roll speed is 115-150 rpm.
Rotary Breaker	Coal is fed to a rotating perforated drum. Small lumps fall through the perforations while larger lumps are raised by radial lifting shelves and then dropped as the drum rotates. Large rocks and foreign material are carried through the drum and discharged out a refuse chute.	Gravity impact breaks the coal as it falls.	Used as primary breakers and cleaners of run-of-mine coal. Very uniform top size, and very low fines production. Slow rotational speed (12-18 rpm). Problems have been experienced with plastic or clay-containing materials that tend to plug the perforations.

(continued)

Table 3.4.2.1 (continued)

Types of Coal Crushers

Crusher	Description	Principles of Operation	Comments
Hammer Mill	Rotating hammers deliver heavy blows with the aid of centrifugal force to the coal while it is in suspension, driving it against a breaker plate until it is sufficiently reduced in size to pass through the discharge. Grate bars may be included in the discharge opening to fix the maximum size.	A combination of impact and attrition is achieved-impact upon initial contact with the hammers, and attrition as the material is caught between the ends of the hammers and the cage bars at the bottom of the crusher.	Among the most versatile crushers available. Exceptionally large capacity for their size. Quantity of fines produced is higher than with most of the other types of crushers. High rotational speed (700-1800 rpm).
Ring Crusher	Similar in design to the hammer mill, with ring-type hammers substituted for the beater-type hammers used in the hammer mill. Initial reduction takes place as the coal is compressed between the rolling-ring hammers and the breaker plate. Final reduction occurs as the lumps are broken between the rolling-ring hammers and the cage-screen bars.	Compression is the primary force acting as the rings roll acros the feed. Intense concentrated pressure cracks and shatters the coal with negligible rubbing action.	Specifically designed to crush bituminous coals for pulverizers and stokers with a minimum of overgrinding and a minimum of fines. Commonly used in place of rotary breakers where floor space is limited.
Impactor	Coal is dropped into the path of high-speed rotating beaters. The shattered particles are driven against the impactor side wall anvils. The cycle is repeated as the particles rebound into the beater circle. The particles continue to shatter under this high-velocity repeated impact-rebound action until they are swept out the open bottom of the impactor.	Impact is the only significant force acting. Shattering power is applied in direct proportion to particle size, thus utilizing maximum power on large fragments and wasting little on finished particles.	High rotational speeds. Overgrinding (fines production) is very low. Production of a high pecentage of closely sized material is possible. A cubed or granular product is produced regardless of rotor speed.

Source: References 1, 2, 7

Table 3.4.2.2

Factors Influencing the Passage of a Mineral Particle through a Screen Opening

(1) Ratio between cross-section of particle and of mesh.
(2) Percentage of screen open area.
(3) Angle of incidence of feed.
(4) Efficiency of spread of feed over screen area.
(5) Kinetic energy of particle approaching screen opening.
(6) Moisture of feed.
(7) Stickiness of particle and of aggregated particles.
(8) Pressure of particles riding above those next to the screen cloth.
(9) Blinding of screen apertures.
(10) Corrosion of screen material.
(11) Electrostatic bunching.
(12) Shape of particle.
(13) Percentage of near-mesh particles in the feed.
(14) Rate of feed.
(15) Thickness of layer.
(16) Tautness of screen.
(17) Shape of screen apertures.
(18) Orbit imparted to particle by screen vibration.

Source: Reference 1

A screen installed ahead of a crusher may serve as a scalping screen to remove oversize lumps, refuse, and trash, or it may permit undersize particles to bypass the crusher and thus minimize fines content. Installed after a crusher, a screen will separate the crushed product into two or more size consists (if necessary) for subsequent processing on an individual basis.

Screens operate either dry (without addition of water) or wet. Sizing below 3/8 in. is practiced primarily with wet screens, where water sprays wash the fine particles through the screen openings and prevent plugging.

Screens may be stationary or activated. Stationary screens, when used to size raw coals, are installed at sufficiently steep angles to permit oversize material to slide over the screen deck by force of gravity alone. They are relatively inefficient and are used primarily for scalping purposes. Activated screens supplement the force of gravity by mechanical and/or electrical means. Their sizing efficiency can approach 100 percent, although 85 to 95 percent efficiencies are commercially acceptable and more cost effective.

Screen surfaces are constructed of abrasion-resistant materials consisting either of parallel bars, punched plates, or wire cloths with square, round, or rectangular openings. Rubber or plastic-lined screen decks and other special constructions are receiving increasing consideration because they offer reduced noise levels and other advantages.

Grinding or Pulverizing[3,4,5]

Grinding or pulverization processes convert crushed coal (top size 1/2 to 3/4 inch) to a fine powder (typically 70 percent through a 200-mesh screen) suitable for feeding to suspension-type furnaces. In the future, finely ground coal may also be utilized in certain coal cleaning processes, slurry pipelines, and coal gasification or liquefaction processes.

As in crushing, the basic operating principles of grinding involve the proper application of stresses to rupture the coal particles in the most efficient manner. Grinding equipment differs fundamentally from crushing equipment in that the mechanical elements touch one another, except to the extent that they are prevented from doing so by the material being ground. Impact, shear, and attrition (abrasive)forces produce large quantities of fines, and are thus preferentially used in grinding equipment.

Also as in crushing, a number of different types of machinery have been found to work well in coal grinding applications. The major types are summarized in Table 3.4.2.3. When integrated into a pulverized coal-fired boiler plant, all of these systems have several elements in common:

1. Feed properties (primarily grindability and moisture content) significantly affect the capacity of a pulverizer. The Hardgrove grindability index (ASTM Standard D-409) measures the relative hardness, or pulverization difficulty, of a coal relative to a standard coal which has an index of 100. A lower index value indicates a more difficult coal to grind, and in turn a lower capacity for a given mill, as illustrated in Figure 3.4.2-1. For low-rank coals, a modified ASTM 409 procedure is used, in which the grindability determination is made at several moisture levels. On some lignites, even this modified procedure has proved inaccurate. On coals without prior experience, grinding a large sample in an actual pulverizer may be necessary to assure that adequate milling capacity is designed into the new installation.[3,5,8]

Figure 3.4.2.1

Effect of Grindability and Fineness on Pulverizer Capacity

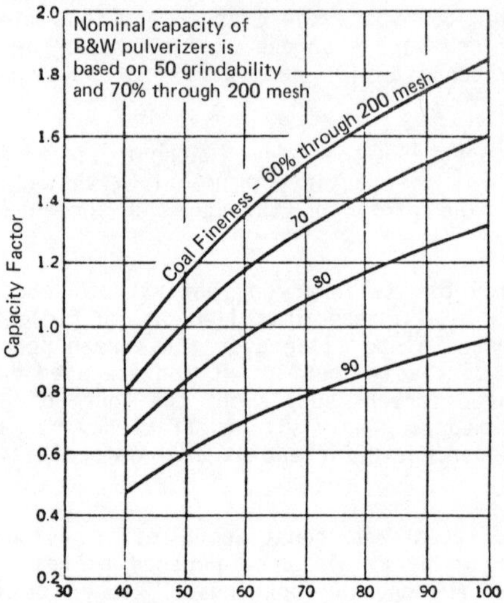

Source: Reference 3

2. Drying of the coal prior to or during pulverization is required to allow pneumatic circulation of the fuel. Most pulverized-coal boilers use in-the-mill drying, in which preheated primary air quickly dries the coal as it is being circulated and ground.[3,10]

3. Classifying and recirculation of the coarse material to the grinding zone is required because it is not feasible to grind all the coal to the desired fineness in a single pass through the grinding ele-

Table 3.4.2.3

Types of Coal Grinders (Pulverizers)

Grinder	Description	Principles of Operation	Comments
Tumbling Mills: Ball Mill	Coal is swept through a large rotating cylinder containing steel balls ranging in size from about 3/4 to 4 inches in diameter. The replaceable steel liner of the cylinder has a wave pattern or other irregular pattern to lift the balls up the side of the cylinder as it rotates. Best grinding is achieved at a rotational speed about 75 percent of critical speed (at which the balls would be held against the wall rather than falling).	Impact and attrition forces are generated by the steel balls, both as they tumble into one another and as they rotate with respect to each other and the cylindrical wall.	Low rotational speed ($\lesssim$75 rpm). Can be operated either wet or dry. Most commonly applied as air-swept grinding systems for producing pulverized power plant feed. For a given capacity, tumbling mills are generally larger, heavier, require more power, and are less efficient in drying the coal than the more modern high- or medium-speed mills. However, they are dependable and require low maintenance.
Rod Mill	Same as ball mill, except that steel rods lying horizontally along the full length of the mill are used instead of balls.	Same basic principles as ball mill, but rod mill relies more on compression and attrition forces than impact.	Rod mills consume less power per unit of work than ball mills, but are not capable of grinding as finely.
Impact Mills: Swing-Hammer Pulverizer	Same basic design as hammer mill described in Table 3.4.2-1. Swing-hammer pulverizer incorporates a cage assembly with closely spaced screen bars to permit reductions to minus 1/3 inch.	Impact and attrition forces are generated by the contact of the rapidly rotating hammers with the coal lumps, and the grinding of the material between the ends of the hammers and the cage bars.	High rotational speed (700-1800 rpm); and high power requirements. Can pulverize coal to minus 14-mesh, suitable for use as pipeline slurry.

(continued)

Table 3.4.2.3 (continued)

Types of Coal Grinders (Pulverizers)

Grinder	Description	Principles of Operation	Comments
Impactor	Same design as described in Table 3.4.2-1. When operated in a closed circuit with air classifier, can pulverize coal to minus 14-mesh or fines.	Free-air impact of coal particles on rotating beaters and side wall anvils.	Somewhat higher capacity than swing-hammer units. Power requirements are high. Susceptible to damage by foreign material.
Roller Mills:			
Bowl Mill	Coal is ground between a rotating bowl and stationary rollers which are held in position by mechanical springs. Centrifugal force feeds the coal between the race and the rollers. Ground coal spills over the top of the bowl and is carried out by circulating hot air. Coal fineness is controlled by adjusting the entrance vanes to the classifier and by adjusting the compression springs to control the pressure of the rollers on the coal.	Abrasion (attrition) is the primary cause of breakage of the coal particles as they rub against each other between the rollers and the bowl.	Medium rotational speed (75-225 rpm). A number of variations on this basic design exist.
Ball-and-Race Mill	Coal is ground by a row of large metal balls that rotate between a spring-mounted stationary upper race, and a rotating lower race (Ball bearing principle). Air circulates the coal through the grinding elements where some of it is pulverized in each pass through the row of balls. Fine particles are carried by the air to the classifier, which returns oversize material to the grinding zone.	Abrasion (attrition) action is applied to the coal particles by the rotation of the balls between the conformed circular races.	Medium rotational speed (75-225 rpm). High load circulation is obtained through the grinding zone, which is very desirable for effective drying and classification.

Source: References 3,4,7

ments. Proper operation of the classifier is crucial because over size particles entering the furnace can cause slagging and emission of unburned carbon.

Chemical Comminution[6]

Chemical comminution has been proposed as a potentially attractive alternative to mechanical comminution plus cleaning of coal. In chemical comminution, coal is exposed to low molecular weight chemicals, such as gaseous or liquid ammonia, at modest temperatures and pressure. The ammonia penetrates the coal along the natural fracture planes and disrupts the forces holding the coal particles together. The impurities and coal particles are separated from each other, with larger particle size distribution than the same coal mechanically ground to achieve the same degree of release. The ammonia has no chemical action on either the coal or impurity fractions, and can be recovered from the fragmented product by washing with water.

The wide variability in the petrographic characteristics (maceral forms and boundaries) and in the distribution of impurities in coal deposits suggests that the response to chemical comminution treatment will be variable depending on the coal. Bench scale tests on 15 coals have shown that bituminous coals are most amenable to chemical fracture, with lignite and anthracite exhibiting less susceptibility to treatment. The fact that coal cleaning (for removal of high pyritic sulfur or ash contents) is not generally practiced on low-rank coals is another indication that this technology is likely to be more useful for eastern bituminous coals.

A 50 tpd pilot plant facility has been proposed to test the essential elements of a chemical comminution process. An accurate assessment of the technical and economic promise of the technology will be possible only after sufficient data from a unit of that type has been obtained.

3.4.2.2 Environmental Control Technology

The primary environmental control requirement associated with coal crushing and grinding operations is the control of dust emissions. This problem is amenable to engineering solutions through the use of conventional technology. The degree of control installed in a particular facility is usually determined by the locally applicable standards.

The design of an in-plant dust control system is a specialized field, with a wide variety of equipment available to meet specific needs. In general the design should include the following:[1]

1. Hoods designed to provide effective dust control at the sources with a minimum air flow through the hoods and minimum power consumption.

2. Ductwork designed to transport the dust-laden air, collected by the hoods, to the air cleaner without settling of dust therein and with a minimum pressure loss.

3. An air cleaner selected which will efficiently remove particles from the air so the effluent in the discharge will conform with air pollution ordinances. Applicable devices include cyclones, baghouses, ESPs, and wet scrubbers.

4. A fan and motor so selected as to provide the necessary air volume at the static pressure (total system resistance) developed by the system.

3.4.2.3 Effects of Low-Rank Coal Properties

Crushing for top size control at the mine, and pulverization for feeding to suspension-fired boilers, are widely practiced on lignites and subbituminous coals. (In fact, these are the only preparation steps generally practiced on low-rank coals in the U.S.) Sufficient experience has been accumulated so that design comminution equipment to accomodate the unique characteristics of low-rank coals is no longer a significant problem. However, the standard laboratory test for grindability (Hardgrove) has been shown to correlate poorly with actual grindability of low-rank coals in pilot-scale or full-scale mills, as discussed below.

The properties of low-rank coals that affect comminution processes are: moisture content, slacking behavior, reactivity, and grindability (as determined in the laboratory and in actual mills). The effects of these properties are briefly described below.

Moisture Content

The high moisture contents of low-rank coals translate into higher feed throughput requirements for given energy production rates. Because this moisture content is largely inherent (surface-mined low-rank coals are not washed for removal of impurities and usually contain little or no surface moisture), the coals actually appear dry and dusty during handling, transportation, and use.

The removal of moisture before or during pulverization has a significant effect on both measured and actual grindability, as discussed below.

Slacking

Lignite particles are fibrous, resilient, and tough; these properties are somewhat less pronounced in subbituminous coals. As moisture evaporates from low-rank coal particles, the surface becomes brittle, and particles spall off without application of physical force. This process is termed "slacking." This structural weakening is beneficial in pulverization if the coal is dried either before or during grinding.[12]

Slacking behavior and the dry, dusty appearance of low-rank coals have led to the popular misconception that a greater percentage of fines is produced from comminution of low-rank coals than high-rank coals. This is important, for example, in fixed-bed gasification processes, where large amounts of coal fines cannot be tolerated. Wide variations in behavior of coals are observed within all ranks, and some coals do produce excessive amounts of fines under certain conditions. However, there are no data that suggest a systematic difference in the fines fractions produced from coals of different rank in crushing and grinding operations.

Reactivity

Low-rank coals, as fired in pc burners, are partially dried and highly reactive. Because of their superior ignition characteristics, low-rank coals need not be ground as finely as bituminous coals. The difference is on the order of 55-65 percent through a 200-mesh screen for lignites, compared to 70 percent for bituminous coal.[3,9]

Hardgrove Grindability Index

The Hardgrove grindability index of North Dakota lignites has been shown to vary in a characteristic manner as moisture is removed from the lignite. As measured in a modified procedure (that directly weighs the fine particle material and allows more satisfactorily for loss of moisture during testing) developed at GFETC, the index indicates that the lignite is relatively easy to grind at its native moisture content of 35-40 percent; becomes more difficult as moisture is removed down to 20-25 percent; then becomes more "grindable" again as even more moisture is removed. This is illustrated in Figure 3.4.2.2.

Little success has been obtained in correlating the Hardgrove grindability index to the actual performance of pilot plant and full-scale pulverizers using North Dakota lignites.[9-13] Contrary to the indication of the index, the actual grindability is lowest at the lignite's natural moisture content, and increases steadily as moisture is removed.

Figure 3.4.2.2

Hardgrove Grindability Index Versus Moisture Content of Two North Dakota Lignites

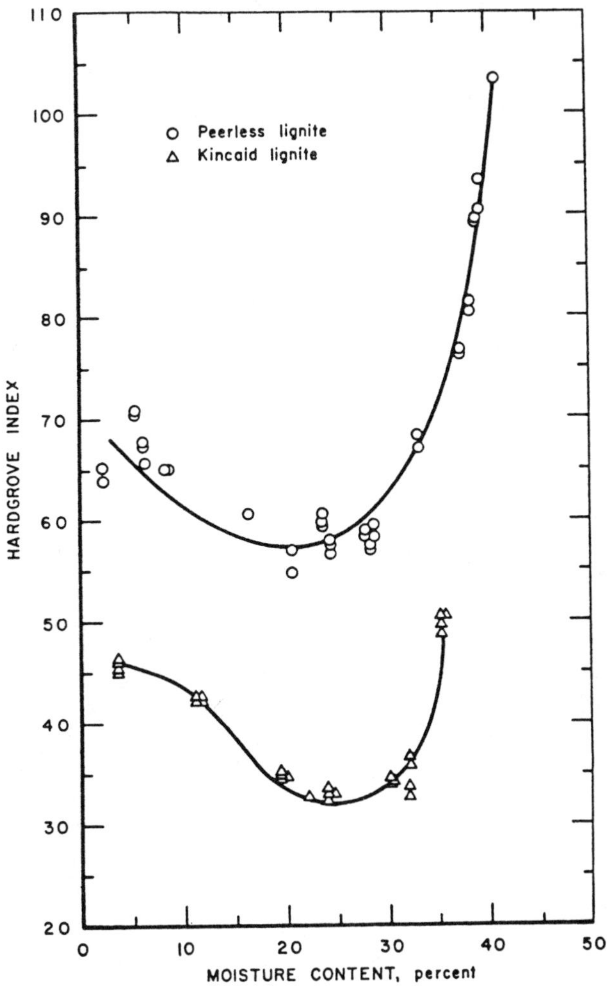

Source: Reference 9

Actual Grindability

Pulverization tests in several different types of pilot-scale mills showed essentially the same behavior: as moisture is removed from the North Dakota lignite, product fineness increases, capacity of the mill increases, and power requirements decrease. Figure 3.4.2.3 shows a compilation of data from a large number of tests at GFETC. The amount of minus-200 mesh material produced per kilowatt-hour is seen to increase dramatically as the lignite is dried. Figure 3.4.2.4 shows a similar compilation of data on the increase in mill capacity as moisture is removed. It should be noted that the effects of moisture removal were determined to be essentially the same when drying occurred before pulverization as when drying occurred in the mill itself.[9-13]

It is reasonable to expect that other low-rank coals behave in a similar manner in pulverizers, although data are unavailable. All existing pc-fired boilers take advantage of the improved grindability of low-rank coals at low moisture contents through the use of in-the-mill drying, using preheated primary combustion air. (For example, lignite as burned typically contains 25-30 percent moisture, versus 35-40 percent as mined.) In no case has it proved economically attractive (yet) to pre-dry the coal and obtain even higher grindability values (as well as less water passing through the combustion chamber). The advantages do not outweigh the costs of a separate drying step.

Figure 3.4.2-3

Pounds of Pulverized Lignite Produced per Kilowatt-Hour Versus Moisture Content of Lignites. Pilot Plant and Commercial Scale Test Results

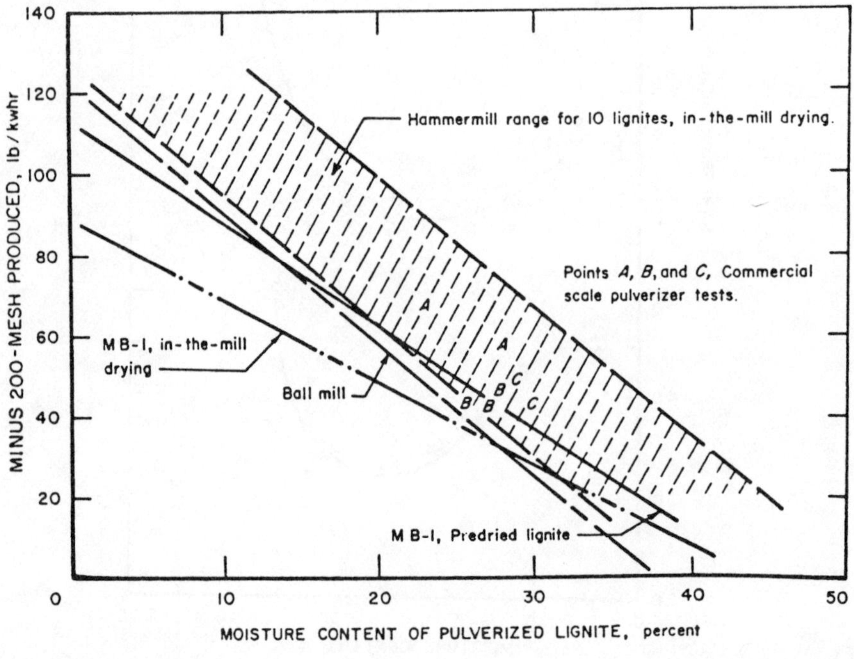

Source: Reference 9

Figure 3.4.2.4

Effect of Drying on Capacity of Pilot Plant Pulverizers

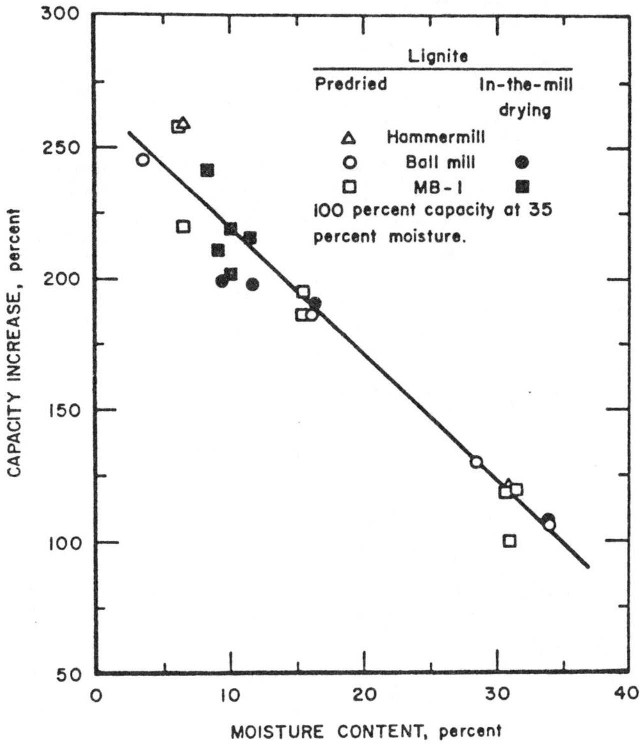

Source: Reference 9

3.4.2.4 Current Status

Low-rank coals are almost exclusively surface mined. Typically, coal is loaded into large trucks or rail cars and hauled to a primary crushing plant at the mine where its top size is reduced for ease in handling and transporting. All of the major types of crushers identified in section 3.4.2.1 are utilized in these operations.

Low-rank coals are also almost exclusively used as fuel in utility (and some large industrial) boilers. A very small fraction of these units are stokers, and utilize sized lumps of coal from secondary crushing operations. The remainder are pc-fired units which have grinding mills integrated into their designs. All of the major types of grinders identified in section 3.4.2.1 are utilized in low-rank coal-fired power plants. Essentially all of these units use preheated combustion air (600-700°F) as the drying and transporting medium.

In general, the industrial designers of crushing and grinding equipment consider the use of present day comminution equipment on low-rank coals to be a question of design adaptation rather than one of fundamental gaps in comminution science. Currently, there is almost no R&D work by DOE or EPRI in progress on crushing and grinding technology; the majority of the development work takes place in the private sector and consists of efforts directed at site-specific applications.

A single federally sponsored program directed at producing a coal grinding handbook has been underway since 1976. This program is being conducted by Kennedy van Saun Corporation and is entitled "Developing/Modifying Coal Grinding Procedures and Equipment to Produce Predictable Size Distributions During Coal Preparation."[14] The major motivation for this program is the need for prepared coal that falls within specific size ranges in order to optimize its utilization in advanced conversion and preparation processes. Perhaps because of this emphasis, the project does not include the testing of lignites, and includes only a single subbituminous coal from Wyoming.

References - Section 3.4.2

1. Leonard, J.W., and D.R. Mitchell (Editors). Coal Preparation, American Institute of Mining, Metallurgical, and Petroleum Engineers, Inc., New York, 1968.

2. Phillips, P.J. Coal Preparation for Combustion and Conversion, EPRI AF-791, May 1978.

3. Babcock & Wilcox Company. Steam/Its Generation and Use, New York, 1978.

4. Spink, K. Comminution Methods and Machinery, Mining and Minerals Engineering, February 1972, pp. 5-24.

5. Frisch, M., and Holder, G.C. Correlation of "Grindability" with Actual Pulverizer Performance, Combustion, June-July 1933, pp. 29-34.

6. Higginson, G.W., and H.R. Maddocks. Using Chemical Comminution to Ease Solids Separation in Coal Liquefaction, presented at the Fifth Annual International Conference on Coal Gasification, Liquefaction,and Conversion to Electricity, University of Pittsburgh, August 1-3, 1978.

7. Roman, G.H. "Coal Crushers and Crushing," Coal Age, February 1967, pp. 108-122.

8. Ellman, R.C., and J.W. Belter. Grindability Testing of Lignites, Bureau of Mines Report of Investigations 5167, November 1955.

9. Ellman, R.C., J.W. Belter, and L. Dockter. Lignite Pulverization: A Status Report, in "Technology and Use of Lignite - Proceedings: Bureau of Mines - University of North Dakota Symposium, Grand Forks, ND, April 27-28, 1967," Bureau of Mines Information Circular 8376, May 1968.

10. Ellman, R.C., J.W. Belter, and L. Dockter. Effects of In-the-Mill Drying on Pulverizing Characteristics of Lignite, Bureau of Mines Report of Investigations 6074, 1962.

11. Ellman, R.C., and J.W. Belter. Pulverizing Lignite in a North Dakota Powerplant, Bureau of Mines Information Circular 8032, 1961.

12. Dockter, L., J.W. Belter, and R.C. Ellman. Pulverizing Lignite in a Ball Mill, Bureau of Mines Report of Investigations 7010, August 1967.

13. Ellman, R.C., L. Dockter, and J.W. Belter. Pulverizing Lignite in a Ring-Roller Mill, Bureau of Mines Report of Investigations 7631, 1972.

14. Luckie, P.T. Developing/Modifying Coal Grinding Procedures and Equipment to Produce Predictable Size Distributions during Coal Preparation, Annual Progress Report for the period October 1977-December 1978, Kennedy Van Saun Corp. Under DOE Contract No. EX-76-C-01-2475, FE-2475-26, January 15, 1979.

3.4.3 Mineral Matter Control

3.4.3.1 Technology Description

The purpose of mineral matter control is to upgrade as-mined coal by reduction or alteration of its mineral matter content in order to obtain improved performance in specific utilization processes. Historically, as practiced on bituminous coals to meet product quality specifications, the technology consists of coal cleaning (beneficiation) processes in which extraneous mineral matter, including some pyritic sulfur, is separated from the coal by washing in water. A number of other types of processes, such as chemical cleaning and ion exchange, are being developed but have not reached commercial status yet. In the future, if any mineral matter control processes are applied to low-rank coals, these developmental processes are perhaps more likely to be used than the traditional physical coal cleaning techniques.

Essentially no mineral matter control is practiced on low-rank coal in the U.S. today. The reasons for this include very low extraneous mineral matter content, high inherent moisture content, and others discussed in Section 3.4.3.3. However, some growth of this technology for low-rank coals may be anticipated, as evidenced by an increasing number of economic tradeoff or scoping studies, small-scale research and development efforts, and references to the problems and opportunities of low-rank coal beneficiation in the literature. Some of the tradeoffs that are being explored include the following:[5]

1. Coal cleaning to remove sulfur, combined with less severe flue gas desulfurization, could reduce the total cost to a utility of meeting the NSPS with certain low-rank coals.[6,7] This might be particularly true in cases where ash-alkali wet scrubbing or dry sorption SO_2 removal processes were unable to consistently achieve 70 percent removal, but were able to do so when combined with coal cleaning; the combination might prove less costly and more reliable than conventional lime or limestone wet scrubbing. One engineering study has estimated that intensive cleaning of Montana Rosebud subbituminous coal could reduce the requirement for SO_2 removal from the flue gas to 58 percent, instead of 70 percent without cleaning.[6] (However cost estimates showed no economic advantage for the combined system under the assumptions used.)

2. Coal cleaning to reduce the quantity and variability of mineral matter (or specifically sodium content), from high-fouling low-rank coals, could reduce the total cost to a utility of fouling and slagging control, particulate removal, and ash disposal. The cleaning technology needs further development, and the cost tradeoffs need to be better quantified, before the potential benefits in this area can be realized in practice.

3. Coal cleaning might be used as a retrofit strategy by utilities or industries seeking emission offsets in either SO_2 or particulate matter.

4. For low-rank coal being transported long distances, coal cleaning might reduce overall shipping and handling costs per BTU. However, moisture reduction (see Section 3.4.4) would be likely to have a more important role than mineral matter reduction, in terms of impact on low-rank coal transport costs.

In addition to these areas currently being explored, there is a possibility that changes in mining practices and depletion of the highest quality reserves will create a need for some low-rank coal cleaning in the future.

The remainder of this section presents brief descriptions of available and developing mineral matter control technology, arranged into two generic areas as follows:

1. Physical Processes - these techniques rely on the differences in physical properties between the coal and the mineral-containing materials. The most important property difference is specific gravity; another is the surface affinity to specific reagents; a third area being explored is the magnetic properties of coal and minerals. Existing preparation plants use physical processes almost exclusively.

2. Chemical Processes - these techniques utilize the chemical characteristics of the coal and its constituents to effect the the removal of undesirable materials. These processes, for the most part, are in developmental stages.

Physical Processes

The vast majority of physical coal cleaning processes rely on the differences in specific gravity between coal and the extraneous minerals mixed with it. In essentially all cases the impurities have higher specific gravities than the clean coal, as illustrated in Table 3.4.3.1. The specific gravity range for coal itself depends largely on the amount of ash bound within the coal particle. Any coal particle, independent of size, contains both inherent and extraneous ash. Inherent ash is bound within the structure of the coal and cannot be removed by physical means. Extraneous ash is either contained as deposits within the coal and is liberated by crushing, or consists of material that has been combined with the coal during the mining process (e.g., overburden). The extraneous ash contained within the coal particle can be removed only to the extent to which the coal is crushed. Thus, the most intensive coal cleaning processes incorporate steps in which the coal is finely ground and the fines are washed.

The standard test to determine the specific gravity distribution of the various components comprising the coal is the washability analysis. This analysis is a laboratory procedure referred to as the float-sink test that determines the weight percentages and chemical compositions of materials collected from liquid baths of different specific gravities. The material that floats at a particular specific gravity is called the yield or product and the material that sinks is the reject. Data on the composition of these fractions usually consist of ash content, sulfur content, and heating value. Washability analyses represent the theoretical limitations of physical coal beneficiation since the laboratory tests are performed under "ideal" batch conditions using pure heavy liquids that are too expensive to use in large-scale continuous processes.

Table 3.4.3.1

Specific Gravities of Coal and Impurities

Material	Specific Gravity
Coal	1.12 - 1.35
Bone coal[a]	1.35 - 1.7
Carbonaceous shale	1.6 - 2.2
Shale	2.0 - 2.6
Clay	1.8 - 2.2
Pyrite	4.8 - 5.2

[a] Bone coal is coal with a high ash content.

A wide variety of processes and equipment is utilized in the many coal preparation plants that separate mineral matter from a large percentage of the bituminous coal mined in this country. Some types of equipment are best suited for cleaning the coarser size fractions of coal; these are briefly described on Table 3.4.3.2. Other types of equipment are designed for fine coal cleaning, and are described on Table 3.4.3.3. As indicated on Table 3.4.3.4, jigs and dense medium vessels are by far the most widely used types of equipment, together accounting for 80 percent of the total coal cleaned in the United States (1973 data).

All of these devices are described in considerable detail in references 1-4, including schematic drawings and photographs, performance curves, design and operating data, and cost comparisons. This detailed information is not included here because of the almost total lack of current low-rank coal applications for these processes.

In addition to these widely used commercial processes, a few coal cleaning techniques are being developed which may have future low-rank coal applications. One is the Otisca process, which uses trichlorofluoromethane (CCl_3F, one form of Freon) as a dense medium in place of the more commonly used suspension of magnetite in water. Claims for the process, based on pilot plant results, include better ash and sulfur removal at a lower cost per ton of coal.[8]

Another novel dense medium process utilizes an air-fluidized bed of magnetite as the dense medium. Tests in a 4-inch diameter unit have shown the basic feasibility of the process, as well as indicating the existence of complex relationships among the process variables. The concept combines the good separating capabilities of high density medium with the advantages of a dry process - advantages which are particularly applicable to western low-rank coals.[9]

A third cleaning process under exploratory development is high gradient magnetic separation (HGMS), which is directed primarily at pyritic sulfur separation. HGMS methods are currently used commercially to remove impurities called colorbodies from Kaolin clay. The dispersed nature of pyritic sulfur in coal combined with its paramagnetic distinction from diamagnetic coal led to the initial investigations of HGMS processes applied to coal cleaning. A large number of HGMS methods have been proposed and are undergoing experimental development. These include both wet and dry processes which use various chemical and physical means to enhance the weak paramagnetic properties of pyrite.[10,12]. Most of the current developmental work is under EPRI sponsorship.

Table 3.4.3.2

Types of Equipment Used for Coarse Coal Cleaning

Equipment	Description	Principles of Operation	Comments
Jig	The coal bed rests in water on a perforated screen plate. The water-filled hutch compartment below the screen is connected to an adjacent vessel where pulses of air pressure are applied (by valve as in the Baum jig, or by plunger). The pulses are transmitted through the water causing an upward motion through the screen. The jig box is divided vertically into compartments separated by weirs which control the flow of the float strata. Float material feeds from one compartment to the next until it passes over a weir into a delivery sluice for dewatering. Sink material (refuse or middling) is withdrawn by an ejector from each compartment. Adjustment of the refuse gate height, the refuse withdrawal rate, and a float control determines the refuse separation.	A bed of coal particles is stratified in water by pulsations. Segregation by specific gravity is achieved through alternate expansion and contraction of the bed. As the expanded bed settles back into the compacted state, the higher specific gravity particles (i.e., the refuse) settle at a faster rate.	Most popular, simplest wet beneficiation vessel. Handles full size range up to 8-inch, but only limited upgrading of fine materials is obtained. Most effective at separating gravities above 1.55 specific gravity, and with coal containing less than 10 percent near-gravity material. Relatively low cost, high capacity per unit (up to 1,000 tph). Requires even feed distribution over entire width of vessel, thoroughly wetted coal, constant feed rate and washability for best performance.
Dense Medium Vessels	Coal is fed to a vessel (cone, drum, or trough) containing the dense medium. Finely-ground and dispersed magnetite in water is the preferred medium because the magnetite can be easily	A large-scale application of the principles of separation by specific gravity used in the laboratory float-sink procedure for preparing washability analyses. Raw coal is immersed in a fluid having a	Affords very good control of sharpness of separation, at specific gravities that can readily be adjusted between 1.3 and 1.8. Handles all size consists above 1/4-inch, with up to 25 percent near-gravity material in the feed.

(continued)

Table 3.4.3.2 (continued)

Types of Equipment Used for Coarse Coal Cleaning

Equipment	Description	Principles of Operation	Comments
Dense Medium Vessels	recovered magnetically and resued. Specific gravity is easly controlled by the concentration of suspended solids. Other media less frequently used are sand in water, organic liquids, dissolved salts (calcium and zinc chloride), and sand in air (fluidized bed). In a cone-type vessel, the feed enters the top near the center, and the float material travels to the periphery and overflows a tangential weir. Sink material is discharged from the bottom of the cone, usually via a central airlift tube. The cone geometry and large volume create stability and long residence time. Trough-type vessels utilize either chain conveyors for removal of the float and sink products, or an oscillating blade for removal of sink material.	specific gravity intermediate between the coal and the reject material. Less than perfect separation is obtained due to practical limits on particle retention time, and the use of magnetite suspensions in water, rather than pure fluids used in the laboratory.	Relatively high operating costs. Requires deslimed feed and a dense medium recovery system.
Concentrating Table	Coal and refuse particles are spread out over a ribbed, tilted, rhomboid-shaped table to which a reciprocating motion in the horizontal direction is imparted. As the feed flows down the table, the combination of forces stratifies the coalbed behind the riffles. The wash water carries the light coal particles over the riffles to the bottom of the table where they are collected as clean coal product. The refuse at the bottom of the bed is moved horizontally behind the riffles to the end of the table.	The differential notion imparted to the table is such that is approaches one end of its travel with greater speed than the other. Thus both the retardation and acceleration are greater at one end than at the other, imparting to the table a conveying action. The basic principle of separation is the difference in specific gravity between coal and refuse particles.	Rarely used for coarse coal cleaning. Poor sharpness of separation and control of cutpoint. Used frequently for fine coal cleaning (see Table 3.4.2.3).

(continued)

Table 3.4.3.2 (continued)

Types of Equipment Used for Coarse Coal Cleaning

Equipment	Description	Principles of Operation	Comments
Hydrocyclone	Coal and water enter the side of the conical vessel tangentially. Sink material (refuse) is accelerated to the outside and bottom of the cone, and exits at the bottom through an annular space created by a central upward blast of air. Float material is carried by the air stream up and out a central tube called a vortex finder. Hydrocyclones have been used for coarser coals, although they are most useful for cleaning coals in the 3/4-inch x 100 mesh range, and then preferably as scalping units ahead of other beneficiation devices.	An autogenous (self-creating) dense medium is established through the high centrifugal acceleration of particles in water. Denser particles are accelerated to the outer diameter of the cyclone. Because of the squat conical configuration used (small L/D), coarsest and heaviest particles tend to orbit in the lower cone region, forming a barrier against the discharge of coal through the apex orifice.	Simple design, low space requirements, low maintenance, no medium recovery system required. Relatively inefficient when used singly. When used ahead of concentrating tables, the combined efficiency is greater than with tables alone, particularly at lower specific gravities. One-third to one-half the feed is clean enough following the hydrocyclone to bypass the table.
Pneumatic Separator (air concentrator)	Coal and refuse particles are stratified by pulsating air as they travel across an inclined surface. A layer of refuse forms at the bottom of the strata, and falls into pockets or wells from which it is withdrawn by screw conveyors. The upper layer (clean coal) is removed at the end of the incline. Dust created by the pulsating air is sucked into an overhead hood and is recovered in a dust collection system (e.g., cyclone and baghouse). The collected dust may be recombined with the clean coal if the final quality is acceptable. Alternatively, it may be burned as fuel or discarded as refuse.	Air tables, or air jigs, utilize air as a separating medium in much the same way as coarse coal jigs utilize water. A dry vibrating table is designed so that its vibratory motion segregates the particles according to specific gravity, shape, and size.	Not widely used commercially, but receiving renewed development attention. Possibly most advantageous for low-rank coals because no water is required and no surface moisture is added to the coal. Unlike wet cleaning processes, dry concentration will not clean wet fine coal efficiently, or eliminate dust as a problem.
Launder	Coal and refuse particles are transported down an inclined surface by flowing water. Stratification of particles occurs parallel to the inclined surface. The higher specific gravity particles (refuse) settle and move more slowly than the clean coal particles due to friction. These units are now considered obsolete.	Boundary layer, frictional forces, and the decreasing velocity of water in a trough from bottom to top are significant in creating the velocity difference between high specific gravity particles (slow) and low specific gravity particles (fast).	Poor sharpness of separation and control of cutpoint. Relatively low cost when used in dual role as washer and conveyor. Not currently used in the U.S.

Sources: References 1-4

Table 3.4.3.3

Types of Equipment Used for Fine Coal Cleaning

Equipment	Description	Principles of Operation	Comments
Dense Medium Cyclone	Dense medium cyclones are generally mounted nearly horizontal with only a sufficient angle to facilitate drainage. As in the hydrocyclone, sink material (refuse) is accelerated to the outside and bottom of the cone, and exits through the apex. Float material is carried up and out the vortex finder.	Dense medium fluid is utilized as in dense medium veseels, but the separation between the coal and the refuse is accelerated by centrifugal force. Because of the forces acting on the magnetite in the medium, the density of the medium in the underflow is heavier than in the overflow. The specific gravity of separation is always higher than the specific gravity of the prepared medium. Extremely high forces acting on the particles (up to 200 G's) increase throughput and efficiency of separation of fine particles.	Generally processes the 1/4-inch x 28 mesh size fraction. The larger units are 28 inches in diameter and have a capacity of over 100 tph. Cone angles are commonly about 20°. Good sharpness of separation with up to 25 percent near-gravity material in the feed. Very high operating costs; heavy medium recovery system required.
Concentrating Table (Deister Table)[a]	Coal and refuse particles are spread out over a ribbed, tilted, rhomboid-shaped table to which a reciprocating motion in the horizontal direction is imparted. As the feed flows down the table, the combination of forces stratifies the coalbed behind the riffles. The wash water carries the light coal particles over the riffles to the bottom of the table where they are collected as clean coal product. The refuse at the bottom of the bed is moved horizontally behind the riffles to the end of the table.	The differential notion imparted to the table is such that it approaches one end of its travel with greater speed than the other. Thus both the retardation and acceleration are greater at one end than at the other, imparting to the table a conveying action. The basic principle of separation is the difference in specific gravity between coal and refuse particles.	Generally processes 3/8-inch by 0 size coal. Tables have a capacity of 10 to 15 tph. Fair sharpness of separation with up to 10 percent near-gravity material in the feed. Low operating costs. Allows isolation of pyrite. Specific gravity adjustment (between 1.6 and 1.8) is accomplished by adjusting the side tilt and longitudinal inclination of the deck.
Fine Coal Jig (Feldspar Jig or Batac Jig)[a]	Basically the same as the standard jig described in Table 3.4.2.2. However, the screen plate is covered by a bed of sized stones, usually feldspar, which prevent all but the finest particles from sifting through. In the Batac jig, air pulsations are produced directly beneath the bed screen instead of in an adjacent chamber. The refuse particles sift down through the larger feldspar particles, while the clean coal is discharged at the end of the jig.	As the beds of feldspar lift with the pulsations, with feed material passing above, the turbulent effect on small particles is reduced. Otherwise, the principles of operation are the same as for the standard jig, relying primarily on specific gravity differences between the coal and refuse.	Fine coal jigs have a considerably smaller capacity than conventional Baum-type jigs, and poor efficiency when processing feed sizes containing many particles smaller than 28 mesh. Fair sharpness of separation when processing feed with up to 10 percent near-gravity material.

(continued)

Table 3.4.3.3 (continued)

Types of Equipment Used for Fine Coal Cleaning

Equipment	Description	Principles of Operation	Comments
Froth Flotation Cell	Feed coal slurry, conditioned with suitable chemical reagents, enters a cell where air is bubbled up through the slurry. Agitation and aeration are obtained by mechanical, hydraulic, or pneumatic systems (Wemco cells use a rotor-stator device, and Heyl and Patterson cells use submerged vortex chambers[a]). Clean coal is skimmed off as a froth (20-30% solids) at the top. Refuse remains suspended in the waters and passes from the cell into a static thickener.	Depends entirely on surface characteristics of coal and refuse, not on specific gravity differences. Flotation relies on the selective adhesion of air to the coal particles and the simultaneous adhesion of water to the refuse particles, as finely disseminated air bubbles pass through the feed slurry. Reagents can be utilized to enhance the hydrophobic property of the coal without diminishing the hydrophilic properties of the mineral solids.	Optimum size range is 28 to 150 mesh. Low volatile coals are easier to float than high volatile coals. Lignites are the least floatable coals, and the froth flotation process is not believed to be applicable to lignites. Pyrites tend to be floated with the clean coal, and thus are not effectively separated. Operating costs are relatively low. Chemical reagents are required.
Hydrocyclone	Coal and water enter the side of the conical vessel tangentially. Sink material (refuse) is accelerated to the outside and bottom of the cone, and exits at the bottom through an annular space created by a central upward blast of air. Float material is carried by the air stream up and out a central tube called a vortex finder. Hydrocyclones have been used for coarser coals, although they are most useful for cleaning coals in the 3/4-inch x 100 mesh range, and then preferably as scalping units ahead of other beneficiation devices.	An autogenous (self-creating) dense medium is established through the high centrifugal acceleration of particles in water. Denser particles are accelerated to the outer diameter of the cyclone. Because of the squat conical configuration used (small L/D), coarsest and heaviest particles tend to orbit in the lower cone region, forming a barrier against the discharge of coal through the apex orifice.	Rarely applied for fine coal cleaning. Multiple stages required. Poor sharpness of separation. High relative operating costs.

Notes:
[a]Reference to a company or product name does not imply approval or recommendation of the product.

Sources: References 1-4

Table 3.4.3.4

Distribution of the Coal Cleaned in the United States By Method of Separation Used

METHOD OF SEPARATION	PERCENTAGE OF COAL PROCESSED	
	1971	1973
Jigs	43	48
Tables	13	12
Launders	2	3
Dense Medium	33	32
Pneumatic	5	0
Flotation	3	5

Source: Reference 3

One of the leading magnetic separation processes, called Magnex, involves chemical pretreatment of the pyrite and mineral matter with (gaseous) iron pentacarbonyl, to enhance their magnetic susceptibilities. Separation of these components (pyrite with some pyrrhotite-like material, and minerals with crystallites of iron) is then accomplished from the coal (which is unaffected by the $Fe(CO)_5$ pretreatment) by conventional magnetic means. This dry process has been successfully tested at the 200 lb/hr pilot plant level on a large number of bituminous coals.[23] Because of the important chemical pretreatment step, the process is usally classified as a chemical coal cleaning process, and thus is mentioned again in the next subsection.

Chemical Processes

Processes which utilize chemical reaction systems to upgrade coal quality are in various stages of development. In general, the objectives of these processes are to remove sulfur, sodium, and/or mineral matter that is bound within the coal structure, and thus not available to be removed by physical separation processes.

In a sense, the SRC-I (Solvent Refined Coal) process might be considered the ultimate chemical coal cleaning process, because it produces an almost ash- and sulfur-free, high-BTU solid fuel from coal. (SRC from Wyoming subbituminous coal contains about 0.2 percent ash, 0.1 percent sulfur, and has a heating value of about 15,000 BTU/lb.) Some minor modifications are required at a coal-fired boiler to handle this prepared fuel, such as larger dust control systems and water-cooled burners.[13] Solvent refined coal is prepared through a liquefaction process whose severity is low enough that the product is a solid at room temperature (see Section 3.5.4). If it is viewed as representing an upper bound on the "reasonable" cost of a chemical coal cleaning process, then the figures of roughly $\$4\text{-}5/10^6$ BTU, or \$120-160/ton of product, should be borne in mind.

Many chemical cleaning processes are currently under development for removal of both organic and inorganic sulfur from coal. The inorganic sulfur compounds occur mainly as pyritic minerals, with small amounts of sulfate minerals such as melanterite, jarosite, and gypsum. These are the sulfur compounds that can be separated from the coal to some extent by physical processes. The organic sulfur forms consist mainly of mercaptans, sulfides, disulfides, and thiophenes. Their removal requires at least a partial breakdown of the organic coal matrix. Chemical mechanisms that can be classified into six major groupings as shown on Table 3.4.3.5. At this time, the reduction mechanism (with hydrogen) has been shown to be the most effective.[14,15]

Table 3.4.3.5

Organic Sulfur Removal Mechanisms

1. Solvent Partition

$$R_1S_xR_2 + X \longrightarrow \left.\begin{matrix}R_1\\R_2\end{matrix}\right> S_x^{+}\ X^{-}$$

2. Thermal Decomposition

$$R_1S_xR_2 \longrightarrow R_1R_2 + S_x$$

$$RCH_2CH_2SH \longrightarrow RCH = CH_2 + H_2S$$

3. Acid-Base Neutralization

$$RSH + OH^- \longrightarrow RS^- + H_2O$$

4. Reduction

$$R_1S_xR_2 + 4H \longrightarrow R_1H + R_2H + H_2S_x$$

$$R_1S_xR_2 + 2R_3H_2 \longrightarrow R_1H + R_2H + H_2S_x + 2R_3$$

5. Oxidation

$$R_1S_xR_2 \xrightarrow[(0)]{} R_1SO_3H + R_2SO_3H \xrightarrow{H_2O} R_1OH + R_2OH + 2H_2SO_4$$

6. Nucleophilic Displacement

$$R_1S_xR_2 + Nu^- \longrightarrow R_1S_xNu + R_2^-$$

$$R_1S_xR_2 + Nu^- \longrightarrow R_1S_{x-1}Nu + R_2S^-$$

$$R_1S_xR_2 + Nu^- \longrightarrow R_1S_x^- + R_2Nu$$

$$R_2S^- + R_1S_{x-1}Nu \longrightarrow R_1S_{x-1}R_2 + NuS^-$$

Source: Reference 14

The technical and economic feasibility of a chemical coal desulfurization process depend primarily on the properties of the reagent. The desulfurization reagent must be selective and not significantly react with other coal components. The reagent should be regenerable and be either soluble or volatile so it can be recovered from the coal matrix. Finally, the reagent should be inexpensive since a portion of it will certainly be lost to either irreversible sorption on the coal matrix or by reaction.[14] Another very important factor is the extent to which the cleaning process removes carbon values (BTU's) along with sulfur and ash. This can be a significant problem with "deep" cleaning processes, which can result in rejecting as much as 40 percent of the total energy value of the coal.[15]

An EPA-sponsored assessment of chemical coal cleaning processes in 1978 identified 29 processes, of which 11 U.S.-developed processes were classified as major processes.[15] These are listed on Table 3.4.3.6, along with some pertinent information on their methodology, sulfur removal capability, stage of development, problems and cost estimates.

As a general rule, the development of all of these processes has focused heavily on bituminous coals, rather than low-rank coals (which also tend to be lower-sulfur coals). Therefore, very little pertinent data on the effectiveness of the different types of processes in desulfurizing low-rank coals are available. It is probably fair to say that processes which remove only pyritic sulfur (such as the first four listed on the table) are of little interest for low-rank coal applications. To illustrate this point, Table 3.4.3.7 shows the sulfur reduction obtained on three western low-rank coals by the JPL chlorinolysis process, which is one of the processes that appears to be quite effective in removing both organic and pyritic sulfur. The total sulfur removal obtained was: 39 percent for the Zap (N.D.) lignite, which contains 1.22 percent sulfur as mined; 14-34 percent for the Carbon, Seam 80 (WY) subbituminous coal, which contains 1.23 percent sulfur; and 64 percent for the Big D Seam (Lewis, WA) subbituminous coal, which contains 3.36 percent sulfur as mined.[21]

These are substantial reductions in coal sulfur content, particularly for the higher-sulfur Washington state coal. However, the amount of relief that would be provided by this type of deep coal cleaning to the FGD SO_2 removal requirement in order to comply with the 1979 NSPS is relatively small. This is illustrated on Table 3.4.3.8, where some assumptions have been made (about coal heating value, complete conversion of coal sulfur to SO_2 in the stack and no loss of coal in the cleaning process) to calculate estimated FGD system requirements, with and without coal cleaning by the JPL process. As shown, cleaning the North Dakota lignite reduces the FGD design severity from 83 percent SO_2 removal to 72 percent. The stack gas SO_2 removal requirements for the Wyoming subbituminous coal are 74 percent without JPL coal cleaning versus 61 percent with cleaning, and for the Washington subbituminous coal are 90 percent without cleaning and 72 percent with cleaning. Thus in each case a fairly rigorous SO_2 scrubbing job is required, despite the substantial reductions in coal sulfur content achieved by the chemical cleaning process.

Some brief additional comments on the chemical coal cleaning processes that were listed on Table 3.4.3.6 are as follows:[15]

1. The Magnex process was briefly described earlier under the subject of magnetic separation. The use of iron carbonyl presents some health and safety difficulties.

2. The Syracuse process was also briefly described earlier (in Section 3.4.2) as a chemical comminution process. After removing the ammonia which chemically comminutes the coal, conventional physical cleaning steps are used to separate the coal from pyrite and ash.

3. The Meyers (TRW) process is the only chemical coal cleaning process developed to the 8 t/d pilot scale level. Thirty-two different coals have been tested at bench scale, including two western coals.

4. The Ledgemont oxidative leaching process has potential materials selection problems due to the presence of corrosive dilute sulfuric acid; and potential disposal problems with lime-gypsum-ferric hydroxide sludge, which may contain leachable heavy metals.

5. The PETC air and steam leaching process also has materials selection problems due to dilute sulfuric acid which is generated by the process, and is highly corrosive at the operating temperatures and pressures.

6. The portion of the GE process which recovers the sulfur values and regenerates the NaOH is still conceptual.

7. The Battelle process has two major technical problems: feasibility of the closed-loop caustic regeneration feature in a continuous process is not yet demonstrated; and the product coal may contain excessive sodium residues, causing severe furnace fouling and slagging problems.

8. The JPL process uses a trichloroethane solvent which is listed by EPA as a priority pollutant.

9. The IGT process is in essence a hydro-pyrolysis process which creates a low-sulfur char product but loses considerable coal to the offgas due to oxidation, volatilization, and gasification. The process will only be cost-competitive if a satisfactory utilization scheme for the offgas can be devised.

10. The KVB process requires oxygen concentrations in the treat gas that exceed the explosion limits for coal dust, thus creating potentially hazardous conditions; some nitrogen may be incorporated into the coal structure from the NO_x emissions from combustion of the clean coal product.

11. The Arco Process has similar materials selection and sludge disposal problems to the Ledgemont and PETC processes.

A different type of chemical coal cleaning process, specifically targeted at high-sodium (high-fouling) low-rank coals, is the use of ion exchange to reduce sodium content. Sodium has been found to be the principal cause of fireside tube fouling at high load operation for North Dakota lignite-fired power plants. Lignites with 0.4 percent or higher Na_2O content on a dry basis (corresponding to about 4 percent Na_2O in the ash) exhibit medium or high fouling rates. Sodium levels as high as 10 percent or more in the ash are also present in some Western subbituminous coals.[25,26]

Removal or reduction of the sodium levels in pilot plant tests has proven to be the most successful method of controlling fouling for lignites. Since the sodium in lignites has been shown to be ionically bound to the uniformly distributed humic acids in the coal substances, physical separation techniques are not effective. Sodium can be removed from the coal, however, by ion exchange techniques. The exchange mechanism allows replacement of the sodium by ions of greater ionic weight, of higher valence, and/or of higher concentration. Therefore, sodium can be replaced by potassium, calcium, iron, magnesium or hydrogen, depending on

Table 3.4.3.6

Summary of Major Chemical Coal Cleaning Processes

Process & Sponsor	Method	Type Sulfur Removed	Stage Of Development (1978)	Problems	Annual Operating Cost, $/Ton Clean Coal[a]
"Magnex", Hazen Research Inc., Golden, Colorado	Dry Pulverized Coal Treated with FE $(CO)_5$ Causes Pyrite To Become Magnetic. Magnetic Materials Removed Magnetically	Up to 90 Percent Pyritic	Bench & 91 KG/Day (200 LB/Day) Pilot Plant Operated	Disposal of S-Containing Solid Residues. Continuous Recycle of CO to Produce FE $(CO)_5$ Requires Demonstration	15.7
"Syracuse" Syracuse Research Corp., Syracuse, N.Y.	Coal is Comminuted by Exposure to NH_3 Vapor; Conventional Physical Cleaning Separates Coal/Ash	50-70 Percent Pyritic	Bench Scale	Disposal of Sulfur Containing Residues	12.0
"Meyers", TRW Inc., Redondo Beach, CA.	Oxidative Leaching Using $FE_2(SO_4)_3$ + Oxygen in Water	90-95 Percent Pyritic	8 Metric Ton/Day PDU for Reaction System. Lab or Bench Scale for Other Process Steps	Disposal of Acidic $FESO_4$ and $CASO_4$, Sulfur Extraction Step Requires Demonstration	18.4[d]
"LOL" Kennecott Copper Co. Ledgemont, MA	Oxidative Leaching Using O_2 and Water @ Moderate Temp. and Pressure	90-95 Percent Pyritic	Bench Scale	Disposal of Gypsum Sludge, Acid Corrosion of Reactors	21.9
"PETC" (DOE) Oxydesulfurization Bruceton, PA.	Air oxidation & Water leaching @ High Temperature and Pressure	~95 Percent Pyritic; up to 40 Percent Organic	Bench Scale 11 KG/Day (25 LB/Day) Continuous Unit Under Construction	Gypsum Sludge Disposal Acid Corrosion at High Temperatures	26.6[b]
"GE" General Electric Co., Valley Forge, PA.	Microwave Treatment Of Coal Permeated with NAOH Solution Converts Sulfur Forms to Soluble Sulfides	~75 Percent Total S	Bench Scale	Process Conditions Not Established. Caustic Regeneration Process Not Established.	16.8

(continued)

Table 3.4.3.6 (continued)

Summary of Major Chemical Coal Cleaning Processes

Process & Sponsor	Method	Type Sulfur Removed	Stage Of Development (1978)	Problems	Annual Operating Cost, $/Ton Clean Coal[a]
"Battelle" Laboratories Columbus, Ohio	Mixed Alkali Leaching	~95 Percent Pyritic; ~25-50 Percent Organic	9 KG/Hr (20 LB/Hr) Mini Pilot Plant and Bench Scale	Closed Loop Regeneration Process Unproven. Residual Sodium in Coal	30.9
"JPL" Jet Propulsion Laboratory Pasadena, CA.	Chlorinolysis In Organic Solvent	~90 Percent Pyritic; up to 70 Percent Organic	Lab Scale But Proceeding to Bench and Mini Pilot Plant	Environmental Problems Conversion of HCL to CL_2 Not Established.	21.0
"IGT" Institute of Gas Technology Chicago, IL.	Oxidative Pretreatment Followed by Hydrodesulfurization at 800°C	~95 Percent Pyritic; Up to 85 Percent Organic	Lab and Bench	Low Btu Yield (<55%) Change of Coal Matrix	40.8
"KVB", Inc., Tustin, CA.	Selective Oxidation in the Presence of NO_x, followed by water or caustic washing	~63 Percent to 89 Percent Totals	Lab Scale	O_2 Concentration Exceeds Explosion Limit for Coal Dust. Nitrogen Uptake by Coal will increase NO_x Emissions	22.5[c]
"ARCO" Promoted Oxydesulfurization, Los Angeles, CA.	Oxygen oxidation and Water Leaching at High Temperature and Pressure	~95 Percent Pyritic ~8-19 Percent Organic	Lab Scale	Gypsum Sludge Disposal Acid Corrosion at High Temperatures	NA[e]

Notes:
[a]Raw coal cost is not included.
[b]1979 cost estimates are $16-20/ton (References 16,17).
[c]KVB's estimate (1978) is $18.50/ton (Reference 18)
[d]1979 cost estimates are $20-25/ton (References 19,20,21)
[e]1979 cost estimates are $27-39/ton for original design, and $17-19/ton for revised design (Reference 24).

Source: Reference 15

Table 3.4.3.7

Laboratory-Scale Desulfurization of Three Western Low-Rank Coals by the JPL Chlorinolysis Process

No. of Runs	Chlorination Time (Min.)	Residual Sulfur Analysis (Wt. %)				Sulfur Removal (%)			Dechlorination Residual Cl (Wt. %)	
		Organic	Pyritic	Sulfate	Total	Organic	Pyritic	Total	Before	After
		PSOC-086, Lignite, Zap, Mercer, North Dakota						6900 BTU/lb (as received)		
RAW COAL		0.63	0.52	0.03	1.22				0.00	
1	30	0.35	0.23	0.17	0.75	44	59	39	--	0.33
1	60	0.32	0.35	0.06	0.73	50	37	39	8.00	--
		PSOC-097, Subbituminous A, Seam 80, Carbon, Wyoming						10,665 BTU/lb (as received)		
RAW COAL		0.84	0.38	0.01	1.23					
1	30	0.70	0.31	0.05	1.06	17	18	14	--	0.28
1	60	0.74	0.05	0.02	0.81	12	87	34	--	0.15
1	120	0.79	0.19	0.06	1.05	5	50	15	--	0.22
		PSOC-240, Subbituminous B, Big D, Lewis, WA						8700 BTU/lb (as received)		
RAW COAL		1.75	1.60	0.01	3.36				0.02	
1	120	0.49	0.68	0.05	1.22	72	58	64	--	0.26

Source: Reference 22

process conditions. Ion exchange studies to remove sodium from high fouling lignites have progressed from experimental efforts, through the pilot plant stage, to a recently completed conceptual design of a large processing plant. The pilot process has been successfully tested with high-sodium lignite to produce a low-fouling product.[27,31]

Table 3.4.3.8

Approximate Reduction in FGD System SO_2 Removal Requirements Obtained by Cleaning Low-Rank Coals using JPL Process

Coal	Zap (ND) Lignite	Seam 80 (Carbon WY) Subbituminous A	Big D (Lewis, WA) Subbituminous B
Sulfur in Coal, wt %	1.22	1.23	3.36
Heating Value of Coal, Btu/lb.	6900	10,665	8700
Uncontrolled SO_2 Emissions, lb/10^6 Btu	3.54	2.31	7.72
NSPS Sulfur Removal Requirement, %	83	74	90
NSPS SO_2 Emission, lb/10^6 Btu	0.6	0.6	0.77
Sulfur Removed From Coal by JPL Process, %	39	34	64
Cleaned Coal SO_2 Emission With no FGD, lb/10^6 Btu	2.16	1.52	2.78
Sulfur Removal Required From Flue Gas by FGD, %	72	61	72

GFETC pilot plant studies investigated the variables affecting the removal of sodium from lignite by ion exchange in a continuous countercurrent unit with a capacity up to 100 lbs/hr of solids.[27] Operational variables investigated were particle size, effectiveness of various cations in the exchange solution, solids and liquid residence time, and multistage reusing of solids and liquids on the sodium content of the product. Results indicate that exchange is most effective with small particle sizes with a concentrated exchange solution. (However, under these conditions, loss of product to waste water and required cleanup are the greatest.) Another finding was that reducing the lignite's moisture content reduces the ion exchange potential, possibly by collapsing and sealing off the capillaries.

Combustion tests were run on lignite whose sodium (Na_2O) content in the ash was reduced from 7 percent to 0.5 percent. Ash deposits on the fireside tubes were greatly reduced and equivalent to those obtained when burning lignite from the same mine having a natural sodium content of less than one percent in the ash. In a commercial application, the sodium would probably be reduced to 3.0 percent Na_2O in the ash, a level at which excessive fouling is not generally observed.

During 1979, the University of North Dakota, under contract to GFETC, completed a conceptual design and economic evaluation of an ion exchange process to remove sodium.[29,30] The design basis utilized commercially available equipment to process 4000 tons per day of lignite and reduce the sodium content in the ash from 8.5 percent (as Na_2O) to 4 percent. The economic evaluation determined the overall cost to process the coal would be $1.30 per ton. This very preliminary estimate is sufficiently encouraging that further work on the process and its economics will be pursued. An important part of this effort will be an evaluation of the costs of boiler downtime due to excessive fouling, as compared with the costs to prepare a non-fouling coal.

In the conceptual design (Figure 3.4.3.1) the lignite feed is crushed to 100 percent through 6 mesh with approximately 18 percent passing 28 mesh. Forty percent of the crushed lignite is bypassed to the end of the process line. The remaining portion is fed to the countercurrent ion exchange system to be contacted with dilute H_2SO_4. Extracted lignite is conveyed to a size separation and rinse step with the plus 28 mesh and the minus 28 mesh material dewatered separately. The treated solids are then combined with the bypassed lignite.

Liquid effluent from the ion exchanger is pumped to a solid-liquid separation step from which recovered solids are returned to the dewatered product and the clear liquid treated by reverse osmosis. From the reverse osmosis step, the purified water is returned to the process and the concentrated brine is disposed of in an evaporation pond.[29]

Ion exchange processing of lignite appears to represent a possible option for eliminating a major problem in its use in combustion applications. It may be expected that eventual commercial processes will incorporate other upgrading features such as moisture reduction by drying and perhaps agglomerating (briquetting) processes to improve handling and storage characteristics.[27]

3.4.3.2 Environmental Control Technology

The environmental control requirements associated with the physical coal cleaning processes may be briefly summarized as follows:[32]

1. Solid waste resulting from separation processes includes coarse refuse from jigs and dense-medium vessels; fine refuse from air tables, dense-medium cyclones, wet concentrating tables and hydrocyclones; sludge from water clarification circuits; magnetite from dense-medium processes (0.5 lb/ton of feed coal); and chemical reagents from froth flotation processes. These processes, applied to medium and fine-sized feed coal streams, generate on the order of 25 percent of their coal feed as waste.

2. Contaminated water from wet separation processes.

   The characteristics of process water are highly dependent upon the characteristics of coal being processed and the particular process or recovery technique utilized in the operation. The principal pollutant present in process water is suspended solids. Some minerals also are present as dissolved solids. Among the major pollutant constituents or parameters identified in effluents from coal preparation plants are:

Acidity or Alkalinity	Total Suspended Solids
Total Iron	Total Dissolved Solids
Dissolved Iron	Sulfates
Ammonia	

Figure 3.4.3.1

Conceptual Flow Diagram of an Ion Exchange Process to Remove Sodium from Lignite

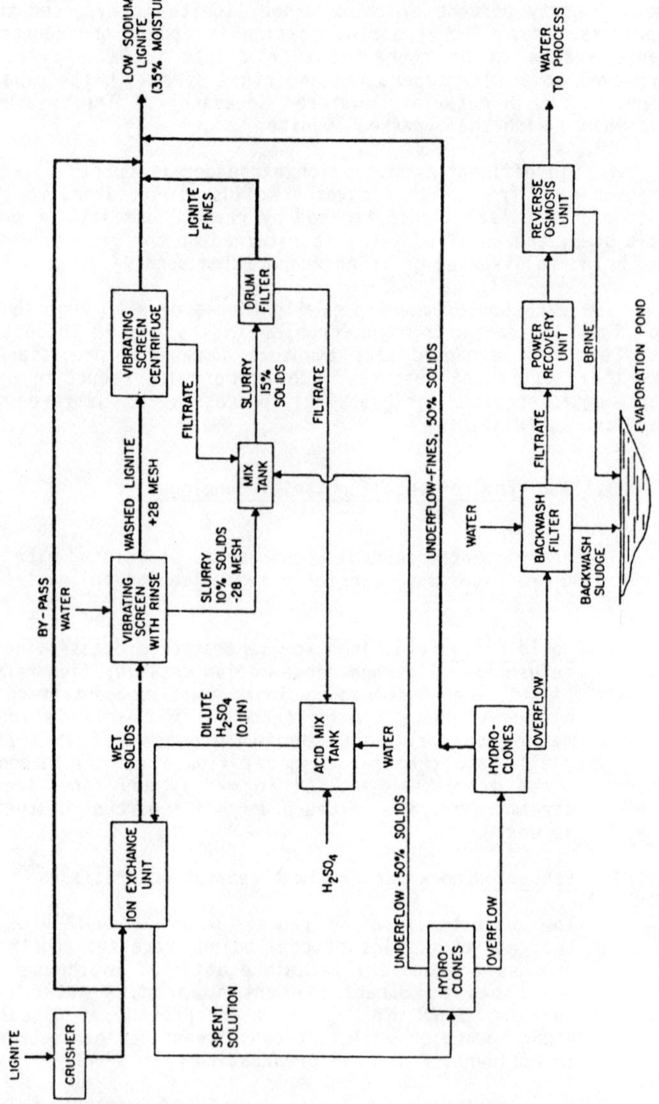

Source: Reference 29

Process water from dense media processes may contain magnetite; water from froth flotation operations may contain potentially toxic or noxious chemical reagents. The quantities of water used in processing range from 180 to 1800 gallons per ton of coal processed. A major portion of the water used in coal cleaning is recirculated.

3. Air Emissions from pneumatic coal cleaning consist of particulates only, because ambient air is used to separate coal from refuse. The quantity and pressure of the air used depends on the size and kind of coal to be cleaned. For pneumatic cleaning of minus 3/8-inch coal, an average volume of exhaust air is about 14,100 cu ft per ton of feed coal. The exhaust air usually picks up about 65 to 70 percent of the minus 48 mesh material in the feed coal, and about 20 percent of minus 3/8-inch coal is smaller than 48 mesh. Therefore, the uncontrolled exhaust air contains about 260 to 280 pounds of dust per ton of feed coal treated or 128 to 138 grains of dust per cubic foot.

4. Recently, the fate of potentially toxic elements in coal during coal cleaning has received special attention. Coal has been found to contain nearly every naturally occuring element. Coal cleaning affects the distribution of these elements between clean coal and refuse portions. Table 3.4.3.9 shows the concentrations of a few selected elements in raw coal, clean coal at 75 percent weight recovery, and in the resulting refuse. The enrichment factor is defined as the concentration of an element in the clean coal (or in the refuse) divided by the concentration of the same element in raw coal. Of 29 elements measured in this way by the Illinois State Geological Survey, all but boron and germanium had higher concentrations in the refuse than in the raw coal. Beryllium is distributed approximately evenly between the clean coal and the refuse.

   Reduction of trace elements is an added benefit of coal cleaning for reducing the environmental pollution from burning coal; however, the concentration of trace elements in the solid waste may increase the potential for environmental contamination from this source.

Table 3.4.3.9

Enrichment Factors in Float-Sink Separation of Illinois Coals

Element	Concentration, ppm (unless otherwise stated)			Enrichment Factor	
	Raw Coal	Clean Coal	Refuse	Clean Coal	Refuse
S	4.4 %	1.6 %	12.9 %	0.36	2.93
As	11.5	1.5	41.0	0.13	3.57
Be	3.0	2.9	3.3	0.97	1.10
Ge	6.7	8.1	2.3	1.21	0.34
Se	2.8	1.3	7.3	0.46	2.61

Source: Reference 33, as cited in Reference 32

Technologies for control of many of the pollutants mentioned are well-known. For example, dust collection devices include cyclones, fabric filters, electrostatic precipitators, and wet scrubbers. Gases and odors can be controlled by wet or dry absorbers. Water treatment technologies for suspended materials include filtration, sedimentation, and flotation. Control of dissolved materials can be achieved by neutralization, absorption, ion exchange, reverse osmosis, freezing, and/or biological oxidation.

One alternative to the use of ponding, or dewatering and disposal, is a selective oil agglomeration process that recovers coal fines from cleaning plant reject waters.[34] Another alternative is the use of completely closed water circuits, which eliminates the need for coal refuse ponds; a new U.S. Steel Corporation preparation plant is designed with a totally closed water circuit.[35]

A potentially critical environmental problem that has not received sufficient attention until recently is the chemistry of refuse disposal piles and the associated groundwater contamination possibilities. In eastern coal disposal sites, it has been found that oxidative degradation of pyrite and marcasite in the refuse produces acidic materials, which can result in highly contaminated leachates with pH's often falling below 2. At that level of acidity, many trace elements readily dissolve. Potential preventive technologies that are being evaluated include: codisposal of the refuse with lime or limestone (which neutralizes the acidity and greatly reduces dissolution); codisposal with subsoils or alkaline waste materials; surface coatings and cementing agents; calcining (an expensive procedure); and water treatment.[36]

The preceding comments apply to the conventional physical coal cleaning processes, with which there is ample operating experience. For the chemical cleaning processes which are under development, some of the same concerns may be applicable. However, these processes will also involve some new and unique problems associated with the disposal of spent chemicals, contaminated with various coal-derived organic and inorganic species. An exhaustive treatment of these potential environmental control problems is not possible at the present time because commercial processes, including disposal or regeneration steps, have not yet been defined. As promising processes reach the large-scale pilot plant and prototype commercial plant testing stages, relevant data will be gathered and control technologies will be developed or applied as necessary.

3.4.3.3 Effects of Low-Rank Coal Properties

The net result of the effects of low-rank coal properties on the applicability of mineral matter control technologies is represented by a simple fact: in the United States today, there are no physical or chemical lignite cleaning plants in operation; there is one physical coal cleaning (jig) plant washing the subbituminous coal being strip mined near Centralia, Washington, which is highly contaminated with rocks, clay, and slate.[2] There are only three reported lignite-washing operations in the world: one in the Federal Republic of Germany and two in the Soviet Union.[37]

Clearly, there is no incentive to clean low-rank coals under current and past market conditions in this country. This is due to a combination of interrelated factors. Low-rank coals are essentially all surface mined, and thus tend to contain very little extraneous mineral matter - which is the primary material removed in the eastern bituminous coal preparation plants. The mineral matter contained in low-rank coals tends to be very finely dispersed, and a substantial fraction is in fact organically bound. Thus little or no benefit is to be gained through the use of common washing practices with these coals. The high inherent moisture content of low-rank coals tends to be a disincentive to utilize wet processes, which add a substantial amount of surface moisture that further dilutes the coal's heating value. Low-rank coals tend to have low sulfur contents, and quite often only a third of the total sulfur is pyritic, which is the most readily separable form. Thus the use of physical or chemical coal desulfurization processes does not tend to have as large a payoff with low-rank coals as it does with high-sulfur bituminous coals.

There are, of course, exceptions to these general statements. Some Gulf Coast lignites tend to have higher separable mineral matter contents than their counterparts further north. The highly contaminated Washington state coals have already been noted; and other similar geologic conditions exist, where highly faulted and fractured strata make mining of "clean" low-rank coal difficult. In addition, economic and market conditions change, as do environmental regulations and other relevant external factors. It is not unreasonable to expect that incentives for removing or altering the mineral matter content of some low-rank coals will begin to appear. For example, as the costs of environmental control technology begin to dominate the capital and operating costs of coal-fired power plants, the potential benefits of coal cleaning will probably become more apparent (even if these benefits relate to such "non-environmental" factors as ash fouling of boiler tubes).

For the purposes of this study, which include the identification of unique technology requirements associated with low-rank coal development, it is useful to assume that some or all of the mineral matter control technologies described in section 3.4.3.1 will have a role in that development. One question that then arises is, to what extent do the unique properties of low-rank coals preclude, enhance, or alter the applicability of those processes? Relevant data to answer this question are quite limited. In general terms, addressing each of the major technology areas in turn, it is apparent that:

1. The physical coal cleaning processes are not likely to be affected in any dramatic way by the properties of low-rank coal. While various low-rank coal properties may render particular equipment inappropriate for a specific application, this will occur with any coal. In general, the maturity of mineral matter control technology is such that differences in coal properties are directly handled by design adaptations or slight modifications in a process. Based on isolated statements found in the literature, low-rank coals apparently are not suitable for treatment in froth flotation cells due to their poor flotability. However, through the use of appropriate flotation reagents it is possible that effective flotation systems for low-rank coals could be established.

2. The chemical coal desulfurization processes are also unlikely to be dramatically affected by the properties of low-rank coals. However, not enough is known about the precise forms of (particularly organic) sulfur in different coals, and how differences that are rank-related (if any) might affect specific processes.

3. The ion exchange process for sodium removal is one instance of a technology that is applicable only to low-rank coals. The incorporation of sodium ions into the salts of humic acids found in low-rank coals is apparently a property found only in the geologically younger coals.

Washability Data on Low-Rank Coals

Not surprisingly, very little data is available on the washability characteristics of low-rank coals. The few sources of such data that are available are briefly summarized here. It should be noted that high moisture content in coal has a considerable effect on float-sink testing. Utmost care and special procedures are required in performing the test work in order to preserve the inherent moisture and avoid the production of misleading data on specific gravity, yield, and ash relationships.[45] Certain low-rank coals, after being float-sunk, can give erroneously high chlorine contents in the float products. If the fractions are immediately washed with benzene after float-sink testing, then lower chlorine values can be obtained.[46]

A 1950 Bureau of Mines report gives washability data on a lignite sample from an underground mine, which has since closed, the Burleigh mine, Wilton, N.D.[38] The initial ash content of this sample was 7.8 percent. The report states:

> "The amount of 1.50 specific-gravity sink material, usually considered to be extraneous high ash matter, is negligible. A rejection of 50 percent of the raw feed as refuse would result in an ash reduction of 1 percent in the clean coal. These studies indicate that the lignitic coal is not amenable to washing except for possible rejection of extraneous material included with the lignite during the mining phase."

The same report contains washability analyses of subbituminous coals from Rosebud, Montana, the Elkol mine, Wyoming (Adaville coal bed, Ham's Fork region), and the Monarch mine, Wyoming (Monarch bed, Powder River region). The Rosebud coal exhibited a small amount of separable heavy mineral matter. The raw coal contained 11.5 percent ash (moisture-free), while the float material at 1.5 specific gravity contained 6.8-8.8 percent ash, and yielded 89-97 percent of the original material, depending on crush size. The Elkol mine coal was extremely low in ash (2.9-3.6 percent), and float material at 1.5 sp.g. was only slightly improved, to 2.5-2.8 percent ash. The Monarch mine coal was a similar case, with 4.5-5.3 percent ash in the raw coal, and 4.2-4.5 percent ash in the float material at 1.5 sp.g. It should be noted that the purpose of this report was to determine if U.S. coals could be cleaned to a level of 2.5 percent ash for use in coal liquefaction; the coals examined were predominantly eastern bituminous coals.

An extensive survey of the sulfur reduction potential of U.S. coals contains washability analyses of 455 coal mine samples, of which 44 are from the western U.S.[39] These 44 samples include lignite, subbituminous, and bituminous coals, but are predominantly bituminous. Sources of the samples were as follows: Arizona-6; Colorado-11; Montana-5; New Mexico-9; North Dakota-1; Utah-8; and Wyoming-4. The samples averaged 8.9 percent ash, 0.68 percent total sulfur (0.23 percent pyritic sulfur), and 12, 437 BTU/lb (moisture-free). Figure 3.4.3.2 shows averaged data for the 44 samples on the ash and sulfur reduction levels obtained as a function of crushing severity. Increased reductions in ash and sulfur contents obtained at finer sizes would require substantial fines handling circuits in the preparation plants.

A review of these data indicated that Rosebud (Montana) subbituminous coal appeared to be a good candidate for washing, when the NSPS was 1.2 lb SO_2 per 10^6 BTU.[40] Washing that 1 percent sulfur coal, which contains about 54 percent of its sulfur as pyritic sulfur, reduced the sulfur content to 0.56 percent, which is equivalent to 0.95 lb $SO_2/10^6$ BTU. As shown in Table 3.4.3.10, essentially no additional ash or sulfur removal is obtained by washing this coal at finer sizes than 1 1/2-inch top size.

A Bureau of Mines report is also available on washability of the subbituminous San Juan Basin coals of New Mexico and Colorado.[41] Eighteen samples were analyzed from 11 core holes. Ash content of the samples ranged from 4.9 percent to 61.3 percent, and averaged 20 percent. All of the samples were substantially upgraded by removal of the sink-1.50 or sink-1.60 sp.g. material. For example, at 1.5 sp.g., all but 6 of the float fractions contained less than 10 percent ash; the average was 9.6 percent.

Figure 3.4.3.2

The Effect of Crushing To 1-1/2 inch, 3/8 inch, and 14-Mesh Top Size on the Reduction of Ash, Pyritic Sulfur, Total Sulfur, and Pounds SO_2 Emission Per Million Btu at Various Specific Gravities of Separation For the Western Region Coals

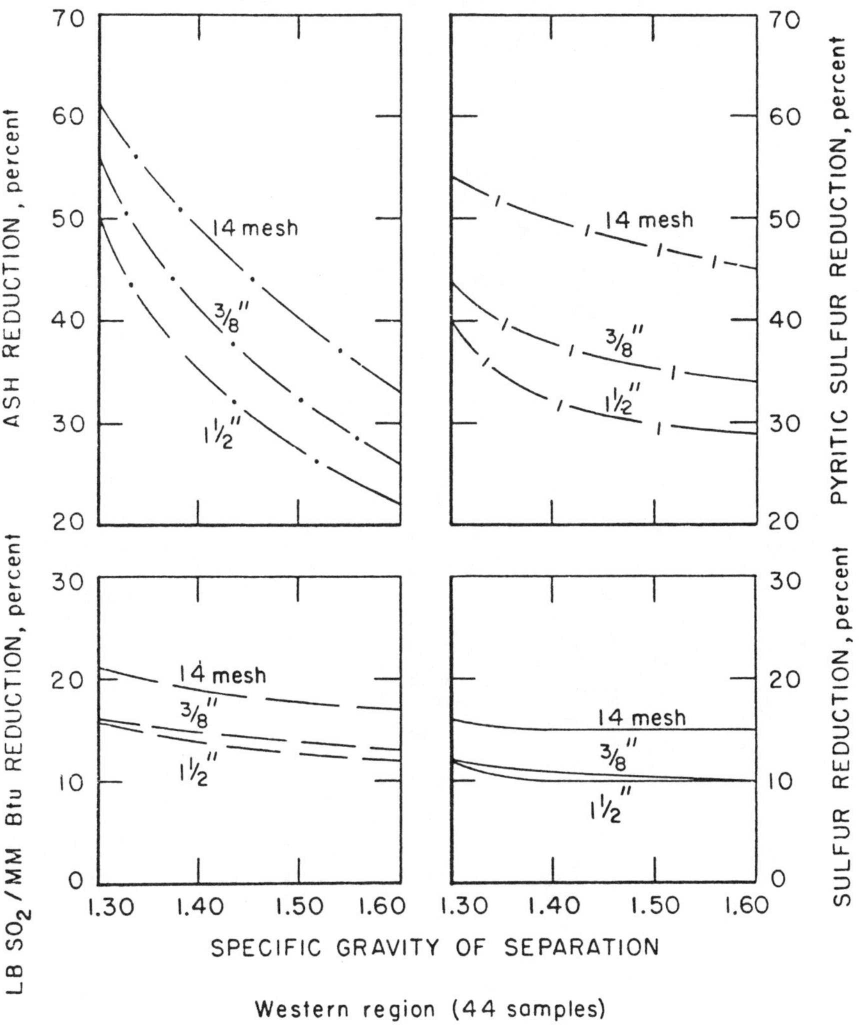

Source: Reference 39

Table 3.4.3.10

Cumulative Washability Data for Rosebud Seam Coal, Rosebud County, Montana

RAW-COAL-DRY BASIS					CLEAN COAL ANALYSES-DRY BASIS							
	% Sulfur					% Recovery			% Sulfur			
% Ash	Pyritic	Organic & Sulfate	Total	Btu	Specific Gravity	Weight	Btu	Ash	Pyritic	Organic	Total	Btu
					1-1/2 in. by 100 Mesh							
					1.30	59.3	61.8	5.9	0.06	0.48	0.54	12,041
					1.40	88.8	91.5	6.8	0.07	0.48	0.55	11,917
					1.60	96.3	98.1	7.8	0.09	0.47	0.56	11,780
9.4	0.53	0.48	1.01	11,560								
					3/8 in. by 100 Mesh							
					1.30	49.8	52.2	4.9	0.06	0.42	0.48	12,178
					1.40	83.8	87.1	5.7	0.08	0.43	0.51	12,068
					1.60	95.4	97.9	6.8	0.09	0.43	0.52	11,917
9.0	0.54	0.44	0.98	11,615								
					14 Mesh by 0							
					1.30	19.7	20.9	4.9	0.08	0.40	0.48	12,178
					1.40	75.1	79.1	5.6	0.10	0.42	0.52	12,082
					1.60	94.5	97.8	7.1	0.10	0.42	0.52	11,876
10.0	0.46	0.42	0.88	11,478								

Source: Reference 39

Recently, the same laboratory has published a report (under DOE and EPA sponsorship) on washability characteristics of Arkansas and Texas lignites.[42] Four channel samples of lignite were obtained from Arkansas and seven from Texas.

Only two of the samples, those collected from Dallas County, Arkansas, could be upgraded to meet the previous EPA new source performance standard (NSPS) of 1.2 pounds $SO_2/10^6$ BTU. The other nine samples, although averaging only a little over 1 percent total sulfur were relativ-ely high in organic sulfur (about 80 percent on the average) and could not be upgraded by washing to meet the standard.

Physical coal cleaning of these samples, however, provided significant reduction of the ash content. The raw coals averaged 17.0 percent ash, ranging from 9.3 percent to 43.1 percent. For the 1 1/2-inch x 100 size composite, the float fraction at 1.30 sp.g. averaged 11.3 percent ash, ranging from 7.1 percent to 19.2 percent. The percent reduction of ash content at this specific gravity ranged from 4 to 55 percent, and on a composite basis averaged 34 percent.

The weight-percent yields obtained at the 1.30 specific gravity of separation for the material crushed to 1 1/2-inches top size were high, except for one sample which yielded 42.8 percent. The other 10 samples ranged from 75.9 percent to 96.7 percent yield. This high yield of float 1.30 material is attributable to the low specific gravity of lignite.

It was demonstrated that much of the sodium can be removed from these Arkansas and Texas lignites by ion exchange. Analyses showed 37 to 91 percent sodium reduction after treatment with a solution containing calcium ions. However, none of these lignite samples contained the high sodium levels associated with severely fouling coals (i.e., more than 4 percent Na_2O in the ash). The Na_2O contents in the ash of the untreated lignite ranged from 0.08 percent to 1.13 percent; in the treated lignite the Na_2O contents in the ash ranged from 0.04 percent to 0.37 percent.

Studies of the North Dakota lignite mineral matter on selected samples at GFETC[43] have shown that mineral elements occur in three distinctive modes--15 to 20 percent occur as mineral matter separable by CCl_4 float-sink techniques, 35 to 70 percent as organically bound elements, and 20 to 40 percent as finely divided nonseparable clay and SiO_2. The float fraction ash has an analysis similar to the total ash. The organically bound mineral elements (Ca, Mg, Na, S, and possibly some Fe and Al) are present primarily as salts of humic acids and are ion exchangeable. Frequently, more than half of the coal sulfur is present as organic sulfur.

Preliminary float/sink tests at GFETC to study the possibilities of washing and float/sink for removing ash, sodium and sulfur from Northern Great Plains lignite indicated that typically about 15 percent of the ash and 30 percent of the sulfur (in the pyritic form) could be removed.[44] The sodium level was not reduced significantly since most of the sodium is bound into the organic coal matrix. Subsequent tests on Western subbituminous coals have indicated similar results. Tests on a Texas lignite from Bryan showed more promising results with up to 57 percent reduction in ash, 35 percent reduction in sulfur and 68 percent reduction in sodium (from 0.38 to 0.11 percent in the dry coal).

A subbituminous coal from San Juan County, New Mexico, experienced a 41 percent reduction in ash (from 22.8 to 13.3 percent and a 16 percent increase in dry heating value. Conversely, the sulfur concentration stayed relatively constant at 0.7 percent in the coal and the sodium in the coal ash increased 70 percent (from 2 to 3.4 percent Na_2O in the ash). The increase in sodium was somewhat offset by an increase in calcium (4.1 to

5.6 percent). Recent tests in the GFETC ash fouling test furnace indicated a slight reduction in the ash fouling potential with the washed coal, but both the raw and washed coals, had high deposit buildup on the probe and walls.

A Texas lignite from Freestone County experienced a 46 percent reduction in ash (30.6 percent to 16.4 percent), a 31 percent reduction in sulfur (1.33 to 0.92 percent) with a corresponding 20 percent increase in the dry heating value. An increase in the sodium in the coal ash from 0.2 to 0.4 percent was somewhat offset by an increase in the calcium from 4 to 6.7 percent. Although the raw and washed samples were obtained and tested at different times, the raw-coal sample is believed to be essentially the same as that used for the washing tests. The recent tests on the washed lignite indicated that the ash fouling potential was still low and the reduction in ash improved the ash deposit problem on the refractory walls.

Minerals concentrated from lignite by float/sink technique using carbon tetrachloride as the suspending medium and separated by hand-picking under a microscope were identified as alumnio-silicates, pyrites, quartz, calcite, and gypsum.[44] Barium, which is a trace element in lignite, appeared occasionally as barite. No minerals were found in which sodium was a major constituent, and only trace amounts of sodium occurred in the ash of the concentrated minerals. It is evident that float/sink treament of pulverized lignite is relatively ineffective for concentrating the finely-divided inorganic minerals in lignite.

3.4.3.4 Current Status

As discussed in the previous section, only one low-rank coal preparation plant is active in the U.S. today. Investigations into a variety of mineral matter control techniques are being pursued, some of which will eventually be utilized in specific cases where they can be justified. Since the objectives and status of these technologies and development efforts have been summarized in the preceding sections, they will not be repeated here.

References - Section 3.4.3

1. Leonard, J.W., and D.R. Mitchell (Editors). Coal Preparation, American Institute of Mining, Metallurgical, and Petroleum Engineers, Inc., New York, 1968.

2. Phillips, P.J. Coal Preparation for Combustion and Conversion, EPRI AF-791, May 1978.

3. Huettenhain, H., J. Yu, and S. Wong. A Technical and Economic Overview of Coal Cleaning, in "Proceedings: Symposium on Coal Cleaning to Achieve Energy and Environmental Goals," Hollywood, Florida, September 1978, pp. 256-297.

4. Killmeyer, R.P., Jr., and R.E. Hucko. Coal Preparation, NSDOE Coal Preparation Laboratory, Pittsburgh, Pa., (undated).

5. Libach, N. "More Coal per Ton," EPRI Journal, June 1979, pp. 6-13.

6. Buder, M.K., et al. Cost Benefit Analysis of Cleaning Two Western Coals for Steam Generation, presented at the Second International Coal Utilization Conference, Houston, Texas, November 1979.

7. Hall, E.H., et al. Physical Coal Cleaning for Utility Boiler SO2 Emission Control, EPA-600/7-78-034, February 1978.

8. Smith, C.D. Coal Cleaning By the Otisca Process, in "Proceedings: Symposium on Coal Cleaning to Achieve Energy and Environmental Goals," Hollywood, Florida, September 1978, pp. 623-636.

9. Weintraub, M., A.W. Deurbrouck, and R.H. Thomas. Dry-Cleaning Coal in a Fluidized Bed Medium, DOE RI-PMTC-4(79), January 1979.

10. Karlson, F.V., K.L. Clifford, W.W. Slaughter, and H. Huettenhain. The Potential of Magnetic Separation in Coal Cleaning, in "Proceedings: Symposium on Coal Cleaning to Achieve Energy and Environmental Goals," Hollywood, Florida, September 1978 pp. 568-598.

11. Ergun, S., and E.H. Bean. Magnetic Separation of Pyrite from Coals, Bureau of Mines R.I. 7181, 1968.

12. Kester, W.M., J.W. Leonard, and E.B. Wilson. "Reduction of Sulfur from Steam Coal by Magnetic Methods," Mining Congress Journal, July 1967, pp. 70-75.

13. Schall, J. Compilation and Assessment of SRC Experience: Data Book, EPRI AF-1019, March 1979.

14. Meyers, R.A. Introduction to Chemical Coal Cleaning, in "Proceedings: Symposium on Coal Cleaning to Achieve Energy and Environmental Goals." Hollywood, Florida, September 1978, pp. 923-933.

15. McCandless, L.C., and G.Y. Contos. Current Status of Chemical Coal Cleaning Processes-An Overview, in Proceedings: Symposium on Coal Cleaning to Achieve Energy and Environmental Goals," Hollywood, Florida, September 1978, pp. 934-959.

16. Morel, W.C. and T.J. George. An Economic Evaluation of the Oxydesulfurization Process for Coal Desulfurization, 3.9 Million Tons Per Year of Cleaned Coal Product. DOE FE/EES-79/8, June 1979.

17. Morel, W.C. and T.J. George. An Economic Evaluation of the Modified Oxydesulfurization Process for Coal Desulfurization, 3.9 Million Tons Per Year of Cleaned Coal Product. DOE FE/EES-79/9, June 1979.

18. Guth, E.D. Oxidative Coal Desulfurization Using Nitrogen Oxides - The KVB Process, in "Proceedings: Symposium on Coal Cleaning to Achieve Energy and Environmental Goals," Hollywood, Florida, September 1978, pp. 1141-1164.

19. Plants, K.D., and T.J. George. An Economic Evaluation of the TRW Gravichem Process - One Unit Train Per Day of Coal, DOE FE/EES-79/10, June 1979.

20. Plants, K.D., and T.J. George. An Economic Evaluation of the Modified TRW Gravichem Process - One Unit Train Per Day of Coal, DOE FE/EES-79/11, June 1979.

21. Santy, M.J., and L.J. Van Nice. Status of the Reactor Test Project for Chemical Removal of Pyritic Sulfur from Coal, in "Proceedings: Symposium on Coal Cleaning to Achieve Energy and Environmental Goals," Hollywood, Florida, September 1978, pp. 960-990.

22. Kalvinskas, J.J., and G.C. Hsu. JPL Coal Desulfurization Process by Low Temperature Chlorinolysis, in "Proceedings: Symposium on Coal Cleaning to Achieve Energy and Environmental Goals," Hollywood, Florida, September 1978, pp. 1096-1140.

23. Kindig, J.K., and D.N. Goens. The Dry Removal of Pyrite and Ash from Coal by the Magnex Process; Coal Properties and Process Variables, in "Proceedings: Symposium on Coal Cleaning to Achieve Energy and Environmental Goals," Hollywood, Florida, September 1978, pp.

24. Beckberger, L.H., et al. Preliminary Evaluation of Chemical Coal Cleaning by Promoted Oxydesulfurization, EPRI EM-1044, April 1979.

25. Gronhovd, G.H., R.J. Wagner, and A.J. Wittmaier. A Study of the Ash Fouling Tendencies of a North Dakota Lignite as Related to its Sodium Content, Transactions, Soc. of Mining Engineers Quarterly, September 1967, pp. 313-322.

26. Sondreal, E.A., P.H. Tufte, and W. Beckering. Ash Fouling in the Combustion of Low-Rank Western U.S. Coals," Combustion Science and Technology, Vol. 16, 1977, pp. 95-110.

27. Paulson, L.E., and R.C. Ellman. Reduction of Sodium In Lignite By Ion Exchange : A Pilot Plant Study, GFETC/RI-79/1, 1979.

28. Paulson, L.E., and W.W. Fowkes. Changes in Ash Composition of North Dakota Lignite Treated By Ion Exchange, Bureau of Mines R.I. 7176, September 1968.

29. Baria, D.N., W.R. Kube, and L.E. Paulson. Conceptual Design of a Commercial Plant for Sodium Removal from Lignite, GFETC, March 1980.

30. Baria, D.N., W.R. Kube, and V. Carver. Conceptual Design of a 1,584,000 Ton per Year Lignite Preparation Facility for Sodium Reduction, prepared by University of North Dakota under DOE Contract No. EW-78-A-03-2101, Draft Report, August 1979.

31. Paulson, L.E., and J.R. Futch. Removal of Sodium from Lignite by Ion Exchanging with Calcium Chloride Solutions, presented at the 179th National American Chemical Society Meeting, Houston Texas, March 23-28, 1980.

32. Lemmon, A.W., G.L. Robinson, and D.A. Sharp. An Overview of Control Technology, in "Proceedings: Symposium on Coal Cleaning to Achieve Energy and Environmental Goals," Hollywood, Florida, September 1978, pp. 793-823.

33. Gluskoter, H.J., et al. Trace Elements in Coal, EPA-600/7-77-064, 1977.

34. Nicol, S.K. Oil Agglomeration for Fine Coal Refuse Treatment, World Coal, August 1979, p. 14.

35. Deurbrouck, A.W. United States Coal Preparation Practices, World Coal, May 1977, p. 24.

36. Wangen, L.E., et al. Control Technology for Coal Cleaning Wastes, in Proceedings of the ASCE/PRC-EPRI Workshop "Solid Waste Research and Development Needs for Emerging Coal Technologies," San Diego, CA, April 23-25, 1979, p. 79.

37. Graham, R.J. Washability Characteristics and their Role in Coal Preparation, presented at the Gulf Coast Lignites Conference, Houston, Texas, May 17-19, 1978.

38. Fraser, T., W.L. Crevtz, and O.T. Barrett. Preparation Characteristics of Some Coals Available for the Synthetic Liquid Fuels Industry, Bureau of Mines Bulletin 495, 1950.

39. Cavallaro, J.A., M.T. Johnston, and A.W. Deurbrouck. Sulfur Reduction Potential of the Coals of the United States; A Revision of RI 7633, Bureau of Mines R.I. 8118, 1976.

40. Becker, D.F., et al. "Beneficiating Rosebud Subbituminous Coal," Power Engineering, December 1977, p. 61.

41. Deurbrouck, A.W. Washability Examinations of Core Samples of San Juan Basin Coals, New Mexico and Colorado, Bureau of Mines R.I. 7608, 1972.

42. Cavallaro, J.A., and A.F. Baker. Washability Characteristics of Arkansas and Texas Lignites : Report of Investigations, EPA-600/7-79-149 (DOE PMTC-5(79)), June 1979.

43. Honea, F.I., S.J. Selle, and E.A. Sondreal. Factors Affecting Boiler Tube Fouling from Western and Gulf Coast Lignites and Subbituminous Coals, NACE Corrosion/79 Conference, Atlanta, Georgia, March 12-16, 1979.

44. Paulson, L.E., W. Beckering, and W.W. Fowkes, "Separation and Identification of Minerals from Northern Great Plains Province Lignite," Fuel, Vol. 51, July 1972, pp. 224-227.

45. Cooper, H.M., E.C. Tarpley, and R.F. Abernethy. Effects of Moisture on Float-and-Sink Testing of Lignite, BuMines R.I. 4184, 1947.

46. Duzy, A.F., et al. Western Coal Deposits; Pertinent Qualitative Evaluations Prior to Mining and Utilization, in "Technology and Use of Lignite, Proceedings "Grand Forks, North Dakota, May 18-19, 1977, pp. 13-42.

3.4.4 Moisture Reduction

3.4.4.1 Technology Description

Reduction of moisture levels in coal can be carried out by either mechanical or thermal processes. Both approaches are used throughout the coal industry, but generally the mechanical techniques are applied to dewatering (the separation of the solid from a slurry for example). Thermal methods are normally applied to drying (for example, to remove the surface moisture from fines or the inherent moisture from coal). With regard to low-rank coals, mechanical techniques might be utilized in the future for the dewatering of coal pipeline slurries and wet physical coal cleaning plant product streams. Thermal processes are the most likely approach to be utilized in reducing the high inherent moisture levels in low-rank coals for transport and other utilization.

Mechanical Dewatering[1,2]

The difficulty of dewatering increases with increasing surface area of the coal particles to be dewatered. The finer the particle size distribution, the greater the surface area available for the adherence of water. Dewatering difficulty also increases as the desired moisture levels are reduced. Numerous devices and techniques have been developed to address both lower desired moisture levels and increased surface area. Figure 3.4.4.1 illustrates the many equipment types used for mechanical dewatering as a function of size consist and final desired moisture levels. The areas of greatest concern are particle sizes less than about 1/4 inch and surface moisture levels below about 2.5 percent.

Figure 3.4.4.1

Dewatering Equipment in Common Use in the Coal Industry
(Particle Size Ranges Handled and Approximate Moisture Produced)

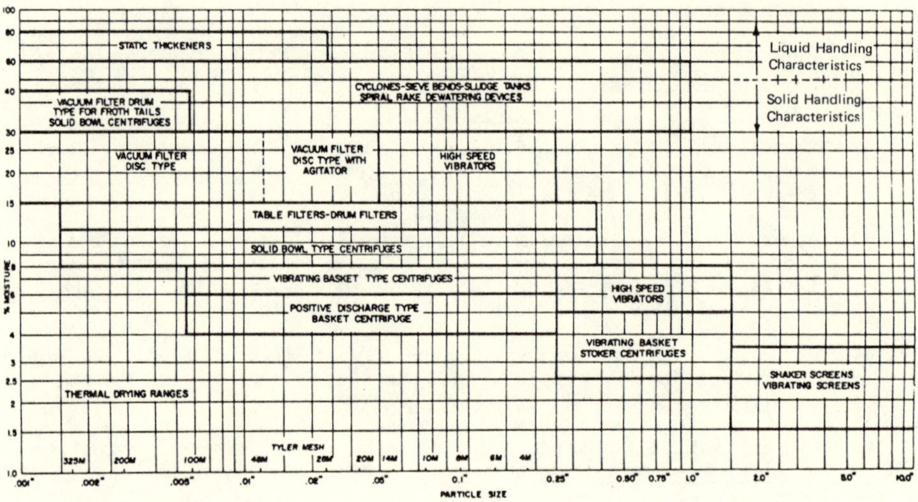

Source: Reference 1

The two potential applications for mechanical dewatering systems for Western low-rank coals will be in wet coal washing plants and in pipeline slurry dewatering stations. Wet preparation plants will require dewatering for the coarse stream; conventional equipment should be adequate for this application. Conventional equipment will perform on the fines streams as well, but the effectiveness of this equipment with low-rank coals is uncertain. Coal slurry pipelines dewatering needs will be for coal fines only. While the Black Mesa coal slurry pipeline currently uses mechanical dewatering adequately, there is a desire to develop more efficient methods.

The most applicable process categories of mechanical drying processes are centrifugation and filtration. Centrifugal type dryers rely on high rotational accelerations to dewater. For a typical 48-inch diameter centrifuge rotating at 250 rpm, the centrifugal acceleration is greater than 40 times the gravitational acceleration, resulting in a correspondingly increased force acting to remove the water from the coal. These devices find application in virtually every wet coal washing plant for dewatering coal from 1-1/2 inch and under. Products from centrifugal dewatering devices tend to be uniform, consistent and easily handled.

While centrifugal acceleration strongly affects the amount of moisture that can be removed from the coal, it is not the final determinant. Characteristics of the machine and the coal can play an important role in the degree to which the coal can be dewatered.

Centrifuges are classified as perforate basket or solid bowl types Perforate basket machines with transport devices are the most common type found in the coal industry today. The vibrating basket machine is the design being used in most new equipment, however.

Perforate basket machines with transport devices have two rotating elements: an outside conical screen frame and an inside solid cone which carries spiral hindrance flights. Both rotate in the same direction, but the screen element moves slightly faster than the cone. The wet coal enters the machine at the top, falls on the apex of the cone and the centrifugal force developed by the rotating cone throws the coal-water mixture against the screen. The water passes through the perforations and is collected in an effluent chamber. The flights spiral downward and, as the screen moves relatively slowly around the flights in the direction of their downard pitch, the solid material is gradually transported to the bottom of the screen cone. Because of the conical shape of the perforated basket, the coal and water are subjected to zones of increasing centrifugal force.

Vibrating basket centrifuges employ vibrating as well as rotating baskets. The vibrating movement causes the solids to flow through the machine, and tends to loosen the material to aid the dewatering process. Because of the low speed generally used in these centrifuges, the product moisture is usually higher than the moisture produced by the transport type machine. However, wear and horsepower are also low and coal degradation is of minor importance.

The vibrating basket centrifuge can be used successfully to dewater stoker size coal with minimum degradation because of its very low operating speeds. It is not unusual for such a machine to produce 2-1/2 to 3 percent surface moisture on 3/4 x 1/4 inch stoker coal. These machines can be fed coal containing 60 percent moisture and single units can be designed to handle up to 150 tph of solids. Surface moisture on well-screened 3/8 inch by 1/2 mm sizes can be as low as six to seven percent.

The solid bowl centrifuge consists essentially of two rotating elements. The outer element is a solid bowl in the shape of a truncated cone. Mounted inside the bowl and rotating at a slightly different speed is a second element, a helical conveyor, which is shaped to follow the contour of the bowl. Feed is introduced through a stationary pipe and flows into the bowl proper. In the bowl the centrifugal forces cause the solids to settle out of the liquid. The solids are conveyed to the small diameter end of the bowl by the helical conveyor. The effluent of the machine flows to the base end of the cone where it is discharged through effluent ports.

One of the principal advantages of this machine is that it can be fed dilute slurries. For example, in Deister table plants no dewatering screens are required between the table and the centrifuge as is necessary when basket type machines are used. The solid bowl machine will, however, require more horsepower because it must accelerate this water load as well as the coal load. A decrease in required horsepower and a decrease in cake moisture can be obtained if a fine opening seive bend is used to dewater and remove fine solids from the feed to the machine.

The moisture contained in the solid bowl product and also the ash content of the cake can be improved by the addition of a water rinse in the machine. By this expedient, fine, clay-like substances are washed from the product and discarded with the effluent. Because of the raking action of the helical conveyor, there is some degradation caused by this machine. It ranks between the positive discharge basket type and the vibrating basket types in degradation, with the vibrating basket type giving the least degradation.

Like centrifuges, filters process a suspension with a high solids fraction and separate the water to produce a compact wet cake. This process is accomplished by placing a filtering medium (cloth, screen, porous medium) in the suspension and applying a suction to draw the water and the solid to the filtering surface. Water passes through the surface but the solids remain on the surface forming the filter cake. The filter cake is removed by reversing the pressure on the filter surface and/or by the use of mechanical scrapers.

Filtration theory has advanced significantly since filters were first used for coal dewatering, but the technology remains more an art than a science. The rate of water removal tends to: a) increase proportionately with the product of filtration area and pressure drop and b) decrease with the viscosity of the suspension, the specific resistance of the cake, and the weight of the cake per unit filtering area. Perhaps the most important element of the solid material is its size consist. In all cases a finer consist reduces the filtrate volume removal rate.

An inspection of the data obtained from numerous installations in the field indicates that the minus 200-mesh fraction plus the amount of clay or slimes in this fraction, exert a marked influence on filtration rate characteristics of coal. It has been found that the best correlating factor representing specific cake permeability is obtained by using the square root of the percent minus 200-mesh fraction times the percent ash in this fraction. Correlations for filtration rate based on this factor, using actual plant results, are very dependable.

There are two filter designs currently in use in coal dewatering applications -- discs and drums. The disc type design employs a succession of filtration discs. The coal/water suspension spills into each successive disc if the system is vertically oriented. In the horizontal position the discs dip into the suspension. The drum filter is a cylinder around which is wrapped the filter cloth. The drum dips into the suspension or the suspension is fed from a top reservoir onto the drum. As the drum rotates the filter cake forms and the cake is removed either by mechanical scraper or pneumatically.

The disc type filter is the accepted design for coal filtration. Perhaps the only advantage of the drum filter is that more complete cake discharge will result in cases where thin cakes are obtained. In coal, however, the cake formed is relatively thick and no trouble is experienced in discharging the cake from the disc as long as proper care is taken in maintaining proper alignment of discs and scraper blades. Maintenance costs are less on the disc and, in general, the disc filter is easier to handle than the drum.

Thermal Dewatering[1,2,11]

Current and projected coal economics have warranted the increased utilization of minus 1/4 inch coal as a feedstock to coal preparation plants. Where subjected to wet processing techniques, these fines tend to retain a higher fraction of moisture due to the associated greater surface area than do the coarse fraction. Thermal dewatering has been applied in many preparation circuits to treat the high moisture fines and bring their moisture levels down to usuable levels. The benefits that can be obtained from thermal drying procedures include:

1. Improved transport economics resulting from
   - Improved heating values
   - Reduced likelihood of freezing problems during adverse weather conditions; and
2. Improved utilization economics due to
   - Increased boiler efficiency since water need not be evaporated during combustion

- Lower throughput requirements due to improved heating value

- Increased pulverizer capacity due to low moisture

Essentially all industrial coal dryers are continuous direct contact units that employ convection as the primary means of heat transfer. Hot gases, generally consisting of combustion products, are brought into direct contact with wet coal on a continuous gas flow - coal feed basis. The hot gases evaporate the water retained on the surface of the coal particle and remove the water in the vapor state.

Thermal drying of coal can be divided into two periods: an initial constant-rate period in which the surface of the coal is entirely wet; and a final falling rate period in which the surface is partially dry. During the initial period the magnitude of the constant drying rate can be improved by increased particle surface area; by increased differences between the vapor pressure of water at the coal surface and the partial pressure of water vapor in the atmosphere; by increased mass velocity of air; by increased differences between the wet and dry bulb temperature of the drying air; by decreased bed bulk density; and by decreased thickness of bedding.

Drying during the final period progresses at a decreasing rate as a result of the decreasing wetted evaporation surface area. The nature of the coal substance and its characteristic release of moisture at various temperature gradients become the major factors in this phase of drying. In addition to these coal-specific factors, drying time during the falling rate period will be reduced by: reduced bed bulk density; reduced average particle diameter; reduced differences between the moisture at the beginning (critical moisture) and end (equilibrium moisture) of the falling-rate drying period; increased bed drying area; increased mass velocity of drying gases and increased differences between particle surface temperature and air temperature.

Different types of convective direct contact coal dryers can be classified as fluidized bed; entrained flow (suspension or flash); multi-louvre; vertical tray or cascade; continuous carrier; or drum.

The fluidized-bed dryer employs hot drying gas (usually air) that is forced through a constriction plate upward through the bed of material to be dried in such a manner that the material is suspended or fluidized above the plate. The fluidized material exists in a state of dynamic equilibrium in which the bed is rapidly mixing in both the horizontal and vertical directions. Corresponding to the turbulent environment are very high heat and mass transfer rates which greatly improve the drying capacity of the unit. The coarse dried material typically discharges from the dryer through a motorized conveyor-airlock. The fines which are suspended in the gas stream are collected in a dry dust collector and are usually recombined with the coarse material discharged from the dryer, producing a product with minimum dust loss.

Entrained flow dryers operate on the entrained flow principle where coal particles are continuously introduced into a turbulent gas stream that carries the particle through a prespecified distance within which the moisture is removed. In one typical design, drying gases are produced by a stoker-fired furnace and directed into the drying column. The drying column and the furnace capacity are carefully matched to ensure that coal particles will be carried up the drying column to the cyclone which separates the coal from the gas. Residence time in the drying column is about 1/2 second. Inlet temperature of the gas is about 1200°F and outlet temperature is approximately 200°F.

The essential feature of fluidized or entrained drying is, therefore, the mixing of fine coal with hot gases in such proportions that the resultant mixture reaches a temperature of 275° to 325°F. The coal must remain in this environment long enough to be heated to about 250°F, at which temperature 90 to 95 percent of the inherent moisture is evaporated. Approximately 50 seconds is required for particles of 1/4-inch by 0 size. The size of the drying chamber and the velocity of the gases traveling through the system are selected to provide the optimum residence time of the coal in the system. Operation of a given drying unit will vary with different coals containing more or less moisture, because the heat required for drying varies almost directly with the moisture in the coal and the capacity of a drying unit varies inversely with the heat required.[4]

The multi-louvre dryer is adapted to handling large capacities and is applicable to those materials requiring rapid treatment, operating on a comparatively short retention time. It is extremely flexible, being effective as a dryer or a cooler, or a combination of the two. The moving element in this machine consists of specially designed flights attached to and carried by two strands of roller chain. These chains are away from the air stream and do not contact the material. The material is carried up in the flights and then flows downward in a shallow bed over the ascending flights. It gradually moves across the dryer, a little at each pass, from the feed point to the discharge point. One advantage of this type of machine is the gentle flowing action of the material, which exposes all particles to incoming air and results in minimum degradation of the particles.

The cascade dryer consists of a series of shelves mounted like stairs which vibrate the coal as it cascades down through them. The dried coal is collected in a conveyor at the bottom for evacuation. Adjustment of shelf pitch provides for variation in speed of travel of the coal. Hot gases, which are ordinarily generated in a coal-fired furnace, are fed to the back of the dryer and are drawn upward through and between the wedge wire shelves. As the coal is fed into the dryer at a determined rate, it is subjected to the temperature best suited to start moisture evaporation. This temperature is controlled automatically at all times through the control circuit. If a small excess of coal of higher moisture percentage enters the dryer, these controls compensate by letting a greater volume of gases flow through the dryer. In a continuous carrier dryer, coal travels through the machine on screen decks and is subjected to both thermal and mechanical drying. Hot gases are drawn through the coal on the screen decks. The gas flow is automatically alternated each second between two sections. This brings the full force of flow on one screen section at a time, resulting in gas pressure so strong that the coal is clamped to the deck and excess water is squeezed from it. When the pressure is relieved and the reciprocating throw of the screen section loosens and moves the coal forward, the hot gases are sucked into the coal bed enveloping each suspended particle of coal. A rapid heat exchange occurs with resultant evaporation of moisture. Each particle of coal is subject to an average of 50 of these drying cycles during its passage through the drier.

The drum-type dryer consists of a solid outer cylindrical shell, and an inner shell composed of full length louvres. This inner shell of overlapping louvres, which supports the bed of material, increases in diameter in the direction of flow. The material gently travels toward the discharge end as the drum slowly revolves, in similar fashion to a kiln. In operation the heat transfer medium, introduced through the louvre openings, permeates the bed and intimately contacts every particle.

In addition to the convective thermal drying processes described above, there are a number of drying processes that utilize conduction and/or radiation heat transfer modes as the method of drying. One example of this, which was discussed in Section 3.4.2 (comminution), is in-the-mill drying that commonly occurs while coal is being pulverized for feeding to a furnace.

Another technique involves transporting the wet coal through a trough by a hollow screw. The heat transfer fluid, usually oil, flows through the screw and usually through an annular space around the trough. Heat conducted to the coal evaporates the moisture which is carried away by a minimal air flow over the coal. The major advantage to this type of drying arrangement is that dust entrainment is minimal since coal is conveyed mechanically rather than pneumatically.

Steam filtration has as its objective the reduction of water content by heating water in the filter cake to lower interstitial surface tension and increase fluidity. This increased flowability of the water facilitates its removal in the filtration process. The process was developed in Europe as a means of increasing the efficiency of a disk filter. It is based on the fact that the application of superheated steam to the filter cake results in significant additional moisture removal. Steam filtration operational and equipment costs have been determined to be strongly competitive with conventional thermal drying systems.

The steam filtration method basically involves the covering of a vacuum filter with suitable hooding, and the application of steam within the hood. The steam condenses, when pulled through the coal cake by the vacuum system and consequently releases latent heat which raises the cake temperature. At the resultant higher cake temperature, the water viscosity is reduced, which greatly facilitates additional water drainage from the filter cake. Several advantages are achieved with steam filtration when compared with conventional thermal drying practices in addition to operational and equipment cost benefits. The advantages include:

1. Fine flotation coal, dried by steam filtration, can be by-passed from a thermal dryer, resulting in a simplified thermal dryer design and operation. The most stringent of final loaded coal moisture levels can be obtained by this approach.

2. By keeping the fine flotation coal out of the thermal dryer, the dustiness of the loaded coal can be eliminated and thermal dryer fire and explosive dangers can be greatly reduced.

3. In some cases, steam filtration of the total product normally dried in thermal dryers is sufficient to attain certain final loaded coal moisture specifications.

Steam Drying[3]

The term steam drying is used to denote water removal brought about by treatment of low-rank coal at pressure with saturated steam. This treatment causes shrinkage, removes water and stabilizes the lump, thus improving its handling and weathering properties. Some evolution of gases occurs due to the thermal destruction of carboxylic acid groups to yield carbon dioxide. Some of the water is removed non-evaporatively, i.e. as a liquid, and some is flashed when the vessel containing the coal is depressurized. If energy recovery is practiced the process is more efficient than evaporative drying and the mass of steam required to remove a unit mass of water is fractional.

The process has been operated commercially in Europe since 1927 in the form of multivessel batch process known as Fleissner drying. At present there are three or four Fleissner plants operating in Eastern

Europe and Turkey, the largest handling about 600,000 tpy of raw coal. Patents have been taken out for Fleissner plant operation in the U.S.A. as well as Europe. The difficulty of handling solids into and out of batch pressure vessels has made the process economically unattractive.

Pilot plant work on the Fleissner process has been conducted at the University of North Dakota and at GFETC.[5,6] Wide variations were noted in release of liquid water from different lignites, and were attributed to possible differences in capillary shrinkage and/or colloidal properties of lignite as a function of temperature. The most recent information to appear is a description of coal thermal properties, the heat transfer rate, and the kinetics of water removal in a steam system for Victorian (Australian) brown coal.[7]

In addition to the beneficial effects on the volume and surface properties of lignite, steam drying has been noted to remove approximately half the Na^+ and Cl from Australian lignite at 200°C, and almost all at 300°C.[3] Thus this type of treatment may be an alternative to removal of sodium from high-fouling coals by ion exchange.

The Koppelman process is a proprietary continuous steam-drying process that has been claimed to produce an upgraded lignite product containing 12,000 BTU/lb that will not reabsorb moisture.[8] A coal-water slurry is pumped into a 1500 psig tubular reactor and the coal is pyrolyzed in steam at 1000°F. Offgas with a heating value of 400-500 BTU/scf is expanded and burned for power production. The dried lignite is discharged through lock hoppers to a cooling system. Tests are reportedly being done in a pilot plant with a continuous input of 180 to 400 lb/hour.

3.4.4.2 Environmental Control Technology

Very little information has been found in the literature on the environmental control problems associated with coal moisture reduction processes. For the mechanical dewatering processes which remove surface moisture, the primary apparent problem is the proper treatment and disposal of the separated water. In all modern coal preparation plants, this water is recirculated to the washing units and does not represent an effluent. The thermal processes drive off the moisture as steam which creates no environmental problem; however, high dust loadings are produced in many of the thermal drying units, which require conventional collection systems such as cyclones, baghouses, and ESPs. The steam drying process would have similar environmental control requirements.

3.4.4.3 Effects of Low-Rank Coal Properties

The properties of low-rank coals that have the greatest impact on technology for moisture reduction are: 1) the high inherent moisture content, and 2) the physical properties of dried low-rank coals.

Moisture Content

Removal of moisture from low-rank coals is a complex process (at the microscopic level) due to the various ways in which water may be held in, and released from the coal. Water in coal may be obtained from (1) decomposition of organic molecules which is sometimes called combined water, (2) surface absorbed water, (3) capillary condensed water, (4) dissolved water and (5) water of hydration of inorganic constituents. There is no method of distinguishing between the amounts of water in coal in these various ways.[9]

Lower rank coals contain more natural bed moisture than higher rank coals; the inherent moisture progressively decreases with rank. As the coalification process proceeded from peat to lignite and coals of higher rank, there was a continual decrease in the capillaries of all sizes in the coal particle but especially in the large ones. Thus, the pore volume decreased progressively from peat to bituminous. A freshly mined lignite loses moisture very rapidly on exposure to air because much of the water is loosely bound. The vapor pressure of this moisture is normal. Upon further loss of moisture, the vapor pressure decreases, apparently from the water beginning to evaporate from the capillaries. During the drying of lignite, a physical change occurs through shrinkage, causing a collapse of some of the capillaries. Thus, the dried lignite cannot absorb as much water as it originally held. This phenomenon of hysteresis has been studied by many and the theory of collapsed capillaries is the most accepted one.[9]

Because of the high moisture content and different drying characteristics of low-rank coals, the material capacity of a given thermal drying system is substantially reduced when compared to the drying of a bituminous coal.[10] The throughput of these devices tends to be limited by the water evaporation rate, and thus the (dry) coal throughput is roughly inversely proportional to the feed coal's moisture content. Other than this, however, no substantial differences or operating difficulties have been noted when low-rank coals have been tested in conventional thermal driers.[11,12,13,14]

Mechanical dewatering will not be directly affected by high inherent moisture contents of low-rank coals, since this moisture is not what is being removed. Mechanical dewatering will be used for dewatering coal water slurries from pipeline transport and from beneficiation plants. The objective will be to reduce the surface moisture to the greatest extent possible. High inherent moisture will indirectly affect this technology by increasing the need for more effective mechanical moisture reduction processes. For example, current mechanical dewatering may reduce a slurry pipeline coal to 15 percent surface moisture. This may be acceptable if the inherent moisture of the coal is 5-10 percent; however, in the case of western low-rank coals where the moisture levels are 25-35 percent, a 15 percent surface moisture may not be acceptable.

Physical Properties of Dried Low-Rank Coal

It was noted above that the process of moisture removal causes physical change in a low-rank coal through shrinkage and collapse of the pore structure. Accompanying these changes are some desirable and undesirable physical properties, which affect the systems required for handling and utilizing dried low-rank coals, and their costs. Also, the properties of dried low-rank coal are reported to be significantly different depending on whether the drying is done by air or by steam.

Air-dried low-rank coals do exhibit a pronounced slacking tendency, and thus any handling of the dried coal reduces the size of particles and creates dust. The extent of this process is related to the degree of drying, the original size of the particle, and the petrographic structure of the coal.[13]

Air-dried low-rank coals are also reported to be more reactive particularly to oxygen, than as-mined coals. Studies at GFETC have found that the reactivity of freshly dried test samples of lignite to oxygen does not vary significantly with the moisture content. This suggests that the degree of moisture reduction is not a major consideration in respect to product reactivity. A major factor in respect to reactivity of lignite under any given sample condition is temperature. A process in which dryer product is cooled and exposed to air was found to be very effective in reducing reactivity towards oxygen. In general, dried subbituminous coal was somewhat more reactive than dried lignite. At elevated temperatures, reactivity of dried subbituminous coal was found to be greater than that of the as-received coal tested at the same temperature.[14]

Reaction with oxygen of void space air produces an initial temperature rise in dried coals during transport or in stockpiles. The rate of heat transfer in dried coals is slow. Little cooling can be accomplished in an enclosed bin or stockpile within reasonable time periods without benefit of a suitable conductant mechanism or procedure.[14]

Dried low-rank coals will regain only a portion of the moisture removed if exposed to saturated air conditions. The lower the initial moisture content of dried coal, the lower will be the final moisture content after exposure. In no case does the moisture content return to original moisture content. A period of 3 to 4 days of continuous exposure is required for the reabsorption process to be completed.[14]

The bulk density of dried low-rank coals without a vibration or settling treatment is greater than that of as-mined coals. A vibration or settling procedure was found to increase bulk density of dried coals but to a lesser degree than that which occurred for the as-mined coal. Results suggest that volume requirements for handling and storage of a given weight of dried low-rank coals are not greatly different than those required for as-mined coals.[14]

With regard to the reactivity of dried low-rank coals, it should be noted that the liquefaction reactivity of these coals is greatly reduced by the collapse of the pore structure that occurs upon air drying. The slurry drying technique utilized in the Exxon Donor Solvent process, in which the moisture is driven from the combined coal/recycle solvent stream in a high-temperature stirred tank, apparently avoids this loss of reactivity.

The properties of steam-dried low-rank coal are somewhat different. In fact, the original purpose of developing the process was to retain the lump size of coal. Steam-dried lignite exhibits some surface cracks but not the light, flaky appearance of air-dried lignite. The exterior surfaces are tougher, and breakage tests have shown considerably less degradation with steam-dried lignite than with air-dried lignite. However, some conflicting results have been reported that appear to indicate significant temperature effects. Steam-dried coal will not reabsorb moisture from the atmosphere to the same extent as the air-dried equivalent.[3]

3.4.4.4 Current Status

No commercial moisture reduction facilities are in operation in this country on low-rank coals, with the exception of the surface moisture reduction steps in the Centralia coal washing operation, and the Parry driers at Sandow, Texas which have been operating for 25 years. A number of pilot and prototype commercial drying tests have been conducted during the past four decades,[3,4,6,11,13] and the prospect of an upgraded, transportable fuel continues to stimulate a low level of activity.

A detailed description of the unique combustion characteristics of dried Texas lignite at the Sandow plant was given in a 1959 paper.[15]

As a result of the extensive development work undertaken by the Texas Power and Light Company and the USBM to adapt the Parry carbonization process to Texas lignite, and through the efforts of Dr. V.F. Parry and others,[4] the Aluminum Company in 1950 made a decision to construct a primary aluminum smelter on the site of one of the Texas lignite fields. A 300-megawatt steam power plant was to supply the smelter, using carbonized lignite as the basic fuel.

The carbonizing process was planned for two stages of development, the first stage consisting of a drying operation to reduce moisture content from the 28 to 35 percent of the raw lignite to approximately 3.6 percent in the dried product. The second stage would carbonize the dried lignite to extract low-temperature tars, while producing lignite char as the power-plant fuel.

The size degradation of lignite upon heating permitted the use of the concept of suspension burning in a conventional slagging-type furnace. The development of this steam-generating unit was primarily centered around handling a product which was originally coarse ground and dried or carbonized to permit degradation to a relatively small size. The high reaction rate of the dried lignite and lignite char, its relatively small particle size, and the possibility of further size degradation on entering the combustion zone permitted the elimination of the conventional pulverizer in the fuel-burning system.

The dried lignite exhibited very high abrasion when transported in suspension. The original pneumatic transport between the driers and storage silo eroded rapidly and was replaced with a mechanical conveyor system. Some unexpected ash fouling and slagging problems were encountered and largely solved through design modifications. From the description in the paper, these problems appear to be typical of those now considered normal in burning as-mined lignite.

References - Section 3.4.4

1. Leonard, J.W., and D.R. Mitchell (Editors). Coal Preparation, American Institute of Mining, Metallurgical, and Petroleum Engineers, Inc., New York, 1968.

2. Phillips, P.J. Coal Prepartion for Combustion and Conversion, EPRI AF-791, May 1978.

3. Stanmore, B.R. The Steam Drying of Lignite - A Review, preliminary unpublished report prepared under contract to the University of North Dakota, 1979.

4. Parry, V.F., et al. Drying and Carbonizing Fine Coal in Entrained and Fluidized State, Bureau of Mines R.I. 4954, April 1953.

5. Oppelt, W.H., R.C. Ellman, O.C. Ongstad, and W.R. Kube. Experiments in Fleissner Drying of North Dakota Lignite, Bureau of Mines R.I. 5122, April 1955.

6. Oppelt, W.H., W.R. Kube, and T.W. Kamps. Drying North Dakota Lignite to 1,500 Pounds Pressure by the Fleissner Process, Bureau of Mines R.I. 5527, 1959.

7. Stanmore, B.R., and A.R. Boyd. "Removal of Water from Brown Coal by Treatment with Steam". Fuel Processing Technology, 1 (1977/1978), pp. 305-313.

8. (Anonymous). "How to Get Water out of Lignite, Wood, and Peat", Chemical Engineering, March 27, 1978.

9. Duzy, A.F., and L.E. Little. Low-Rank Coals-Moisture Reduction, presented to the 27th Power Technology Conference, Omaha, Nebraska, June 16, 1971.

10. Luckie, P.T., and E.A. Draeger. The Thermal Drying of Western Coals, presented at the AIME/SME Fall Meeting, Salt Lake City, Utah, September 10-12, 1975.

11. Hand, J.W. "Drying of Western Coal", Mining Congress Journal, May 1976, pp. 30-32.

12. Ziesing, G.F., et al. Drying of Low-Rank Coal for MHD Application, Symposium on Technology and Use of Lignite, Grand Forks, N.D., May 1979, DOE GFETC/IC-79/1, 1979, pp. 459-495.

13. Ellman, R.C., L.E. Paulson, and S.A. Cooley. Commercial-Scale Drying of Low-Rank Western Coals; Part I. Rail Shipment Test Observations, in "Technology and Use of Lignite; Proceedings of a Symposium," Grand Forks, North Dakota, May 14-15, 1975, pp. 312-340.

14. Paulson, L.E., S.A. Cooley, and R.C. Ellman. Shipment, Storage, and Handling Characteristics of Dried Low-Rank Coals, in "Technology and Use of Lignite; Proceedings" Grand Forks, North Dakota, May 9-10, 1973.

15. Caracristi, V.Z., and H.D. Mumper. Combustion of Crushed, Dried Texas Lignite and Char in Steam-Power Boilers, ASME Paper 59-SA-36, presented at the Semi-Annual ASME Fuels Division Meeting, St. Louis, Mo., June 14-18, 1959, 15 pp.

3.4.5 Briquetting and Pelletizing

3.4.5.1 Technology Description

Briquetting and pelletizing are processes which convert a low-grade solid fuel into one of superior quality. Lignite or subbituminous coal can be converted from a somewhat friable material with a high water content into a hard, compact briquette or pellet with an increased heating value and better handling and storage properties. (Similarly, bituminous coal fines can be converted into lumps of acceptable size.) The coal may be carbonized (pyrolyzed) prior to briquetting to yield smokeless coke briquettes of still higher quality; or the briquettes made from raw coal may be carbonized for the same purpose. A binder may or may not be required to give the briquette adequate physical strength, depending on the properties of the coal.[1]

The process of briquetting[a] which is applied to many materials in addition to coal, consists of applying pressure to a mass of particles (with or without the addition of a binder) to form a compact or agglomerate. The important and complex physical/chemical interactions involved are the cohesive forces between solids, the adhesive properties of binders, and the rheological behavior of the coal particulate masses and agglomerates. The process has been utilized commercially for a least 240 years, and has been developed to a high level of maturity largely through empirical practice rather than the application of basic physical and chemical theory. The age and maturity of the technology are indicated by the fact that the classical textbook[2] on the subject was written by G. Franke in Germany in 1910 (a 1930 edition is also available).[1]

[a]The term "briquetting" is used throughout this section as a general descriptor for the various coal agglomeration processes, including pelletizing and extrusion.

Because the technology is so well developed and so widely reported in the literature (244 older references are cited in Reference 1 (1953) and 59 recent references are cited in Reference 4 (1979), for example) only a very brief overview of the technology is included here. An attempt has been made to limit the information in this review to that which is pertinent to U.S. low-rank coals. However, this is difficult because of the international scope of the literature and the technology, as well as the fact that the literature specifically dealing with briquetting of U.S. coals is quite sparse. Because of the largely empirical nature of the technology, the optimal techniques, product qualities, and costs of briquetting a specific coal can only be determined through intensive testing of that coal.

An additional reason for limiting the description of briquetting technology in this study is that the production and sale of low-rank coal briquettes in this country are not limited by lack of adequate technology, but rather by market demand for the products. Briquettes are manufactured and used in other areas of the world where adequate or superior substitutes (coking coal, petroleum products, or natural gas) are not locally available at competitive prices. To the extent that these types of market forces develop in this country in the future, known briquetting technology will be applied to coals of various ranks and the demand will be satisfied, as is the case today.

The essential steps in the briquetting process are (not necessarily in the order listed):

1. Crushing and grinding to a typical top size of about 4 mm, with about 50 percent below 1 mm. (The presence of excessively fine material is undesirable.)

2. Drying to an optimum moisture content, which is generally between 12 and 18 percent for low-rank coals. It is very important that the moisture be distributed evenly through the coal particles. (In the pelletizing process, the drying step is applied to the pellets (greenballs) which are manufactured at 40-42 percent moisture.[5])

3. Mixing and heating of the coal and the binder (if any), which is the most important step in the process. European plants predominantly use coal pitch as the binder; American plants generally use bitumen (asphalt) derived from petroleum. The fraction of binder varies, but is generally 5-10 percent of the product by weight.

4. Tempering the mixture (generally, cooling) to the optimum temperature for pressing, which is generally 100-150°C for low-rank coals. This step also allows time for the moisture to become evenly distributed throughout the coal particles (i.e., moisture is transferred from the larger particles to the smaller ones).

5. Pressing the mixture to form the briquette, and cooling and loading the briquettes. The Exter press which is an extrusion device, is used to produce more than 90 percent of the world's output of brown coal briquettes. Many different types of presses are used for bituminous coal briquetting, but the rotary-table and the double-roll types are the most common.

The higher the rank of coal, the harder it becomes, and the more difficult it is to briquette without a binder. To summarize a great deal of literature, brown coal and bituminous coal are members of an essentially continuous series, and the transition in properties from one to the other is a gradual one. It is considered unlikely that the nature of the cohesion between brown coal particles is fundamentally different from that between bituminous coal particles.

The methods which succeed in making hard briquettes are those which aim to bring the particles close together. With bituminous coal, which is harder than brown coal, this is achieved by the adoption of one or more of the following conditions: (1) the use of high pressures; (2) the introduction of sheer strain under load; and (3) the preparation of the coal by fine grinding, by addition of a binder, and by briquetting at the appropriate temperature so that it is in its optimum state of plasticity.

3.4.5.2 Environmental Control Technology

The primary environmental control requirement in a coal briquetting plant is the control of dust emissions in the raw coal storage, drying, comminution, and handling systems. Also, a small water effluent stream may be created in the drying step which requires treatment prior to discharge.

3.4.5.3 Effects of Low-Rank Coal Properties

Brown coal briquettes are generally manufactured without the addition of binder.[1] However, the lignites and subbituminous coals of this country are somewhat higher in rank than the foreign brown coals, and most investigators have found that U.S. low-rank coals produce superior briquettes when a binder is used.[3] Essentially all commercial briquette manufacturers in this country utilize binder; the preferred type of binder is bitumen (or asphalt) derived from petroleum.[1] In 1943, Parry[3] reported that "coal is being briquetted to some extent without binder in the United States and Canada," but that:

> "Binders are used in virtually all American briquetting plants. By employing binders in the raw-coal briquetting mixes, it is possible to employ moderate compacting pressures that contribute to lower maintenance costs than do high pressures. Probably this accounts for the limited work on the briquetting of low-rank coals at high pressure without binder.
>
> Low-rank coals briquetted with binder have better weathering properties than raw coal because of the sealing action of the binder, which repels water and retards loss of moisture. The ability to withstand transportation shock is also generally improved. Achievement of desirable burning properties depends to some extent upon the choice of binder, but even the best binders do not eliminate deterioration in the fuel bed. During combustion the binder should not contribute much additional smoke to that produced by the coal, and it should be of such a nature as to retard disintegration of the briquet into fine coal during combustion."

In his investigation of briquetting of raw and steam-dried (Wyoming) subbituminous coal, Parry[3] also manufactured some briquettes for comparison purposes from Pocahontas No. 4 (bituminous) coal. Although the briquettes manufactured from the subbituminous coal (both raw and steam-dried) were of satisfactory quality for commerical use, Parry concluded that there were:

> "... fundamental differences between briquets made from the two ranks of coal. The results indicate that it is impossible to make briquets from subbituminous coal equal to those made from

low-volatile bituminous, even when more binder is employed. Any investigators attempting to make briquets from lignite or subbituminous coal should remember this. Briquets made from Pocahontas coal improve with age, whereas subbituminous-coal briquets deteriorate; furthermore, so-called good subbituminous-coal briquets do not hold up while burning."

In a more recent (ongoing) experimental effort sponsored by DOE's (formerly Bureau of Mines') Pittsburgh Mining Technology Center, satisfactory pellets have been produced from North Dakota lignite.[4, 5] Binders have been used in all tests; the best quality pellets have been made with an asphalt binder (actually an emulsion of 25 percent asphalt, 4 percent starch, and 71 percent water). The ingredients (90-95 percent raw lignite or char, 5-10 percent asphalt binder) are crushed, ground, mixed, and then pelletized in a 24-inch diameter by 10-inch deep inclined balling disc at 40-42 percent moisture. The greenballs are nominally 1-inch diameter, and are dried at 110°C to an optimum moisture between 10 and 16 percent. Extensive mechanical strength and simulated weathering tests have been performed with encouraging results. Significant reported findings from this first phase of the project were as follows:[4, 5]

1. Pellets of good mechanical strength and weathering resistance can be made from lignite and lignite char with asphalt emulsion binders. Satisfactory pellets may be made with asphalt levels as low as 5 percent. That conclusion, however, is subject to the establishment of standards of quality for lignite pellets.

2. The mechanical properties of the lignite pellets are enhanced when the moisture content of the pellets is maintained at maximum permissible levels.

3. Gilsonite is a potentially effective binder; but, under the conditions employed in this program, yielded more brittle pellets than did the asphalt binder. This may be due to the gilsonite having a higher softening temperature or higher degree of penetration to realize its full potential as a binder.

4. Pellets made from lignite char appear to be equivalent in mechanical properties to pellets made from lignite. In terms of weathering resistance the char pellets appear to be somewhat better than the raw lignite and appear to have great potential as a smokeless fuel.

5. A fine grind, such as -200 mesh that is required for iron ore, is not required for pelletizing lignite. A satisfactory size consist appears to be -10 mesh with about 50 percent within the -10 +50 mesh range and 20 percent at -200 mesh.

6. Drying lignite pellets with about 40 percent greenball moisture first increases and then decreases compression strength upon drying to zero percent moisture. Maximum strength is not at zero percent moisture. This is unique because in most pelletizing applications the highest strength is achieved at or near zero percent moisture. Pellets at 10 percent moisture are significantly stronger than pellets at zero percent moisture.

7. Slow drying generally results in stronger pellets than quick drying.

8. Lignite tars produced by pyrolyzing lignite at temperatures within the range 700 to 1000°F are not good binders for lignite pellets. A commercial grade of lignite tar obtained

from a North Dakota plant likewise is not a good lignite binder.

9. In addition to pelletizing, lignite can be agglomerated by briquetting and by extrusion.

   In terms of product quality and processing cost, extrusion appears to be slightly better than briquetting.

The second phase of the project, currently underway, consists of:[5]

- Confirmation of laboratory results on successively larger scale pelletizing equipment.
- Exploration of optimum lignite crushing and pellet drying equipment.
- Production or larger quantities of product for evaluation and to demonstrate consistency and repeatability of results.
- Based on data from the pilot work, the design of a 4,000 T/D lignite pellet plant.
- Analysis of the capital and operating costs of the 4,000 T/D plant together with the results of the pilot scale work.

3.4.5.4 Current Status

Briquetting of low-rank coals has been a widely used commercial technology since the 19th century. In 1919, over 32 million tons of coal briquettes were produced worldwide (almost all in Europe), approximately 2/3 of which were produced from brown coal. In 1958, worldwide production had grown to almost 94 million tons (including 955,000 tons in the U.S.), with the proportion of brown coal briquettes still about 2/3 of the total. At that time, East Germany was by far the largest producer, with over 53 million tons of brown coal briquettes.[1]

The market for this product is primarily the small solid fuel user, who requires a stable, easy-to-handle substance that burns cleanly and (preferably) without smoke. For example, 80 percent of the 1958 briquette production in Germany went to the domestic market, and the remainder was sold to industry.[1]

In the United States at the present time, there are only two known commercial briquetting operations using low-rank coals. Husky Corporation manufactures charcoal briquettes from lignite in North Dakota for home barbeques. The FMC Corporation has a briquetting operation in Kemmerer, Wyoming, in which briquettes are made from subbituminous coal char for metallurgical use.

The process scheme in the Husky briquetting plant at Dickinson includes carbonization in two Lurgi carbonizers, followed by pyrite separation, grinding, mixing, briquetting, drying, and bagging.[6] During the winter (prior to 1964), pitch and asphalt binder were added in the mixing step to produce fuel briquettes. The Komarek Greaves briquettes press handled the wet mix at a rate of 18 tons per hour. During the rest of the year, barbecue briquettes were made by mixing the char with starch binder and water. Since that time, the market for fuel briquettes decreased to the point where the plant now produces barbecue briquettes only.

The FMC coke process is a continuous process for producing form coke of uniform size and quality that is well-suited for metallurgical uses from either coking or non-coking coals.[7]

In 1960 a demonstration plant was erected near Kemmerer, Wyoming, approximately one mile from the Elkol mine. The plant produces about 200 TPD of coke from mine run coal.

The product from this plant has been tested in electric elemental phosphorus and calcium carbide furnaces and in both pilot and commercial pig iron blast furnaces. The pilot blast furnace test was made during a joint venture arrangement with U.S. Steel Corporation.

During the early operation of the plant numerous mechanical and some process scale-up problems were experienced. These problems were solved, proving the process operationally sound.

Although a plant of this size is too small for economic operation for merchant coke, the plant is currently being operated to supply part of FMC's coke demand for phosphorus. Conditions peculiar to this phosphorus plant make the operation justifiable.

Ground coal is pyrolyzed in a fluidized bed reactor at successively higher temperatures under controlled atmospheric conditions. The purpose of the various steps is to destroy any coking and agglomeration tendency, to expel tar which is recovered for use as binder, and to produce a calcinate with less than 3 percent volatile matter.

The briquettes formed from calcinate and air blown tar are heat treated in an oxidizing atmosphere at about 450°F on a traveling grate. The reaction is a combination of polymerization and dehydrogenation reactions that result in a binder carbon in the finished coke having about the same chemical reactivity as the calcinate and, therefore, permitting the coke to be consumed in any application without disintegration of the briquettes.

The cured briquettes are devolatized in a shaft kiln at a temperature of about 1600°F to reduce the volatile matter content to about 3 percent. The remaining volatile content in the finished coke consists of hydrogen, carbon dioxide, carbon monoxide and nitrogen. The coke contains no condensables.

The overall residence time of the coal in the process amounts to about six hours, excluding intermediate storage between the pyrolysis and the briquetting section.

More than fifty different coals from throughout the world have been evaluated in FMC's bench-scale or pilot plant equipment. These coals have ranged from anthracite to lignite. With anthracite it is necessary to use a binder from an outside source and with lignite the reactivity of the calcinate is so great that very special precaution has to be taken in curing and coking steps. However, with medium volatile bituminous coals to high volatile subbituminous coals FMC has demonstrated that adequate tar can be obtained to produce a strong form coke.

<u>References - Section 3.4.5</u>

1. Rhys Jones, D.C. "Briquetting," in Lowry, H.H. (editor), <u>Chemistry of Coal Utilization; Supplementary Volume,</u> John Wiley & Sons, Inc. New York, 1963, pp. 675-753.

2. Franke, G. <u>Handbuch der Brikettbereiting,</u> Euke, Stuttgart, 1910. (trans. by Loutsberry, Chas., Griffin Co., Ltd., London, 1916.) (as cited in reference 3).

3. Parry, V.F., and J.B. Goodman. Briquetting Subbituminous Coal, Bureau of Mines R.I. 3707, June 1943.

4. Goksel, M.A., and L. Valentyik. Production and Evaluation of Lignite Pellets, presented at the Institute of Briquetting and Agglomeration 16th Biennial Conference at San Diego, California, August 6-8, 1979.

5. Grant, A.J., and R.E. McKeever. Pelletizing and Drying of Lignite, presented at "Coal Technology '78", Houston, Texas, October 1978.

6. Henderson, C. and S.V. Margolin. Processing of Lignite to Commercial Products, in Technology and Use of Lignite: Proceedings of Bureau of Mines - University of North Dakota Symposium, Grand Forks, N.D., April-May 1963, BuMines IC 8234, 1964, pp. 40-47.

7. _______, FMC Coke Process, FMC Corporation brochure, 1970, on file at GFETC.

3.4.6. Storage

3.4.6.1 Technology Description

Coal is stored at several points between the mine and its eventual use: at the mine loading terminal, at rail/barge transhipping terminals and at the offloading terminal, which is often a powerplant. Coal storage acts as a buffer against various potential forces that cause fluctuations in supply such as:

1. Transporation Modes - larger shipments are delivered less frequently and from greater distances. Disruption of a single shipment could stop plant operation without proper storage.

2. Adverse Weather - severe weather conditions may cause interruption in transport of coal to user. Sufficient storage allows plant to continue operating.

3. Labor Strikes - effects of mining or transporation industry strikes can be minimized via satisfactory long-term storage.

4. Price Fluctuations - while this element may not affect continuity of supply, storage capabilities can help allow purchase of coal at best prices.

In general, tranport mode and weather-related problems can be addressed in a matter of days, thus requiring only minimal storage. However, the storage capacities required to adequately address the problems of labor disputes are significantly greater than the capacities required by common operational problems.

The amount of coal that is stored is strictly an economic decision. In the case of a powerplant, the decision must be made between the cost of storing the coal and the potential cost of shutting down the plant and purchasing electricity elsewhere. The cost of storing coal has several elements including the cost of the raw coal, cost of storage facilities and equipment, cost of product degradation and the opportunity cost of capital tied up in the coal. Cost elements of shutting down the plant and purchasing electricity elsewhere depend on a large number of plant factors as well as the availability and spot price for electricity when it is needed.

Coal can be stored in stockpiles exposed to the elements or it can be stored in silos or bins completely or at least partially protected from the environment. Stored coal is termed either "live" or "dead" storage. Live storage refers to the portion of the stored coal that is used within a relatively short time to smooth out supply variations. Dead or long-term storage is coal that is only used as a source of supply in the event of emergency supply problems.

The objective of any coal storage system is to minimize to the greatest extent possible product degradation and spontaneous combustion. Product degradation results from oxidation, weathering, slacking and windage loss or any combination of the above. Stockpiling techniques are formulated to minimize these occurrences.

Spontaneous combustion results from local oxidation rates exceeding heat dissipation rates so that local temperatures exceed ignition points. Techniques to avoid spontaneous combustion are aimed at sealing the interior of the pile from available oxygen. These techniques are generally similar to those that address product degradation.

Stockpiles

There are three basic types of coal stockpiles; conical-shaped, wedge-shaped, and kidney-shaped. The most common stockpile is the conical-shaped configuration. The pile is stocked by a fixed cantilevered boom conveyor that feeds a telescoping chute. The purpose of the chute is to minimize dust emissions. The pile can be constructed on flat ground and the coal used as dead storage; or it can be constructed above an underground reclaiming facility, in which case part of the pile is used as live storage and part as dead storage.

One of the major disadvantages of the conical-shaped pile is that only about 20 percent of the total can be used as live storage when a single reclaim chute at the center is used. Live storage can be increased to about 55 percent by placing a reclaim device across the entire diameter of the pile rather than only at its center.

Generally, compaction procedures are not practiced with this type of storage pile.[1] Since the pile is generally used in the part live, part dead storage configuration, spontaneous combustion does not tend to be a problem. For low-rank coals, this pile configuration may well present spontaneous heating problems, especially if it were used for dead storage only. Where a pile is formed by allowing unsized coal to fall, roll or slide, a natural size segregation occurs. The larger sizes fall to the outside and bottom, and the fines collect in the interior and the top. Air moves easily through the coarse lumps and begins reacting with the fines in the interior. With highly reactive coals, there is a tendency to experience local "hot spots" where the heat generated by the oxidation reaction is greater than the local heat dissipation.

The wedge-shaped stockpile has the greatest capacity of any configuration. There are basically two types of wedge-shaped stockpiles: 1) that used for short-term storage (i.e. part live, part dead), and 2) that used for long-term storage (i.e. all dead). In the first type of stockpile, relatively little compaction is carried out because nearly 60 percent of the coal is live storage. Another use for the wedge-shaped pile with multiple reclaim chutes is the storage of different grades of coal along the length of the pile. By selective opening and closing of the reclaim chutes various coal blends can be obtained. Use of this type of pile is found primarily in shipping terminals where large quantities of coal must be reclaimed in short time intervals to load a unit train or barge.

The second type of wedge-shaped stockpile is generally employed by the end user for long-term dead storage. These storage piles are carefully compacted and sealed for protection against deterioration and spontaneous combustion. This is accomplished by spreading the coal in approximately 1-foot deep layers at a time, and thoroughly packing each layer to eliminate air spaces. The top is slightly crowned and symmetrical to permit even runoff of water. With bituminous coals, the sides and top are covered with a 1-foot compacted layer of fines and then capped with a 1-foot layer of sized lump coal. It is not practical to seal low-rank coal piles with coarse coal, since it will weather and slack to small size in a short period of time. Instead, the top and sides of the pile are compacted with slack-size coal. Drift (snow) fences are often installed on the piles to avoid drifting of fines, and the top and sides of the pile are periodically trimmed.[9]

Regardless of the coal stockpiled, the use of material, such as asphalt, for airtight sealing is not recommended. Capping methods described above have proved effective. Sealing with asphalt or road-tar coating may be conducive to self-ignition in those areas near breaks that could cause a "stack effect." Also, the materials are an additional expense, and they may prove difficult to handle or pulverize when the stockpile coal is used.[9]

A third type of stockpile is the kidney-shaped pile. This type of pile is built by a radial stacker that rotates about a fixed point. The kidney-shaped stockpile can be built above a reclaim facility that maximizes the live storage portion of the pile.[1] Another method of reclaiming is through the use of a rotary bucket stocker-reclaimer. This device is essentially a radial stacker with a rotary bucket attached to the end of the boom. For reclaiming, the bucket is activated and coal is recovered from the pile and deposited on the conveyor belt whose direction has been reversed. While this type of stockpile can be used for dead storage, this is generally not done because the capital cost of a radial stacker cannot justify long periods of idleness.

Enclosed Storage

It is sometimes desirable to use an enclosed bin or silo rather than an open stockpile. The advantages of enclosed storage are:

1. reduced fugitive dust emissions

2. reduced product degradation

3. reduced spontaneous combustion

4. reduced handling problems - especially frozen coal.

Enclosed storage facilities are used primarily by shipping terminals to speed loading procedures. Coal storage bins and silos are constructed of either steel or concrete although, in the large sizes, steel structures have not proven as economical as concrete. The dimensions of concrete silos have responded to demands for larger capacities and to developments in the state of the art in construction techniques. The jump form is used when building silos up to 45 ft. in diameter, whereas silos up to 70 ft. in diameter use slipforms. The capacity of a 70 ft. silo, depending on its height, is 10,000 to 15,000 tons.. Storage silos can be classified as either mass-flow bins, or funnel-flow bins. In mass-flow bins, all the solid is in motion whenever any of it is drawn out; the solid is utilized in a first-in, first-out sequence.

Most existing bins are of the funnel-flow type. The solid flows toward the outlet though a channel extending upwards from the feeder or grate. The channel expands from the outlet to a circular shape and is surrounded by non-flowing solid. As the solid flows out, layers of the non-flowing solid fall into the dropping channel. Such a pattern leads to a first-in, last-out sequence of flow because the solid which was first deposited at the bottom of the bin around the channel does not discharge until the bin is finally emptied out. Funnel-flow binds are acceptable for coarse, free-flowing, chemically stable solids which do not segregate.

3.4.6.2 Environmental Control Technology

Control of fugitive dust emissions from coal storage facilities is accomplished as a routine part of the operation. Dust control depends on the use of proper compaction techniques, covered or hooded transfer points, water or oil sprays, and other such techniques.

3.4.6.3 Effects of Low-Rank Coal Properties

Several properties associated with low-rank coals may have a distinct impact on the storage of these coals. These properties are: 1) high reactivity; 2) size consist; and 3) moisture content. Despite the difficulties imposed by these coal properties, both lignite and subbituminous coals can be stored for long periods provided proper procedures are followed in preparing and maintaining the stockpile.[2] It is in the "live" storage or short-term storage of coal, where compaction and sizing practices are not used, that low-rank coal properties can have their greatest impact.

High Reactivity

The strong tendency of low-rank coals to oxidize in comparison to bituminous coals can result in extensive product degradation, and possibly in spontaneous combustion, during storage. Figure 3.4.6.1 shows the tendency of low-rank coals to oxidize at higher rates than coals of higher rank. These curves were generated under laboratory controlled conditions. As shown, the low-rank coals heat considerably faster. In addition (for all coals), the rate of oxidation (slope of the curve) increases with increasing temperature.

Figure 3.4.6.1

Heating Rates of Coal in Storage

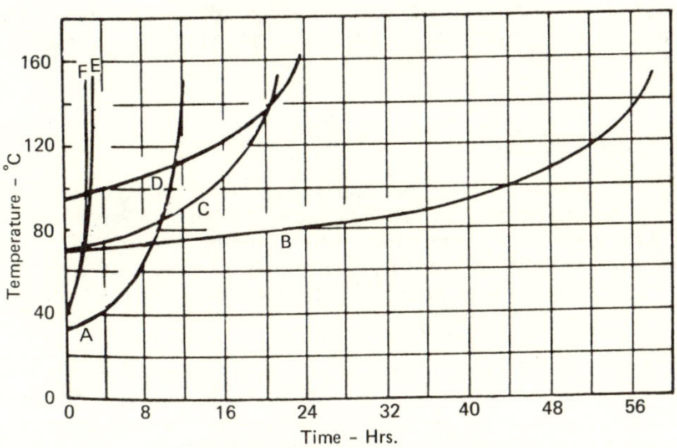

A = Alaska Subbituminous C
B = Pittsburgh Seam (High Vol. A Bit.)
C = Elkhorn No. 3 (High Vol. A Bit.)
D = Eagle Seam (High Vol. A Bit.)
E = North Dakota Lignite
F = Wyoming Subbituminous B

Source: Reference 3

Thus if adequate oxygen is supplied to the low-rank coal surface, oxidation will take place at rates appreciably higher than those found for bituminous coals. If oxygen is supplied to the coal surface and the heat generated by the exothermic reaction is not dissipated at an equal rate, the local temperature will begin to rise resulting eventually in spontaneous combustion.[1]

Size Consist

The slacking nature of low-rank coals tends to exacerbate the potential fugitive dust emissions and the product degradation via surface oxidation resulting from newly exposed surface area.

The dustier character of low-rank coals increases the concerns that must be given to dust control. Fugitive dust emissions were minimized in one long-term storage test by constructing a drift fence perpendicular to the prevailing wind direction.[2] This approach may or may not prove adequate for future storage sites depending on local emission regulations. Dust emissions are a particularly important element of live storage, since the coal is generally scheduled for more movement and handling. One attempt at reducing dust emissions used No. 6 oil as a spray coating and experienced significant reductions in dust generation.[4]

An additional potential problem resulting from the slacking tendency of low-rank coals is that oxidation can be increased by the generation of new surface area. If the slacking occurs in a region where oxygen is readily available to the coal surface, oxidation will undoubtedly result. The tendency of a coal to self-heat has been shown to be directly proportional to its exposed surface area.[5]

Moisture Content

Spontaneous heating in storage piles is believed to be affected by moisture content of the coal, although specific functional relationships have not been formulated. One laboratory investigation showed that oxidation rates between 25 and 95°C are maximinum near the equilibrium moisture content of about 20 percent for lignite dried in an inert atmosphere. The rates at 5 and 36 percent coal moisture were about half that at 20 percent moisture. The heat of condensation of enough water to raise a piles moisture content from 3 to 4 percent can also cause a 30°F rise in temperature in the pile.[5]

3.4.6.4 Current Status

Storage of western low-rank coals has been approached by industry in a very positive fashion. Although some technical problems exist as discussed in the preceding section, the understanding of their basic mechanisms is sufficient to design and operate satisfactory storage facilities. A few examples of typical low-rank coal storage systems follow:

The 400 MWe San Miguel Power Plant in Atascosa County, Texas, was scheduled for initial operation in December 1979. Live storage is contained in a 14,000 ton stockpile and in two mass flow-type concrete silos with a capacity of 28,000 tons, or about two days operating requirements. The silos have been carefully designed to eliminate common flow and storage problems such as bridging, rat-holing, and spontaneous combustion. Dead storage will contain 250,000 tons, sufficient for about 18 days coal supply.[6]

The 440 MWe Big Stone Plant in South Dakota will receive lignite via unit train consisting of specially designed covered gondola cars. Figure 3.4.6.2 shows a simplified schematic of the unloading and storage configuration. As shown, live storage will be in a pit with an integral reclaim system. Dead storage sufficent for 30 days operation (or 265,000-tons) can supply either the live storage or the plant directly.[7]

The Coal Creek Station is a two-unit 1100 MWe minemouth plant in Central North Dakota. About 125,000 tons of lignite will be kept in live storage piles in addition to another 50,000 tons live storage in three storage silos. Dead storage of 1 million tons (about 60 days supply) will be maintained for emergency supply interruptions. Both storage piles will use combination stacker/reclaimers under normal circumstances. However, emergency recovery hopper systems are installed for use in extreme circumstances.[8]

References - Section 3.4.6

1. Landers, W.S., and D.J. Donavon. "Storage, Handling, and Transportation," in Lowry, H.H. (editor, Chemistry of Coal Utilization, Supplementary Volume, John Wiley and Sons, New York, 1963, pp. 296-311.

2. Ellman, R.C., et al. "Long-Term Storage of Lignite at Garrison Dam, Riverdale, North Dakota," U.S. Bureau of Mines, RI 7037, October 1967.

3. Leonard, J.W., and D.R. Mitchell, (Editors). Coal Preparation, American Institute of Mining, Metallurgical, and Petroleum Engineers, Inc., New York, 1968.

4. Ellman, R.C., L.E. Paulson, and S.A. Cooley. Commercial Scale Drying of Low-Rank Western Coals; Part I. Rail Shipment Test Observations, in "Technology and Use of Lignite; Proceedings of a Symposium," Grand Forks, North Dakota, May 14-15, 1975, pp. 312-340.

5. Sondreal, E.A., and R.C. Ellman. Laboratory Determination of Factors Affecting Storage of North Dakota Lignite: Computer Simulation of Spontaneous Heating, U.S. Bureau of Mines, RI 7887, 1974.

6. Gee, L.S. Utilization of Low-Grade Lignite, "Gulf Coast Lignites Conference," Houston, Texas, May 17-19, 1978.

7. Johnson, O.B., and R.F. Middleton. Big Stone Plant: Design Features and Fuel Handling, in "Technology and Use of Lignite; Proceedings," Bureau of Mines, I.C. 8650, Grand Forks, North Dakota, May 9-10, 1973, pp. 34-48.

8. Hickock, W.W., et al. Coal Creek Station Steam Generators, in "Technology and Use of Lignite; Proceedings," GFERC/IC-77/1, Grand Forks, North Dakota, May 18-19, 1977, pp. 43-73.

9. Babcock & Wilcox Company. Steam/Its Generation and Use, New York, 1978.

3.4.7 Blending

The blending of two or more coals to achieve a specific end product is commonly practiced. For utility applications, low-sulfur western coals have been combined with high-sulfur eastern bituminous coals to produce a blend whose sulfur content meets prevailing regulations. The steel industry is perhaps the most familiar with coal blending to produce higher

Figure 3.4.6.2

Loading, Unloading, and In-plant Handling of Lignite At the Big Stone Plant, South Dakota

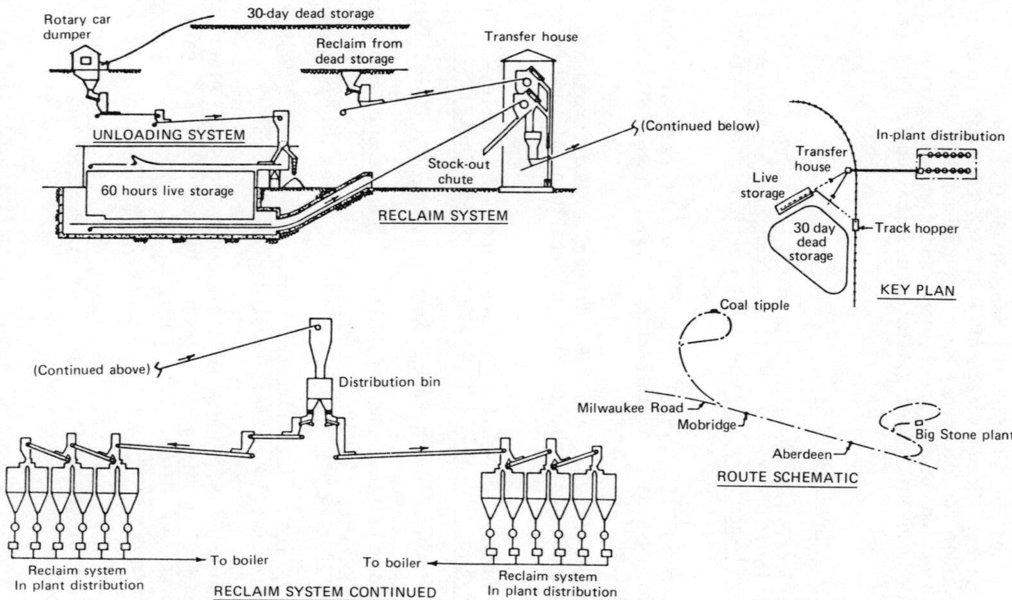

Source: Reference 7

quality coke. While blending may be practiced for a number of reasons, the general objective is to impart some characteristic(s) to the end product by blending two or more coals that do not separately possess that characteristic. Various characteristics or qualities that might be desired as a product of blending might be minimum variability in physical or chemical properties such as:

1. Minimum property variation

2. Mineral matter (sulfur, ash or sodium)

3. Grindability

4. Heating Value

5. Ash-fusion temperature or slag viscosity

Coal blending is accomplished by an organized control of coal mining, monitoring and recovering so as to produce the desired end product. Equipment that is used for blending is comprised of standard stacking and reclaiming systems currently used extensively in storage yards. The best illustration of a modern blending operation is given through example.[1]

The Navaho mine in the Four Corners area of New Mexico supplies all coal to the Arizona Public Service Company's Four Corners Power Plant. The strip mine, owned and operated by Utah Construction and Mining Company, supplies the plant annually with approximately 2-1/2 million tons of subbituminous coal that contains an average heat content of 9,000 Btu/lb. Because of conditions within the mine, the quality of the run-of-mine coal can vary between 7,000 and 10,200 Btu/lb. This variation in calorific content would cause serious operational and economic problems to the power plant and had to be avoided.

The answer was found in an extensive, fairly sophisticated, and highly automated coal blending system which handles all coals, after they are crushed at the mine to a -3/4 inch product. The blending system includes ten separate storage piles arranged so that each completed pile is approximately 800 ft. long, 90 ft. wide, 32 ft. high, and accommodates 30,000 tons of coal. A pile is built by means of a double-wing stacker which continually travels the length of the pile while depositing in horizontal layers a continuous stream of coal received from the mine over the conveyor system. To control the quality of blended coals, a mine engineer at the mine schedules the two loading shovels in the pits in an effort to have one shovel in a face where the calorific value is higher than average and the other shovel in a face where it is lower than average. As a pile is being built, a running inventory is kept of the grade. If the grade varies too much from the 9,000 Btu/lb average, the engineer can revise the loading schedule or direct the coal into another pile. Once a pile has been built, it can be scheduled for reclaiming at the convenience of the plant. Because of climatic conditions and the reactive nature of the coals, the practice has been to reclaim each pile not later than two weeks after its completion. This practice has circumvented the problem of fires caused by spontaneous combustion, yet it allows for a "live" storage capacity of up to 240,000 tons of coal. This storage capacity represents a ten-day supply if all generating units operate 100 percent.

The reclaiming is accomplished by either one of two bridge-type bucketwheel reclaimers, each equipped with two 25 ft. diameter bucketwheels supported from and transversing on an inverted triangular-shaped truss that is carried by propelling trucks moving on rails paralleling each side of the blend piles. All movements of the reclaimer, such as wheel rotation, traverse, and advance, are automated and adjustable over a wide range.

The coal delivered to the Four Corners Power Plant must meet specifications covering: minimum calorific value, maximum volatile matter, maximum ash, maximum moisture, maximum alkalis, maximum grindability, and maximum ash-fusion temperature. Experience has shown that regulation of the Btu content will cause all other specifications to fall substantially within specified limits. Carefully kept records between 1964 and 1969 indicate that the average fluctuation of daily heating value of the blended coal was only about 1.7 percent (150 Btu/lb) with some rare excursions up to 5.5 percent.

Another example of coal blending to meet user specifications occurs in Hanna, Wyoming, where Energy Development Company's new 180 tph preparation plant cleans the entire production of an adjacent underground mine and subsequently blends the cleaned product with surface mined coal. The cleaning and blending operation improves the underground coal by increasing its heating value by up to 500 Btu/lb and decreasing the ash content to 1.2 percent or less. The product is then shipped to Iowa Public Service Company where it makes up 30 percent of their coal needs.[2]

References - Section 3.4.7

1. Phillips, P.J. Coal Preparation for Combustion and Conversion, EPRI AF-791, May 1978.

2. Jackson, D. "Wyoming Gets Its First Prep Plant," Coal Age, November 1978, p. 81.

3.4.8 - SELECTED REFERENCES

The following publications are particularly informative on the subjects of coal preparation, handling, and storage in general, or on the effects of low-rank coal properties on those processes. Many other publications, as listed under References at the end of each section, provide data on specific subjects within these technical areas.

1. Leonard, J.W., and D.R. Mitchell (Editors). Coal Preparation, American Institute of Mining, Metallurgical, and Petroleum Engineers, Inc., New York, 1968.
   This is a comprehensive text and reference manual on all aspects of coal preparation technology, with individual chapters written by recognized experts in the field. Both theory and practice are covered in detail, supported by extensive use of photographs, tables, and charts. Essentially all the data and discussions pertain to the use of bituminous coals.

2. Phillips, P.J. Coal Preparation for Combustion and Conversion, EPRI AF-791, May 1978, 364 pp.
   This report covers the technology of steam coal preparation by stressing topics of particular interest to the electric utility industry. It covers the full scope of coal processing, from mine face to post-combustion stack gas clean-up, and provides essential information for assessing the potential contribution of physical (as opposed to chemical) coal beneficiation to a utility's fuel procurement and utilization strategy. A methodology is presented for quantifying direct costs corresponding to six different levels of preparation, ranging from mere rubbish removal from ROM coals to intensive beneficiation of crushed and sized coals in prepared media. Some data are provided relative to low-rank coal preparation.

3. Ellman, R.C., J.W. Belter, and L. Dockter. Lignite Pulverization: A Status Report, in "Technology and Use of Lignite - Proceedings: Bureau of Mines - University of North Dakota Symposium, Grand Forks, ND, May 27-28, 1967," Bureau of Mines Information Circular 8376, May 1968, pp. 29-39.

The history of commercial lignite pulverization is reviewed, and comments are made concerning past and present problems. Research at the Grand Forks Coal Research Station (now Energy Technology Center) is summarized, including grindability studies and pilot plant scale as well as commercial sized equipment tests. Variations in pulverization within and between seams are noted. The techniques of predrying or increasing the degree of in-the-mill drying increase the capacity of pulverizers and reduce power requirements.

4. Rogers, S.E., and A.W. Lemmon, Jr. (Editors). Proceedings: Symposium on Coal Cleaning to Achieve Energy and Environmental Goals, Hollywood, Florida, EPA-600-7-79-098a and b, September 1978.
The proceedings document a total of 49 presentations covering the physical and chemical coal cleaning programs of EPA, DOE, the Electric Power Research Institute, and numerous industrial organizations; European and Soviet plans for the future; and problems of ongoing operations. The proceedings include the following topics: coal characteristics, coal cleaning overview, physical coal cleaning technology, environmental assessment and pollution control technology, and chemical coal cleaning technology. The first three topics are covered in Volume I; the last two, in Volume II.

5. Paulson, L.E., and R.C. Ellman. Reduction of Sodium in Lignite By Ion Exchange: A Pilot Plant Study, GFETC/RI-79/1, 1979, 50 pp.
The report documents the results of pilot plant tests on the variables in removing sodium from lignite by ion exchange. A continuous processing vessel with countercurrent flow of lignite and treating solution was used to simulate a possible commercial technique. Data show that sodium was reduced in 1/2 by 0-inch lignite from 7 percent (as Na_2O in ash) to 3-4 percent by contact with a $CaCl_2$ solution for several hours; sulfuric acid was also used successfully as the treating solution. Boiler fouling tendencies using ion-exchanged lignite were markedly reduced.

6. Duzy, A.F., et al. Western Coal Deposits; Pertinent Qualitative Evaluations Prior to Mining and Utilization, in "Technology and Use of Lignite, Proceedings" Grand Forks, North Dakota, May 18-19, 1977, pp. 13-42.
Coal exploration programs and analytical techniques for evaluation of Western U.S. coal deposits are discussed. Included in the programs are drilling density, desirable analyses, testing of coals and impurities, potential coal beneficiation, and problems associated with evaluations for efficient utilization. The discussion of potential coal beneficiation provides an excellent overview of the problems specific to low-rank coals.

7. Paulson, L.E., S.A. Cooley, and R.C. Ellman. Shipment, Storage, and Handling Characteristics of Dried Low-Rank Coals, in "Technology and Use of Lignite; Proceedings," Grand Forks, North Dakota, May 9-10, 1973, Bureau of Mines IC 8650, 1974, pp. 49-75.
The report documents a program in which rail car quantities of lignite and subbituminous coal were dried, transported, and then stockpiled for several years. Characteristics of the dried coals are described, and techniques for proper shipment, storage, and handling of the dried coals are defined.

8. Rhys Jones, D.C. "Briquetting," in Lowry, H.H. (Editor), Chemistry of Coal Utilization; Supplementary Volume, John Wiley & Sons, Inc., New York, 1963, pp. 675-753.
This is a comprehensive article on the technology of coal briquetting as practiced throughout the world. The different techniques utilized for bituminous coals and brown coals are described in detail. A small amount of information is included on the briquetting of U.S. low-rank coals.

Processing and Utilization

3.5.1 Conventional Combustion

3.5.1.1 Introduction and Summary

The direct combustion of coal currently accounts for the largest consumption of coal in the United States, and will continue to do so for some time to come. The primary use of this energy source is for steam generation, which is used directly and indirectly in industrial processes, and by utilities for electric power generation.

The three primary methods of burning coal are pulverized coal firing, cyclone firing and stoker firing. Pulverized coal furnaces are in widest use among utilities and represent the most modern design for this application. Cyclone furnaces were introduced because of their ability to burn coals having low ash fusion temperatures and to recover a high percentage of the coal ash as slag instead of allowing it to escape the combustion section and form deposits on boiler tube surfaces. The high operating temperatures required in cyclone furnaces favor the formation of nitrogen oxides and, largely for this reason, no cyclone units have been installed in recent years. Stoker firing is generally limited to smaller applications of less than 100,000 lb/hr steam. All three combustion techniques represent established technology and are therefore referred to as methods of conventional combustion.

Environmental control technology is a major area of importance in conventional combustion. A considerable portion of the capital cost of new electric utilities is devoted to systems which maintain emission controls over stack gases, wastewater, solid waste and fugitive emissions. Stack gas cleaning methods are oriented toward limiting emissions of particulate matter, sulfur compounds and nitrogen oxides.

The special properties of low-rank coals influence virtually all aspects of direct combustion. Of primary importance are the high moisture levels (low heating values) and alkaline contents typical in low-rank coals. High sodium levels in low-rank coal ash exacerbate the ash fouling of boiler tubes by creating hard, tenacious deposits. The sodium tends to volatilize in the high temperature zone of the furnace and acts as a fluxing agent for other ash constituents, causing them to melt and deposit on tube surfaces. Coals having high sodium contents also tend to produce a fly ash fraction rich in sodium sulfate and having a very small particle size.

The high silica levels found particularly in Texas lignites cause or aggravate a number of problems. The abrasive characteristics of silica accelerate erosion rates of coal feeding systems and furnace burners. High-silica coals promote rapid deterioration of fabric filters. When combined with high sodium levels, high silica contents can result in hard, massive deposits on boiler heat transfer surfaces.

Some minerals found in low-rank coals are beneficial to power plant operation. High calcium and magnesium levels tend to mitigate the effects of sodium fouling. In addition, the low sulfur levels typically found in low-rank coal reduce the extent of flue gas desulfurization problems, especially when coupled with the high alkalinity of low-rank coal fly ash. Alkaline fly ash scrubbing systems have been installed on a number of commercial utility boilers using low-rank coals. Disposal of the scrubber sludge may be a problem due to the tendency of some of the materials to be leached when contacted with ground waters.

The higher reactivity of low-rank coals aids combustion but can be a problem because of its tendency to spontaneously ignite during transport or storage. Another useful characteristic of low-rank coals during combustion is that they do not agglomerate as do many higher rank coals. This allows a larger particle size while still being able to assure complete burnout. The inherent high resistivities of some low-rank coal fly ashes make collection of these ashes with electrostatic precipitators difficult, although coals with high sodium levels produce ash with much lower resistivities. These problems, together with the 1979 NSPS, are forcing utilities to lean towards use of baghouses, fly ash conditioners, or new concepts in control techniques to meet particulate standards.

The high moisture content of low-rank coals necessitates the use of hot primary air for drying in the mills prior to combustion. High surface moisture also can cause problems in the coal feed systems.

A major disadvantage of low-rank coal is its low heating value. This increases the tonnage of coal pulverized and burned for a given power output as compared to higher-rank coals. Hardware changes, such as more and larger mills for pulverization and larger furnaces for combustion, are required. Additionally, the handling capacities for coal transport and stack gas cleanup must be correspondingly larger. Retrofit of boilers from higher-rank coals to lower-rank coals usually means a significant derating of output.

Currently, there are over 29,000 megawatts of generating capacity in the U.S. based on low-rank coals. The vast majority of these plants are pulverized-coal-fired with a very small number of cyclone and stoker units. Most of these plants utilize electrostatic precipitators for their particulate removal requirements, and those that require sulfur removal primarily use limestone wet scrubber systems. Most planned units or those presently under construction are pulverized-coal-fired units. Their modes of environmental controls have shifted somewhat, however. A much larger number of baghouses will be seen and sulfur removal techniques are beginning to shift to ash alkali and dry scrubber systems. The use of low-rank coals for electric power generation will double by the mid-to-late 1980's.

From study of the special problems relating to combustion of low-rank coals, a number of key issues have been identified for RD&D work. These issues have been subdivided into two sections; those that pertain to the combustion process and those that pertain to environmental control technology.

Key Issues in Combustion Processes

1. Increase overall boiler availability by increased understanding of ash fouling and slagging.

2. Substitute coal for oil by direct ignition of coal.

3. Improve stoker furnaces for small scale operations.

Ash Fouling and Slagging

Ash fouling of heat transfer surfaces is the most serious operating problem of boilers fired on low-rank Western U.S. coals.[1] Problem coals cause rapid ash deposition which can force repeated unscheduled shutdowns. During difficult operating periods, the capacity of the boiler can be lowered by as much as 10 to 20 percent, the thermal efficiency reduced by 10 percent, and 1 percent of the steam generated may be used to operate on-line cleaning devices. The number of operating days per year can be reduced by 10 percent. The economic penalty for forced outage of a large boiler is very severe, amounting to over one hundred thousand dollars per day in lost revenue for a 500 MW unit. These costs have prompted extensive research on ash fouling, starting in the 1950s and continuing today. Extensive work has been done on high-fouling coals in the United States, Canada, Australia, England, and Germany. Substantial progress has been made in understanding ash fouling. However, the phenomena involved are complex, and more research is required to achieve control of the problem.

The ash fouling process is determined mainly by the properties of the coal ash. However, the instantaneous fouling rate is influenced by numerous operating variables, not the least of which is the cleanliness of the boiler resulting from previous operation under fouling conditions. To the observer, the progress of fouling often seems erratic because all the determining factors cannot be closely followed.

For a constant boiler load and uniform fuel quality, the course of fouling after a complete boiler cleanup is typically represented by a slowly accelerating fouling rate. Deposits are cyclically built up and partially removed by soot blowers, but there is a slow net accumulation. As accumulation proceeds, temperatures in the boiler are increased. The fouling rate in turn increases along with the temperature, so that the process is accelerated and fouling is moved up and out of the furnace proper and through the convection sections of the boiler. At some point, the temperature of the surface of the ash-laden furnace walls may exceed the fusion temperature of the ash, and heavy slagging will occur.

Any furnace variable that affects the burning rate of the coal can also influence fouling. For example, a coarse grind in a pulverized coal fired boiler will cause the larger coal particles to continue burning as they are carried up through the furnace, thereby increasing both temperature and fouling in the convection section. Changes in the tilt of burners, the amount of excess air, and the air/fuel distribution can have similar effects, which raise potential concerns in regard to the changes in boiler design that are proposed for reducing NO_x formation.

The factors that control the proportion of ash appearing as fly ash will also influence fouling, particularly in wet bottom or cyclone boilers. Occurrence of low slag viscosity in a cyclone burner would, for example, increase the carry through of coal particles that are burned in suspension, causing higher dust loadings and higher temperatures at the furnace outlet.

Boiler load has a very pronounced effect on the rate of fouling, and dropping load is the usual practice for extending boiler operation after severe fouling has commenced. Intensive on-line cleaning is also used in an endeavor to improve the condition of the boiler at low load. If severe fouling becomes a chronic problem, a generating unit may be derated to provide more dependable service below its nameplate rating. Derating is a very costly solution, since it results in idle capacity for the turbine-generator and all auxiliaries to the boiler.

There are a number of methods available or being developed to control ash fouling. These include:

- Boiler design
- Soot blower design
- Restriction of sodium level in the coal by selective mining, blending and upgrading
- Use of additives

The general approach to improving the reliability of boilers intended for high fouling coals is to, in effect, derate the boiler when it is designed, thereby matching a conservatively designed and thus more expensive boiler to the inherently more reliable turbine-generator and auxiliaries. Boilers for fouling coals are designed with considerable furnace height and with a large furnace volume (low volume heat release rate) so as to allow ample time for burnout of suspended coal particles and to produce a low gas temperature at the furnace exit. A conservative volume heat release rate for a high-fouling lignite is 7200 Btu/hr-ft^3 about half that for a low-fouling coal. Furnace exit gas temperature would be about 1900°F at rated load. Other special features include ample spacing between burners, wide tube spacing and shallow tube bank depth in the convection section, steeply sloping floors under the superheater pendants for shedding deposits into the main furnace, and large numbers of soot blowers. The inclusion of these special design features has been reported to add about 15 percent to the capital cost of the steam generator.[40]

On-line cleaning devices (soot blowers) have been improved and the numbers installed greatly increased.[41] Since the capital and operating costs of these devices are substantial, it is not uncommon for boilers to be built with a minimum of soot blowers, and for more to be added later as required. Soot blowers are normally located to act on furnace walls, convection surfaces, and the air heater. Steam blowers of fixed and retractable design for removing hard deposits are designed for steam pressures up to 250 psig (17.0 atm) and with relatively large nozzles for high impact. Air blowers are also used, particularly in arid regions where water costs would be prohibitive. Since about 1968 in the U.S., water blowers pioneered in Germany and Australia have been installed on furnace walls to provide improved removal of slag by imparting thermal shock as well as inertial impact. Water blowers have been very effective, but they are used only where needed because of a reduction in the expected life of tubes resulting from repeated thermal shock.

The rate of ash fouling for U.S. lignite and subbituminous coal has been found to be predominantly a function of the sodium level. Fouling increases sharply as sodium content in the ash increases. The percent of ash in the coal is also a major factor in fouling. High calcium content in the ash has a favorable effect, acting to decrease fouling. Because of these effects, various means to reduce the sodium content of the coal can be used, such as selective mining of low-sodium portions of the coal seam, blending with low-sodium coal, or removal of sodium and/or ash by chemical cleaning of the coal.

It has been determined that fouling can be reduced from a high level to a low level by washing high-sodium coal with hard water to exchange calcium for sodium. Further research on a laboratory scale is not required to prove this remedy, but substantial development would be required to establish the practicality and economic feasibility of a commercial process. Serious problems can be envisioned in dewatering fine coal (particularly lignite) and in disposing of or treating large volumes of sodium-laden wash water.

The near-term emphasis in research on remedial methods will be directed toward a search for an effective boiler additive for reducing fouling. Additives containing calcium and magnesium have been tested in the past. Calcium and magnesium compounds are expected to act by reducing the fluxing tendency of the matrix parent. Aluminum compounds which react with sodium to form high-melting products will also be tested in an attempt to tie up sodium in a harmless form.

Direct Spark Ignition Using Low-Rank Coal

Growing concern over oil usage in the country has increased interest in the possible use of pulverized coal for ignition systems in coal-fired plants rather than the present oil method. There is a potential for significant savings of oil if coal spark ignition systems are adopted. A base load plant could save 18,300 gallons of oil per year. A cycling plant could save about 234,100 gallons per year and a two-shift plant, as much as 1,816,000 gallons per year. It is unclear whether low-rank coals could be utilized in this spark ignition role. With proper upgrading, however, they may well be able to be used in this manner. Work should be done to examine their usefulness in this area and establish any treatment that may be required.

Small Scale Stoker Furnaces

Very little work has been done in the past 20 years to develop or improve small stoker furnaces to burn low-rank coals. There are, however, many potential applications for these small furnaces in residential, commercial, and small industrial facilities. It may therefore be beneficial to develop and demonstrate the use of these small stoker furnaces.

In addition to furnace development, fuel improvement for these systems is desirable. Development of a high grade fuel that burns efficiently and incorporates an absorbent (such as limestone) to control sulfur dioxide emissions will allow implentation of these sytems in an environmentally acceptable manner. This may entail refining briquetting or pelletizing techniques to meet the requirements for these furnaces.

Key Issues in Environmental Control Technology

- "Integrated systems assessment" of environmental control techniques for low-rank-coal-fired power plants
- Performance of ESP on low-rank coal
- Analysis of fabric filter applications and problem areas (e.g., fabric life, flow distributions, air to cloth ratio).
- Collection of data on SO_2 removal efficiency of ash-alkali scrubbing for various coals, and on other factors such as scaling and oxidation state
- Effectiveness of spray dryers or dry sorbent injection
- Collection of data on emissions of hydrocarbons and trace elements from power plants
- Evaluation of new particulate removal devices or concepts

- Opportunities for utilizing solid waste (as an alternative to disposal)
- NO_x control techniques
- Evaluation of chemical treatment processes for fixation and disposal of FGD sludge
- Minimization of powerplant water requirements

<u>Integrated Systems Assessment of Environmental Control Techniques</u>

Environmental control systems are becoming the dominant problem in both designing and operating coal-fired power plants. EPRI has estimated that approximately 40 percent of the capital cost of a typical new coal-fired unit is used to meet environmental control requirements for air, water, and solid wastes; by 1985, ECT's share could rise to 50 to 60 percent.[38] The tendency to add control systems in series (e.g., combustion modification for NO_x, scrubber for SO_2, ESP for particulate) adds to the complexity of plants, and is likely to have a compounded negative effect on unit availability, heat rate, permissible rate of load change, complexity of startup and shutdown procedures, and other plant operations. EPRI has initiated a research project on integrated environmental control that is intended to optimize and simplify overall powerplant designs.[39] Testing of skid-mounted pilot-scale cleanup systems arranged in various configurations will be conducted at the Arapahoe test facility. Initial studies will consider interface and compatibility problems among various units, such as:

a. Effects of ammonia carryover from postcombustion NO_x control processes - e.g., air preheater deposition/corrosion; conversion of SO_2 to SO_3; NH_3 stack emissions and potential visible plume; baghouse blinding; effects on fly ash resistivity; impact on SO_2 scrubber chemistry and solid wastes.

b. Alternative process configurations for particulate and SO_2 control - e.g., baghouse or ESP downstream of wet or dry SO_2 scrubber.

c. Use of cooling tower blowdown as scrubber makeup, and use of solid wastes from cooling tower sidestream treatment as SO_2 scrubber reagent makeup.

The EPRI project is illustrative of some of the kinds of R&D required to develop integrated environmental control systems for low-rank coals. The overall problem is very complex, for two reasons: (1) the properties of low-rank coals affect the selection, design, and operation of individual control units in unique ways that have not been completely characterized or embodied in optimal systems; and (2) as new or improved individual control techniques are developed, the criteria for effectiveness of upstream or downstream units can change significantly. For example, selection of a spray dryer or dry sorbent injection system for SO_2 control might suggest the selection of a baghouse for particulate control because of the efficient gas-solid contact obtained (as opposed to an ESP which is designed to minimize flue gas-particulate contact).

The "systems integration" function that is required to deal with this complex issue is illustrated in Figure 3.5.1.1. Examples of individual environmental control technologies are shown at the top of the

Figure 3.5.1.1

Integrated Systems Assessment

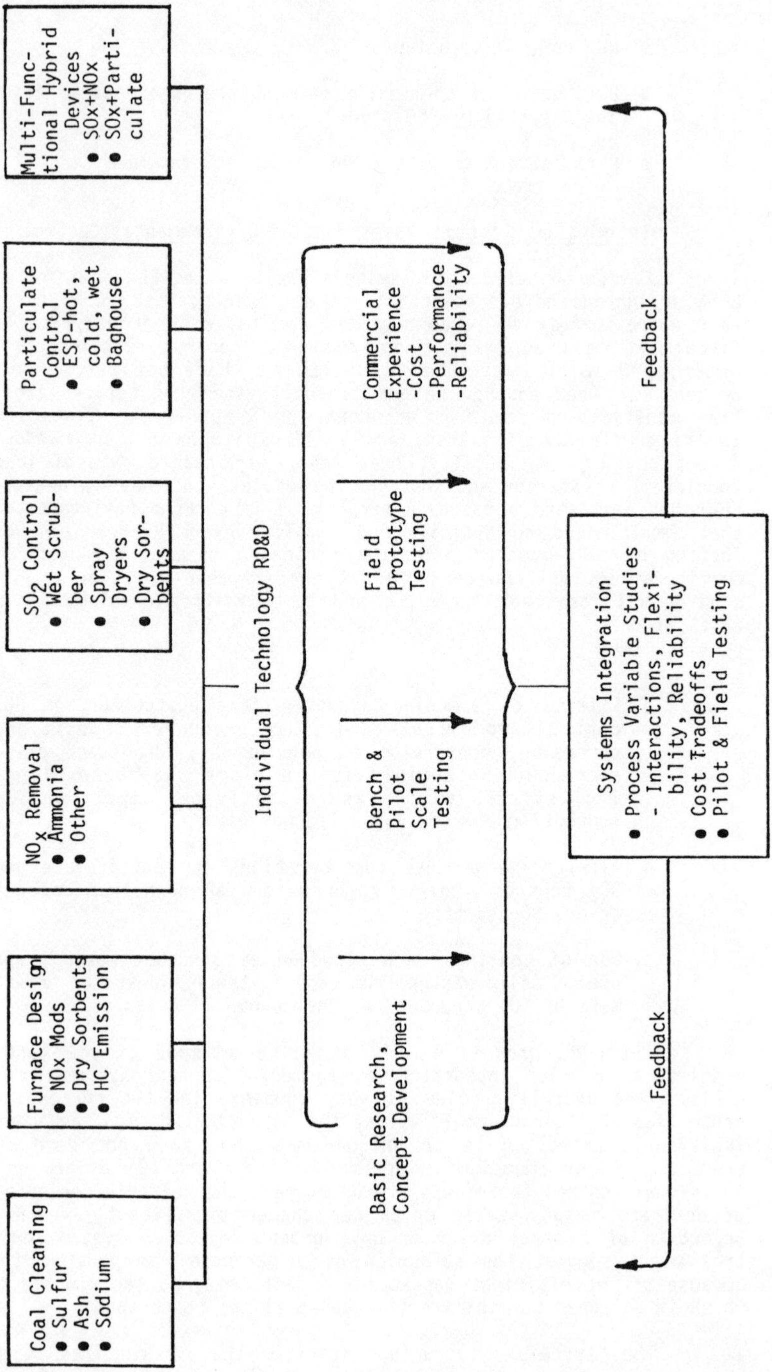

flowchart. These technologies are the subjects of a continuous array of RD&D work, ranging from basic scientific investigation and new concept development to statistical evaluation of commercial performance. In general, each technology is pursued within relatively narrow bounds in order to optimize performance of some specific function such as SO_2 removal from flue gas. Interactions with other system components are not excluded from consideration; however, they are definitely not the main thrust of the research effort.

The integrating function consists of a more limited array of technology evaluation and testing, from the point of view of optimizing the overall system. This involves the identification of both synergistic effects and incompatibilities created by certain combinations of systems. Between these extremes, the objective is to evaluate the overall system in terms of process variable studies and cost tradeoffs. This is supported by pilot and field testing of combined sytems to the extent that data cannot be reliably derived from operations of these individual units by themselves. The output from the systems integration work represents feedback to the individual RD&D projects, primarily through the setting of new or revised objectives or performance targets for individual components.

Performance of Electrostatic Precipitators

Control of particulate emissions has become one of the most difficult environmental regulations to meet. Initially, cyclone systems were used to collect particulate matter from combustion sources. At the time, the combustion units were relatively small and particle sizes from the units were fairly large. The cyclones operated at a suitable efficiency for these size distributions to meet their objectives, which were primarily to protect induced draft fans from erosion. As combustion units increased in size, pulverized coal systems became predominant. Electrostatic precipitators (ESP's) then came into use and were effective on most eastern coals, but some western coals presented a problem due to high resistivities and fine particles produced from low-melting alkali metals. These early ESP's were low-cost units, underdesigned by today's standards, and were intended only to obtain better collection efficiency than cyclones and to meet local objections.

During the 1970's, much more stringent design criteria were imposed on ESP's by the Clean Air Act emissions standards. The current NSPS of 0.03 lb/million Btu is presenting a major problem to utilities. Electrostatic precipitators have a difficult time collecting high-resistivity, submicron-size particles, making their usefulness somewhat questionable.[14]

There are five major techniques available for overcoming the high resistivity fly ash problem. They are:[14]

- Brute force
- Flue gas conditioning
- Source conditioning
- Operating at elevated temperatures
- Operating at depressed temperatures

The high resistivity of fly ash reduces the effectiveness of ESP's by disrupting the electrical conditions within the unit. The brute force method counters this effect by use of a sufficiently large precipitator to provide the desired result. The advantage to this method is that no particular or unusual operating procedures are required for the precipitator.[14] It is not useful when a problem is occurring at an existing station and it is difficult to determine the exact size requirements before actual operation begins.

Flue gas conditioning is a technique in which a chemical is introduced into the gas stream to interact with the ash to reduce its resistivity. One such conditioning agent is sulfur trioxide. The use of conditioners has met with mixed results.[13] Public Service Company of Colorado has had considerable success with sulfur trioxide;[15] however, they warn that its effectiveness is very dependent on the type of ash and should only be used in retrofit operations. Tests with ammonia and triethylamine as conditioning agents to improve collection efficiency have been inconclusive.[16] The lime content of the ash appears to be one influential factor but additional research is required to fully characterize the problem. One must also consider the environmental impact of the conditioning agent.[13]

Source conditioning is a technique utilized to modify the chemical composition of the fly ash. The objective of this approach is to reduce the chemical durability of the ash and/or increase the number of available charge carrying ions.[14] Sodium is an example of a source conditioning agent. Very little work has been conducted in this area and uniform conditioning of the flue gas stream may be a problem.

Another option is to change the operating temperature of the ESP. At typical stack temperatures (300°C), the ash resistivity is at a maximum and can be lowered either by cooling the flue gas before the precipitator or by placing the ESP in a hotter region such as before the air preheater. The preferred option has been the latter. At hot-side temperatures, resistivity is lowered primarily by means of sodium ions. However, the precipitator may also experience a loss in efficiency because of electrical characteristics similar to the back corona effect.[17] For coals low in sodium as well as low in sulfur, use of hot-side techniques may not solve loss of collection efficiency problems associated with variations in coal types. Two other problems develop when this technique is applied: the volume of gas handled is about 50 percent greater than cold-side, and thermal expansion problems are increased.[14]

The resistivity can also be reduced by operating at reduced temperatures. Tests sponsored by the Montana Power Company[18] showed the resistivity to drop by a factor of about 30 between operation of their ESP at 300°F and at 200°F.

From this discussion it is evident that the most effective method for improving ESP collection efficiencies is going to be dependent on the chemical characteristics of the fly ash. It will therefore be necessary to determine which constituents of a given fly ash are limiting with respect to available control techniques. It is, however, still not clear whether any of these methods will consistently produce removal efficiencies sufficient to meet 1979 NSPS. Some units have met these standards and shown good results.[19]

Performance of Fabric Filters

Recent revisions in the particulate control standards are beginning to shift control techniques from the use of precipitators to fabric filters (or baghouses) especially when dealing with high-resistivity western coals. A study by the Chemical Engineering Department at Manhattan College[20] concluded that when the sulfur content of the coal drops below 1 percent, baghouses are more economical than electrostatic precipitators. A similar study by EPRI[21] showed the economics for baghouses to clearly surpass ESP units at 1979 standards and in fact improve if future standards are tightened. This is due to the rising cost on ESP units as more stringent standards are imposed while the cost of baghouses remain relatively constant due to their inherent high efficiencies.

Experience with baghouses in the utility industry has been somewhat limited, however, and some mixed results have been noted. The Cameo Station in the Public Service Company of Colorado System and the Martin Drake Plant in Colorado Springs have both had relatively good results with their baghouse units.[22] Success stories have been reported elsewhere at Holtwood, Kramer and Sunburg stations.[23] These were relatively small units; some problems have been observed at larger ones. The Harrington Station in Texas has experienced problems, both with high pressure drops due to cleaning problems and with bag deterioration.[24] Similar bag problems have occurred at the Montecello plants in Texas. Bags expected to last 2 years had at one point been replaced twice within an 8-month period.[25] This is thought to be due to the high silica and abrasive nature of the Texas lignite.

These examples point out two areas requiring further attention. One will be advanced cleaning methods to handle the very large total quantity of particulate matter collected in large power plants. The other area will be development of fabrics capable of dealing with the high abrasive mineral content of ash from some western coals.

Ash-Alkali Scrubbing for SO_2 Removal

Traditionally, the control of particulate and SO_2 emissions have been treated separately with particulate controlled by electrostatic precipitators or baghouses and SO_2 controlled by wet scrubber systems. Wet scrubbers operate with a reactive alkali medium, usually a lime or limestone slurry, and precipitate sulfur out of the flue gas stream as a sulfite or sulfate. A characteristic of western coals is that they contain high levels of alkali such as calcium, magnesium and sodium. This fact initiated the possible use of these alkalis inherent in the flue gas stream of plants fired by western coals to perform the sulfur scrubbing, eliminating or reducing the need for an external alkali source.

This idea was developed and these alkali species can be leached out of the fly ash stream for use in flue gas desulfurization wet scrubbers. At present, there are nearly 2,600 MW of generating capacity in the western United States that utilize either fly ash or fly ash supplemented with lime or limestone. An additional 3,500 MW of western generating capacity which will use alkaline fly ash are being planned or are in various stages of construction.[26]

The use of fly ash alkalis benefits the overall environmental control system by reducing the volume of particulate matter to be controlled. By passing the flue gas stream first through a venturi scrubber to leach out the alkalis, then through a sprayer system to scrub the sulfur, the particulate load in the stream is greatly reduced. At its cooled temperature, the flue gas is now prepared for a cold-side precipitator. This is only one of a number of system designs for use in ash alkali systems, but most provide a number of overall economic benefits:[27]

- Elimination or reduction of alkali costs
- Reduction in initial capital investment
- Reduction in waste solids handling and disposal cost
- Improved system reliability

Whether addition of lime or limestone as a supplement to the ash alkali will be required depends on its ability to remove sulfur to meet

NSPS. The 1979 standards require 70 percent removal of sulfur for most low sulfur western coals (see Section 3.5.1.2 - "Flue Gas Desulfurization"). Most plants at this time either use lime or limestone as supplement or have it available if required.[26] Removal efficiencies from these plants have varied from about 50 percent into the upper 90's; however, their ability to remove 70 percent consistently without supplemental alkali is questionable.

Spray Dryers and Dry Sorbent Injection

The use of a spray dryer system for flue gas desulfurization is an attractive alternative to wet systems for a number of reasons. Spray dryers are very simple and have been used for many years in the chemical and food industries. In these systems an alkali slurry such as lime is pumped into a drying chamber as a fine atomized mist. The sulfur dioxide present in the flue gas is then absorbed into the water droplets and reacts with the alkali. The water droplet then vaporizes in the hot flue gas leaving the reacted particle to be collected by a particulate control device.

The major advantages proposed for the spray dryer system compared to wet scrubbers are:[29]

- Reduced waste handling
- No scaling or plugging problems
- Low cost construction
- Significantly lower operation and maintenance requirements
- Reduced energy requirements
- Lower particulate loading
- Lower water consumption

Spray dryer scrubbing has been demonstrated in Japan for a number of years using caustic soda as the reactive material.[30] Application in the United States has been slow due to lack of abundant alkali material. There are only a few geographic areas where reactive material, such as nahcolite, is available. In late 1977, however, a pilot spray dryer scrubber was installed at the Neal Station of Basin Electric Power Cooperative for the purpose of investigating the spray dryer scrubber concept with lime and other alkali materials.

The tests at Neal Station concentrated on the use of lime as the reactive material. Promising results were obtained with SO_2 removal efficiencies climbing well above 90 percent. It was observed that the gas temperature leaving the reactor has a strong influence on the removal efficiency. The closer this gas temperature is to the saturation temperature of water, the better the removal. Unfortunately the station lacks the flexibility for changing conditions and feeds readily and more pilot plant data would be useful.

Three spray dryer units are being installed in large low-rank coal utility plants presently: the Otter Tail Power Company Coyote Station, the Basin Electric Company Antelope Valley Station and the Laramie River Station.

More data are required before the usefulness of these systems can be completely assessed. Spray dryer systems have received considerable attention up to now but investigations with packed beds and granular filters would also be useful. Further analysis into the effect of adsorbent particle size, injection temperature, injection technique, and alkali material should be conducted to more fully understand the problem.

Another concept that has been experimented with in the past is the injection of dry sorbents into the furnace or into the flue gas stream upstream of a particulate collector. Reaction kinetics tend to limit SO_2 sorption, and the injected dry material can cause operating problems in the plant. However, this method does have promise as a retrofit technique that could be used effectively in certain special situations.

Effects of Trace Elements, POM and Radiological Elements Emitted From Coal-Burning Power Plants

Little information is known about the environmental and health effects caused by trace elements in atmospheric pollution. Lim[31] writes about trace elements from coal combustion: "Coal combustion is neither the only nor the most important source of trace metals . . . but the extent to which coal combustion may be raising the total health hazard in the environment ought to be known." Lim cites some possible effects of trace elements as:

- Deposition of trace elements and their compounds in plant equipment may reduce the overall efficiency of the plant.
- Trace elements may interfere with catalytic activity in coal conversion.
- Some trace elements are toxic to humans, animals and plants when they exist above certain concentrations. Others may be carcinogenic. Therefore, a potential danger to the environment may exist as a result of the effluent streams.

Studies in these areas would be beneficial to ascertain the nature and extent of these problems.

The formation and transformation of polycyclic organic matter (POM) from coal combustion was studied by Natusch.[32] This study showed that adsorption of vapor phase POM will be preferentially concentrated in particles whose aerodynamic size falls in the range which can remain airborne for several days and which is capable of being deposited in the pulmonary region of the human respiratory system when inhaled. Further work should be done to determine the health hazards of inhalation of POM.

McBride, Moore, Witherspoon and Blanco[33] have shown (using a theoretical model) that radiation doses from airborne effluents of a coal-fired plant may be greater than those from a nuclear plant. The major pathway of exposure for the radioactivity in the emissions was ingestion of contaminated foodstuffs. It was concluded that the public health significance of the computed doses was relatively minor compared to the health effects associated with airborne releases of nonradioactive material (particulate, NO_x, SO_2 and so on).

A major problem encountered in dealing with organic and trace element emissions is the lack of accurate sampling and analytical procedures to determine the exact nature and extent of the problem. Techniques must be developed to both monitor the emissions and verify that monitoring is providing an accurate characterization of the emissions. Many present practices have produced questionable results and there have yet to be developed techniques to determine many of the trace element and organic constituents.

Advanced Particulate Removal Systems

To meet 1979 NSPS for particulate control, advanced designs are being developed having high collection efficiencies for small particle sizes and high resistivities. The Environmental Protection Agency formed a branch in 1973 to investigate new devices and designs to assess their usefulness in advanced particulate control.[34] They have tested a number of devices and some have shown excellent results even for very small particle sizes. Most of these are hybrids between electrostatic and wet scrubber designs.

Most electrostatic devices may be broken into two groups: those that isolate the charging process and those that do not.[17] The purpose of isolating the charging process is to use high fields or AC fields for more effective charging. Collection of particles is accomplished in a separate zone where the current can be kept small to avoid back-corona problems. This approach has been called the two-stage precipitator. Changes in precipitator design not associated with separate charging sections usually involve modification of the electrodes. Some of these devices use a third, nondischarging electrode to maintain the field for collection while the voltage through the collected dust layer has time to decay before reaching back-corona levels.

Although many of these devices show promise, the testing must be conducted in the field on low-rank coals before their usefulness to fly ash collection can be accurately assessed.

Opportunities for Utilizing Solid Waste

Increased interest is being focused on utilizing fly ash waste from coal-burning power plants. Ash is hardly ever recognized as a mineral raw material but recent studies have been oriented toward classification of ash constituents.

Manz[35] is doing work at the University of North Dakota to quantify the physical-chemical properties of ash in an effort to expand its use. At present, ash is being used in the following ways:

- Mixed with cement
- Partial replacement of cement in concrete
- Stabilizer for road bases, parking areas, etc.
- Fill material for roads, construction sites
- Filler for asphalt mixes
- Oil well cementing

The use of fly ash for these purposes is proceeding at a relatively low level. However, the future will almost certainly see its expanded use.

Investigations into the feasibility of making bricks from fly ash has involved fly ash and clay mixtures. In the proper proportions the mixture can be extruded from a die as a stiff mud. Most brick plants currently use this process to produce bricks, therefore the utilization of fly ash as another ingredient would not involve much capital investment.

Other work being done on low-rank coal fly ash utilization is:

- The Texas Transportation Institute, Texas A&M University, College Station, Texas, has completed a study involving the production of a synthetic aggregate from lignite fly ash.

- Manitoba Hydro, Winnipeg, Manitoba, Canada, under the direction of K.A. Lenz, Concrete Engineer, conducted a thorough laboratory investigation of both a lignite and subbituminous fly ash for possible replacement of cement in concrete to be used in a hydro dam project.

Continuing work in these and other areas may provide beneficial alternatives to fly ash dumping. Many possible applications still require considerable further development, however, and research funding should be continued.

NO_x Control

Nitrogen oxides are produced by reaction of oxygen with both nitrogen present in combustion air and nitrogen inherent in the fuel source. There are a number of combustion modification techniques available to reduce NO_x production in the furnace. They accomplish these reductions generally in one of two ways:[36] (1) by making less oxygen available in the burner zone where fuel nitrogen is volatilized, or (2) by causing combustion to be extended over a longer time and space so that heat losses reduce the peak temperature.

It appears at this point that meeting 1979 NSPS with pulverized coal combustion will not be a problem for western coals by using combustion modifications and low NO_x burners.[37] Advanced burners should be developed as standards tighten and forecasts indicate it will be late in the 1980's before alternative control techniques such as flue gas treatment will be required.

Chemical Treatment and Disposal of FGD Sludge

There are two primary solid waste streams resulting from coal-fired utility plants, fly ash from particulate control devices and scrubber sludge from SO_2 removal systems. The primary methods of disposal currently in use are ponding and sludge fixation with subsequent landfill or mine disposal.[29] Chemical fixation has been employed in many sludge operations to attempt to provide an environmentally acceptable disposal method by lowering permeability, increasing strength properties, and reducing compressibility. Fixation techniques have been successful to some extent; however, problems have been encountered in trying to correlate laboratory data with field results.

Ultimately, it would be desirable to develop a method to predict for each type of waste:[38]

- The materials handling problems associated with sludge and sludge/ash mixtures

- The long-term physical behavior of such material dispersed in the field

- The effect that leachate from such materials may have on surface and groundwater

The most pertinent disposal criterion is the EPA's "Alternative Waste Management Techniques for Best Practical Waste Treatment." This criterion is restricted to impact on the quality of groundwater and surface water affected by waste material, and to the structural quality of the waste materials as it relates to the actual disposal site. This means groundwater concentrations shall be limited to the maximum levels given in the National Interim Primary Drinking Water Regulations or the United States Public Health Service drinking water standards, whichever is less.

Most sludge studies deal with sludge produced by lime or limestone systems. However, the chemical characteristics of these are much different than those encountered in fly ash alkali systems. Also these sludges vary for the different coal sources, operating conditions and supplemental alkalis used in each case. Much work is still required to characterize these types of sludge. Work is also required into determining the best mode of ultimate disposal, whether it be underground mines, landfill or other techniques.

Minimization of Powerplant Water Requirements

One major problem facing expanded low-rank coal utilization in western states is adequate water supplies for plant operation. Methods to reduce water requirements are presently being developed and some are being applied at the present facilities. Two major water consumers in plant operations are cooling water for the steam supply system and now wastewater from wet scrubber systems. Use of dry cooling towers is one method for reducing the volume of cooling water required for routine plant operations.

A new source of water consumption is flue gas desulfurization systems. Previously, most western plants could avoid this water drain by use of "compliance coal" that required no sulfur removal system. The 1979 NSPS will, however, require sulfur scrubbing on all new plants. This new drain on already limited water resources may cause significant problems in some areas. This will increase emphasis on the development of spray dryers and dry sorbent injection systems. Rapid development in these areas may be crucial to expanded use of coal-fired systems in western states.

References - Section 3.5.1.1

1. Sondreal, E.A., P.H. Tufte, and W. Beckering, Ash Fouling in the Combustion of Low-Rank Western Coals, Combustion Science and Technology, 1977, Vol. 16, pp. 95-110.

2. Gronhovd, G.H., A.E. Harak, and P.H. Tufte, Ash Fouling and Air Pollution Studies Using a Pilot-Plant Test Furnace, 1969 Lignite Symposium, U.S. BuMines IC 8471, 1970, pp. 69-88.

3. Honea, F.I., S.J. Selle, and E.A. Sondreal, Factors Affecting Boiler Tube Fouling from Western and Gulf Coast Lignites and Subbituminous Coals, Paper No. 73, International Corrosion Forum, Atlanta Georgia, March 12-16, 1979.

4. Basin, J.A, Experience with High Sodium Lignites, paper presented at University of Newcastle, August 1979.

5. Baker, B.K., and W.H. Gardiner, Modification of Ash Behavior in Lignite Fired Boilers, Symposium on Technology and Use of Lignite, Grand Forks, N.D., May 1977, GFERC/IC-77/1, pp. 74-81.

6. Lundberg, R.M, Combustion of Western Coal, American Institute of Chemical Engineers, San Francisco, CA, November 25-29, 1979, Paper No. 146.

7. Boelter, W.A., R.C. Austin, R.C. Herrick, K.D. Pollins, Conversion of St. Clair Power Plant to Subbituminous Coal, 2nd International Coal Utilization Conference, Houston, Texas, November 1979.

8. Moore, G.F. and R.F. Ehrler, "Western Coal Laboratory Characterization and Field Evaluation of Cleaning Requirements," Journal of Fluids Engineering, ASME, August 1973.

9. Honea, Frank, GFETC, Personal communication concerning field trip to visit various Texas utilities.

10. Honea, F.I., S.J. Selle, E.A. Sondreal, The Effects of Overfire and Low Excess Air on NOx Emissions and Ash Fouling Potential for a Lignite-Fired Boiler, for presentation at the American Power Conference, Chicago, Illinois, April 24-26, 1978.

11. Hall, R.E., David G. Lachapell, Status of EPA's Coal-Fired Utility Boiler Field Test Program for Controlling Air Pollutant Emissions, Proceedings, Second NO_x Control Technology Seminar, July, 1979, prepared by Electric Power Research Institute.

12. Johnson, S.A., P.L. Cioffi, T.M. Sommer, M.W. McElroy, The Primary Combustion Furnace System - An Advanced Low-NOx Concept for Pulverized Coal Combustion, Proceedings, Second NO_x Control Technology Seminar, July 1979, prepared by Electric Power Research Institute.

13. Abbott, J.H. and D.C. Drehmel, Control of Particulates from Combustion, presented at Third National Conference on the Interagency Energy/Environment R&D Program, U.S. EPA, June 1 and 2, 1978.

14. Nichols, G.B. and R.E. Bickelhaupt, Electrostatic Collection of Fly Ash from Western Coals: Some Special Problems and the Approach to Their Solution, Southern Research Institute, Birmingham, Alabama.

15. Green, G.E., Operating Experience with Particulate Control Devices, Public Service Company of Colorado, April 1973.

16. Southern Research Institute, Flue Gas Conditioning for Enhanced Precipitation of Difficult Ashes, FP-910, October 1978.

17. Drehmel, D.C., Recent Developments in Particulate Control for Coal-Fired Power Plants, 72nd Annual Meeting of AIChE, November 25-29, 1979.

18. Berube, D.T., Low Gas Temperature Solution to High Resistivity Ash Problems, Symposium on Electrostatic Precipitator Technology, Birmingham, Alabama, February 24, 1971.

19. Young, D.G. and K.N. McConnell, Precipitator Operation on a Western Fuel, presented at the 2nd International Coal Utilization Conference and Exhibition, Houston, Texas, November 6-8, 1979.

20. Glaubinger, R.S., "High Rating for Baghouses," Chemical Engineering, February 11, 1980.

21. Sterns-Roger Incorporated, Economics of Fabric Filter Versus Precipitators, EPRI FP-775, June 1978.

22. Drehmel, D.C., Recent Developments in Particulate Control for Coal-Fired Power Plants, presented at 72nd Annual Meeting of AIChE, November 25-29, 1979.

23. Frederick, E.R., "Fabric Filtration for Fly Ash Control," JAPCA 29 (1): pp. 81-85, January 1979.

24. Perry, J. and G. Cammack, A Case Study of Utility Baghouses, presented at 2nd International Coal Utilization Conference and Exhibition, November 6-8, 1979.

25. Honea, F.I. and E.A. Sondreal, Ash Properties and Combustion Characteristics of Gulf Province Lignites, presented at SME-AIME Fall Meeting, October 1979.

26. Ness, H.M., P. Richmond, G. Eurick, and R. Kruger, Power Plant Flue Gas Desulfurization Using Alkaline Fly Ash from Western Coals, presented at USEPA Flue Gas Desulfurization Symposium, Las Vegas, Nevada, March 5-8, 1979.

27. Johnson, C., "Fly Ash Removes SO_2 Effectively from Boiler Flue Gases," Power Engineering, June 1979.

28. Tufte, P.H., E.A. Sondreal, K.W. Korpi, and G.H. Gronhovd, Pilot Plant Scrubber Tests to Remove SO2 Using Soluble Alkali in Western Coal Fly Ash, Symposium on Technology and Use of Lignite, Grand Forks, N.D., May 1973, BuMines IC 8650, 1974, pp. 103-133.

29. Ness, H.M., S.J. Selle, and O.E. Manz, Power Plant Flue Gas Desulfurization for Low-Rank Western Coals, Symposium on Techology and Use of Lignite, Grand Forks, N.D., May 1979, DOE GFETC/IC-79/1, pp. 117-146.

30. Hurst, T.B., Dry Scrubbing: A New Concept for Flue Gas Desulfurization, presented at 2nd International Coal Utilization Conference and Exhibition, November 6-8, 1979.

31. Lim, M.Y., Trace Elements from Coal Combustion - Atmospheric Emissions, IEA Coal Research, May 1979.

32. Natusch, David F.S., Formation and Transformation of Polycyclic Organic Matter from Coal Combustion, Progress Report, Contract #EE-77-5-02-4347.

33. McBride, J.P., R.E. Moore, J.P. Witherspoon, R.E. Blanco, "Radiological Impact of Airborne Effluents of Coal and Nuclear Plants," Science, December 8, 1978, Vol. 202, Na 4372.

34. Abbott, J.H., Advanced Techniques for Fine Particulate Control, Particulate Technology Branch, EPA, Research Triangle Park, N.C.

35. Manz, Oscar E., Utilization of Lignite and Subbituminous Ash, Symposium on Technology and Use of Lignite, Grand Forks, N.D., May 1973, BuMines IC 8650, 1974, pp. 204-219.

36. Selle, S.J., F.I. Honea, and E.A. Sondreal, Direct Use of Lignite for Power Generation, presented at IGT New Fuels and Advances in Combustion Technology Conference, March 26-30, 1979.

37. Mason, H.B., K.G. Salvesen, A.B. Shimizu, N. Concion, Utility Boiler NOx Emission Characterization, presented at Second NO_x Control Technology Seminar, EPRI FP-1109-SR, July 1979.

38. Yeager, K.E., C.R. McGowin, and S.B. Baruch, Potential Impact of R&D Programs on Environmental Control Costs for Coal Fired Power Plants, Fossil Fuel Power Plants Department, Electric Power Research Institute, June 1979.

39. Giovannia, D., "Integrated Environmental Control," Fossil Fuel and Advanced Systems Division R&D Status Report, EPRI Journal, May 1979, p, 43.

40. Nelken, H.H., and J.G. Altman. Design Considerations for Modern Utility Boilers, Presented at the American Power Conference, Chicago, Illinois, 1975.

41. Buckley, W.R., R.F. Ehrler, J.R. Hayman, and G.F. Moore, Select From Air, Steam and Water For Removing Boiler Fireside Deposits, Power, July, 1974, pp. 19-26.

Selected References - 3.5.1.1.1

The references and abstracts listed below are considered comprehensive sources of information on their selected topics. The list is provided only as a recommended starting point for research into these various areas.

1. Sondreal, E.A., P.H. Tufte, and W. Beckering, Ash Fouling in the Combustion of Low-Rank Western Coals, Combustion Science and Technology, 1977, Vol. 16, pp. 95-110.
This paper gives a complete review of experimental, theoretical and operating data concerning ash fouling of boiler tubes by low-rank coals. The results of many years of research into this problem at the Grand Forks Energy Technology Center are summarized.

2. Lundberg, R.M., Combustion of Western Coal, American Institute of Chemical Engineers, San Francisco, CA, November 25-29, 1979, Paper No. 146.
As a representative of Commonwealth Edison Co., the author describes incentives behind the company's shift from high sulfur Illinois coal to western coals, and the changes required in the power production facilities to accommodate the new coal. Provides an excellent utility viewpoint

3. Honea, F.I., S.J. Selle, E.A. Sondreal, The Effects of Overfire and Low Excess Air on NO_x Emissions and Ash Fouling Potential for a Lignite-Fired Boiler, for presentation at the American Power Conference, Chicago, Illinois, April 24-26, 1978.
Some uncertainties remain as to the effect on the operation of lignite-fired powerplants of efforts to reduce NO_x emissions using currently available technology. The uncertainties are greatest for units burning high-sodium Northern Great Plains lignites. These units experience aggravated boiler-tube ash fouling problems during periods of high coal sodium content. In an effort to study the effects of NO_x control on ash fouling potential, the Grand Forks Energy Research Center (GFERC) of the Department of Energy conducted two weeks of testing on a tangentially fired boiler burning a high-fouling North Dakota lignite. Ash fouling rates were studied under conditions of both overfire air and low excess air. The report presents experimental results and examines the tradeoffs between NO_x production and ash fouling.

4. Nichols, G.B. and R.E. Bickelhaupt, Electrostatic Collection of Fly Ash from Western Coals: Some Special Problems and the Approach to Their Solution.
The electrostatic precipitator is the primary air pollution control device for removing particulate material from effluent gas streams from coal-fired power boilers. The operation of this device is dependent upon three steps in the process of collection: particle charging, particle collection, and the removal and disposal of the collected material. These three steps must be performed at near optimum conditions for the efficient operation of the device.

Although electrostatic precipitation theory is not covered, factors affecting the problems are discussed together with ideas for potential solutions.

5. Drehmel, D.C., Recent Developments in Particulate Control for Coal-Fired Power Plants, 72nd Annual Meeting of AIChe, November 25-29, 1979.

The need to limit both SO_x and particulate emissions has provided a double challenge to conventional coal-fired power plant control technology. Lower particulate emissions require more efficient devices. Lower SO_x emissions achieved with low sulfur coal combustion are coupled with poor electrostatic precipitator performance. To solve these problems, possible solutions include use conditioning, "hot-side" precipitators, or novel precipitators. Examples of novel precipitators are the Buell Trielectrode Electrostatic Precipitator, the University of Denmark Pulse Generator/Precipitator, and the Cold Electrode Electrostatic Precipitator. Also under development are two stage precipitators using novel charging sections such as the Southern Research Precharger, the APS High Intensity Ionizer, and the University of Tokyo Boxer Charger. As an important alternative to electrostatic precipitators, baghouses have shown general success in controlling coal-fired powerplants. The various control technologies are discussed in relation to their preferred application in different situations.

6. Sterns-Roger Incorporated, Economics of Fabric Filter Versus Precipitators, EPRI FP-775, June 1978.

The relative economics of electrostatic precipitators and fabric filters were studied for a variety of U.S. coals. The economic selection in each case was found to be dependent on the particulate emission limit to be met, the physical and chemical characteristics of the coal and fly ash, and the bag replacement schedule of the fabric filter. The studies were based on a 500-MW (net) pulverized-coal-fired powerplant provided with a well-designed fly ash collector system of high reliability.

7. Ness, H.M., P. Richmond, G. Eurick, and R. Kruger, Power Plant Flue Gas Desulfurization Using Alkaline Fly Ash from Western Coals, presented at USEPA Flue Gas Desulfurization Symposium, Las Vegas, Nevada, March 5-8, 1979.

A characteristic of western coals is that they contain high levels of alkali such as calcium, magnesium, and sodium. These alkali species can be leached from powerplant fly ash for use in flue gas desulfuriztaion (FGD) wet scrubbers in lieu of lime or limestone. At present, there are nearly 2,600 MW of generating capacity in the western United States that utilize either fly ash or fly ash supplemented with lime or limestone. An additional 3,500 MW of western generating capacity which will use the alkaline fly ash are being planned or are in various stages of construction. This report describes the western alkali ash FGD systems.

8. Tufte, P.H., E.A. Sondreal, K.W. Korpi, and G.H. Gronhovd, Pilot Plant Scrubber Tests to Remove SO_2 Using Soluble Alkali in Western Coal Fly Ash.

Sulfur dioxide emission from powerplants has become the subject of much recent legislation and extensive engineering study. The most fully developed process for reduction of sulfur emissions is the lime-throwaway wet-scrubbing process. The present study utilizes an adaptation of the basic principles of lime and limestone scrubbing in that the necessary alkali is the calcium oxide in fly ash derived from western coals. The process under consideration combines the known capability of wet scrubbing to remove a high percentage of fly ash from flue gas with the potential of the alkali in some coals to remove a significant portion of the SO_2 in the flue gas. Fly ash alkali could be augmented with lime (for additional SO_2 removal) if required to meet emissions regulations.

9. Manz, Oscar E., Utilization of Lignite and Subbituminous Ash. The United States and Canada are on the threshold of making significant progress in the utilization of lignite and subbituminous ash. Total ash production is growing rapidly, and there is no indication that the growth rate will diminish during the next few years.

   The paper examines production rates for various ash materials, possible uses, and test and analysis results.

3.5.1.2 Technology Description

The primary method for coal utilization at this time is through direct combustion. Direct firing in utility and industrial processes is a well-known technology encompassing many applications including steam boilers, kilns, furnaces, and heaters.

There are two basic methods of burning coal: in suspension or in a fuel bed. In fuel bed firing, coal is fed onto a grate and the type of burning is determined by the direction of flow of the fuel and air. This type of device is known as a stoker. Fuel beds have historically provided the most economical method of firing in almost all industrial boilers rated at less than 100,000 pounds of steam per hour.[1]

Suspension burning occurs in pulverized-coal (pc) and cyclone furnaces, and fluidized-bed combustors (FBC). These types of firing systems result in thorough mixing of coal particles and air for rapid release of energy. A pc-fired furnace burns finely pulverized coal fully entrained in the turbulent flow of combustion air. A cyclone furnace burns 1/4-inch by zero crushed coal by swirling it with the primary air into horizontal cylindrical burners. The larger coal particles in the cyclone furnace are trapped and burned in the molten layer of slag lining the walls of the burner, and smaller particles burn by entrainment. FBC units burn coal in a boiling bed of noncombustible particles suspended by an upward flow of air but not entrained out of the combustor. Pulverized-coal and cyclone systems are generally economical when capacities are in excess of 100,000 pounds of steam per hour.[1]

3.5.1.2.1 Pulverized Coal

The function of a pulverized-coal system is to pulverize the coal, deliver it to the fuel-burning equipment, and accomplish complete combustion with a minimum of excess air. This method is used primarily in utility steam boilers and to some degree in large industrial applications. The coal must be pulverized to a size small enough for combustion to be completed before reaching the cooler section of the furnace.[a] This type of firing allows combustion of virtually all types of coal.

Coal Pulverization

The extent and type of coal pulverization is influenced by a number of factors including:

- reactivity
- moisture content
- heating value
- grindability
- abrasion potential

---

[a]Pulverized coal specifications usually require 60 to 90 percent by weight to pass through a 200 mesh screen in a standardized test.

The delivered coal size determines the need for crushing before pulverization. Coal can have a top size no greater than 2 inches to ensure smooth passage through the feeders to the pulverizer.

During pulverization, a stream of hot air is used to dry the coal, remove the fines, transport the coal to the burners and supply primary air for combustion. Excessive moisture can limit pulverizer throughput by exceeding the drying capacity of the hot air source. Also, an excess of surface moisture coupled with a large percentage of fines impedes coal flow in the conveyor systems, bins, and coal feeders.[2]

The heating value of the coal will directly determine the quantity of fuel that must be pulverized and burned. For a specified unit energy output, the lower the heating value, the greater the quantity of pulverized coal required. This effect can be so severe that lower rank coals usually require both larger, and more, mills in order to maintain the necessary throughput.[3]

Pulverizer capacity is also affected by variations in the grindability of coals. The Hardgrove grindability index is a measure of the amount of fines produced by a standard laboratory pulverizer operation. Table 3.5.1.1 shows grindability and other coal preparation factors for selected coals of various ranks. The grindability index must be considered in the design of the milling system. The grindability is affected by a number of factors including moisture, mineral makeup and other physical properties.[8]

Table 3.5.1.1

Pulverizer Requirements
(Nominal 600-MW Unit Qf = 5400 x 10^6 Btu/hr)

	Eastern Bituminous	Midwestern Bituminous	Sub-Bituminous	Texas Lignite	Northern Plains Lignite
Hardgrove Grindability	55	56	43	48	35[a]
Energy content (Btu/lb) (moist, mineral-matter-free)	10,500-14,000	10,500-14,000	8,500-10,500	6,000-8,500	6,000-8,500
Typical mill selections[b]					
Number required	6	6	6	6	7
Nominal capacity (T/hr)	50	63	85	92	100
Primary air temperature for drying coal (°F)	525	640	725	750	750

[a]Modified Hardgrove grindability.

[b]Mill selection based on one full spare with remaining mills at 0.9 x new capacity.

Source: Reference 2.

On low-rank coals, a modified Hardgrove procedure has been developed to help predict the actual pulverizer performance. In this modified procedure, several prepared samples are dried to different moisture levels and the grindability index is determined for each level. The grindability value selected is the one at the moisture level expected of the coal around the grinding elements. On some lignites, this modified procedure is inaccurate. Grinding samples in an actual pulverizer may be necessary to assure that adequate milling capability is designed into new installations.[4]

The mill must deliver coal to the fuel nozzle at the proper particle sizing and moisture condition at tonnages required to produce the desired heat input. Varying degrees of fineness are required with different ranks of coal to assure flame stability and to minimize unburned carbon losses. The degree of pulverization required for each fuel represents a balancing of physical and chemical influences to produce the desired results.

Furnace Design

Proper sizing of the furnace is one of the most important design considerations for a steam generating unit. The arrangement of burners, location and extent of convective heat transfer surfaces, and number of sootblowers are all influenced by coal properties. A properly proportioned furnace must be used to ensure proper retention time for the gaseous combustion products. The furnace outlet temperature at the entrance to the convective section must be below the critical fouling temperature of the lowest quality coal to be burned to prevent slagging and fouling on convective surfaces yet still sufficiently high to afford good heat transfer from the flue gas to the steam.[4]

In the design of a boiler, important parameters are the combustion heat release rate per unit of furnace volume and per unit area of radiant surface, which are dependent on coal characteristics. To achieve good plant operability, the furnace size for the design coal must be established based on the combined effects of the heating value, moisture content and ash properties. Figure 3.5.1.2 shows relative furnace dimensions for typical coals from various regions, and indicates the order of magnitude penalty that must be paid to effectively burn high-fouling low-rank coal.

After the furnace has been sized properly, consideration must be given to the arrangement of the superheater, reheater and economizer heating surfaces located in the upper portion of the furnace and in the convection pass. In locating the various sections within the unit, a proper balance must be sought to maintain a temperature difference to transfer heat from gas to steam without raising metal temperatures to a level that will promote wastage of the tubes. The corrosion properties of low-alloy boiler tubes typically limit steam temperature to 1050 or 1100°F. Also, the velocity of the flue gas must not exceed reasonable limits in order to minimize tube erosion from fly ash. Depending upon the ash quantity and quality, the velocity is generally designed for the range of 60 to 70 feet per second. For coals yielding a heavy loading of erosive ash, the velocity may be limited to 50 ft/sec or less.[5]

Burner type and firing configuration are also important design parameters. There are two commonly used types of firing configurations for large utility boilers as shown in Figure 3.5.1.3. In horizontal firing, as the name implies, the coal-air mixture is blown horizontally into the furnace. One of the most common methods of combustion in this mode is the horizontally opposed design. Here coal-air mixtures are blown in on

Figure 3.5.1.2

Effect of Coal Rank on Furnace Sizing

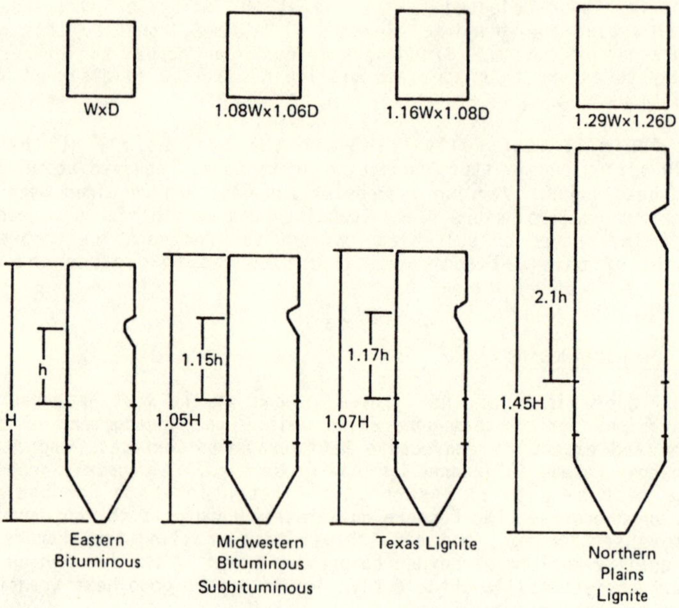

Source: Reference 2.

Figure 3.5.1.3

Two Commonly Used Suspension Firing Methods

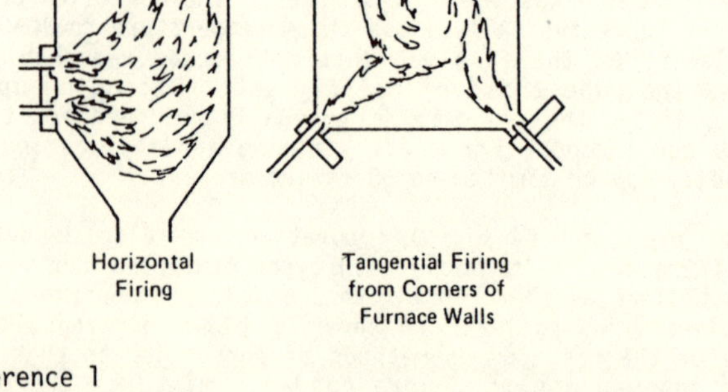

Source: Reference 1

opposite sides of the firebox and impinge on one another near the center of the furnace. This type of combustion produces high flue gas temperatures and high carbon burnout but has the disadvantage of producing high NO_x levels due to the high flue gas temperatures.

In tangential firing, burner nozzles project the streams of coal and air along a line tangential to a small circle, lying in a horizontal plane, at the center of the furnace. With this type of firing, combustion is rapid and flame lengths are long.[1] Temperatures tend to be lower in these furnaces, causing NO_x production levels to be less than other pc-fired furnaces and cyclone units.

Combustion Characteristics

The two properties of coal that have a direct influence on combustion rate and carbon burnout are the agglomerating properties and particle reactivity.

Agglomerating coals are those that soften and melt when heated to the extent that they transform into a molten mass. Subbituminous coals and lignite are nonagglomerating (the particles do not go through a melting stage) and consequently do not require as fine a pulverization to assure burnout since surface area to mass ratios do not decrease during combustion as with the agglomerating coals.[2]

Differences in combustion reactivity from one rank to another have been explained by changes in the level of organically bound oxygen. Organically bound oxygen is considered to be that which is an inherent part of the coal structure exclusive of water and mineral matter. As rank decreases, organic oxygen increases by a factor of 5 or 6. Some feel[3] that, when the fuel is heated, a portion of this oxygen becomes available for the oxidation process. Lower rank coals are therefore more reactive than bituminous and do not require the same degree of fineness to ensure complete combustion.[3]

Slagging and Fouling Properties

The quantity and nature of the mineral matter in coal determine the degree to which the heat transfer performance in boilers is reduced as a function of time. Accumulation of slag and fouling deposits, accelerated surface wastage by corrosion and erosion, and particulate emissions are all a result of mineral matter content. Some of the ash properties of concern to slagging, fouling, and corrosion are:

- Fusibility temperatures
- Base-to-acid ratio

Suggested mechanisms for the transfer of ash material from the gas stream to the tube surface include: (1) vapor phase diffusion, (2) thermal diffusion of small particles, especially in the formation of the inner white layer, (3) electrostatic attraction, and (4) inertial impaction. Because the major buildup occurs on the leading surface of boiler tubes, inertial impaction is evidently the prime mode of deposition in forming the bulk of the deposit, which is comprised of the outer sinter layer. Retention efficiency, which is influenced by particle size, geometry, and melting behavior, is obviously as important as the frequency of impaction. Fluid dynamic forces tend to counteract inertial forces and thereby tend to keep the particle moving along the flue gas stream-lines around the boiler tube. As a practical application of this, it has been suggested that finned tubes might be useful in altering the fluid dynamics in a way that would be favorable to reduced fouling.

Strength and cohesion in the fly ash deposit may be determined by a combination of van der Waals' forces, liquid film effects, retention in a liquid or solid matrix, or retention by geometric shape, for example by whiskers. The occurrence of a partial melt phase within the deposit depends on the fusion temperatures of the many individual particles of fly ash, the flue gas temperature, the boiler tube temperature, and the extent of prior ash deposition. The melting behavior of fly ash particles depends on the distribution of their analysis and the phase melting diagram accounting for all ash constituents. Some phase diagrams are available for ash slag and for ceramic materials. Statistical correlations of the ASTM fusion temperature versus ash composition may also have some value in predicting the melting behavior of individual fly ash particles. In the presence of a melt phase, liquid phase diffusion may also play a role in determining deposit growth and hardness.[13]

Despite the numerous factors which affect the fouling process, it is essential, when looking for the controlling factor, to return to the fact that the deposition process for Western coals is determined overwhelmingly by the role of sodium in fluxing deposits. It is instructive to compare fouling deposits produced by a high sodium coal and a low fouling deposit produced by a low sodium coal. High sodium deposits have a continuous melt phase that envelops and connects particles into a strongly bonded network. Low sodium deposits have no continuous melt phase, and the particles are held together by weak particle-to-particle surface bonding. It is because of this important difference that the investigation of the melt phase and the reactants which produce it have been major objectives of the study of fouling mechanism at GFETC. Hereafter the continuous melt phase will be termed "matrix," and the fluxable subfraction of fly ash from which it is produced will be termed "matrix parent."

The organically bound sodium in Western coals is partially volatilized in the combustion process. In tests at GFETC, reheating either fly ash or laboratory ash has resulted in substantial loss of sodium above 2400°F. Volatilization from the organic structure of the coal would be expected to occur more easily than revolatilization from glassy fly ash particles. The specific chemical species of sodium existing in the high temperature zone is not known; thermodynamic calculations may not be predictive because of the likelihood of "frozen equilibrium" over the very short period spent in the high temperature region of the flame. The sodium may exist for a short time as atomic Na or Na_2O; these react to form NaOH because of the presence of water vapor; and at lower temperatures, reaction with CO_2, SO_2, O_2, and SO_3 occur to form Na_2CO_3 and Na_2SO_4. Sodium sulfate is the predominant specie in the presence of the sulfur oxides.

The sodium that is not volatilized in the flame is retained in the complex ash particle produced from non-volatilized inherent ash. The portion of sodium retained and not volatilized is not known, but it is believed to be significant. Below 1800°F, the volatilized sodium is condensed into the entrained fly ash in a manner which causes the finer fraction of the fly ash to be substantially enriched in sodium. Sodium enrichment also occurs in deposits which are formed at lower temperatures rather than higher temperatures. In the end, the roles of the volatilized sodium and the retained sodium are the same, since both react to flux the ash and worsen fouling.

The form of the sodium entering the combustion process has not been found to significantly alter the severity of fouling in burning Western coals. Changes in level of sodium by ion exchange (which alters the amount of Na held on the coal structure) and by addition of $NaC_2H_3O_2$, NaOH, NaCl, Na_2CO_3, or Na_2SO_4 have had essentially the same effect on the severity of fouling. The water soluble sodium in lignite does not correlate with fouling.[13]

Slag tap furnace designs (such as cyclone furnaces) operate at temperatures in excess of the ash fusion temperatures and depend on the molten ash particles to form liquid slag deposits on the furnace walls. An equilibrium slag deposit is formed as the molten substance flows to the furnace tap hole. Continuous tapping of the slag occurs. This design is beneficial in reducing the amount of particulate in the flue gas stream. As much as 70 percent of the coal ash can be removed by way of the slag tap while the remainder exits as entrained particulate.[4] No new orders for wet-bottom pc-fired or cyclone furnaces have been made recently due to the design's inherently high NO_x production and other operating problems. It appears the utility preference will remain with the dry-bottom pc-fired units.

Industrial Process Heating from PC-Fired Units

Currently, pulverized coal firing in the process industries is essentially limited to cement and lime calcining kilns. Other process heating applications that fire pulverized coal on a limited basis are copper smelting, ceramic kilns, and glass melters.

The major piece of heating equipment in the cement and lime calcining industries is the rotary kiln. The rotary kiln is a refractory-lined horizontal cylinder rotating at about 1 revolution per minute. These kilns range up to 300 feet in length and up to 25 feet in diameter. The cylinder is slightly inclined to maintain product flow through the length of the kiln. Firing takes place at the discharge end. Fuel is introduced through a burner pipe and combustion takes place in the kiln as it mixes with combustion air. Flame control is not considered to be an important element in product quality control and as such is not given the priority afforded it in boiler furnace design. The important element of cement calcining is the time-temperature history in the kiln.

Product quality and equipment life are important factors in the design of pulverized-coal-fired calcining kilns. Coal property ranges recommended by users are shown in Table 3.5.1.2.

Table 3.5.1.2

Calcining Kiln Fuel Quality Limitations

Coal Property	Range	Impact
Total sulfur	1.5% maximum	Product quality deterioration
Ash fusion temperature	2500° F minimum	Product quality deterioration
Exit gas oxygen content	0.5-1.5%	Product quality deterioration
Na_2O, Vanadium, SO_3	Low	Refractory attack

3.5.1.2.2 Cyclone Furnace

The cyclone furnace was developed in the mid-1940's by Babcock and Wilcox Company as a high-temperature, high-turbulence combustion device that operates separately from the heat transfer sections of the boiler. The high-temperature turbulent, slagging environment promotes complete combustion prior to entering the heat transfer sections of the boiler.

Although the furnace was originally developed for low fusion temperature central Illinois coal, it has been applied successfully to all ranks of coal. The current primary application of the cyclone furnace is with western low-rank coals. It is applicable to coals having a slag viscosity of less than 250 poise at 2600° F provided the ash analysis does not indicate excessive formation of iron or iron pyrites. Cyclone firing, where applicable, is considered to have the following advantages:[4]

- Reduction of fly ash content in the flue gas
- Savings in the fuel preparation, since only crushing is required instead of pulverization
- Reduction in furnace size

A cyclone furnace is a water-cooled, refractory-lined horizontal cylinder in which crushed coal is combusted. Air enters the cylinder tangentially and imparts a whirling (cyclonic) motion. Coal (95 percent-4 mesh) and primary air are introduced at the burner end of the cylinder. The coal particles are entrained in the high velocity stream and thrown against the furnace wall by centrifugal force where they are held in the slag layer. The high-velocity tangential (secondary) air supplies combustion oxygen to the coal particles and removes the products of combustion. Molten slag drains to the bottom of the furnace and is discharged. Gaseous products of combustion flow from the discharge end of the furnace directly into the radiant heat transfer section of the boiler, as shown in Figure 3.5.1.4.

Figure 3.5.1.4

Cyclone Firing Method

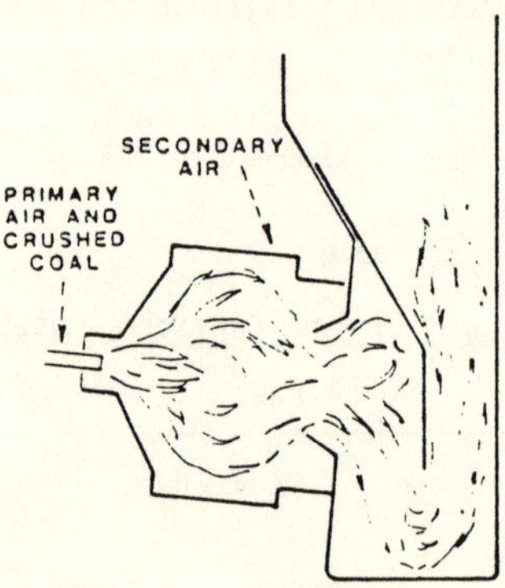

Source: Reference 1

In contrast to pulverized coal furnaces, the burner region heat release rates (and consequently local temperatures) in cyclone furnaces are extremely high. In particular, the temperature is high due to relatively low heat absorption rates in the furnace itself. Furnace temperatures in the 3000° F range are sufficient to fuse most coal ash on the refractory walls of the cyclone furnace. As with wet-bottom pulverized coal firing, cyclone operation reduces the quantity of fly ash carryover as compared with typical pulverized coal furnaces.[6,10]

Cyclone furnaces, unlike pulverized-coal furnaces, have limited design flexibility with respect to major alteration of furnace volume and heat transfer rates. Accordingly, the application of a cyclone furnace to a specific coal is dependent on the properties of the coal. The most important fuel characteristics are shown in Table 3.5.1.3.

Table 3.5.1.3

Coal Characteristics for Suitability to Cyclone

Characteristics	Range
Ash	6-25% wt.
Volatile Matter	15% wt. minimum
$\left(\frac{Fe_2O_3}{CaO + MgO}\right)$ Ash vs. Sulfur	Variable (see Figure 3.5.1.5)
Slag Viscosity	250 poise @ 2600° F maximum

Source: Reference 6

Ash content is important to provide a layer of slag on the furnace walls. Volatile content must be high enough (>15 percent) in order to maintain a rapid combustion rate. Moisture content is also an important parameter, but the maximum allowable level is variable depending on coal rank, secondary air preheat, and preparation equipment.

Coals high in sulfur or having a high ratio of iron to calcium plus magnesium are unsuitable to the cyclone furnace because they have a tendency to form iron or iron sulfide, both of which adversely affect the cyclone. Figure 3.5.1.5 shows the suitability range based on the tendency to form iron and iron sulfide.

Figure 3.5.1.5

Coal Suitability to Cyclone Furnaces Based On Tendency to Form Iron and Iron Sulfide

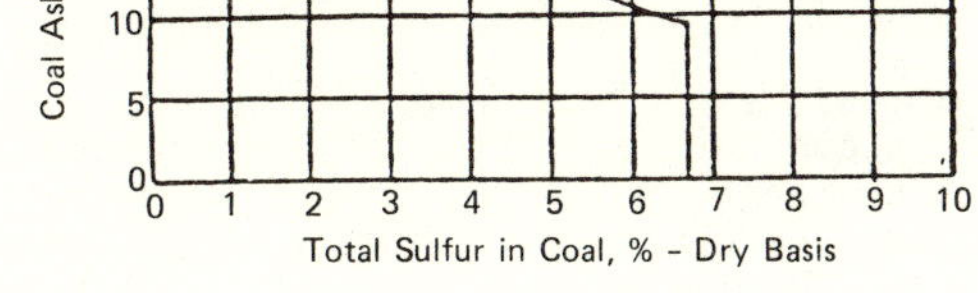

Source: Reference 6

Another important coal characteristic is the viscosity of the slag. The viscosity must be sufficiently low so as to permit slag flow at normal furnace operating temperatures. Slag will just flow on a horizontal surface at a viscosity of 250 poise. Slag viscosity has been shown to be primarily a function of the silica content of the ash but is also influenced by Fe_2O_3, CaO, MgO and other metallic oxides.

3.5.1.2.3 Stoker[1,4]

In a stoker furnace, coal is placed on a grate in the high temperature region of the furnace.' As air is forced up through the coal bed on the grate, the fresh coal is heated, volatiles in the coal are distilled off, and a coke or char is left on the grate. The coke or char then burns to form carbon dioxide and carbon monoxide, leaving ash material at the bottom of the bed.

Gaseous volatile matter distilled from the coal and carbon monoxide produced by partial combustion of the coke are burned above the fuel bed with unconsumed primary air and with secondary air which is injected above the fuel bed. Approximately 40 to 60 percent of the total heat liberated in the furnace is produced by combustion of these gases.[1]

A successful installation requires the selection of the correct type and size of stoker for the fuel to be used and the desired rate of heat release. Where applicable, stokers exhibit a good operating range, the capability of burning a wide range of solid fuels, and low power requirements. Almost any coal can be burned on some type of stoker.

Stokers can be divided into two general classes depending on the direction from which raw coal reaches the fuel bed: (1) overfeeds, in which the fuel comes from above, and (2) underfeeds, in which it comes from below. The overfeed group includes spreader, chain and vibrating grate stokers. Types of underfeed stokers are single retort and multiple retort.

The grate area required for a given heat release in a stoker is determined from allowable fuel burning rates established by experience as shown in Table 3.5.1.4. These rates are based on using coal suited to the stoker type in each case.

Table 3.5.1.4

Maximum Allowable Fuel Burning Rates

Type of Stoker	Btu/sq ft-hr
Underfeed - Single Retort	425,000
Underfeed - Multiple Retort	600,000
Water Cooled Vibrating Grate	400,000
Chain Grate and Traveling Grate	500,000
Spreader - Stationary & Dumping Grate	450,000
Spreader - Traveling Grate	750,000
Spreader - Vibrating Grate	400,000

Source: Reference 4

The spreader stoker is most generally used in a capacity range up to 100,000 lb of steam per hour. It responds rapidly to load swings and can burn a wide range of fuels. Underfeed stokers of the single retort, ram-feed and side ash discharge type are used primarily for space-heating and for small industrial boilers supplying less than 30,000 lb of steam per hour. Larger size underfeed stokers of multiple retort, rear ash discharge type have been largely displaced by spreader stokers and by water cooled vibrating grate stokers in the intermediate size range. Chain and traveling grate stokers also are gradually being displaced by the spreader and vibrating grate types. Characteristics of the various types of stokers are listed in Table 3.5.1.5.

Table 3.5.1.5

Stoker Characteristics

Design/Operating Parameter	Spreader	Chain and Traveling Grate	Underfeed
Quick response to load change	Excellent	Fair	Fair
Minimization of carbon loss	Fair	Fair	Fair
Prevents coal segregation	Fair	Poor	Poor
Utilizes wide variety of coals	Excellent	Poor	Poor
Burns extremely fine coals	Poor	Poor	Poor
Permits smokeless combustion at all loads	Poor	Good	Good
Minimizes fly ash discharge	Poor	Good	Good
Minimizes maintenance	Good	Good	Fair
Minimizes power consumption	Good	Good	Good
Handles ash easily	Excellent	Good	Fair

Source: Reference 1

Spreader Stokers[1,4]

Spreader stokers are widely used in industry today because they are capable of burning a variety of coals ranging from eastern bituminous coal to lignite. These stokers throw coal into the furnace over the fuel bed with a uniform spreading action, which permits burning of the fine fuel particles. The larger pieces, that cannot be supported in the gas flow, fall to the grate for combustion in a thin, fast-burning bed. This method of firing allows quick response to load fluctuations because ignition is almost instantaneous when the firing rate is increased, and the thin fuel bed can be burned out rapidly when it is desired to decrease the rate. Turndown capability normally extends from maximum capacity to 20 percent of full load, but minimum load can be designed for as low as 12.5 percent of maximum.

Underlying the active fuel bed is a layer of ash. This, together with the flow of air through the grate, serves to keep metal parts at allowable operating temperatures. For this type of stoker, combustion air can be preheated to 300-500°F for increased efficiency without creating a grate maintenance problem.

A typical spreader stoker installation consists of feeder-distribution units (in widths and numbers as required to distribute the fuel uniformly over the entire grate), specifically designed air metering grates, forced draft fans for both undergrate and overfire air, dust collecting and reinjecting equipment, and combustion controls to coordinate fuel and air supply with load demand.

The size consistency of coal fed to a spreader stoker has a direct bearing on boiler efficiency and on the tendency of the installation to discharge particulates. Coal segregation is a problem with any type of stoker, but the spreader stoker can tolerate a small amount of segregation because the feeding rate of the individual feeder-distributors can be varied. Size segregation, where fine and coarse coals are not distributed evenly over the grate, produces a ragged fire and poor efficiency.

When boiler load is relatively constant, and the fuel has 30-40 percent volatile matter, coal ranging from 0 to 3/4 inch is recommended, with not more than 20-30 percent passing through a 1/4-inch round hole screen. The remainder should be fairly well distributed between 1/2 inch and 3/4 inch at the stoker hopper.

When load varies significantly or drops substantially, a range from 0 to 3/4 inch is recommended, with at least 40 to 45 percent passing through a 3/4-inch screen. When a relatively high combustion rate is maintained, coal with size up to 1-1/4 inch maximum is permissible. A consistently good size distribution between 1-1/4 top size and 3/4-inch screenings must be maintained.

The suspension burning produced by a spreader stoker, while it helps combustion efficiency, has the disadvantage of causing a greater carryover of particulate matter in the flue gas than occurs with other types of stokers. Because much of the carryover is unburned carbon, installation of a carbon reinjection system is necessary to return the unburned fuel to the furnace. Reinjection of this material can generally increase furnace efficiency by 2-3 percent.

Chain or Traveling Grate[1,4]

These stokers are constructed of assembled links, grates or keys, joined together in an endless belt arrangement that passes over sprockets or return bends located at the front and rear of the furnace. Coal is fed from a hopper onto the moving assembly and enters the furnace, where it is heated by radiation from the furnace gases. Hydrocarbon and other combustible gases are driven off by the distillation (pyrolysis) of the coal. The fuel bed is ignited and continues to burn as it moves along; the bed grows progressively thinner as combustion continues. At the end of the grate travel, ash falls off the end of the grate into the ash pit. Although there are structural differences, the operation of the chain grate and other traveling grate types is similar. Generally, these stokers use furnace arches to improve combustion by reflecting heat onto the fuel bed. However, a chain grate stoker has been developed which eliminates the need for a furnace arch by utilizing two rows of overfire air jets located in the front wall. These air jets are effective in completing the combustion of the volatile gases over the first two compartments. These compartments maintain heat release rates higher than is permissible with an arch.

Chain and traveling grate stokers can burn a wide variety of fuels including peat, lignite, subbituminous, free-burning bituminous, anthracite and coke breeze. When burning bituminous coal, proper size distribution is important so that there is passage of sufficient air through the fuel bed.

On a natural draft stoker, the coal should be sized to pass through 1-1/4- to 2-inch screens. Forced draft systems require smaller coal sized to pass through a 3/4-inch to a 1-inch round hole screen. Approximately 25 to 30 percent of the feed should be under 1/4 inch.

Because of their high content of volatile matter, subbituminous coals ignite easily and burn freely. For that reason, the maximum sizing can be somewhat larger than for bituminous coals. Since the average moisture content of subbituminous coal is between 18 and 23 percent, tempering to achieve the desired heat release rate when burned may not be necessary, unless considerable moisture loss occurs during transportation and storage.

Even though lignite has a very high moisture content, this fuel burns well on mass burning stokers. The maximum size for this fuel should be 1-1/4 inches, with all the fines resulting from crushing left in the fuel. Lignites, having a high moisture content ranging from 36 to 40 percent, are more difficult to ignite and the maximum size should be in the 3/4-inch to 1-inch range.

Underfeed Stokers[1,4]

As the name implies, underfeed stokers introduce coal into the fuel bed from below. The coal is pushed along in a feed trough, or retort, by a reciprocating ram. Under pressure from the ram, the coal rises in the retort and spills over onto the bed at either side of the trough. As the fuel rises in the retort, it is subjected to heat from the burning fuel above, and volatile gases are distilled off. The gases are mixed with air introduced through tuyeres, which are openings in the grate section adjoining the trough. The volatile mixture burns as it passes upward through the incandescent zone, sustaining ignition of the rising fuel. Burning continues as the upcoming raw coal continually forces the fuel bed to each side. Combustion is completed by the time the coal reaches the side dumping grates. Ash is intermittently discharged to shallow pits, quenched and removed through doors at the front of the stoker.

There are two types of underfeed stokers: single retort and multiple retort. Single retorts utilize different methods of feeding coal and different grate designs. Feeding methods include the reciprocating ram mentioned earlier, and a retort with a sliding bottom having pusher blocks for advancing coal into the retort. Single retorts generally utilize moving grates to provide fuel-bed agitation and to assist movement of the coal to the dump grates at the side. Because of the agitated bed, these designs handle highly caking coals well.

Multiple retort stokers consist of a series of inclined feeding retorts extending from the rear of the boiler, with tuyere sections between them. Rams push coal into the retort and up into the fuel bed. Incoming coal gradually pushes its way up under the fire and secondary pushers move the whole mass slowly to the rear where combustion is completed and the ash drops into the ash pit. The multiple retort stoker was a logical extension of the single retort design and works best with caking coals. However, very few units are now sold and the method has been displaced by newer stoker methods.

The size of the coal furnished to underfeed stokers has a marked effect on their capacity and efficiency. The most desirable size consists of 1-1/4 inch by zero nut, pea and slack in equal proportion. A reduction in the percentage of fines helps to keep the fuel bed porous and extends the range of use to coals with a high free-swelling index.

Water-Cooled Vibrating Grate Stokers[1,4]

The water-cooled vibrating grate stoker was originally designed and used in Europe. It has had wide acceptance in the United States since its introduction in the middle 1950's because of its simplicity, low fly ash carryover, and very low maintenance.

In a vibrating grate stoker, the entire structure is supported by a number of flexure plates allowing the grid and its grate to move freely in a vibrating action that conveys coal from the feeding hopper onto the grate and gradually to the rear of the stoker. Ashes are automatically discharged to a shallow or basement ash pit.

Vibration of the grates is intermittent, and the frequency of vibration is regulated by a timing device. Timing is regulated by the automatic combustion control system to conform to load variations, synchronizing the fuel feeding rate with the air supply.

The water-cooled vibrating grate stoker is suitable for burning a wide range of coals from bituminous to lignite. Even with coals having a high free-swelling index, the gentle agitation of the fuel bed tends to keep the bed porous without the formation of large agglomerates. A well-distributed, uniform fuel bed is maintained without blow holes or thin spots.

The furnace design for this stoker should include water-cooled walls to prevent slag formation adjacent to the stoker. A rear arch extending over approximately one third of the stoker length directs the gases forward to mix with the fuel-rich volatile gases released in the ignition zone. A short front arch is adequate for most bituminous fuels. The use of high pressure air jets - from 27 to 30 inches of water (gauge) - through the front arch provides turbulent gas mixing and promotes combustion. In rare cases, with extremely low-volatile fuels, some refractory facing of the front water-cooled arch may be desirable to increase the temperature over the ignition section.

Burning rates of these stokers vary with different fuels but, in general, the maximum heat release rate should not exceed 400,000 Btu/sq ft-hr. In this range, carbon carryover is held to a minimum.

Water-cooling of the grates makes this stoker especially adaptable to multiple-fuel firing, as a shift to oil or gas does not require special provision for protection of the grates. A normal bed of ash left as a cover gives adequate protection from furnace radiation.

The strategic placement of burners in this type of furnace configuration may, in many cases, permit operation with a bare grate without exceeding safe limits for metal temperature.

References - Section 3.5.1.2

1. Roe, D.E., A.W. Karr, K.R. Lemmerman, J.E. Sinor, Solid Fuels for U.S. Industry, Volume II, March 1979.

2. Hensel, R.P. and D.A. Harris, Properties of Solid Fuels and their Impact on Boiler Design and Performance, ASME-IEEE-ASCE Joint Power Generation Conference, Dallas, Texas, September 10-13, 1978.

3. Hensel, R.P., Western Coals: Properties That Influence Boiler Design, Coal Technology, 1979, Volume 2, Houston, Texas, November 6-8, 1979.

4. Steam: Its Generation and Use, Babcock and Wilcox, New York, NY, 1978.

5. Harris, D.A., Effect of Various Fuels on Furnace Design, paper presented at REA Generating Conference, Owensboro, Kentucky, June 26-29, 1972.

6. Lowry, H.H. (editor), Chemistry of Coal Utilization - Supplementary Volume, John Wiley & Sons, Inc., New York, 1963.

7. Environmental Control Implications of Generating Electric Power from Coal, Technology Status Report, Volume II, Argonne National Laboratory, December 1976.

8. Duzy, A.F., M.P. Corriveau, R. Byrom, and R.E. Zimmerman, "Western Coal Deposits Pertinent Qualitative Evaluations Prior to Mining and Utilization," Technology and Use of Lignite, Grand Forks, ND, May 18-19, 1977.

9. Honea, F.I., S.J. Selle, and E.A. Sondreal, "Factors Affecting Boiler Tube Fouling From Western and Gulf Coast Lignites and Subbituminous Coals," N.A.C.E. Corrosion/79 Symposium, Atlanta, Georgia, March 13, 1979.

10. Gronhovd, Wagner, and Wittmeir, American Power Conference, 1966.

11. Handbook of Chemistry and Physics, 52nd edition, edited by R.C. Weast, published by the Chemical Rubber Company, Cleveland, Ohio.

12. Sondreal, E.A. and R.C. Ellman, "Fusibility of Ash From Lignite and Its Correlation with Ash Composition," Grand Forks Energy Research Center, RI-75-1, 1975.

13. Sondreal, E.A., P.H. Tufte, and W. Beckering, Ash Fouling in the Combustion of Low-Rank Western Coals, Combustion Science and Technology, 1977, Vol. 16, pp. 95-110.

Selected References - 3.5.1.2.4

The references and abstracts listed below are considered comprehensive sources of information on their selected topics. The list is provided only as a recommended starting point for research into these various areas.

1. Hensel, R.P. and D.A. Harris, Properties of Solid Fuels and their Impact on Boiler Design and Performance, ASME-IEEE-ASCE Joint Power Generation Conference, Dallas, Texas, September 10-13, 1978.

   Since the world has recognized that reserves of gas and oil are inadequate to meet our long-term demands for energy, fuel prices have continued to rise higher and higher. Consequently, there has been a renewed and increased emphasis on the use of lignite, subbituminous and bituminous coals for the generation of power. Although coals represent over 65 percent of all U.S. energy reserves, other materials considered as a potentially supplemental source for heat energy are municipal wastes, wood and forest products, peat and anthracite. While solid wastes and wood may be classified as solid fuels for electric generation, they are usually considered as fuels to be fired supplemental to coal. Since coals are and will continue to be the main source of solid fuels, this discussion is confined to coals and the application of coal properties to furnace design and operation.

2. Sondreal, E.A., P.H. Tufte, and W. Beckering, Ash Fouling in the Combustion of Low-Rank Western Coals, Combustion Science and Technology, 1977, Vol. 16, pp. 95-110.

   This paper gives a complete review of experimental, theoretical and operating data concerning ash fouling of boiler tubes by low-rank coals. The results of many years of research into this problem at the Grand Forks Energy Technology Center are summarized.

3. Duzy, A.F., M.P. Corriveau, R. Byrom, and R.E. Zimmerman, "Western Coal Deposits Pertinent Qualitative Evaluations Prior to Mining and Utilization," Technology and Use of Lignite, Grand Forks, ND, May 18-19, 1977.

   The purpose of this paper is to demonstrate how coal deposits in the western United Sttes are (or should be) evaluated, with the objective being focused on their efficient utilization. Evaluation of a coal deposit for physical development of a coal mine includes a complete coal exploration program, together with a mining feasibility study; each of these areas, including other than quality considerations, would require a very lengthy paper and, thus, will be discussed only briefly for purposes of illustration.

4. Sondreal, E.A. and R.C. Ellman, "Fusibility of Ash From Lignite and Its Correlation with Ash Composition," Grand Forks Energy Research Center, RI-75-1, 1975.

   Fusibility of ash from Northern Great Plains lignites was correlated with the oxide constituents in a 10-component analysis. The correlations developed established methods and guidelines for predicting the fusibility of ash in these coals and adjusting it for boiler fuel use or other conversion processes.

Compared to ash from bituminous coals, lignite ash has higher proportions of Ca, Mg, S, and Na. Indices useful for predicting ash-softening temperature of bituminous coals are not satisfactory for lignites. The need for improved indices for lignites was a principal justification for this study.

3.5.1.3 Environmental Control Technology

Increasingly stringent environmental standards are causing control technologies to become a major element in the design specifications and cost of new coal-fired power plants. These changing performance standards (see Table 3.5.1.6) are forcing utilities to attack the problem on an overall plant scale, integrating furnace design, particulate control, SO_2 scrubbing and other control measures together to produce a reliable system that meets the standards at minimum cost.

The primary areas of environmental concern are:

- Stack gas cleaning
- Solid waste management
- Wastewater effluents
- Fugitive emissions of particulate matter

Recently, increased attention has also been given to trace element and organic emissions.

The following sections discuss standards for performance, control technologies, and design strategies to meet proposed or existing requirements.

3.5.1.3.1 Stack Gas Cleaning

Gaseous emissions from coal-fired plants can be broken into four major categories:

- Particulate
- Sulfur dioxide
- Oxides of nitrogen
- Trace element and organic emissions

The first three areas are presently covered by New Source Performance Standards and hence have received the bulk of the development work and operating experience. Trace element and organic emissions are receiving increasing interest due to concern over possible carcinogenic and other toxic effects. Very little work has been performed in this area, however, and all implications are not completely understood. Also, this study does not address the problem of CO_2 emissions and the environmental effects it may cause.

Particulate Control

Control of particulate emissions from coal-fired plants is a major problem due to the composition, mineral matter content of the coal, and

Table 3.5.1.6

New Stationary Source Performance Standards
For Electric Utility Steam Generating Units, December 1971 and June 1979

Pollutant		December 1971 Standards	June 1979 Standards
SO_2:	Emission limit	1.2 lb/10^6 BTU	1.2 lb/10^6 BTU (based on a 30-day rolling average)
	Reduction in potential emissions		90% when emissions are 0.6 lb/10^6 BTU or greater
			70% when emissions are less than 0.6 lb/10^6 BTU
Particulate Matter:	Emission limit	0.1 lb/10^6 BTU	0.03 lb/10^6 BTU
	Opacity of emission	20%	20% (based on a 6-minute average)
			Based on a 30-day rolling average:
NO_x:	Emission limits	(i) 0.70 lb/10^6 from the combustion of coals except lignite.	(i) 0.50 lb/10^6 BTU from the combustion of subbituminous coal, shale oil, or any solid, liquid, or gaseous fuel derived from coal.
		(ii) 0.80 lb/10^6 BTU from the combustion in a cyclone furnace of any fuel containing more than 25 percent, by weight, lignite which has been mined in North Dakota, or Montana.	(ii) 0.80 lb/10^6 BTU from combustion in a slag tap furnace of any fuel containing more than 25 percent, by weight, lignite which has been mined in N. Dakota, S. Dakota, or Montana.
		(iii) Combustion of a fuel containing more than 25 percent, by weight, coal refuse is exempt from the NO_x standards and monitoring requirements.	(iii) Combustion of a fuel containing more than 25 percent, by weight, coal refuse is exempt from the NO_x standards & monitoring requirements.
		(iv) 0.60 lb/10^6 BTU from the combustion of lignite except as stipulated in (ii) above.	(iv) 0.60 lb/10^6 BTU from the combustion of any solid fuel not specified in (i), (ii), or (iii) above.

Notes: 1. These standards apply to electric utility steam generating units capable of combusting more than 73 megawatts heat input (250 million BTU/hour) of fossil fuel.

2. A major difference between the December 1971 and June 1979 NSPS is in compliance testing. The June 1979 NSPS require continuous stack monitoring and a 30-day rolling average for the SO_2 and NO_x emissions. The December 1971 required that emission monitoring only be performed at the beginning of plant operation and thereafter when the EPA deemed it necessary.

high removal requirements. This problem is aggravated by the presence of combustion-generated fine aerosols resulting from condensation.

Development of a particulate control startegy is dependent on several factors: total particulate loading, size distribution of particles, and degree of removal required. Table 3.5.1.7 provides a general size distribution for particulate matters from the primary coal-fired furnace types,[1] which is plotted in Figure 3.5.1.6. Because of the differences in feed size requirements, stokers produce larger fly ash particles than pc-fired units, with correspondingly lower carryover rates. In addition to the effects of furnace design, the total loadings and size distributions vary somewhat with the rank of coal. Figure 3.5.1.7 shows how size distributions can vary for different types of coal.[2]

Table 3.5.1.7

Fly Ash Characteristics and Size Distributions for Various Furnace Types

	Underfeed Stoker	Spreader Stoker	Pulverized Coal Dry Bottom	Pulverized Coal Slag Tap	Cyclone Furnace
Carry-over fraction of ash in coal,%*	20-30	30-80	55-85	45-55	10-30
Combustible Content, %	10-50	20-60	5-10	5-10	5
Dust loading, lb. 1000 lb. gas**	0.5-4	1.5-10	6-9	4-6	0.6-1.5
Typical size, cumulative percent Microns					
<10	7	11	40		86
<20	15	23	70		91
<44	30	42	80		95
<74	38	56	97		98
<149	57	73	100		100

* Includes combustible.

** Based on coal of 15% ash content. Dust in gas leaving furnace before any collectors. Corrected to 50% excess air.

Source: Reference 3

The degree of removal required is dependent on the federal or state regulations controlling particulate emissions. Until June 1979, the Federal standard was 0.1 lb/10^6 Btu heat input (~99.0 percent removal[a]); the current New Source Performance Standard is 0.03 lb/10^6 Btu heat input (~99.7 percent removal[a]). (This June 1979 NSPS applies to new or modified electric utility steam generators capable of combusting more than 250 million Btu/hr of fossil fuel.[4]) New Mexico has a specific regulation limiting fine particulate of less than 2 microns in diameter to less than 0.02 lb/10^6 Btu.

[a]These removal efficiencies assume a particulate loading of 10 lb/10^6 Btu at the furnace outlet.

Figure 3.5.1.6

Fly Ash Size Distributions
For Various Furnace Types
(plotted from Table 3.5.1.6)

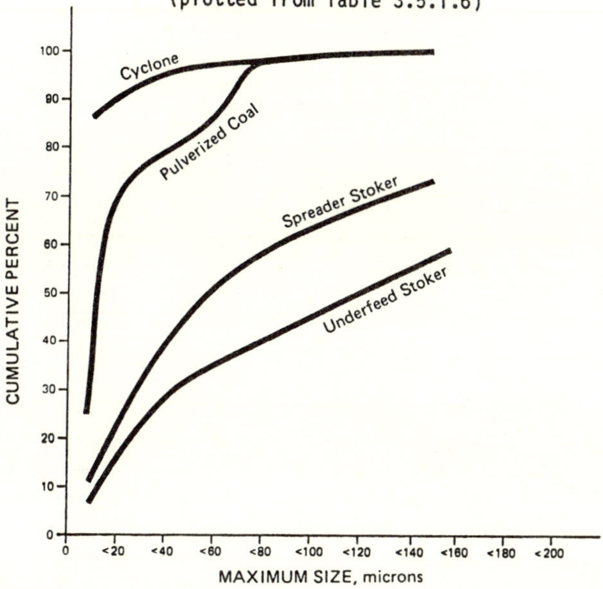

Source: Reference 3.

Figure 3.5.1.7

Inlet Particle-Size Distribution For Ash From Four Coals

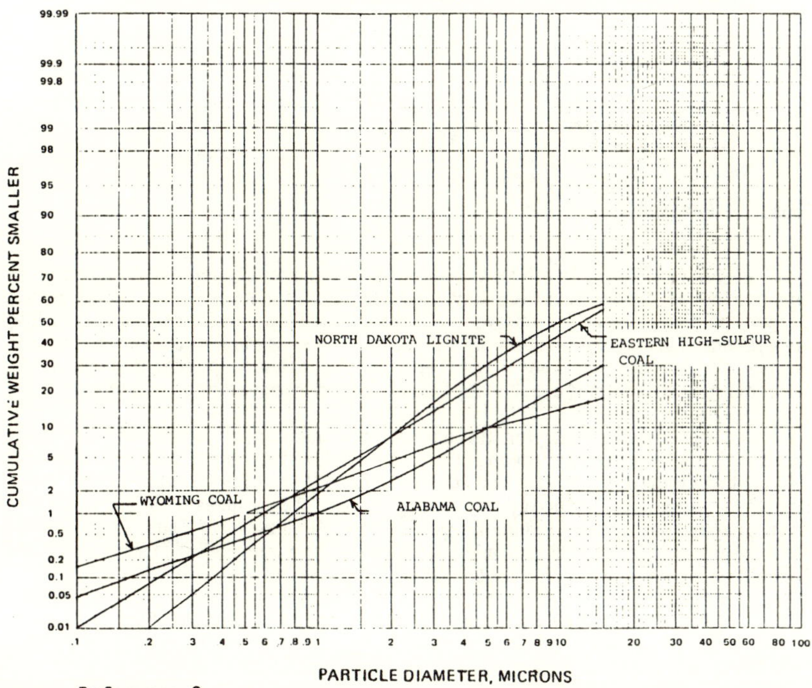

Source: Reference 2

Once required removal efficiencies, particulate loading and size distributions have been established, the specific control device can be selected. Figure 3.5.1.8 gives the approximate removal efficiency as a function of particle size for an assortment of particulate control devices.

Figure 3.5.1.8

Particulate Removal Efficiencies
For Various Control Devices

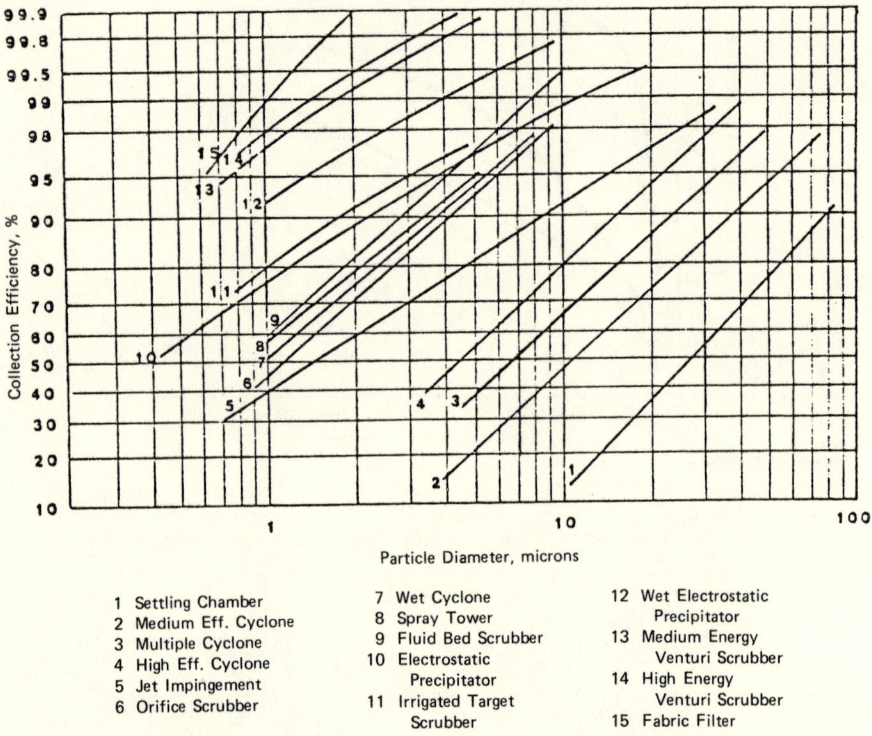

Source: Reference 5

In small scale industrial boilers where stoker furnaces are used, the choice of control device may be made from a number of candidates due to larger particle size. However, in large utility boilers (pulverized-coal or cyclone furnaces) where particle sizes are small, the choice of control device is limited to fabric filters, electrostatic precipitators (ESP's), or venturi scrubbers to achieve the required removal requirements.

The final selection is based on a number of additional characteristics such as pressure drop, flue gas velocity, and cost. Table 3.5.1.8 indicates typical values of these parameters for a number of control devices. High gas velocity requirements and large pressure drops generally discourage the use of venturi scrubbers in favor of the other devices. This means large utilities are faced with the choice of fabric filters or ESP's for their particulate control systems.

Short descriptions of the various control devices follow.

Table 3.5.1.8

Operational Parameters for Various Particulate Control Devices

Equipment	Pressure drop, in H.O.	Gas Velocity, 1000 ft/min	Installed cost, $/cfm
Cyclone collector	3.5-5.0	- -	0.35-1.05
Fabric Filter			
Shaker type	2.5	0.002-0.006	0.75-1.25
Reverse Jet	3.5	0.018-0.028	1.00-1.50
Precipitator	0.9	0.2-0.5	1.00-3.50
Venturi Scrubber			
Low energy	5-15	6[1]; 0.6[2]	0.75-1.50
Medium energy	15-20	9[1]; 0.6[2]	1.00-2.00

1. At throat

2. Through cyclone separator

Source: Reference 3

Electrostatic Precipitators

Electrostatic precipitators utilize electrostatic forces to remove particulate matter from the flue gas stream. The efficiency of collection is directly dependent on the resistivity of the particles. Many western low sulfur coals produce fly ash with higher resistivity than eastern coals under comparable conditions. Figure 3.5.1.9 illustrates the effect of temperature on resistivity. Various strategies to deal with this problem have been developed and will be discussed below.

Single-Stage Precipitator. The single-stage electrostatic precipitator is made up of a number of positively charged, vertical, parallel plates. Between each pair of plates is a discharge cathode made up of a series of fine wires hung vertically 6 to 12 inches apart. Gas passing between the plates is ionized by the intense electric field created between the positively charged plates and the negatively charged wires. Particles are then ionized by the gas and migrate toward the plates as the gas proceeds through the precipitator. Agglomerated particulate matter attached to the walls is shaken free by small hammers called "rappers." Fly ash settles to a dust hopper located below the precipitator and is removed by a screw conveyor or rotary valve. Factors which influence the precipitator performance are:

- ratio of plate area to gas flow
- time particle is exposed to electric field
- strength of field
- gas viscosity
- particle resistivity
- sulfur content of the fuel
- temperature

- carbon content of the particulate
- particulate loading
- ability to remove particles from collector plates

The primary design parameter for ESP's is the "specific collection area" (SCA) defined as the square feet of plate area per 1000 actual cubic feet per minute (ACFM) of flue gas. A graph of SCA versus particulate emissions limit is shown in Figure 3.5.1.10 for various coals.[5]

Figure 3.5.1.9

Typical ASME Laboratory Resistivity Curves as a Function of Temperature and Humidity for Fly Ash from a North Dakota Lignite Having One Percent Sulfur

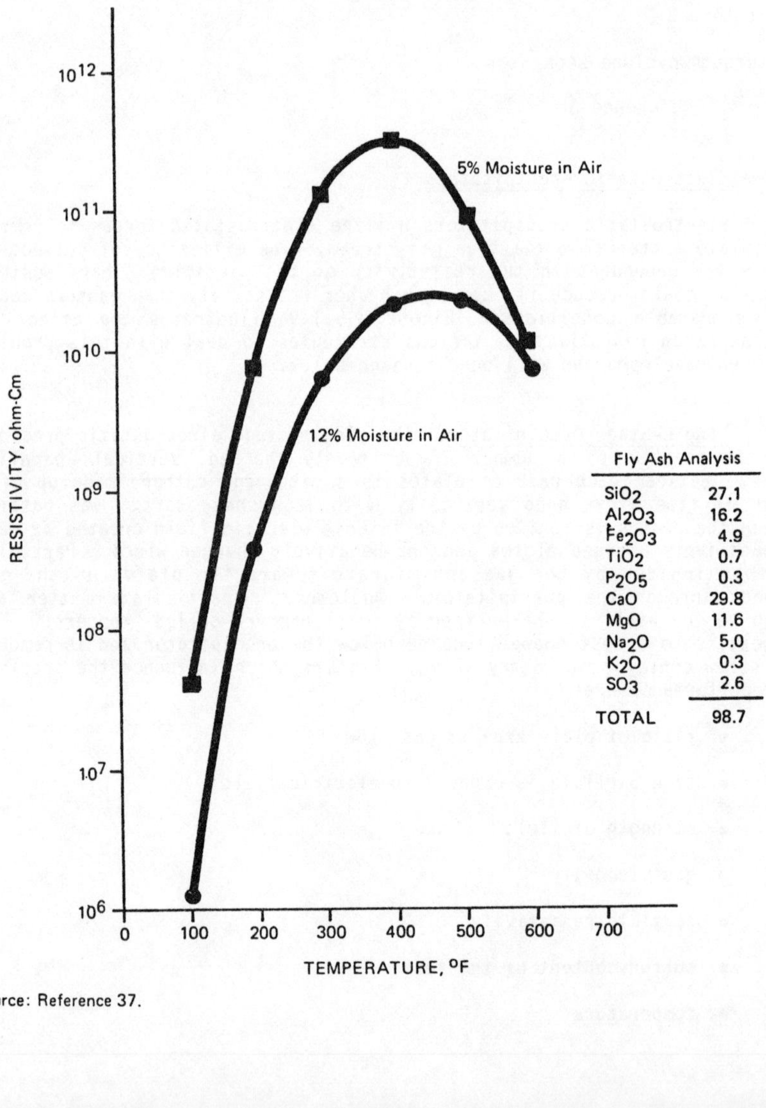

Source: Reference 37.

Figure 3.5.1.10

Collecting Area Requirements For Electrostatic Precipitators

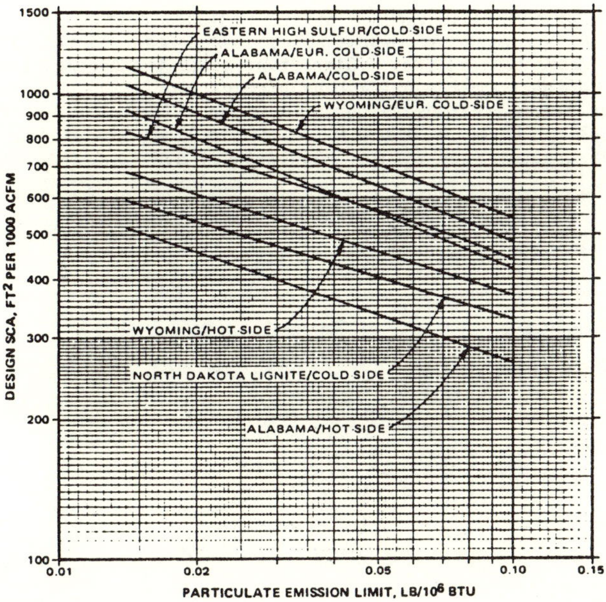

Source: Reference 2

Two-Stage Precipitators. This precipitator is identical to the single-stage precipitator except that ionization is accomplished in a separate stage. The ionization section can be similar to the conventional wireplate precipitator with a higher voltage for efficient ionization, or it can be similar to the high-intensity ionizer shown in Figure 3.5.1.11. In this figure, the ionization step precedes three conventional collector fields. The ionizer array acts as a diffuser plate, with each perforation containing an intense radial corona discharge. Each element contains a cathode disc concentrically contained within an anode venturi throat. The venturi provides a high degree of energy recovery with a pressure drop of approximately 0.5 inches of water. The ratio of cathode to throat area (~0.2) leads to particle charges 3-5 times higher than in conventional ionizers.[4]

The higher resistivity fly ash from some western coals causes a deterioration in both the charging and collection processes. Consequently, the two-stage precipitator can operate efficiently only when sufficient collector plate is also provided.

Hot-side versus Cold-side Precipitation. Utilities have the option of locating their ESP before or after the air preheater. The changes in location alter the temperature of the flue gas stream which in turn greatly affects the resistivity of the fly ash (refer again to Figure 3.5.1.9). A typical cold-side condition is a flue gas temperature of 284°F and corresponding resistivity of 4×10^{12} ohm-cm. This contrasts with typical hot-side conditions of 743°F and a resistivity of 8×10^{9} ohm-cm.[2] Thus, locating the precipitator after the economizer but before the air preheater will result in stable, highly efficient operation because of the high temperatures. The precipitator in this situation can be designed for relatively high current densities and will provide stable operation over a relatively wide range of fuel compositions. Hot-side precipitators are used in some plants burning high-resistivity (low-rank) coals.[2]

The primary disadvantage of the hot-side precipitator is the need to handle an increased gas volume (50 to 100 percent higher) due to operation at elevated temperatures. Other than that, the precipitator is nearly identical to the normal single-stage precipitator. The design must allow for greater thermal expansion during startup. Other considerations which increase the capital and operating costs include insulation, higher heat losses, and greater wear (higher maintenance) due to increased gas flow.[5]

Figure 3.5.1.11

HIGH-INTENSITY IONIZER

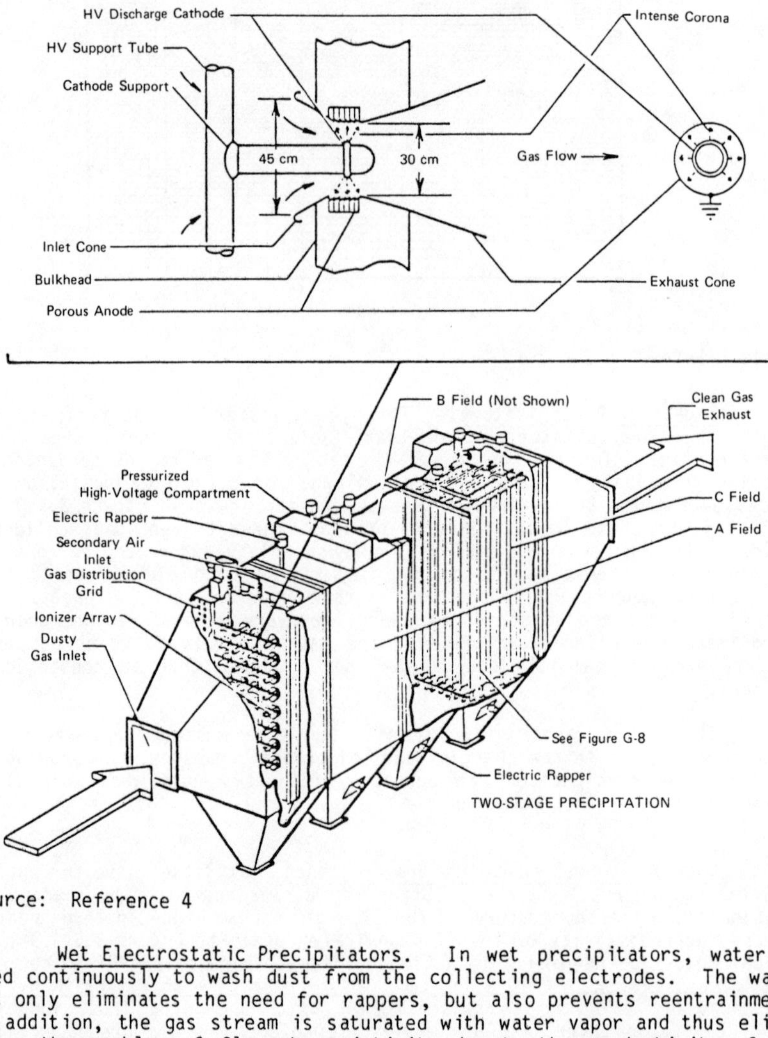

TWO-STAGE PRECIPITATION

Source: Reference 4

<u>Wet Electrostatic Precipitators</u>. In wet precipitators, water is used continuously to wash dust from the collecting electrodes. The water not only eliminates the need for rappers, but also prevents reentrainment. In addition, the gas stream is saturated with water vapor and thus eliminates the problem of fly ash resistivity due to the conductivity of the

water vapor. Consequently the wet electrostatic precipitator may be an attractive method of collecting high resistivity fly ash. The major disadvantage is cost. Wet precipitators must be fabricated from corrosion resistant materials. This may increase the cost to 2.5 times that for dry precipitators. The corrosion problem arises from the reaction of the water with the chemicals in the gas stream. The disposal problems associated with the fly ash slurry will, in most cases, increase the capital and operating costs as well.[5]

Fly Ash Conditioners. Many chemical agents have been investigated to determine their effectiveness as fuel or flue gas additives. The objective is to reduce fly ash resistivity and thus increase collection efficiency.

Test results have been variable. In some instances, the additive is effective but the cost is prohibitive (e.g., phosphorus pentoxide). The most commonly used additive is sulfur trioxide (SO_3). Sulfuric acid, which forms SO_3 upon injection, SO_3, and sodium carbonate all reduce fly ash resistivity. Vanadium and iron oxides supposedly catalyze the oxidation of SO_2 to SO_3, thus increasing SO_3 content in the flue gas. The mechanisms by which other additives increase fly ash removal efficiencies have not been reported. In some cases a fly ash conditioner can be found that will significantly reduce particulate emissions when burning low sulfur western coal. Testing is likely to be required to determine costs and expected efficiency. Some additives may have undesirable environmental side effects.[5]

Fabric Filters

Fabric filters use tubular-shaped fabric bags to filter out particulate matter contained in the flue gas. Filtration is efficient and normally results in removal efficiencies greater than 99 percent. The efficiency of each individual bag, however, varies with time as dust accumulates on the bag surface. After a fixed period of time, a section of the baghouse, containing a number of individual bags, is isolated from the gas stream. Each bag is cleaned by one of the methods described in the following sections. When cleaning is complete, the section is returned to service and another section is removed for cleaning. This cleaning cycle continues until every bag has been cleaned. The cleaning sequence can be operated manually when needed or it can be automated to function continuously. An illustration of a baghouse is given in Figure 3.5.1.12.

Figure 3.5.1.12

Baghouse Schematic

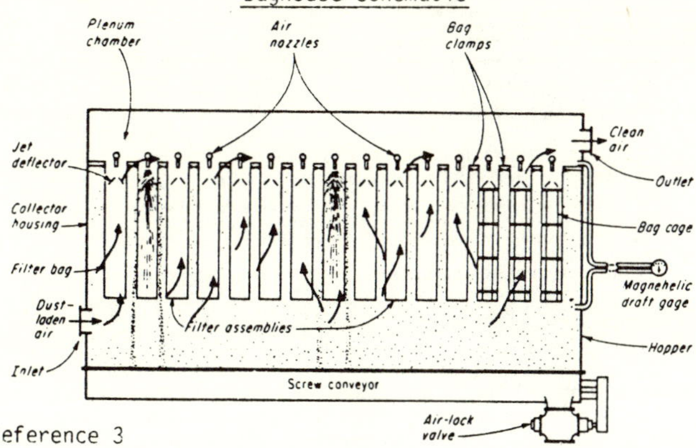

Source: Reference 3

The major design parameters for baghouses include the following (typical values for these parameters on a 500-MW plant are given in parentheses):[2]

- inlet gas volume (1.97×10^6 afcm)
- gross air-to-cloth ratio (1.81)
- cleaning method (reverse-air)
- fabric (Teflon-coated fiberglass)
- total cloth area (1,087,000 ft^2)
- normal operating temperature (300°F)
- maximum operating temperature (550°F)
- normal operating pressure drop (4 in. of H_2O)
- cleaning cycle time (30 minutes)

Baghouse costs remain essentially constant as a function of required removal efficiency due to their inherent high collection efficiency. Electrostatic precipitator costs, on the other hand, increase significantly as collection efficiency requirements increase. Figure 3.5.1.13 illustrates this relationship. Baghouses appear to become the economic choice when permissible emission levels drop much below the 0.1 lb/10^6 Btu level.[2]

Figure 3.5.1.13

Economic Comparison of ESP's and Baghouses for Two Low-Rank Coals

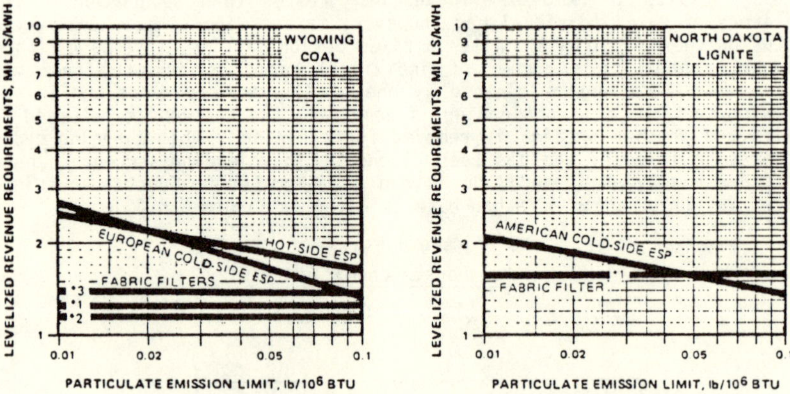

LEVELIZED REVENUE REQUIREMENTS FOR COLLECTION SYSTEMS ON 500 MW (net) POWER PLANTS, BASE YEAR 1978

*1. 20-COMPARTMENT FABRIC FILTER WITH 2-YEAR BAG-REPLACEMENT.
*2. 20-COMPARTMENT FABRIC FILTER WITH 4-YEAR BAG-REPLACEMENT.
*3. 40-COMPARTMENT FABRIC FILTER WITH 2-YEAR BAG-REPLACEMENT

WIDTH OF LINES REFLECTS RELATIVE ACCURACY OF THE ESTIMATES

Source: Reference 2

Because the emission standards have only recently been placed below this level, only a few large, modern pulverized-coal units are presently in operation with a baghouse and therefore no accurate reliability record is available (see Section 3.5.1.5). Many problems with fabric filter life have been noticed in these plants.[29]

The major cleaning methods for baghouses are briefly discussed below:

Air or Mechanical Shaking. Bags can be cleaned by mechanically shaking the frame on which the bags are mounted or by blowing air between the rows to wind-whip each bag. This is an efficient, low-cost cleaning method which normally is reliable.

Bubble Cleaning. In this method, a large pulse of air is added to the top of the bag. The bag takes the shape of a bubble which rapidly moves down the bag like a wave. This rippling action flexes and cleans the bag as it progresses downward. The major disadvantage of this method is the large quantity of compressed air required. A similar method can be applied to bags where the normal flow is outside-in (frames are required to support the bag). In this situation, an ejector is used to create a pressure wave in the bag. This can be done without isolating the bag from the incoming gas stream.

Reverse Air Cleaning. Here, the bags are cleaned by reversing the air flow after the bags are sealed off from the gas stream. This is an efficient method but requires a large capacity to handle the recycled gas.

Sonic Cleaning. Sonic generators can be used to release the dust from the bags but reverse air or shaking is usually required in addition for efficient cleaning. This method also may develop complications from governmental regulations relating to noise.

Mechanical Collectors

Mechanical collectors use centrifugal, inertial, or gravitational forces to separate particulate matter from the gas stream. Cyclones use centrifugal forces and can be purchased singly or in multiple units containing a number of smaller cyclones. The inertial collector effects an abrupt change in the direction of flow, causing the momentum of the particulate matter to separate it from the flue gas. The settling chamber uses gravitational forces to accomplish particulate removal.

Single Cyclones. In the single cyclone, gas enters tangentially to a cylindrical or conical vessel. As the gas is forced to travel in a spiral path, centrifugal forces are applied to the particulate matter. These particles move to the outside wall where their speed is reduced and they fall to the bottom for removal. The clean gas is removed from the center of the cyclone through a standpipe. Dust is removed by a screw conveyor or rotary valve. In other types of cyclones, the vortex is created by vanes located at the top of the cone. Here, gas spins as it passes through the vanes and particles are removed by a process similar to that described above.

The centrifugal force on a particle varies directly according to the mass of the particle and the square of the velocity, and inversely with the radius of the collector. Thus larger particles, having more mass, are removed with a higher efficiency than lighter particles. Efficiency can be increased by increasing the velocity or by decreasing the diameter of the cyclone. The latter, however, reduces the capacity. Efficiency also varies directly with the dust loading. Mechanical collectors are most effective for a particle size range above 10 microns. With particles less than 10 microns, the efficiency drops below 90 percent. Cyclones are not used for the high efficiency collection of fly ash that is currently required.

Multiple Cyclones. Multiple cyclones incorporate several small diameter cyclones into a single intake and discharge manifold. This enables the parallel operation of a number of high efficiency cyclones at higher flow rates than with any single cyclone. Careful design is required in these systems to ensure equal dust loadings to each cyclone. This is necessary for high efficiency, and to prevent backflow, plugging or re-entrainment from the dust bin.

Cyclones also can be used in series to increase the efficiency above that obtained from a single cyclone. This is an effective method to improve efficiency without increasing the velocity (which may break down dust agglomerates).

Inertial Collectors. Inertial separation devices cause sudden changes in the direction of the gas stream. The inertia of the particles causes them to impinge on a target or into a collection area. Only the larger particles are removed in this way. Smaller particles deflect with the flue gas and are reentrained. Performance data are scarce, and must be obtained from pilot tests or vendor's data. In general, the efficiency of inertial collectors is comparable to that for high efficiency cyclones. Pressure drop can be as high as 6 inches of water.[5]

Settling Chambers. Gas velocity is a major factor influencing the particulate loading from a boiler. As the gas velocity increases, an increased quantity of larger particles is entrained in the gas stream. In a similar manner, when the gas velocity is reduced, the large particles settle out and can be removed from the gas stream. Particulate control devices have been based on this principle to settle large size particles. The gas stream is passed through an enlarged section of ductwork which allows cinders and large dust particles to settle. These devices, once called cinder traps, are not highly efficient and must be used in conjunction with other more advanced control systems. Certain boiler designs lend themselves to a higher rate of settling, or they can be supplied with an enlarged economizer section which acts as a settling chamber. Some applications have settling changers just before electrostatic precipitators to reduce the load. The efficiency is 30 to 40 percent for particles with sizes less than 45 microns. Particles greater than 45 microns can be removed with efficiencies of 75 percent or more.[5]

Wet Particulate Scrubbers

Wet scrubbers remove particulate matter from the gas stream by contacting the gas with water. Some gaseous pollutants can be removed simultaneously with the fly ash. This aspect will be discussed in a later section concerning SO_2 removal.

The mechanism for particulate removal is primarily inertial impaction (i.e., impaction on water droplets or on wetted surfaces). The exit gases from all wet scrubbers are saturated with water and cooled below the inlet temperature. In general, these gases must be reheated to provide buoyance for dispersion by tall stacks and to protect stack linings. Wet scrubbers also require treatment and/or disposal of the scrubber solution to remove suspended solids and treat any chemicals leached from the fly ash.

Capital costs for wet scrubbers are generally lower than other particulate control methods with the exception of mechanical collection. A wet scrubber may cost only 25 percent as much as an electrostatic precipitator. However, the operating and maintenance costs are much higher, as high as five times that for a precipitator. This is due to several factors.[5]

Scrubbers have the disadvantage of a large pressure drop (thus requiring more energy than electrostatic precipitators) for efficient collection of fine particles. They also introduce the necessity of handling a wet fly ash slurry and recirculating the scrubber liquid. Scaling, corrosion, plugging, and mist entrainment in the effluent gas are typical problems.[6]

Five general types of scrubbers are currently in use. Each has been fabricated in a variety of designs by a number of different manufacturers. The general characteristics of each type of wet scrubber are described below.

Venturi Scrubbers.[5] In the venturi scrubber, dust-laden flue gas is contacted with water and forced through a converging section in the ductwork. The high velocities in the venturi throat atomize the liquid and result in intimate gas-liquid contact. Gas from the throat area is decelerated in the diverging section to allow the wetted particles to agglomerate. Entrained water and fly ash are removed from the gas stream in a cyclonic separator.

High particulate removal efficiencies can be accomplished with the venturi scrubber. However, an extremely high energy consumption is required to obtain these efficiencies. Pressure drops of up to 100 inches of water are required for high efficiency (99 percent plus) collection of submicron particulate. Some manufacturers have designed specially shaped diverging sections to maximize the removal efficiency for a given pressure drop and energy consumption.

Many different scrubber designs have been developed to date. One unique design uses a deck of parallel metal pipes. Liquid spray and inlet gas are directed toward the deck which functions like a series of parallel venturi slots as the gas passes between the pipes. Other designs allow for adjustment in the throat size by placing a cone-shaped plug in circular throats or by providing dampers on either side of rectangular throats. Some manufacturers also vary the method of liquid introduction; spraying, swirling around the walls, and injection on the top of the cone have all been used.

Spray Scrubbers.[5] Spray scrubbers are the simplest type of wet scrubber available, and come in almost any shape. Scrubbing liquor is sprayed into the gas stream by the use of high pressure spray nozzles. Some designs use splash plates to finely disperse the liquid in the gas stream. Countercurrent, cocurrent and cross flow configurations have all been used.

The efficiency of the spray scrubber depends on the relative velocities of the liquid and gas. Cocurrent scrubbers have the lowest relative velocities and are effective only for the removal of larger particles. Countercurrent designs create maximum relative velocities and efficiencies, up to 90 percent for particles 10 microns and larger. Most spray scrubbers are relatively low in capital and operating costs.

Moving Bed Scrubbers.[5] Packed bed towers provide good removal efficiencies because of the high ratio of liquid surface area to gas volume. However, packed beds cannot be used for particulate removal because of the tendency for packing to plug up with ash. The moving bed scrubber avoids this problem because of the fluidized action of the bed. The bed is made up of low density plastic spheres contained between two retaining grids. Flue gas velocity, usually kept between 7 and 13 ft per second, is a critical factor in maintaining bed mobility. Velocities higher or lower than this will allow the bed to rest on either retaining grid and thus decrease the efficiency. Scrubbing liquor is sprayed from the top down through the moving bed. Particulate matter is removed from the gas stream by impaction on the wetted spheres. Several stages can be provided for high efficiency, if the application warrants it.

Moving bed scrubbers generally are effective on particles down to 1 micron in size. Some designs can provide efficiencies of 99 percent for particles 2 microns and larger. Scrubbers of this type have been used on utility boilers with 95 to 98 percent efficiency. Maintenance costs, however, may be high due to the erosion of the plastic spheres by the fly ash. Northern State Power uses a venturi prescrubber in front of its marble bed scrubber to try to reduce wear problems.

Impingement Scrubbers. Impingement scrubbers use impaction on a wetted plate as the primary mechanism for particulate removal. Gas is generally forced through an opening and impacted on a wetted plate positioned perpendicular to the gas flow. These "impingement baffles" are positioned side by side to form a tray similar to a valve tray in distillation, or they can be positioned vertically in a variety of configurations. Recirculating liquor is sprayed onto the impingement tray and captures the particulate matter which contacts the baffle. Efficiencies can range from 80 to 99 percent depending upon the particle size, the number of impingement stages, and the pressure drop. One design provides 90 to 99.5 percent efficiencies for 0.5 micron particles when operating at a pressure drop of 30 inches of water.

Orifice Scrubbers.[5] Orifice scrubbers use a self-induced spray curtain to remove particulate matter from the gas stream. The spray is formed when flue gas is forced through a partially submerged orifice. This operation produces a high degree of gas-liquid contact but the atomization effect is not as high as that for the venturi scrubbers. Fly ash, impacted upon the fine liquid droplets, is removed from the gas stream in a mechanical type collector or a mist eliminator. Many manufacturers have also developed scrubbers of this type which generate the spray curtain in a specially designed entrainment channel. Pressure drops for both types of orifice scrubbers can range up to 15 inches of water. Particulate removal efficiencies are lower than for other high energy scrubbers, but are still acceptable for particles greater than 1 micron in size.

Novel Particulate Control Devices[7]

In recent years because of the need to collect smaller particles to achieve higher collection efficiencies, hybrid devices have been developed utilizing principles of "conventional" methods. The cost of conventional devices capable of collecting fine particles rises significantly due to need for increased size and energy requirements. This has also provided incentive for development of advanced systems.

Many of the advanced concepts or devices combine the principles of cyclones and electrostatic precipitators to achieve very high collection efficiencies. Others use electrostatically charged fabric filters to capture submicron size particles. Figure 3.5.1.14 shows particle penetration[a] versus diameter for many of these novel devices tested by the EPA. The major disadvantage to all of these advanced systems is high cost but as regulations become more stringent, their viability increases.

Experimental testing of new particulate concepts has yielded favorable results, but indicates that much work still remains in the field.

Figure 3.5.1.14

Penetration vs. Particle Size for Novel Devices tested by EPA

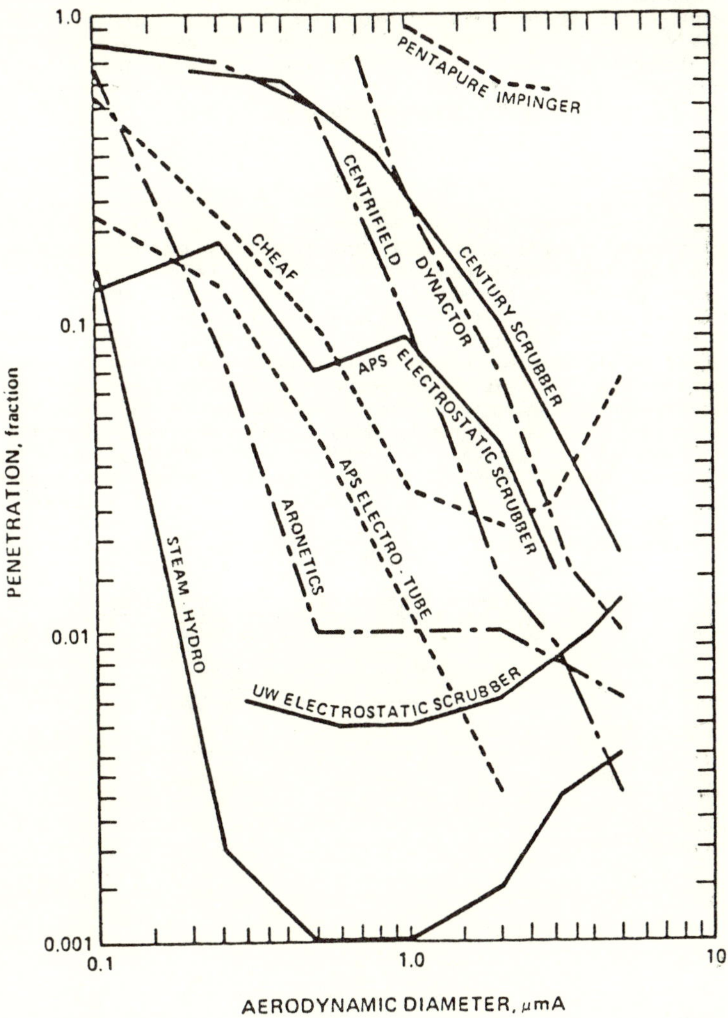

Source: Reference 7

[a]Penetration is defined as the fraction of the particulate matter that remains in the flue gas stream at the device outlet; penetration (fraction) = 1 - collection efficiency (%)/100.

One pilot scale (0.47 m^3/sec) investigation of a three-electrode particle precharger achieved charging levels in high resistivity dusts which were attainable only on low and moderately resistive dusts in conventional systems.[38] In a separate laboratory scale study, a range of operating temperatures (38 to 343°C) was used to determine the effectiveness of particle charging, especially at hot-side ESP temperatures, in positive and negative corona discharge. Electrode geometries and electrical operating conditions utilized were typical of full scale ESP's. Experiments were conducted at fine particles, with radii between 0.3 and 1.5 microns. At comparable voltages and currents for positive and negative corona discharges, the ratio of the values of negative to positive charge for radii between 0.6 and 1.3 microns was shown to increase from about 1 to 2 as temperatures were raised from 37 to 343°C. A mathematical model of the charging setup (using ionic charging theory) showed good agreement with all positive charging data, but good agreement with negative charging data only at temperatures below 37°C. Differences between measurements and model predictions are consistent with the postulation of free electron charging at high temperature (negative corona).[39]

Flue Gas Desulfurization

Sulfur dioxide (SO_2) is generated when the sulfur contained in coal is burned. Until June 1979, the federal standards regarding SO_2 emissions from coal plants required emissions to be less than 1.2 lb/10^6 Btu.[4] Some western coals have a sulfur content sufficiently low to meet this standard with no SO_2 control systems. This coal is referred to as "compliance coal." The June 1979 New Source Performance Standard for SO_2 has altered this situation such that all utility coal-fired plants will be required to remove from 70 to 90 percent of the SO_2 from the flue gas. The exact removal efficiency required is a function of the sulfur content of the received coal (see Figure 3.5.1.15). The curved portion of the line represents the maximum allowable emission of 0.6 lb/10^6 Btu, which applies at all removal levels between 70 and 90 percent. The maximum emission of 1.2 lb/10^6 Btu remains in force for high sulfur coals not included on the graph.

Figure 3.5.1.15

New Source Performance Standards for SO_2 Emissions from Electric Utility Steam Generators

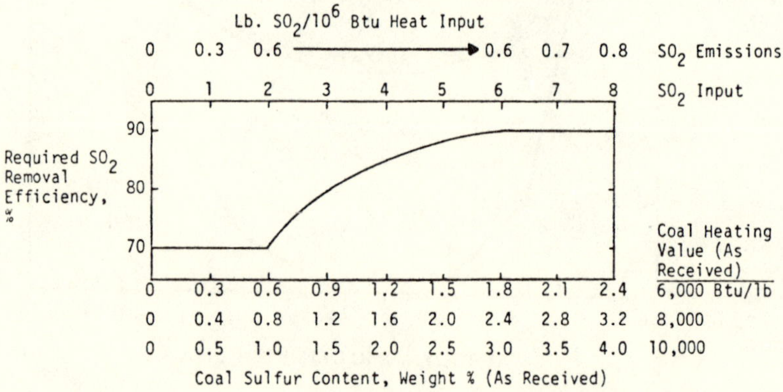

Two general types of control have been used over the past 50 years: wet and dry flue gas desulfurization (FGD) processes. The dry processes

consist of a gas/solid contacting system where the SO_2 is absorbed by the solid matter. Dry lime/limestone injection, and adsorption on activated carbon, finely ground charcoal, coke and silica gel have all been tested and applied to boilers to varying degrees. The wet FGD processes have had by far the greatest use to date in utility and industrial boiler applications. Eight processes have been used on a commercial scale in the United States. In all of these processes, SO_2 is absorbed in a water slurry or solution and reacts to form sulfurous acid (H_2SO_3). This acid in turn is neutralized by an alkali contained in the system. All processes are classified into one of two categories, nonregenerable or regenerable. Nonregenerable processes generate a sulfite/sulfate waste product which requires disposal. The regenerable processes regenerate the alkali and produce one of several by-product sulfur compounds which can be marketed. A list of the major throwaway and regenerable processes is given in Table 3.5.1.9. The following sections will discuss these various processes.

Table 3.5.1.9

Sulfur Dioxide Removal Processes

Nonregenerable	Regenerable
Lime Slurry	Activated Carbon
Limestone Slurry	Adsorption
Ash Alkali	Magnesium Oxide
Aqueous Sodium	Wellman-Lord
Aqueous Ammonia	Copper Oxide Adsorption
Double Alkali	
Chiyoda 101 and 121	
Citrate	
Aqueous Carbonate	
Catalytic Oxidation	

Source: Reference 5

Wet Scrubbing Processes

Lime/Limestone Wet Scrubbing[5]

In most existing applications there are three sections to the lime slurry process, an electrostatic precipitator (ESP) and two stages of lime scrubbing. Particulate matter is removed in the ESP and in the first stage venturi scrubber. SO_2 is removed primarily in the second stage scrubber, but a considerable amount (up to 20 percent) is removed in the first stage scrubber. Various scrubber designs have been successfully used in both the first and second stages. With some systems, the electrostatic precipitator can be eliminated by high efficiency particulate scrubbing in the first stage. Several new units are coupling spray towers with ESP's or baghouses for SO_2 removal.

Lime (CaO) is slaked (reacted with water) to form calcium hydroxide

($Ca(OH)_2$), and is used in the scrubber recirculation system. SO_2 dissolves in the lime slurry (pH of 6.0 to 8.0) and forms sulfurous acid (H_2SO_3). This acid dissociates and reacts with the lime slurry according to reaction 1. The sulfurous acid can also be oxidized by any dissolved oxygen (reaction 2) and react with the lime slurry according to reaction 3 forming a calcium sulfate precipitate (gypsum). SO_2 also can be oxidized in the gas phase to SO_3 which would react similarly to form gypsum.

$$Ca(OH)_2 + H_2SO_3 \longrightarrow CaSO_3 \cdot 2H_2O \quad (1)$$

$$H_2SO_3 + 1/2O_2 \longrightarrow H_2SO_4 \quad (2)$$

$$Ca(OH)_2 + H_2SO_4 \longrightarrow CaSO_4 \cdot 2H_2O \quad (3)$$

Cleaned gas from the scrubber is reheated (typically to 175` F) for buoyancy, prevention of corrosion and acid rain, and vented through tall stacks. The reheat system can be eliminated but this may create a visible water plume from the stack and an undesirable "acid rain" in cold weather. The reheat step detracts from the overall plant efficiency. In the case of low sulfur western coals, for which 70 percent SO_2 removal is required, another alternative is available. It will be possible to scrub one portion of the flue gas stream to a higher removal efficiency, allowing the remaining portion to bypass the scrubber such that the combined stream will still meet the removal requirements. The hot bypass stream will provide the reheat and hence reduce scrubbing costs.

Calcium sulfite/calcium sulfate slurry from both scrubbers is thickened and the resulting sludge can be disposed of in a pond, or else chemically fixed and disposed of in a landfill. Sludge also can be oxidized in an air-blown reactor to form an environmentally acceptable disposal sludge or hopefully to form by-product gypsum for use in the production of wallboard.

Lime slurry scrubbing without sludge oxidation is applicable only to plants which have an environmentally acceptable sludge disposal system. This process can be designed to meet a wide range of flow rates from 1200 scfm for small industrial users to over 3.5 million scfm for large utilities. SO_2 loadings as high as 4000 ppm have been handled. Currently, only one application of the lime slurry process is operational in the industrial sector and one other is planned. The operational unit handles 280,000 acfm of flue gas. SO_2 removal efficiencies for the lime slurry process depend on several variables:

- Reactivity of the scrubbing liquid
- Degree of gas-liquid contact
- Liquid to gas ratio
- Gas residence time
- The number of scrubber stages

High efficiencies can be realized by appropriately adjusting the above factors. In most installations operating today, the efficiency has ranged between 80 and 98 percent.

A major contributor to low overall efficiency is scrubber maintenance, where the scrubber must be bypassed to facilitate repair or cleanout. This availability problem can be overcome at some expense by installing extra modules and transferring the load.

The limestone slurry process is similar to the lime slurry process in that particulate matter can be removed either in an ESP or simultaneously with SO_2 removal in the first stage wet scrubber. The chemistry is also similar to that for the lime process since both are calcium based.

Limestone, containing as much as 95 percent $CaCO_3$ and varying amounts of $MgCO_3$, is crushed in a wet ball mill and pumped as a slurry to the absorber recycle system. Pilot studies have been conducted to investigate the use of benzoic acid in the dissolution of the limestone. SO_2 dissolves in the limestone slurry (pH=5.8-6.4) and reacts with the dissolved $CaCO_3$ according to reaction 1. $CaCO_3$ will also react with the oxidized sulfur compounds (H_2SO_4) to form gypsum as shown in reaction 2.

$$CaCO_3 + H_2SO_3 \longrightarrow CaSO_3 \cdot 1/2H_2O + CO_2 + 1/2H_2 \quad (1)$$

$$CaCO_3 + H_2SO_4 \longrightarrow CaSO_4 \cdot 2H_2O + CO_2 \quad (2)$$

Stack gas reheat and sludge disposal requirements are basically the same as for the lime slurry process. Typically the sludge in the limestone process has a better settling rate than that in the lime slurry process.

Unlike the magnesium in lime, the higher the magnesium content in limestone ($CaCO_3$), the lower the reactivity for sulfur removal. Some developers suggest the use of a catalyst to increase absorptivity, reduce scaling, and to produce an easily oxidized sludge.

Efficiencies and the application range for the limestone slurry process are similar to the lime process. Low availability caused by scaling in the scrubber and the resulting maintenance is the major item of concern. Several industrial applications of the limestone process are currently operational. These systems are used for boiler operation, and handle as much as 320,000 scfm of flue gas.[30]

The major advantages of these systems are:[8]

- Relative simplicity
- No product to market
- Process can tolerate fly ash in system

The limestone system has the added advantage of more development and operating experience than any other system; it is also the least expensive system at moderate removal efficiencies. However, the lime slurry process provides better control over SO_2 removal efficiency through pH adjustments.

The major disadvantages of the lime/limestone wet scrubbing systems are:[8]

- Large quantities of solid waste require disposal
- Stack gas reheat is desirable
- Mudding occurs at wet/dry interfaces
- Very high liquid-to-gas ratios are required

The limestone system has also faced scaling and corrosion problems in the scrubber, pumps, and reheater tubes. Tables 3.5.1.10 through 3.5.1.16 summarize existing and planned lime, limestone and lime/limestone processes in utility and industrial applications.

Table 3.5.1.10

Utility FGD Systems
Lime Processes
--Operational--

Utility	Unit		Location	Capacity MW	Startup Date	Supplier[1]	Design Removal % Part.	SO_2	Scrubber product disposal
Allegheny Power	Pleasants	1	Belmont,W.Va.	519	3/79	B&W	99.55	90.00	landfill
Big Rivers Electric*	Green	1	Sebree, Ky.	242	10/79	AAF	--	--	--
Columbus & Southern Ohio Electric	Conesville	5	Conesville, Oh.	411	1/77	UOP	99.60	89.50	landfill
Columbus & Southern Ohio Electric	Conesville	6	Conesville, Oh.	411	6/78	UOP	99.60	89.50	landfill
Duquesne Light	Elrama	1-4	Elrama, Pa.	510	10/75	CMCO	99.00	83.00	landfill
Duquesne Light	Philips	1-6	South Height, Pa.	410	7/73	CMCO	99.00	83.00	landfill
K.C. Power and Light	Hawthorn	3	Kansas City, Mo.	90	11/72	CE	99.00	70.00	pond
K.C. Power and Light	Hawthorn	4	Kansas City, Mo.	90	8/72	CE	99.00	70.00	pond
Kentucky Utilities	Green River	1-3	Central City, Ky.	64	9/75	AAF	99.50	80.00	lined pond
Louisville Gas & Electric	Cane Run	4	Louisville, Ky.	188	8/76	AAF	99.00	85.00	lined pond
Louisville Gas & Electric	Cane Run	5	Louisville, Ky.	190	12/77	CE	99.00	85.00	lined pond
Louisville Gas & Electric	Mill Creek	3	Louisville, Ky.	442	8/78	AAF	99.00	85.00	pond
Louisville Gas & Electric	Paddy's Run	6	Louisville, Ky.	72	4/73	CE	99.00	90.00	pond
Pennsylvania Power	Bruce Mansfield	1	Shippingport, Pa.	917	12/75	CMCO	99.80	92.10	landfill
Pennsylvania Power	Bruce Mansfield	2	Shippingport, Pa.	917	12/77	CMCO	99.80	92.10	landfill
Utah Power & Light	Hunter	1	Castle Dale, Ut.	360	5/79	CMCO	99.50	80.00	pond
Utah Power & Light	Huntington	1	Price, Ut.	366	5/78	CMCO	99.50	80.00	landfill
				6199					

Notes:
1. Suppliers: UOP - Air Correction Division, UOP
CMCO - Chemico
CE - Comubstion Engineering
AAF - American Air Filter
B&W - Babcock & Wilcox

Source: Reference 30

*Updated from Reference 5 based on Reference 33.

Table 3.5.1.11

Utility FGD Systems
Lime/Limestone Processes
--Operational--

Utility	Unit	Location	Capacity MW	Startup Date	Supplier	Design Removal % Part.	SO_2	Scrubber product disposal
TVA	Shawnee 10A	Paducah, Ky.	10	4/72	UOP	--	--	--
TVA	Shawnee 10B	Paducah, Ky.	10	4/72	Chemico	--	--	--
			20					

Utility FGD Systems
Lime/Limestone Processes
--Under Construction--

Utility	Unit	Location	Capacity MW	Startup Date	Supplier	Design Removal % Part.	SO_2	Scrubber product disposal
Louisville Gas & Electric	Mill Creek 1	Louisville, Ky.	358	4/81	Combustion Engineering		--	--
Louisville Gas & Electric	Mill Creek 2	Louisville, Ky.	350	4/82	Combustion Engineering		--	--
			708					

Utility FGD Systems
Lime/Limestone
--Planned--

Utility	Unit	Location	Capacity MW	Startup Date	Supplier	Design Removal % Part.	SO_2	Scrubber product disposal
Central Maine Power	Sears Island 1	Sears Island, Me.	600	11/87	--	--	--	--

Source: Reference 31

Table 3.5.1.12

Utility FGD Systems
Limestone Processes
--Operational--

Utility	Unit	Location	Capacity MW	Startup date	Supplier	Design Removal % Part.	Design Removal % SO_2	Scrubber product disposal
Alabama Electric	Tombigbee 2	Leroy, Al.	179	9/78	PPS	99.30	85.00	lined pond
Alabama Electric	Tombigbee 3	Leroy, Al.	179	6/79	PPS	99.30	85.00	lined pond
Arizona Electric Power	Apache 2	Cochise, Az.	195	8/78	RC	99.50	85.00	pond
Arizona Electric Power	Apache 3	Cochise, Az.	195	4/79	RC	99.50	85.00	pond
Arizona Public Service	Cholla 1	Joseph City, Az.	119	10/73	RC	80.00	92.00	pond
Arizona Public Service	Cholla 2	Joseph City, Az.	350	4/78	RC	99.70	75.00	pond
Central Illinois Light	Duck Creek 1	Canton, Il.	378	9/76	RS/ENV	99.80	85.00	lined pond
Colorado Ute Electric Assoc.*	Craig 2	Craig, Co.	447	8/79	PPS	--	--	--
Indianapolis Power and Light	Petersburg 3	Petersburg, In.	532	12/77	UOP	99.30	85.00	pond
Kansas City Power and Light	La Cygne 1	La Cygne, Ks.	874	2/73	B&W	99.50	80.00	unlined pond
Kansas Power and Light	Jeffrey 1	Wamego, Ks.	540	8/78	CE	99.00	80.00	pond
Kansas Power and Light	Lawrence 4	Lawrence, Ks.	125	1/76	CE	98.90	73.00	pond
Kansas Power and Light	Lawrence 5	Lawrence, Ks.	420	11/71	CE	98.90	52.00	pond
Salt River Project*	Coronado 1	St.Johns, Az.	280	11/79	PK	--	--	--
South Carolina Public Service Authority	Winyah 2	Georgetown, S.C.	140	7/77	B&W	99.40	69.00	pond
South Mississippi Electric	R.D.Morrow 1	Hattisburg, Ms.	124	8/78	RS/ENV	99.60	85.00	landfill
South Mississippi Electric	R.D.Morrow 2	Hattisburg, Ms.	124	6/79	RS/ENV	99.60	85.00	landfill
Southern Illinois Power Coop	Marion 4	Marion, Il.	184	5/79	B&W	99.60	89.40	landfill
Springfield City Utilities	Southwest 1	Springfield, Mo.	194	4/77	UOP	99.70	80.00	landfill
Tennessee Valley Authority	Widows Creek 8	Bridgeport, Al.	550	5/77	TVA	99.50	80.00	pond
Texas Utilities	Martin Lake 1	Tatum, Tx.	595	4/77	--	99.40	--	landfill
Texas Utilities	Martin Lake 2	Tatum, Tx.	595	5/78	RC	99.40	--	landfill
Texas Utilities	Martin Lake 3	Tatum, Tx.	595	2/79	RC	99.40	--	landfill
Texas Utilities	Monticello 3	Mt.Pleasant, Tx.	800	5/78	CMCO	99.50	74.00	landfill
			8714					

Notes:

1. Suppliers: PPS - Peabody Process Systems
   RC - Research Cottrell
   RS/ENV - Riley Stoker/Environeering
   UOP - Air Correction Division, UOP
   PK - Pullman Kellogg
   B&W - Babcock and Wilcox
   CE - Combustion Engineering
   TVA - Tennessee Valley Authority
   CMCO - Chemico

Source: Reference 31

*Updated from Reference 5 based on Reference 33.

Table 3.5.1.13

Utility FGD Systems Lime Processes
-Under Construction-

Utility	Unit	Capacity MW	Supplier	Startup Date
Allegheny Power	Pleasants 2	519	Babcock & Wilcox	9/80
Big Rivers Electric	Green 2	242	American Air Filter	6/80
Cincinnati Gas & Electric	East Bend 2	650	Babcock & Wilcox	10/81
East Kentucky Power Coop	Spurlock 2	500	ADL/Combustion Equipment Co.	10/80
Louisville Gas & Electric	Mill Creek 4	495	American Air Filter	7/81
Pennsylvania Power	Bruce Mansfield 3	917	Pullman Kellogg	5/80
Utah Power & Light	Hunter 2	360	Chemico	6/80
		3683		

Utility FGD Systems Lime Processes
-Contract Awarded-

Utility	Unit	Capacity MW	Supplier	Startup Date
Allegheny Power	Mitchell 33	300	Chemico	8/82
Arizona Public Service	4 Corners 4	755	United Engineering	1982
Arizona Public Service	4 Corners 5	755	United Engineering	1982
		1810		

Source: Reference 31

Table 3.5.1.14

Utility FGD Systems
Limestone Processes
-Under Construction-

Utility	Unit	Capacity MW	Supplier	Startup Date
Arizona Public Service	Cholla 4	126	Research Cottrell	6/80
Assoc. Electric Coop	Thomas Hill 3	670	Pullman Kellogg	1/82
Basin Electric Power Coop	Laramie Riv. 1	600	Research Cottrell	4/80
Basin Electric Power Coop	Laramie Riv. 2	600	Research Cottrell	10/80
Colorado Ute Electric Assoc.	Craig 1	447	Peabody Process Systems	1/80
Commonwealth Edison	Powerton 51	450	UOP	11/79
Hoosier Energy	Merom 2	441	Mitsubishi Inter.	1/82
Indianapolis Power & Light	Petersburg 4	530	Research Cottrell	10/83
Kansas Power & Light	Jeffrey 2	490	Combustion Engineering	6/80
Salt River Project	Coronado 2	280	Pullman Kellog	1/80
San Miguel Electric Coop	San Miguel 1	400	Babcock & Wilcox	7/80
Sikeston Bd. of Mun. Util.	Sikeston 1	235	Babcock & Wilcox	1/81
S. Carolina Public Service	Winyah 3	280	Babcock & Wilcox	5/80
Springfield Water, Light & Power	Dallman 3	205	Research Cottrell	7/80
TVA	Widows Creek 7	575	Combustion Engineering	10/80
Texas Power & Light	Sandow 4	382	Combustion Engineering	7/80
		6711		

Utility FGD Systems
Limestone Processes
-Contract Awarded-

Utility	Unit	Capacity MW	Supplier	Startup Date
Hoosier Energy	Merom 1	441	Mitsubishi Inter.	4/81
Houston Lighting & Power	W.A. Parish 8	512	- -	11/81
Lakeland Util.	McIntosh 3	364	Babcock & Wilcox	10/81
Pub. Serv. of Indiana	Gibson 5	650	Pullman Kellogg	/82
S. Carolina Public Service	Winyah 4	280	American Air Filter	5/81
S. Western Electric Power	Henry Perkey 1	720	UOP	12/83
TVA	Paradise 1	704	Chemico	6/82
TVA	Paradise 2	704	Chemico	3/82
Texas Mun. Power Agency	Gibbons Creek 1	400	Combustion Engineering	1/82
Texas Utilities	Martin Lake 4	750	Research Cottrell	/85
		5525		

Source: Reference 31

Table 3.5.1.15

Utility FGD Systems
Limestone Processes
-Planned-

Utility	Unit	Capacity MW	Supplier	Startup Date
Central Illinois Light	Duck Creek 2	416	--	1/84
Northern States Power	Sherburne 3	860	Combustion Engineering	5/84
Seminole Electric	Seminole 1	620	--	6/83
Seminole Electric	Seminole 2	620	--	6/85
Texas Power & Light	Twin Oaks 1	750	--	8/84
Texas Power & Light	Twin Oaks 2	750	--	8/85
NY State Elec. & Gas	Somerset 1	870	--	6/84
Pacific Gas & Electric	Montezuma 1	800	--	6/86
Pacific Gas & Electric	Montezume 2	800	--	6/87
Salt River Project	Coronado 3	280	--	/84
		6766		

Source: Reference 31

Table 3.5.1.16

SUMMARY OF COMMITTED LIME/LIMESTONE SYSTEMS FOR U.S. INDUSTRIAL BOILERS AS OF MARCH 1978

Process	Vendor	Company/ Location	New or retrofit	Size of FGD unit scfm	Size of FGD unit MW	Characteristics	Startup mo/yr
Lime	Koch Engineering	Armco Steel Middletown, OH	R	84,000	42	Coal 0.8% sulfur	1975
Lime	Carborundum Environ. Systems, Ltd.	Carborundum Abrasives Buffalo, NY	R	30,000	15	Coal 2.2% sulfur	1980
Lime/ Limestone	Research Cottrell-Bahco	Rickenbacker Air Force Base Columbus, OH	R	55,000	27	Coal 3.6% sulfur	1976
Limestone		Dupont Texas	N	320,000	160	Coal 0.5% sulfur	1982

Source: Reference 30

Ash Alkali Wet Scrubbing

The fly ash alkali process is an effective way to utilize the alkaline content of the fly ash produced from the combustion of low sulfur western coal. This low sulfur level coupled with the relatively low removal efficiencies (70 to 80 percent) usually attained by the fly ash alkali process make an acceptable combintation for many coals even under the 1979 NSPS (refer again to Figure 3.5.1.15). However, the low efficiencies have restricted the use of the process in the past to "near compliance" coals.

The fly ash alkali process uses the alkaline content of the fly ash to remove SO_2 from the flue gas. Lime or limestone is used as a supplement in case the fly ash alkali provides insufficient SO_2 removal to meet emission standards. Flue gas from the boiler is contacted with recycle water (pH 2.8-4.5), typically in a venturi scrubber. Here the fly ash is removed and the alkaline content, consisting of compounds such as Na_2O,

MgO, and CaO, is leached out. SO_2 is absorbed in the recycle liquor and oxidized to form sulfuric acid; little sulfurous acid is formed. This is due to the high oxygen-to-sulfur ratio encountered with the combustion of western coal. Sulfuric acid in turn reacts with the hydrated alkaline species according to the following reactions:

$$Ca(OH)_2 + H_2SO_4 \longrightarrow CaSO_4 \cdot 2H_2O \qquad (1)$$

$$Mg(OH)_2 + 5H_2O + H_2SO_4 \longrightarrow MgSO_4 \cdot 7H_2O \qquad (2)$$

$$2NaOH + H_2SO_4 \longrightarrow Na_2SO_4 \cdot 2H_2O \qquad (3)$$

Clean gas from the scrubber is passed through a mist eliminator and is typically reheated to 175° F before discharge. Fly ash removed in the scrubber settles out in the bottom of the scrubber and is pumped as a slurry to the thickener. Slaked lime or fine ground limestone is added to react with any residual H_2SO_4 and to adjust the pH for scale prevention. Thickener sludge is disposed of in a pond, or else dewatered, fixed and disposed of in a landfill. Overflow from the thickener is recycled to the scrubber.

The potential usefulness of the ash alkali process for a specific coal depends primarily on the molar ratio of calcium oxide and other alkaline materials in the ash to coal sulfur content. For a coal containing 7.5 percent ash, with 25 percent calcium oxide equivalent in the ash, the calcium oxide is chemically equivalent to slightly more than 150 percent of a 0.7 percent coal sulfur content. Some lignites have an alkali-to-sulfur mole ratio of several hundred percent. The amount of fly ash alkali available and utilized in a wet scrubber depends on the coal ash content, method of firing (pc versus cyclone), and scrubber operating conditions; however, there are many western low-rank coals with ample fly ash alkali available to react with sulfur dioxide in a wet scrubber.[10]

In the ash alkali scrubbing process, operating conditions such as liquid-to-gas (L/G) ratio and pH have a direct impact on fly ash alkali utilization and supplementary alkali requirements. As an example, for any given SO_2 removal requirement, as the L/G is increased, the operating pH in the system can be decreased somewhat while still meeting the outlet SO_2 concentration requirement. The utilization of the alkaline fly ash increases as the operating pH of the system decreases by encouraging the dissolution of fly ash components. Hence a tradeoff exists because reducing L/G lowers the cost but reduces the fly ash utilization.[9]

The fly ash alkali process, because of its simplicity, can be retrofitted to some existing plants. However, scaling and low efficiency may be a problem.

As an alternative to the process described above, fly ash can be removed in an electrostatic precipitator and stored for use in the FGD process. In this case both the fly ash and any lime used would be added to the thickener. This system can be applied easily to existing plants which have particulate removal systems already in operation. With the fly ash being removed in a separate system, the absorber can be optimized to increase SO_2 removal efficiencies and to lower pressure drop.

Thirteen applications of the ash alkali process are currently operating in coal-fired power plants (see Table 3.5.1.17). Some utilities consider their systems to be particulate removal systems only, removing as little as 15 percent of the SO_2; SO_2 emissions are sufficiently low that no FGD system is needed. With proper design, removals of over 85 percent of the SO_2 have been demonstrated at favorable alkali to sulfur ratios. Low availabilities caused by scaling within the scrubber and mechanical problems have resulted in unacceptable performance on some units; however, chemical scale formation can be controlled.

Table 3.5.1.17

FGD Wet Scrubbers Utilizing Alkaline Western Fly Ash

Utility/Station	Capacity	Status	FGD Design	Alkali Source
Northern States Power Co. Sherburne County	Unit 1 - 700 MW Unit 2 - 700 MW Unit 3 - 700 MW	In-service In-service Planned -- 1984	Venturi-marble bed Venturi-spray tower	Subbituminous fly ash - limestone
Minnesota Power & Light Co. Clay Boswell	Unit 3 - 1 MW Unit 4 - 500 MW	Pilot plant Under construction	Venturi-spray tower	Subbituminous fly ash - lime
	Unit 1 - 500 MW	Planned -- mid-1980's	Undetermined	
Montana Power Co. Colstrip	Unit 1 - 360 MW Unit 2 - 360 MW Unit 3 - 700 MW Unit 4 - 700 MW	In-service In-service Planned Planned	Venturi-spray tower	Subbituminous fly ash - lime
Square Butte Electric Coop. Milton R. Young	Unit 2 - 440 MW	In-service	Spray tower	Lignite fly ash - lime
United Power Association-Cooperative Power Association Coal Creek	Unit 1 - 550 MW Unit 2 - 550 MW	Under construction Under construction	Spray tower	Lignite fly ash - lime

Source: Reference 10

This process is ideal for remote plants where access to alternative alkali sources is difficult. Large storage areas are not required for raw materials and less area is required for ponding and disposal as compared to other throwaway processes. The settling properties of the sludge are excellent, but the trapped water may contain high concentrations of dissolved salts derived from the ash. An increase in the sodium or magnesium concentration in the scrubber liquor has been shown to increase SO_2 removal by a double alkali type mechanism.

Aqueous Sodium Process

The aqueous sodium process is the most widely used process for flue gas desulfurization in the industrial sector. The process uses a solution of sodium carbonate or sodium hydroxide to remove SO_2 from flue gases. The sodium carbonate can be obtained from raw trona (~60 percent sodium carbonate equivalent), from commercial soda ash, or from a waste liquor containing soda ash. Sodium hydroxide can be obtained from a waste caustic stream or purchased. The latter is the most expensive source of the alkali. This process is the simplest and most reliable of the nonregenerable FGD processes.

Hot flue gas from the boiler passes through a mechanical type dust collector or an electrostatic precipitator to remove most of the fly ash (75 percent plus). The gas enters a venturi scrubber which contacts the gas with an aqueous sodium solution. Here the remaining fly ash and some of the SO_2 is removed according to reactions 1 and 2. The residual SO_2 is absorbed in a tower containing a single sieve tray. Gas from the tower is passed through a mist eliminator and typically is reheated before it is discharged to the atmosphere.[5]

$$Na_2CO_3 + SO_2 \longrightarrow Na_2SO_3 + CO_2 \qquad (1)$$

$$2NaOH + SO_2 \longrightarrow Na_2SO_3 + H_2O \qquad (2)$$

New scrubber solution is made by mixing trona with water. The solution is then clarified to remove inert materials contained in the ore. Sodium carbonate under certain conditions may crystallize out of concentrated solutions, thereby plugging pumps and piping. Downtime because of these "frozen" lines can be averted by operating at reduced concentrations or by heat traced piping. Waste liquor from the scrubber recycle systems is continuously bled off (5 percent) and neutralized before disposal in a lined pond. As an alternative, the disposal solution can be oxidized by aeration to form sodium sulfate (Na_2SO_4). This solution, with no oxygen demand, can then be disposed, treated, or evaporated to produce byproduct sodium sulfate. The major advantages of this process are:[8]

- Extremely simple
- No scrubber scaling
- Low liquid-to-gas ratio in scrubber
- No product to market
- Process can tolerate fly ash in system
- Corrosion and erosion are minimal
- High removal efficiency is possible
- Response time of process to changes in SO_2 feed is rapid

The aqueous sodium process is especially applicable to boilers where low SO_2 and particulate emissions are required. It also is well suited to applications that currently produce a waste effluent containing caustic or soda ash in significant quantities. The removal efficiency for the process can approach 99 percent.

The disadvantages are:

- Na_2CO_3 or NaOH is an expensive raw material
- Soluble sodium salts require disposal
- Stack gas reheat is desirable

At present there are over 100 sodium-based scrubbers operating commercially on industrial boilers, totaling an equivalent of over 1000 MW. In addition, four utility boilers totaling 925 MW (Ried Gardner 1, 2 and 3, in Moapa, NV, and the Jim Bridger Station in Rock Springs, WY) use the aqueous sodium process. Flows vary from 8,000 to 723,000 scfm. SO_2 concentrations to the scrubber are reported to range between 150 and 2000 ppm, although higher concentrations could be handled efficiently.[5] These data are tabulated in Tables 3.5.1.18 and 3.5.1.19.

Aqueous Ammonia Process

The aqueous ammonia process uses a solution of ammonium sulfate ($(NH_4)_2SO_4$), ammonium sulfite ($(NH_4)_2SO_3$), ammonium bisulfite (NH_4HSO_3), and ammonium hydroxide (NH_4OH) to remove particulate and SO_2 from boiler fuel gases. The process was developed by replacing the water in a particulate scrubber with ammonia-laden process water. SO_2 is readily absorbed and reacts with the ammonium salts according to reactions 1, 2, and 3; SO_2 can be oxidized to form SO_3 which reacts similarly to form $(NH_4)_2SO_4$. The ammonium sulfite also can be oxidized to the sulfate form as shown in reaction 4.

$$(NH_4)_2SO_3 + SO_2 + H_2O \longrightarrow 2\ NH_4HSO_3 \quad (1)$$

$$2NH_4OH + SO_2 \longrightarrow (NH_4)_2SO_3 + H_2O \quad (2)$$

$$NH_4HSO_3 + NH_4OH \longrightarrow (NH_4)_2SO_3 + H_2O \quad (3)$$

$$(NH_4)_2SO_3 + 1/2O_2 \longrightarrow (NH_4)_2SO_4 \quad (4)$$

Gas is contacted with recirculating absorbent in a venturi scrubber. Entrained solution is removed in a cyclonic separator equipped with a mesh or chevron type mist eliminator. Scrubber liquor from the venturi and the separator is combined in a reaction tank where the pH is adjusted with fresh ammonia liquor. The pH is a critical control parameter, because a low pH produces a dense "blue" plume due to the gas phase reaction of SO_2 and ammonia.[5] Stack gas reheat can be used to eliminate the water vapor plume normally associated with wet FGD processes.

A bleed stream is removed from the reaction tank to maintain a constant water and ammonia balance. This stream can be oxidized and filtered to form a pure ammonium sulfate fertilizer for resale as a solution or after crystallization.

The aqueous ammonia process is well suited to boiler applications where an ammonia waste stream is available at low cost. A higher cost ammonia source can be used if the by-products can be marketed. However, because of the environmentally unacceptable blue plume, only 14 ammonia units have ever been installed, mostly in sugar processing plants.[30]

Table 3.5.1.18

Utility FGD Systems
Sodium Carbonate Processes
--Operational--

Utility	Unit	Location	Capacity MW	Startup Date	Supplier[1]	Design Removal % Part.	Design Removal % SO_2	Scrubber product disposal
Nevada Power	Reid Gardner 1	Moapa, Nv.	125	4/74	ADL/CEA	97.00	90.00	pond
Nevada Power	Reid Gardner 2	Moapa, Nv.	125	4/74	ADL/CEA	97.00	90.00	pond
Nevada Power	Reid Gardner 3	Moapa, Nv.	125	6/76	ADL/CEA	99.00	85.00	pond
Pacific Power & Light	Jim Bridger	Rock Springs, Wy.	550	9/79	UOP	--	--	--
			925					

Notes:
1 Suppliers: ADL/CEA - Arthur D. Little/Combustion Equipment Assoc.
UOP - Air Correction Division, UOP

Source: Reference 31

Table 3.5.1.19

Performance Data for Operating Industrial Sodium Scrubbing Systems

Installation/ Location	Sorbent	....Fuel.... Type*	% S	Start-Up Date	No. of FCD Units	SO_2 Inlet (ppm)	Percent Removal	Waste Disposal
Alyeska Pipeline Valdez, Alaska	NaOH	O	<0.1	6/77	1	150	96	oxidation/dilution
American Thread Martin, NC	Caustic waste	C	1-1.5	1973	2	500	70	pond
Belridge Oil McKittrick, CA	NaOH	O	1.1	6/78	2	500	90	wastewater treatment
Canton Textiles Canton, GA	Caustic waste	C	0.8	6/74	1	500	70	pond/waste treatment
Chevron Bakersfield, CA	Na_2CO_3	O	1.1	7/78	3	700	90	pond/waste treatment
FMC Green River, WY	Na_2CO_3	C	1	5/76	2	800	95	pond
General Motors Dayton, OH	NaOH	C	0.7-2.0	9/74	2	$1.43/10^6$**	86	clarify/adjust pH/ to sewer
General Motors Pontiac, MI	NaOH	C	0.8	4/76	2	–	–	combine with ash/landfill
General Motors St. Louis, MO	NaOH	C	3.2	1972	2	2,000	90	oxidize/neutralize/discharge
General Motors Tonawanda, NY	NaOH	C	1.2	6/75	4	$1/10^6$**	90	combine with ash/landfill
Georgia Pacific Orosett, AK	Caustic waste	B,C,O	1.5-2	7/75	1	500	80	to city sewers
Getty Oil Bakersfield, CA	Na_2CO_3	O	1.1	6/77-12/78	6	600	90-96	pond
Great Southern Cedar Springs, CA	Caustic waste	B,C,O	1-2	1975	2	1,000	85-90	ash pond
ITT Rayonier Fernandina, FL	NaOH	B,O	2-2.5	1975	2	1,200	80-85	to paper process
Kerr-McGee Trona, CA	Na_2CO_3	O	0.5-5	6/78	2	–	98	pond
Mead Paperboard Stevenson, AL	Na_2CO_3	O	1.5-3	1975	1	1,500	95	to paper process
Mobil Oil San Ardo, CA	Na_2CO_3/NaOH	O	2-2.5	1974	28	1,500	90	pond
Nekoosa Papers Ashdown, AK	NaOH	C	1-1.5	2/76	2	600	90	waste treatment
Northern Ohio Sugar Freemont, OH	NaOH	C	1	10/75	2	–	–	pond
St. Regis Paper Cantonment, FL	NaOH	B,O	<1	1973	1	–	80-90	clarification/aeration
Texaco San Ardo, CA	NaOH	O	1.7	11/73	32	1,000	73	pond/wells/softening and reuse
Texasgulf Granger, WY	Na_2CO_3	C	0.7	9/76	2	860	90	pond

*B is bark, C is coal and O is oil.
**Pounds per British thermal units.

Source: Reference 30

Double Alkali Process[5,6]

The double alkali process is sometimes classified as a regenerable FGD system. Even through the scrubbing liquor is regenerated, a by-product sulfur compound is not produced. Because of this and because a large volume of waste sludge is generated, this process is more accurately classified as a nonregenerable process. The process uses a sodium hydroxide (NaOH)/sodium sulfite (Na_2SO_3) solution to cool the gas and to absorb SO_2. SO_2 absorbed in the liquor reacts with the sodium compounds according to reactions 1 and 2. Sodium sulfate (Na_2SO_4) and other sulfate compounds also can be formed by the oxidation of any of the sulfur species. Clean gas from the scrubber is passed through a mist eliminator and is reheated prior to discharge.

$$2NaOH + SO_2 \longrightarrow Na_2SO_3 + H_2O \quad (1)$$

$$Na_2SO_3 + SO_2 + H_2 \longrightarrow 2\ NaHSO_3 \quad (2)$$

Scrubbing liquor from the recycle system, containing sodium salts and possibly some unreacted sodium hydroxide, is combined with a number of other streams and sent to the regeneration section of the plant. Here the liquor is reacted with lime or limestone to precipitate sulfur compounds and to restore the alkaline content of the liquor for recycle. Lime reacts with sulfite in the liquor according to reactons 3 and 4, and limestone according to reacton 5. Sulfate reacts in a similar manner.

$$Ca(OH)_2 + 2NaHSO_3 \longrightarrow Na_2SO_3 + CaSO_3{\cdot}1/2H_2O + 3/2H_2O \quad (3)$$

$$Ca(OH)_2 + Na_2SO_3 + 1/2H_2O \longrightarrow 2\ NaOH + CaSO_3{\cdot}1/2H_2O \quad (4)$$

$$CaCO_3 + 2NaHSO_3 + 1/2H_2O \longrightarrow Na_2SO_3 + CaSO_3{\cdot}1/2H_2O + CO_2 + H_2O \quad (5)$$

The concentration of sulfate in solution depends on the degree of sulfite oxidation. However, because of the solubility relationship, the concentration of sulfate at steady state rises sharply for relatively small increases in oxidation. If the relative oxidation of sulfite exceeds 15-25 percent, the concentration of sulfate will be too high for efficient operation. This can be corrected by purging the solution or by increasing the pH with lime to precipitate gypsum (similar to reaction 4). With lime the system can be operated over a wider pH range than with limestone. Systems which have high oxidation and require regeneration to NaOH (low concentrations of Na_2SO_3) are called "dilute." Those systems which have low oxidation and regenerate to Na_2SO_3 are termed "concentrated." Concentrated systems usually have an active sodium (Na+ associated with SO_2 absorption) concentration greater than 0.15 molar.

Precipitated solids are removed from the regenerated liquor in a clarifier. Some of the solids are recycled back to the regeneration section to reduce the possibility of scale formation in the reaction tank. Recycle liquor for the dilute system contains a considerable amount of calcium, and must be treated with Na_2CO_3 (trona or soda ash), Na_2SO_3 or CO_2 to precipitate the calcium. This softening process eliminates scaling in the scrubber. The concentrated system has a high concentration of Na_2SO_3 which prevents high calcium concentrations. Thus the reactor clarifier is not needed. Solids from the regeneration process are dewatered, washed to recover the alkali content, and disposed of in a landfill. Handling properties of the sludge are substantially different from lime or limestone sludge because it is a granular material and does not require fixation.

To use the double alkali process, the plant must have a suitable sludge disposal system. The process is currently being used on several industrial boilers having flow rates ranging from 10,000 to 640,000 CFM with SO_2 loadings reported from 800 to 2,000 ppm. Three utility applications are currently under construction. These range in size from 250 to 575 MW.

The advantages of this process are basically the same as the sodium process. The major disadvantages are:[8]

- Process is complex and requires many additional pieces of equipment relative to lime/limestone scrubbing
- Large quantities of solid waste require disposal

- Stack gas reheat is desirable
- Two or three separate solids handling systems are required
- Sodium losses occur and must be made up

SO_2 removal efficiencies for the double alkali process, similar to the aqueous sodium process, have been very high. The system can be designed for removal efficiencies in excess of 90 percent because the sodium alkali scrubbing agent is very reactive. Availability of the double alkali process (the fraction of time the system is operational) has been reported to be greater than 90 percent in most installations. U.S. utility and industrial double-alkali systems are listed in Tables 3.5.1.20 and 3.5.1.21.

Table 3.5.1.20

Utility FGD Systems
Dual Alkali Processes
--Operational--

Utility	Unit		Location	Capacity MW	Startup Date	Supplier[1]	Design Removal % Part.	Design Removal % SO_2	Scrubber product disposal
Central Illinois Public Service	Newton	1	Newton, Il.	617	9/79	B/E	99.50	95.00	--
Louisville Gas & Electric	Cane Run	6	Louisville, Ky.	288	4/79	ADL/CEA	--	95.00	pond
Southern Indiana Gas & Electric	A.B. Brown	1	West Franklin, In.	265	3/79	FMC	99.50	85.00	landfill
				1170					

Notes:

1. Suppliers: B/E - Buell/Envirotech
ADL/CEA - Arthur D. Little/Combustion Equipment Assoc.
FMC - FMC

Source: Reference 31

Table 3.5.1.21
Full Scale Industrial Double-Alkali Systems in the U.S.

System Operator & Location	System Application	Vendor or Developer	Size[c] (MW, Equivalent)	Active Alkali	Start-up Date
FMC Modesto, CA	Reduction kiln	FMC	10 (gas rate) 30 (regen.)	Conc.	Dec. 1971
General Motors Parma, OH	Coal-fired industrial boilers	General Motors	64	Dilute	March 1974
Caterpillar Tractor Co. Joliet, IL	Coal-fired industrial boilers	Zurn Industries	30	Dilute	Sept. 1974
Firestone Pottstown, PA	Coal-fired industrial boilers (demonstration)	FMC	4	Conc.	Jan. 1975
Gulf Power Co. Scholz Plant, Sneads, FL (Southern Company Services)	Coal-fired utility boiler (prototype)	CEA/ADL	20	Conc.	Feb. 1975[a]
Caterpillar Tractor Co. Mossville, IL	Coal-fired industrial boiler	FMC	70	Conc.	Oct. 1975
Caterpillar Tractor Co. Morton, IL	Coal-fired industrial boiler	Zurn Industries	19	Dilute	Jan. 1978
Caterpillar Tractor Co. East Peoria, IL	Coal-fired industrial boilers	FMC	100	Conc.	April 1978
Caterpillar Tractor Co. Mapleton, IL	Coal-fired industrial boiler	FMC	65	Conc.	(Jan. 1979)[b]
ARCO Polymers Monaca, PA	Coal-fired industrial boilers	FMC	118	Conc.	(June 1980)[b]
Grissom Air Force Base Bunker Hill, IN	Coal-fired industrial boilers	Neptune/ Airpol	15	Conc.	(Sept. 1979)[b]
Chansler Oil Bakersfield, CA	Oil-fired industrial boiler	FMC	35	Conc.	(Mar. 1979)[b]

a. System ceased operation in July 1976
b. Projected start-up date
c. Size based on a conversion of 2000 scfm per MW equivalent

Source: Reference 30

Chiyoda 101 and 121 Processes[6]

Chiyoda International has been involved in the development of two flue gas desulfurization systems. The CT-101 process is based on sulfuric acid, and was favorably tested on a pilot scale at the Scholz Power Plant in 1975-76. The CT-101 process is utilized in a few plants (up to 300 MWe) in Japan. More recently, Chiyoda has engaged in the development of the CT-121 process, a limestone-based system.

Chiyoda 101

In this Japanese process the SO_2 is absorbed in dilute (2-5 percent) sulfuric acid containing iron salts to catalyze oxidation. The oxidation is completed in a separate tower through which a flow of air is maintained. A slip stream of acid is withdrawn and treated with limestone to precipitate gypsum. The scrubber and the oxidizer in this process are appreciably larger than scrubbers used in other FGD processes and larger pumps must be used for the high liquid/gas ratio that is required.

Advantages

1. No scrubber plugging or scaling
2. Waste product is dry gypsum
3. Process can tolerate fly ash in system
4. Process is insensitive to limestone reactivity
5. Good particulate removal ahead of SO_2 absorber leads to high quality gypsum

Disadvantages

1. Extremely high liquid-to-gas ratios are required
2. High power requirements
3. Scrubber and oxidizer are extremely large
4. Product gypsum is dewatered in a centrifuge which is a high maintenance item
5. Acid resistant materials are required
6 Process has not been demonstrated on a coal-fired boiler
7. Stack gas reheat is desirable
8. Large volumes of dilute H_2SO_4 must be handled
9. Magnesium and chloride ions may require a purge stream

Chiyoda Thoroughbred 121 System

Background

Chiyoda International started bench scale development work on the

CT-121 Process in 1975 at the Kawasaki R&D Center. Testing was performed in a 650-scfm pilot plant, and conceptual design studies for commercial operations were completed. A prototype unit began operation on August 30, 1978 at the Scholz Power Plant of Gulf Power in Chattahoochee, Florida. Equipment for the prototype unit was modified from the CT-101 plant that had been previously operated at that station.[31]

The CT-121 process utilizes limestone, and the overall chemistry is similar to conventional limestone systems. Unique features are that all chemical and process steps are carried out in one vessel, and that the sorbed SO_2 is completely oxidized to sulfate (gypsum).

Process Description

The CT-121 process uses a jet bubbling reactor to absorb, oxidize, and neutralize SO_2 from flue gas. The reactor is a single vessel, consisting of flue gas inlet and outlet, air inlet, limestone slurry inlet, and gypsum slurry outlet. Air and mechanical agitation are also provided. A simplified flow diagram for this process is shown in Figure 3.5.1.16.

Figure 3.5.1.16

Process Flow Diagram of Prototype Plant, Gulf Power Co., Sholz Station

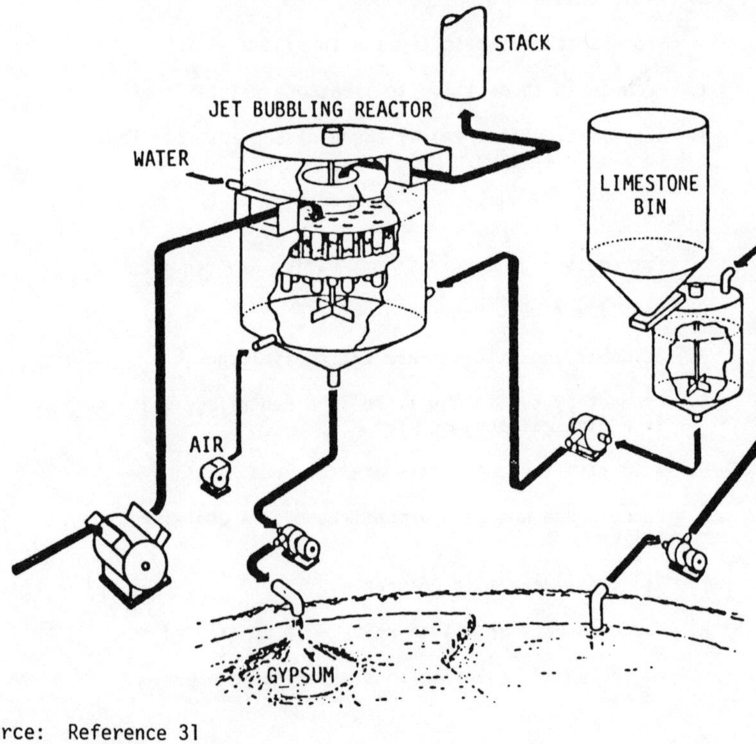

Source: Reference 31

Flue gas enters the reactor through an array of vertical spargers which are submerged 4 to 16 inches below the liquid surface. Gas velocity of 15-65 ft/sec entrains the liquid and generates a bubbling or froth layer. SO_2 is absorbed in the froth layer to form calcium sulfite which is oxidized to calcium sulfate. The froth layer provides good mass transfer for this SO_2 absorption reaction to occur. Cleaned flue gas is demisted and exhausted to the atmosphere.[30,31]

Immediately below the "jet bubbling zone" through which the flue gas passes, the liquid is moderately agitated by air bubbling and a stirrer. Air is introduced to the reactor through spargers at a rate 2-3 times higher than that required by stoichiometry. A gypsum crystal concentration of 10 to 20 weight percent is maintained in this portion of the reactor to provide surface area for crystal growth. The gypsum settles to the bottom of the tank and a slurry containing 10 to 25 percent solids is continuously drawn off. The slurry is pumped to a gypsum stack where the solids settle out by gravity, and the supernatant liquor (stack overflow) is pumped back to the reactor.[30,31]

Chemistry

One overall reaction describes the system:

$$SO_2 + CaCO_3 + 1/2O_2 + 2H_2O \longrightarrow CaSO_4 \cdot 2H_2O + CO_2$$

Operating Status

The prototype unit at the Scholz Power Plant was operated for ten months starting on August 30, 1978.[32] The FGD system is designed to handle 50 percent of the flue gas from either boiler, or the equivalent of 23.5 MWe. Flue gas from the boiler passes through an ESP designed for 99.5 percent particulate removal, then enters the Chiyoda reactor at a rate of 53,000 acfm. The stainless steel reactor is 24 ft in diameter. Scrubbed gases pass through a vertical, two-stage, double pass chevron mist eliminator and exit through a 76-ft fiber-reinforced plastic stack at 130°F (no reheat is provided).[31] A detailed report on the test results is scheduled for release (by EPRI) in mid-1980.[32]

The plant operated over 97 percent of the time during the 10-month test period, and did not experience scaling or other "chemical" problems. The major portion of the downtime was used for modifications to the fan and air compressors. The gypsum product was found to be 96 percent pure by weight, and stacked very well. In the opinion of the Southern Company Services, Inc., principal investigator on the test project, the CT-121 process should be considered as a viable competitor to the current commercial processes. However, its ability to be cost-competitive with other systems at very low SO_2 inlet concentrations would have to be examined for each case individually.[32]

Effluents

The by-product from the CT-121 Process is gypsum, which presumably could be utilized in the cement-finishing process. If this were the case, the need for solid waste disposal would be eliminated. Alternatively, gypsum has suitable physical properties for disposal by landfilling. A closed-loop water system eliminates liquid effluents.

Advantages

- Ready availability of limestone as the raw material and good market for gypsum product.
- Relative simplicity of the process should pay off both in cost and reliability

Disadvantages

- Lack of commercial operating experience causes uncertainty as to overall viability and costs
- Ability of process to follow load could be limited by sparger performance

Citrate Process

In the Citrate process developed by the Bureau of Mines, flue gas is cooled to between 110° and 150°F. SO_2 is absorbed in a buffered solution containing sodium citrate, citric acid and sodium thiosulfate. The absorbed SO_2 is reacted with H_2S at about 150°F and atmospheric pressure to precipitate elemental sulfur and regenerate the solution for recycle. The sulfur is separated by oil flotation and melted. The H_2S for the reduction is made by reacting two-thirds of the recovered sulfur with natural gas and steam. Since thiosulfate ion depresses oxidation of bisulfite, only small amounts of sodium sulfate (which must be removed in a bleed stream) are formed.

Aqueous Carbonate Process

In this process the SO_2 is absorbed by aqueous sodium carbonate in a spray dryer. The dry solids are mixed with carbon (char or coal) and are blown by air into molten sodium carbonate (1830° F) where reduction to sodium sulfide occurs. The reduction mixture is quenched, filtered, and carbonated (with off-gas from the reducer) to evolve hydrogen sulfide. The sodium carbonate solution is recycled and H_2S is converted to sulfur in a Claus reactor. The various steps of the process have been tested only separately.

Catalytic Oxidation (Cat-Ox)[6]

In this process, flue gas from which particulate matter has been effectively removed is passed at 840° F through a vanadium pentoxide catalyst bed to oxidize the SO_2 to SO_3. The gas is then cooled (430° F) and the SO_3 is absorbed in a circulating stream of (80 percent) H_2SO_4. In a new installation, a hot electrostatic precipitator (outlet 840° F) would be used, and heat would be recovered by installing the air preheater of the boiler after the catalyst bed. Retrofit of the process to an existing unit requires extensive modifications and duct work and is not considered to be practical.

Advantages

1. No solid waste other than fly ash
2. No raw material requirements
3. Simple process equipment
4. Part of operating cost can be recovered by sale of H_2SO_4
5. No scaling or erosion problems in system

Disadvantages

1. Must have a sure market for H_2SO_4

2. Acid quality is low

3. More suitable for new installation due to process operating temperature

4. Fly ash causes catalyst fouling

5. High-temperature particulate removal is required

6. Installation on new unit necessitates treating flue gas between the economizer and air heater

7. Expensive materials of construction in acid recirculation loop

8. The process has not yet been successfully demonstrated on a large scale

Magnesium Oxide Process[5]

The magnesium oxide process is a regenerable FGD process currently being used exclusively on utility boilers. Flue gas is treated in an electrostatic precipitator to remove most of the fly ash. The remaining fly ash is removed in the first stage wet scrubber. Here, the gas is cooled and humidified by direct contact with recycle water in a venturi scrubber. Fly ash is removed from the recycle water by a thickener and disposed of in a landfill.

Particulate-free flue gas is next contacted with magnesium oxide slurry in another venturi scrubber. The reaction of MgO with water is shown in Reaction 1. SO_2 is absorbed and reacts according to Reaction 2. SO_2 in the flue gas also can be oxidized to form SO_3, which reacts to form $MgSO_4$.

$$MgO + H_2O \longrightarrow Mg(OH)_2 \qquad (1)$$

$$Mg(OH)_2 + SO_2 + 2H_2O \longrightarrow MgSO_3 \cdot 3H_2O \qquad (2)$$

A bleed stream is removed from the magnesium oxide recycle system, centrifuged and dried to remove the water of hydration. The dried crystals containing magnesium sulfite, sulfate and some unreacted oxide are usually sent to a sulfuric acid plant for regeneration to MgO and SO_2. Regeneration can be carried out onsite or at some distance from the absorption system. In regeneration, accomplished thermally (1800-2200° F) in a rotary kiln, magnesium sulfite reacts according to Reaction 3. Coke is added to reduce the sulfate; this is shown in Reaction 4. The SO_2 generated here can be used to produce sulfuric acid or elemental sulfur. The MgO is recycled to the absorption section of the plant.

$$MgSO_3 \xrightarrow{\text{heat}} MgO + SO_2 \qquad (3)$$

$$MgSO_4 + 1/2C \xrightarrow{\text{heat}} MgO + SO_2 + 1/2CO_2 \qquad (4)$$

By maintaining larger inventories of MgO, extended outages of the regeneration facility can be tolerated. If another company or plant is used for regeneration, care must be taken to ensure the long-term availability of these facilities before the MgO process is considered. Plants have been built, only to discover that the regeneration facilities have discontinued service.

The magnesium oxide process has been applied to smelter off-gases, sulfuric acid plant effluents, Claus plant effluents and utility boiler flue gases. However, the experience with coal-fired boilers is somewhat limited. Industrial non-boiler applications have ranged in size from 28 to 162 MW equivalent, with SO_2 loadings from 1,500 to 25,000 ppm. One utility magnesium oxide process has been in operation since 1975 (120 MW) at Eddystone, Pennsylvania, three more contracts have been awarded (150, 240 and 334 MW) at Eddystone and Phoenixville, Pennsylvania, and one is planned (600 MW at TVA's New Johnsonville, Tennessee facility). These plants are summarized in Table 3.5.1.22.

SO_2 removal efficiencies for the magnesium oxide process are usually higher than lime or limestone systems despite the similarity in chemistry. This is due to the increased reactivity of MgO over lime or limestone. Two utilities - Boston Edison and Potomac Electric and Power - have reported efficiencies of 90 percent. Industrial nonboiler applications have reportedly ranged from 90 to 99.5 percent efficiency.

Past installations have exhibited relatively low reliability of performance (low availabilities). Frequent equipment and processing problems have been encountered. However, this should be improved with time due to the low potential for scale formation with MgO as compared to lime or limestone systems.

Advantages

1. No solid waste other than fly ash
2. Part of operating cost can be recovered by sale of H_2SO_4 if market is available
3. No scaling in system
4. High-quality concentrated acid is produced
5. High SO_2 removal efficiency is possible
6. Oxidation can be tolerated

Disadvantages

1. Must have a sure market for H_2SO_4
2. Fly ash must be kept out of regeneration system
3. Must operate acid plant
4. Auxiliary fuel is required for dryer and calciner
5. Complex solids handling operations
6. Stack gas reheat is desirable
7. Losses and slight deactivation of MgO by repeated regeneration
8. High liquid to gas ratio in scrubber

Wellman-Lord Process[5]

The Wellman-Lord process is a regenerable FGD process which uses a sodium-based alkaline solution for removing SO_2 from combustion gases and other sulfur-laden process effluents. The gas stream is treated in a wet venturi scrubber to remove particulate, and to cool and humidify

Table 3.5.1.22

Utility FGC Systems
Magnesium Oxide Processes
--Operational--

Utility	Unit	Location	Capacity MW	Startup Date	Supplier[1]	Design Removal % Part.	Design Removal % SO_2	Scrubber product
Philadelphia Electric	Eddystone 1A	Eddystone, Pa.	120	9/75	UE	99.00	90.00	sulfuric acid, resalable

Utility FGD Systems
Magnesium Oxide Processes
--Contract Awarded--

Utility	Unit	Location	Capacity MW	Startup Date	Supplier[1]	Design Removal % Part.	Design Removal % SO_2	Scrubber product
Philadelphia Electric	Cromby	Phoenixville, Pa.	150	6/80	UE	--	--	--
Philadelphia Electric	Eddystone 1B	Eddystone, Pa.	240	6/80	UE	--	--	--
Philadelphia Electric	Eddystone 2	Eddystone, Pa.	334	6/80	UE	--	--	--
			724					

Utility FGD Systems
Magnesium Oxide Processes
--Planned--

Utility	Unit	Location	Capacity MW	Startup Date	Supplier[1]	Design Removal % Part.	Design Removal % SO_2	Scrubber product
Tennessee Valley Authority	Johnsonville 1-10	New Johnsonville, Tn.	600	1982	TVA/UE	--	--	--

Notes:
1. Suppliers: UE - United Engineers
TVA/UE- Tennessee Valley Authority/United Engineers

Source: Reference 31

the gas stream. The wet scrubber discharges directly to an absorber with two or more scrubbing stages. Here SO_2 is absorbed in a sodium sulfite solution according to Reaction 1. Reaction 2 shows the oxidation which takes place in the absorber. A considerable amount of the sulfite can be

$$Na_2SO_3 + H_2O + SO_2 \longrightarrow 2NaHSO_3 \qquad (1)$$

$$Na_2SO_3 + 1/2O_2 \longrightarrow 2NaHSO_4 \qquad (2)$$

oxidized, depending upon the absorber type and the amount of oxygen in the gas stream.

The sodium sulfite/bisulfite/sulfate liquor from the absorber is sent to an evaporator with a small bleed stream being sent to a chiller crystallizer for sulfate control. In the crystallizer, the bleed stream is cooled and the sodium sulfate (Na_2SO_4) crystals which form are removed from the slurry and dried for sale or disposal. This system controls the buildup of the non-reactive sulfate ion.

The bulk of the absorber solution is combined with cool sulfite liquor from the sulfate purge section and with hot recycle slurry from the evaporator. This hot slurry is sent to the evaporator where the major portion of the absorbed sulfur dioxide and water are driven off according to the reverse of Reaction 1. A double effect evaporator also can be used to significantly reduce the operating costs (up to 45 percent reduction in steam requirements). Residual sodium bisulfite is regenerated in a dissolving tank according to Reaction 3. Soda ash can be used for makeup as well. Reaction 4 corresponds to this system.

$$NaHSO_3 + NaOH \longrightarrow Na_2SO_3 + H_2O \qquad (3)$$

$$Na_2CO_3 + 2NaHSO_3 \longrightarrow 2Na_2SO_3 + H_2O + CO_2 \qquad (4)$$

Both alternatives replace sodium values lost in the sulfate purge. The sodium sulfite solution, containing some sodium bisulfate, is recycled to the absorber. The pure sulfur dioxide stream released from the evaporator can be used to produce high grade sulfur, sulfuric acid or liquid sulfur dioxide.

Two systems have been developed by Davy Powergas as alternatives to the sulfate crystallization described above. These systems use either natural gas or coal to chemically reduce the sodium sulfate. The purge stream would thus be regenerated and returned to the process. With either system, chemical makeup would be reduced to a minimum and sodium sulfate would not be produced, thus eliminating the marketing or disposal problems.

Surge tanks can be added to the system to store enriched absorber liquor and regenerated solution. This eliminates downtime in the absorber when the regeneration system is down for maintenance. An arrangement like this will facilitate remote operation of the regeneration system.

The Wellman-Lord process can be applied to any flue gas containing SO_2. It is equally applicable to nonferrous smelters, sulfuric acid plants and Claus plants. The maximum size of a single absorber which Davy Powergas recommends is 300,000 scfm of flue gas. Larger applications can be handled with multiple units. Current installations of the Wellman-Lord process range in size from 30,000 to 78,000 cfm for sulfuric acid plants, 30,000 cfm for Claus plants, and 89,000 to 435,000 cfm for oil-fired boilers. Several coal-fired utility boilers are operating with Wellman-Lord units or are in the construction stage. These range in size from 115 MWe to 534 MWe. SO_2 loadings have ranged from 900 to 13,000 ppmv. Tables 3.5.1.23 and 3.5.1.24 summarize industrial and utility applications of the Wellman-Lord process.

Table 3.5.1.23

Utility FGD Systems
Wellman Lord Processes
--Operational--

Utility	Unit	Location	Capacity MW	Startup Date	Supplier[1]	Design Removal % Part.	Design Removal % SO_2	Scrubber product
Northern Indiana Public Service	Dean H.Mitchell 11	Gary, In.	115	7/79	DPG	98.50	90.00	resalable elemental Sulfur
Public Service of New Mexico	San Juan 1	Waterflow, N.M.	361	4/79	DPG	99.50	85.00	resalable elemental Sulfur
Public Service of New Mexico	San Juan 2	Waterflow, N.M.	350	8/79	DPG	99.50	85.00	resalable elemental Sulfur
			826					

Utility FGD Systems Wellman Lord Processes
--Under Construction--

Utility	Unit	Location	Capacity MW	Startup Date	Supplier[1]	Design Removal % Part.	Design Removal % SO_2	Scrubber product
Delmarva Power & Light	Delaware City 1-3	Delaware City, De.	180	4/80	DPG	--	--	--
Public Service of New Mexico	San Juan 3	Waterflow, N.M.	534	12/79	DPG	--	--	--
Public Service of New Mexico	San Juan 4	Waterflow, N.M.	534	1/82	DPG	--	--	--
			1248					

Note: 1. Supplier: DPG - Davy Powergas
Source: Reference 31

Table 3.5.1.24 Summary of Operating Industrial Wellman-Lord Systems in the U.S.

Company/location	Completion date	Feed gas origin	Gas flow 1000 Nm^3/hr	Gas flow (scfm)	Design SO_2 concentration, ppm	Disposition of SO_2
Olin Chemical* Paulsboro, New Jersey	July 1970	Sulfuric acid plant	76	(45,000)	in 6,000 out 500	Recycle to acid plant
Std. Oil of California El Segundo, California	September 1972	Claus plants	51	(30,240)	in 10,000 out 250	Recycle to Claus plant
Allied Chemical Calumet, Illinois	November 1972	Sulfuric acid plants	51	(29,850)	in 2,700 out 250	Recycle to acid plants
Olin Chemical Curtis Bay, Maryland	May 1973	Sulfuric acid plants	133	(78,046)	in 4,000 out 250	Recycle to acid plants
Std. Oil of California Richmond, California	August 1974	Claus plant	51	(30,000)	in 10,000 out 250	Recycle to Claus plant
St. Oil of California El Segundo, California	January 1975	Claus plant	51	(30,000)	in 10,000 out 250	Recycle to Claus plant

* Plant operation suspended as of January 1, 1976.

Source: Reference 30

Operating efficiencies for the Wellman-Lord process, similar to those for the aqueous sodium process, are high compared to the lime or limestone processes. Systems can be designed for removal efficiencies in excess of 95 percent because the sodium scrubbing agent is very reactive. Availability of the Wellman-Lord process has been reported to be greater than 97-98 percent in most installations.

Advantages

1. Sulfur is the most environmentally acceptable by-product
2. Relatively small amount of waste to dispose of relative to throwaway processes
3. High removal efficiency is possible
4. No scaling in system
5. Low liquid-to-gas ratio in scrubber
6. Sulfur can either be marketed or stored

Disadvantages

1. Reducing agent is required
2. Purge stream is required and causes high sodium makeup
3. Fly ash must be kept out of regeneration system
4. Process is expensive
5. Process is not demonstrated on a coal-fired boiler
6. Stack gas reheat is desirable
7. Purge stream is composed of soluble salts that are very difficult to dispose of in an environmentally acceptable manner
8. Corrosive process environment requires expensive materials of construction
9. High steam requirements

Wet Scrubber Devices[5]

The scrubber is the most critical component in wet sulfur dioxide removal systems. It brings the flue gas into contact with the absorbent liquid or slurry. Scrubbing devices break into two major groups. One type removes both particulate matter and SO_2 simultaneously while the other type removes only the gaseous pollutant (SO_2). Of the five principal types of scrubbers, venturi, spray and moving bed scrubbers (discussed in the particulate removal section) fall within the first category. Packed towers and tray columns plug easily and hence require removal of solids prior to treatment. Scrubber selections must be made on the basis of performance in the following areas:

- Performance characteristics of the scrubber to remove SO_2 and particulate
- Scaling tendency of the system

- Physical characteristics of the absorbing media
- Physical properties and flow rate of the flue gas

Venturi Scrubbers. The venturi scrubber used for SO_2 absorption is identical to that described under particulate control. As the gas and liquid pass through the venturi throat the liquid is atomized into fine droplets. This dispersion increases the surface area available for mass transfer between the gas and the liquid. Downstream of the throat, the velocity decreases and the droplets, not containing dissolved SO_2, coalesce. As the velocity decreases, much of the entrained liquid settles from the gas stream. Any remaining liquid is removed in a cyclonic separator or another type mist eliminator.

The venturi scrubber minimizes plugging because the liquid, containing high concentrations of suspended solids, is dispersed into and flows with the gas stream. In addition, scale formation can be minimized because of the "open area" provided for gas flow. Wet-dry interfaces, however, have caused serious operational problems. The mass transfer rate, although somewhat lower than that for packed or plate towers, can be effective for gaseous pollutants which react chemically with the scrubbing liquid. The amount of mass transfer area provided by this scrubber is relatively insensitive to changes in the gas flow rate.

Spray Scrubbers. Open spray towers operated at low pressures (such as the unit at Square Butte Electric) have minimum plugging tendencies and high removal efficiencies at suitable L/G ratios. If high pressure sprays are used to increase the dispersion, plugging of the spray nozzles becomes a problem. However, the main advantage of this design is low P operation, which more than compensates for the possible higher L/G ratio. Once-through systems are unacceptable because of the environmental effects created from the water imbalance. Alternative designs have been developed to overcome these problems. One such design uses a specially designed rebound surface to atomize the spray. Increased mass transfer is experienced because of the contact with the finely divided water droplets. Venturi effects as the gas passes through the rebound zone also may be responsible for the high efficiencies.

Moving Bed Scrubbers. In moving bed scrubbers, the particulate- and SO_2-containing flue gas is contacted with liquid in a region containing captive, movable, solid spheres (see particulate control section for details). Gas velocity through the system is sufficient to cause the spheres to become mobile and move about in a random manner. The wetted surfaces within the bed provide sufficient mass transfer area for SO_2 absorption. Tendency for plugging and the buildup of particulate matter in the bed is small. However, scaling may be a problem due to the precipitation of calcium salts on the retaining grids, on the walls, and at wet/dry zone interfaces. In addition, the plastic spheres may wear at an increased rate for systems with higher suspended solids concentrations than would be present for particulate control alone.

Packed Towers. Packed bed towers have been used for decades for mass transfer in the chemical process industries. In most applications, the scrubbing liquid is allowed to drain downward over porous packing materials such as Raschig Rings, Pall Rings, Berl Saddles, etc., in a vertical tower. The gas to be scrubbed is introduced at the bottom of the tower and flows upward through the packing elements. The entire liquid film covering the packing is thereby contacted by the upward-flowing gas. Mass transfer in the tower is greatly enhanced by the increase in the interfacial area provided by the wetted packing.

The SO_2 concentration in the gas will decrease as the gas travels upward; the SO_2 content of the liquid will increase as it travels downward. In principle any degree of SO_2 removal is possible with proper specifications of tower dimensions, absorbent purity, and flow rate.

Packed beds generally are not used with high particulate levels in the flowing streams, as the immobile solid packing is susceptible to plugging. Under typical operating conditions, the pressure drop in a packed column will be on the order of one inch of H_2O per 10 inches of packed height.

Tray Columns. Tray scrubbers are characterized by horizontal mass transfer devices called trays or plates. The scrubbing liquid flows transversely across the top of the tray, while the gases being scrubbed bubble upward through small openings in the tray. In desulfurization, the SO_2 is transferred from the interior of the gas bubbles through the interfacial surface of the bubble, and into the surrounding liquid. The depth of the liquid on the tray is fixed by an overflow weir. Trays are commonly mounted inside the tower in such a way that the direction of the liquid flow alternates as it flows across each plate in succession down the tower. The simplest tray design is a sieve tray, in which there is a pattern of holes in the tray through which the vapor passes. Bubble-cap trays include inverted caps with slots on the sides, through which the vapor must pass.

As in the case of the packed tower, the bulk gas flow is countercurrent to the liquid. In principle, any degree of absorption is possible with the proper choice of the number of trays and the absorbent purity and flow rate. In general, tray columns are used as "after-scrubbers" operated on clean water to prevent carryover of corrosive mist which is high in dissolved solids. Carryover of such mist results in high corrosion rates for downstream mist eliminators, and an increase in particulate emissions.

Dry Flue Gas Desulfurization

A recent trend in the West has been the use of dry systems for sulfur dioxide control including char-bed units, spray dryer units and completely dry injection units. Dry FGD technology is especially appropriate for western regions where water is scarce and where low sulfur coal predominates. Advantages of dry scrubbing include the elimination of flue gas reheat requirements, and the collection of reacted sorbents in a fabric filter or electrostatic precipitator, resulting in a dry waste material for disposal. Additionally, dry methods of FGD would be attractive alternatives for utilities requiring retrofit applications. There would be a lower water requirement, no complex chemical equipment to operate, and capital equipment expenditures would be minimal.[11]

The Environmental Protection Agency is encouraging the development of dry scrubbing systems. The June 1979 NSPS for SO_2 emissions was tailored to encourage the use of dry systems for low sulfur coals.[4]

A summary of the advantages of dry scrubbing versus wet scrubbing is given below.[13]

Waste Handling. A completely dry injection system has no sludge handling equipment, which is usually troublesome and has a record of high maintenance. Wet scrubbers require thickeners, centrifuges or vacuum filters, and sludge-fly ash blenders in order to obtain a dry product. The product from a dry scrubber can be handled with conventional dry handling systems used for fly ash. For the coals tested, the dry product appears to handle as well as the fly ash.

Wet/Dry Interface. Scaling and plugging is common in wet scrubbers at wet/dry interfaces and on scrubber packing materials and demisters. In the dry system, the interface point occurs in suspension; only dry powder makes contact with the walls. There are no packed beds or demisters in a dry scrubber.

Materials of Construction. Wet scrubbers require expensive alloy materials or coatings for protection for corrosion and erosion. The dry system can use low carbon steel for vessels and ductwork. The I.D. fans can be safely located just ahead of the stack without fear of fan corrosion and imbalance.

Operations. It is estimated that considerably fewer operations personnel will be required for the dry system. Wet scrubbers have proven to take considerable manpower for operations and maintenance. The dry scrubber offers flexibility of operation. Feed rates can be immediately adjusted with little concern for pH control. Turndown capability for a dry scrubber is on the order of 10:1. Wet scrubber modules usually must be left in service at low loads to recirculate slurry. In the dry system, modules and/or atomizers can be removed from service quickly and easily as the load varies.

Maintenance. Wet systems have inherently high maintenance costs with slurry handling equipment recirculating abrasive materials at high pressures and volumes. The dry system operates with low pressures and low material volumes. Liquid to gas ratios for a spray dryer system are about 0.2 to 0.3 gallons per 1000 actual cubic feet compared to about 40-100 for a wet scrubber. The atomizer, which is probably the highest wear item in the system, can be removed and replaced quickly. The elimination of dewatering equipment reduces the maintenance expense considerably from that for a wet scrubber.

Energy Requirements. The dry system requires approximately 25 to 50 percent of the energy required for a wet system.

Particulate Collection. The gas volume to the particulate collector is reduced to below that leaving the air heater as a result of the spray down in the dry scrubber. The gas temperature to the particulate collection device remains constant at all loads. The spray dryer conditions the fly ash with the added moisture resulting in a lower resistivity ash in the precipitator. Although the scrubber reactants produce additional particulate, the increased humidity and lower temperatures have a positive effect on precipitator operation. Pilot test results have indicated that the increased humidity does not adversely affect baghouse operations, and the baghouse benefits from the reduced gas volume due to reduced temperatures.

Water Consumption. The water requirements for a dry system are much less than for a wet system. The spray-dryer scrubber at the Laramie River Station Unit 3 will use about 50 percent of the amount required for the wet scrubber on Unit 1 or Unit 2. Low quality water such as cooling tower blowdown or ash water may be used in the spray dryer. Only a small quantity of treated water, about 20 percent of the total requirement, is needed for lime slaking.

Four major types of processes are being applied to dry systems:

- Lime

- Sodium
- Activated Char
- Copper Oxide

Lime Process

Dry lime systems utilize lime chemically in the same way as described earlier for wet scrubbing. Current applications use a spray dryer system. A slurry containing concentrated lime is pumped to a drying chamber. The slurry is atomized to a fine mist by the use of high-speed rotary disks or high-pressure spray nozzles. The flue gas sulfur dioxide is then absorbed into the water droplets and reacts with the lime. The amount of water added is about 0.3 gal/acf and is evaporated when contacted with the hot flue gas. A diagram of a spray dryer system is shown in Figure 3.5.1.17. The advantages of this system over the wet system are:[14]

- Low capital investment compared to a wet scrubber plus ESP
- No flue gas reheat requirements
- Water chemistry and pH controls typical of wet scrubbers are eliminated
- Minimal corrosion and erosion, and no sulfate/sulfite scaling
- Lower energy requirements
- Dry sludge-material for disposal

Figure 3.5.1.17

Typical Spray Dryer FGD System

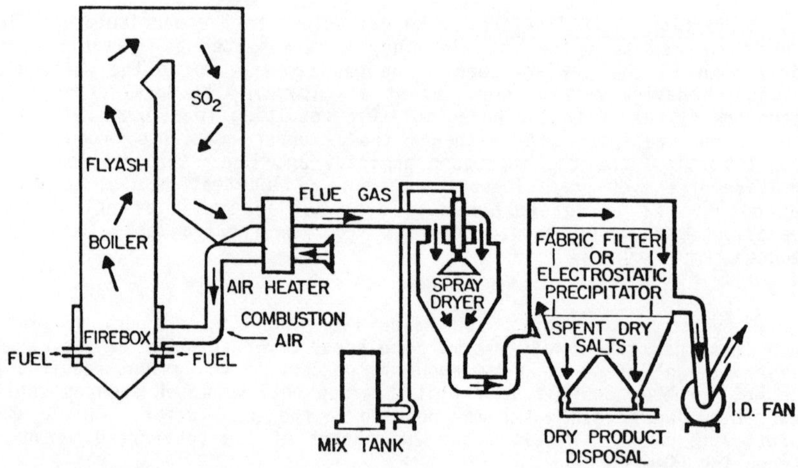

Source: Reference 13

Sodium Processes

Five sodium compounds have been proposed for use in dry alkali systems:[13]

- Manufactured sodium bicarbonate ($NaHCO_3$)
- Trona ($Na_2CO_3 \cdot NaHCO_3 \cdot 2H_2O$)
- Sodium carbonate monohydrate ($Na_2CO_3 \cdot H_2O$)
- Soda ash (Na_2CO_3)
- Nahcolite (natural occurring $NaHCO_3$)

The most promising of these compounds is nahcolite which is a widely occurring mineral in the Rocky Mountain states. In its pure form, it is a white crystalline mineral which dissociates at 518° F according to the reaction:

$$2NaHCO_3 \xrightarrow{\text{heat}} Na_2CO_3 + CO_2 + H_2O$$

It is known, however, that over extended periods of time the mineral will decompose at lower temperatures. The resulting soda ash will undergo an absorption reaction with sulfur dioxide according to:

$$Na_2CO_3 + SO_2 + 1/2O_2 \longrightarrow Na_2SO_4 + CO_2$$

An additional benefit of using nahcolite is that it has been observed to absorb NO_2 moderately well by the reaction:[12]

$$3NO_2 + Na_2CO_3 \longrightarrow 2NaNO_3 + NO + CO_2$$

This reaction indicates that only two-thirds of the nitrogen dioxide can be removed since one out of three moles are converted to nitric oxide. The advantages of the lime process also apply to nahcolite but pilot studies indicate SO_2 removal efficiencies for nahcolite to be higher. The pilot work has been performed using both countercurrent and fixed-bed systems.[12]

Activated Carbon Process (Foster-Wheeler Char Adsorption)

In this process, SO_2 in the flue gas (at 300° F) is adsorbed on char, where in the presence of air and moisture it is converted to sulfuric acid, which is strongly retained in the interior pore system of the char pellets. (Oxides of nitrogen are also adsorbed to some extent.) As it becomes saturated, char is removed and heated (in an inert atmosphere) to 1200° F by mixing it with sand at 1470° F. Under these conditions the sulfuric acid is reduced to SO_2, the nitrogen oxides dissociate, and some char is oxidized to carbon dioxide. In the sulfur recovery section, the (concentrated) SO_2 is passed through a vessel containing crushed coal at 1290-1470° F, where it is reduced to sulfur. Gaseous sulfur is condensed from the effluent and the tail gas is recycled for removal of residual SO_2.

Advantages

1. Dry process - no stack gas reheat is required
2. Sulfur is the end product
3. High removal efficiency is possible
4. Some fly ash can be tolerated in the adsorber

5. SO_2 is reacted directly with coal to form sulfur

6. Sulfur can be either marketed or stored

7. Has potential for NO_x removal

Disadvantages

1. Chemical and physical attrition of char makes the process expensive

2. SO_2 adsorption and reduction steps have not been demonstrated beyond the pilot scale

3. Hot screening is required to separate sand from char

4. Thermal energy is required for regeneration

5. Moving bed is relatively complex to operate

Shell Flue Gas Desulfurization Process (SFGD)

The SFGD process, which is being licensed in the United States through Universal Oil Products Corporation, concentrates the SO_2 by dry reaction with supported copper oxide pellets and regeneration with hydrogen. The SO_2 concentration in the regenerated offgas is increased by removing water and impurities in an absorption/stripping system. The concentrated SO_2 is then reduced to elemental sulfur in a Claus plant.

Advantages

1. Dry processs - no stack gas reheat is required

2. Sulfur is the end product

3. High removal efficiency is possible

4. Minor waste disposal problem

5. Sulfur can be either marketed or stored

Disadvantages

1. More suitable for new installation due to the high reaction temperature required for SO_2 removal

2. Hydrogen is required both for regeneration of the reactant (CuO) and reduction of SO_2 to H_2S

3. Complicated process unit operations with numerous steps

4. High fly ash loading may cause fouling of the reactors

5. Reactor design is complicated

6. Large diameter valves are required both upstream and downstream from reactors

7. Process is not demonstrated on a coal-fired boiler

8. Installation on new unit necessitates treating flue gas between the economizer and air heater

9. Expensive materials of construction are required in the SO_2 concentration steps

Nitrogen Oxide Control

Combustion modification and flue gas treatment are two approaches applicable to conventional combustion processes for control of nitrogen oxides (NO_x) emissions. The former suppresses the formation of the NO_x in the furnace through the alteration of combustion conditions, and the latter destroys the NO_x after they have been formed. Major emphasis in the United States has been on combustion modification, since it is the less costly way of complying with current federal emission regulations.

The June 1979 New Source Performance Standard for NO_x emissions will have a significant impact on boiler selection for new facilities. These regulations concerning low-rank coals are:

1. 210 ng/J (0.50 lb/million Btu) heat input from combustion of subbituminous coal, shale oil, or any solid, liquid, or gaseous fuel derived from coal;

2. 340 ng/J (0.80 lb/million Btu) heat input from the combustion in a slag tap furnace of any fuel containing more than 25 percent, by weight, lignite which has been mined in North Dakota, South Dakota, or Montana;

3. 260 ng/J (0.60 lb/million Btu) heat input from the combustion of any solid fuel not specified in (1) or (2).

NO_x formed in combustion processes originates from two sources: (1) oxidation of molecular nitrogen in the combustion air via a thermal process to form "thermal NO_x"; and (2) conversion of nitrogen chemically bound in the fuel to "fuel NO_x." The relative contributions of these two sources have a very important bearing on the effectiveness of different combustion-modification techniques. For pulverized coal firing, bench-scale studies have shown that the fuel NO_x can be responsible for more than 75 percent of the total NO_x production, suggesting that combustion-modification techniques that suppress only thermal NO_x (e.g., water injection) will not be able to achieve sufficient reduction of NO_x produced in coal-fired boilers. Combustion-modification techniques that have been actively investigated in coal-fired utility boilers, therefore, include only low excess air (LEA) firing, staged combustion, flue gas recirculation (FGR), combinations of these techniques, and new burner designs.[6]

Design of new boilers for NO_x reduction has reversed the traditional trend toward smaller furnaces with more intense combustion. Large combustion chambers can give more effective cooling and more flexibility in combustion control.[8]

Combustion Modifications

Several methods can be used to alter the combustion system and thus reduce the NO_x formation. These techniques are effective as a result of lower peak flame temperatures, reduced availability of oxygen in the flame or altered residence time/temperature profiles in the combustion zone. NO_x formation is a kinetically limited and high temperature sensitive reaction. Table 3.5.1.25 provides a summary of combustion modification techniques for coal-fired utility boilers.[6]

Low Excess Air Firing. Low excess air firing reduces the amount of excess air being fed to the combustion zone and thus reduces the oxygen

Table 3.5.1.25

Summary of Combustion Modification Techniques for Control of NOx Emissions from Coal-Fired Utility Boilers

Technique	Principle of Operation	NOx Emission Level NO_2 basis, 3% O_2[b]	NOx Control Effectiveness[c]	Limitations	Applications
Low Excess Air Firing (LEA)	Reduces oxygen level in primary flame zone and suppresses thermal and fuel NOx	Wall firing 450-600 ppm Tangential 350-450 ppm	20 to 30%[d]	Unburned hydrocarbons; CO emissions at low level of excess air; increased fouling	Routine use in utility boilers
Staged Combustion (SC)	Biased firing or overfire air ports reduces oxygen level in primary flame zone (first stage)	Wall firing 350-450 ppm Tangential 250-350 ppm	30 to 45%	Fouling of convective* section; poor first stage; ignition; soot formation; load reduction	Retrofit of existing boilers and design of new units
LEA Plus SC	Combination of the above two	Wall firing 350-400 ppm Tangential 250-300 ppm	40 to 50%	Limitations with SC plus corrosion on wall tubes	Retrofit of existing boilers and design of new units
Flue Gas Recirculation (FGR)	Recycled flue gas reduces the temperature and oxygen concentration of flame zone	Wall firing 500-700 ppm	10 to 20%	Limited effectiveness in coal-fired boilers	None expected for coal-fired boilers
New Burner Designs	Controlled mixing of fuel and air to suppress fuel and thermal NOx	Wall firing 350-450 ppm	40 to 50%	Unknown	Under development

[b]NOx emission levels reported as NO_2, based on 3% excess O_2. Values cited are nominal for average unit capacity and operating conditions; the range of available data is much wider than the values listed.

[c]Expressed as functional reduction relative to baseline NOx emission levels of 500-900 ppm for wall-fired utility boilers and 400-600 ppm for tangential-fired boilers.

[d]For reduction of 10 to 15% in excess air.

*Not observed on North Dakota lignites.

Source: Reference 6

available for fuel NO_x formation. The amount of combustion air also affects the peak flame temperature. Flame temperature increases as the excess air is decreased. This tends to negate somewhat the effects of the reduced oxygen content in the combustion zone. This is one of the most widely applied techniques for NO_x control.

Maximum combustion efficiency for coal normally requires 20 to 30 percent more air (excess air) than the stoichiometric amount depending on the boiler type and the properties of the coal burned. Field studies have shown that, if the excess air is reduced by 10 percentage points (e.g., to 15 percent from 25 percent), NO_x emissions will generally be reduced about 20 percent. This relationship varies only slightly with boiler type.[6]

One must consider, however, that as excess air levels decrease, CO, hydrocarbons and particulate concentrations will increase. The tradeoff for these emissions and NO_x is very dependent on fuel characteristics and boiler type. Variations in air concentration may also affect wall-tube corrosion and ash fouling.

Nevertheless, firing with low levels of excess air is typically the first technique deployed in a control program and is normally included even when other control techniques are used. The lowest excess air level achievable in combination with other techniques, such as staged combustion, is usually higher than when low excess air is the only control employed.[6]

The effects of overfire and excess air on NO_x formation and ash fouling were investigated by researchers at the Grand Forks Energy Technology Center.[34] The test program involved runs at the Hoot Lake Station of the Otter Tail Power Company (Fergus Falls, Minnesota) on a tangentially fired (high fouling N.D. lignite fuel) boiler. Several significant conclusions were reached:

1. "The reduction in overall excess air to the boiler reduced the NO_x level by a small amount but increased the potential for ash fouling significantly. Aside from the probe tests, severe running slag on furnace walls and tube fouling were observed in the boiler during tests at low excess air. Potential problems with increased slag formation make this technique undesirable, especially for dry bottom boilers."

2. "The NO_x level was reduced without modification of the boiler by using the lower three burners and was reduced further by changing the damper positions from the normal to an overfire air conditon. A total reduction of about 36 pct from 400 ppm to 225 ppm was achieved."

3. "The burner selection appears to influence the ash fouling potential as measured by the probe deposit rate. The deposit rate is lowest for operation with the three bottom burners."

4. "The change in damper positions from normal to overfire air conditions does not appear to increase the ash fouling potential as measured by the probe deposit rate."

5. "Changes in the burner tilt angle can modify the NO_x level and ash fouling potential but do not change the relative advantages of burner selection and damper position."

6. "The test results on probe deposit rates do not correlate well with the flue gas temperatures obtained by HVT, based on the strong relationship between fouling rate and temperature observed in previous test programs (2). Since the probe weights are corroborated by observations made on the boiler, some of the temperature measurements may be in error."

The study recommended that additional tests be made at other tangentially fired powerplants using similar marginally high fouling lignite, and on a cyclone type boiler. It was also recommended that secondary overfire air and improved burner concepts be tested at the Hoot Lake facility.

Staged Combustion. In this process, some of the combustion air is diverted from the initial combustion zone at the burner and injected into the combuston zone farther downflame. The injection points are usually called "overfire ports," "sidefire ports" or "NO_x ports" depending upon the location. This modification causes the combustion process to be fuel rich initially, and slows combustion in the initial combustion zone. The injection of cool combustion air downflame allows for complete combustion and cools the flue gas below the nitrogen oxides formation temperature of 3300°F more rapidly. This control method is inexpensive and effective, but fireside deposits, corrosion, and slagging may create problems in the combustion zone.

The effectiveness of NO_x control by staged combustion is highly dependent on the fuel/air stoichiometry in the burners of the primary stage. Studies have shown that this technique is capable of reducing NO_x emissions from coal-fired utility boilers by as much as 45 percent. The extent of NO_x reduction is limited, however, by poor flame stability, convective section fouling, and excessive formation of unburned hydrocarbons. Recent studies in wall-firing utility boilers by TVA[15,16] have shown that severe corrosion of wall tubes may occur in locally reducing regions that result from staged combustion. In addition, the number of available burners, the limited capacity for varying the fuel rate to individual burners, and the availability of overfire air ports for air addition above the burners place constraints on the use of staged combustion in retrofit applications. Nevertheless, staged combustion has been regarded as the most successful method for NO_x control, and overfire air systems have been included in the design of new coal-fired boilers to meet current NO_x emission standards.[6]

In addition to limiting the formation of thermally generated nitrogen oxides, staged combustion also reduces the conversion of fuel-bound nitrogen. The mechanism for the conversion is not highly temperature sensitive, so a combustion modification which reduces only the bulk gas temperature will not greatly limit the conversion. Staged firing, however, reduces the available oxygen in the combustion zone near the burner, and is effective in limiting the conversion of fuel-bound nitrogen.[5]

Biased Firing. Biased firing is an economical method of implementing staged combustion. In this process, the fuel is shut off to one or more burners while the air flow is maintained at the same rate. The technique can be applied only to multiple burner furnaces and cannot be used on stoker fired boilers. The major difficulty in applying this procedure is in selecting the best burner to turn off and the best settings of excess O_2, air register, etc. which will allow satisfactory operation at reduced NO_x levels without smoke or carbon monoxide formation. In some cases, taking burners out of service forces the unit to a lower load because the fuel system on the remaining in-service burners cannot handle the increased fuel flow necessary to maintain full load.

With a burner out of service, the total nitrogen oxide emissions have been reduced up to 54 percent. An advantage of this type of combustion modification is that the boiler efficiency is relatively unaffected. However, fireside deposits, corrosion and slagging may be a problem.[5]

Flue Gas Recirculation. Flue gas recirculation has been successful in reducing the total nitrogen oxides concentration in flue gases by 10 to 40 percent, especially with pulverized coal boilers. The only effect on boiler efficiency is the work required in the recirculation fan. Recirculation of flue gas into the combustion zone reduces the flame temperature, and thus the formation of thermal NO_x. The lower oxygen concentration in the combustion gases also may reduce the formation of fuel NO_x. Flue gas recycle to secondary air in pulverized coal boilers has resulted in reductions in NO_x concentrations of up to 30 percent. The recirculation rate corresponding to this reduction was 30 percent.

The method or location of injecting the flue gas is important. There are several methods for adding flue gas to the burner. It can be mixed with the combustion air, or a separate passage can be provided.

In some boilers, flue gas recirculation may be uneconomical for large NO_x reductions. In some cases, capital cost may be high. This combustion modification is best applied to large, new boilers.

In general, flue gas recirculation is considered to be less effective as an NO_x control method than is LEA firing or staged combustion. Furthermore, the high installation cost of ducting and the energy penalty (~0.5 percent) attributed to recirculation fans make this method rather unsuitable for coal-fired boilers.[17]

New Burner Design and Tuneup. Advanced burner design is an alternate approach to reducing formation of both thermal and fuel NO_x by means of controlled mixing of the fuel and air. With modified burner designs, the basic NO_x control principles underlying the staged combustion and GFR techniques can be incorporated internally in the furnace design, thereby avoiding some of the operational problems normally associated with the use of these techniques in retrofit situations. In the long run improved burner design will probably replace the practice of external combustion modifications and achieve NO_x levels considerably below that of present-day burners.

A burner, once in operation, must be properly maintained to ensure optimum efficiency and low NO_x formation. Burner maintenance or tuneup consists of adjusting the excess air to the proper level for each load, adjusting the burner registers to give a hard, bright flame and replacing worn parts in the burner itself. The major objective in burner tuneup is to alter the fuel and air mixing patterns to provide as much aerodynamically staged mixing as possible without additional air injection downflame. The low NO_x configuration that one hopes to achieve is a long, narrow flame where the fuel and air mix gradually over the entire flame length. Such flames can be achieved by reducing the swirl of the secondary air and by changing the angle at which the fuel is injected into the secondary air stream. In the normal operating range of a burner with variable air swirl, decreasing the swirl usually decreases NO_x.

At very high swirl settings, NO_x may again decrease if the flame changes to a widely flaring shape which is more effectively cooled by the walls and by entrainment of cooled combustion products from within the furnace.

Improvements in thermal efficiency as well as reduced pollutant emissions can be obtained economically by proper maintenance of the burner if, for example, the overall excess O_3 can be lowered by improved fuel/air mixing. New burner designs are available which advertise low emissions as well as versatile operation.

Flue Gas NOx Removal Treatment Processes. Removal of nitrogen oxides from boiler flue gas is considerably more difficult than preventing their formation by combustion modifications. The problems inherent in this type of system are that large amounts of gas must be handled, and the NO_x is present in dilute concentrations.

Catalytic Reduction Processes. Gases containing small amounts of NO_x can be reacted with a reducing gas (e.g., ammonia) in the presence of a catalyst. Because SO_2 in the combustion gas tends to poison the catalyst SO_2 must be removed by other means before the gas contacts the catalyst. Catalytic reduction is the most effective process for removing NO_x; however, catalyst regeneration costs are prohibitive.

Absorption Processes. Absorption processes for the simultaneous removal of NO_x and SO_2, oxidation/reduction processes and reduction processes, have been developed by the Japanese. In the oxidation/reduction process, NO is oxidized to NO_2 by ozone or chlorine dioxide (ClO_2) in the gas phase and absorbed in a sodium sulfite solution or a lime/limestone slurry, containing a catalyst. The reduction process uses a sulfite solution containing EDTA (ethylene diamine tetracetic acid) and a ferrous compound to absorb NO_x from the flue gas. Each system progresses through a variety of complex reactions to produce by-products such as N_2, NH_3, gypsum or any one of a number of sulfur or nitrogen containing salts. Removal efficiencies have been reported up to 80 percent for NO_x removal and 98 percent for SO_2.

The major problem with the oxidation/reduction process is the high cost of ozone. Chlorine dioxide is cheaper, but the chlorine complicates the process. The major problem of the reduction process is the high liquid to gas ratio and the large scrubber required for absorption. In general, these processes are still developmental and have not yet been shown to be economically feasible.

3.5.1.3.2 Solid Wastes and Sludges

The major solid waste problem from coal combustion results from fly ash collected in particulate control devices and bottom ash or slag. Solid waste also results from dry scrubber systems used for sulfur dioxide removal. Sludge from wet scrubber systems can be a significant disposal problem.

Disposal of Ash and Dry Scrubber Waste

Disposal of solid waste by landfill (the time-honored method) involves potential pollution of surface and subsurface water. Leaching tests on ash samples from existing power plants suggest that runoff and seepage from ash dumps may exceed established water quality criteria with respect to concentrations of various chemical species, including toxic trace elements. Other problems associated with disposal by landfill are structural instability, wind erosion, and the unsightliness of the ash pile. However, these wastes are not classified as "hazardous" under the latest version of the Resource Conservation and Recovery Act, and are not regulated by any federal agency. The EPA is currently reviewing these wastes and is expected to establish a policy for them in the near future. Individual states may have regulations covering these wastes, independent of federal law.

Leaching by percolation of rain, melted snow, or surface runoff represents the greatest pollution threat posed by the dumps. In fact, it is a threat wherever ashes are exposed to rain or surface water. However, ash that is hauled away and put to use in embankments, road and base construction, and the like (where it is spread or mixed with soil, sand, gravel, and/or stabilized by cementitious reactions) will yield a leachate that is more readily attenuated by the underlying or adjoining soil. Moreover, in such application, the elements being leached will probably be exhausted in some reasonably short time.

The leachability of a given species in the ash will be determined largely by its solubility. The resulting solution is usually alkaline due to the effect of free lime, although some ashes, especially from eastern coals, produce acidic leachates. These acidic liquors often exceed stream criteria with respect to sulfate, total iron, zinc, lead, cadmium, and manganese, but they are usually acceptable with respect to copper.[11]

There are two primary methods for prevention or reduction of groundwater contamination: (1) exclusion or diversion of all or part of the surface flow and subsurface water, and (2) maintenance of a suitable system of subsurface and trench drains.[12] For protection of ash dumps from erosion and liquefaction, the main control methods are: (1) properly designed drainage and filter system to ensure fine material is not carried off, (2) suitable vegetative cover and slope protection, and (3) good compacting for wastewater control.[18]

Reclamation has not been practiced much in the past, but with increasing concern for the environment greater care will have to be given to site selection; management of the landfill, including monitoring for potential water pollution; and functional and aesthetic restoration of the dump after it is filled.

Ash does not readily support plant growth. This is due partly to lack of the necessary nutrients (a need that can be supplied by application of fertilizers) and partly to the presence of inhibiting elements such as boron. Yet natural colonization of an ash surface by vegetation often takes place. The first plant to appear is usually a moss, which can cover a moist ash surface within six months and which effectively binds together the ash particles to inhibit erosion. A common weed is usually the next colonizer to appear, and this can be followed by a range of other plants.[19]

A variety of trees (alder, honey, locust, spruce, poplar and willow) are tolerant to ash conditions, but their growth is not sufficiently vigorous to sustain commercial timber production. A number of shrubs can also be grown on ashes. Thus it should be possible to landscape an ash disposal site with these tolerant species.

Several other methods of disposal have been examined. Direct deposition of ashes in underground mines is a possible disposal option. Filling of deep mines with powerplant ash has been used as a mine subsidence prevention technique.[21] The feasibility of this approach is currently being investigated by the Bureau of Mines for disposal of lime/limestone scrubber sludges.

Another possible method of ash disposal would be the utilization of ashes in the rehabilitation of areas which have been destroyed by strip mining, quarrying, and the like. It has been suggested that in these cases the cost of ash disposal could be shared by the utilities, mining companies, land developers, and local governments.[18]

Ash Utilization

Commercial use of ashes is an attractive alternative to disposal by dumping or landfill. First, it represents a means of alleviating the ecological impact resulting from disposal of ashes in lagoons or land sites. Second, to the extent that the ashes are sold or given away to be put to some practical use, it reduces disposal costs. Third, there are technical as well as economic advantages to the user who utilizes ash as a raw material, e.g., for structural compositions, pavings, etc.

The construction industry currently consumes the largest amount of fly ash. Table 3.5.1.26 shows the commercial utilization of fly ash in 1975. Other markets are presently being added and developed. The major areas of use are:

- Concrete and cement
- Aggregates for lightweight concrete
- Time- and cement-stabilized fly ash pavements
- Asphalt pavements
- Road fill
- Slag as anti-agent
- Sand-blasting grit
- Load-bearing fill
- Brick

Table 3.5.1.26
Commercial Utilization of Ash in the U.S. in 1975[a]
(Kilotons)

Use	Fly Ash	Bottom Ash	Boiler Slag
Type 1-P Cement	225	70	36
Partial Substitute for Cement	945	- -	- -
Lightweight Aggregate	90	35	- -
Stabilization & Roads	450	525	72
Filler in Asphalt Mix	135	- -	- -
Ice Control	- -	280	54
Blast Grit & Roof Granules	- -	420	864
Misc.	180	350	414
Ash Removal at no Cost to Utility[b]	1,080	875	270
Ash Utilized[b] from Disposal Sites	1,395	945	90
1975 Total Utilized	4,500	3,500	1,800

[a] Compiled by the National Ash Association and Edison Electric Institute.

[b] Specific end use not known.

Source: Reference 6

Other potential uses of fly ash or bottom slag include:

- Agricultural soil amendment
- Component of grout
- Production of mineral wood
- Filler or extender in a variety of materials
- Extinguishing burning spoil piles
- Source of chemicals (sulfur, alumina, iron oxide, etc.)

Researchers at the University of North Dakota in Grand Forks have examined the suitability of using lignite and subbituminous fly ash in a variety of applications.[35] Mixtures of 25 percent ash and 75 percent clay have produced excellent extruded brick. Although the lignite fly ashes investigated display pozzolanic properties (cementitious behavior in the presence of water and lime), there is considerable variation in the chemical and physical compositon of the various samples examined. Further, none of the fly ashes comply with the present ASTM specification of Fly Ash and Raw or Calcined Natural Pozzolans for Use in Portland Cement Concrete. However, excellent results have been obtained by various researchers when fly ashes with similar properties (to those tested at the University of North Dakota) were used to replace part of the cement in concrete. Since 1970 there has been a concerted effort to have lignite and subbituminous fly ash accepted by ASTM for use in portland cement concrete. Supporting data have been gathered from many sources, including the University of North Dakota, in the form of chemical and physical specifications. For ashes which do not meet the ASTM code, performance data have been submitted which show lignite fly ashes producing higher compressive strengths in many cases than did specification bituminous fly ash. Specific information regarding the physical morphology of the fly ash, its chemical properties and quantative performance as a concrete additive is available in reference 35.

Wet Scrubber Sludge

Several throwaway flue gas desulfurization (FGD) processes generate solid wastes containing varying amounts of calcium sulfate ($CaSO_4$), calcium sulfite ($CaSO_3$), fly ash (up to 70 percent dry weight), scrubbing liquor (typically 30 to 60 percent), and a small amount of unreacted lime or limestone. The physical characteristics of this sludge can vary a great deal. Some sludges in the form of a slurry can be disposed of in a pond. Others appear to be a dry solid, but after a rain storm or upon shaking, these sludges can become fluid. Because of this, sludges from lime and limestone scrubbing systems must be stabilized or "fixed" to give them good, long-term mechanical properties and to improve their resistance to chemical leaching. Several methods have been developed to eliminate the physical problems encountered with these sludges.

The lime/limestone slurries withdrawn from the reaction tank usually have a solids content of about 10 percent. In some cases they are pumped directly to the settling pond, but usually they are routed to a clarifier for preliminary dewatering. This not only reduces the acreage required for the pond, but can also increase the solids content of the slurry to a point where chemical treatment is practical. Flocculants are sometimes added to increase the rate of settling and decrease the size of the clarifier, but they do not significantly increase the solids content of the underflow.[22] In a number of more recent FGD installations, the clarifier underflow is further dewatered by centrifugation or vacuum filtration to facilitate chemical stabilization.

There are two principal methods of chemical treatment of FGD wastes, oxidation to gypsum and fixation. Oxidation to gypsum has been very useful in Japan because gypsum is marketable there. In the United States however, very little attention has been given to this process because a market does not exist. It does appear that it may become cost effective as an approach to an environmentally acceptable method of disposal.

The fixation processes operate in different ways, utilizing fly ash, lime, silicates, and/or polyvalent metal ions to form low grade concretes. In general, the objective is to solidify the sludges by promoting pozzolanic and other cementitious reactions between lime compounds in the additives, alumina/silica in the fly ash, and calcium-sulfur compounds in the sludge. The sludge in its final form will have low permeability and good mechanical properties, and will not degrade with time.

The primary methods of sludge disposal are similar to those for fly ash. The first option is ponding. Sludge in the form of slurry or nearly dry solids can be sent to lined disposal ponds to contain the leachate and allow the solids to settle while the water evaporates. Experience shows, however, that this method results in ponds that appear dry, but may at any time liquefy and behave like quicksand. The solids do not compact themselves, leaving a hazard that must be monitored indefinitely.[5]

Sludges can also be dewatered and sent to a landfill. However, the sludges can exhibit thixotropic properties within the landfill as well, and in wet conditions they can form a mud or slurry with poor mechanical properties. Soluble matter in the sludge also can be leached from the landfill.

Physical stabilization is the third option. Dewatered sludge may be mixed with a dry material such as fly ash and transported to a landfill. This process brings the sludge to a state where it is non-thixotropic and nonplastic; but due to a nonrigid matrix these properties can return if water infiltrates the landfill. Leaching may still be a problem in this system.

Other possible options include disposal in undergorund mines or possibly ocean disposal after the wastes are treated. The alternatives are presently under investigation by EPA.[6]

There are a number of possible beneficial uses of FGD wastes. The use of lime/limestone wastes as a filler material and source of sulfur is being investigated by TVA.[23] The use of natural gypsum as a calcium source for peanuts and soybeans has been well established. Use as an ingredient in road-base and paving materials, artificial aggregates and lightweight concretes is being investigated. Use in cement manufacture and conversion of wastes to elemental sulfur are also being evaluated.

Major deterrents to utilization of lime/limestone scrubber wastes are their variable physical and chemical properties, high transportation costs, dewatering requirements for many applications, and their inability to compete economically with other materials. Although some applications appear to be technically and economically feasible, the potential usage is so limited that only a small fraction of the expected sludge production could be marketed. At least for the near term, disposal of wastes by ponding or landfill will continue to be the major alternatives.[24]

3.5.1.3.3 Wastewater Effluents

Liquid effluents may be broken down into three categories:

- Blowdown and equipment-cleaning waste
- Solids-handling water
- Coal-cleaning effluents

Table 3.5.1.27 provides a general breakdown of wastewater sources by class. Most facilities recycle wastewater streams using the cleanest stream in areas requiring relatively clean makeup and the dirtiest streams in the ash transport recycle system.

Table 3.5.1.27

Classes of Various Waste Water Sources

Class	Source
High Volume	Nonrecirculating main condenser cooling water
Intermediate Volume	Nonrecirculating house service water Blowdown from recirculating main cooling water system Nonrecirculating ash sluicing systems Nonrecirculating wet-scrubber air pollution control systems
Low Volume	Clarifier water treatment Softening water treatment Evaporator water treatment Ion exchange water treatment Reverse osmosis water treatment Condensate treatment Boiler blowdown Boiler tube cleaning Boiler fireside cleaning Air preheater cleaning Stack cleaning Miscellaneous equipment cleaning Recirculating ash sluicing systems Recirculating wet-scrubber air pollution control systems Intake screen backwash Laboratory and sampling streams Cooling tower basin cleaning Rad wastes Sanitary system Recirculating house service water Floor drainage Miscellaneous streams
Rainfall Runoff	Coal pile drainage Yard and roof drainage Construction activities

Source: Reference 6

Cooling towers accumulate mineral deposits due to evaporation and the addition of corrosion inhibitors. It is therefore necessary to discharge, either continuously or intermittently, a fraction of the circulating water (usually in the range of 0.3 to 1.0 percent of the flow). This blowdown contains high concentrations of suspended and dissolved solids and sometimes significant amounts of residual chlorine. It may also contain chromium, zinc, phosphate, or other corrosion inhibitors and biocides in excess of effluent standards. A recent study shows the best treatment (technically and economically) of the blowdown to be lime-soda softening (or related chemical precipitation).[25] The water can then be reused as cooling tower makeup (since makeup water generally requires pretreatment by softening anyway), reused as scrubber makeup or discharged. Boiler blowdown contains mainly a high level of solids and can be similarly treated. Cooling tower and boiler blowdown are also often used for ash sluicing water, and then evaporated. However, reusing this sluicing water is more economical.

Future environmental regulations are likely to have a significant impact on wastewater disposal systems. The main problems with ash pond effluents are suspended solids and trace elements. It is likely that ash pond effluents will have to be recycled or reused in other ways to meet the EPA BATEA requirements of 1983.[6]

3.5.1.3.4 Fugitive Emissions

Fugitive air emissions consist of gaseous and particulate pollutants which are released in small quantities from the plant in general, and not from specific uniform openings within the plant. Most fugitive dust emissions generated from the handling of fine, dry ash can be controlled by the use of enclosures to contain the dust or by water sprays to collect and/or agglomerate the fine particles. Fugitive emissions from the landfill area can be controlled by water sprays to hold down dust, or by spraying the surface with chemicals to form a coating which resists wind erosion. Other methods include covering the site with a daily earth cover, revegetating the area, or using shrubs and other plants as windbreaks.

Fugitive liquid emissions arise from leaks around pumps, piping, and other process equipment. In addition, ponds and spills in and around the plant can allow liquid to migrate into the groundwater. Ash quench water or transport water can be effectively contained in the ash pond by lining the pond with any one of a number of plastic liners, clay liners or other bulk materials such as asphalt or concrete. Leaks from process equipment can be controlled by good maintenance practices or by collecting and recycling the effluent.[5]

References - Section 3.5.1.3

1. Lowry, H.H. (editor), Chemistry of Coal Utilization - Supplementary Volume, John Wiley & Sons, Inc., New York, 1963.

2. Stearns-Roger Incorporated, Economics of Fabric Filters Versus Precipitators, EPRI FP-775, June 1978.

3. "Power from Coal," Special Report, Editors of Power, March 1974.

4. Federal Register, Volume 44, No. 113, Monday, June 11, 1979.

5. Roe, D.E., A.W. Karr, K.R. Lemmerman, J.E. Sinor, Solid Fuels for U.S. Industry, Volume II, March 1979.

6. Environmental Control Implications of Generating Electric Power from Coal, Technology Status Report, Volume II, Argonne National Laboratory, December 1976.

7. Abbott, J.H., Advanced Techniques for Fine Particulate Control, Environmental Protection Agency, Research Triangle Park, North Carolina.

8. Rosenberg, H.S., R.B. Engdahl, J.M. Genco, S.G., Bloom, D.A. Ball, and J.H. Oxley, Status of Stack Gas Technology for SO_2 Control, Part II, EPRI-209, August 1975.

9. Arthur D. Little, Inc., Application of Scrubbing Systems to Low Sulfur/Alkaline Ash Coals, EPRI FP-595, December 1977.

10. Ness, Richmond, Eurick, and Kruger, Power Plant Flue Gas Desulfurization Using Alkaline Fly Ash From Western Coal, GFETC, Flue Gas Desulfurization Symposium, Las Vegas, Nevada, March 5-8, 1979.

11. Ness, H.M. and S.J. Selle, Power Plant Flue Gas Desulfurization for Low-Rank Western Coals, 1979 Lignite Symposium, Grand Forks, N.D., May 30-31, 1979.

12. Knight, J.H., The Use of Nahcolite for Removal of Sulfur Dioxide and Nitrogen Oxides from Flue Gas, Superior Oil Company, October 1977.

13. Evaluation of Dry Alkalis for Removing Sulfur Dioxide from Boiler Flue Gases, Bechtel Corporation, EPRI FP-207, October 1976.

14. Goblirsch, G.M. and E.A. Sondreal, Low-Rank Coal Atmospheric Fluidized Bed Combustion Technology, GFETC, 1979, Lignite Symposium, May 30-31, 1979.

15. Moulton, L., Bottom Ash and Boiler Slag, Third International Ash Utilization Symposium, Pittsburgh, March 13-14, 1973.

16. Faber, J.H., U.S. Overview, Fourth International Ash Utilization Symposium, St. Louis, March 24-25, 1976.

17. Hyland, E., Factors Affecting Pozzolan Marketing, Third International Ash Utilization Symposium, Pittsburgh, March 13-14, 1973.

18. Vogel, G.J., W.M. Swift, J.C. Montagna, J.F. Lenc, and A.A. Jonke, Recent ANL Bench-Scale Pressurized Fluidized-Bed Studies, Proc. of the Fourth International Conference on Fluidized-Bed Combustion, McLean, Virginia, pp. 21-44, December 9-11, 1975.

19. Barber, E.G., Land Reclamation and Environmental Benefits of Ash Utilization, Third International Ash Utilization Symposium, Pittsburgh, March 13-14, 1973.

20. Barber, E.G., Win Back the Acres: The Treatment and Cultivation of PFA Surfaces, Brochure of the Central Electricity Generating Board, Great Britain, (no date).

21. U.S. Environmental Protection Agency, "Standards of Performance," Federal Register 36, No. 147, 1971.

22. Haas, J.C., and W.J. Lombardi, Landfill Disposal of Flue Gas Desulfurization Sludge, Proc. Sixth Symposium on Coal Utilization, National Coal Assn./Bituminous Coal Research, Inc., Coal Conf. and EXPO III, Louisville, KY, pp. 207-220, October 19-21, 1976.

23. Leo, P.P. and J. Rossoff, Control of Waste and Water Pollution from Power Plant Cleaning Systems; First Annual R&D Report, EPA-600/7-76-018, U.S. Environmental Protection Agency, Washington, D.C., October, 1976.

24. Jones, J.W., Environmentally Acceptable Disposal of Flue Gas Desulfurization Sludges: The EPA Research and Development Program, Proc. Symposium on Flue Gas Desulfurization, Atlanta pp. 887-927, November 1974.

25. Williams, S.L., Processing of Cooling Tower Blowdown for Reuse without the Production of a Liquid Waste Stream, paper presented at the Third National Conference on Complete Water Reuse, Cincinnati, Ohio, June 27-30, 1976.

26. Bradley, W.J., et al., Effect of Subbituminous Western Coal Ash on AFBC Power Generating Costs, AIChE Conference, Houston, Texas, April 1979.

27. Goblirsch, G.M. and E.A. Sondreal, Fluidized Combustion of North Dakota Lignite, Ninth Biennial Lignite Symposium at Grank Forks, N.D., May 1977, GFETC/IC-77/1.

28. Hulett, L.D. et al., Trace Element Measurements at the Coal-Fired Allen Steam Plant: Particle Characterization, Coal Utilization Symposium, Louisville, KY, October 22-24, 1974.

29. Personal Communication, Frank I. Honea, Grand Forks Energy Technology Center.

30. Dickerman, J.C. and K.L. Johnson, Technology Assessment Report for Industrial Boiler Applications: Flue Gas Desulfurization, EPA-600/7-79-178i, November 1979.

31. Laseke, B.A., Jr., Electric Utility Steam Generating Units - Flue Gas Desulfurization Capabilities As of October 1978, EPA-4050/3-79-001, January 1979.

32. Personal Communication, Randall E. Rush, Southern Company Services, Inc., Birmingham, AL, March 24, 1980.

33. Personal Communication, B.A. Laseke, PEDCo, Inc., Cincinnati, Ohio, March 1980.

34. Honea, F.I., et al., The Effects of Overfire Air and Low Excess Air on NO_x Emissions and Ash Fouling Potential for a Lignite Fired Boiler, Presented at the American Power Conference, Chicago, Illinois, April 24-26, 1978.

35. Manz, O.E., Utilization of Lignite and Subbituminous Ash, Symposium on Technology and Use of Lignite, Grand Forks, N.D., May 1973, BuMines IC 8650, 1974, pp. 204-219.

36. Personal Communication, John Sudol, EPA San Francisco Operations, June 2, 1980.

37. Selle, S.J., P.H. Tufte, and G.H. Grondhovd, A Study of the Electrical Resistivity of Fly Ashes from Low-Sulfur Western Coals Using Various Methods, Bureau of Mines, U.S. Dept. of the Interior, Grand Forks, N.D., presented at the 65th Annual Meeting of the Air Pollution Control Association, Miami Beach, FL, June 18-22, 1972.

38. Pontius, D.H., P.V. Bush, and W.B. Smith, Electrostatic Precipitators for Collection of High Resistivity Ash, EPA Contract No. 68-02-2193 prepared by the Southern Research Institute, Birmingham, AL, August 1979.

39. McDonald, J.R., M.H. Anderson, and R.B. Mosley, Charge Measurements of Particles Exiting Electrostatic Precipitators, prepared for the EPA under Contract No. 68-02-2610 by Southern Research Institute, Birmingham, AL, April 1980.

Selected References - 3.5.1.3.5

1. Stearns-Roger Incorporated, Economics of Fabric Filters Versus Precipitators, EPRI FP-775, June 1978.
An abstract for this reference will be found in Section 3.5.1.1 reference abstracts.

2. Environmental Control Implications of Generating Electric Power from Coal, Technology Status Report, Volume II, Argonne National Laboratory, December 1976.
An assessment of "Environmental Control Implications of Generating Electric Power from Coal" was carried out for the Division of Environmental Control Technology in the ERDA Office of Environment and Safety by Argonne National Laboratory and a number of subcontractor companies. This report is the first in a series of reports issued under the program and represents efforts from inception of the program in March 1976 through December 1976. The primary emphasis in this initial report is on the characterization and evaluation of existing and near-term technologies for coal utilization. Environmental regulations and the health effects of pollutants are also reviewed.

3. Arthur D. Little, Inc., Application of Scrubbing Systems to Low Sulfur/Alkaline Ash Coals, EPRI FP-595, December 1977.
The objectives of this study were to collect and evaluate available information on low sulfur western coals which produce alkaline fly ash and the behavior of such fly ash in scrubbing systems; and to determine whether sufficient data are available to develop guidelines and specifications for future scrubbing systems based upon fuel and ash analyses and boiler design information. Data are reviewed and summarized from five pilot plant operations and 19 full-scale operating systems on about 4,100 MW of boiler generating capacity fired with low sulfur western coals. Data analysis performed as part of the study concentrated on unique aspects of fly ash in scrubbing systems and approaches to predicting ash behavior based upon fuel and ash characteristics.

4. Ness, Richmond, Eurick, and Kruger, Power Plant Flue Gas Desulfurization Using Alkaline Fly Ash From Western Coal, GFETC, Flue Gas Desulfurization Symposium, Las Vegas, Nevada, March 5-8, 1979.
A characteristic of western coals is that they contain high levels of alkali such as calcium, magnesium, and sodium. These alkali species can be leached from power plant fly ash for use in flue gas desulfurization (FGD) wet scrubbers in lieu of lime or limestone. At present, there are nearly 2,600 MW of generating capacity in the western United States that utilize either fly ash or fly ash supplemented with lime or limestone. An additional 3,500 MW of western generating capacity are being planned or are in various stages of construction which will use the alkaline fly ash. This report describes the western alkali ash FGD systems.

5. Evaluation of Dry Alkalis for Removing Sulfur Dioxide from Boiler Flue Gases, Bechtel Corporation, EPRI FP-207, October 1976.
This report involves a review and evaluation by Bechtel of the Wheelabrator and Superior documents on their investigations of particulate and dry SO_2 scrubbing processes by injection of alkali powder, particularly nahcolite, into the furnace or the boiler flue

gases. Also included is a review of the earlier work at the Mercer Station of the Public Service Electric and Gas Company of New Jersey.

The principal points covered in the evaluation are the chemistry, mining, transportation, waste disposal, and economics of the several processes reviewed. Data reflecting the influence of nahcolite and other dry alkalis upon the reduction of NO_x emissions are also included. Finally, a recommendation is made as to what additional bench-scale and prototype investigations now need to be conducted to fully demonstrate the technology.

6. Honea, F.I., et al., The Effects of Overfire Air and Low Excess Air on NO_x Emissions and Ash Fouling Potential for a Lignite Fired Boiler, Presented at the American Power Conference, Chicago, Illinois, April 24-26, 1978.
A reference abstract for this reference may be found in Section 3.5.1.1 reference abstracts.

7. Manz, O.E., Utilization of Lignite and Subbituminous Ash, Symposium on Technology and Use of Lignite, Grand Forks, N.D., May 1973, BuMines IC 8650, 1974, pp. 204-219.
An abstract for this reference is given in Section 3.5.1.1 reference abstracts.

3.5.1.4 Effects of Low-Rank Coal Properties

The unique properties of low-rank coal influence all aspects of the direct combustion process from milling to firing to environmental control.

These effects are discussed, where relevant, throughout the preceding and following sections on direct combustion and ECT. In this section, an attempt is made to summarize the major points through a short discussion of the various effects of each "unique property" in turn. Unless otherwise noted, the comments below refer to pulverized coal-fired steam generators.

High Sodium Content

The high sodium content which is characteristic of some low-rank coals (particularly North Dakota lignites) has several important effects on the performance of boilers and environmental control systems. The most notable effect is the severe ash fouling of heat transfer surfaces that occurs with high-sodium coals. As discussed in detail in Section 3.5.1.1, high-sodium ash fouling deposits have a continuous melt phase that envelops and connects particles into a strongly bonded network.[1,2] Design and operating modifications practiced with high-fouling low-rank coals in effect derate the boiler.[3] High calcium and magnesium levels present in some low-rank coals tend to suppress the fouling somewhat; this finding has led to some experimentation with the use of additives.[4]

High sodium content tends to reduce the high resistivity of low-rank coal fly ash, which aids collection in electrostatic precipitators. However, there is a tendency for this high sodium fly ash to cake in the hoppers.[5] In some cases, addition of small amounts of sodium as a fly ash conditioning agent can improve ESP performance without causing a large enough increase in sodium content to affect fouling.

High-sodium coals tend to produce a very fine particulate fraction rich in sodium sulfate, which is of concern because of the increasing attention being paid to control of the fine particle emissions.

The possibility of leaching of sodium salts from ash disposal sites is another area of concern in which data are lacking and regulatory scrutiny is increasing.

High sodium content in fly ash can cause undesirable reactions when the fly ash is utilized as an additive in portland cement concrete.[8]

Alkaline Ash

Low-rank coal ash is typically high in the alkali constituents calcium, magnesium, sodium, and potassium. One effect of this high alkalinity is that partial removal of SO_2 occurs by retention on the fly ash in a pc-fired boiler. Field studies have shown that sodium is the alkali constituent that has the greatest effect on SO_2 emitted from lignite-fired boilers.[12]

The ash alkali wet scrubbing process makes considerably better utilization of the alkali in low-rank coal fly ashes. In this process, alkali compounds are leached out of the fly ash to be used as sorbent in lieu of lime or limestone. Nearly 2600 MW of generating capacity utilize fly ash, or fly ash supplemented with lime or limestone, in wet scrubber systems.[13]

The problems mentioned above with regard to sodium content and its effects on fly ash disposal or utilization also apply to alkaline ash in a general sense.

Low Heating Value

The low heating value of low-rank coals significantly increases the quantity of fuel required for pulverization and burning for a given steam or electric power production rate. The number and size of the mills must be larger to handle the desired throughput.[3] The size of the furnace must also be larger than for higher rank coals to maintain the same energy output.[6] In addition, essentially all ancillary process equipment (stack gas handling and cleaning, ash handling and disposal, etc.) must be sized for the larger volumes of their respective streams. Conversely, in cases where boilers are converted from high-rank coal to low-rank coal firing, significant derating of output is usually required.

High Moisture Content

The high moisture content of low-rank coals necessitates a high primary air temperature for drying in the mills for pulverization. It also limits the throughput to the pulverizer depending on the capacity of the hot air source. Excessive surface moisture will also tend to impede the coal flow in conveyor systems, bins and feeders.[6] In northern regions with severe winter weather, high moisture content can cause freezing (thawing) problems in coal delivery, storage, and handling systems.

Grindability

The low grindability of low-rank coal causes the need for the capacity of the mills to be increased to meet sizing requirements for pulverized coal.[3] Modification of the standard Hardgrove grindability test has been required to obtain more reliable data on high moisture coals. Increased slagging at the Leland Olds Power Station was attributed to the low grindability of their lignite.[7]

Silica Content

High silica content in coal (typical of Texas lignites) can cause or aggravate a number of problems. High-silica ash is very abrasive and can cause erosion of coal feeding systems and burners. At the Monticello Power Plant in Texas, high silica ash has caused rapid deterioration of baghouse fabric filters. When high silica content is coupled with high sodium levels, massive deposits can form on heat transfer surfaces.[9] The silica content also has a strong influence on the viscosity of the slag.

Reactivity

The degree of reactivity (for combustion) in coals can be explained in part by the level of organically bound oxygen. Low-rank coals have significantly more oxygen than coals of higher rank. As the coal is heated during combustion, dissociation of this oxygen from the organic matrix occurs, leaving reactive sites for combustion. Due to the greater number of such active sites in low-rank coals (because of the higher oxygen content), they do not require the same degree of fineness as bituminous coals to insure complete combustion in pulverized coal systems.[3] However, the high reactivity of low-rank coals also requires special handling and precautions to avoid oxidation or spontaneous combustion in coal delivery, storage, and handling operations at the plant.

Agglomerating Properties

Low-rank coals do not go through a softening and melting stage when heated and hence do not tend to agglomerate into molten masses. As such, the same degree of fineness is not required to assure burnout compared to the bituminous coals since particle surface area will not tend to increase due to agglomeration.[6]

High Ash Resistivity

A major problem encountered in electrostatic precipitation of particulate matter from some low-sulfur western coals is the inherent high ash resistivity. Collection efficiencies are drastically reduced compared to collection of higher sulfur coal ash. Conversely, ESP's must be sized with much larger specific collecting areas and/or located on the "hot side" of the air preheater in low-rank coal-fired plants, or use additives.

Low Sulfur Content

The 1971 NSPS allowed combustion of some low-rank coals with no SO_2 control systems due to their low sulfur content. New (June 1979) standards now require a minimum of 70 percent removal, forcing the application of control devices.[11] The low sulfur content also is a major cause of the low fly ash resistivity hindering ESP performance as discussed above.

Low sulfur content combined with the high ash alkalinity of low-rank coals presents unique opportunities for SO_2 removal in both pc and FBC systems with minimal or no addition of makeup sorbents such as limestone.

References - Section 3.5.1.4

1. Sondreal, E.A., P.H. Tufte, and W. Beckering, "Ash Fouling in the Combustion of Low-Rank Western U.S. Coals," Combustion Science and Technology, 1977, Vol. 16, pp. 95-110.

2. Gronhovd, G.H., R.J. Wagner and A.J. Wittmaier, "A Study of the Ash Fouling Tendencies of a North Dakota Lignite as Related to Its Sodium Content," Quarterly, Trans. SME, September 1967, pp. 313-322.

3. Hensel, R.P., Western Coals: Properties That Influence Boiler Design, Coal Technology, 1979, Volume 2, Houston, Texas, November 6-8, 1979.

4. Baker, B.K. and W.H. Gardiner, "Modification of Ash Behavior in Lignite Fired Boilers," Symposium on Technology and Use of Lignite, Grand Forks, N.D., GFERC/IC-77/1, 1978, pp. 74-81.

5. Selle, S.J. and L.L. Hess, Factors Affecting ESP Performance on Western Coals and Experience with North Dakota Lignites, Symposium on Particulate Control in Energy Processes, EPA-600/7-76-010, September 1976.

6. Hensel, R.P. and D.A. Harris, Properties of Solid Fuels and Their Impact on Boiler Design and Performance, ASME-IEEE-ASCE Joint Power Generation Conference, Dallas, Texas, September 10-13, 1978.

7. Hillier, L.V., A 212-MW Lignite-Fired Generating Station at Center, North Dakota, Bureau of Mines, University of North Dakota Symposium, Grand Forks, North Dakota, April 27-28, 1967.

8. Manz, O.E., "Utilization of Lignite and Subbituminous Ash," Symposium on Technology and Use of Lignite, Grand Forks, N.D., May 1973, BuMines IC 8650, 1974, pp. 204-219.

9. Sondreal, E.A., G.H. Gronhovd, and W.R. Kube, "Research and Development Relating to Lignite Use in Power Production" (Preprint), presented at Gulf Coast Lignite: Geology, Utilization and Environmental Effects, Austin, Texas, June 2,3 and 4 1976, 32 pp.

10. Lowry, H.H. (ed.), Chemistry of Coal Utilization - Supplementary Volume, John Wiley & Sons, Inc., New York, 1963.

11. Federal Register, Volume 44, No. 113, Monday, June 11, 1979.

12. Gronhovd, G.H., P.H. Tufte, and S.J. Selle, "Some Studies on Stack Emissions from Lignite-Fired Powerplants," Symposium on Technology and Use of Lignite, Grand Forks, N.D., May 1973, BuMines IC 8650, 1974, pp. 83-102.

13. Ness, H.M., S.J. Selle, and O.E. Manz," Power Plant Flue Gas Desulfurization for Low-Rank Western Coals," Symposium on Technology and Use of Lignite, Grand Forks, N.D., May 1979, GFETC/IC-79/1, 1979, pp. 117-146.

3.5.1.5 Current Status

The primary use for lignite and subbituminous coal in the United States today is for electric power generation. Lists of the existing and planned generating facilities utilizing low-rank coal are given in Tables 3.5.1.28 through 3.5.1.32. All lignite-fired power plants at this time are found in Texas and Fort Union Region. Most of these are minemouth facilities. Subbituminous coal-fired generating stations are scattered through the midwest and western United States.

A condensed breakdown of plant location versus operating characteristics is shown in Table 3.5.1.33. In all regions, pulverized coal-fired furnaces dominate utility choice both in presently operating plants and those planned or under construction. Most stoker furnaces are old and operate at low power levels. There are some cyclone furnaces in the midwest and Fort Union area but NSPS for NO_x is discouraging further development. The major method employed for SO_x control at the present time is limestone wet scrubbing, although ash alkali wet scrubbing is utilized in a few Fort Union Region and midwestern plants. Several dry scrubbing systems are planned for new utility boilers. Particulate control in the immediate future will continue to be dominated by electrostatic precipitators although a significant number of baghouses have been planned for new facilities.

Industrial use of low-rank coal at the present time is very limited. Lignite applications are largely limited to process steam generation.[23] Alcoa Aluminum uses three tangentially fired boilers at their Sandow plant in Rockdale, Texas. The American Crystal Sugar Company utilizes stoker furnaces in five of its six locations (four of the five located in North Dakota, one in California). The subbituminous coals have wider applications although total tonnage as compared to electrical generation is still small. Some cement, glass and metal industries can utilize subbituminous coal, as can industries requiring process steam.

Table 3.5.1.28

Lignite-Fired Electric Power Plants in the Fort Union Region

Station		Location	Owner	Start-Up	Size (MW)	Fuel	Furnace Type	Transport Method	Environmental Controls
Lewis & Clark		Sidnev. MT	Montana-Dakota Utility Co.	1958	50	Lignite	Tangential	BN RR	Mech. Dust Collector, Tall Stack Flooded Disc Scrubber
Hoot Lake	1	Fergus Falls, MN	Otter Tail Power	1949	7.5	Lig. (N.D.)	Stoker	BN RR	ESP
	2	Fergus Falls, MN	Otter Tail Power	1959	66	Lig. (N.D.)	Tangential, P.C.	BN RR	ESP
	3	Fergus Falls, MN	Otter Tail Power	1972	79	Lig.	Pulverized Coal	BN RR	ESP
Ortonville	1	Ortonville, MN	Otter Tail Power	Old	15	Lig. (N.D.)	Stoker	Milw. RR	ESP
	2	Ortonville, MN	Otter Tail Power	6-1972	21	Lig. (N.D.)	Spread-Stoker		ESP
Moorhead		Moorhead, MN	Moorhead Public Service Dept.	1969	25	Lig. (N.D.)	Stoker	BN RR	Mech. Dust Collector, ESP, 250 Ft. Stack
Beulah		Beulah, N.D.	Montana-Dakota Utility Co.	1949	13.5	Lig. (N.D.)	Stoker	Minemouth	Mech. Dust Collector
Heskett		Mondan, N.D.	Montana-Dakota Utility Co.	1963	100	Lig. (N.D.)	Stoker	Minemouth, BN RR	Mech. Dust Collector, ESP 300 ft. Stacks
Leland Olds	1	Stanton, N.D.	Basin Elec. Power Coop.	1966	215.7	Lignite (ND)	Horizon. Opp.	Minemouth	ESP
	2	Stanton, N.D.	Basin Elec. Power Coop.	1975	440	Lignite (ND)	Cyclone		ESP
W.J. Neal		Velva, N.D.	Basin Elec. Power Coop.	1960	38.5	Lignite (ND)	Tangential		ESP
F.P. Wood		Grand Forks, N.D.	Minnkota Power Coop.	1950	21.5	Lignite (ND)	Stoker	RR	Dust Collector
M.R. Young	1	Center, N.D.	Minnkota Power Coop.	1970	234.6	Lignite (ND)	Cyclone	Truck	ESP, Ash Alkali and Limestone Scrubber Spray tower
	2	Center, N.D	Minnkota Power Coop.	1977	435	Lignite (ND)	Cyclone		ESP, Ash Alkali and Limestone Scrubber Spray tower

(continued)

Table 3.5.1.28 (Continued)

Lignite-Fired Electric Power Plants in the Fort Union Region

Station	Location	Owner	Start-Up	Size (MW)	Fuel	Furnace Type	Transport Method	Environmental Controls
Stanton	Stanton, N.D.	United Power Assoc	1966	172.0	Lignite (ND)	Front	BN RR Unit Train	ESP
Acme	Sheridan, WY	Montana-Dakota Utility Co.	Old	8.0	Lignite	Stoker	(To Be Closed)	
Antelope Valley 1	Bismarck, N.D.	Basin Elec. Power	4-1982	440	Lignite	Tangential, P.C.	Minemouth	Dry SO_2 Scrubber and Baghouse
2	Bismarck, N.D.	Basin Elec. Power	11-1983	440	Lignite	Tangential, P.C.		Dry SO_2 Scrubber and Baghouse
Coyote	Bealuh, N.D.	Montana-Dakota Utility Co. Montana Power Co.	5-1981	410	Lignite	Cyclone	TK, Minemouth	Dry Scrubber, Baghouse
Big Stone	Milbank, S.D.	Montana-Dakota Utility Co.	5-1975	428	Lignite	Cyclone		ESP
Coal Creek 1	Underwood, N.D.	United Power Assoc.	7-1979	550	Lignite	Pulverized Coal	Conveyor	Spray Tower - ash, baghouse
2	Underwood, N.D.	United Power Assoc.	11-1980	550	Lignite	Pulverized Coal	Conveyor	Alkali and Lime, baghouse

Sources: References 7, 8, 12, 13, 16 and 23

Table 3.5.1.29

Subbituminous Coal-Fired Electric Power Plants in the Western Region

Station		Location	Owner	Start-Up	Size (MW)	Fuel	Furnace Type	Transport Method	Environmental Controls
Cholla	1	Joseph City, AZ	AZ Public Service Co.	Exist.	116	Sub. (N.M.)	Pulverized Coal	RR, Truck	Wet Scrubber Collector
	2			Exist.	235	Sub. (N.M.)	Pulverized Coal	RR, Truck	Wet Scrubber Collector
Four Corners	1	Fruitland, N.M.	AZ Public Service Co.	Exist.	175	Sub. (N.M.)	Pulverized Coal	Minemouth,RR,TK	Wet Scrubbers and ESP, Ash Alkali
	2	Fruitland, N.M.		Exist.	177	Sub. (N.M.)	Pulverized Coal	Minemouth,RR,TK	Wet Scrubbers and ESP, Ash Alkali
	3	Fruitland, N.M.	So.Ca.Edison P.S. of N.M.	Exist.	220	Sub. (N.M.)	Pulverized Coal	Minemouth,RR,TK	Wet Scrubbers and ESP, Ash Alkali
	4	Fruitland, N.M.	El Paso Elec.	Exist.	800	Sub. (N.M.)	Pulverized Coal	Minemouth,RR,TK	Wet Scrubbers and ESP
	5	Fruitland, N.M.	Tucson Gas & Elec.	Exist.	800	Sub. (N.M.)	Pulverized Coal	Minemouth,RR,TK	Wet Scrubbers and ESP
Navajo	1	Page, AZ	Salt River Proj.	Exist.	750	Sub. (AZ.)	Pulverized Coal	RR, Truck	ESP, no SO_2
	2	Page, AZ	Salt River Proj.	Exist.	750	Sub. (AZ.)	Pulverized Coal	RR, Truck	ESP, no SO_2
	3	Page, AZ	Salt River Proj.	Exist.	750	Sub. (AZ.)	Pulverized Coal	RR, Truck	ESP, no SO_2
Osage	1	Osage, WY	Black Hills Power & Light	6-1985	100	Sub. (WY.)	Stoker	BN, RR	ESP
Tri State				7-1988	350	Sub.		RR	
Apache	2	Cochise, AZ	AZ. Elec. Power Corp.	1-1979	175	Sub. (N.M.)	Pulverized Coal	RR, Truck	ESP
	3	Cochise, AZ		4-1979	175	Sub. (N.M.)	Pulverized Coal	RR, Truck	ESP
Coronado	1	St.Johns,AZ	Salt River Proj.	6-1979	350	Sub. (N.M.)	Riley-Stoker Turbo	RR	ESP,Pullman-Kellogg Scrubber
	2	St.Johns,AZ	Salt River Proj.	4-1980	350	Sub. (UTAH)	Riley-Stoker Turbo	RR	ESP,Pullman-Kellogg Scrubber
	3	St.Johns,AZ	Salt River Proj.	1986	350	Sub. (UTAH)	Riley-Stoker Turbo	RR	ESP, or Baghouse
Cholla	3	Joseph City, AZ	AZ Public Service Co.	6-1980	242	Sub. (N.M.)	Pulverized Coal	RR	
	4	Joseph City, AZ		6-1981	347	Sub. (N.M.)	Pulverized Coal	RR	
Springerville	1	Springerville,AZ	Tucson Gas & Elec.	5-1985	330	Sub. (N.M.)	Pulverized Coal	RR	Dry Baghouse for Particulate
	2	Springerville,AZ		6-1987	330	Sub. (N.M.)	Pulverized Coal	RR	Dry Scrubber for SO_2

(continued)

Table 3.5.1.29 (Continued)

Subbituminous Coal-Fired Electric Power Plants in the Western Region

Station		Location	Owner	Start-Up	Size (MW)	Fuel	Furnace Type	Transport Method	Environmental Controls
Dave Johnston	1	Glenrock, WY	Pacific Power & Light Co.	4-1972	330	Sub. (WY)	P.C., Tangential	Truck	ESP
	2	Glenrock, WY	Pacific P&L Co.		330	Sub. (WY)	P.C., Tangential	Truck	ESP
	3	Glenrock, WY	Pacific P&L Co.		330	Sub. (WY)	P.C., Tangential	Truck	ESP
	4	Glenrock, WY	Pacific P&L Co.		330	Sub. (WY)	P.C., Tangential	Truck	Venturi Scrubber, Ash Alkali
Valmont		Boulder, CO	Public Service of Colorado	11-1971	118	Sub. (CO,WY)	Pulverized Coal	Union Pacific RR and Truck	Mechanical Separ. ESP, Limestone Scrubber
Arapahoe	1	Denver, CO	Public Service of Colorado	9-1973	112	Sub. (CO)	Pulverized Coal	C&S RR	Mechanical Collector, ESP, Scrubber
	3	Denver, CO	Public Service of Colorado	5-1979	44	Sub.	Pulverized Coal		Baghouse
Clark	1	Canyon City, CO	Southern CO.Power	8-1978	16.5	Sub.	Stoker		Baghouse
	2	Canyon City, CO	Southern Co.Power	10-1978	22		Stoker		Baghouse
R.D. Nixon		Fountain, CO.	City of Colorado Springs	Late 1979	200	Sub.	Pulverized Coal		Baghouse
Martin Drake		Colorado Springs, CO	City of Colorado Springs	9-1978	85	Sub.	Pulverized Coal		Baghouse
Nucla	1	Nucla, CO.	CO. UTE Elc. Assoc.	12-1973	13	Sub.	Stoker		Baghouse
	2	Nucla, CO.	CO. UTE Elc. Assoc.	2-1974	13	Sub.	Stoker		Baghouse
	3	Nucla, CO.	CO. UTE Elc. Assoc.	5-1974	13	Sub.	Stoker		Baghouse
Colstrip Station	1	Colstrip, MT	Montana Power Co.	Exist.	358	Sub.(MT)	Tangential, P.C.	Minemouth	Ash Alkali Venturi Scrubber
	2			Exist.	358	Sub.(MT)	Tangential, P.C.	Minemouth	Ash Alkali Venturi Scrubber
Laramie River	1	Wheatline, WY	Basin Elec. Power Coop	10-1980	174	Sub.(WY)	Pulverized Coal	RR	ESP, Wet Limestone Scrubber
	2	Wheatline, WY	Basin Elec. Power Coop	10-1980	133	Sub.(WY)	Pulverized Coal	RR	ESP, Wet Limestone Scrubber
	3	Wheatline, WY	Basin Elec. Power Coop			Sub.(WY)	Pulverized Coal	RR	ESP, Spray Dryer

Source: References 2, 3, 4, 6, 7, 13 and 22

Table 3.5.1.30

Lignite-Fired Electric Power Plants in the Gulf Coast Region

Station		Location	Owner	Start-Up	Size (MW)	Fuel	Furnace Type	Transport Method	Environmental Controls
Sandow	1	Rockdale, TX.	Alcoa & Texas P&L Co.	1954	3x120	Lignite	Tangential	Truck	Cold ESP, no SO_2 Control
	2			1981	575	Lignite	P.C., Split Tang.	Minemouth	ESP and Limestone Scrubber
Big Brown	1	Fairfield, TX.	TX. Utilities Co.	1971	575	Lignite	P.C., Split Tang.	Minemouth	ESP, no SO_2
	2			1972	575	Lignite	P.C., Split Tang.	TK and RR	ESP, no SO_2
Monticello	1	Mt.Pleasant, TX.	TX. Utilities Co.	1974	575	Lignite	P.C., Split Tang.	Minemouth	Cold ESP, Baghouse, No SO_2
	2			1975	575	Lignite	P.C., Split Tang.	TK & Elec. RR	Cold ESP, Baghouse, No SO_2
	3			1978	750	Lignite	P.C., Horizontal		ESP, Limestone Spray Tower
Martin Lake	1	Tatum, TX.	TX. Utilities Co.	1977	750	Lignite	P.C., Split Tang.	Minemouth	ESP, Limestone Packed Absorber Tower
	2			1978	750	Lignite	P.C., Split Tang.	TK & Elec. RR	ESP, Limestone Packed Absorber Tower
	3			1979	750	Lignite	P.C., Split Tang.		ESP, Limestone Packed Absorber Tower
	4			1983	750	Lignite	P.C., Split Tang.		ESP, Limestone Packed Absorber Tower
San Miguel	1	Christine, TX.	S.TX. & Medina Elec. Coop Brazos Elec. Power	1980	400	Lignite	P.C., Horizontal	Truck	ESP, Limestone Scrubber
	2			1985	400	Lignite	P.C., Horizontal	Minemouth	ESP, Limestone Venturi Tray
Forest Grove		Athens, TX.	TX. Utilities Co.	1982	750	Lignite	P.C., Horizontal	Minemouth, TK	ESP, Limestone Scrubber
Gibbons Creek	1	Carlos, TX.	TX. Municipal Power Agency	1981	400	Lignite	P.C., Tangential	Minemouth	ESP, Limestone Scrubber
	2			1984	400	Lignite		Truck	
Twin Oak	1	Bald Prairie, TX.	TX. Utilities & Alcoa	1984	750	Lignite	P.C., Tangential	Minemouth	ESP and Limestone Scrubber
	2			1985	750	Lignite	P.C. Tangential	Truck	ESP

(continued)

Table 3.5.1.30 (Continued)

Lignite-Fired Electric Power Plants in the Gulf Coast Region

Station		Location	Owner	Start-Up	Size (MW)	Fuel	Furnace Type	Transport Method	Environmental Controls
Mill Creek	1	Oak Hill, TX.	TX. Utilities Co	1986	750	Lignite		Minemouth	
	2				750	Lignite		Truck	
Unnamed	1	Harrison Co.TX.	Southwestern Elec. Power Co.	Mid-1980's	640	Lignite		Minemouth	
	2				640	Lignite			
Oak Knoll	1	Oletha, TX	TX. Utilities Co	1980's	750	Lignite		Minemouth	
	2			1980's	750	Lignite			
Big Cajun 3	1	Threeport,LA	Cajun Elec. Power	1-1985	540	Lignite	Tangential, P.C.	Minemouth	Dry Scrubber with Baghouse or ESP with Wet Scrubber
	2		Cajun Elec. Power	1-1989	540	Lignite	Tangential, P.C.	Minemouth	
Big Cajun 4		Threeport,LA	Cajun Elec. Power	1-1986	540	Lignite	Tangential, P.C.	Minemouth	
Dolet Hills			Central Louisiana Electric Co.	3-1986	640	Lignite			No Dolet Hills Plant
Lovelady	1		Gulf States Util.	9-1985	600	Lignite	STILL UNDER CONSTRUCTION		
	2		Gulf States Util.	9-1987	600	Lignite	STILL UNDER CONSTRUCTION		
Pirkey		Marshel, TX	Southwestern Elec Power Co.	1985 3-1984	640	Lignite	Front & Rear Burners, P.C.	Minemouth, TK	ESP Limestone Scrubbers
Karnack			Southwestern Elec Power Co.	1-1988	295	Lignite	STILL UNDER CONSTRUCTION		

Sources: References 1, 5, 10, 11, 12, 15, 19

Table 3.5.1.31

Subbituminous Coal-Fired Electric Power Plants in the Midwestern Region

Station		Location	Owner	Start-Up	Size (MW)	Fuel	Furnace Type	Transport Method	Environmental Controls
Aurora		Aurora, MN	MN Power & Light	6-1971	2 x 58	Sub. (MT)	Pulverized Coal	BN RR Unit Train	Wet Scrubbers, Ash Alkali
Columbia	1	Portage, WI	Wisconsin P&L Co. & Madison G&E Co.	Exist.	279	Sub. (WY)	Pulverized Coal	RR P&P RR	*
Stateline	2	Hammond, IL	Commonwealth Edison	1929	140	Sub. (MT)	Pulverized Coal	RR	
	3		Commonwealth Edison	1955	190	Sub. (MT)	Pulverized Coal	RR	ESP
	4		Commonwealth Edison	1962	318	Sub. (MT)	Cyclone	RR	ESP
Waukegan	6	Waukegan, IL	Commonwealth Edison	1952	88	Sub(WY,MT)	Cyclone	RR	ESP
	7	Waukegan, IL	Commonwealth Edison	1958	328	Sub(WY,MT)	Pulverized Coal	RR	Hot Precipitation
	8	Waukegan, IL	Commonwealth Edison	1962	358	Sub(WY,MT)	Pulverized Coal	RR	ESP
Crawford	8	Chicago, IL	Commonwealth Edison	1961	323	Sub. (MT)	Pulverized Coal	Barge, RR	ESP
Fisk	19	Chicago, IL	Commonwealth Edison	1959	338	Sub. (MT)	Pulverized Coal	Barge, RR	ESP
Joliet	6	Joliet, IL	Commonwealth Edison	1959	335	Sub. (WY,MT)	Cyclone	Barge, RR	ESP
	7	Joliet, IL	Commonwealth Edison	1965	535	Sub. (WY,MT)	Pulverized Coal	Barge, RR	ESP
	8	Joliet, IL	Commonwealth Edison	1966	535	Sub. (WY,MT)	Pulverized Coal	Barge, RR	ESP
Will County	1	Romeoville, IL	Commonwealth Edison	1955	141	Sub. (MT)	Cyclone	RR	Limestone Scrubber, ESP
	2	Romeoville, IL	Commonwealth Edison	1955	163	Sub. (MT)	Cyclone	RR	ESP
	3	Romeoville, IL	Commonwealth Edison	1957	257	Sub. (MT)	Pulverized Coal	RR	ESP
	4	Romeoville, IL	Commonwealth Edison	1963	515	Sub. (MT)	Pulverized Coal	RR	ESP
Sherbourne Co. Generating Plant	1	Becker, MN	Northern States Power Co.	Exist.	680	Sub. (MT)	Tangential, P.C.		Limestone Venturi Rod Scrubber/fly ash alkali
	2			Exist.	680	Sub. (MT)	Tangential, P.C.		Limestone Venturi Rod Scrubber/fly ash alkali

(continued)

Table 3.5.1.31 (Continued)

Subbituminous Coal-Fired Electric Power Plants in the Midwestern Region

Station		Location	Owner	Start-Up	Size (MW)	Fuel	Furnace Type	Transport Method	Environmental Control
Ottumwa	1	Chillicothe,IA	Iowa Elec L&P Iowa-IL G&E	1-1981	100	Sub. (WY)	Tangential, P.C.	RR	ESP, No scrubbers
	2		Iowa P&L Iowa Public Serv.	1-1981	125	Sub. (WY)		RR	(PLANS PENDING)
	3		Iowa-So. Ut. Co.	1-1981	225	Sub. (WY)		RR	
Carroll	1	Carroll, IA	Iowa Public Ser.	5-1981	5.1	Sub. (WY)	Stoker	RR	Mechanical Precipitation
	2			5-1981	5.1	Sub. (WY)	Stoker	RR	
Eagle Grove		Eagle Grove, IA	Iowa Public Serv.	5-1981	7.7	Sub. (WY)	Stoker	RR	Mechanical Precipitation
Hawkeye	1	Storm Lake, IA	Iowa Public Serv.	5-1981	7.3	Sub. (WY)	Pulverized Coal	TK, RR	ESP
	2	Storm Lake, IA	Iowa Public Serv.	5-1981	9.7	Sub. (WY)	Pulverized Coal	TK, RR	ESP
Louisa	1		Iowa IL G&E Iowa Public Serv.	1-1983	641	Sub.		RR	
Sherborne	3	Becker, MN	Northern States Power Co.	5-1984	800	Sub. (MT)	Pulverized Coal	RR	Venturi Marble bed, fly ash alkali-limestone
Clay Boswell	1	Cohasset, MN	Minn. P&L	5-1973	350	Sub. (MT)	Pulverized Coal	RR BN	Wet Scrubbers, baghouse
	2	Cohasset, MN	Minn. P&L	Pilot	1	Sub. (MT)	Pulverized Coal	Unit Train	Venturi Spray tower : fly ash alkali-lime
	3	Cohasset, MN	Minn. P&L	1-1981	500	Sub. (MT)	Pulverized Coal		
Pleasant Prarrie	1		Wisconsin Elec. Power Co.	7-1980	580	Sub.	*	RR	
	2			4-1982	580	Sub.		RR	
Edgewater	5	Sheboygan, WI	Wisconsin P&L Co.	6-1983	200	Sub. (IL)	Cyclone	RR	ESP
Weston	3	Rothschild, WI	Wisconsin Public Service	3-1982	300	Sub. (WY)	Pulverized Coal	RR	ESP
Independence	1	Newark, AR	Arkansas P&L	1-1983	700	Sub. (WY)	Tang., P.C.	RR	ESP, no scrubbers, 1000 ft. Stacks
	2	Newark, AR	Arkansas P&L	1-1985	700	Sub. (WY)	Tang., P.C.	RR	ESP, no scrubbers, 1000 ft. Stacks

(continued)

Table 3.5.1.31 (Continued)

Subbituminous Coal-Fired Electric Power Plants in the Midwestern Region

Station		Location	Owner	Start-Up	Size (MW)	Fuel	Furnace Type	Transport Method	Environmental Control
White Bluff	1	Redfield, AR	Arkansas P&L	6-1980	700	Sub. (WY)	Tang., P.C.	RR	ESP, no scrubbers, 1000 ft. Stacks
	2			5-1981	700	Sub. (WY)	Tang., P.C.	RR	ESP, no scrubbers, 1000 ft. Stacks
Iatan	1	Weston, MO	Kansas City P&L	3-1980	650	Sub. (WY)	P.C. Front & Rear Burners	RR	ESP
	2			3-1988	650	Sub.	STILL UNDER CONSTRUCTION		
Fossil Unit			Minnesota P&L	11-1986	500	Sub.		RR	
Jeffery Energy Ct.	1	St. Mary's, KS	Kansas P&L Co.	6-1980	700	Sub. (WY)	Tangential, P.C.	RR	ESP SE Wet Sprayer Scrubber
	3		Kansas P&L Co.	6-1982	700	Sub. (WY)	Tangential, P.C.	RR	ESP SE Wet Sprayer Scrubber
	4		Kansas P&L Co.	6-1984	700	Sub. (WY)	Tangential, P.C.	RR	ESP SE Wet Sprayer Scrubber
Northeastern	3	Oologah, OK	Public Service of Oklahoma	6-1979	450	Sub. (WY)	P.C., Tangential	RR	ESP, 600 ft. Stack
	4	Oologah, OK	Public Service of Oklahoma	6-1980	450	Sub. (WY)	P.C., Tangential	RR	ESP, 600 ft. Stack
Hugo			WEFA	4-1981	376	Sub.			
Gentleman		Sutherland, NB	Nebraska Public Power District.	1979	600	Sub. (WY)	P.C. Radiant Boiler	RR	ESP, No Scrubbers
Neal	1	Salix, IA	Iowa Pub. Serv.	5-1979	293	Sub. (WY)	Pulverized Coal	RR	ESP
	2	Salix, IA	Iowa Pub. Serv.	5-1979	100	Sub. (WY)	Pulverized Coal	RR	ESP
	3	Salix, IA	Iowa Pub. Serv.	5-1979	120	Sub. (WY)	Pulverized Coal	RR	ESP
Hodgett, J.P.		Elmo, WI	Daryland Power Coop	11-1979	350	Sub. (WY)	P.C. Front/Rear Burner	RR	

(continued)

Table 3.5.1.31 (Continued)

Subbituminous Coal-Fired Electric Power Plants in the Midwestern Region

Station		Location	Owner	Start-Up	Size (MW)	Fuel	Furnace Type	Transport Method	Environmental Control
Fox Lake	3	Sherburn, MN	Interst. Power Co	Exist.	85	Sub. (MT)	Pulverized Coal	RR	ESP
Bay Front	1	Ashland, WI	Lake Superior Dist. Power Co.	Exist.	6	Sub. (MT)	Stoker	RR, Barge	Mechanical Collector
	2		Lake Superior Dist. Power Co.	Exist.	6	Sub. (MT)	Stoker	RR, Barge	Mechanical Collector
	3		Lake Superior Dist. Power Co.	Exist.	6	Sub. (MT)	Stoker	RR, Barge	Mechanical Collector
	4		Lake Superior Dist. Power Co.	Exist.	22	Sub. (MT)	Pulverized Coal	RR, Barge	Mechanical Collector
	5		Lake Superior Dist. Power Co.	Exist.	22	Sub. (MT)	Pulverized Coal	RR, Barge	Mechanical Collector
	6		Lake Superior Dist. Power Co.	Exist.	30	Sub. (MT)	Pulverized Coal	RR, Barge	Mechanical Collector
Syl Laskin	1	Aurora, MN	Minnesota P&L	Exist.	55	Sub. (MT)	Pulverized Coal	RR	Wet Scrubbers
	2			Exist.	55	Sub. (MT)	Pulverized Coal	RR	Wet Scrubbers
Kramer	1	Bellevue, NB	Nebraska Public Power District	Exist.	23	Sub.(WY,CO		RR	Mechanical Collectors, Baghouse
	2	Bellevue, NB	Nebraska Public Power District	Exist.	23	Sub.(WY,CO		RR	Mechanical Collectors, Baghouse
	3	Bellevue, NB	Nebraska Public Power District	Exist.	23	Sub.(WY,CO		RR	Mechanical Collectors, Baghouse
	2	Bellevue, NB	Nebraska Public Power District	5-1977	36				

(continued)

Table 3.5.1.31 (Continued)

Subbituminous Coal-Fired Electric Power Plants in the Midwestern Region

Station		Location	Owner	Start-Up	Size (MW)	Fuel	Furnace Type	Transport Method	Environmental Control
Sheldon	1	Hallam, NB	Nebraska Public Power District	Exist.	105	Sub.(UT,WY)	Cyclone	RR	ESP
	2		Nebraska Public Power District	Exist.	115	Sub.	Cyclone	RR	ESP
Mitchell		Mitchell,S.D.	Northwestern Public Service	Exist.	7.5	Sub. (WY)	Stoker	RR	Mechanical Dust Collectors
North Omaha	1	Omaha, NB	Omaha Public Power District	Exist.	75.6	Sub. (WY)	Pulverized Coal	RR	ESP
	2		Omaha Public Power District	Exist.	102.1	Sub. (WY)	Pulverized Coal	RR	ESP
	3		Omaha Public Power District	Exist.	102.1	Sub. (WY)	Pulverized Coal	RR	ESP
	4		Omaha Public Power District	Exist.	131.2	Sub. (WY)	Pulverized Coal	RR	ESP
	5		Omaha Public Power District	Exist.	218.6	Sub. (WY)	Pulverized Coal	RR	ESP
Elk River	1	Elk River, MN	United Power Assoc.	Exist.	12.1	Sub. (MT)	Stoker	RR	Baghouses
	2	Elk River, MN	United Power Assoc.	Exist.	11.9	Sub. (MT)	Stoker	RR	Baghouses
	3	Elk River, MN	United Power Assoc.	Exist.	26	Sub. (MT)	Pulverized Coal	RR	Baghouses
Council Bluffs	1	Council Bluffs IA	Iowa Power and Light	Exist.	46.6	Sub. (WY)	Pulverized Coal	RR	ESP
	2			Exist.	90.6	Sub. (WY)	Pulverized Coal	RR	ESP
	3			Exist.	303.5	Sub. (WY)	Pulverized Coal	RR	ESP

(continued)

Table 3.5.1.31 (Continued)

Subbituminous Coal-Fired Electric Power Plants in the Midwestern Region

Station		Location	Owner	Start-Up	Size (MW)	Fuel	Furnace Type	Transport Method	Environmental Control
Lansing	4	Lansing, IA	Interstate Power	Exist.	260	Sub.(MT,WY)	Pulverized Coal	RR, Barge	ESP
Boone	1	Boone, IA	Iowa Elec. L&P	Exist.	9	Sub.(MT)	Pulverized Coal	RR	ESP
	2			Exist.	18	Sub.(MT)	Pulverized Coal	RR	ESP
Iowa Falls	4	Iowa Falls, IA	Iowa Elec. L&P	Exist.	8.5	Sub.	Stoker	RR, Truck	Mechanical Collector
Prairie Creek	1	Cedar Rapids, IA	Iowa Elec. L&P	Exist.	19.5	Sub.(CO)	Stoker	RR	ESP & Mechanical Collector
	2		Iowa Elec. L&P	Exist.	19.5	Sub.(CO)	Stoker	RR	ESP & Mechanical Collector
	3		Iowa Elec. L&P	Exist.	49	Sub.(CO)	Pulverized Coal	RR	ESP & Mechanical Collector
	4		Iowa Elec. L&P	Exist.	132	Sub.(CO)	Pulverized Coal	RR	ESP & Mechanical Collector
Sixth St.	1	Cedar Rapids, IA	Iowa Elec. L&P	Exist.	8	Sub.(CO)	Pulverized Coal	RR	ESP & Mechanical Collector
	2		Iowa Elec. L&P	Exist.	3	Sub.(CO)	Pulverized Coal	RR	ESP & Mechanical Collector
	4		Iowa Elec. L&P	Exist.	18	Sub.(CO)	Pulverized Coal	RR	ESP & Mechanical Collector
	6		Iowa Elec. L&P	Exist.	8	Sub.(CO)	Pulverized Coal	RR	ESP & Mechanical Collector
	7		Iowa Elec. L&P	Exist	18	Sub.(CO)	Pulverized Coal	RR	ESP & Mechanical Collector
	8		Iowa Elec. L&P	Exist	30	Sub.(CO)	Pulverized Coal	RR	ESP & Mechanical Collector
Southerland	1	Marshalltown, IA	Iowa Elec. L&P	Exist.	31.5	Sub.(CO)	Pulverized Coal	RR	ESP
	2		Iowa Elec. L&P	Exist.	31.5	Sub.(CO)	Pulverized Coal	RR	ESP
	3		Iowa Elec. L&P	Exist.	80.0	Sub.(CO)	Pulverized Coal	RR	ESP
Maynard	6	Waterloo, IA	Iowa Public Serv.	Exist.	22.8	Sub.	Pulverized Coal	RR	ESP
	7			Exist.	51.8	Sub.	Pulverized Coal	RR	ESP
Columbia	4	Columbia, MO	City of Columbia Water & Light	1979	16.5	Sub.	Stoker		Baghouse
	5			1979	22	Sub.	Stoker		Baghouse

Sources: References 9. 17, 18, 20, 21

Table 3.5.1.32

Subbituminous Coal-Fired Electric Power Plants in the Gulf Coast Region

Station		Location	Owner	Start-Up	Size (MW)	Fuel	Furnace Type	Transport Method	Environmental Control
Coletta Creek	1	Goliad, TX	Central P&L Co.	12-1979	550	Sub. (CO)	P.C., Tangential	RR, (Barge from So. Africa, trial)	ESP, no scrubbers
	2		Central P&L Co.	1-1988	640	Sub.	STILL UNDER CONSTRUCTION		
Lake Diversion		Oklaunion, TX	West TX Utility	1-1987	640	Sub.	STILL UNDER CONSTRUCTION		
Big Cajun 2	1	Baton Rouge, LA	Cajun Elec. Pow.	1-1980	540	Sub. (WY)	P.C., Horizontal	RR, water	ESP, no scrubbers
	2	Baton Rouge, LA	Cajun Elec. Pow.	1-1981	540	Sub. (WY)	P.C., Horizontal	RR, water	ESP, no scrubbers
	3	Baton Rouge, LA	Cajun Elec. Pow.	1-1983	540	Sub. (WY)	P.C.	RR, water	ESP, no scrubbers
R.S. Nelson	5	Lake Charles, LA	Gulf States Ut.	3-1984	540	Sub. (WY)	P.C., Tangential	RR	Hot ESP
	6			3-1982	540	Sub. (WY)	P.C., Tangential	RR	Hot ESP
Welsh	2	Welsh, TX	Southwestern El. Power Company	Open	540	Sub. (WY)	P.C.	RR	ESP
	3		Southwestern El. Power Company	3-1982	540	Sub. (WY)	P.C., Front & Rear	RR	ESP
Clark	2	Amarillo, TX	Southwestern Pub Service Company	6-1978	350	Sub.	P.C.		Baghouse
	3	Amarillo, TX	Southwestern Pub Service Company	1980	350	Sub.	P.C.		Baghouse

Source: Reference 19

Table 3.5.1.33

U.S. Low-Rank Coal-Fired Electric Power Plants[a]

Fuel:	Fort Union Lignite		Gulf Coast Lignite		Western Subbituminous Coal					
Plant Location:	Fort Union Region		Gulf Coast Region		West		Midwest		Gulf Coast	
Plants, Capacity:	No.	MWe	No.	MWE	No.	MWe	No.	MWe	No.	MWe
Operating Plants (1979)	19	2,920.3	9	5,660	29	8,145.9	75	11,179.4	3	1,440
Furnace:										
PC	7	1,171.2	9	5,660	23	7,718	53	9,675.9	3	1,440
Stoker	8	211.5	-	-	6	427.5	11	133.5	-	-
Cyclone	4	1,537.6	-	-	-	-	7	1,265	-	-
Unknown	-	-	-	-	-	-	4	105	-	-
Wet scrubber:										
Limestone	3	719.6	4	3,000	3	580	3	1,501	-	-
Ash-alkali	3	1,219.6	-	-	6	1,618	3	1,476	-	-
Spray dryer	-	-	-	-	-	-	-	-	-	-
Particulate removal:										
ESP	14	2,328.8	11	5,660	17	6,672	49	8,810.4	2	1,090
Baghouse	1	550	2	1,150	8	406.5	8	156.5	1	350
Mechanical	5	210	-	-	9	2,965	21	482	-	-
Unknown	-	-	-	-	-	-	4	805	-	-

[a]This table is a condensation of the information presented in Tables 3.5.1.27 through 3.5.1.31.

(continued)

Table 3.5.1.33[a] (Continued)

Fuel:	Fort Union Lignite		Gulf Coast Lignite		Western Subbituminous Coal					
Plant Location:	Fort Union Region		Gulf Coast Region		West		Midwest		Gulf Coast	
Plants, Capacity:	No.	MWe	No.	MWE	No.	MWe	No.	MWe	No.	MWe
Plants Under Construction and Announced	4	1,840	23	13,850	11	2,706	27	11,611.9	9	4,870
Furnace:										
PC	3	1,430	12	7,035	7	1,556	15	6,917	7	3,590
Stoker	-	-	-	-	3	800	3	17.9	-	-
Cyclone	1	410	-	-	-	-	1	200	-	-
Unknown	-	-	11	6,815	1	350	8	4,477	2	1,280
Wet scrubber:										
Limestone	1	550	10	5,745	3	857	4	4,650	-	-
Ash-alkali	1	550	-	-	-	-	7	3,707	-	-
Spray dryer	3	1,290	2	1,080	2	330[d]	-	-	-	-
Particulate removal:										
ESP	3	1,428	11	5,745	6	1,107	8	4,210	6	3,240
Baghouse	4	1,840	2	1,080	2	680	-	-	1	350
Mechanical	-	-	-	-	-	-	-	-	-	-
Unknown	-	-	11	6,755	3	939	10	4,182.1	2	1,280

[a]This table is a condensation of the information presented in Tables 3.5.1.27 through 3.5.1.31.

Several industries in Texas are planning expansions to utilize lignite reserves in the state.[15] Other industries plan on shipping other western coals to the state for use. Most of the industries will utilize the fuel for process steam generation in petrochemical plants. Some smaller companies are investigating the uses of western coals and lignite in brick, cement, and lime kilns. Activated carbon is currently made from lignite at several Texas locations. Texas has traditionally been a "natural gas state," and only recently has lignite been considered for industrial uses.

A number of foreign generating facilities utilize low rank coals. Table 3.5.1.34 lists some of those in commercial operation using conventional combustion.

Table 3.5.1.34

Data on Selected Foreign Low-Rank Coal-Fired Electric Power Plants

Country	Station	Initial Start-up Date	Largest Unit (MWe)	Total Capacity (MWe)	Feedstock
Australia	Yallourn	1954 1964	20 120	546	Brown coal - 67% moisture 1800-3800 Btu/lb.
	Hazelwood	1964	200	1600	Brown coal - 67% moisture 1800-3800 Btu/lb.
	Yallourn W	1972	350	350 (1050 to be added)	Brown coal - 67% moisture 1800-3800 Btu/lb.
	Morwell	1958	- -	25	
West Germany	Niederaubern		600		Brown coal - 52-57% moisture 2960-4600 Btu/lb.
	Neurath		600		Brown coal - 52-57% moisture 2960-4600 Btu/lb.
	Essen		600	3600 final configuration	Brown coal - 57% moisture 2970 Btu/lb.
Turkey	Tunebilec		150		Coal - 22-24% moisture 3600 Btu/lb.
Greece	Megalopolis		125		Coal - 64% moisture 1600 Btu/lb.
	Ptolemais IV		300		
Greece	Aliveri				Lignite - 30% moisture 4850 Btu/lb.
Spain	Puentes				Brown coal - 45% moisture 3100 Btu/lb.

Source: Reference 14

References - Section 3.5.1.5

1. Personal communication, Tanny, R., Texas Power and Light Company, September 19, 1979.

2. Personal communication, Plant Supervisor, Navajo Plant, Salt River Project, September 19, 1979.

3. Personal communication, Herrman, R., Arizona Electric Power Coop, September 20, 1979.

4. Personal communication, Kimble, R., Salt River Project, September 25, 1979.

5. Personal communication, Mewth, H., Texas Utilities Company, September 27, 1979.

6. Personal communication, Gin, D., Tucson Gas & Electric Company, September 27, 1979.

7. Personal communication, Prete, L., Department of Energy, Statistics Division, September 26, 1979.

8. Personal communication, McConnon, S., United Power Association, September 25, 1979.

9. Personal communication, Benedict, W., Commonwealth Edison Company.

10. Personal communication, Mathews, D., Texas Municipal Power Agency, September 28, 1979.

11. Schwartz, M.S., Combustion and Utilization Characteristic of Texas Lignites, Gulf Coast Lignites Conference, Houston, Texas, May 17-19, 1978.

12. Stephens, R.H., et al., Standard Support and Environmental Impact Statement for Standards of Performance: Lignite-Fired Steam Generators, EPA, March 1975.

13. Nehring, R., B. Zycher, and J. Wharton, Coal Development and Governmental Regulation in Northern Great Plains: A Preliminary Report, National Science Foundation, R-1981-NSF/RC, August 1978.

14. Morrison, G.F., Combustion of Low Grade Coal, IEA Coal Research, London, R.N. ICTIS/TR02, June 1978.

15. Integrated Assessment of Texas Lignite Development: Volume I, Texas Energy Advisory Council, Report No. ES-011, April 1979.

16. Gronhovd, G.H. and E.A. Sondreal, Technology and Use of Low-Rank Coals in the U.S., Seminar on Technologies for the Utilization of Low Calorie Fuels, Varna, Bulgaria, April 20-22, 1976.

17. Personal communication, Scott, W., Department of Energy, Atlanta Regional Office, September 21, 1979.

18. Personal communication, Krefft, H., Department of Energy, Chicago Regional Office, September 27, 1979.

19. Personal communication, Johnson, J., Department of Energy, Fort Worth Regional Office, September 20, 1979.

20. Personal communication, Cigal, G., Department of Energy, New York Regional Office, September 21, 1979.

21. Personal communication, Rinder, H., Department of Energy, Chicago Regional Office, September 20, 1979.

22. Personal communication, Cone, M., Department of Energy, San Francisco Regional office, September 19, 1979.

23. Standards Support and Environmental Impact Statement, Volume 1: Proposed Standards of Performance for Lignite-Fired Steam Generators, EPA-450/2-76-0030a, Emission Standards and Engineering Division, Environmental Protection Agency, Research Triangle Park, North Carolina, December 1976, 190 pp., pp. v-20 - v-21.

3.5.2 Fluidized-Bed Combustion

3.5.2.1 Introduction and Summary

Fluidized-bed combustion technology is being developed as an alternative to the conventional pulverized coal boiler equipped with stack gas scrubbing systems. The two major applications for this technology will be industrial steam/heat generation (3.5 quads by the year 2000) and electric power generation (2 quads by the year 2000). These estimates represent a market penetration fcr new capacity of 50 percent for industrial applications and 15 percent for power generation between 1980 and 2000.[1]

The advantages offered by fluid-bed combustion over the conventional pulverized coal approach may be summarized as follows:

- Utilization of high sulfur coals without pretreatment or flue gas desulfurization
- Reduced combustion temperatures resulting in lower NO_x emissions
- Increased combustion efficiency
- Reduced possibility of tube fouling because of the lower combustion temperature
- Reduced excess air requirements
- Reduced combustor size/increased heat transfer to the working fluid
- Fuel versatility
- Compact modular unit, package construction for units smaller than 50,000 lb/hr
- Easy to handle by-product material

Generally, these advantages are due to the intense turbulence in the bed, and to the comparatively long solid residence times in the bed without a long linear flow path requirement. The use of low-rank coals in fluid-bed combustors offers additional advantages:

- Sulfur capture via the inherent alkaline nature of low-rank coal ash, without added sorbents
- Tolerance of a wide variation in coal properties as is typical of low-rank coals

However, the successful utilization of low-rank coals in fluid-bed combustors will depend upon the resolution of several problems, some of which are common to all FBC applications and some which are unique to low-rank coals. Those unique to low-rank coals arise primarily from the properties and composition of the coal, and the impact of these characteristics on materials and operating characteristics. In addition, the waste materials pose possible disposal problems because the alkali sulfates in the ash are water soluble and are thus prone to leaching upon contact with ground waters.

In determining which problems might be the primary impediments to utilizing the low-rank coal resource, a series of key issues were identified which warrant continuing attention. Many of these issues are interrelated because they arise from some inherent characteristic of the coal. For example, low-rank coals sometimes occur with high inherent alkali content which permits substantial "self control" of sulfur. However, the high alkali also contributes to waste disposal problems, hot corrosion, and potential gas stream corrodents which could damage turbines in pressurized applications. The key issues are summarized in tabular form in Table 3.5.2.1 and a discussion of each is contained in the following subsection.

Table 3.5.2.1

Key Issues for Fluidized Bed Combustion

- Design Configuration of FBC Optimized for Low-Rank Coal
- Sulfur Retention by Inherent Alkali in Low-Rank Coals
- Agglomeration of Solids in Fluidized Bed
- Properties of Limestone-Deficient, Ash-Rich Fluidized Bed
- Materials Problems and Selection for Low-Rank Coal FBC
- Hot Gas Cleanup and Turbine Reliability for Pressurized FBC
- Coal and Sorbent Feeding and Distribution
- NO_x Control
- Temperature, Gas, and Solids Distribution in Low-Rank Coal FBC
- Disposal of Spent Solids From Low-Rank Coal-Fired FBC's
- Heat Transfer Data
- Effects of High Reactivity on FBC Design

Design Configuration of FBC Optimized For Low-Rank Coal

Design of industrial and utility scale AFBC facilities will require intelligent selection of operating parameters, materials, and process design parameters. Sensitivity to changes in materials and operating parameters will also have to be known so that the tradeoffs associated with first cost and operating costs can be evaluated, and an optimum design evolved for the particular low-rank coal of interest. While substantial data has been and will continue to be gathered for higher rank coals burned in fluid bed units, there is substantially less information to guide the designer who wishes to construct a low-rank coal fired plant.

Low-rank coal work done to date has been primarily confined to small scale pilot installations. Considerable experimentation with six inch units has given fairly good predictive data for SO_2 control that can probably be extrapolated to much larger scale installations. However, the predominance of wall effects in such units makes the data for NO_x formation and aspects of bulk bed behavior (such as values of heat transfer coeffficient) unreliable and probably misleading.

Data taken from 18-inch units is probably suitable for use in scaling up to moderate industrial units, possibly 4 foot by 4 foot in size. For this level of scale up, the data gathered thus far would be an adequate guide with regard to operating parameters using coals tested to date. Within these restrictions, the probable "best" operating conditions for low-rank coals are as follows:

Bed Temperature	1500-1600F
Superficial Velocity	6-7 feet per second
Excess Air	15-20 percent
Coal Feed Size	1/4 inch
Sorbent	Alkaline ash plus Limestone, especially with high sodium coals

Sulfur Retention by Inherent Alkali in Low-Rank Coals

Sulfur dioxide control arising from inherent alkali in the ash has been demonstrated to be good for several coals. Figure 3.5.2.1, developed at the Grand Forks Energy Technology Center, shows that for a collection of ten different western coals operating with inherent alkali, most are capable of meeting the 1971 new source performance standard, while a small percentage are potentially capable of meeting the proposed new source performance standard of 0.2 pounds of SO_2 per million Btu of heat input. Figure 3.5.2.2 shows that for at least one Wyoming coal ash reinjection can contribute modest increases in the degree of sulfur control as compared with inherent alkali alone.

For the cases where inherent alkali alone is insufficient to achieve the desired level of control, addition of sorbents has been demonstrated to give improved sulfur capture compared to the inherent alkali alone. Figure 3.5.2.3 illustrates the effects of adding limestone, nahcolite, and trona to a North Dakota coal, while Figure 3.5.2.4 illustrates results obtained from a variety of western coals, and compares these with retention data expected from an Illinois #6 coal. The data suggest that the alkali in the coal is nearly as effective in the capture of sulfur as the alkali contained in sorbents.

Figure 3.5.2.1

Sulfur Dioxide Emissions Versus Alkali-to-Sulfur Ratio

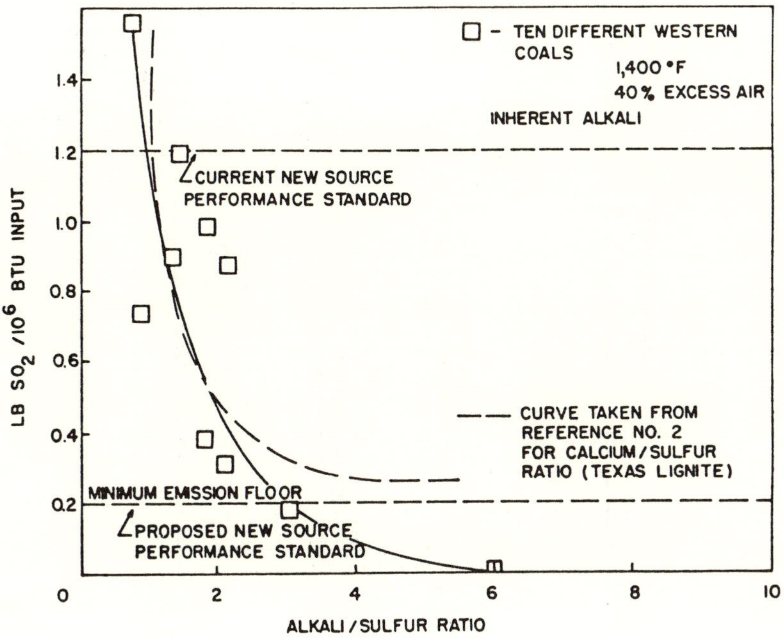

Source: Reference 1

Figure 3.5.2.2

Sulfur Retention by the Inherent Ash Alkali and Additional Limestone During the Fluidized-Bed Combustion of Wyodak, Wyoming, Subbituminous Coal

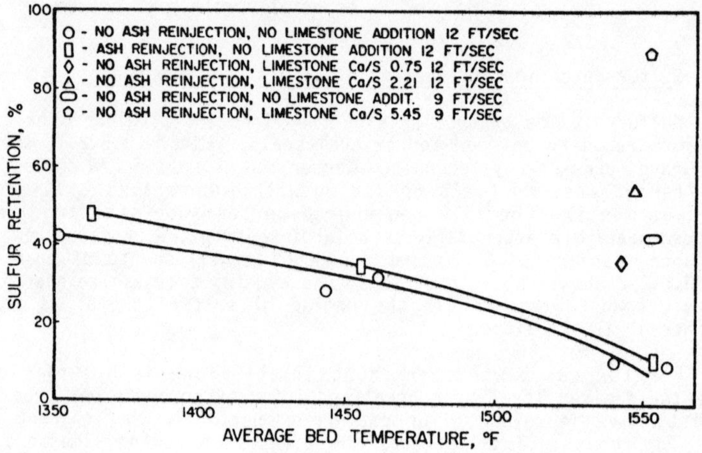

Source: Reference 2

Figure 3.5.2.3

Effects of Sorbent Type on Sulfur Retention

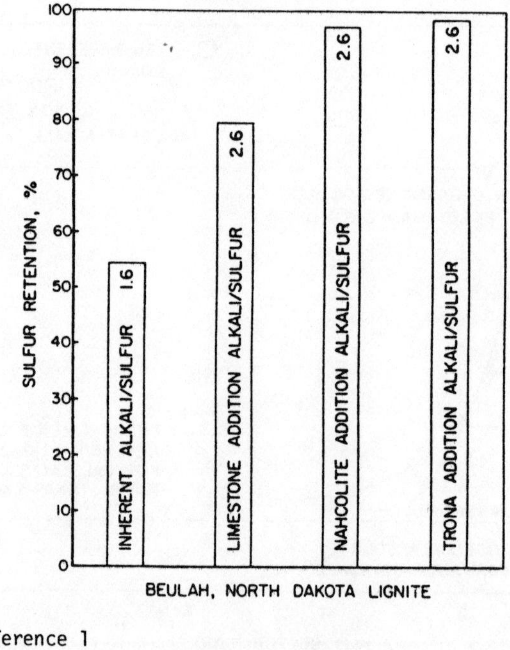

Source: Reference 1

Figure 3.5.2.4

Percent Sulfur Retention as a Function of Total Alkali-to-Sulfur Ratio

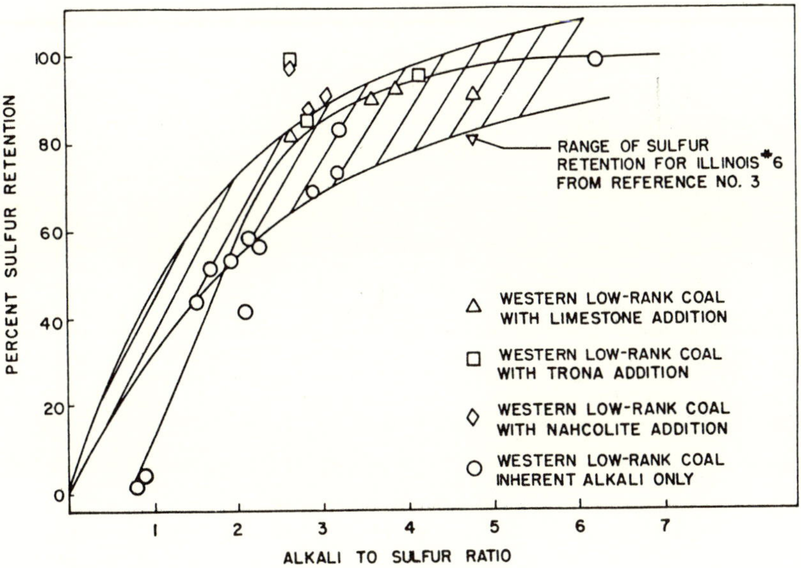

Source: Reference 1

Agglomeration of Solids in Fluidized Bed

One difficulty arises in operating an FBC with a sand bed which relies on inherent alkali for sulfur capture. In cases where the sodium content of the ash is more than 3 percent, the sand bed has demonstrated a tendency to agglomerate. The agglomerates formed either float in the bed, or stick to both cooled and uncooled surfaces somewhat in the manner of boiler tube fouling. The origin of the agglomerates is unclear, but appears to be due to sodium aluminosilicate eutectics having low melting points which form from the combined presence of coal ash and the sand bed. Some level of control can be achieved by the addition of limestone sorbents or the presence of Al_2O_3 which are believed to suppress the silica concentration in the bed. It has also been noted that the bed tends to preferentially retain sodium ash, and further changes in the character of the sand have been noted (possibly a coating of the individual sand grains) which may contribute to the changes in bulk heat transfer coefficient that have been observed.

Properties of Limestone-Deficient, Ash-Rich Fluidized Bed

As indicated above, the properties of a low-rank coal ash bed are somewhat different from the ash-sorbent-coal bed used in eastern coal fired FBC's. Both the composition and bulk behavior of the bed can be expected to be different with possible impact on operating conditions, material lifetime, and plant operating performance.

Results from tests of about 100 hours have shown that bed composition gradually changes, particularly in the case of a bed that begins as sand. Silica content declines, and sodium content rises with time. In addition sand in the bed takes on a different appearance and the rheological properties of the fluidized medium may change as well. Changes in heat transfer coefficient as a function of time have also been noted. No tests have been reported thus far that indicate what the "steady state" character of such a bed might be, and it may be that steady state is not achieved for hundreds of hours of operation.

Materials Problems and Selection for Low-Rank Coal FBC

Corrosion characteristics of the ash bed can also be expected to differ from those resulting from FBC of eastern coals. High alkali content combined with variability in the feed material and possible high erosivity could cause combined erosion/corrosion effects that are more severe than those experienced with eastern coals. The complexity of corrosion mechanisms together with corrosion dependency on minor constitutents in the coal ash make extrapolation of corrosion rates from existing data difficult.

Corrosion tests have been performed for eastern coals using a wide variety of materials under varying conditions, and a greater level of materials understanding is being attained for these coals. Part of what has been learned covers the nature of the testing procedures themselves, and a clearer understanding of the risk associated with extrapolating from test data has been obtained. Virtually no comparable testing has been accomplished for low-rank coals.

Hot Gas Cleanup and Turbine Reliability for Pressurized FBC

Pressurized FBC technology offers the possibility of superior process efficiency compared to atmospheric units. In the operation of pressurized systems, exhaust gases from combustion of coal are expanded through gas turbines to extract mechanical power. Contaminants in the gases can lead to attack and failure of highly stressed turbine components operating at high temperatures. Control of these contaminants is essential to the success of the technology.

Vaporized alkali metal cations (particularly sodium and potassium) present in the hot gas appear to be the most corrosive agents. These chemicals react with SO_2 to form sulfites and sulfates which condense on turbine blades when the combustion gases cool during expansion. Severe corrosion and early blade failure are the results of this phenomenon. Low-rank coals have substantially higher levels of alkali than eastern coals, and thus one would expect evolution of substantially more sodium sulfate than would occur with other coals.

Early hot corrosion work was performed at Combustion Power Company where it was noted that corrosion rates in testing were less than expected. Research indicated that alkali sulfates might be controlled by other components contained in the ash. Clay added to the bed, particularly kaolin, combined with sodium to form a fine solid particulate of feldspar, which could be removed in standard particulate control equipment. Because some low-rank coals contain kaolin or similar aluminum silicate minerals in the ash, it may be that these same coals may be able to self-control the generation of sodium in the gas stream in much the same way that they self control sulfur emissions. However no data has been located which demonstrates either the presence or absence of this effect.

Laboratory scale experiments performed by Argonne National Laboratory determined that sodium and potassium were transported in the chloride form, and reacted with SO_2 to form the corrosive sulfate when temperatures were reduced as would happen in a turbine.[3] Thus control of the chloride should yield a reduction in sulfate.

In the Argonne work simulated flue gas doped with sodium and potassium chloride vapor was passed through packed beds of diatomaceous earth, kaolin clay, and activated bauxite. Experiments showed that sodium and potassium chloride were captured with approximately the same efficiency. Figure 3.5.2.5 shows a summary of some experimental results. Similar tests were conducted to examine capture of alkali sulfates with these sorbents, and lab scale tests at atmospheric and elevated pressures showed the sorbents to be effective on sulfates as well.

Figure 3.5.2.5

Effect of Sorbent Bed Temperature On Sorption Capacity As a Function of Experimental Time. Experimental Data for 880°C Was Adjusted to be on the Same Basis as the 800°C Data

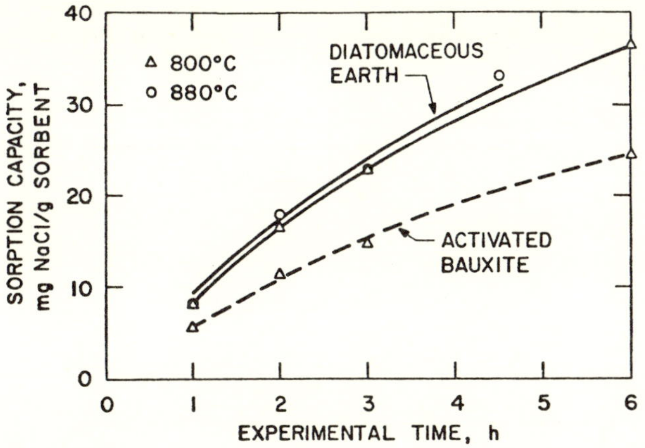

Source: Reference 3

Coal and Sorbent Feeding and Distribution

Large-scale, long-duration feeding of solids into the fluidized bed and subsequent ash-sorbent removal from the system pose substantial problems. The problems are common to all coal-fired FBC systems, and the vast majority of work performed in these areas is applicable to FBC's operating with low-rank coals. Possible special problems which might arise with low-rank coals include corrosiveness, abrasiveness, and unique ash handling and disposal problems not shared (at least in degree) by other coals.

NO_x Control

NO_x emissions from coal fired FBC installations are potentially a problem which could impede advancement of the technology. NO_x emissions arise from two sources: fixation of atmospheric nitrogen (a high temperature phenomenon) and conversion of nitrogen bound in the fuel. Low temperature operation in fluid beds makes conversion of fuel bound nitrogen the dominant source of NO_x in coal fired FBC's. The high nitrogen content of some low-rank coals suggests that NO_x emissions could impede the utilization of the resource.

Some work has been accomplished by GFETC in which NO_x emissions were parametrically studied for several western coals at atmospheric pressure. This work is summarized in Figures 3.5.2.6 and 3.5.2.7 which show the trends in NO_x emissions as a function of temperature and percent excess air. No comparable data was found for western coals in pressurized systems.

Figure 3.5.2.6

Nitric Oxide Emission Level as a Function Of Average Bed Temperature

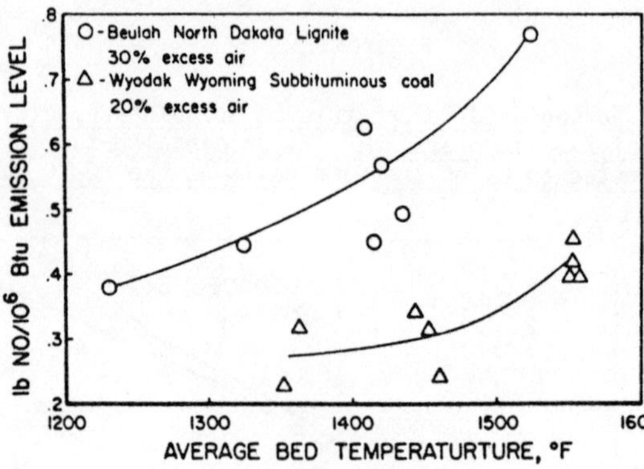

Source: Reference 4

Figure 3.5.2.7

Nitric Oxide Emission Level as a Function of Excess Air

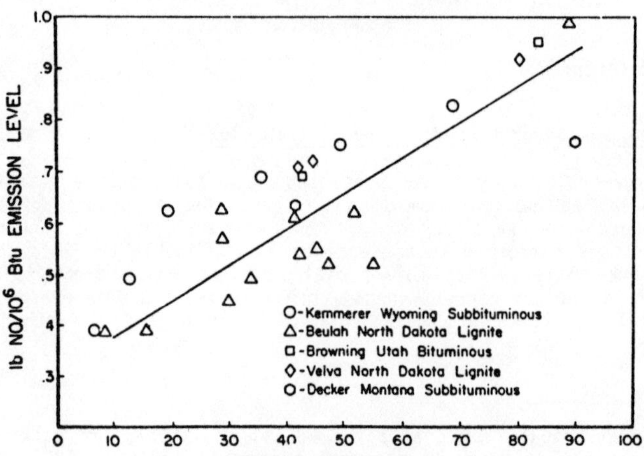

Source: Reference 4

Past work with pressurized FBC systems burning eastern coa has shown that NO_x emissions are lower when the coal is burned at pre: sure. The reason for this is unknown, but it is speculated that N(formed in the bed is locally reduced to oxygen and nitrogen by the presen of carbon monoxide, and this effect is enhanced by pressure. Pressuriz work has been done almost exclusively with eastern coals, but since weste coals tend to be higher in fuel bound nitrogen, the effects of N(suppression at pressure become of great interest in low-rank coal util zation.

Temperature, Gas, and Solids Distribution in Low-Rank Coal FBC

An understanding of the variations in temperature, gas composition, solids concentration, and other basic parameters is useful in obtaining a better understanding of the fluid bed. The information can be used to determine optimum feed port location, optimize utilization of sorbents, and improve combustion efficiency.

Basic studies in this area should be directed toward the unique properties of low-rank coals and their effects on operation of a fluidized bed. These include high coal reactivity, alkaline ash, agglomeration of particles, and capture of SO_2 on the ash.

Disposal of Spent Solids from Low-Rank Coal Fired FBC's

All fluid bed combustors operating with coal produce large amounts of dry, solid wastes. Possible environmental impacts associated with the disposal of these wastes need to be studied. The high inherent alkali content of many low-rank coals suggests that the waste products will be high in alkali sulfate content, materials which have potentially high solubility in water. This could in turn have substantial impact on the nature of disposal practices.

Some work has been done and is continuing under the sponsorship of both EPRI and EPA. Methodologies are being devised to define the problems of leachate contamination of ground water, chemical migration through the ground, and stabilization of landfills. Much of the work performed with scrubber sludge may also be applicable. However, most of the work has focused on wastes generated from eastern coals; virtually no data exist on FBC wastes from low-rank western coals.

Substantial research is needed to determine the character of the low-rank coal waste material, with and without the use of sorbents such as limestone, nahcolite, and trona. The research should build on past work with eastern coal wastes and scrubber sludge experience, but should emphasize those aspects of the problem unique to low-rank coals including:

- high alkali content
- solubility in water
- variability in ash composition
- interaction with underlying soils that may be unique to western locations

Heat Transfer Data

Some observers have noted a deterioration in average heat transfer coefficient with the passage of time. It is not clear why this occurs, but the effect has been repeatedly observed, and would be expected to have substantial impact on the design of fluid bed systems.

Work in which deterioration of heat transfer coefficient has been noted has occurred primarily at Combustion Power Company with some similar work performed at GFETC. It is currently believed that the changes are due to shifts in the character of the bed with the passage of time. It has been noted that the bed preferentially retains sodium, and that with sand beds, the character of the sand changes with the sand taking on a coated appearance. Also agglomeration effects with high sodium coals have been noted. These effects would be expected to impact on the behavior of particle-to-tube contact with subsequent effects on bulk heat transfer coefficient.

Most of the data are from relatively short term tests, and it is difficult to know what the long-term equilibrium character of the bed and the average heat transfer coefficient will be. Examination of these effects will require long term testing in systems large enough to give bulk bed behavior typical of full scale installations. Bed composition, emission levels, and heat transfer coefficient need to be monitored as a function of time in an effort to determine what the "steady state" levels will be. This testing could be performed concurrent with corrosion and coal-sorbent evaluation tests, and other than the cost of added chemical analysis and added instrumentation, there should be little marginal increase in cost required to obtain this information.

Effects of High Reactivity on FBC Design

The high reactivity of some low-rank coals suggests that changes in fluid bed design may be possible that take advantage of this property. High reactivity effects could possibly increase volumetric heat release rates with subsequent reduction in equipment size and cost.

Past work done primarily at GFETC suggests that volumetric heat release is probably constrained by heat transfer, and that it is unlikely that heat transfer rates can be increased much because of higher coal reactivity. The reaction of coal particles in the bed tend to be diffusion limited rather than kinetic rate limited with the result that high reactivity has little effect on these particles. Kinetic effects are important only for small coal particles, and for these reactivity effects become more important.

References - Section 3.5.2.1

1. Goblirsch, G.M. and E.A. Sondreal. Low-Rank Coal Atmospheric Fluidized-Bed Combustion Technology, Symposium on Technology and Use of Lignite, Grand Forks, N.D., May 1979, DOE GFETC/IC-79/1, 1979, pp. 75-108.

2. Goblirsch, G. Fluidized-Bed Combustion of Wyodak Anderson Canyon Subbituminous Coal from Wyoming, DOE GFETC In-House Report in Support of the Burns and Roe, Inc. 600 MW AFBC Conceptual Design Program, 1978.

3. Johnson, I., G.J. Vogel, S.H.D. Lee, et al. Support Studies in Fluidized-Bed Combustion, Quarterly Report, Argonne National Laboratory, Argonne, Illinois, 1978.

4. Goblirsch, G.M. and H.M. Ness. Fluidized-Bed Combustion Tests on Lignites and Subbituminous Coals at the Grand Forks Energy Technology Center, for presentation at the DOE Symposium on Environmental Control Activities, November 28-30, 1978, Washington, D.C.

3.5.2.2 Technology Description

Fluidization is a phenomenon which occurs when a porous bed of solid particles is contacted with a stream of liquid or gas moving at sufficient velocity to cause movement of the solid particles. Within a range of fluid velocities (defined by particle density and size, and fluid viscosity and density), the fluid-solid system behaves in a way which is in certain respects qualitatively and quantitatively similar to that of an agitated fluid; hence the name "fluidized bed."

The behavior of a bed of solids in contact with a flowing fluid varies through several different phases as the fluid velocity changes.

Fluid first enters the bottom of the bed at very low velocities, passing through the spaces between the solid particles. As fluid velocity is increased, the pressure drop experienced by the fluid increases, as shown by segment OA in Figure 3.5.2.8.[1,a]

Figure 3.5.2.8

Pressure Drop in Fluidized Beds

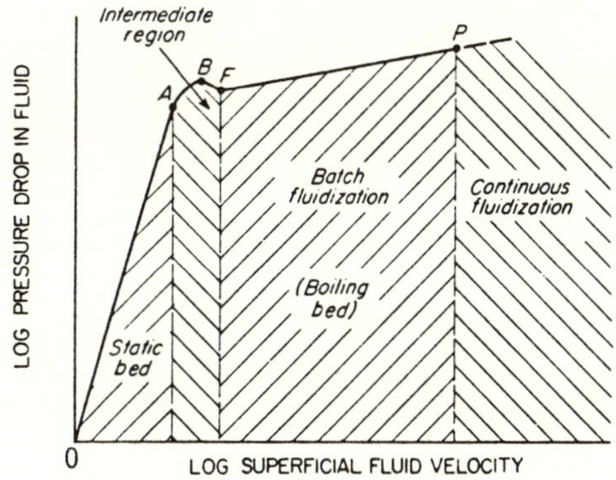

Source: Reference 1

When the gas pressure drop across the bed offsets the force of gravity on the particles, incipient fluidization occurs (point A in Figure 3.5.2.8) and is accompanied by a slight expansion of bed volume without breaking contact between any of the solid particles. The increased porosity resulting from bed expansion results in a lower rate of increase of pressure drop with fluid velocity (segment AB). At point B, the bed is in the maximum state of expansion while still having all of the solids in contact. A further increase in fluid velocity causes particle separation and the onset of true fluidization. As the particles separate, a slight decrease in pressure drop may occur which is soon reversed, marking the batch fluidization regime. From point F onward, the solid bed particles move rapidly in random motion, and resemble a boiling liquid. However, since the gas velocity between particles in the bed is considerably higher than the gas velocity above the bed, particles which are thrown above the bed immediately return. This condition exists until the superficial gas velocity is increased sufficiently to entrain bed particles out of the vessel (point P). This condition can be described as the simultaneous flow of two phases, and is referred to as continuous fluidization.

The principles of fluidization can be applied to the combustion of coal. In a broad sense however, fluid-bed combustion is neither restricted to combustion in a dense phase classical fluidized bed nor to the combustion of a specific fuel. Although the combustion of different fuels can be achieved efficiently in a hot bed of fluidized solids at very high temperatures (over 3000°F if construction materials allow), the main advantages of fluid-bed combustion are most apparent at low combustion temperatures. Under these conditions, stable ignition of the injected fuel can always be assured by the high heat capacity of the fluidized solids and the high rate of heat transfer which occurs from bed material to incoming fuel and combustion air. Lower combustion temperatures shift the equilibrium away from NO_x formation, allow sulfur capture by bed materials and promote NO_x decomposition under reducing combustion conditions.[2]

[a]References for Sections 3.5.2.2 - 3.5.2.5 are listed at the end of Section 3.5.2.5.

The composition of the bed material in a fluidized-bed combustor is somewhat variable, but always contains a large fraction (up to 98 percent) of a non-combustible material such as limestone, dolomite or coal ash. This material functions as a sulfur sorbent and a heat transfer medium, allowing uniform temperature control of the bed. A small amount of coal (generally 2-3 percent) comprises the rest of the bed material. This bed is then fluidized by an air stream which provides the oxygen for combustion.

Fluidized-bed combustion can be designed for a variety of operating modes, configurations, temperatures and pressures. Figure 3.5.2.9 describes the differences in reactors as a function of gas velocity through the bed, and can be compared to Figure 3.5.2.8. The figure also introduces the concept of slip velocity, the difference in the mean gas and solids velocities. It can be seen that starting with the classical fluidized-bed (boiling bed regime in figure 3.5.2.8), the slip velocity constantly increases until a significant amount of bed material becomes entrained. In the extreme of the transport reactor (continuous fluidization), the solids velocity asymptotically approaches the gas velocity in two phase cocurrent flow.

Figure 3.5.2.9

Fluidized Bed Reactor Types
As A Function of Gas Velocity

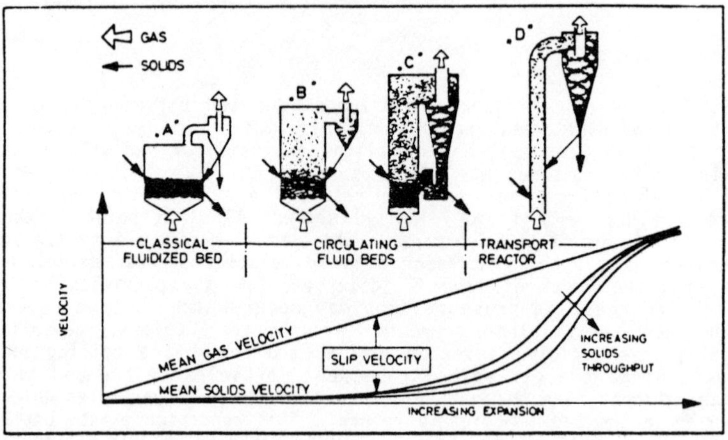

Source: Reference 2

Within the regimes of the classical fluidized and circulating fluid beds, several alternate configurations exist for contacting the combustion air with the bed material:

- downflow operation
- upflow operation
- multi-solids fluidized bed
- cyclonic fluidized bed

These four configurations are presented in Figure 3.5.2.10.

Figure 3.5.2.10

Fluidized-Bed Configurations
For Classical and Circulating Regimes

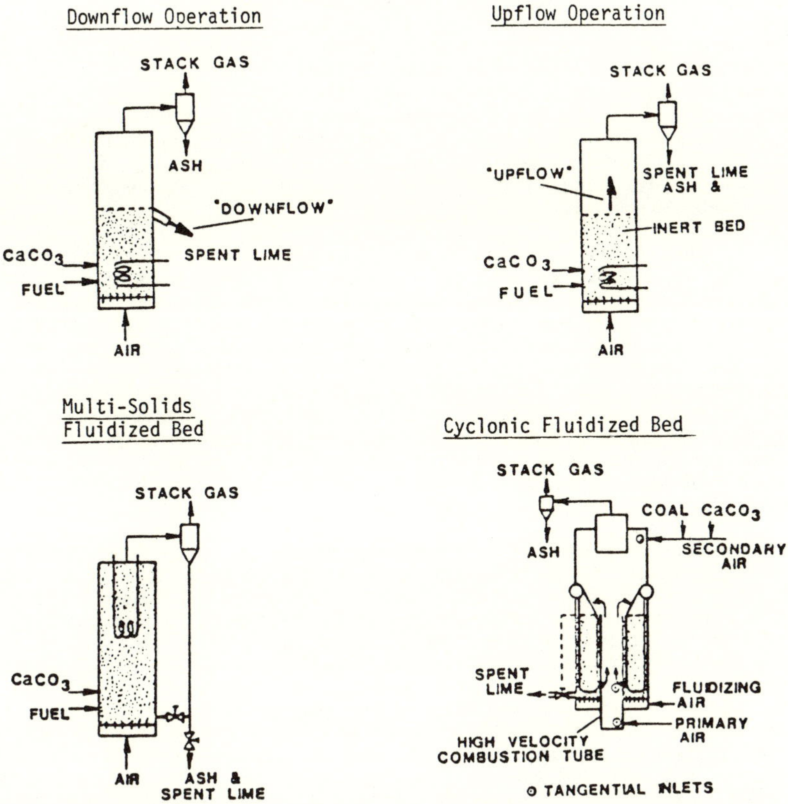

Source: Reference 9

When operated in a "downflow" fashion the feed materials (limestone and fuel) are fed into a precalcined bed of lime. Gas velocities are generally low enough so that little of the lime is elutriated. Only ash from the coal is carried out overhead. The solid waste product, consisting of spent lime ($CaSO_4$ + CaO) and some ash, is removed by displacement with fresh feed and discharged through an overflow pipe. Constant bed level is thus automatically maintained. Actually, material can be discharged controllably from any size opening below bed level by means of a differential pressure regulator actuating a valve in the discharge line. The cooling surface can be immersed directly in the bed to extract the necessary amount of heat for maintaining bed temperature at the desired level.

When operated in an "upflow" fashion the bed consists of a coarse inert material. Velocities are somewhat higher, and the feed materials are finer, so that all the spent lime as well as the ash are carried out with the fluidizing gas. This mode of operation is suitable if the gas-solid reactions are relatively fast because the solids retention time is relatively short, particularly for a one-bed system.

Multi-Solids Fluidized Bed Combustion is a recent development by Battelle Memorial Institute. In this system, a conventional fluidized bed using a high specific gravity material forms the base bed. Above this bed is a recirculating entrained bed of fine particles of the same or different material. Coal and finely ground limestone are fed to the dense bed as in conventional systems. During operation, the coal is burned in the dense bed and SO_2 is absorbed at the same time. Heat released is transferred to the recycled entrained bed material as it passes through the dense bed and steam is generated by tubes located in the region above the dense bed. Tests to date have been performed at velocities up to 3 to 4 ft/sec and claims of satisfactory SO_2 removal with Ca/S ratios as low as 1.4 have been made. Detailed data are considered to be proprietary by Battelle. The ability to follow load over a 5:1 range is a feature claimed by this unit.

The cyclonic fluidized bed is another high unit capacity device patented by Babcock and Wilcox. The unit is constructed with a central high velocity cyclonic combustion chamber where char and limestone are fired with primary air. This high velocity mixture then encounters a low velocity region at the top of the device where coal and additional air are injected tangentially. Because of the low velocities in this section, unburned coal, char and limestone fall out of the gas stream and are distributed to the perimeter of the unit due to the swirling motion. Here, at high temperature, they enter an annular fluidized bed which contains the steam generation tubes. The bed does not overflow due to the constant removal of material from the bottom for injection into the high velocity combustion chamber.

Operating pressure in a parameter which has a significant impact on the integration of the combustor in a utility power generation or industrial steam generation system. Atmospheric pressure fluid-bed combustion (AFBC) is suitable for raising steam in boiler tubes submerged within the bed itself, as well as from the hot combustor offgas, this latter part similar to the operation of a conventional pulverized coal fired boiler system. Pressurized fluid-bed coal combustion (PFBC) is also well suited for steam generation within the bed and downstream of the combustor, but has the additional potential of power generation by expansion of the hot pressurized gas. Exploitation of the full advantage offered by PFBC systems requires the use of a hot gas clean-up system for particulate removal and turbine blades which are capable of withstanding higher solids loadings than are currently experienced by modern day gas turbines. Hot gas clean-up systems and advanced turbine blade technology are both in the developmental stage, and for this reason, PFBC technology is farther from commercialization than is AFBC technology.

Atmospheric Fluid-Bed Combustion

A schematic diagram of an atmospheric pressure fluidized bed boiler is presented in Figure 3.5.2.11. The mixture of crushed limestone (or other non-combustible) and coal has an initial particle size of about 1/8 inch (for the classical boiling bed combustor). Before air is admitted to the bed, auxiliary burners heat the bed to approximately 1100°F, which is sufficient to sustain combustion when the fluidizing combustion air is admitted. As the coal burns, the bed attains its normal operating temperature of 1500 to 1600°F.

Saturated steam is generated in tubes submerged in the bed. Some designs also call for steam generation in the water walls, or these may instead act as multiple downcomers providing water for the in-bed tubes. Steam superheating is generally carried out in the freeboard section of the device (the open space above the bed), because the extremely high rates of heat transfer which occur within the bed would raise tube metal temperatures to excessive levels.[b] Some designs call for economizer coils (feedwater preheating) to also be placed in the freeboard.

Since the air velocity is set to achieve fluidization of these coarse particles, some carryover of fine particles (produced through abrasion or shrinkage due to combustion) may occur. This phenomenon will result in a loss of efficiency if the elutriated fines contain unburned carbon, and hence some mechanism is needed for collection and recycle of this material to extinction. This function is accomplished by the primary cyclone in Figure 3.5.2.11.

Figure 3.5.2.11

Atmospheric Fluidized Bed Combustion Boiler

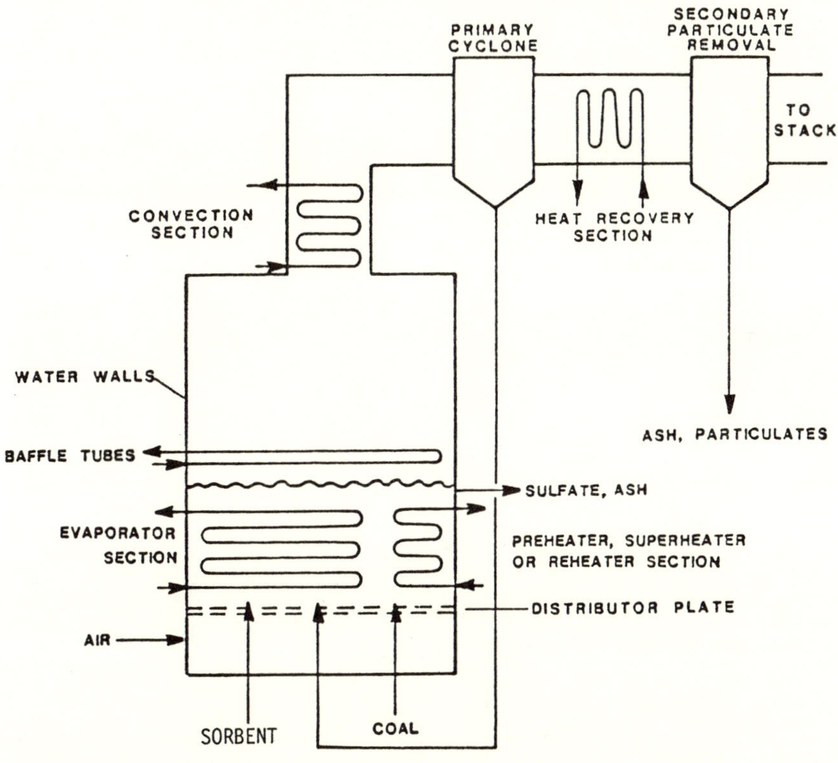

It is desirable to maintain reasonably high superficial gas velocities to reduce the size and cost of the AFBC unit. To resolve the problem of unburned carbon loss, some designs incorporate a carbon burn-up cell (CBC)in place of recycling the primary underflow to the combustor as discussed above.[c] These CBC units typically operate at higher temperatures than the combustor, and at lower gas velocities (approximately 2000°F, 5 ft per second). A study by Pope, Evans, and Robbins indicates that overall combustion efficiency can be increased to 99 percent or more for the combined AFBC-CBC system.[3]

---

[b]Steam generation in submerged tubes is allowable providing that sufficient liquid, usually 50 percent, is continuously available to absorb latent heat, keeping the tubes cool.

[c]Recycle of small particles to the bed may be ineffective in assuring complete carbon burnout due to the rapid rate at which they are elutriated from the bed.

Babcock and Wilcox[4] have estimated that the percentage of sulfur in the elutriated coal dust in the combustor offgas is approximately 50 percent of the percentage of unburned coal. At a combustor carbon conversion efficiency of 85 percent, approximately 93 percent of the sulfur is converted to SO_2, the remaining 7 percent leaving the combustor unburned.[5] This solid material is then collected and charged to the CBC. Due to the high operating temperature of the cell, data suggest that this sulfur may not be captured by either partially sulfated limestone or fresh limestone injected into the CBC.[5] The resulting 7 percent sulfur bypass of the system would require that a greater percentage of sulfur be captured in the combustor to comply with emission regulations. This would be achieved by increasing the Ca/S feed ratio, resulting in a much greater consumption of limestone overall. However, experimental studies indicate that a certain type of limestone (Greer) or dolomite can function effectively even when partially sulfated.[5] This suggests that partially spent bed material could be charged to the CBC unit, giving the CBC unit a significant role in meeting SO_2 emission objectives. However, the high operating temperatures of the burn-up cell produce much greater concentrations of "thermal" NO_x than produced at the lower operating temperatures of the fluid bed. EPRI data indicate that NO_x levels emitted from CBC units may approach 800 ppm;[d] mixing this gas with the main flue gas would raise the overall NO_x level, but would probably still be within the present environmental standard. NO_x levels may therefore be a concern if future standards are tightened.

The relationship of an AFBC unit to its auxiliaries in a power generation plant is shown in Figure 3.5.2.12. Coal is delivered to the plant and screened to separate particles of the proper size for fluidization from coal which will require crushing. Stored in silos, coal and limestone are withdrawn and blown into the bed as required. Coal feed rate is determined by the steam demand; limestone is fed to the bed in response to increases in the level of SO_2 contained in the combustor off gas.

Figure 3.5.2.12

Flow Diagram of a Fluidized-Bed Boiler System

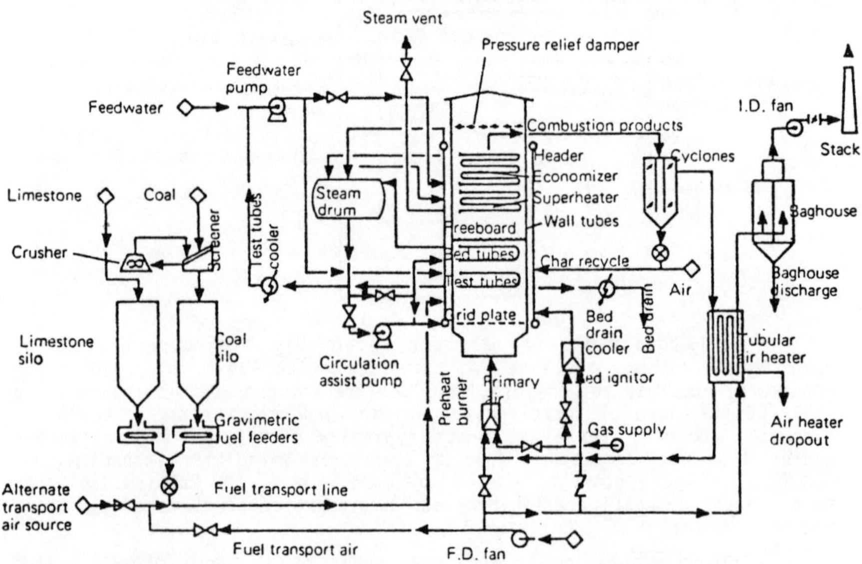

Source: Reference 7

---

[d]This may be compared to NO_x levels of approximately 350 ppm for fluid bed units, and a current federal standard of approximately 525 ppm.

Primary air (combustion air) is provided to a plenum under the bed distributor plate. Preheating of the air may be accomplished by either an auxiliary burner, heat exchange with flue gases, or a combination of both methods.

As noted earlier, particulate matter recovered from the primary cyclones is either recycled to the bed or a carbon burnup cell. Material collected from downstream cyclones or baghouses is removed as waste, and is usually attractive as a landfill material. Other landfill material is obtained when the bed is drained due to a buildup of solids inventory resulting from solids recycle or sorbent (limestone or dolomite) addition.

The remaining systems shown in Figure 3.5.2.12 are common to existing coal fired power plants.

Pressurized Fluidized-Bed Combustion

A considerable amount of effort is being focused on the development of fluid bed combustors which operate under pressurized conditions (3 to 10 atmospheres). In addition to retaining all of the advantages of AFBC operation, PFBC also offer:

1. Smaller boiler size for a given heat duty. This is possible due to the fact that, for a given duty, required bed area is inversely proportional to operating pressure.

2. Combustion efficiency within the bed is such that a properly designed unit should eliminate the need for solids recycle.

3. Using dolomite, sulfur retention is improved with pressure. However, the opposite is true of limestone.

4. NO_x emissions are further reduced by pressure.

Exploitation of the full advantage of PFBC units requires the use of a combined gas turbine/steam turbine cycle such as that shown in Figure 3.5.2.13. In addition to steam which is generated in the bed, the hot combusted gases can be expanded in a gas turbine after suitable cleanup (removal of particulate matter. Additional steam is raised from the gas turbine exhaust. The figure also shows a regeneration system for the sulfur sorbent used in the bed.

Aside from the PFBC unit itself, two other components are of critical importance to the efficient operation of the system. The solids separation system shown is of paramount importance in protecting the turbine system from large quantities of erosive materials. It must perform this task in a pressurized environment of high temperature, in the presence of significant quantities of high speed erosive particles. Second, the turbine blades must be capable of withstanding a higher solids loading than is common for present day machines, despite a proper functioning hot gas clean-up system. Both of these items are the focus of intense development work.

Although concern has been addressed over erosion problems faced by PFBC gas turbines, several indications suggest that the problem may not be as severe, for an equal solids loading, as that faced by a conventional gas turbine arrangement. The ash produced by the fluid bed is of a fluffy consistency, unlike the vitreous slags produced in some pc boilers, and has been described as "non-erosive."[8] Gas turbines operating off of PFBC generated gas will also be aided by the ability of the bed to retain corrosive agents such as alkali and heavy metals, and sulfur compounds.

Figure 3.5.2.13

Pressurized FBC Boiler with Sorbent Regeneration

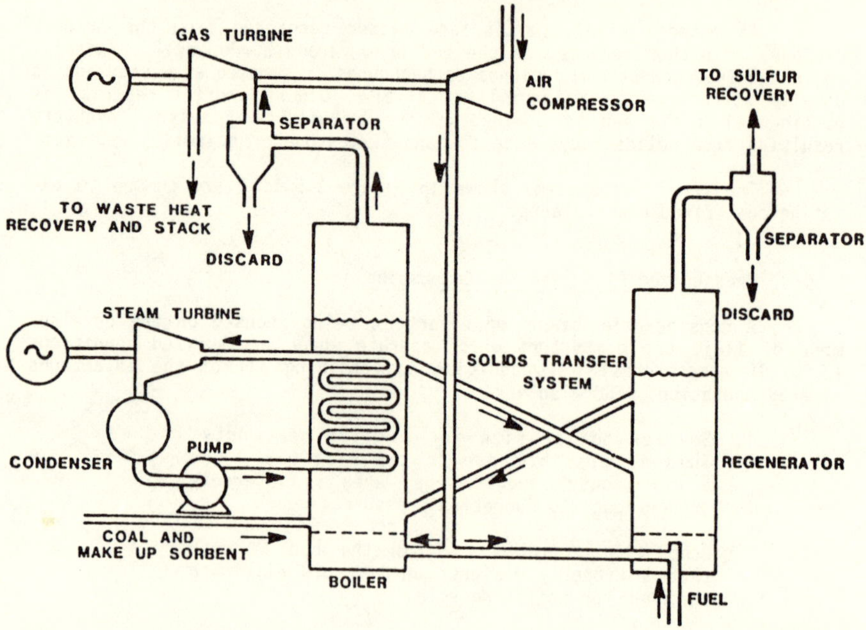

The use of combined gas and steam cycles provides the PFBC power generation system with a 3 to 4 percent efficiency advantage over AFBC systems, which is the principle driving force behind their development. This economic incentive also applies in a similar way to combined cycle coal gasification systems. Although sulfur removal differs from that used in PFBC, overall conversion efficiencies of coal to electricity are similar, and the economics for the two systems lie in a narrow band within the accuracy of engineering estimates. Because of the major role of the gas turbine in PFBC systems, this approach is primarily suited to large electric utility use and not to smaller industrial users whose primary need is steam generation for process use.

Fast Fluidized Beds

While pressure is an important process variable in determining the application of a fluidized bed system, gas velocity through the bed is important in the combustor design, and operation in the "circulating" fluidized regime (where slip velocity is greatest) may provide advantages not available from batch fluidization. The next step, combustion in continuous fluidization, more closely resembles the operation of a conventional puvlerized coal burner, and loses many of the advantages offered by combustion in a classical or circulating fluidized state.

Through their many years of industrial experience with fluidized bed technology, the German firm Lurgi Chemie und Huttentechnik GmbH is familiar with the concept of burning coal in a circulating bed combustor. Figure 3.5.2.14 is a schematic drawing of the Lurgi approach for two stage coal combustion.

Figure 3.5.2.14

Two-Stage Combustion in a Lurgi Circulating Fluid Bed

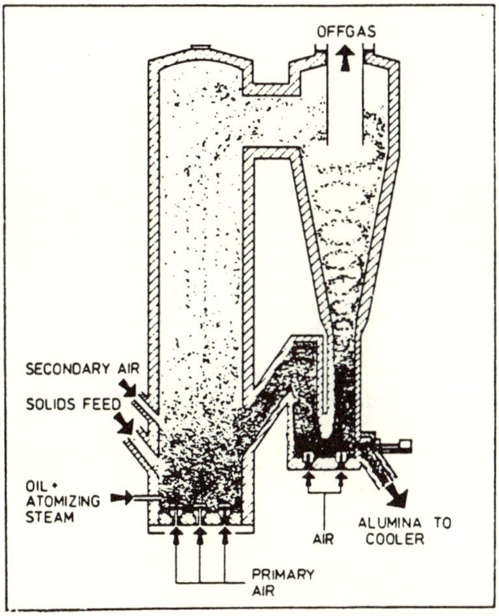

Source: Reference 2

According to Lurgi, this technique can be used advantageously in low temperature combustion of finely grained solid fuels, especially high ash containing coals and carbonaceous shales.[2] The desired feed size range for a device of this type is somewhere between that of pulverized fuel boilers and that generally specified for classical (boiling type) fluidized beds. By pneumatically injecting the coal feed into the lower section of the bed (which is free of tube bundles to avoid erosion), combustion is initiated under conditions which do not favor NO_x-formation from fuel bound nitrogen.[2] NO_x which is formed is eventually decomposed again in the presence of low percentages of residual carbon in the bed material.[2] In the upper part of the circulating bed where the walls are constructed as tube walls for steam raising, combustion with a minimum of excess air is completed. The circulating bed may consist mostly of ash, finely ground limestone or any other similar material.

One of the advantages of the circulating bed concept is the high rate of heat transfer available due to the smaller particle sizes and higher gas velocities. Lurgi indicates that such an arrangement may allow for better load following capabilities, since changes in gas velocity will greatly affect the heat available for steam raising.[2]

The desulfurization reaction when burning coal with limestone occurs at faster reaction rates, greater limestone utilization and higher overall sulfur capture with fine particle sizes. High degrees of sulfur capture with low Ca/S ratios (less than 1.5) can be expected, and the same effect is true for NO decomposition on carbon.[2]

Internal feasibility studies at Lurgi show that there is a potential for large power generation facilities based on the circulating bed concept, but not approaching the pulverized fuel boiler upper capacity range. However, there seems to be an interesting range of application in the industrial boiler field, delivering process heat.[2]

3.5.2.3. Environmental Control Technology

One of the principal incentives for pursuing fluidized-bed combustion is the ease of controlling environmental pollutants common to conventional coal combustion systems. The pollutants of primary concern are SO_2, oxides of nitrogen (NO_x), particulate matter, trace elements and solid waste.

The design of fluid-bed combustion systems to meet environmental factors will depend on a variety of factors:

1. Application - (electric utility, industrial, cogeneration, etc., including load requirement specifications)

2. Plant location - (as it affects availability of various sulfur sorbents, and variation in fuel heating value, sulfur and nitrogen contents, and trace constituents)

3. Environmental requirements - (specifically air emission standards and solid wastes)

4. Combustor design - (e.g. temperature control by heat transfer surface, excess air or circulating solids)

Thus, each FBC system under consideration is optimally a custom design which is compatible with specific site limitations and will function within process and environmental constraints.

SO_2 Control

Control of sulfur dioxide produced during the combustion process with limestone, dolomite or coal ash (in the case of low-rank coals) in the bed has been experimentally demonstrated. However, this is only one option for controlling SO_2 emissions. Table 3.5.2.2 lists this and other control methods.

Table 3.5.2.2

Options for Increased SO_2 Control In Fluid-Bed Combustors

- Increase Feed Rate of Fresh Sorbent (expressed as calcium-to-sulfur ratio, Ca/S)

- Increase Gas Residence Time in Bed By
  - reducing gas velocity
  - increasing bed depth

- Reduce Sorbent Particle Size (may require a reduction in gas velocity)

The first option increases the number of active sorbent particles in the bed, thereby increasing the overall SO_2 removal rate. In addition, however, this option will result in a significant net energy loss due to endothermic calcination of the sorbent:[e]

$$\text{heat} + CaCO_3 \longrightarrow CaO + CO_2$$

[e]Sorbent calcination is necessary for adequate sorbent performance in the bed.

Factors of secondary importance are increased capital investment for larger sorbent receiving, storage, handling and feeding facilities, larger spent sorbent handling, processing, storage and disposal facilities, greater particulate control requirements, and increased environmental impact of spent sorbent.

The second option, increasing gas residence time, is effective but has limitations. Lowering gas velocity requires that bed cross sectional area be increased to maintain a constant heat rate (overall, this has a small effect on plant capital cost). Bed depths in AFBC are probably limited to about 6 ft of expanded bed depth due to significant auxiliary power consumption for primary air compression and because of the limited discharge pressure available from simple fans.[10,f]

The third option, finer particle sizes (down to about 500 microns) may require bed operation at lower gas velocities and/or greater recycle of the material collected from the primary cyclones. This would have a minor cost impact.[10]

The effect of FBC operating conditions on sorbent requirements is shown in Figures 3.5.2.15 and 3.5.2.16, where the calcium to sulfur ratio is expressed as that required to achieve 90 percent removal with the two different limestone types (Grove and Carbon). The estimates through AFBC system using no sorbent enhancement options, and operating at 815°C with 70 percent excess air.[10] In applying the curves to low-rank coals, it must be remembered that no effect has been assumed for sulfur absorption by alkaline ash. The estimates also assume that the carbon burnup cell releases an insignificant amount of SO_2. Figure 3.5.2.17 gives sulfur removal performance based on Carbon limestone for a bed operating under two different sets of conditions, as predicted by a Westinghouse kinetic model.[10] The effect of finer particle size and lower gas velocity in reducing the Ca/S requirement for a given removal percentage is clearly evident.

Particulate Control

The lack of data on the behavior and physical properties of low-rank coal ash in fluidized-bed combustion systems may present a serious problem in specifying particulate removal equipment. Furthermore, it is likely that such information, for any coal, must be derived from experience with a large scale system.[6]

For AFBC systems conventional technology exists that may be capable of controlling particulate emissions at the 0.03 lb/MM Btu level. This includes hot ESP (750°F) or fabric filters. Cyclones have been proposed for carbon collection, but these primary devices would be inadequate for fine fly ash collection, and may even prove to be inadequate for carbon collection. However, ESP systems may be faced with serious problems. For hot side ESP's, it is reported that alkali salts in fly ash are a major reason for the high performance of these units. The low operating temperatures in the fluid bed, and the tendency to retain alkali metals may therefore retard ESP performance. For low-rank coals, higher alkali metal concentrations in the ash may counteract this trend. Cold side ESP systems may have collection problems due to 1) excessive carbon resulting from poor combustion; and 2) the high ash resistivity from low SO_3 levels in the flue gas. This last point may be especially true for low-rank coal fired boilers operating with limestone or dolomite sorbents.

---

[f]References for sections 3.5.2.2 - 3.5.2.5 are listed at the end of section 3.5.2.5.

Figure 3.5.2.15 Ca/S Molar Feed Required to Maintain 90% Sulfur Removal in AFBC With Grove Limestone

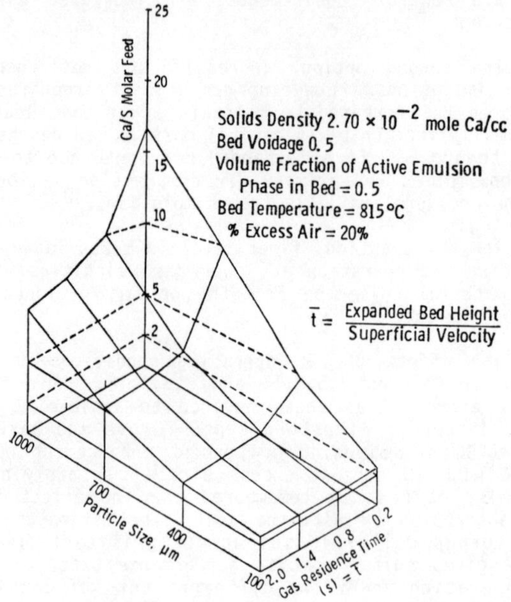

Source: Reference 10

Figure 3.5.2.16 Ca/S Molar Feed Required to Maintain 90% Sulfur Removal in AFBC With Carbon Limestone

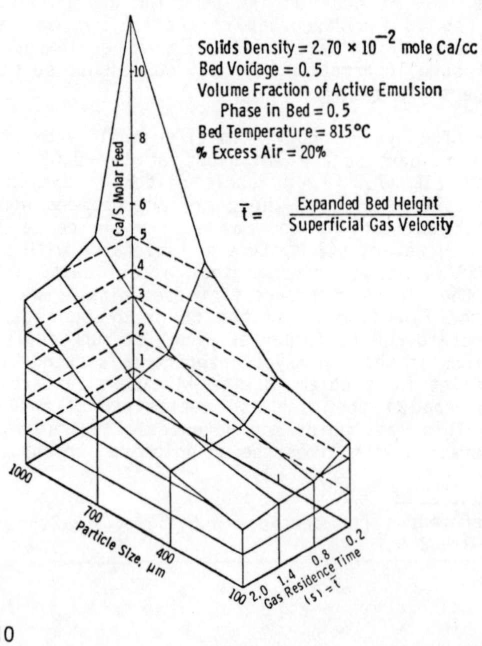

Source: Reference 10

Figure 3.5.2.17

Sulfur Removal Performance For Typical Sorbents (Projected Using Westinghouse Kinetic Model)

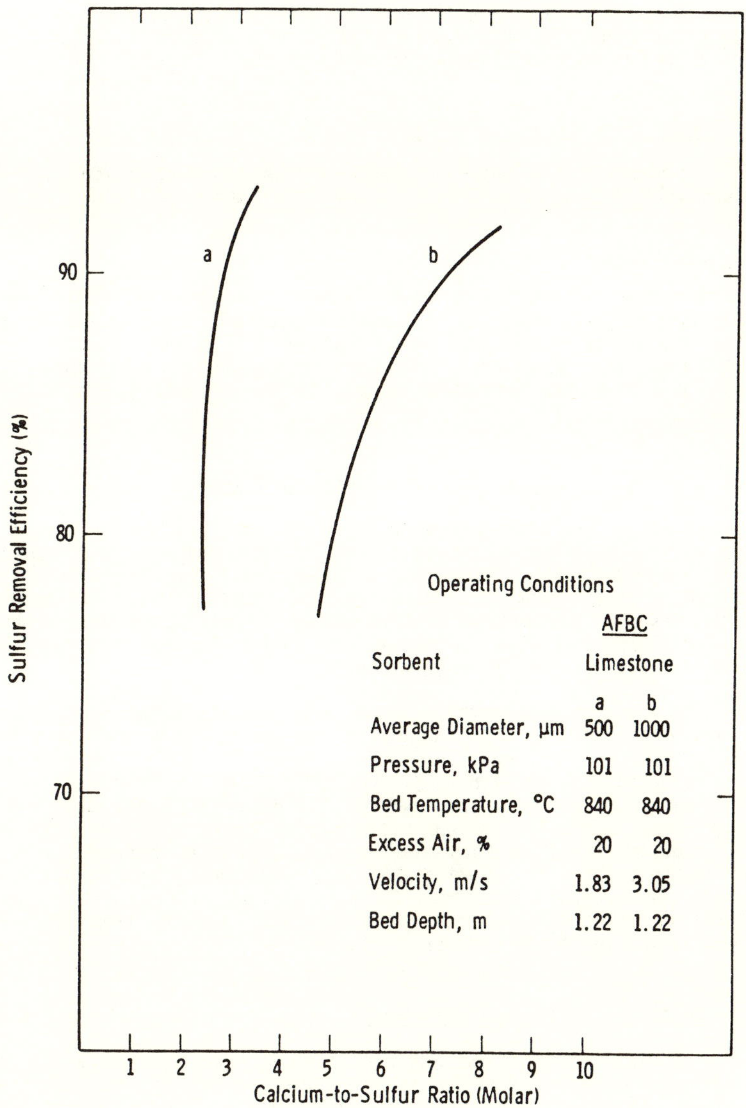

Source: Reference 10

Nitrogen Oxides

Due to the low operating temperatures, the main sources of NO_x emissions from fluid-bed combustors is fuel bound nitrogen and not "thermal" NO_x produced in the combustion zone.

In a statistical evaluation by Westinghouse,[10] it was estimated that AFBC emissions would meet a 0.6 to 0.7 lb NO_2/MM Btu standard. The

1979 standard for lignite and subbituminous coals is 0.6 and 0.5 lb. NO_x/MM Btu, respectively. Whereas for the control of sulfur oxides and particulates, the control techniques and relationships to operating and design parameters are relatively clear, the emission of nitrogen oxides is largely uncontrolled. The relationship between emissions and operating and design parameters in FBC units is not clearly understood at present, but it is expected that continuing research will yield a sufficient understanding of the problem to meet existing emission standards.

Trace Elements

Inorganic trace emission studies by Argonne conclude that FBC emissions will be similar to those of conventional boilers.[6]

Solid Wastes

Disposal of the solid waste (spent limestone and coal ash) may be best handled by using it as a landfill material. However, the presence of lime (CaO) may complicate this solution; the lime produces high pH levels, in the leachate. Additional work is required to solve the disposal problem, but limestone regeneration may be an attractive alternative to disposal.

3.5.2.4 Effects of Low-Rank Coal Properties

Although fluidized-bed combustion is easily adaptable to a wide variety of fuels, the specific technical problems and economic incentives will differ importantly depending on such fuel characteristics as the sulfur and nitrogen content, the amount and composition of the inorganic content, and the properties of the ash residues. Low-rank coals are particularly distinguished by their low sulfur content and the unique behavior of low-rank coal mineral matter.

The process data available for low-rank coals has been generated by a number of pilot scale AFBC units, located at the Grand Forks Energy Technology Center (bed cross sectional area: 2.2 ft^2), the Morgantown Energy Technology Center (1.8 ft^2), Combustion Power Corporation (7.1 ft^2), and Fluidyne Engineering Corporation (17.8 ft^2). These data address low-rank coal sulfur retention and emission characteristics, NO_x emissions, combustion efficiency and bed agglomeration tendencies during fluidized-bed combustion. During 1980, EPRI sponsored North Dakota lignite burns at the Babcock and Wilcox 36 ft^2 AFBC at Alliance, Ohio, and at the Lurgi circulating bed facility in Germany. Results from these tests were not available at the time of this writing.

Sulfur Retention and Emissions

The alkaline nature of low-rank coal ash makes possible the use of this ash as a sorbent for sulfur dioxide released during the combustion of these coals. Although coal ash may be inferior to dolomite or nahcolite as a sulfur sorbent, the low sulfur contents of low-rank coals provides an acceptable alkaline-to-sulfur ratio for this use. Bituminous coals are unsuitable for this due to both their non-alkaline ash, and high sulfur contents.

Sulfur retention tests have been performed at GFETC on North Dakota Lignite, Montana and Wyoming subbituminous coals, and a Utah bituminous coal, and the following data abstracted from reference 11.[a] These

[a]References for Sections 3.5.2.2 - 3.5.2.5 are listed at the end of Section 3.5.2.5.

tests cover a wide range of coal compositions, with sulfur ranging from 0.4 to 1.4 percent, and the inherent alkali-to-sulfur ratio from about 0.8 to 6.0. Data are available for both inert bed materials (alundum and silica) and for beds containing active sorbents (limestone, dolomite, and nahcolite). Texas lignites have been tested in the METC AFBC test unit.

The most important factor in determining sulfur retention is the molar ratio of total alkali to sulfur found in the coal and added sorbent. The total alkali is calculated on the basis of the calcium, magnesium, and sodium in the coal ash, and any alkali from added sorbents. This ratio can predict the percentage sulfur retention over a very wide range of coals and additional sulfur sorbents, as indicated in Figure 3.5.2.4 (see section 3.5.2.1). The shaded region in the figure represents the range of sulfur retention for Illinois No. 6 coal containing approximately 4 percent sulfur burned in a limestone bed at Ca/S feed ratios from 0 to 4.0.[11] The keyed data points represent lignites and subbituminous coals burned either in an inert bed (with the inherent coal alkali as the only sorbent) or in beds containing some limestone, nahcolite, or trona (the latter two are naturally occurring sodium carbonate/bicarbonate minerals). The beds containing additional sorbents were not at equilibrium, and sulfur retention may increase if the bed is allowed to become rich in sorbent material. It is important to note that most of the data points for the low-rank coals fall within the shaded region determined for the high-sulfur Illinois coal -- which leads to the conclusion that the inherent alkali in the coal and the two sodium sorbents all performed about as effectively as limestone. This approximation is not exact, but it does illustrate the predominant effect of the stoichiometric ratio in determining sulfur retention.

The principal operating variables affecting sulfur retention are the average bed temperature, superficial gas velocity, and air-to-fuel ratio, for which data have been reported previously.[11,12,13] The effects of bed temperature and velocity, which are the important design variables, are shown in Figures 3.5.2.18 and 3.5.2.19. The retention of sulfur on the inherent coal ash alkali in North Dakota lignite is maximum at an average bed temperature of 1400°F without ash reinjection and at 1500°F with ash reinjection (at a superficial velocity of 7 ft/sec). The lower temperature for the maximum observed without ash reinjection is attributed to the temperature elevation above the bed average in the region of the burning coal particles where the sulfur capture by the inherent alkali takes place.

Figure 3.5.2.18

Percent Sulfur Retention As A Function of Average Bed Temperature

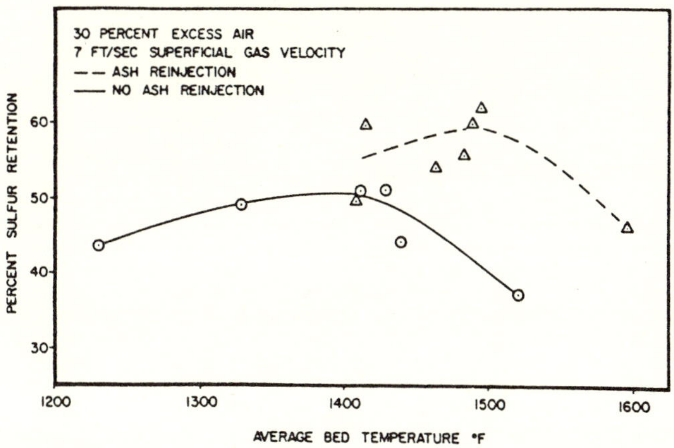

Source: Reference 11

Figure 3.5.2.19

Effects of Superficial Gas Velocity on Sulfur Retention

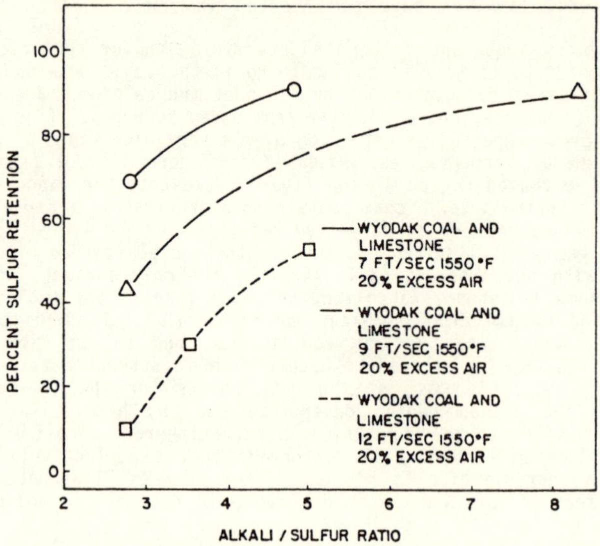

Source: Reference 11

The effect of temperature can be different, depending on other operating conditions and the specific coal burned. For example, a Wyodak, Wyoming, subbituminous coal exhibited a continuously decreasing sulfur retention from about 40 percent to 10 percent as the temperature was increased from 1350° to 1550°F at a superficial velocity of 12 ft/sec.[11]

The parametric effect of superficial gas velocity on sulfur retention is shown in Figure 3.5.2.19 for Wyodak subbituminous coal burned at 1550°F and 20 percent excess air. The sulfur retention versus alkali relationship is lowered by about 15 percent between 7 and 9 ft/sec, and is further reduced by 20 percent between 9 and 12 ft/sec.

The effect of an increase in air-to-fuel ratio is generally to raise sulfur retention.[12] However, the data obtained to date show considerable scatter and are not well correlated overall. It is reasonable to expect that the effect of excess oxygen could differ widely depending on the oxidation state of intermediate and final products of reaction. The different compounds and forms in which the several alkaline elements are contained in the coal and additives may affect the sulfur sorbence of these compounds at differing excess air levels. The oxidation of sulfite to sulfate ($SO_3^=$ to $SO_4^=$) may also be catalyzed by major or trace elements present in the ash.

A comparison of different sorbents for a single coal under identical operating conditions is shown in Figure 3.5.2.3 (see section 3.5.2.1). The inherent alkali-to-sulfur ratio of 1.6 for Beulah, North Dakota, lignite resulted in a sulfur retention of 55 percent. Three different additives were then tested at an alkali-to-sulfur add ratio of 1.0 (total alkali-to-sulfur ratio of 2.6). Limestone was the least effective of the additives tested, resulting in a sulfur retention of just under 80 percent. The sodium sorbents were more effective, resulting in retentions of 96 percent and 98 percent, respectively, for nahcolite and trona. Some other coals have been tested at two different add ratios of nahcolite. One Montana subbituminous coal with an inherent alkali-to-sulfur ratio of about

3.0 to 1.0 was tested at ratios of 0.5 and 0.9, and showed retentions of 87 and 90 percent, respectively. Sodium sorbents may offer the logistical advantage of shipping and handling a very small amount of additive material, which is an important consideration in choosing a sorbent for a low-sulfur coal, particularly in geographical regions where limestone is not locally available. The principal problem with sodium sorbents is the production of a soluble waste, sodium sulfate.

In Figure 3.5.2.1 (see section 3.5.2.1), the emission of SO_2 in lb/MM Btu is compared for ten Western lignites and subbituminous coals [data points and solid curve (1)] and Texas lignites [dashed curve (2)]. Again the retention/emission of sulfur is well correlated with the alkali-to-sulfur ratio, with essential agreement between the Western coals and the Texas lignites over most of the range. The leveling off indicated for the Texas lignites at about 0.25 lb SO_2/MM Btu for high stoichiometric ratios may be device-specific, or the result of the operating conditions under which the data were taken.

Emissions of Nitrogen Oxides

The emission of nitrogen oxides (NO_x) in AFBC of low-rank coals is determined primarily by bed temperature and percent excess air and, to a lesser (and undetermined extent), by fuel nitrogen content. The nitrogen content of Northern Great Plains lignites averages about 1.0 $\pm$ 0.3 (moisture- and ash-free); that for Texas lignites 1.3 $\pm$ 0.4; and that for Montana subbituminous coals 1.5 $\pm$ 0.2. No definitive study has been performed on the mechanism of NO_x formation/reduction for low-rank coals, but the current assumption would be that the small variation in total fuel nitrogen content is not a significant factor in determining the emission level.

The correlations of NO_x emissions with average bed temperature is given in Figure 3.5.2.6 (see section 3.5.2.1). For Beulah, North Dakota, lignite (1.0 pct maf N content), the NO_x emission was doubled from 0.4 to 0.8 lb NO_2/MM Btu as the bed temperature was increased from 1230°F to over 1500°F at 30 percent excess air. For Wyodak, Wyoming, subbituminous coal (1.2 pct maf N content), the NO_x emission increased from about 0.3 to 0.4 lb NO_2/MM Btu between 1350°F and 1550°F; these lower levels of NO_x are attributed to the lower excess air (20 pct) at which these tests were performed.

The effect of excess air on NO_x emissions is shown in Figure 3.5.2.7 (see section 3.5.2.1) which presents data for five Western coals (lignite, subbituminous, and bituminous) at a bed temperature of 1400°F. Discounting experimental scatter, the data indicate a direct linear correlation of NO_x versus percent excess air, which increases the emission from 0.5 to 0.9 lb NO_2/MM Btu between 10 percent and 90 percent excess air.

The 1979 NSPS for NO_x emissions is 0.6 lb NO_2/MM Btu for lignite in pc fired boilers (0.8 lb/MM Btu for North Dakota lignite in cyclone burners), and 0.5 lb/MM Btu for subbituminous and higher rank coals. The majority of the data in Figures 3.5.2.6 and 3.5.2.7 at bed temperatures below 1400°F and excess air levels below 30 percent meet these standards.

Combustion Efficiency

Combustion efficiency in AFBC can be a problem, more so than in conventional firing, because of the lower temperatures and the elutriation loss of carbon particulate at the higher velocity. The solution to the problem can be approached either by capturing unburned particles and burning them in a separate fluidized bed operating at a lower velocity and higher temperature (carbon burnup cell), or by employing recycle of the collected carbon particulate back into the main bed. The extended contact time of particulates and SO_2 achieved by either method improves sulfur retention marginally on the inherent coal alkali as well.

Combustion efficiencies for low-rank coals are shown in Figure 3.5.2.20 as a function of average bed temperature. The solid curve represents data obtained with Western low-rank coals at velocities ranging from 7 to 12 ft/sec without recycle. The dashed line is for data obtained with the Beulah lignite while recycle of ash was taking place. The upper curve is from data obtained by Morgantown Energy Technology Center on Texas lignite at 3 to 4 ft/sec and ash recycle. At 7 to 12 ft/sec the combustion efficiencies range from 92 to 96 percent, associated with the increase in bed temperature from 1300° to 1550°F.

Figure 3.5.2.20

Low-Rank Coal Combustion Efficiency Versus Average Bed Temperature

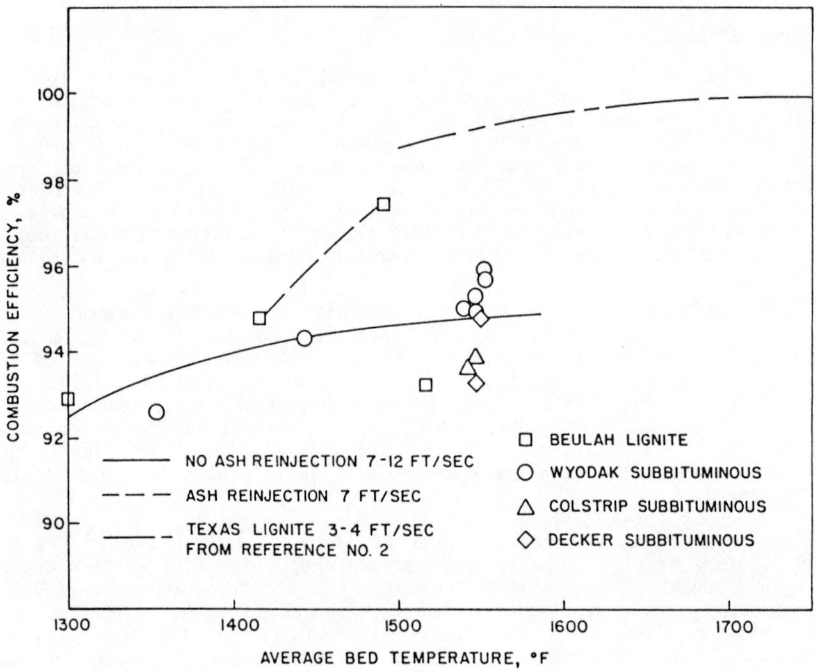

Source: Reference 11

In general, the combustion efficiencies shown for low-rank coals are higher than those observed for bituminous coals under similar conditions, which reflects a higher reactivity for the lower-rank coals. In systems with ash recycle, close to complete carbon utilization can be obtained.

High Sodium Content

One problem which could present a major drawback to the full utilization of lignite in AFBC is the observed tendency for the sodium to form agglomerations of bed material. After the combustion of the high-alkali Beulah, North Dakota, lignite in GFETC's 0.2 ft^2 combustor at a test condition of 1500°F for approximately 12.5 hours, fused pieces of bed material appeared in the removed bed material. Some of these agglomerations appear to have been formed on cooling tubes or thermo-couple wells. This agglomerated bed material later broke off and fell to the bottom of the bed. The largest agglomeration was approximately 1 1/2 inches long, 1/4 inch wide and 3/8 inch deep, and was apparently formed on a cooled tube.[11]

High magnification photomicrographs of bed agglomerates showed two distinct surfaces on opposite sides of some particles, one surface being fused completely and the other only partially fused. This indicates that agglomerates forming on cooled tubes showed a fused consistency on the exterior (bed side) surface, and only a partially fused surface on the tube-side.

An analysis of the partially melted material next to the tube surface was made using the scanning electron microscope. This analysis of the inner surface, as presented in Table 3.5.2.3 shows a substantial enrichment of sodium in the deposit as compared to the coal ash analysis (24 wt pct Na_2O on the tube surface versus 6 pct Na_2O in the ash).

An analysis of a random location on the surface of the outer fused area is shown in Table 3.5.2.4. This outer surface has only a slight enrichment in sodium, which would suggest that the deposits formed on the tube are associated with the condensation of sodium and sulfur as sodium sulfate. The sodium material is volatilized out of the material as it grows away from the cooled surface and is fused.

Table 3.5.2.3

Surface Analysis of Single Particles From Tube Side of Agglomeration

	Pct.
Na_2O	24.3
MgO	2.8
Al_2O_3	3.4
SiO_2	3.2
P_2O_5	0.3
SO_3	48.7
K_2O	4.7
CaO	9.3
Fe_2O_3	3.3
TiO_2	0.0
TOTAL	100.0

Table 3.5.2.4

Surface Analysis of Fused Side of Agglomeration

	Pct.
Na_2O	8.7
MgO	4.8
Al_2O_3	6.7
SiO_2	8.7
P_2O_5	0.2
SO_3	30.1
K_2O	0.4
CaO	30.1
Fe_2O_3	9.3
TiO_2	1.0
TOTAL	100.0

Source: Reference 11

A very large agglomeration was formed around the lower portions of the 7 ft^2 CPC combustor after a shakedown test with the high sodium Beulah lignite.[11] However, this agglomeration was formed under an unusual operating condition, and may not represent what would happen during normal operation. The combustor was operating at about 1800°F when an electrical malfunction caused the combustor to automatically shut down. When the combustor was opened after the run, the large agglomerations were found, which were very likely formed at least in part when the hot bed was slumped.

Bed agglomeration problems in low-rank coals may be alleviated by limestone addition. In this event, the injection rate of limestone would depend on bed properties rather than SO_2 removal requirements.

3.5.2.5 Current Status

Fluidized bed technology is currently evolving from the research and development stage to the demonstration project phase. Commercially available industrial-size units are now emerging into the market place.

Four pilot AFBC units are reported to have operated on low-rank coals. The Grand Forks Energy Technology Center has the largest data base on the Northern Great Plains lignites and subbituminous coals. The other units are at the Morgantown Energy Technology Center (METC), Combustion Power Company in Menlo Park, California, and the Fluidyne Engineering Corporation of Minneapolis, Minnesota. Other significant FBC units may be classified as developmental, demonstration or commercial:

Developmental Systems

- Babcock & Wilcox Ltd. (Renfrew Scotland) - retrofit of conventional spreader stoker boiler.
- Babcock & Wilcox Research Center (Alliance, Ohio) - 6' x 6' FBC Development Facility steam generator under contract with the Electric Power Institute (EPRI).
- Battelle Sub-Scale Experimental Unit System - preliminary program to Battelle Demonstration Project.

Demonstration Projects

- Battelle Columbus Laboratories - steam generator for space heating.
- Combustion Engineering Corp. - steam generator for space heating and power generation.
- Exxon Research and Engineering Co. - crude oil heater for petroleum refinery.
- Fluidyne Engineering Co. - system to provide hot air for space heating and industrial heating.
- Georgetown University - steam generator for space heating and possible power generation.
- Rivesville Allegheny Power System - electric power generation.

Commercial Systems

- British Babcock - State of Ohio Program - retrofit of a 60,000 lb/hr stoker fired boiler at an Ohio State psychiatric hospital. Completed in 1979.
- Johnston Boiler Co. - developed under license from Combustion Systems Ltd. (a group composed of three United Kingdom companies) using Renfrew Boiler data.
- Riley Stoker Corp. - to be manufactured and marketed under agreement with B&W Contractors (Subsidiary of B&W Ltd.) - 50,000 to 500,000 pounds per hour steam capacity. This agreement was very recently announced; therefore a description of the Riley Stoker design is not included in this presentation.

The test devices which have burned low-rank coals will be described first.

GFETC

The data developed at GFETC have come mainly from a 0.2 ft^2 atmospheric fluidized-bed combustor. The unit is constructed from schedule 40, type 316 stainless steel, and uses a flat drilled plate distributor. Heat is removed from the bed by vertical heat transfer tubes in the bed, which are cooled with a flow of water. The nominal fuel feed rate to this combustor is 15 lb/hr of lignite at 7 ft/sec superficial gas velocity and 20 percent excess air.

GFETC is also operating a 2.25 ft^2 experimental AFBC unit (Figure 3.5.2.21). The combustor is a refractory-lined steel shell with a removable door which provides for easy access to the combustor. Horizontal water-cooled tubes are installed through the access door allowing the surface area of the heat removal system to be easily varied to provide for different operational conditions. The area of the heat exchange surface in the bed can be varied from 0 to 38.34 ft^2 by increments of 1.6 ft^2. The nominal fuel feed rate to this combustor is 180 lbs/hr of lignite at 7 ft/sec superficial gas velocity and 20 percent excess air. The combustor system has been designed to operate over a wide range of conditions, as listed below:

Average bed temperature	1300 - 2000°F
Superficial gas velocity	3 - 12 ft/sec
Excess air	10 - 50 percent
Ash reinjection (pct of cyclone catch)	0 to 100 percent

This AFBC uses a flat-drilled plate distributor which is interchangeable with other plates for different operation conditions. The unit also has complete gas analysis and operational control systems dedicated to its operation. The systems control the percent of excess air (air-to-fuel ratio), average bed temperature (through air preheat temperature modulation), superficial gas velocity, and ash reinjection rate. The exhaust gases are continually monitored for O_2, CO_2, CO, NO_x, SO_2, total HC, and opacity. These and about 70 other operational and functional parameters are measured and recorded automatically every hour. On-line data processing is available with the use of a mini-computer which also provides system alarm and process display data.

To date over a dozen tests have been run in this unit, and have shown excellent correlation with data obtained in the 6 inch device. Data from this unit are also in agreement with data taken from the CPC combustor (described later) for the same coals.

Figure 3.5.2.21

2.25 Sq Ft Fluidized-Bed Combustor at GFETC

Source: Reference 11

MERC

The Morgantown Energy Technology Center (METC) has two 18-inch diameter AFBC's in the pilot plant at Morgantown, West Virginia; unit 2 is shown in Figure 3.5.2.22. The bed surface of each of these units is 1.8 ft^2. Unit 1 is a steel-shelled, refractory-lined unit with a conical-shaped distributor plate. The nominal coal feed rate to the combustor is about 110 lbs/hr. The unit has in-bed and freeboard cooling tubes to control bed and flue gas temperatures. The approximate range of operating parameters of Unit 1, in which lignite has been burned, are:

Average bed temperature	1300 - 1800°F
Superficial gas velocity	2.5 - 6 ft/sec
Excess air	Substoichio-metric - 60 percent
Ash reinjection (pct of cyclone catch)	0 or 100 percent

This unit uses six individually water-cooled hairpin tubes of 0.6^2 each, for a total of 3.6 ft^2 in the bed for primary cooling. The bed temperature control is obtained by taking out or bringing into service any number of these tubes. The heat transfer tubes are made from 1/2 inch schedule 160, type 316 stainless steel, because they must withstand bed temperatures without cooling water flow. The flue gas from this combustor is monitored for O_2, CO_2, CO, NO_x, SO_2, and total HC on-line. The Morgantown Center has the most AFBC performance data on Texas lignites and coal refuse.

Combustion Power Corporation

Combustion Power Compnay, Inc. (CPC) of Menlo Park, California, is performing combustion tests on Beulah, North Dakota, lignite under contract to GFETC. The CPC combustor is 36 inches in diameter, for a bed surface area of 7.1 ft^2. This unit is also a steel shell with refractory lining. The nominal feed rate is 500 lbs/hr of coal. Figure 3.5.2.23 is a photograph of the unit at the CPC facility in Menlo Park. The distributor plate used for these tests is a CPC design and is of a modified bubble cap design. The range of operational conditions under which the series of lignite tests is being performed are:

Average bed temperature	1300 - 1800°F
Superficial gas velocity	3 - 9 ft/sec
Percent of excess air	10 - 50 percent
Ash reinjection (pct of cyclone catch)	0 or 100 percent

The unit normally uses about a 5 ft expanded bed, which leaves a freeboard of approximately 13 ft. Ash collected in the cyclone separator may be fed back to the bed if desired. The cyclone was designed to collect 90 percent of all the particulate greater than 5 microns. Vertical cooling tubes have been installed in the combustor for the lignite test program. Tubes of several different configurations are available and are selectively installed to control operating conditions before the run begins. The surface of these in-bed heat exchangers can be varied from about 2.3 to 43.7 ft^2. On-line flue gas analysis equipment is available to measure concentrations of O_2, CO_2, CO, SO_2, NO_x, and total HC.

Figure 3.5.2.22

1.8 Sq Ft Atmospheric Fluidized-Bed Combustor At Morgantown Energy Technology Center

Source: Reference 11

Figure 3.5.2.23

7.1 Sq Ft Atmospheric Fluidized-Bed Combustor
At Combustion Power Company, Inc.

FluiDyne

FluiDyne Engineering Corporation of Minneapolis has performed tests on North Dakota lignite for a private company. They have tested lignite in both an 18 inch square combustor and a 40 x 64 inch combustor. A photograph of the Fluidyne 17.8 ft^2 AFBC is shown in Figure 3.5.2.24. The range of conditions under which coal has been tested in this unit are:

Average bed temperature	1350 - 1650°F
Superficial gas velocity	2 - 6 ft/sec
Excess air	20 - 200 percent
Ash reinjection (pct of cyclone catch)	100 percent

The average coal feed rate to the bed is about 800 lbs/hr, but varies considerably with operational conditions. This is an air-cooled unit and is, therefore, somewhat different in performance, and is particularly well suited for testing industrial process air requirements. The bed depth for most of the testing that has been done was 35 inches, but recently tests with deep beds up to 54 inches have been conducted. FluiDyne has tested both an electrostatic precipitator and a baghouse on this unit. Flue gas analysis equipment is available to measure O_2, CO_2, CO, SO_2, and NO_x on-line.

Figure 3.5.2.24

Fluidized Bed Combustor Designed For Air Heating
FluiDyne Engineering Corporation

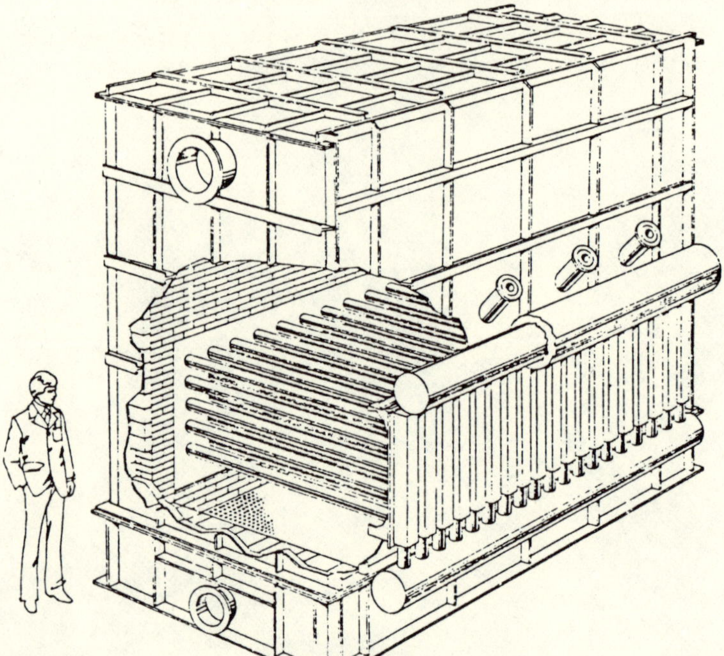

One important feature of the FluiDyne system is its distributor plate. Its proprietary design is based on small air nozzles and small diameter air holes uniformly distributed over an air distribution grid plate. A plate of this design is reported to have been subjected to more than 150 startup-shutdown cycles with no evidence of thermal deformation.

Important operating parameters for the FluiDyne combustor are nominally as follows:

Bed Temperature	1465°F
Bed Depth	35-54 inches
Freeboard Height	5 ft
Excess Air	30%
Superficial Air Velocity	3.6 ft/sec.
Heat Transfer Coefficients	
Tubeside Fluid:	
Liquid	50 Btu/hr ft²°F
Low pressure air	20 - 25
High pressure air	30 - 40
Flyash Recycle (screw feed)	40% (to bed)
Solids Feeding System	Pneumatic
Injection points	combustor walls beneath bed surface
Operational Flexibility	
Air Flow:	1,000 - 12,500 lbs/hr
Coal	150 - 500 lbs/hr
Limestone	50 - 180 lbs/hr
Turndown ratio	3.3:1 (1.65 to 5.5 MM Btu/hr)
Combustion Efficiency	96.3%

The other principal FBC units which have been tested are as follows:

<u>Babcock and Wilcox, Ltd.</u>[14]

Although there are many experimental FBC systems around the world, the B&W unit at Renfrew, Scotland was the first one large enough to permit realistic optimization studies to be conducted on the scale-up engineering problems associated with boiler operation and control.

The B&W Ltd. system was originally a spreader stoker water-tube boiler which was converted to a fluidized-bed combustor, Figure 3.5.2.25 and recommissioned in May, 1975. Over 5,000 hours of satisfactory operation have been accumulated with this system. The unit has a bed area of 100 square feet with a steam generating capacity of about 45,000 lb/hr. All tests have been conducted with coal, but the boiler has recently been

Figure 3.5.2.25

Babcock Renfrew Fluidized Boiler Retrofit

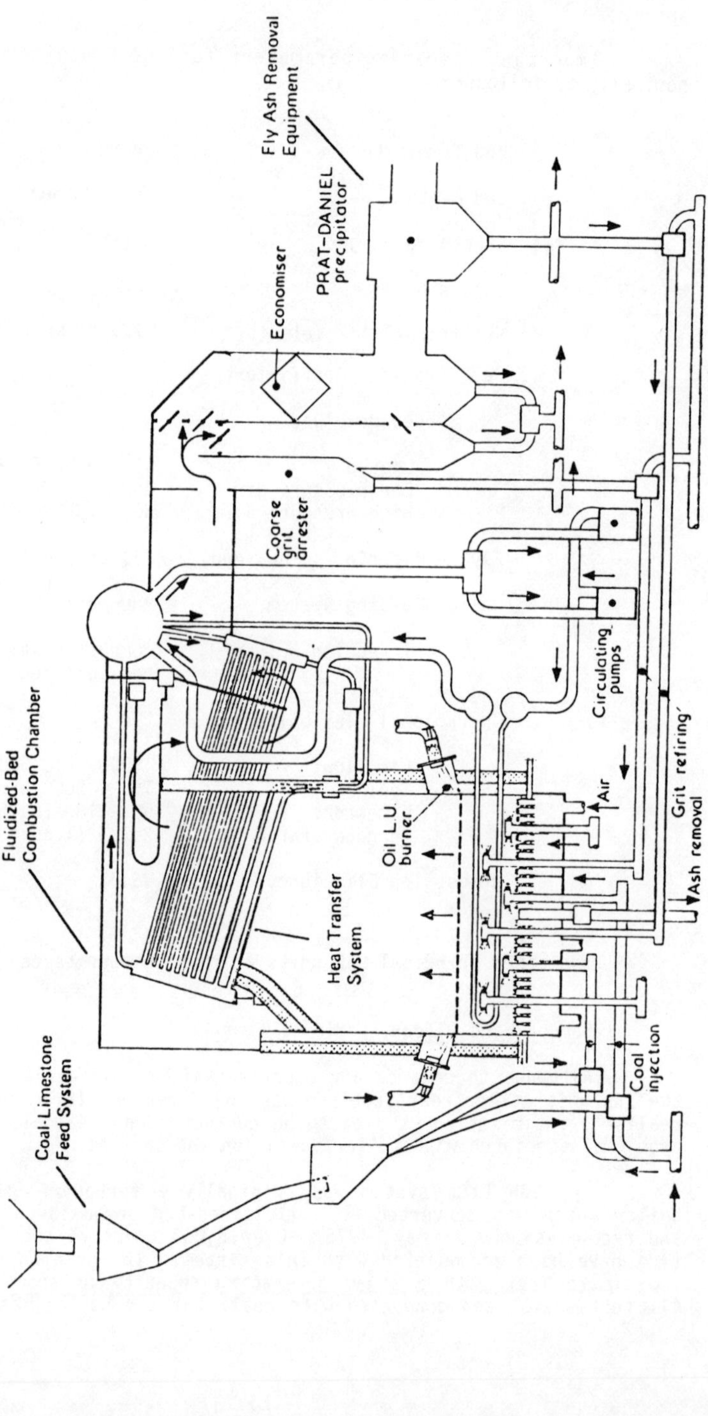

converted for combined coal/oil/gas firing. The studies conducted with this facility have included: startup, load control, temperature uniformity, fuel size and feeding arrangements, fluidizing velocity, particle elutriation and carbon losses, heat transfer rates, corrosion and erosion, sulfur retention and NO_x generation. The data obtained with this system were used in the design of the Johnston Boilers discussed later.

The operating parameters for this system are:

• Bed temperature	1550 - 1650°F
• Calcium to sulfur ratio	6:1 max
• Fluidized-bed depth	3 ft
• Superficial air velocity	4 - 10 fps
• Excess air	Variable
• Flyash recycle	100% greater than 295 microns particle size
• Solids feeding	Pneumatic
location	Under bed
• Operational flexibility	
turndown ratio	2:1

The B&W Ltd unit is distinguished by its proprietary design which separates the functions of bed support from air injection and maintains a low plate temperature.

Babcock and Wilcox Research FBC

This unit was built by B&W under contract with the Electric Power Research Institute.[14] The main furnace structure of this system is an atmospheric pressure water jacket with fireside refractory lining. It is a 6 x 6 foot unit with a steam generating capacity of 10,000 pounds per hour at 150 psig (saturated). It is large enough to help bridge the gap between the bench scale/pilot scale boilers now in operation or proposed and large scale semi-commercial demonstration units. The system will also generate about 3000 pounds per hour of 1000°F superheated steam in addition to the saturated steam. The saturated steam will be used for heating purposes at the Research Center but the superheated steam will be vented.

This program will obtain data to improve combustion efficiency, investigate the effects of hot bed oxidizing/reducing environment on metals, and provide heat transfer data. The effect of multiple coal feed points on combustion efficiency will also be evaluated.

A schematic of the device is shown in Figure 3.5.2.26, showing the combustor and auxiliaries.

Steam generation tubes are located both in the bed and in the freeboard space. The tube bank located in the bed utilizes 1 1/2" O.D. tubes designed for three types of heat transfer: (1) boiler water, (2) superheating and (3) two rows of tubes operating at 1000 psig utilizing sub-cooled water and instrumented to obtain heat transfer data.

The upper (freeboard) steam section is designed to produce additional 150 psig steam while cooling the combustion gas to 900°F. The steam drum is divided to separate the steam produced by the in-bed tubes from that produced by the freeboard tubes. Flow through the circuit is produced by forced circulation.

Figure 3.5.2.26

Babcock & Wilcox Alliance Research Center
Fluidized Bed Combustion Development Facility

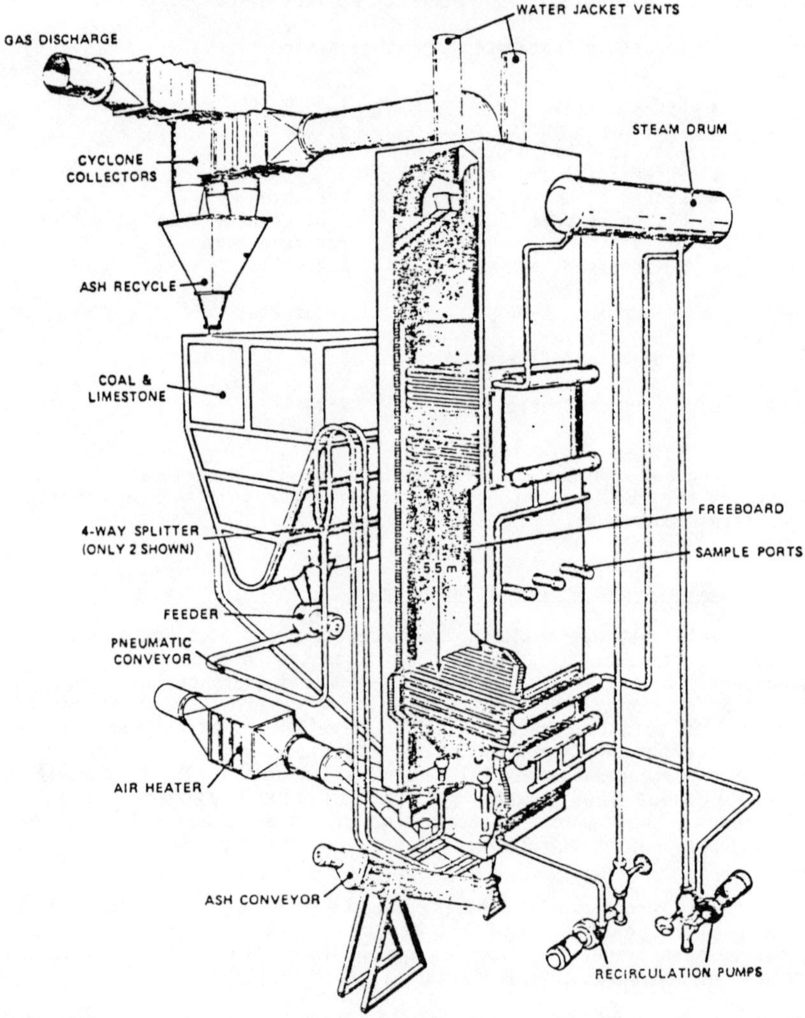

Output control is achieved by the number of beds in fluidization. The bed is divided into four quadrants, each with its own air and coal/limestone supply. Four separate air ducts feed the plenum chamber. Each duct has a separate damper and venturi flow meter for control and measurement.

The original distributor plate was made of calendered woven nichrome wire mesh. This plate was unsatisfactory because of plugging due to dust in the incoming air stream, plugging due to fines in the bed and warpage due to high temperatures. The present plate is 1/16 inch thick stainless steel sheet perforated by 0.093 inch diameter holes and has an open area of 2.0 percent. This plate is sandwiched between two stainless steel grids. The plate is designed for 10 inches water pressure drop at an air velocity of 8 feet per second. Although some plugging, leaking of bed material into the plenum, and plate warpage have been encountered, these problems have not greatly hampered facility operations. This plate has withstood temperature excursions better than the woven wire plate. The use of a refractory covered, bubble-çap plate, successfully tested on a smaller unit, is being considered for the 6' and 6' system.

Four cyclone separators are mounted at the furnace exit to collect particulates escaping the furnace. Material collected by these cyclones can be recycled to the fluidized-bed via the coal feed system or removed from the unit by the ash handling system.

The flue gas from the cyclones passes through a large venturi flow meter and is then cooled to 400°F before it enters the induced draft fan and passes out the stack.

The bed (ash) removal system consists of five drain pipes which extend from the bed through the distributor plate and the windbox into the boiler room basement. Each ash pipe has a separate shutoff valve activated by an air cylinder. Presently the bed depth and the rate of bed removal can be controlled by the pressure drop across the fluidized bed. Ash from these drains drops into a cooling conveyor which carries the material out of the boiler room basement to an ash hopper.

Nominal operating parameters of the system are:

• Bed temperature	1600°F
• Calcium to sulfur ratio	5:1 maximim
• Bed depth	2.3 - 4 ft
• Freeboard height	18 ft
• Superficial air velocity	5 - 10 fps
• Excess air	5 - 30 percent
• Heat transfer coefficient	45 - 50 Btu/hr ft²°F
• Flyash recycle	none in current tests
• Solids feeding	pneumatic injection
location	under bed
• Operational flexibility	
turndown ratio	4 - 5:1

<u>Battelle</u>[14]

The Battelle Multi-Solids FBC demonstration unit, conceptually shown in Figure 3.5.2.27 will generate 40,000 lb/hr of saturated steam at 100 psig for use in an existing space heating system.

The Battelle system is unique because it employs two fluidized beds: one bed for the combustion process and a second bed for heat transfer, which contains the steam tubes. The separation of the combustion process from the heat transfer process permits each to be optimized for maximum efficiency. A dense solid material, Ohio River river-bed pebbles, is fluidized with the coal and limestone in the combustion bed, while a lighter material (sand) continuously circulates from the combustion bed to the heat transfer bed where it transfers its heat to the steam tubes, and then returns to the combustion bed. Several of the advantages claimed for this process are:

- Erosion and corrosion of the boiler tubes minimized by the low gas velocities (1-2 fps) in the external boiler.

- The external boiler may be designed to maximize heat transfer to the steam tubes.
- Rapid shutdown by stopping air, coal and limestone feed. Bed does not require cooling following shutdown so can therefore retain high temperature to facilitate rapid restart.
- Improved calcium utilization and sulfur capture.

Figure 3.5.2.27

Battelle MS-FBC With External Heat Exchanger

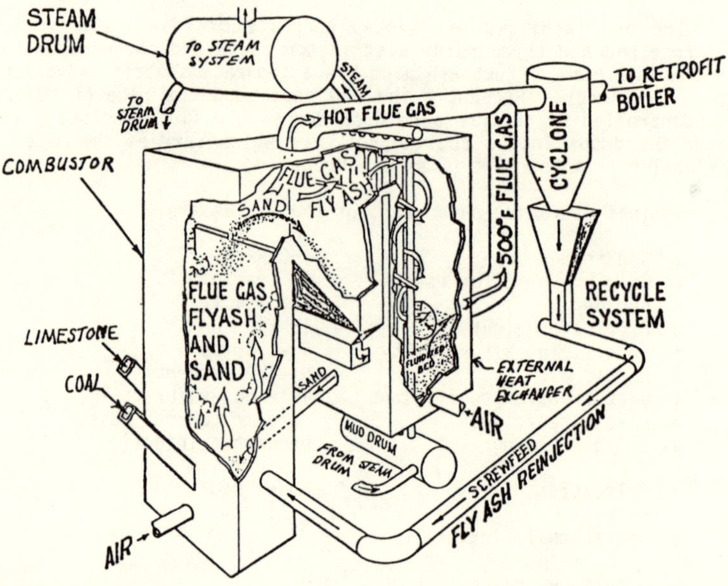

Source: Reference 14

The Battelle system has no provision for coal crushing or drying. The system is capable of burning 1-1/4" x 1/4" coal with varying moisture content. The limestone will be -325 mesh. The beds will be initially charged with the sand and pebbles, which will remain in the system and replenished periodically as required by operational losses. The sand make-up loss will be approximately 22 pounds per hour.

Coal and limestone will be gravity fed through the side walls beneath the bed surface by water-cooled air-purged drop pipes. The air purge will prevent back-up of bed material into the pipes. The pipes will be fed by screw conveyors from storage bins.

Flyash will be recycled by means of a screwfeeder into the combustion bed. Figure 3.5.2.27 shows the Battelle FBC system. As previously mentioned, this design is unique since the heat transfer system is located in a secondary (hot sand) bed which is separate from the primary or combustion bed. Tubes are positioned both in the hot sand bed and in the freeboard space of the heat transfer chamber. Tubes within the bed are vertical. Flow through the water circuits is by forced circulation. Approximately 27,000 pounds of steam per hour are generated in the secondary bed.

The hot flue gases are then ducted to an existing boiler in which an additional 13,000 pounds of steam per hour are generated.

The system uses horizontal sparger tubes located at three levels in the lower portion of the combustor. Air, when emitted from these tubes, which are perforated on their undersides, fluidizes that portion of the bed above the tubes. Air flow to each of the three layers of tubes may be independently controlled.

A cyclone dust collector is used to remove large size particulate matter (flyash) from the flue gas stream. The flyash is recycled to the combustion bed. The gas stream from the cyclone system will pass through a baghouse filter for final particulate removal.

Relevant operation variables for the Battelle system are:

• Bed temperature	1650°F
• Bed depth	4.75 ft
• Freeboard height	31 ft
• Superficial air velocity	30 - 40 fps
• Excess air	20 percent
• Heat transfer coefficient	45 - 50 Btu/hr ft^2°F
• Flyash recycle (screw feeder)	20 percent
• Solids feeding	Gravity drop
location	in bed
• Operational flexibility	
turndown ratio	3:1
(Coal, air, limestone and and all variable)	

Combustion Engineering

The Combustion Engineering demonstration plant will generate 50,000 lb/hr of superheated steam at 365 psig and 560°F for space heating use at the U.S. Navy Great Lakes Training Station, Illinois. The unit will be an addition to an existing power plant for space heating and power generation. The purpose of this program is to develop a packaged FBC steam generator which may be built and shipped as a modular unit. It must be able to meet all environmental standards without the use of auxiliary SO_2 removal equipment, and must be competitive with other coal-fired units. A cross section through the Combustion Engineering FBC System is shown in Figure 3.5.2.28. The general arrangement of the system is shown in Figure 3.5.2.29.

Coal and limestone will be stored in bins, metered by weigh belt feeders and mixed in a hopper. However, the coal, prior to release to the mixing hopper will be blended with recycled ash in a screw conveyor and then discharged to the hopper. A pneumatic feed system will deliver the coal to any of the individual bed segments. The coal-limestone mixture will be injected pneumatically through horizontal nozzles located in the bed just above the distributor plate. The nozzles will be fed from below the plate.

This system uses a three drum design (2 mud drums - 1 steam drum), with boiler tubes in an "A" configuration having the horizontal tubes at a 10° slope as shown in Figure 3.5.2.29. The front and rear walls of the boiler are water-cooled. The superheater tubes are completely immersed in the fluidized-bed. The turndown will be controlled by varying the fluidized bed area, fuel feed and air feed rates.

Figure 3.5.2.28
Combustion Engineering-FBC Cross Section
U.S. Navy Great Lakes Training Station

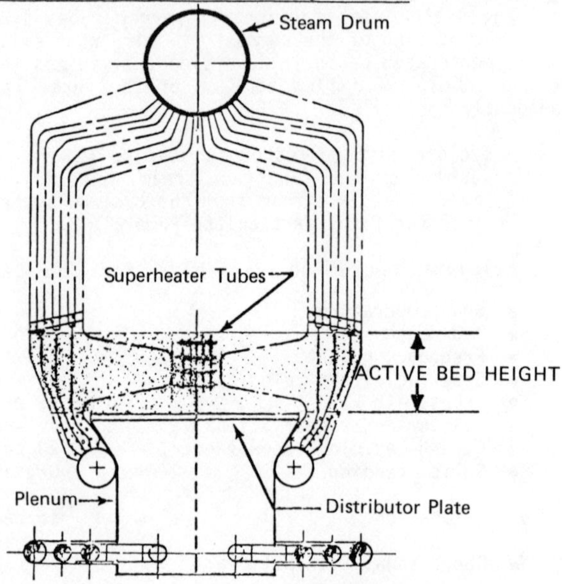

CROSS SECTION through AFBC INDUSTRIAL BOILER

Source: Reference 14

Figure 3.5.2.29
Combustion Engineering Steam Generation System
U.S. Navy Great Lakes Training Station

GENERAL ARRANGEMENT

Fluidized Bed Demonstration Boiler

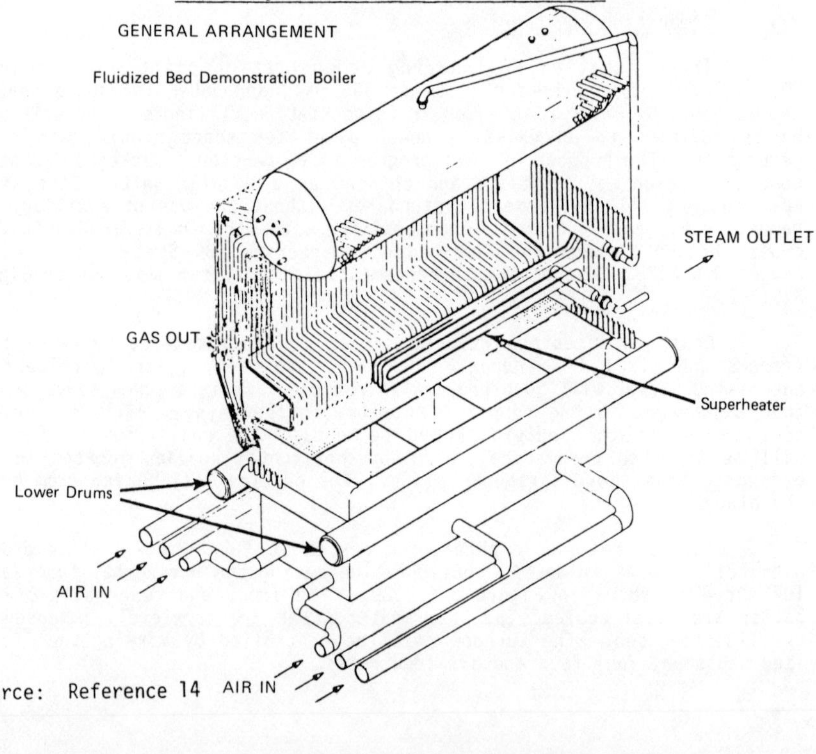

Source: Reference 14

The bed is segmented into six areas, each having an independently controlled air supply. The total bed area of 140 square feet is divided into four zones of 17.5 square feet each and two zones of 35 square feet each. The number of beds in operation provides the primary output control. Output may also be regulated by varying the temperature in each zone from 1400°F to 1700°F by air and fuel flow control. It is anticipated that a wide turndown ratio may be achieved, the actual values of which will be established as part of the demonstration program.

The flue gases pass through a cyclone for flyash removal. The removed flyash is transported by a pneumatic system to a separator, from which it falls into a storage silo. Approximately 85 percent of the flyash is blended with the coal stream for recycle through the combustion bed. A baghouse house is used for final gas particulate removal.

Nominal operating parameters for the C-E System are:

• Bed temperature	1400 - 1700°F
• Bed depth	3 ft
• Freeboard height	9 ft
• Superficial air velocity	7 fps
• Excess air	20 percent
• Heat transfer coefficient	50 Btu/hr ft^2°F
• Flyash recycle	85 percent
• Solids feeding	pneumatic injection
location	below bed

<u>Exxon</u>[14]

The purpose of this program is to extend the fluidized-bed combustion state-of-the-art technology presently used for steam generation to the preheating of crude oil to 600°F or higher prior to distillation or other refinery processing.

Two areas of technology peculiar to oil process heater application which differ from boiler applications are the affects of tube size and bundle configuration on heat transfer and the problem of oil degradation which results in coke deposits on the interior walls of the heat transfer tubes.

Indirect fired process oil heaters usually have tubes which are 4 to 8 inches in diameter in contrast to the 1 to 2 inch diameter tubes used for steam generation. The Exxon program will establish the affects of these larger tubes on fluidization characteristics and will define the optimum or acceptable configuration of the tube bundle immersed in the fluidized-bed. The parameters controlling hydrocarbon coking and the affects of the coke laydown on heat transfer through the heater tubes will also be investigated.

The Exxon project consists of a research and development program, Phase I, (completed in 1979) and a demonstration program, Phase II.

Phase I includes visualization studies now completed, using a cold flow two dimensional fluidized-bed unit, and hot flow studies using a Process Stream Coking Unit and a High Temperature Heat Flux Unit. These units will utilize both electric and propane fired heat sources. The latter unit will also be fitted for coal firing. The flow plan for the High Temperature Heat Flux Unit is shown in Figure 3.5.2.30.

If the tests of Phase I are successful, the High Temperature Heat Flux Unit will be installed in an Exxon refinery for the demonstration studies. This unit has a bed area of 9 square feet. When fired with coal, the anticipated thermal output capacity will be 10 - 15 x 10^6 Btu/hr with a coal consumption of about 1,560 pounds per hour. It is planned to burn coal having a sulfur content of 3 percent. Coal fired design parameters have not yet been established.

Figure 3.5.2.30

Exxon Flow Plan FBC Heat Flux Unit

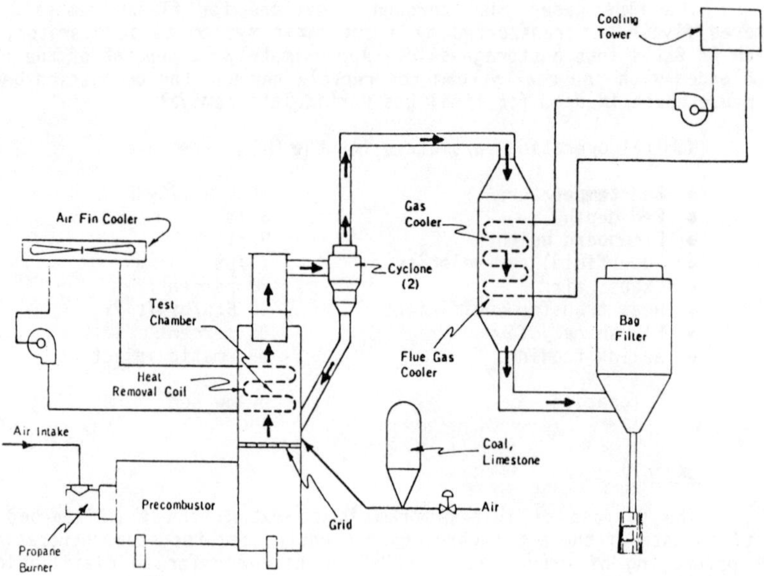

Georgetown University

The Georgetown demonstration facility, shown schematically in Figure 3.5.2.31, has a steam generating capacity of 100,000 lb/hr at 675 psig. The steam system has been integrated into the existing university power plant which provides steam for space heating and cooling. The purpose of this program is to demonstrate the use of FBC to burn high sulfur coal in an environmentally acceptable manner in a congested urban location without auxiliary SO_2 removal equipment. The FBC unit and all auxiliary equipment will be totally enclosed in a building which is to be aesthetically compatible with the surrounding university complex. The system can also produce steam up to 625 psig for possible power generation.

Coal is stored in three 346 ton capacity bunkers, transported by elevator to a crusher if not delivered presized, then fed by conveyor to a weigh scale which feeds a short term storage (20 ton) bin. Coal is delivered to the combustor by two stoker feeds located in the sidewalls above the bed. The stokers distribute the coal uniformly over the bed area.

Limestone is stored in a 350 ton bunker and transported pneumatically to 5 ton bins. From here the limestone is metered by weigh scales to two 4 inch diameter pipes which feed the combustor by gravity just above the bed surface.

Both of the stoker feeders and one of the limestone injection pipes is located on the side wall. The remaining limestone feed pipe is located on the front wall near the side wall which contains the coal feeders.

Figure 3.5.2.31

Fluidized-Bed Steam Generator Georgetown University
100,000 Lbs/Hr 675 PSIG Design Pressure Saturated Steam

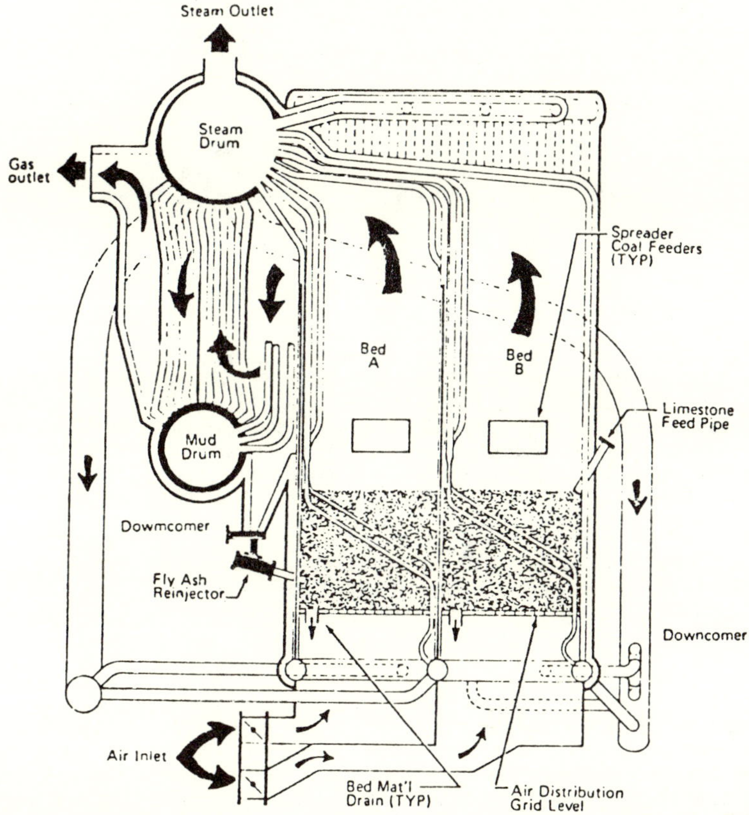

Source: Reference 14

The Georgetown system includes three parallel steam generating circuits which consist of a convection boiler bank, immersed bed tubes and water walls. A water-tube wall also divides the bed segments. Flow through the circuits is by natural circulation. The steam, generated at 675 psig, is reduced to 275 psig for use in the existing university power plant steam system.

Control of the system is achieved by using the divided bed, each segment of which may be independently controlled to provide an overall turndown ratio of 4:1. Each bed is provided with a separate stoker spreader for coal feed and one four inch diameter limestone feed pipe.

Mechanical (cyclone) separators at the boiler bank inlet and outlet remove flyash from the flue gas. The flyash is reinjected directly into the fluidized-bed. A baghouse will remove fine particulate matter from the flue gas. Spent bed material will be gravity fed to screw conveyors for cooling prior to disposal.

Nominal operating parameters for the Georgetown system are:

• Bed temperature	1600°F
• Bed depth	4.5 ft
• Freeboard height	8 ft
• Superficial air velocity	6 - 8 fps
• Excess air	20 percent
• Heat transfer coefficient	40 Btu/hr ft²°F
• Flyash recycle	78 percent
• Solids feeding	Stoker spreader
location	combustor walls, freeboard
• Operational flexibility	
turndown ratio	4:1

Rivesville[15]

Early in 1970, the consulting engineering firm of Pope, Evans, and Robbins approached the Allegheny Power System management with a proposal to participate in a fluidized-bed combustion program. The Rivesville station, owned by the Monongahela Power Company and located in northern West Virginia, was selected as the site for the demonstration project. Initial operation began in 1976, and within one year after start-up, the boiler nameplate rating had been achieved for a short period and sulfur capture demonstrated.

The Rivesville FBC operates at a minimum of three cell steam flow of 125,000 pounds per hour at 1200 psig and 900°F. Peak load stream is generated at 1250 psig and 940°F, and represents a turndown ratio of approximately 3:1.

Since 1978, the unit has experienced serious operational difficulties. These problems have caused major set-backs in the testing schedule, which originally called for a series of tests on low-rank coals, but were never carried out. A major milestone was reached early in 1980 when the unit successfully completed a 200 hour demonstration run. Current plans call for demolition of the unit if FY-81 funds are not received; otherwise, the unit will be operated for six months and then dismantled.

British Babcock - State of Ohio[14]

This commercial installation was initiated as a result of the twin crises encountered in Ohio during the severe 1977-78 winter: acute fuel shortages and a conflict between the EPA SO_2 regulations and Ohio's vast reserves of high sulfur coal which has in the past supported the state's industrial base.

This unit is a retrofit of a 60,000 lbs/hour boiler at the Central Ohio Psychiatric Hospital in Columbus. The project was contracted by the Ohio Energy and Resource Development Agency with Woodhall-Duckham (USS) Ltd., a British Babcock subsidiary.

The plant is designed to accept a wide range of coals and limestone and is intended to demonstrate, on an industrial scale, the reliability and operating economics of a high sulfur coal fired fluidized-bed combustion boiler under a variety of operating conditions. It is closely based on the results obtained with the Babcock Renfrew system. Design details are not available.

Johnston Boiler

The Johnston Boiler Co. commercial units, Figure 3.5.2.32, were developed under technology and patent license with Combustion Systems Limited, a joint British organization. The design is based on the results obtained with the B&W Ltd. Renfrew tests. The Johnston commercial units are packaged modular systems available in nine sizes. The steam generating capacity of these units range from 2500 lb/hr (3×10^6 Btu/hr) to 50,000 lb/hr (60×10^6 Btu/hr). These units are designed to operate on a wide variety of fuels including natural gas, propane, No. 2 and No. 6 fuel oils and wood in addition to all grades of coal.

The Johnston units are segmented into three zones, each of which is capable of independent operation. This provides a primary turndown ratio of 3:1. Additional optional control may be provided by varying the fuel and air flow to each segment over a range of 2:1. Thus, by combining the two modes of control an overall turndown ratio of 6:1 may be achieved.

Tests, conducted to determine potential turndown ability, indicated a possible range of 10:1, but with efficiency penalties. Johnston hopes, however, to achieve this high turndown capability without efficiency losses in future designs.

The Johnston Boiler Units can be operated with coal having a maximum size of 1 1/4 inches. Coal and limestone feed hoppers are located above the combustor, integral with the FBC unit. Variable speed screw feeders transport the coal and limestone to gravity feed drop pipes, which pass through the combustor sidewall at a point just above the bed surface. Unburned carbon is recycled by pneumatic injection through the combustion chamber wall beneath the bed surface.

Steam is generated by an integral firetube heat exchanger (Figure 3.5.2.32). Although heat transfer tubes are not immersed within the bed, all surfaces of the combustion chamber are lined with water tubes, as shown in the figure. Water tubes are also used for bed dividers and baffles in the freeboard space. Cyclone dust collectors remove flyash which is recycled directly to the bed. Ash and spent bed material is pneumatically transported from the bed for disposal.

Operating parameters are as follows:

• Bed temperature	1600°F
• Fluidized bed depth	3 ft
• Freeboard height	6 ft
• Superficial air velocity (variation due to changing cross sectional area)	12 fps (bottom) 6 fps (top)
• Excess air	25 percent
• Heat transfer coefficient	7500 Btu/hr ft^2
• Flyash recycle	Variable
• Solids feeding	Gravity drop
location	
• Operational flexibility	
turndown ratio	6:1

The following tables summarize operating parameters for some of the fluid-bed combustion systems described above. Table 3.5.2.8 lists foreign FBC projects.

Figure 3.5.2.32
Johnston Boiler System

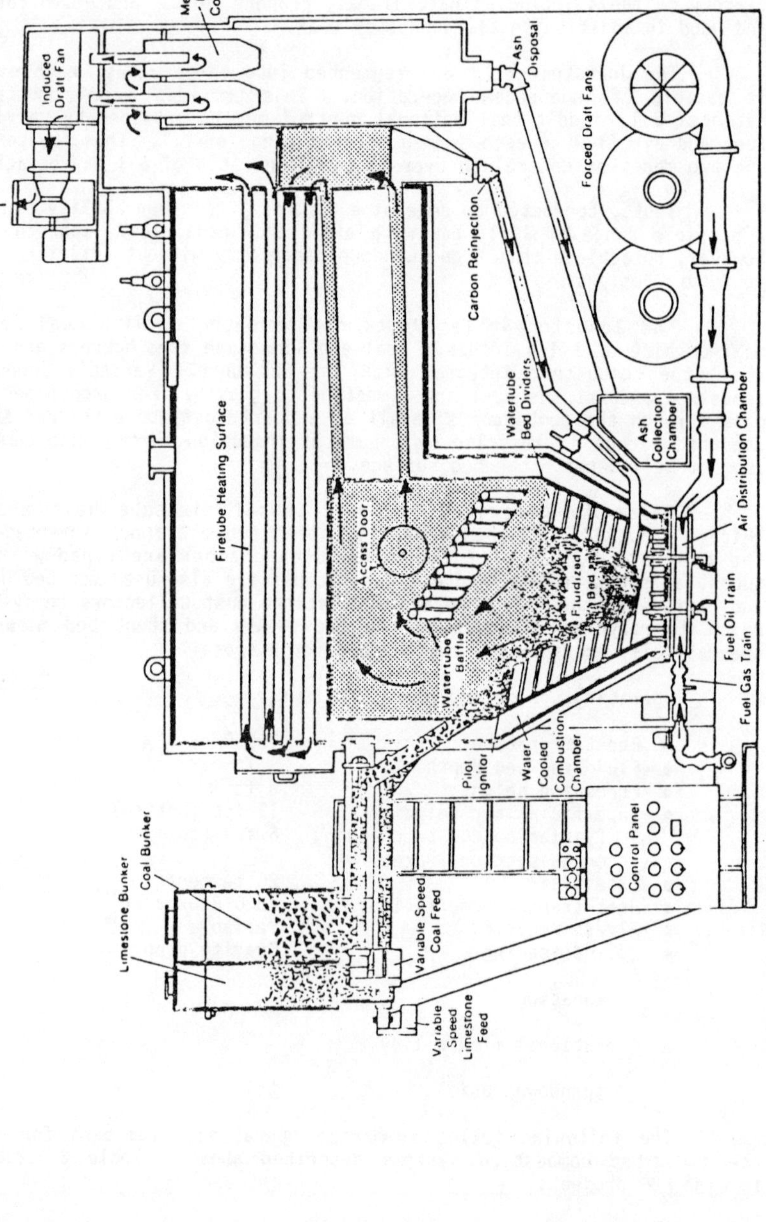

Source: Reference 14

Table 3.5.2.5

FBC Units
General Description

System	Status of Start-Up Date	Steam Cap. or Heat Output	Nominal Bed Area, Sq. Ft.	Coal Type	Coal Sulfur Cont.	Feed Size (Inches)	Rate (Lb/Hr)	Sorbent Type-Feed Rate (Lb/Hr)
B&W Ltd.	R&D Operational	45,000 Lb/Hr	100	Various-	3.5-5%	95%-1/8	-	Limestone
B&W Res. Cent.	R&D Operational	10,000(Sat.)Lb/Hr	36	Ohio No. 6-	3.5-4%	1/4 x 0	1,880	Limestone-600
Battelle	Demo-1979	40,000 Lb/Hr	15	Illinois No. 6-	High	1-1/4 x 1/4	4,700	Limestone-2240
Comb. Eng.	Demo-1980	50,000 Lb/Hr	140	Midwest Bitum.	3.5%	1/4 x 0	6,710	Limestone-3260
Exxon	Demo-1979	15 x 10^6 Btu/Hr	9		3%		1,562	Limestone
Fluidyne	Demo-1979	30 x 10^6 Btu/Hr	17.8	Illinois No. 3-	3.6%		500	Dolomite-180
Georgetown	Demo-1979	100,000 Lb/Hr	212	East Bitum.	3.29%	1-1/4 x 3/8	9,565	Limestone-3133
Johnston Boiler	Commercially Available	10,000 Lb/Hr 50,000 Lb/Hr	12-24 to 60-120 Bed Bottom and Top Areas	All Grades-	High-	1-1/4 x 0	263- 5,260	Limestone- Approx. 50% of Coal Feed, (Highly Variable)

Source: Reference 14

Table 3.5.2.6 Nominal Combustion Chamber Parameters

	Bed Temp, °F	Calcium/ Sulfur Ratio (Molar)	Fluidized Bed Depth, Ft.	Freeboard Height, Ft.	Superficial Air Velocity Ft/Sec	Excess Air %	Heat Trans. Coefficient, Btu/Hr Ft^2°F	Flyash Re-Cycle, % of Coal Feed	Distributor
B&W Ltd.	1,550-1,650	6:1 MAX	3	15	4-10	Exp. Var.	Exp. Var.	Exp. Var.	Separate Bed Support & Air Injection Plates
B&W Res. Cntr.	1,600	5:1 MAX	2.3-4.0	18	5-10	5-30%	45-50	Exp. Var. (0% MIN)	Proprietary Design
Battelle	1,650	3:1	4.75	31	30-40	20	45-50	20	Sparger Tubes
Comb. Eng.	1,400-1,700	4:1	3	9	7	20	50	85	Perf. Plate 0.1" Dia. Holes Proprietary Design
Fluidyne	1,465	1.1:1	3.5	5	3.6	30	20-40	40	Proprietary Design
Georgetown	1,600	3:1	4.5	8	6-8	20	40	78	Perf. Plate Prop. Design, 0.1" Holes 3/4" Pitch
Johnston Boiler	1,600	2:1-3:1	3	6	12 (Bed Bottom) 6 (Bed Top)	25	7500 Btu/Hr-Ft^2 (Nominal Overall Rate)	Operating Variable	Horizontal Air Injection Proprietary Design

Source: Reference 14

Table 3.5.2.7 Output Control (Turndown)

System	Turndown Ratio	Segments	Air Flow Lb/Hr	Coal Feed Lb/Hr	Limestone Feed Lb/Hr
B&W Ltd.	2:1	Multi-Variable No.	Experim. Variable	Experim. Variable	Experim. Variable
B&W Res. Cntr.	4-5 Objective	4	27,900 MAX	3000 MAX	1000 MAX
Battelle	3:1	Single	Variable (Not Specified)	4700 MAX	2240 MAX
Comb. Eng.	To Be Determined	6	Variable	6710 MAX	3200 MAX
Fluidyne	3.33 ($1.65-5.5 \times 10^6$ Btu/Hr)	Single	1,000-12,500	150-500	50-180
Georgetown	4:1	2			
Johnston Boiler	3:1 STD. 6:1 Optional	3	Variable	Variable	Variable

Source: Reference 14

Table 3.5.2.8

Foreign FBC Projects

Country	Location	Start-up	Feedstock	Process	Project	Product
West Germany		Spring 1980	Coal	PFBC	Gas turbine cycle with a pressurized fluid bed combustor	3 MWe
West Germany	Essen	1972	Coal	AFBC	16"x32" unit, burns 100 lbs/hr - test unit	
West Germany	Saarberg	1980	Coal	AFBC	Power Plant	
New South Wales	Camden	October 1977	Coal	FBC	Fluidized Bed Pilot Plant; 2.2 tons/hr	
New South Wales			Brown Coal		1.3 $ft.^2$ FBC Test Unit	
Czechoslovakia	Bechovice	1958	Low Quality Brown coals-can be used with lignite	AFBC	Commercial Plant - 25 MWe	
Finland	Kajaani	April 1977	Coal	FBC	Commercial unit- $25MW_{th}$ 10 ft. diameter bed	Heating only
France	Grenoble	1968	Coal		Commercial 25MWe	
Morocco	Casablanca	1968	Coal		Commercial 60MWe	
India	Trichy	June 1976		AFBC	Test Unit, 6 ft diameter, 22 tons/day 20,000 pph steam	
India	Hardwar	Under Construction	Coal	PFBC	660 lb/hr	

References - Section 3.5.2.2 - 3.5.2.5

1. McCabe, W.L. and J.C. Smith. Unit Operations of Chemical Engineering, second edition, p. 172, published by the McGraw-Hill Company, 1967.

2. Reh, L. "Fluid Bed Combustion in Processing, Environmental Protection and Energy Supply," presented at the International Fluidized-Bed Combustion Symposium of the American Flame Research Committee, Boston, Massachusetts, April 30-May 1, 1977.

3. Robison, E.B., et al. Study of Characterization and Control of Air Pollutants From a Fluidized-Bed Combustion Unit - The Carbon Burnup Cell, EPA Contract CPA 70-10, Pope, Evans and Robbins, February 1972.

4. Lange, H.B. and C.L. Chen. SO_2 Absorption in Fluidized-Bed Combustion of Coal - Effect of Limestone Particle Size, EPRI Project No. RP 719-1, The Babcock and Wilcox Company, Alliance, Ohio, March 1977.

5. Snyder, R.B., et al. "The Role of a Carbon Burnup Cell in Reducing SO_2 Emissions From Fluidized-Bed Coal Combustion Plants," presented at the 1977 meeting of the Western States Section of the Combustion Institute, October 17-18, Stanford, California, Paper No. 77-30.

6. Summary Evaluation of Atmospheric Pressure Fluidized-Bed Combustion Applied to Electric Utility Large Steam Generators, prepared for EPRI by the Babcock and Wilcox Company, Barberton, Ohio, October 1976.

7. Berman, I.M. "Fluidized-Bed Combustion Systems: Progress and Outlook," Power Engineering Magazine, November 1979.

8. McKenzie, E.C. "Burning Coal in Fluidized Beds," Chemical Engineering, August 14, 1978.

9. Roe, D.E., A.W. Karr, K.R. Lemmerman, and J.E. Sinor. Solid Fuels For U.S. Industry, Volume II, March 1979.

10. Keairns, D.L., R.A. Newby, and N.H. Ulerich. "Design of Fluidized-Bed Combustion Systems to Meet Environmental Standards," presented at the Fifth Annual International Conference on Commercialization of Coal Gasification, Liquefaction, and Conversion to Electricity, University of Pittsburgh, August 1-3, 1978.

11. Goblirsch, G.M. and E.A. Sondreal. "Low-Rank Coal Atmospheric Fluidized-Bed Combustion Technology," Symposium on Technology and Use of Lignite, Grand Forks, N.D., May 1979, DOE GFETC/IC-79/1, 1979, pp. 75-107.

12. Goblirsch, G.M. and E.A. Sondreal. "Fluidized Combustion of North Dakota Lignite," Ninth Biennial Lignite Symposium at Grand Forks, N.D., May 1977, GFERC/IC-77/1.

13. Goblirsch, G.M., R.W. Fehr, and E.A. Sondreal. "Effects of Coal Composition and Ash Reinjection on Sulfur Retention Burning Lignite and Western Subbituminous Coals." Proceedings of the Fifth International Conference on Fluidized-Bed Combustion, Volume II, Washington, D.C., December 1977.

14. Kapp, G.S. and W. T. Harvey, Jr. "Fluidized-Bed Combustion: Where Are We Today, A State-of-The-Art Summary," presented at the Kentucky Industrial Coal Conference, University of Kentucky, Lexington, Kentucky, May 2-3, 1979

15. McCarthy, H.T. "The Rivesville Installation From the View of the Monongahela Power Company," presented at the Fifth International Conference on Fluidized Bed Combustion, December 12-14, 1977, Washington, D.C.

3.5.2.5.1 - Selected Reference Abstracts For Fluidized-Bed Combustion

A broad overview of information on fluidized bed combustion of coal may be gained from the Tables of Contents of the Proceedings of the Annual International Conferences on Fluid Bed Combustion sponsored by EPRI, EPA, and TVA. The papers contained in these proceedings are well organized, brief, and cover key areas of interest. In addition the proceedings contain good overview papers for those desiring only a brief introduction to the topic.

In the specific area of low-rank coal use in fluid beds, the amount of literature is remarkably small. However, several papers are available that can give the reader a quick appreciation of the various topic areas unique to low-rank coals. These are listed with comments below:

1. "Fluidized Bed Combustion Systems: Progress and Outlook," Power Engineering Magazine, November 1979.
   Excellent overview with a review of some key projects. Small reference to lignite fuels.

2. McGowin, C.R., et al. Conceptual Design of a Gulf Coast Lignite-Fired Atmospheric Fluidized Bed Power Plant, presented at the Annual Meeting of the American Institute of Chemical Engineers, San Francisco, California, November 25-29, 1979.
   A moderately detailed 35 page paper presenting designs and economic comparisons of a nominal 530 megawatt net AFBC plant and comparable lignite fired pulverized coal plant operated as minemouth systems. Summarizes current thinking on utility installations.

3. Goblirsch, G.M. and E.A. Sondreal. Low-Rank Coal Atmospheric Fluidized Bed Combustion Technology, Symposium on Technology and Use of Lignite, Grand Forks, N.D., May 1979, DOE GFETC/IC-79/1, 1979, pp. 75-107.
   Describes some DOE test facilities, and summarizes effect of sulfur retention with inherent alkali for several low-rank coals. Brief discussion of agglomeration with photographs and results of chemical analysis.

4. Johnson, I., G.J. Vogel, S.H.D. Lee, et al. Support Studies in Fluidized Bed Combustion, Quarterly Report, Argonne National Laboratory, Argonne, Illinois.
   The Argonne reports summarize work in a variety of topic areas, but those reports beginning in mid-1978 include information on turbine corrodent studies, specifically removal of alkali metal compounds that would damage turbomachinery used in pressurized FBC installations.

3.5.3 Gasification

3.5.3.1 Introduction and Summary

The incentives for using low-rank coals as feedstocks for coal gasification systems depend upon a wide variety of factors. Although political and regulatory influences are important, their influence is secondary in comparison to questions of a technical or economic nature. The range of application of coal gasification technology varies from low pressure, low-Btu gas generation to high pressure pipeline gas designs. These systems employ fixed, fluidized and entrained flow gasifiers blown with air or oxygen (and in some cases, hydrogen or steam) which operate in non-slagging, agglomerating or slagging modes with or without the aid of catalysts.

In reviewing the application of low-rank coals to the various gasifiers commercially available or under development, it is clear that testing of low-rank coals in these units has been far from exhaustive. For the many reasons outlined in this report, low-rank coals are good gasification feedstocks, and in many cases better overall than bituminous coals. Because of their suitability to gasification, demonstrations of gasifier performance may benefit from the inclusion of low-rank coals in the test program, a practice which has not been common to date.

Fixed-bed gasifiers may be operated in either slagging or non-slagging modes. Non-slagging fixed-bed gasifiers are in wide commercial use and therefore represent the lowest risk systems available today. The slagging version of the fixed-bed design, although not commercial, represents an improvement in throughput (by a factor of three), thermal efficiency and reduced steam consumption.

Fluidized-bed systems have been in industrial use for many years, although the Winkler design is the only commercial coal gasifier available. Fluidized-bed gasifiers offer the possibility of operating with a sulfur sorbent directly in the bed, thereby reducing the requirement for downstream H_2S removal. Ash removal in fluidized-bed gasifiers generally requires that the ash be in an agglomerated state.

Entrained flow gasifiers are operated in a slagging mode, and their high temperature, cocurrent flow assures that no higher organics remain in the raw gas. As a rule, these systems have the highest throughput of the three configurations (fixed, fluidized, and entrained) and are also the largest consumers of oxidant per unit weight of coal gasified.

Environmental control issues for coal gasification address air, water and solid waste pollutants. The lower sulfur content of low-rank coals will reduce the cost of meeting sulfur emission standards relative to the gasification of higher rank coals. Liquid phase pollutants result from raw gas and ash quenching, and process condensate blowdown. The degree of cleanup required will depend to a great extent upon the gasifier design, with fixed-bed units generating the dirtiest effluents. Low-rank coals may be expected to produce greater amounts of phenols than higher-rank coals, exacerbating the water treating problems. Solid wastes from gasification plants may also pose environmental concerns, primarily due to the potential for water leaching of hazardous components. Finally, trace element emissions may occur in air, water or solid waste emissions and although mass emission rates are small, the potential health hazards may be significant.

Low-rank coals have physical and chemical characteristics which significantly affect their performance as coal gasification feedstocks relative to bituminous coals. The non-caking property of lignite and subbituminous coals significantly simplifies the operation of fixed-bed gasifiers, and is also beneficial in the operation of fluidized-bed gasifiers. Although entrained flow reactors handle both types of coal successfully, the non-caking feature can also reduce complications associated with pneumatic solids handling in gasification systems where high temperatures are encountered during solids transfer operations.

High moisture contents present in low-rank coals act to dilute gasifier energy production, result in large volumes of waste liquor, and in extreme cases, may be too high for use in fixed-bed slagging gasifiers.

The higher reactivity in gasification reactions displayed by lignite and subbituminous coals may mean higher allowable reactor space velocities (and hence higher throughputs), higher carbon conversion and/or lower gasification temperatures.

The higher content of volatile matter (ratio of volatile matter to fixed carbon) in low-rank coals, and in particular the higher content of phenols, means a significantly different wastewater treating system will be required for plants based on fixed-bed gasifiers.

The ash properties of low-rank coals are significantly different than that found in other coals. This ash has been shown to be an important catalytic agent in promoting gasification reactions, and displays different temperature viscosity behavior than other coal ashes. Its chemical reactivity with certain refractory materials has caused problems in test experiments, and will continue to be an issue when selecting compatible refractory linings for gasifier walls and hearths.

Low-rank coals display lower sulfur contents than eastern coals, which may offer unique opportunities for their use in small installations without environmental control equipment. The resulting lower content of H_2S in raw gas streams reduces the requirements imposed on acid gas recovery systems, thereby lowering capital and operating costs. Fugitive sulfur emissions are also expected to be lower as a result of lower initial concentrations in the feed coal.

These unique low-rank coal properties as applied to the various types of gasification systems raise a number of key issues which impact the utilization of the resource:

Key Issues for Gasification

A. High Moisture Content

1. Effect on gasifier operation
2. Effect on wastewater treating system
3. Effect on ability to minimize plant water consumption
4. Effect on feeding system requirements
5. Effect on process yields

B. Fines Problems

1. Effect on gasifier selection
2. Effect on design of feeding system

C. High Coal Reactivity

1. Effect on gasifier throughput
2. Effect on optimal gasifier design

D. Non-caking Characteristics

1. Effect on choice of gasifier type

E. Volatile Matter Content

1. Effect on wastewater treating

F. Ash or Slag Charactertistics

1. Define important characteristics of low-rank coal ash and slag

2. Role in catalysis of gasification reactions
3. Effects on refractory linings in slagging gasifiers
4. Effects on downstream equipment and disposal requirements
5. Effects on processes using added catalysts
6. Slag viscosity

G. Low Sulfur Content

1. Effect on choice of acid gas removal system
2. Effect on gas cost
3. Advantages for use in small gasifier installations

H. Higher Oxygen Content

1. Effect on process yields
2. Effect on reaction rates and reactor throughput
3. Effect on product slate and marketability

The following is a discussion of these key issues and an assessment of the processes which show the best applicability to the gasification of low-rank coals.

A. High Moisture Content

Effect on Gasifier Operation

The high moisture content of some lignites may cause problems in slagging fixed-bed gasifiers when levels are sufficiently high to quench the reaction or cause operational difficulties. As the moisture level increases, the heat produced from the gasification/combustion reactions in the lower region of the gasifier will be insufficient to vaporize the moisture in the incoming feed. The resultant drop in bed temperature may stop the flow of molten slag, shutting down the unit. Tests at Grand Forks Energy Technology Center indicate that the maximum allowable moisture content for successful operation of their pilot scale fixed-bed gasifier is between 37 and 40 percent.[1] In an entrained flow gasifier, a similar problem may result if the reaction temperature drops below the required temperature for ash fusion. Non-slagging gasifiers do not face this problem, since slag removal occurs in the solid phase. In fact, the Lurgi dry ash gasifier requires large amounts of steam injection to prevent slagging of the ash. The fluid bed should reach its operational limit at some level between the slagging fixed-bed and dry ash gasifiers since ash is generally recovered in an agglomerated form.

Effect on Wastewater Treating System

Although high moisture levels in low-rank coals increase volumes of wastewater to be treated, the extent of this problem in comparison to higher rank coals is not completely clear. Depending upon the wastewater treatment process involved, the concentrated liquors from bituminous coal gasification may have to be diluted before treatment can be effective. However, treatment of a large volume of dilute contaminants also poses difficulties, particularly when removal must be made to very low levels. Therefore, the effect of high moisture contents in the feed coal depends upon the wastewater treatment system used, the extent to which water is recycled, and the ultimate level of treatment required. In circumstances where concentrated liquors from bituminous coals do not have to be diluted, the large volumes of wastewater generated from low-rank coal gasification will be more costly to treat, from both capital and operating viewpoints.

Minimization of Plant Water Consumption

Current process designs for coal gasification plants reflect their need for large quantities of fresh water and their production of large amounts of dirty water. As such, the siting of these plants in locations where water supplies are minimal or uncertain may be subject to a host of legal, social, political and economnic questions. In particular, the arid West and some possible plant sites in the East are subject to local and temporal water shortages. In all regions, the discharge of water which has been contaminated by coal conversion operations is constrained by environmental regulations.

Due to the fact that the transportation economics of coal are sensitive to the high moisture contents, the commercialization of coal gasification technologies using low-rank coals is likely to first occur as minemouth installations. If coal gasification is to be a viable enterprise in these arid regions, then the use of water for evaporative cooling must be minimized, process effluents must be recycled, and unusual water sources such as municipal waste or brackish groundwaters may have to be utilized.

The production of gaseous fuels from coal may be regarded in part as a hydrogenation process in which water acts as a hydrogen source. However, the difference in hydrogen to carbon ratio of products versus feed represents only one consumer of water. Other net consumers of water are blowdown from power generation, and evaporation due to wet cooling tower operation. In addition, water is required for mining, coal preparation, and ash disposal, and these must therefore also be considered along with the plant water requirements.

Although it is not feasible to completely eliminate the plant requirement for fresh water,[2] it is at the discretion of the plant designer to reduce water consumption to very low levels. The use of techniques such as air cooling, waste water purification and recycle, and collection and utilization of water from high moisture coals is however, not without added cost. The largest consumer of plant water is evaporative cooling, the quantities depending mainly upon the overall plant energy efficiency. In other words, heat which is not recovered in a product must by some means be rejected to the environment. An appreciable fraction of this waste heat will be lost directly to the atmosphere, and the remainder may be disposed by forced air convective cooling or evaporative cooling, depending on temperature level and process requirements, economics (air cooling is often indicated for streams above 140°F) and water availability. In water rich areas, typically 25 to 65 percent of the total unrecovered heat is absorbed in water evaporation, while in arid sectors or areas of high water cost only 10 to 30 percent of the unrecovered heat would be used in this manner.[3]

In all estimates of water consumption, reuse or recycling of water is assumed, and the required water treatment facility is integrated into the coal gasification facility. For the highest level of treatment required to minimize the consumption of water, the complete water treatment cost including amortized capital costs should be in the range of \$0.02 to \$0.10/10^6 Btu in the product fuel.[3] This cost is not likely to exceed five percent of the sale price of the product fuel from a coal gasification plant.[3] A less extreme approach using both wet and dry cooling which is designed to reduce (by one-half to two-thirds) but not minimize plant water consumption can be carried out at cost not likely to exceed one percent of the sale price of the product fuel.[3]

For a commercial-sized (250 million SCFD of pipeline gas) coal gasification plant designed to reduce water consumption, the actual requirement (not considering uses for mining and disposal) may be of the order of 4 million gallons per day, as compared to the commonly quoted range of 9 to 40 million gallons per day.[4] If recovery of moisture from the raw gasifier effluent is used as a method to further reduce fresh water consumption, the effect of using a high moisture coal feedstock can be substantial. For example, a commercial plant of this size will consume approximately 28,000 tons per day of lignite having a moisture content of 35 percent.[5] This represents over two million gallons per day of water, the recovery of a substantial portion of which could represent a significant offset in the plant's fresh water requirements.

From the standpoint of water conservation, it is fortunate that high moisture coals are present and available as coal gasification feedstocks in arid regions of the country. Although the utilization of their moisture to reduce fresh water consumption imposes added costs, these costs together with costs for increased use of air cooling and additional water purification and recycle may not be excessive or prohibitive. The issue is therefore one of engineering design and economic tradeoffs which will differ for each individual case according to technical, economic, environmental and political constraints.

Effect on Feeding System Requirements

The high moisture content of low-rank coals causes a reduced energy density in the slurry as compared to equal amounts of higher rank coals. An appreciable fraction of the moisture content of the low-rank coals is bound and hence this water is not useful in the slurry process but will be detrimental to the energy balance in the gasifier. Correction of this problem will require a higher ratio of coal to water in the slurry which may require special drying or other preparation techniques. This higher coal to water ratio will further aggravate already troublesome pumping problems and seal erosion in equipment.[6]

Effect of Moisture Content on Process Yields

The high moisture content of low-rank coals requires that greater quantities of coal be fed to the gasifiers to achieve the same process energy input which are obtained from higher-rank coals. The greater solids handling requirement and the greater volume of gases and liquids generated as a result of the increased coal feeding rate result in larger process equipment from coal handling through waste heat recovery and water treating (although acid gas removal requirements may be greater for certain applications due to larger amounts of CO_2 which result from lignite gasification). This effect is expected to be most pronounced for lignite and to a lesser extent for subbituminous coal.

In certain cases, the additional water vapor present in the raw gas may partially offset the reduced yield of fuel gases anticipated from a low-rank coal feedstock. For example, for synthesis gas uses and in the manufacture of SNG, shift conversion is carried out to produce additional hydrogen from carbon monoxide generated in the gasifier. Water vapor is required as a reactant and is introduced by injection of steam in cases where raw gas water vapor pressure is too low. This advantage is not applicable when the desired product is a medium- or low-Btu gas, since shift conversion is not required in these cases. In fact for low-Btu gas production the water vapor may lower the product gas heating value to unacceptable levels for certain end uses, particularly if entrained flow gasifiers are used. Due to their high operating temperatures, these gasifiers produce only CO and H_2 as fuel components of the product gas, and in air blown configurations have a maximum heating value of approximately 160 Btu/SCF (which is rarely achieved).

Significant amounts of water vapor or other diluents can reduce this heating value to low levels (although gas turbines can be designed to function reliably on gas of 100 Btu/SCF or greater heating value). Gasifiers requiring water slurry feeds or high rates of steam injection will compound the effect and may necessitate product gas drying by a series of low-temperature knock out drums or a glycol drying system.

A clearer definition of the incentive for using lignite and subbituminous coals as gasification feedstocks could be gained from an engineering/economic analysis of the performance of the two feedstocks as compared to bituminous coal in plants which have been modified to take advantage of the properties of each feedstock.

B. Fines Problems

Fines are produced during crushing and handling of all types of coal. In addition, slacking or decrepitation is a process whereby the surface moisture of the coal evaporates, causing cracking and crumbling of the surface. This occurs particularly in lignites and can be the principal source of fines if long-term exposure to weather is allowed.

Effect on Gasifier Selection

Fines are a particular problem to fixed-bed gasifiers.[7] This problem occurs because of the need to maintain the proper gas flow which in turn requires a well-controlled particle size distribution (plus 1/4 inch). Introduction of fines near the top of the gasifier results in their entrainment out of the bed. Fines can also be a problem in fluid-bed systems if required gas velocities are sufficiently high to cause carryover of the fines. This may result in erosion, fouling or plugging in downstream equipment and a thermal efficiency reduction if the fines contain unreacted carbon. Contrary to these difficulties, entrained flow gasifiers require a pulverized coal feed.

Effect on Feeding System Design

Fines can be a major problem in water slurry feed systems. In addition to making the slurry flow thixotropic, a high fines content increases the slurry viscosity resulting in increased pumping power requirements. Fines have also caused erosion problems in pump seals and have lead to rapid deterioration of moving parts. In lock hopper systems, abrasion due to fines has been a problem. This is compounded by high silica levels present in some Western coals. New dry pumping systems being developed will likely encounter similar abrasion problems.

C. High Coal Reactivity

Effect on Gasifier Throughput

The rate at which coal can be converted in the gasifier is a function of several parameters, including the reactivity of the coal. It has been shown that low-rank coal chars can display over 100 times greater reactivity towards gasification than chars from higher-rank coals.[8] Due to this property of low-rank coals, it is possible to process these coals at higher space velocities than common for higher-rank coals. This translates into lower capital costs in the gasification sector, due either to smaller allowable sizes for the gasifiers, or a fewer number of units required.

This property of low-rank coals has particular application to dry ash fixed-bed units. In this case, the low operating temperatures required to maintain the dry ash condition also significantly slows the gasification rate in comparison to slagging operation. Thus, highly reactive low-rank coal feedstocks show an advantage over bituminous coals in this area. Due to the increase in reactivity with temperature, this difference is much less noticeable at slagging temperatures.

The importance of maximizing coal throughput in the gasifier is limited by the relatively small fraction of the total plant capital cost accounted for by the gasifiers (approximately 10 percent[5]). In addition, considerations of mechanical reliability suggest that high mass flux devices may encounter more operational difficulties or a more frequent maintenance schedule. It may be necessary therefore to place added emphasis on operational reliability as a part of reactor design efforts on high throughput devices, particularly those associated with electric power generation.

Effect of High Reactivity on Optimal Gasifier Design

It may be possible to take advantage of the high reactivity of low-rank coals with a gasifier which has been specially designed for this purpose. For example, in the hydrogasification of coal to produce methane, the reaction equilibrium is shifted toward methane at low temperatures and high pressures. Using a highly reactive coal may allow good reactor conversions in a short residence time reactor. Reactor design considerations should then address reactor temperature profile, pressure gradients, and the degree of mixing of the reactants. Reactor geometry and injection method and orientation will be important factors in determining the completeness of mixing and conversion efficiency in the reactor.

Since the problems and advantages associated with the coal feedstock may vary considerably according to the rank of coal, it may be desirable to explore optimal reactor designs for individual coals in specific applications. In some cases it may be possible to achieve a substantial improvement in performance with only minor modifications of an existing design. It is likely that a laboratory study to define the relationship between reactor design and performance for low-rank coals would yield information applicable to other coals, and would represent a significant increase in the current knowledge in this area.

D. Non-Caking Characteristics

The non-caking characteristics of low-rank coals make them especially valuable as feedstocks to certain types of gasifiers. This property allows a simplifed coal preparation system and alleviates some of the severe operational problems found in fixed and fluidized-bed gasifiers with higher-rank coals. Caking coals generally require pretreatment or, for fixed bed gasifiers, mechanical stirrers to prevent coal agglomeration. These methods are not required for low-rank coal systems and hence costs are somewhat reduced.

In the case of fluid bed systems, coal particles are diluted by char, ash, or other inert materials (such as a sulfur sorbent) to prevent agglomeration. Although caking coals can be used to some extent due to the reduced coal concentration in the bed, low-rank coals may be used at any concentration without concern for caking problems.

Since particle interaction is minimized in the gas stream for entrained flow systems, caking tendencies are not a problem within the reactor. However, pneumatic conveying systems operating at high temperatures (such as in a char recycle loop) may be very sensitive to nozzle plugging with caking coals. Again, low-rank coals hold an advantage in this application.

E. Volatile Matter Content

In fixed-bed gasifiers, volatile matter is driven off in the upper regions of the reactor. Hence, the heating value of the raw gas produced from low-rank coals can be enhanced by the presence of light hydrocarbon gases. In addition, devolitilization of oils, phenols, napthas and other components in low-rank coals, will require different separation and treatment systems from those associated with higher-rank coal gasification because of differences in composition and quantity.

In entrained flow systems, where gasification temperatures are very high, partial oxidation of all tars and oils occur and hence the need for complex gas liquor treatment facilities is eliminated.

Effect on Wastewater Treating

There are two contaminants in most gasification wastewater that present major problems, phenols and ammonia. Phenols are one of the most toxic constituents of the wastewater and are generally removed by one of three methods; recovery, degradation, or a combination of these methods. Recovery processes include distillation, solvent extraction, crystallization and activated carbon adsorption. These methods concentrate the phenols for further use. Degradation techniques include bio-oxidation, ozonation, incineration and activated carbon adsorption with thermal regeneration. The technique selected for use is dependent primarily on the concentration level and cleanup requirement.

Ammonia is typically a by-product of the coal conversion process and its formation is favored by low temperature, high pressure reactor conditions. Ammonia levels in the water may reach 3600 mg/l to 14,000 mg/l or more, far above the allowable discharge concentration. The presence of ammonia can be damaging to gasification equipment metallurgy as well as to the environment. It can also inhibit phenol removal when certain solvent extraction processes are used.

Recovery of ammonia is economically advantageous in that anhydrous ammonia or ammonium sulfate are readily salable items. The recovery is complicated however by the presence of CO_2 and H_2S as well as the phenols. In some cases, ammonia may be stripped off with the acid gases. There are a number of proprietary processes available for ammonia recovery.

F. Ash or Slag Characteristics

Important Characteristics of Low-Rank Coal Ash and Slag

The unit operations in coal gasification plants place a number of requirements on the chemical and physical properties of the coal mineral matter. Within the gasifier, important ash characteristics are temperature viscosity behavior including ash softening and initial deformation points, corrosive and erosive action on refractory surfaces and other gasifier internals, and catalytic effects on gasification rates. Ash properties may also affect downstream equipment and disposal requirements.

Role of Inherent Catalysis in Gasification Reactions

Reaction rates in coal gasification are dependent upon several factors, including temperature, pressure, presence of catalysts, composition of reactants and physical properties of the coal (number of active sites, porosity, etc.). In particular, gasification rates are strongly dependent upon temperature. The use of coals having catalytically active mineral matter is most advantageous in lower temperature gasifiers. The exploitation of those coals having inherent catalytic mineral matter may be a way to enhance the technical and economic viability of proposed gasification processes.

Researchers at Ford Motor Co.[9] have found that the rates of steam gasification of low-reactivity coals and graphite can be significantly enhanced by the addition of ash from highly reactive lignites. However, it was not possible to distinguish clearly between a catalytic lowering of the activation energy for the reaction, and an increase in the number of the gasification sites. An understanding of the mechanism of catalytic activity would help to choose optimal reactions for a given coal in a specific process, and may also identify which coals available in the resource are most valuable with respect to use in leading gasification projects. An increase in demand for low-rank coals may also arise as part of blending schemes where processors are seeking enhanced reactivity of higher-rank feedstocks.

Effects on Refractory Linings in Slagging Gasifiers

Some coal gasifiers rely on an interior lining of refractory material to limit heat losses and preserve the structural integrity of the vessel during high temperature operation. If these gasifiers are operated in a slagging mode, the refractory liners may be subject to corrosive attack by the molten mineral matter in the coal which may reduce the life of the liner to unacceptably low time periods. Replacement of refractory materials requires shut down of the reactor (and a proportionate decrease in plant output if another unit is not brought on line) and can be overall a rather costly procedure.

Some low-rank coals have high fraction of calcium in their mineral matter, which has been shown to be corrosive to silicon carbide refractory liners. However, low-rank coals are also generally lower in iron, which is also a corrosive agent in certain cases (lower iron contents also reduce the two-phase slag problem encountered in fixed-bed gasifiers).

To insure a reasonable service life from refractory liners in slagging gasifiers, an effort must be made to establish chemical compatability between liner and slag. This can only be assured after a laboratory or pilot plant test program has identified the agents present in low-rank coal mineral matter which are corrosive to those refractory liners in question at slagging conditions over extended time periods.

Effects of Low-Rank Coal Mineral Matter on Downstream Equipment and Disposal Requirements

Ash particles escaping from the gasifier in the raw gas effluent may have a corrosive or erosive effect on downstream process equipment. However, it is expected that this problem will differ more according to the type of gasifier than to the rank of coal fed to the gasifier.

Disposal of mineral matter from coal gasifiers may be complicated by the possibility of contamination of groundwaters with harmful constituents by water leaching of the buried material. This issue is not a concern associated with slagging gasification and the problems arising from dry ash disposal are similar to those encountered by electric utilities in the disposal of fire box and fly ash (for a discussion of this issue, see Direct Combustion).

Effects on Processes Using Added Catalysts

Enhancement of coal gasification reaction rates may be accomplished by the addition of a catalyst material to the coal before or during gasification. The possibility exists for the mineral matter inherent in low-rank coal to react with these added catalysts in such a manner as to: a) increase their effectiveness in promoting gasification; b) inhibit their catalytic effect, or c) reduce the availability of the catalyst for recovery and recycle.

Depending on the cost of the added catalysts, the process design may specify a catalyst recovery step, or may allow the added catalyst to be disposed along with the ash. Low-rank coals have shown some tendency towards cation exchange with potassium in Exxon's Catalytic Gasification Process. Since the potassium catalyst is expensive, a fairly high recycle is necessary to maintain attractive process economics. The exchange of potassium for sodium and calcium in low-rank coal makes a proportion of the added catalyst inaccessible to the normal catalyst recovery system, and is ultimately disposed of along with the coal ash. In addition, catalyst recycle loops may require a purge stream to prevent buildup of exchanged ions, thus representing another pathway for catalyst loss.

The chemical interaction of low-rank coal mineral matter with added catalysts to produce enhanced or inhibited reaction rates is a potentially important and catalyst specific topic. Certainly these possible interactions and their effect on gasification rate should be well characterized before proceeding past a laboratory scale development program.

Slag Viscosity

Ash or slag removal from the gasification zone is a topic of prime importance due to its effect on gasifier operation. In a dry ash fixed-bed gasifier, the ash is removed from the gasifier through a slowly rotating grate. Fluidized-bed devices rely on a continuing removal of solid inert material (often a mixture of ash and sorbent) to maintain proper bed height. Slagging gasifiers require a certain minimum slag viscosity to assure proper slag flow through the tap hole and into the quench vessel.

In order to assure proper operation of the gasifier, the consistency of the ash or slag must fall within predetermined limits at the gasifier operating temperature.

Prediction of the behavior of the slag under operating conditions can therefore be an important tool in assuring proper operation. In cases where the viscosity of the slag is not low enough for proper slag flow, fluxing agents may have to be added. Low-rank coals in certain cases may have lower fluid temperatures than bituminous coals.

Depending upon the type of molten slag discharge used, slag viscosity will take on varying levels of importance. Where the slag is removed by intermittent tapping, as in BGC device, the flow properties are less important than in the Koppers-Totzek gasifier, for example, where slag flow is continuous. In gasifiers which operate at lower temperatures, slag consistency is also important. The agglomerating ash concept requires that ash particles adhere to each other at bed temperatures so that segregation can be achieved by virtue of agglomerate particle size and density. Dry ash removal systems would be fouled by a sticky or fluid ash.

For several coals which have been studied to date, initial attempts are being made to predict the viscosity-temperature relationship of low-rank coal slags by using a correlation developed at the Grand Forks Energy Technology Center.[10] Successful operation of a gasifier requires a prior knowledge of ash or slag behavior, particularly when faced with a varying coal feedstock. Researchers at the National Bureau of Standards have made progress in this area by investigating the effect on viscosity of varying concentrations of chemical compounds found in coal ash.[11] Although their work was conducted for application to MHD development, minor extrapolations of their data may yield information relevant to coal gasification studies.

In light of the difficulty in determining a coal slag viscosity by experiment, the correlative work done so far should be extended to other coals and other process conditions. If possible, it should also be broadened to include predictions for other physical properties as a function of mineral matter composition, temperature and the presence of other chemical agents (such as fluxes).

G. Low Sulfur Content

Effect on Choice of Acid Gas Removal System

Sulfur removal from the gasification product stream is required to meet environmental regulations but is also required to prevent catalyst poisoning and corrosion of process equipment. The low sulfur content of low-rank coals translates into a reduced requirement for removal to meet environmental standards and satisfy process needs. However, significant volumes of CO_2 are produced from the gasification of low-rank coal and in cases where non-selective acid gas removal is required the increase in size and operating requirements imposed by the additional CO_2 more than offsets the decreased volume of sulfur compounds.

Systems removing both the H_2S and the CO_2 are referred to as "non-selective" while those that preferentially separate these compounds are referred to as "selective." Most absorption systems rely on direct gas-liquid contact with the acid gases being absorbed into the solution. A number of aqueous agents are available as the absorbing fluid. An example of a non-selective absorber is a potassium carbonate system while some amine-based aqueous systems operate selectively.

One problem faced in non-selective treatment systems is disposal of H_2S from the absorber once the synthesis gas stream has been cleaned. One option is the Claus process where the H_2S is first thermally oxidized to SO_2 and the SO_2 is reacted again with H_2S to produce elemental sulfur. However, the Claus process requires a minimum level of H_2S in the feed gas, and may not be applicable in some low-rank coal based gasification plants. Other sorbents used for selectively removing sulfur include activated carbon, iron oxide, zinc oxide or molecular sieves, although sulfur levels are generally too high to justify use of these systems, even when low-sulfur coals are used as gasification feedstocks.

Effect on Gas Cost

As mentioned earlier, acid cleanup from the gasification of low-rank coals is economically attractive due to the low sulfur content but the increased CO_2 production is a negative aspect of their use. A useful comparison would be to evaluate the engineering/economic tradeoff between reduced sulfur levels and increased CO_2 production in low-rank coals.

Applications of Low-Sulfur Coal to Small Gasifier Installations

The low-sulfur content of Western coal may be advantageous for use in small gasifiers where controlling regulations are much less stringent. For units with power inputs less than 75 MW, the North Dakota sulfur emission standard is 3.0 $lbs/10^6$ Btu. For lignite with a heating value of 6500 Btu/lb, all coal with less than 2 percent sulfur will meet emission standards. This will include all of North Dakota's lignite. With no sulfur removal requirement, the economics of these smaller units become much more attractive, and could become a major base for expansion of low-rank coal utilization.

H. High Oxygen Content

Effect on Process Yields

The high oxygen content of low-rank coals affects the process yields in much the same way as the high moisture content in that it has a dilution effect. The same external oxygen input is required for the gasification of low-rank coal as for the higher rank but much of the low-rank coal is already partially oxidized and hence the energy content of the product gas is lower. Since the energy content of the raw gas is decreased, the throughput must be increased to produce an equivalent energy content in comparison to the high-rank coal.

Effect of High Oxygen Content on Reaction Rates and Reactor Throughput

The reactivity of the low-rank coals is higher than bituminous coals. In the gasification process, as discussed earlier, a significant amount of CO_2 is produced. This freed oxygen opens reaction sites in the coal which increases the reactivity and hence reduces the required operating temperature.

Effect of High Oxygen Content on Product Slate

In fixed-bed gasifiers, many of the high volatile compounds may be captured as condensates. Some of these compounds will be partially oxygenated and may provide useful chemical feedstocks.

Assessment of Processes with Best Applicability to Low-Rank Coals

The chemical properties and physical characteristics of low-rank coals differ significantly from those associated with higher-rank coals, and therefore present advantages and disadvantages when considered as feedstocks for gasification processes. Many of the properties of low-rank coals which are most significant to coal gasification apply to a greater or lesser extent to all gasifier types. Some properties however, apply only to specific gasifier configurations or end uses for the product gas.

The properties of low-rank coals which impact the choice of gasifier or process application are the following:

- High moisture content
- Low sulfur content
- High oxygen content
- Volatile matter content
- High coal reactivity
- Non-caking properties
- Slag viscosity and chemical behavior

Due to the extremely wide variety which can be observed in coals of any rank, it is not always possible to make general statements regarding the behavior of low-rank coals as a group. In addition, a property which may be an advantage in one gasifier or in producing a particular gas product may be of no advantage whatsoever in another circumstance, and may indeed be detrimental. Therefore, these characteristics of the resource should be evaluated with respect to their application to specific cases.

Depending upon the conditions of pressure and type of oxidant chosen, coal gasification reactors can be operated to produce a gas suitable for different uses. Gasification at high pressures produces a raw gas which may be used for combined cycle power generation, chemical synthesis, or transmitted via pipeline for use as an industrial fuel gas. Conversely, low pressure gasification could be considered for combined cycle power generation only if the product gas were compressed prior to combustion, a costly process alternative. Low pressure gas may also be restricted in its use as a feedstock for chemical synthesis, and certainly cannot be transmitted long distances for use as a fuel gas.

The use of oxygen as the oxidizing agent produces a medium-Btu gas which may be used as fuel, or a chemical synthesis feedstock. Air blown gasification results in a low-Btu raw gas. Either gas may be used to produce methane, although most systems require a medium-Btu feed for shift conversion and methanation.

The gasification process applications which should be evaluated relative to use with low-rank coals are: low-Btu gas (for use as an industrial fuel and as a utility power generation fuel), medium-Btu gas (for use as a chemical synthesis feedstock and fuel gas), and high-Btu gas (derived from shift conversion and methanation or hydrogasification).

Low-Btu Gasification

Fuel gas generation is one objective of air blown coal gasification. It is primarily oriented at firing utility and industrial boilers on-site, and therefore can be generated at low pressures. Because of the structure of current environmental regulations, low-rank coals may find particular application to firing industrial boilers (new or retrofit) or in other industrial applications which qualify for a higher sulfur emission rate as a result of being smaller than a certain maximum size. For example, in North Dakota, energy consumers smaller than 75 MM Btu/hr have an allowable emission rate of 3 lb SO_2 per MM Btu fuel charged, compared to a much more stringent standard for larger facilities. Although large users of energy are required to achieve a lower emission rate and will therefore require environmental control equipment regardless of coal fired, low-rank coals are advantageous due to the fact that less sulfur will ultimately be removed.

Because the facilities which could use low-rank coals without sulfur recovery equipment are small energy consumers, it is likely that capital cost will be a major factor in the decision to gasify coal instead of burning oil or gas (if indeed a choice is available). Unlike the case of a large coal gasification complex where the capital cost of the gasification equipment is of the order of 10 percent of the plant cost, the gasifier cost will be a significant capital item for a small industrial user. In addition, many potential users of this technology may have limited plot area for installation of a gasification system and associated coal handling equipment. These considerations imply that gasifier throughput would be an important criterion in selecting a system, because high mass flux (entrained) gasifiers are likely to have a lower capital cost and require less plot area per unit of rated coal capacity than either fixed or fluidized bed gasifiers.

The high operating temperatures found in entrained flow gasifiers may create boiler tube fouling problems when low-rank coals are used as feedstocks. However, the high inherent reactivity of low-rank coals would allow lower temperature operation while still maintaining a high throughput capability. This property could be further enhanced by catalyst addition. Maintenance of proper slag flow characteristics could be assured by the addition of fluxing agents.

Combined cycle gas turbine systems are most cost effective with a pressurized (15-20 atm) gasifier. The high moisture content of low-rank coals could prove to be less of a problem in this system if hot gas cleanup is employed, although it is emphasized that this is still a developmental technology. In this case, water vapor which would normally be condensed and require treatment remains in the vapor phase at high pressure and is available for power generation. Hot gas cleanup also reduces the problems associated with recovery of unreacted organics from fixed-bed gasifiers, since these components are burned to carbon dioxide and water vapor in the power generation step. The difficulties of implementing hot gas cleanup systems could be reduced by using a fluidized-bed gasifier operating with sulfur sorbent. Removal of sulfur compounds is required prior to the gas turbines due to their effect on turbine blade life. Large amounts of water and undesirable organics are condensed when fixed-bed gasifiers are used in conjunction with a cold gas cleanup system.

Medium-Btu Gasification

Medium-Btu gases are derived from oxygen blown gasification and range in heating value from approximately 300 to 600 Btu/SCF depending upon the hydrocarbon content of the raw gas. Principal process applications include fuel gas production (including combined-cycle power generation and boiler firing) and use as a chemical feedstock for synthesis of methanol, formaldehyde, acetic acid, gasoline and other end uses. All applications of medium-Btu gas production benefit from the low sulfur content of low-rank coals.

Fuel gas production can occur at atmospheric or elevated pressures, the latter case applying to long-distance pipeline transmission, chemical synthesis or gas turbine firing. Medium-Btu fuel gas generation at atmospheric pressure is restricted to on-site use as boiler fuel. This system is less likely to be implemented by small users due to the requirement for oxygen, although certain retrofit applications may require a medium-Btu gas due to derating or equipment modification problems anticipated with low-Btu gas. If a high heating value is required, low-rank coals are well suited to use in a fixed-bed gasifier due to their higher ratio of volatile matter to fixed carbon. Devolatilization of the coal in the upper region of the gasifier results in the release of high heating value volatile matter.

As in air blown combined cycle systems, oxygen blown combined cycle power plants using hot gas cleanup might face fewer problems due to the handling of large volumes of wastewater and the loss in power production due to moisture condensation.

Gasification of low-rank coals in fixed-bed gasifiers may be attractive if a high heating value fuel gas is desired, due to the high production of hydrocarbons in the vapor phase. Due to the low initial levels, sulfur recovery may be accomplished to adequate levels by using a fluidized bed gasifier employing a sulfur sorbent without additional acid gas scrubbing.

The generation of medium-Btu gas for use as a chemical feedstock generally requires a high degree of sulfur recovery. The high moisture content of low-rank coals is not a drawback in this case, since water vapor is required for shift conversion downstream. Since stoichiometric quantities of carbon monoxide and hydrogen are required, the presence of organic compounds in the raw gas from fixed-bed (and to a lesser extent from fluidized bed) gasifiers is not advantageous and may be harmful depending upon the type of catalyst used.

High-Btu Gas

Generation of a gas product having a heating value of 900 to 1000 Btu/SCF has been proposed to offset the current use of natural gas in the residential, commercial and industrial sectors. The primary component, methane, can be produced in several ways including shift conversion and methanation of medium-Btu gas, catalytic coal gasification and direct hydrogasification of coal.

SNG production with a medium-Btu feed gas by shift conversion and methanation takes advantage of the high inherent moisture content of low-rank coals to achieve the shift conversion. Other considerations relating to low-rank coals as feedstocks are also identical to those for pressurized medium-Btu gasification to produce synthesis gas.

The reaction equilibrium to produce methane from a mixture of CO, H_2, H_2O and carbon is shifted toward methane at lower temperatures. Since reaction rates are a strong function of temperature, coals which display a natural catalytic activity toward gasification (such as low-rank coals) are desirable feedstocks from this viewpoint. The use of added catalysts to further accelerate reaction rates at low temperatures is complicated by possible interaction of the catalyst with the inherent coal mineral matter. In the case of the Exxon Catalytic Gasification process, this results in an ion exchange between potassium from the added catalyst and several cations present in the raw coal, making catalyst recovery difficult and therefore adversely affecting process economics.

Hydrogasification is a method of producing SNG which has been explored in entrained flow gasifiers in the United States (IGT and Rockwell) and extensively by Rheinbraun in Germany using a fluidized bed to gasify lignite. The novel feature of the Rheinbraun process is the integration of a high temperature oxygen blown Winkler gasifier which produces high pressure hydrogen (and carbon monoxide) in a separate fluidized bed. The high oxygen contents of low-rank coals detract from the system economics by consuming hydrogen. However, high moisture contents tend to shift the equilibrium in the opposite direction, and the high reactivity of low-rank coals towards hydrogasification make them attractive feedstocks overall.

References - Section 3.5.3.1

1. Ellman, R.C., L.E. Paulson, and D.R. Hajicek. Slagging Fixed-Bed Gasification Project Status at the Grand Forks Energy Technology Center, Symposium on Technology and Use of Lignite, Grand Forks, N.D., May 1979, DOE GFETC/IC-79/1, 1979, pp. 236-277.

2. Personal Communication, Dr. Ronald Probstein, Massachusetts Institute of Technology, February 26, 1980.

3. Probstein, R.F., and H. Gold. Water in Synthetic Fuel Production - The Technology and Alternatives, Water Purification Associates, The MIT Press, Cambridge, MA, 1978.

4. Tetra Tech, Inc. Energy From Coal - A State-of-the-Art Review, ERDA Report No. 76-7, U.S. Government Printing Office, Washington, D.C., 1976.

5. Fodor, R.J. Overview of the Great Plains Gasification Associates Coal Gasification Project, presented at the 1979 Lignite Symposium at the University of North Dakota, Grand Forks, North Dakota, May 30-31, 1979.

6. Fluor Engineers and Constructors, Inc. Coal Liquefaction Design Practices Manual, U.S. Department of Commerce, PB-257 541, July 1976.

7. Schora, F.C. The Challenge of Solids Handling in Coal Gasification, Illinois Institute of Gas Technology.

8. Walker, P.L. Gasification Rates Related to Coal Properties, presented at the conference on Scientific Problems of Coal Utilization at West Virginia University, May 23-25, 1977.

9. Otto, K., L. Bartosiewicz, and M. Shelef. "Catalysis of Carbon-Steam Gasification by Ash Components from Two Lignites," Fuel, 1979, Vol. 58, February.

10. Schobert, H.H. Petrochemistry of Coal Ash Slags 2. Correlation of Viscosity with Composition and Petrographic Class, ACS Division of Fuel Chemistry, Symposium on the Properties of Coal Ash, May 24-June 2, 1977, Montreal,

11. Capps, W. Coal Slag Properties Related to MHD, Presented at the Conference on High Temperature Sciences at ANL, April 4-6, 1977.

3.5.3.2 Technology Description

Coal gasification is a process in which coal is converted to a chemical synthesis gas or gaseous fuel product by reaction with air, oxygen, steam, or mixtures of these gases. In addition, the use of hydrogen as the principal gaseous reactant is termed hydrogasification. Conversion processes can be classified according to the type of oxidant used (air or oxygen), the pressure at which gasification takes place (atmospheric or pressurized), the type of gasifier employed (fixed bed, fluidized bed, or entrained flow), and the physical state of the ash in the gasifier (slagging, non-slagging, or agglomerating).

The gases produced by coal gasification vary in chemical composition and potential use according to the method used to produce them. Generally, the gas products are categorized by their heating value, which is directly determined by the chemical constituents in the gas, as shown in Table 3.5.3.1.

Table 3.5.3.1

Typical Gas Products Resulting From Coal Gasification (Dry Basis)

	Heating Value Btu/SCF,HHV*	Major Gas Components	Source
High-Btu Gas	900-1000	CH_4 (>90%)	Oxygen-blown gasification followed by methanation or direct hydrogasification of carbonaceous material
Medium-Btu Gas	300-550	H_2,CO,CH_4	Oxygen-blown gasification with acid gas removal. Methane formed under high pressure equilibration or by catalytic methanation
Low-Btu Gas	≤200	N_2(~50%),H_2, CO, CO_2, CH_4, (3%)	Air blown gasification

* High heating value in Btu per standard cubic foot, as measured at 60°F and one atmosphere pressure.

Low- and medium-Btu gasification plants have been used extensively in the past. One form of low-Btu gas known as "town gas" was in prevalent use as a residential fuel gas throughout Europe, Britain, and the United States in the first half of the century. Conversion to natural gas took place in the United States in the 1930's and 1940's, while the last units to operate were shut down in parts of Britain and Europe in the 1960's. No commercial medium-Btu and virtually no low-Btu gasification plants currently (1980) exist in this country.

High-Btu (pipeline quality) gas has never been produced from coal in large quantities. However, medium-Btu synthesis gas is still produced at many locations in Europe, Britain, and South Africa. The two commercial scale plants in South Africa are based on technology developed in Germany before WW II for the production of liquid fuels from coal via synthesis gas. Numerous developmental programs are currently underway to address applications of low-, medium-, and high-Btu gasification, and are sponsored by both government and private organizations.

The need to supplement natural gas supplies and to displace a portion of the domestic and foreign oil consumed in the United States provides most of the incentive for coal gasification projects currently under consideration. One such example calls for an industrial fuel gas (either low- or medium-Btu) which could be used in place of oil or gas to fire process heaters or steam generators. High-Btu gasification plants will provide a source of pipeline quality gas for residential, commercial, and industrial use. Medium-Btu synthesis gas can be used as a basic chemical feedstock to plants producing a variety of industrial chemicals such as ammonia, methanol, and formaldehyde.

Proposals have been made for new combined cycle electric power plants based on coal gasifiers.[1,a] Here, hot low- or medium-Btu gas from the coal gasifiers is cooled to produce high pressure steam. After sulfur removal and final particulate cleaning, the cold fuel gas is fed to gas turbines where it is burned with excess air and expanded for power generation. An alternate design calls for hot gas cleaning of the raw gasifier effluent, which is then passed directly to gas turbines at high temperature. The hot turbine exhaust is cooled against steam generation and vented as flue gas. Steam raised here and elsewhere in the plant is expanded in steam turbines for additional power generation. Based on currently available gas turbines (2000°F inlet temperature), combined cycle power plants are thermodynamically more efficient than conventional coal-fired power plants based on steam as the only working fluid.[2] Incorporation of hot gas cleanup systems is expected to result in a further efficiency increase, but considerable development is still required for these devices.

A typical coal gasification process is depicted in the block diagram of Figure 3.5.3.1. Several of the product options discussed above are shown on the figure with dashed lines. One observation that can be made for any of the alternative process configurations is that the gasifier itself represents a small fraction of the total equipment (or cost) in the plant. This implies that further optimization of the gasification process itself may not have much of an impact on net gas cost unless the change is one which significantly affects upstream or downstream processing requirements.

In the generalized example, coal is transported from the storage area to the coal preparation section, where grinding, pulverizing, cleaning, size classification, and gasification pretreatment (for caking coals only) may be performed. In the gasification step, the prepared coal is charged to a reaction vessel, which may be of fixed bed, fluidized bed or entrained flow design. This step produces either a low- or medium-Btu gas, depending on the oxidant (air or oxygen) selected for gasification. The hot gas leaving the reactor may carry with it particles of unreacted coal or char, as well as solid, molten, and vaporized ash species.

The next process operation may be one of three alternatives, depending on the composition and solids and liquids content of the raw gas, and its end use. Some designs call for direct water quench of the gasifier effluent, thereby cooling and humidifying the gas. Other systems, such as for combined cycle power generation, cool the hot gas by

[a]References for sections 3.5.3.2 - 3.5.3.6 are listed following section 3.5.3.6.

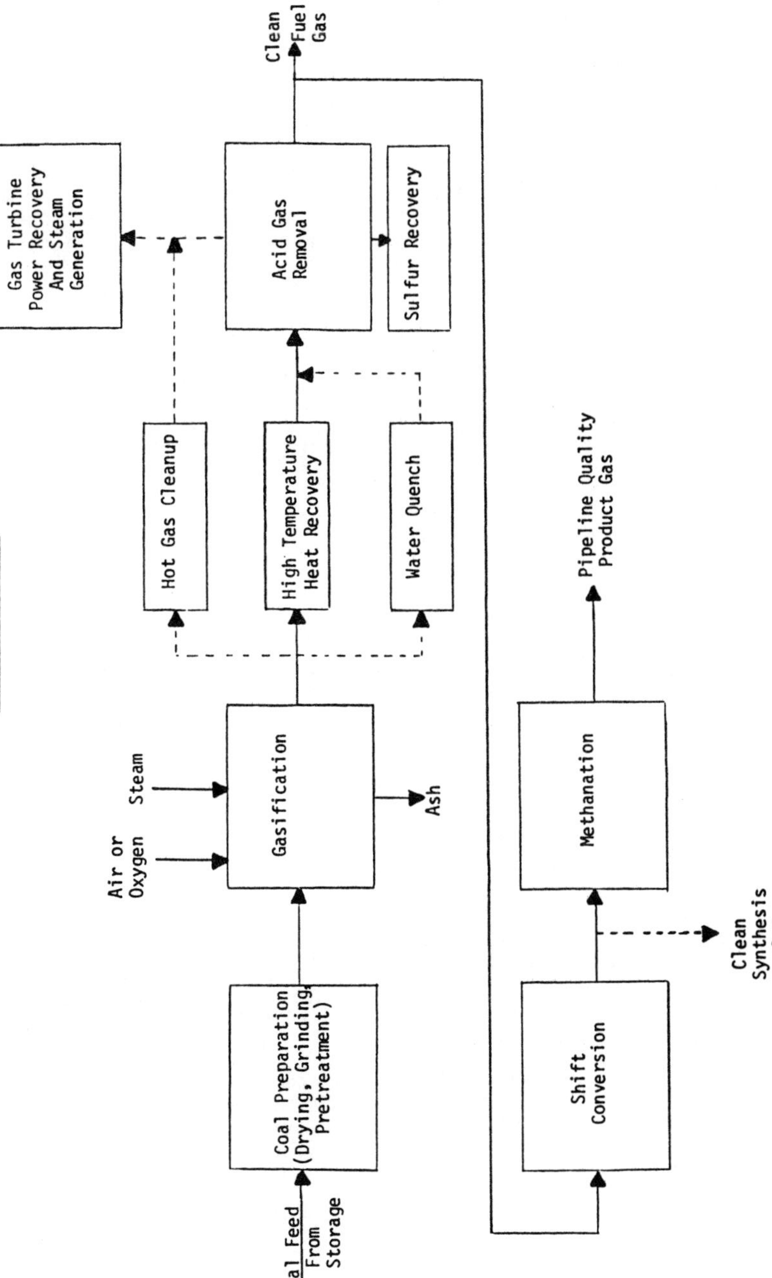

Figure 3.5.3.1
General Block Diagram
Coal Gasification

generating high pressure steam (for power generation) or by oxidant preheating for gasifier use (especially in air blown systems). Because of difficulties with high temperature heat exchange equipment handling dirty gas streams, designs using first a quench followed by waste heat recovery have been proposed. A third possibility under development involves particulate and/or sulfur removal from the gas at high temperatures. Demonstration of this concept may allow higher thermodynamic efficiencies for certain processes.

Acid gas removal usually takes place after the raw gas has been cooled. In this step, sulfur compounds are removed (in processes not having hot gas clean-up) along with partial and sometimes complete removal of carbon dioxide. This step produces a fuel or synthesis gas which is environmentally acceptable, and will not poison downstream catalysts sensitive to sulfur compounds. Fouling and corrosion of metal surfaces will also be reduced as a result of acid gas removal.

The clean gas produced in acid gas removal may be used in a variety of ways. It may be expanded in a gas turbine for power generation (although the combustion cans of turbines intended for methane-rich fuels will require modification for some coal-derived fuel gases), the exhaust gas of which is normally cooled against steam production for additional power generation. Alternatively, the clean fuel gas may be used as an industrial fuel gas.

Production of pipeline quality gas or chemical feedstock requires additional processing. Shift conversion is a catalytic process which increases the hydrogen content of the CO-containing process stream by the water gas shift reaction:

$$CO + H_2O \longrightarrow H_2 + CO_2.$$

This technique is used to adjust the H_2/CO ratio to a level appropriate to the downstream use of the gas. For example, methanol can be produced by reaction over a copper based catalyst at typically 800 psi:

$$2H_2 + CO \rightleftharpoons CH_3OH$$

which consumes hydrogen and carbon monoxide in a ratio of 2:1. For methane production, hydrogen is consumed at a rate of 3 to 1 over a nickel catalyst by the predominant reaction

$$3H_2 + CO \rightleftharpoons CH_4 + H_2O.$$

The actual gas composition used may vary somewhat from these exact molar ratios to optimize reaction rates or account for other production reactions. In the case of high-Btu gas production, the high hydrogen content may be obtained by several stages of shift conversion separated by CO_2 removal units.

After the desired degree of shift conversion, the hydrogen-rich gas is used either as a chemical feedstock in the production of methanol, ammonia, formaldehyde or other products, or in the methanation process. Methanation involves the catalytic formation of methane and water vapor from the reaction of hydrogen with CO and CO_2 (if any):

$$CO + 3H_2 \longrightarrow CH_4 + H_2O \text{ and}$$

$$CO_2 + 4H_2 \longrightarrow CH_4 + 2H_2O.$$

Dehydration of the product gas produces a clean fuel with a heating value similar to that of natural gas.

When synthetic natural gas (SNG) or medium-Btu gas for combined cycle power generation is the desired product, it may be argued that gasification should be carried out at elevated pressure. This is because the medium-Btu gas must be at high pressures for gas turbine use or SNG pipeline transmission, and it is more economical to obtain this pressure by oxidant compression rather than product gas compression due to the volumes of gas involved.

When choosing between air and oxygen blown systems, use of the gas is again the deciding factor. Production of chemical feedstocks or methane is usually done from medium-Btu gas (from an oxygen blown gasifier) since use of low-Btu gas would require removal of nitrogen from the product gas, and may retard reaction rates in the synthesis step. Medium-Btu gas is also favored for atmospheric pressure retrofits to steam-boilers due to the boiler derating which is experienced when firing low-Btu gas. However, low-Btu gas may still be used in this application if the size of the boiler facility is not large enough to justify an oxygen plant, and can accept a significant (20-40%) derating of the boiler. Low-Btu gas is well suited for use as an on-site source of process heat or as a fuel for combined cycle power generation. Due to compression requirements for gas transmission, low-Btu gas is not suited for pipeline transport over long distances.

3.5.3.3 Gasifier Types

Although there are number of ways to classify the different gasifiers which are currently available or under development, they are generally categorized into three different areas according to the method of contacting reactants. Fixed bed (also known as gravitating or moving bed), fluid bed, and entrained flow types are being considered for all applications discussed in section 3.5.3.2.

Fixed bed contacting is the oldest of the three techniques and has applications in many other process industries in addition to coal gasification. Fixed bed coal gasification results in efficient utilization of the available reaction volume due to the countercurrent flow of fuel and gaseous reactants resulting in essentially complete carbon conversion. However, problems arise if swelling or caking coals are used in these gasifiers. This has been one of the major motivations for developing the fluidized bed and entrained flow types.

Fluidized bed gasification offers enhanced flexibility and control over fixed bed designs. Swelling coals do not present as much of a problem due to the severe bed agitation; however, highly caking coals often cannot be used without pretreatment or a higher char recycle. As with fixed bed operation, fines result in particulate carryover from fluid beds using crushed coal as a feedstock, but pulverized coal fluid bed gasifiers are less susceptible to this problem. Excessive carryover of fines will result in decreased conversion efficiencies unless carbon can be recovered from the entrained ash and recycled to the bed. Ash removal in fluid beds is critical, and high ash coals find less application in this type of gasifier.

Entrained flow gasifiers can use all ranks of the coal as feedstock, but gasification efficiencies can be sensitive to coal reactivity due to short residence times in the gasifier. This is generally not a problem because, at the high operating temperatures of these devices, reaction rates are extremely high. Atmospheric pressure entrained flow gasifiers use standard pulverized coal as fuel, and pressurized versions use either a coal slurry (in water or oil), or a pneumatic injection system for coal feeding to the gasifier. Entrained flow gasifiers are generally of the slagging type.

The three gasifier types are shown schematically in Figure 3.5.3.2, along with typical temperature profiles for each gasifier. Table 3.5.3.2 compares the fixed bed, fluid bed, and entrained flow gasifiers on a variety of key operating parameters.

Figure 3.5.3.2

Gasifier Types and Temperature Profiles

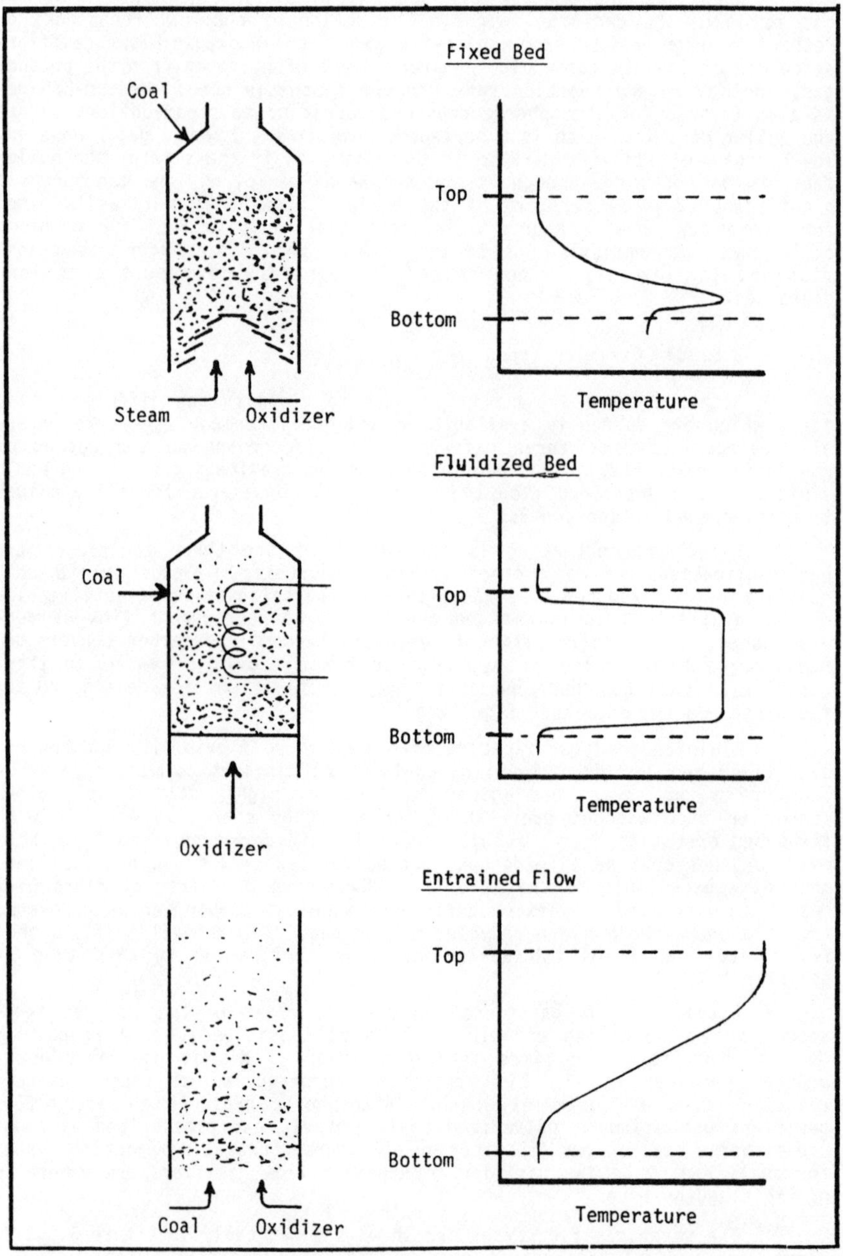

Table 3.5.3.2[e]

Performance Comparison of Leading Commercial and Developmental Coal Gasifiers

Performance Parameter	FIXED BED GASIFIERS				FLUIDIZED BED GASIFIERS			ENTRAINED FLOW GASIFIERS				
	Lurgi	Dry Ash[3]	BGC Slagger[4]	GFETC Slagger[15]	CO_2 Acceptor[5]	U-GAS[3]		Foster-Wheeler[1]		Texaco[7]		Koppers-Totzek[1]
	Air Blown	Oxygen Blown	Oxygen Blown	Oxygen Blown	Air Blown	Air Blown	Oxygen Blown	Air Blown	Oxygen Blown	Air Blown	Oxygen Blown (slurry feed)	Oxygen Blown
Coal Type	Illinois	Illinois	Illinois	Lignite	Lignite	Illinois	Illinois	Illinois	Illinois			
Gasifier Outlet Temperature, °F	955	1080	820	340	1500	1660	1550	1700	1700	2300-2600	2300-2600	2700
Gasifier Outlet Pressure, psig	25	25	25	200	150	25	25	360	360	600	600	25
Air or Oxygen Consumption, lb/lb MAF[b] Coal	2.56	0.54	0.53	0.45	2.6	3.44	0.71	2.86	0.61	1.08	0.86	0.9
Oxidant Feed Temperature, °F	342	321	214	950[f]	390	900	480	800	335	1000	300	480
Steam Consumption, lb H_2O/lb MAF[b] Coal	1.65	2.58	0.34	0.25	1.1	0.65	0.59	0.15	0.62	0.009	0	0
Steam Injection Temperature, °F	590	590	620	950[f]	1200	800	800	453	453	N/A	N/A	N/A
Gasifier Cold Gas Efficiency[c]	90.9	91.2	96.4	88	77[d]	83.3	89.4	77	N/A	78	94	75
Approximate Coal Residence Time	1 hr.	1 hr.	15 min.	45 min.		1 hour		5 sec		100 milliseconds		1 sec
Approximate Gas Residence Time		5 sec		5-10 sec.		2-5 sec		5 sec		100 milliseconds		1 sec

a Lignite coal feed
b Moisture and ash free
c Cold gas efficiency = higher heating value of product gas at 60°F (inc. liquid hydrocarbons, if any) x 100 / higher heating value of coal fed to gasifier
d Cooled, cleaned product gas
e References for sections 3.5.3.2 - 3.5.3.6 are listed following section 3.5.3.6
f This is the steam-oxygen mixture temperature entering the gasifier

Among the air blown gasifiers, the U-Gas fluid-bed type has the highest consumption of air, followed by the Foster-Wheeler design. The extremely low air consumption of the Texaco design is possible since steam is not being used as a moderator.

Steam rate is highest in the Lurgi dry ash gasifiers, and is lower in the slagging gasifier regardless of bed type.

Gasifier outlet temperature is a parameter which in some cases can be used to determine the content of phenols, tars, oils and unreacted hydrocarbons in the raw gas. In cases of countercurrent flow between coal and product gas, sensible heat is lost by the gas in heating and devolatilizing the incoming coal. These unreacted components are, therefore, never exposed to high temperature regions in the reactor where they would be cracked to simpler molecules. Thus, gasifiers having gas outlet temperatures of 1500°F and above generally have significantly reduced contents of condensable materials in the raw gas.

Among fixed bed gasifiers, cold gas conversion efficiencies are on the average higher than those achieved by the other gasifier types, for several reasons. First, coal residence time in fixed bed devices is longer than for any other type, resulting in essentially complete carbon conversions in the reactor. In addition, since a large amount of raw gas sensible heat is used for drying and devolatilizing the incoming coal, the raw gas temperature leaving the gasifier is generally lower than for other types of gasifiers. Thus, less heat is "lost" when calculating the cold gas efficiency based on a product gas temperature of 60°F. Finally, the large amounts of steam fed to the fixed bed unit do not appear in the efficiency calculation, and would represent a significant decrease in the quoted gasifier efficiency if included.

Although there is a considerable variation in coal residence times between the various gasifiers, the gas residence times are all somewhat similar, with the exception of the Texaco entrained flow gasifier.

When using the table for comparative purposes, it is extremely important to keep in mind that data for the individual cases often is based on different coals as feedstocks, different operating pressures, and different oxidant and steam injection temperatures. These data can significantly affect the apparent performance of the gasifier.

Fixed Bed Gasification

Fixed bed gasification refers to a constant depth fuel bed that is supported by a grate or other means. The coal moves slowly from the top of the bed through a gasification zone and residue is removed at the lower extremity of the bed. The bed is maintained at constant depth by the addition of coal to the top of the bed as residue is removed at the bottom and coal is consumed in the gasification process. The general layout of the fixed bed gasifier is shown in Figure 3.5.3.3.

The figure shows a cross section of the fuel bed resting on a grate. The gasifying medium (air or oxygen and steam) flows countercurrent to the flow of coal, and feed coal is added at the top of the unit. In a dry bottom unit, ash lies directly on the grate which serves to insulate the grate from high temperatures in the gasification zone. In the gasification zone above the grate, the hydrogen and carbon monoxide components of the raw gas are formed by partial oxidation, steam decomposition and water gas shift reactions. The high temperature gases created in this zone rise to partially devolatilize the coal, forming the higher hydrocarbons, tar and methane components of the gas. Above this zone, the slightly cooled gases preheat and dry the incoming coal.

Figure 3.5.3.3

Dry Bottom-Fixed Bed Gasifier

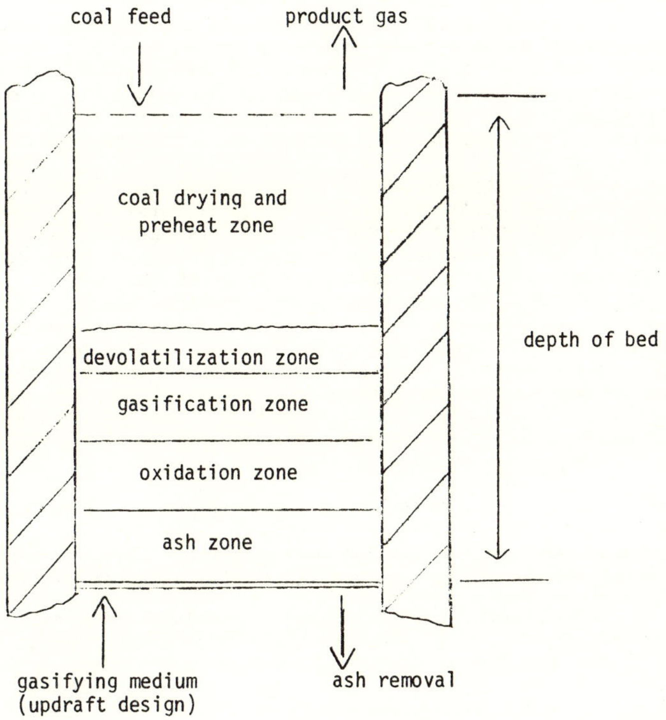

Due to the mechanical motion of the grate, the temperature of gasification, the velocity of the gas passing through the bed, and the limited time for reaction, the most important determinants of the design of a fixed bed gasifier are:

- amount of steam required for gasification (lb/lb coal)
- amount of oxidizer required for gasification (lb/lb coal)
- coal size consist
- caking behavior
- ash fusion temperature
- reactivity

Steam has a dual function in coal gasification. It promotes the formation of hydrogen through the water gas shift reaction ($H_2O + CO \rightleftharpoons CO_2 + H_2$) and by the oxidation of carbon ($H_2O + C \longrightarrow H_2 + CO$) In its other role, steam quenches the gasification reaction due to its high heat capacity and the endothermic nature of the above reactions. Its degree of participation in either of these roles depends upon the temperature of the steam relative to the gasification zone temperature, and the presence or absence of catalysts.[27] In the case of dry bottom gasifiers, the maximum operating temperature is set by the fusion point of the coal ash. Thus, the amount of steam required to maintain the gasification temperature below the ash fusion point will be an important coal characteristic for any given gasifier.

Oxidant consumption determines the heating value of the cold raw product gas since the amount of oxygen consumed in the reaction is inversely proportional to the amount of combustibles in the raw gas. The oxygen content of the coal influences the content of oxidized species in the product gas, and hence high oxygen contents in coal are undesirable because of the lower quality gas produced from such coals.

Because of the importance of proper gas flow rate, the coal feed for fixed bed gasifiers must have well-controlled (plus 1/4 inch) particle size distribution, containing a minimum of fines. This will insure maximum gas-solids contacting, minimum pressure drop and minimum fines carry-over. The caking tendency of the coal is also important, since caking coals must undergo pretreatment before use if adequate bed agitation is not provided. Surface oxidation of the coal has been shown to be effective against caking, as has blending with noncaking coals. If a coal has only moderate caking tendencies, mechanical stirring of the bed may obviate pretreatment.

Slagging (Wet Bottom) Fixed Bed

As an alternative to dry ash operation, fixed bed gasifiers may be designed for slagging gasification. Schematically, the slagging fixed bed gasifier can be represented by Figure 3.5.3.3, with the exception of molten, instead of dry, ash removal at the bottom of the bed.

Maintenance of the slagging condition requires combustion zone temperatures of approximately 2500-3000°F for most coals. Some low-rank coals are expected to require combustion temperatures at the high end of this scale, due to their higher slag viscosity, although slag properties depend on ash chemical composition which exhibits a wide range of variability. Similar to the non-slagging version, endothermic gasification, devolatilization and drying processes take place above the combustion zone reducing the outlet gas temperature to the order of 400-1000°F, depending on the coal moisture content.

Since steam is used in the dry bottom gasifier largely as a reaction moderator to keep bed temperatures below the ash fusion temperature, steam consumption in the slagging fixed bed gasifier is greatly reduced. Because the volume of gas passing through the reactor is much lower per ton of coal gasified, coal throughput is greatly increased (about fourfold) in a slagging gasifier as compared to a dry bottom unit of the same diameter. In addition to increasing the efficiency of gasification and lowering costs for steam consumption, the reduced steam usage drastically lowers the volume of condensed waste liquor normally associated with the cooling of the raw gas downstream of the gasifier.

Slagging fixed bed gasifiers have been operated with caking coals in short experimental tests at Westfield, but this has been made possible by the addition of mechanical devices which break up agglomerates. Use of caking coals in these gasifiers still presents difficulties. Coal sizing is still required, but fines may be fed to a limited extent (less than 10 percent below 1/8-inch for Lurgi) if they are injected with the steam at the bottom of the gasifier. However, entrainment of fines still remains a problem if fines are present in the upper portion of the bed.

In summary, the primary advantages of slagging fixed-bed gasifiers are their relatively high throughput, low steam consumption, low volume of gas liquor produced, and the more rapid reactions that occur at higher temperature (important for non-reactive coals). The major problem areas requiring further development are related to materials selection and mechanical reliability of the slagging section. No suitable refractories have been found that can withstand corrosive attack at the slag taphole for extended periods; however, a water-cooled metal hearth plate developed at GFETC may prove successful.

Fluidized Bed Coal Gasification

Fluidized beds have been a part of industrial technology for nearly half a century. The catalytic cracking of petroleum hydrocarbons in fluidized beds represented a major breakthrough in refining technology during the 1940's and has played a major part in the processing of crude oil ever since. The technique has also been successfully used fcr a wide variety of other process applications, including chemical reaction, absorption (including drying), and coating processes. Fluidized bed principles were first applied to coal gasification with the advent of the Winkler process in the 1920's.

The general principles of fluidization and fluid bed operation were described in the chapter on fluidized bed combustion, and these comments are also applicable to the use of fluidization for coal gasification (see Figure 3.5.3.4). Several advantages may be attributed to the use of fluidized bed gasification, relative to coal gasification in fixed bed or entrained flow reactors:

- High tolerance to a wide range of feedstock compositions
- High specific gasification rate resulting from high heat and mass transfer rates
- Sulfur removal may be accomplished by adding limestone or dolomite to the bed
- Gasification temperature is high enough to eliminate tar formation, but low enough to avoid clinker formation
- High product uniformity results from turbulent environment

However, fluidized bed coal gasification also implies several disadvantages:

- Gas velocities required to sustain fluidization in the bed are often high enough to result in a substantial carryover of fines. This may result in erosion, fouling or plugging of downstream equipment and a loss in thermal efficiency if the fines contain unreacted carbon.
- Acceptable carbon conversions require a high ash content in the bed, lowering the rate of gas production per square foot.
- Turndown capability may be limited due to the requirement of maintaining fluidizing gas velocities.
- Unlike fixed bed gasifiers, the gas leaving the reactor is at the same temperature as the reaction zone. Therefore, sensible heat in the hot gas is not utilized for drying or devolatilizing and must instead

be either quenched or processed through waste heat recovery equipment.

Figure 3.5.3.4

Fluidized Bed Gasifier

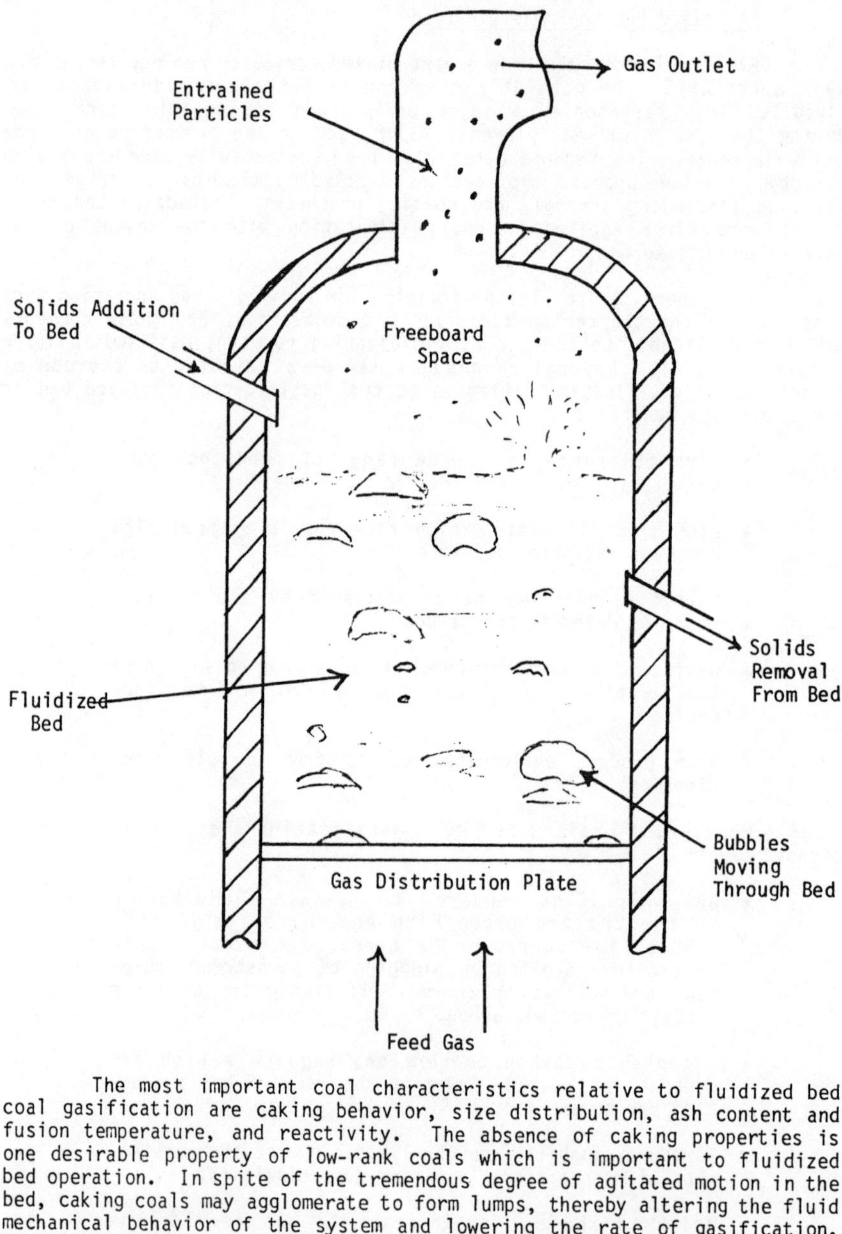

The most important coal characteristics relative to fluidized bed coal gasification are caking behavior, size distribution, ash content and fusion temperature, and reactivity. The absence of caking properties is one desirable property of low-rank coals which is important to fluidized bed operation. In spite of the tremendous degree of agitated motion in the bed, caking coals may agglomerate to form lumps, thereby altering the fluid mechanical behavior of the system and lowering the rate of gasification. Coal particle size distribution is also an important factor, since proper fluidization of the bed requires a wide range of particle sizes. Fines carryover may occur in any fluidized bed, and will result in a loss of efficiency if carbon-bearing particles are not recovered from the overhead gas stream and returned to the bed. Fluidized bed coal gasifiers using puvlerized coal may not require fines recovery equipment, since fluidization velocities will be low relative to gasifiers using crushed coal, thereby reducing the amount of fines carryover.

The ash fusion temperature is a concern in fluid bed gasifiers for several reasons. In gasifiers where a sulfur absorbent such as dolomite or limestone is added, the gasification temperature may be chosen to optimize sulfur recovery and ash properties at the bed temperature must be compatible with the ash removal mechanism. Maximization of gasification hot gas thermal efficiency occurs as gasification temperatures increase, but minimizing volatile ash carryover by lowering bed temperatures is a competing influence. Effective ash removal is in general a problem and high ash coals are undesirable for this reason. Coal reactivity determines the hold-up time in the reactor, and therefore for coals of low reactivity, larger gasifiers will be required for equal coal consumption rates, as compared to more reactive (such as low-rank) feedstocks.

Low-rank coals thus display advantages to higher rank coals due to their higher reactivity and noncaking character, and may show advantages or disadvantages with respect to their use in specific gasifiers owing to the different physical properties of low-rank coal ash.

Fluidized bed coal gasification systems may be considered under two separate categories; integral processes which produce high-Btu gas, and isolated gasifiers proposed for low- or medium-Btu gas production. Table 3.5.3.3 summarizes these differences:

Table 3.5.3.3

Fluidized Bed Gasification Systems

Integrated, High-Btu Systems	Low- or Medium-Btu Gas Producers
Hygas	Winkler
Synthane	U-Gas
Exxon Catalytic	Westinghouse
CO_2 Acceptor	DOE Fast Fluid Bed

The Winkler gasifier is the only fluidized bed gasification system which can be considered commercially available.

Entrained Flow Gasification

Entrainment of solid particles by liquids or gases is a well known technique for contacting chemical reactants or bringing about physical change. Entrained flow coal gasification processes may involve upflow, downflow, horizontal flow or cyclone (swirling) flow of the cocurrent solid and gaseous phases, where the solids may move at velocities equal to or less than that of the entraining gas medium. The entrained upflow arrangement shown in Figure 3.5.3.5 is a special case of the fluidized bed configuration, known as a fast fluid bed or lift pipe reactor.

Because of the fact that all solid particles are carried along by the gas stream in entrained flow coal gasification, gas velocities must exceed the terminal velocity of the largest particles (in upflow configurations). Pulverized coal is generally used in entrained flow gasifiers, which requires lower gas velocities for entrainment and less time for complete reaction due to the larger surface area available for reaction per unit weight. Faster particle consumption and lower particle speed give a combined effect of reducing the size of the gasifier as compared to what would be required for crushed or larger coal particles. Size consistency therefore, is not a major concern except that a maximum size limitation be observed.

Figure 3.5.3.5

Entrained Flow Gasifier

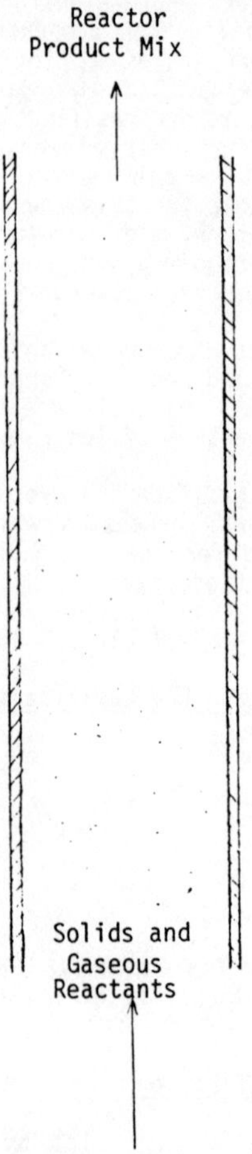

Particle-particle interactions in the reaction zone are minimized because of the high void volume in the gasifier. Ash agglomeration or caking tendencies are therefore unimportant in the gasifier, although caking coals may still adversely affect the operation of certain coal feeding systems (pneumatic systems in particular). Ash fusion temperature of the coal is important since it determines the slagging or non-slagging behavior of the coal.

Although entrainment implies a relatively dilute (with respect to coal) process stream, the high reaction rates brought about by coal pulverization make high mass throughouts feasible and therefore high energy output per unit volume may be achieved in entrained flow designs. However, entrained flow gasification conversion efficiencies may be sensitive to the rank of coal used in the process. For example, it has been estimated that the Koppers-Totzek process can obtain conversion efficiencies of greater than 98 percent when using lignite coals, about 85 percent for subbituminous coals, and a maximum of 60 percent for anthracite.[9] This decreased conversion is probably due to the lower reactivity of higher rank coals acting in concert with the short residence times characteristics of entrained flow gasifiers. In addition, entrained flow gasifiers are also relatively large consumers of oxidizer per pound of fuel, as was indicated on the comparison chart (Table 3.5.3.2).

3.5.3.4 Environmental Control Technology

There are four specific areas for environmental concern in coal gasification:

- gaseous emissions
- liquid effluents
- solid waste disposal

In general, sources of atmospheric pollutants are coal dryers, coal gasifiers, sulfur plant tail gas, cooling towers, process furnaces, and carbon dioxide rejected from acid gas removal. Water effluents consist of gas liquors process condensate blowdown, ash and slag quench water, and rain runoff from coal storage. Gas liquors can contain high concentrations of tars, oils, phenols, naphthas, sulfur and nitrogen compounds, particulate matter, and dissolved solids. Solid wastes arise primarily from coal cleaning, ash recovery, and sludge from waste water treatment. As part of each of these streams, trace element emissions result from gas cleanup system volatiles, oil products, and leaching from solid waste.

Gaseous Emissions

The prominent gaseous pollutants of coal gasification are summarized in Table 3.5.3.4. Assuming the absence of leaks, gasifier emissions may occur when coal is charged to the reactor or when slag or ash is withdrawn. Most of the emissions which arise from this source are raw product gas.[10,a]

Table 3.5.3.4

Gas Emissions From A Typical SNG Plant

Stream	Million std. cu. ft/day	Relative Volume
Synthetic Natural Gas	250	1.0
Coal Dryer Vent Gas	100	0.4
CO_2 From Acid Gas Removal	275	1.1
N_2 From Oxygen Plant	425	1.7
Claus Plant Tail Gas	85	0.3
Flue Gas From Utility Furnace	770	3.1
Air From Cooling Tower	55,000	220.0
H_2O Evaporated in Cooling Tower	1,270	5.1

Source: Reference 10

Several different acid gas removal systems are under consideration for use in coal gasification systems. Each of these however, has its own unique problems in meeting the objective of high sulfur removal, producing a concentrated H_2S stream for sulfur recovery, and producing a clean

[a]References for sections 3.5.3.2 - 3.5.3.6 are listed following section 3.5.3.6.

CO_2 vent stream. It is particularly important that contaminants in the rejected CO_2 stream be kept to a minimum, since the flow rate of this effluent is quite large.

Although the cooling towers process only water, it has been shown that they are a major source of emissions in other industries.[16] Water passing through cooling towers also circulates through heat exchangers, gasifier cooling coils, acid gas removal reboilers, and other process operations where equipment leaks can contaminate the water with appreciable amounts of sulfur compounds, oil, particulates, and trace elements. These contaminants may be stripped out and carried off as airborne pollutants while passing through the cooling tower. In coal gasification, high process pressures of the order of 500 to 1000 psi increase the leakage potential. However, the main problem will result from the fact that partially cleaned wastewaters would be used as make-up water for cooling towers, resulting in evaporation of residual contaminants present in those waste water streams. Aside from cooling water pollutants, the water itself may form mists or plumes of water droplets that are carried out of the tower. These mists can form deposits on equipment, buildings, roads and vehicles and can cause road icing during winter months.

The lower sulfur content of low-rank coals is expected to result in a decreased level of atmospheric sulfur emissions relative to gasification of higher sulfur bituminous.

Liquid Effluents

The major sources of liquid waste streams in coal gasification plants result from quenching raw gas and ash, and from process condensate blowdown.

Gas quench liquor will be present primarily in raw gas streams resulting from fixed bed gasification processes, since fluidized and entrained flow gasifiers substantially consume these components. As soon as the raw gas temperature is lowered, either by direct quench or heat exchange, liquid components of the vapor phase begin to condense. This liquor is removed in knockout drums between successive stages of cooling. If a direct water quench is used, the condensate collected after gas cooling will consist largely of water and will be determined by the process, the coal moisture content and the make-up/blowdown rate for low-moisture coals. The other components most likely to be in this stream are the following:

Tar	Ammonia
Tar Oil	Hydrogen Sulfide
Naphtha	Organic Sulfur Compounds
Phenols	Hydrogen Cyanide
Particulates	Trace Elements

Concentrations of these components will depend upon their raw gas concentration and the method and degree of quenching. Ash which is washed out of the raw gas stream is separated from the quench liquor in a settling tank, but such separation techniques are not completely effective. A complex water treatment system will be required to remove these contaminants and produce discharge water meeting standards set by the Federal Water Pollution Control Act (state standards also apply, and are at least as stringent as Federal law). Of the possible components of the gas quench liquor, phenols require the most complete processing. Extensive dephenolization treatment systems, such as Phenosolvan, Koppers Light Oil Extraction, and Barrett, Jones and Laughlin, may have to be used in large gasification plants. In many cases, dephenolization may be followed by biological oxidation treatment, the effect of which is shown in Table 3.5.3.5 (data taken from pilot plant and commercial coal gasification units).

Table 3.5.3.5

Some Contaminants in Gasifier Sour Water, Parts/Million

Item	Gas Liquor	To Biox	From Biox
Phenol	2,000-4,000	100-500	0.1-0.3
Fatty Acids	--	500-1,500	9
Ammonia	8,000-11,000	200-1,000	5-10
Thiocyanates	5-10	1	0.1
Fluoride	--	56	6
BOD	--	2,500	75
COD	10,000-20,000	1,100	82

Source: Reference 10

Water used in the ash quench tank is generally recycled gas quench liquor, which is used for both ash cooling and transport. However, blowdown streams from other process units may be used. Thus, the ash quench water will contain ash particulates and any water leachable substances present in the ash quench tank, in addition to its contaminants which were present before being charged to the quench vessel. This water may be sent to evaporation ponds for thickening,but air emissions may result from the volatile components. Alternatively, a settling tank may be used to separate an ash slurry from the liquor. The slurry will have a solids loading of approximately 25 percent to 35 percent, the composition of the solids being essentially the coal ash and any unreacted carbon discharged from the gasifier. Water recovered from the ash settlers is suitable for recycle to process condensate or gas quench liquor streams. The dewatered ash or ash slurry is a waste product requiring ultimate disposal. Ash slurry may be combined with dry ash prior to disposal.[17]

Solid Waste Disposal

The major solid waste streams in coal gasification plants requiring disposal are coal cleaning refuse, ash or slag from gasification, waste water sludge, and possibly solids from evaporation of cooling tower blowdown liquors or other final wastewater disposal techniques used to achieve a zero liquid discharge. For coal cleaning refuse, thorough planning will be required for disposal methods such as land fill with revegetation, due to the extremely large volumes involved (as much as 800 acre-ft/year for a large gasification plant). Coal ash recovered from gasification and utility plant areas require similar consideration. Wastewater sludge recovered from the biological oxidation unit, however, may cause odor problems unless appropriate cautions are taken. Owing to its high surface area, char recovered from gasification may be useful for activated carbon, but will ultimately require disposal. Preliminary bench scale tests have shown that it will remove over 90 percent of the phenols present in wastewater.

Leaching of salts and trace metals from solid waste disposal dumps into ground waters must be evaluated for each proposed gasification site. Leaching behavior of low-rank coal ash and slag will be considerably different than that for higher rank coals. Monitoring may be required to insure that minerals such as magnesium sulfate, sodium salts, calcium chloride, sulfur compounds and trace metals do not contaminate water sources (chlorine contents of low-rank coals are low except for some Texas lignites). The situation bears some resemblance to acid mine water, and such background experience should be taken into consideration.[18]

Trace Elements

Coal contains a wide variety of trace elements, many of which are toxic (such as Hg, As, Pb, Cd, Be and others) and have an appreciable volatility under the high temperature conditions present in gasification. Metal cations present in raw coal may be reduced to the metallic state by carbon or other reducing agents in the gasifier, and in the case of low boiling point elements such as mercury, arsenic, cadmium, selenium, and zinc, will be carried out of the gasifier at temperatures above 1700°F. Metal halides are also relatively volatile. Reactions of metals with carbon monoxide to form carbonyls of iron, nickel and cobalt, and formation of arsenic, antimony, selenium hydrides also represent sources of volatile toxic compounds. Most alkali metal salts have appreciable vapor pressures at gasification conditions as well. Table 3.5.3.6 shows the estimate volatility of trace elements present in coal.[19] For a 12,000 ton/day gasification plant, the data show the amount of trace elements subject to carryover in wastewater, tar, and solid effluents streams as well as in gas vent and product gas streams. Past studies have shown that as much as 90 percent or more of the chlorine in the feed coal is ultimately released in gas streams as hydrogen chloride.[18] However, the chemical form and process stream in which most of the other components are found remains unknown.

Table 3.5.3.6

Estimated Volatility of Trace Elements

	Typical Coal parts/million	% Volatile*	lb/day**
Cl	1,500	90+	32,400
Hg	0.3	90+	6
Se	1.7	74	30
As	9.6	65	150
Pb	5.9	63	89
Cd	0.8	62	12
Sb	0.2	33	2
V	33	30	238
Ni	12	24	69
Be	0.9	18	4
Zn	44	e.g.10	106
B	165	e.g.10	396
F	85	e.g.10	204
Cr	15	nil	nil

*Volatility based mainly on gasification experiments (18) but chlorine taken from combustion tests, while zinc, boron, and fluorine taken at 10 % for illustration in absence of data.

**Estimated amount volatile for 12,000 ton/day of coal to gasification.

Source: Reference 10

Environmental Control Equipment

Atmospheric Control Techniques

Many processes for the control of atmospheric emissions find application to coal gasification as well as direct combustion. A complete discussion of these processes may be found in section 3.5.1.3 under Direct Combustion.

Acid gas removal systems are not found in coal direct combustion plants; they are required for removal of H_2S and CO_2 from gasification process streams. Air quality regulations place limitations on sulfur emission levels. In addition to meeting environmental standards, sulfur removal may also be required to prevent catalyst poisoning in certain processes, or to protect process equipment from corrosion.

Carbon dioxide faces no current atmospheric environmental regulations and in most cases does not cause severe corrosion. Due to the large CO_2 volumes resulting from coal gasification, CO_2 absorption greatly increases the cost of acid gas removal, which is a major capital item in coal conversion plants. Its removal is usually justified in cases where product gas heating value is important or where the gas is being transmitted long distances and will require compression. Carbon dioxide is also removed before synthesis gas is charged to processes in which the presence of CO_2 would limit reaction conversions by acting as a diluent or participant in a reverse or competing reaction. A potential beneficial use of CO_2 is in enhanced oil recovery.

In processing situations requiring removal of both carbon dioxide and sulfur compounds, the acid gas removal system is referred to as "nonselective." Conversely, systems which recover sulfur compounds preferentially are termed "selective."

Several different techniques exist for acid gas removal from gas streams. Some processes rely on gas-liquid contacting for either physical or chemical removal, or gas-solid contacting which may use activated carbon as a physical adsorbent, iron oxide or zinc oxide as chemical recovery agents, or molecular sieves for selective recovery of desired chemical species.

Gas-Liquid Contacting Systems

In general, most absorption systems (both selective and nonselective) which are based on gas-liquid contacting have several major features in common. Figure 3.5.3.6 is a simplified flow sheet which can be used to represent either a selective or nonselective acid gas removal system in its most basic form. Raw gas enters the absorber and passes up through the vessel, contacting the absorbing medium countercurrently. Acid gases contained in the raw gas are absorbed in the solution.

Figure 3.5.3.6

Generalized Acid Gas Removal System

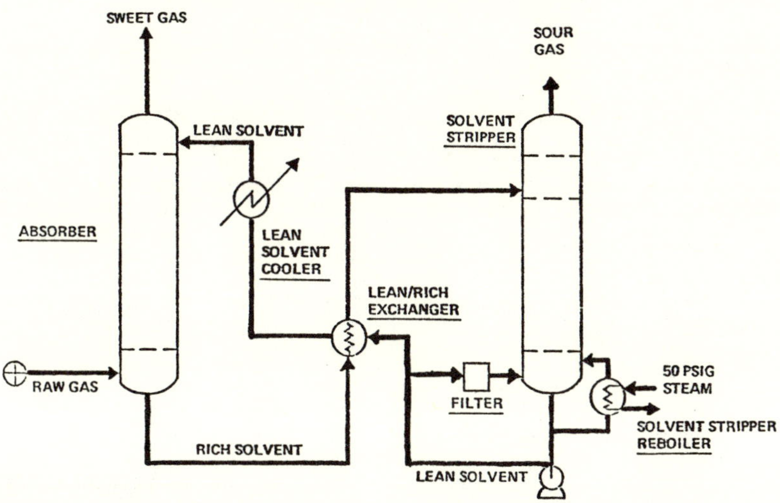

Absorbing solution rich in dissolved acid gas is heated by heat exchange with lean solvent before it is stripped of acid gas in the stripper. Acid gases released in the stripper are sent to the sulfur recovery system. Lean solvent from the bottom of the stripper is recycled through the heat exchanger and back to the absorber to be used again.

Many modifications to the configuration in Figure 3.5.3.6 are possible, and are aimed at increasing sweet gas purity, H_2S concentration in the acid gas stream, reducing column size, solvent circulation rate or stripping steam usage.

Acid gas removal systems using aqueous solutions of potassium carbonate and proprietary additives are usually operated for nonselective gas absorption. Amine-based aqueous systems are generally operated selectively. In addition to these types, there exist a variety of other chemical compounds that are used for acid gas removal. However, selective or nonselective operation may be achieved with any type of system, and is determined by the difference in absorption rate between H_2S and CO_2 (H_2S is usually more absorptive) and the gas residence time in the absorber. Table 3.5.3.7 lists and compares the major acid gas removal processes on the basis of several key performance parameters. Caution should be exercised when using Table 3.5.3.7 to compare utility requirements, since feed and product gas conditions may differ as noted in the Table.

One problem faced by a nonselective system when cleaning a coal derived producer gas (containing approximately 7 percent CO_2 and 0.5 percent H_2S) is the disposal of the H_2S contained in the acid gases which have been removed from the producer or synthesis gas stream. For acid gas streams containing at least 15 percent H_2S, the Claus process is widely used to produce high purity elemental sulfur which may be sold as a by-product.

The Claus process involves a thermal oxidation stage where H_2S is converted to SO_2.

$$H_2S + {}^3/{}_2O_2 \longrightarrow H_2O + SO_2 + \text{Heat} \qquad (1)$$

The SO_2 so formed reacts with H_2S to produce elemental sulfur:

$$2H_2S + SO_2 \rightleftharpoons 3S + 2H_2O + \text{Heat} \qquad (2)$$

Since reaction (2) is reversible, the presence of sulfur limits the conversion of H_2S. Vapor phase elemental sulfur is removed by condensation, and the remaining gases are heated and passed through one or two reaction chambers where reaction (2) is promoted catalytically by a bauxite or alumina catalyst in the converter.

Because the first step involves the flame oxidation of hydrogen sulfide, H_2S must be present in high enough concentrations to support a flame front. Generally, 15 mole percent is recognized as the lower limit, although some operations have succeeded at 7 mole percent using special techniques. Acid gas removal systems which produce acid gas streams lower than 15 percent H_2S (because of dilution with CO_2 and other gases) may require a different sulfur recovery system. Since amine-based systems are generally designed for H_2S selectivity, they will produce sufficient H_2S in the acid gas stream for Claus processing, if the raw gas stream is concentrated enough in H_2S.

Gas-Solid Contacting Systems

Activated Carbon Process

Activated carbon adsorption[11] is a low temperature process

Table 3.5.3.7

Acid Gas Removal Processes

Performance Parameter	Amine Based Systems				
	Shell ADIP	Dow Chemical MDEA	Eickmeyer CATAMINE	Black, Sivalls, Bryson SULFIBAN	Shell SULFINOL
State of Development	Commercial	Commercial	Commercial	Commercial	Commercial
Absorber Operability Limits					
Temperature, °F	100-140	80-125	100-250	100	65-125
Pressure, psia	0-1000+	0-1000+	0-1000+	0-1000+	0-1000+
Purification Limits for H_2S	<5 ppm	<4 ppm	<5 ppm	~ 1 ppm	<1 ppm
Principal Absorbent(s)	di-isopropanolamine (aqueous)	methyl-diethanolamine (aqueous)	proprietary	mono-ethanolamine (aqueous)	di-isopropanolamine (aqueous)
Physical/Chemical Absorption	Chemical	Chemical	Chemical	Chemical	Physical/Chemical
Typical Utility Requirements					
For Feed Gas Composition H_2S/CO_2 mol %	0.5%/7.0%	0.6%/10.0%	0.5%/7.0%	1.7%/9.7%	0.46%/4.9%
and Product Gas Composition H_2S/CO_2 mol %	10 ppm/3.6%	50 ppm/3.3%	10 ppm/5.2%	2 ppm/0.0%	3 ppm/.05%
Steam Consumption lb/MM SCF Feed Gas	7900	10,700	7500	49,000	10,000
Cooling Water Gal/MM SCF Feed Gas	N/A	N/A	50,000 (30°F rise)	161,000	N/A
Electricity kwh/MM SCF Feed Gas	38	15	25	50	60
Solvent loss lb/MM SCF Feed Gas	Mechanical leakage only	0.5	N/A	1.6	N/A

(continued)

Table 3.5.3.7 (continued)

Acid Gas Removal Processes

Performance Parameter	Carbonate Based Systems: Benfield (Hot Carbonate)	Carbonate Based Systems: Eickmeyer CATACARB	Other Processes: Lurgi RECTISOL	Other Processes: Allied Chemical SELEXOL	Other Processes: STRETFORD
State of Development	Commercial	Commercial	Commercial	Commercial	Commercial
Absorber Operability Limits					
Temperature, °F	60-280	60-290	-100 to 0	20-100	80-120
Pressure, psia	High Pressure Operation Favored	High Pressure Operation Favored	200-3000	500-1000	0-100
Purification Limits for H_2S	< 1 ppm	< 1 ppm	< 0.1 ppm	< 1 ppm	< 1 ppm
Principal Absorbent(s)	Potassium carbonate (aqueous)	Potassium carbonate (aqueous)	Methanol	Polyethylene glycol-dimethyl ether	ADA (anthra-quinone disulfonic acid), Sodium metavanadate (aqueous)
Physical/Chemical Absorption	Chemical	Chemical	Physical	Physical	Chemical
Total Utility Requirements					
For Feed Gas Composition H_2S/CO_2 mol %	0.5%/7.0%	0.5%/7.0%	1%/5.5%	0.5%/35.0%	1.0%/0.2%HCN
and Product Gas Composition H_2S/CO_2 mol %	10 ppm/0.01%	10 ppm/0.01%	< 0.1 ppm/0.1%	< 0.1 ppm/11.0%	1 ppm/0.0% HCN (no CO_2 removal)
Steam Consumption lb/MM SCF Feed Gas	8700	11,100	2550	3000	730
Cooling Water Gal/MM SCF Feed Gas	30,000	40,000	121,000	35,000	2160
Electricity kwh/MM SCF Feed Gas	90	55	550	900	19
Solvent loss lb/MM SCF Feed Gas	N/A	N/A	40	0.5	15

developed in the 1920's and is currently in wide commercial use in the United States. However, it is not practical for application to primary sulfur recovery in coal gasification processes, due to its limitation on feed gas sulfur content. Operation of the process calls for absorber bed regeneration after the bed becomes saturated with sulfur compounds. Process streams containing more than approximately 30 ppm of sulfur compounds (low-rank coals may produce sulfur concentrations approximately 100 times greater) require regeneration too frequently to justify use of the process.

Iron Oxide Process

Another low temperature cleaning system, iron oxide has been used as chemical absorbent for gases containing hydrogen sulfide. Feed gas is allowed to flow over a bed of hydrated ferric oxide, in which hydrogen sulfide is removed by oxidation:

$$2Fe_2O_3 + 6H_2S \rightleftharpoons 2Fe_2S_3 + 6H_2O$$

After most of the iron oxide has been sulfided, atmospheric oxygen regenerates the oxide, producing elemental sulfur.

$$2Fe_2S_3 + 3O_2 \longrightarrow 2Fe_2O_3 + 6S$$

This cyclic process is repeated until an excessive pressure drop develops in the bed due to the accumulation of sulfur. At this point the bed material is removed and replaced with fresh oxide, and in some instances the sulfur is recovered. In the manufactured gas industry, iron oxide is mixed with wood shavings as a support. These are placed in trays inside large "dry boxes". Continuous revivification in-situ is accomplished by bleeding a small air stream into the feed gas.

The iron oxide process is not as limited with respect to feed gas H_2S concentration as is the activated carbon process. The latest installations of iron oxide absorbers are limited to gas streams having H_2S concentrations in the range of 160 to 1200 ppm, still to low for primary acid gas removal in coal gasification plants.

Zinc Oxide Process[11]

Zinc oxide is also in wide commercial use as a low temperature sulfur absorbent, but faces limitations similar to those for activated carbon and iron oxide processes. Although these systems are not suitable for primary acid gas absorption in coal gasification, they may find application as sulfur guards to protect methanation catalysts or other sulfur sensitive processes.

Molecular Sieve Adsorption[11]

Molecular sieves belong to a class of chemical compounds known as zeolites, and function in separation processes by physical adsorption of molecules within the interstices of the molecular matrix. Adsorption in molecular sieves is a highly selective process, and is determined by the pore size of the matrix. Molecular sieves are in wide commercial use, particularly for sweetening natural gas, petroleum and petrochemical process streams.

Of the acid gas removal processes based on solid-gas contacting molecular sieves are the best suited for handling gas streams with high H_2S concentrations. Streams of up to one mole percent hydrogen sulfide are currently being desulfurized at low temperatures down to as low as 4 ppm H_2S with this method. However, higher concentrations require shorter cycle times between absorption and regeneration as with the other gas-solids contacting processes previously discussed. This drawback is compounded by the fact that molecular sieves having the proper pore size for adsorption of H_2S and other sulfur compounds will also adsorb H_2O, leaving the gas stream completely dry. In coal gasification processes, the raw gas stream usually contains at least a few percent water vapor, which would require frequent regeneration of large beds of molecular sieve material.

Hot Gas Clean-up Systems

Still a development concept, hot gas clean-up systems are proposed for high temperature removal of sulfur compounds, and in some cases particulate matter, from coal gasifier effluent streams. The proposed concepts are usually based on fluidized bed or entrained flow contacting of the solid sulfur absorbing material (iron oxide, for example) and the raw gasifier effluent in one vessel, and regeneration of the sulfur absorbent in a separate vessel, with continuous solids transfer between the two. One study projected a substantial improvement in thermal efficiency and economic performance for a Lurgi coal gasification combined-cycle power plant fitted with hot gas cleanup over conventional cold gas cleaning methods.[12] However, the same study showed only a marginal incentive for hot gas cleaning when applied to the Foster-Wheeler entrained flow gasifier in a combined cycle plant. The increase in performances for the Lurgi system with hot gas cleanup was attributed to the presence of large amounts of steam and tar, which were not quenched and removed as in the cold cleaning system. Other gasifiers requiring low steam injection rates and producing small quantities of tars might thus be expected to show only a slight advantage for hot gas cleaning. The study's conclusions were based on the assumption that a number of significant technical problems with hot gas cleaning systems could be overcome.

Wastewater Treatment Systems

The wastewater stream that presents unique treatment problems in coal gasification plants is foul condensate produced when coal reactor product streams are cooled. Wastewaters arising from other sources are similar to wastewaters from coal fired power plants and therefore do not require unique treatment systems. For example, methanation reactor condensate is clean enough to be used as boiler feed water after minor treatment.

Wastewater treatment methods are designed for phenol recovery, ammonia stripping, biological oxidation, residual sulfur removal, and cooling tower water control. Composition of the wastewater stream is largely determined by coal rank, process type, and amount of water or steam in the reactor effluent. The initial wastewater treatment step is usually solids and oils removal, after which the water treating sequence may take either route depicted in Figure 3.5.3.7, or may be interrupted at any point and the partially treated water returned for process use.

Phenol Recovery

Phenols constitute one of the most toxic of all wastewater constituents associated with coal conversion. Most of the experience in phenol recovery to date has been associated with petroleum processing and coke oven plants. One important research need is therefore to compile a data base adequate to design an optimal phenol recovery system for coal conversion.

Biological oxidation may be used to prepare phenol-contaminated wastewaters for use in cooling towers. Phenols exert a BOD of approxi-

Figure 3.5.3.7

Sequence of Wastewater Treatment Operations Depending on Water Composition

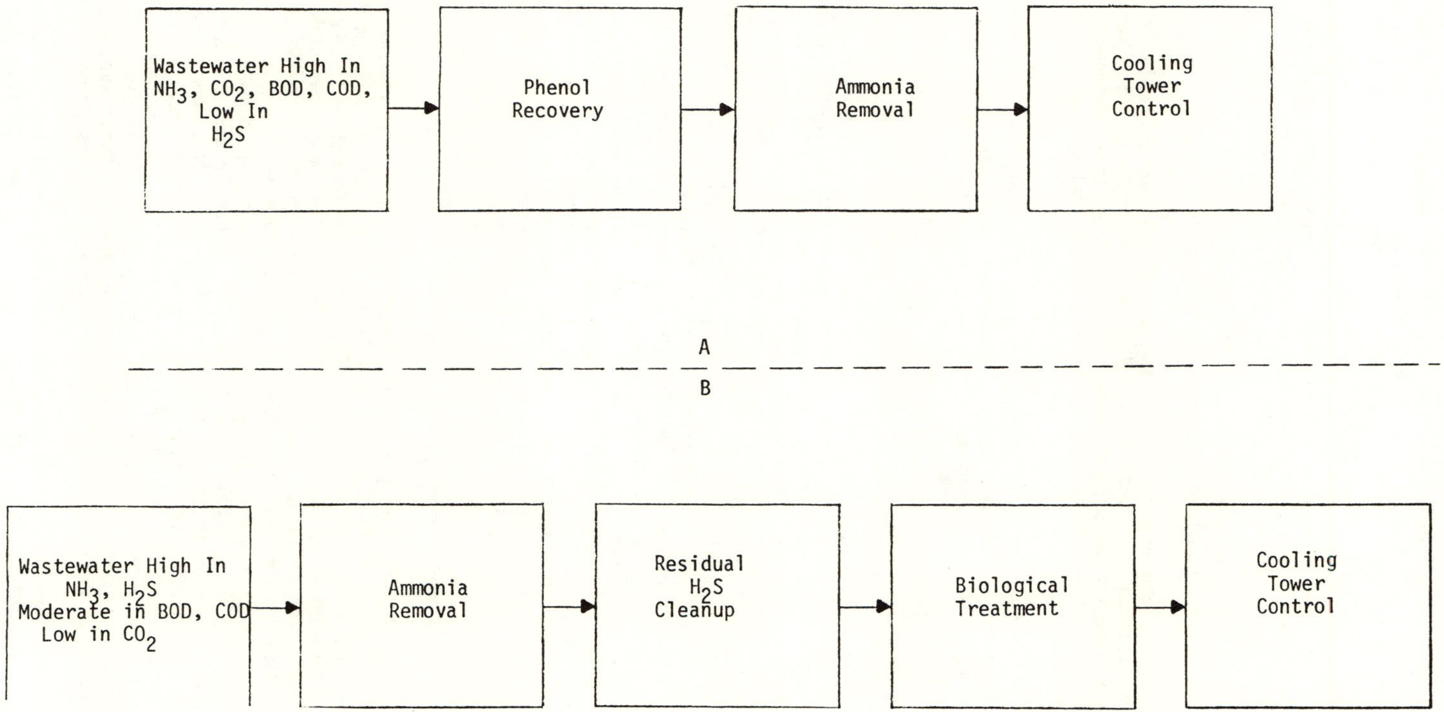

mately 2 lb. oxygen/lb phenol and can be reduced to the order of 100 mg/l by bi-ox processes.[13] The suitability of this concentration depends on the reuse application of the water. Reuse will be necessary to meet the zero liquid discharge requirement.

The alternatives for phenol recovery from wastewaters usually can be classified as one of the following types:

1. Recovery
2. Degradation
3. Combined recovery and degradation

Recovery processes include distillation, solvent extraction, crystallization, and activated carbon adsorption and synthetic polymer adsorption with solvent regeneration. When these methods are used, the phenol components of wastewater streams are collected and concentrated for further use.

Degradation techniques include bio-oxidation, ozonation, incineration, and activated carbon adsorption with thermal regeneration. When degradation techniques are applied, no recovery of phenols occurs.

Recovery may also be supplemented by degradation for ultimate disposal.

Process economics are dependent upon phenol concentration in wastewater streams, and the total volume of wastewater treated. As an example, the cost of solvent extraction is virtually independent of phenol concentration in the inlet stream giving solvent extraction an advantage over other techniques at high contamination levels. However, when high removal efficiencies are required, capital costs for solvent extraction increase sharply, making it unlikely that this method would be used alone for treating dischargeable effluent. Biological oxidation is one process which can be used to supplement solvent extraction, and is economic at inlet stream Phenol concentrations of 1000 mg/l or below.[13] At intermediate concentrations, absorption with solvent regeneration may be used economically, as a back up to solvent extraction, or by itself. Ozonation and activated carbon adsorption find application only for extremely low pollutant levels, and hence are unsuitable for primary wastewater treatment in coal conversion processes. Other methods, such as incineration, are usually impractical for coal conversion.

Ammonia Recovery

Depending upon conditions within the coal conversion reactor, nitrogen present in coal and introduced with oxidents may react with hydrogen to form ammonia, NH_3. Ammonia formation is favored by low temperatures and high pressures in the reactor, and conversions of up to 60 percent of influent nitrogen have been reported for the Synthane process. In low temperature conversion processes such as Hygas or Lurgi gasification, ammonia concentrations in raw gas quench condensates may range from 3,600 mg/l to 14,000 mg/l or more.[13] These levels are orders of magnitude above those found in common wastewaters (raw sewage may contain 35 mg/l, for example) and are far beyond acceptable levels for direct discharge. High temperature reactors such as the Texaco, Combustion Engineering, and Foster-Wheeler designs are relatively clean gasifiers and may not require ammonia recovery processes as part of their wastewater treatment schemes. Coal nitrogen in these gasifiers appears primarily as N_2.

Ammonia must also be limited in recycled process waters, since it can accelerate bacterial growth, corrode copper tubing, deactivate zinc corrosion inhibitors, and escape as an airborne containment from cooling towers. In plants where process condensate is recycled for coal slurrying or as a reactor feed, steady state condensate ammonia levels may be excessively high and therefore must be controlled at some level, as is also the case for cooling tower operation. This level must be determined for each process and coal, and should be the subject of a research study. An optimum ammonia level exists for biological oxidation processes; concentrations above approximately 1800 ppm cause inhibition of oxidation reactions, and a minimum concentration is required to provide bacterial nutrients.[13] Commonly, a phenol to nitrogen ratio of 14 to 1 is used.

Choice of sequence for the processing operations of phenol and ammonia removal depends on several considerations and will vary depending on the coal conversion process in question. If phenol is to be recovered by solvent extraction, ammonia may or may not cause problems, depending on the nature of the solvent used. However, the presence of phenol in the ammonia stripper will introduce phenol in the stripper overhead, which can accelerate catalyst deactivation in sulfur recovery processes downstream.[13] Of course, high reflux ratios in the ammonia stripper may be used successfully to limit phenol concentrations overhead, but this imposes added costs and would not normally be practiced unless other circumstances dictate this mode of stripper operation. Thus, upstream phenol separation may be desirable, but low phenol concentrations may dictate ammonia recovery first followed by biological oxidation for phenol degradation.

The problem of ammonia recovery is complicated by the presence of CO_2, H_2, as well as phenols in coal conversion process condensates. Recovery of ammonia in a salable form such as anhydrous ammonia or ammonium sulfate is a motivating factor in water treatment design, due to the relatively high value of this commodity.

Ammonia and acid gases may be stripped from wastewaters simultaneously followed by ammonia separation and recovery from the gas phase. Alternatively, the acid gases may be stripped with ammonia held in the aqueous phase by reflux or water wash followed by ammonia concentration in a second stage.

The Phosam-W process may be used for anhydrous ammonia recovery. Although the design is a proprietary process of USS Engineers and Consultants (U.S. Steel Corporation), a general description of the processing sequence is available. The four major components of the system are: 1) a wastewater stripper in which H_2S, CO_2, NH_3, and any other volatile components are recovered in a gas phase, after which the wastewater is available for continued processing or use; 2) an absorber which selectively absorbs ammonia from the mixed gas phase, and allows the acidic components to pass on for further treatment; 3) a "Phosam" stripper which provides recovery of the ammonia absorbing solution for reuse, and produces an aqueous ammonia stream; and 4) an ammonia fractionater with a total condenser where aqueous ammonia is reduced to anhydrous ammonia product.

Fertilizer grade ammonium sulfate may be produced from the ammonia present in gasifier wastewater. After ammonia is recovered from a condensate stream, it is contacted with sulfuric acid to form ammonium sulfate. Although this may be one way to utilize both the ammonia and sulfuric acid by-products which may be produced, current market economics for ammonia and ammonium sulfate do not justify this practice (ammonium sulfate is generally cheaper than ammonia).

A proprietory process for ammonia recovery by distillation only is offered by Chemie-Linde-Lurgi (CLL). Although the details are unknown, the design consists of two columns. The first tower strips off acid gases with steam while ammonia is held in solution. The second fractionator concentrates the ammonia to either a 30 percent solution or all the way to anhydrous ammonia.

Biological Oxidation

Biological wastewater treatment requires the presence of bacteria, organic matter, and usually oxygen, and is a process in which these bacteria bring about the reaction of dissolved oxygen and organic matter. Bacterial growth takes place, and organic matter is consumed and converted into smaller residual molecules such as carbon dioxide or methane. This results in an aqueous phase which has had most of its organic content eliminated, and a suspended separable sludge consisting of living and dead bacteria. Approximately half the weight of BOD (biological oxygen demand) is converted to weight of sludge.

Several processes involving biological oxidation have been developed for other treatment needs, and in most cases are not applicable to coal conversion wastewater treatment due to the sensitivity of the bacteria to coal derived chemicals, or differing treatment objectives.

Bacterial conversion in the absence of oxygen (anaerobic treatment) results in the production of methane. Although this approach has the advantage of not requiring large energy inputs for oxygenation or aeration, evidence is lacking that anaerobic oxidation can be easily integrated into coal conversion wastewater treatment. Similarly, bacterial conversion of ammonia to nitrate compounds has been demonstrated, but in most cases is probably too costly for conversion of the high ammonia concentrations found in coal conversion condensate streams. In this process, water which has had its ammonia converted to nitrates is denitrified anaerobically in another vessel to produce gaseous nitrogen which again is accomplished by bacteria. Nitrification must be carried out in the presence of oxygen but at very low concentrations of carbonaceous compounds. The reverse is true of denitrification, and methanol is sometimes added to provide the required carbon content.

Although most of the organic content of coal conversion wastewaters is biodegradable (molecules that are not biodegradable are referred to as biorefractory), high concentrations of organic biodegradables or ammonia are toxic to bacteria. As an example, phenol is readily biodegradable at low concentrations, and is usally kept below 500 mg/l.

For most wastewaters, the bacterial concentration resulting from a one-pass reactor is not high enough to assure an adequate reaction rate. Recycle of a portion of the reactor effluent sludge maintains a high bacterial concentration in the reactor and is known as the activated sludge process. The trickle filter concept achieves high bacterial levels with a different approach by growing the organisms on a solid surface and passing the wastewater in a thin film over the solid substrate. Bacterial agents in biological oxidation will show a varying reactivity to organic compounds, depending upon strain. Sludge obtained from municipal waste treatmnent plants consists of ordinary soil bacteria, which will display a lower efficiency at treating phenolic wastewaters than will sludge obtained from coke oven wastewater treatment, which has been acclimated to phenols. Bacteria which have been mutated deliberately for phenol consumption are available and seem to be more efficient than other types.

The Air Activated Sludge (AAS) Process

The activated sludge process is widely used in municipal waste processing plants and can be applied to sour process condensate streams as shown in Figure 3.5.3.8. The liquid wastewater feed must come from a hold-up tank or storage pond to minimize fluctuations in flow rate and concentration to the bioreactor. Since phosphorus is an essential bacterial nutrient, it must be added along with the wastewater stream. The bioreaction process is accomplished in one or more stirred basins arranged in series, and is such that reactor effluent composition is very nearly the same as the composition of the reactor contents at any given time. Good distribution of feedwater to the reactor may be necessary to avoid high local concentrations which may be toxic to the bacteria in the bioreactor.

Figure 3.5.3.8

Coal Conversion Wastewater Treatment by the Activated Sludge Process

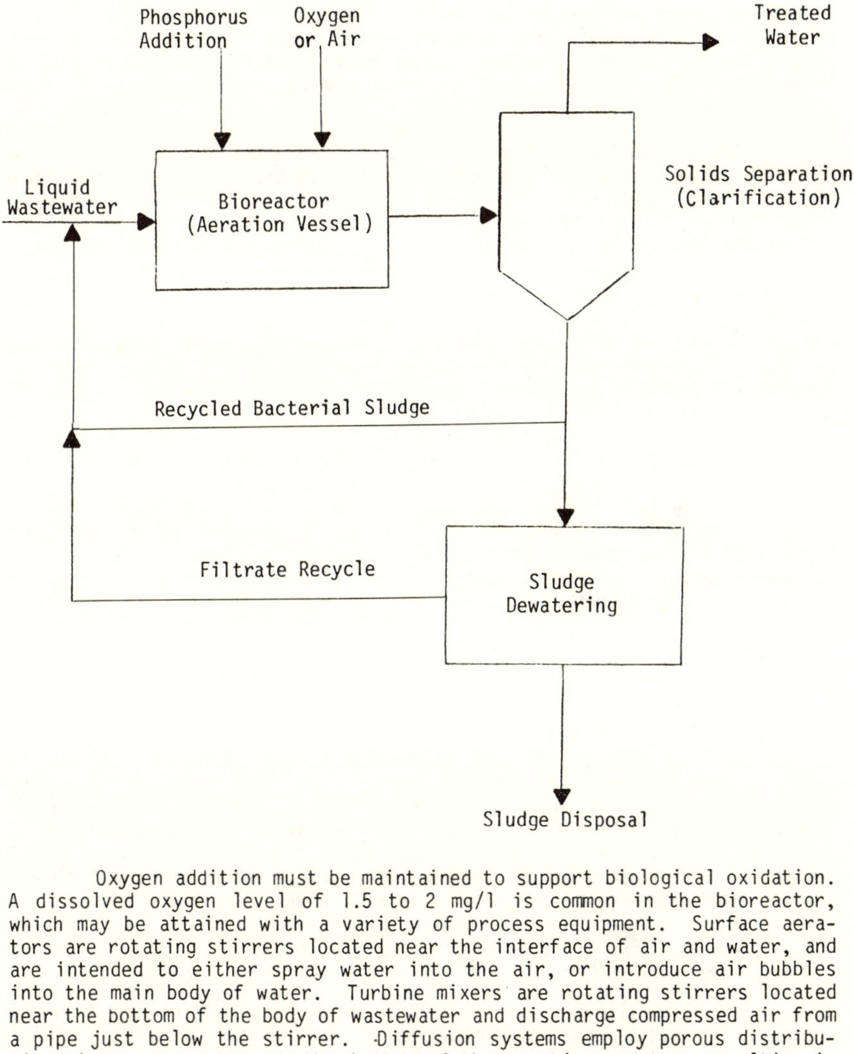

Oxygen addition must be maintained to support biological oxidation. A dissolved oxygen level of 1.5 to 2 mg/l is common in the bioreactor, which may be attained with a variety of process equipment. Surface aerators are rotating stirrers located near the interface of air and water, and are intended to either spray water into the air, or introduce air bubbles into the main body of water. Turbine mixers are rotating stirrers located near the bottom of the body of wastewater and discharge compressed air from a pipe just below the stirrer. Diffusion systems employ porous distribution pipes or spargers at the bottom of the aeration reactor, resulting in fine gas bubbles rising throughout the aqueous mass.

The sludge separation step is an essential part of the activated sludge process. Clarification serves two needs; removal of suspended matter from the treated water stream, and thickening of the solid sludge for recycle of a sufficiently high concentration of biologically active sludge.

High purity oxygen may also be used for oxygen enrichment of wastewaters instead of air. In addition to the requirement of a source of oxygen, several differences arise when the activated sludge process is operated in this mode. Since oxygen is present in the gas phase at a higher partial pressure, its concentration in the aqueous phase may be maintained at a higher level than with air usage. Assuming the reaction rate is dependent somewhat on oxygen concentration in the wastewater, this together with the staged reactor configuration often used (intermediate between a perfectly mixed reaction system and a plug flow arrangement) will result in a smaller reaction volume required for similar conversions. However, high levels of phenols or other toxic components in the feed will have a smaller volume in which to be mixed, and therefore the risk of excessively high local concentrations is greater than in the well mixed reaction system.

The oxygenation and mixing requirement diminishes from the initial to final stages, allowing a reduction in capacity for these two parameters. This reduction in mixing and agitation due to oxygenation is conducive to flocculation and agglomeration of the flocs which aids in the clarification step downstream.

Trickle Filtration

Unlike the air activated or high purity oxygen activated sludge processes, trickle filter processes rely on the thin film contacting of wastewaters on solid bacterial substrates. As such, a liquid (wastewater) instead of a gas (air or oxygen) is pumped resulting in lower operating costs. Trickle filters will generally not give as clean an effluent stream as will an activated sludge process. In wastewaters with high organic loadings, the fraction of the BOD converted may be as low as 40 percent. Use of trickle filter treatment together with an activated sludge process is an established practice, and in the case of coal conversion, a trickle bed treatment system would most likely be placed in front of the activated sludge process to act as a roughing filter. Although this configuration has been used successfully on other wastewater streams, its use associated with coal conversion has not been reported and is therefore regarded as experimental.

Cooling Tower Control

Cooling towers operating in coal conversion plants handle process and raw waters from a variety of sources, and are faced with a unique set of problems which arise from the mixture of different waters used. Optimization of the costs involved in wastewater treatment requires a tradeoff between the cost of ultimate water treating and the resulting normal cooling tower problems, and no wastewater treating and the following categories of operational and maintenance problems.

Scale Formation

Since cooling towers function by evaporation these air-water contacting devices act as concentrators for dissolved salts in their feedwaters, sometimes to the point of precipitation. Scale formation will lead to a loss in heat transfer rates for other equipment. Treated process concentrate may have a lower dissolved salt content but higher levels of BOD, COD, and ammonia than raw river water, suggesting that cooling tower control will be highly specific to recycle rates of treated water and treated and raw water compositions.

Scale formation and deposition may be controlled by limiting the concentration of certain chemical species, particularly those whose solubility decreases with increasing temperature. These salts will then tend to drop out of solution and adhere to heat exchanger surfaces. The anions and cations which must be controlled are those of calcium (Ca^{2+}), magnesium

(Mg^{2+}), carbonate (CO_3^{2-}), phosphate (PO_4^{3-}) and dissolved silicon dioxide (SiO_2). The presence of phosphate is quite uncommon in the cooling waters of industries other than coal conversion, but arises here due to its addition in biological oxidation (where it is not all consumed) and in the cooling tower water itself to encourage biological oxidation.

The relatively high concentration of carbon dioxide in coal conversion wastewaters leads to an increase in pH due to formation of HCO_3^- ions. During passage through the cooling tower, CO_2 may be driven off, causing formation of CO_3^{2-} ions and subsequent precipitation. This may be prevented by the addition of small amounts of sulfuric acid. Water softening by lime or soda ash will precipitate calcium ions and carbonate alkalinity, and can be used to control calcium sulfate precipitation. The phosphate anion may also be removed with lime and the water softener may be operated to remove magnesium cations as the hydroxide which has the added benefit of partial silica adsorption. Even though precipitation may occur, scale formation may be limited by the addition of chemicals which retard crystal growth, such as acrylate and acrylamide polymers. Despite the use of any process, the total dissolved solid level will increase without the use of some blowdown in the cooling water circuit. Droplet loss in the cooling tower constitutes one form of blowdown, although for systems having high TDS levels, droplet settlement may cause damage to foliage and property, and is thus not an acceptable form of blowdown. In general, designers must be sensitive to possible odor and health problems which may arise from trace organics emitted from cooling towers operating on treated wastewater streams.

A balance between successful bio-oxidation of contaminating organics and an over-population of bacteria must be maintained in the cooling tower water. Chlorination has been used for bacterial control, but is not recommended for coal conversion processes due to the possible release of toxic chlorinated phenols and other organics. Other non-chlorine biocides are available for control, such as acrolein.

Filtration, rather than clarification, is suggested for fouling control by suspended solids in the circulating water. This is because clarification requires the addition of chemicals to promote flocculation, although it can handle a higher suspended solids in the feed water.

Limitation of Corrosion

Both metallic and nonmetallic corrosion inhibitors are available. For cases where the surface contacted by the cooling water is not particularly hot (usually the case in coal conversion plants), nonmetallic inhibitors are preferred. For higher temperature service applications, corrosion resistant alloys should be used. In particular, copper alloys are to be avoided due to the presence of ammonia in coal conversion process streams.

Solid Waste Control Techniques

The methods used to control and dispose of the solid wastes from coal gasification except for bio-oxidation processes are similar to those used for direct combustion and will not be repeated here.

Bio-oxidation processes are not found in direct combustion operations, and produce a solid material requiring disposal. The procedure involves transfer of the sludge together with its supernatant liquor from the bi-ox unit to a digester vessel, where some of the sludge is converted to a combustible gas. After digestion is complete, the remaining sludge may be dewatered, dried and disposed of in a sanitary landfill. The procedure may be complicated by the presence of leachable trace contaminants, which would then require disposal in a class 1 site.

3.5.3.5 Effects of Low-Rank Coal Properties

The properties associated with coal rank have significant influence on gasification behavior, process selection, and process design. The predominant effects are on the gasifier itself, but properties of the coal do affect many of the other unit operations as well. Although there are both positive and negative factors involved, low-rank coals are basically preferable feedstocks for the commercially available gasification processes. In fact, a major portion of the ongoing R&D effort in gasification process development is motivated by the inherent difficulty of gasifying caking bituminous coals. Nevertheless, there are some unique problems associated with low-rank gasification, which are identified below.

Caking Properties

Bituminous coals become plastic and agglomerate to form coke as they pass through the temperature range 750-950°F. Coal particles tend to adhere to each other in this state, decreasing the surface area for reaction, and in severe cases plugging up the reactor. Low-rank coals do not have this property. Depending on the type of gasifier, this allows the low-rank coals to be processed in a simpler, less expensive coal preparation and reactor system.

The caking problem is most severe in the fixed bed gasifier type. Some pretreatment step to reduce or eliminate the caking tendency may be used; oxidation of the coal particle surfaces at 700-750°F is one pretreatment step that works for some caking coals. The Slagging BGC-Lurgi fixed-bed process, with a mechanical stirrer in the proper reactor zone, has been operated successfully on nontreated caking coals in short-term tests at Westfield, Scotland. Lignite and subbituminous coals do not require anti-caking pretreatment or stirrers (although drying, size reduction, and classification may still be necessary).

In fluid bed reactors, the caking problem can be controlled by keeping the coal particles so diluted (with char, ash, or other inert materials) that agglomeration is prevented. For example, in the two-stage pressurized process operated by Westinghouse for DOE, a high ratio (up to 100:1) of recycle solids to coal feed is maintained to control agglomeration of the coal. This degree of dilution is not necessary when operating fluid bed reactors on low-rank coals.

Entrained flow reactors can handle caking and non-caking coals equally well because particle interaction is minimized in the flowing gas stream.

Moisture Content

The very high inherent moisture content of low-rank coals acts primarily as a diluent in the gasification process; less carbon is available for reaction per pound of feed compared to lower moisture content (high-rank) coals. Predrying of the coal requires considerable heat and causes particle size degradation (generation of fines). Coal fines can be accepted as feed to fluid bed or entrained flow gasifiers, and thus a predrying step is feasible in conjunction with these gasifiers if the economics justify it. Fixed bed gasifiers require a feed free of fines and thus are not well suited to pre-dried low-rank coals due to their tendency to decrepitate.

Gasification of high-moisture coal in fixed bed units results in large volumes of waste liquor. The moisture and volatile matter are evaporated in the upper portion of the gasifier due to the countercurrent flow temperature profile. This results in a requirement for large gas liquor separation and waste water treating sections.

Some U.S. lignites have so much moisture that they cannot be used directly as feed to slagging fixed bed gasifiers. For example, the moisture content of Gascoyne lignite (greater than 40%) exceeded the limit for which successful operation could be obtained in the GFETC slagging fixed bed gasifier. The heat produced from the gasification/combustion reactions in the lower region of the gasifier was found to be insufficient to vaporize the moisture in the incoming feed coal. As a result the temperature of the fuel bed gradually decreased and slag flow stopped.[15,a]

The Texaco entrained flow gasifier utilizes a slurry feed system. In order to obtain adequate coal concentrations in the slurry, drying of high-moisture lignite prior to the slurry preparation step may be required. The Koppers-Totzek gasifier also requires that high-moisture lignites be dried.

Coal Feed Properties

Fixed bed gasifiers require lump particle sizes with a minimum of fines. This imposes a design requirement on fixed bed, low-rank coal-fed plants that a suitable use for the large quantity of fines can be found. The approach taken by the Great Plains Gasification Associated project is a good example. In that project, the gasification plant will be integrated with a pc-fired power plant. Screened lignite (plus 1/4 inch) will be fed to the Lurgi gasifiers at a rate of 28,400 tons per day; the 15,600 tons per day of undersize coal will go to the power plant.[30] A similar proposal by Fluor Corporation would involve the integration of a slagging Lurgi (fixed bed) gasifier with a Texaco (entrained-bed) gasifier. The plus 1/4 inch portion of the coal would be fed to the Lurgi unit; the minus 1/4 inch portion would be slurried with phenolic liquor (recovered downstream of the Lurgi unit) and fed to the Texaco gasifier.[7] Other proposed designs call for briquetting of the coal fines, or injection into the bottom of the gasifier along with the steam.

For gasifiers which utilize a slurry feed system (e.g., high pressure entrained flow systems such as Texaco), the properties of low-rank coal may present problems. Coal slurries may be comprised of either oil or water as the slurry agent. For higher rank coals in water slurries, the solids concentration is limited by the requirement that the slurry be pumpable. In the case of lignite, however, the solids concentration is limited by the high moisture content of the coal, or in the case of dried lignite, by the high reabsorption of moisture. Some drying processes may remove moisture in a way that limits reabsorption and thus allows a more concentrated slurry to be formed. Such a claim is made for the Koppelman Process, which involves the heating of lignite in water under pressure (1000°F, 1500 psi).[29]

Reactivity

Lignite and subbituminous coals are in general more reactive in gasification reactions than higher rank coals. For the gasification step, this may mean higher reactor space velocities (higher throughputs), higher carbon conversions, and lower gasification temperatures.

This factor might be expected to have more importance in the gasifiers which involve shorter gas-solid contact or residence times, namely the entrained flow and fluidized bed types. For example, it has been estimated that the Koppers-Totzek (entrained flow) process can obtain conversion efficiencies of greater than 98 percent when using lignite, about 85 percent for subbituminous coal, and a maximum of 60 percent for anthracite.[3]

[a]References for section 3.5.3.2-3.5.3.6 are listed following section 3.5.3.6.

The CO_2 Acceptor process is an example of a process designed specifically for highly reactive low-rank coals, and is only operable with that type of feedstock.

Volatiles

In fixed-bed gasifiers, liquid materials such as tars, oils, phenols, naphthas, and other components are produced largely through devolatilization mechanisms. The chemical composition of these substances produced from low-rank coals is significantly different from the devolatilization products of higher rank coals.[9] Therefore, separation, treatment, and utilization of these by-products will be different in plants using low-rank coals than in plants using high-rank coals.

In high-temperature gasification processes such as the entrained flow systems, tar and oil are consumed, thus eliminating the need for complex gas liquor treatment facilities.

Ash Characteristics

The ash of low-rank coals has significantly different chemical and physical properties from high-rank coal ash, which can affect gasifier design and performance in a number of ways. Alkaline components of low-rank coal ash may act catalytically to promote gasification reactions, and may explain in part the higher reactivity attributed to lignite and subbituminous coals. In addition, these ashes will display a different temperature-viscosity relationship than higher rank coal ash and therefore require different gasification temperatures to maintain slagging or non-slagging conditions. Behavior of these ash materials toward refractory materials and sulfur capture in the gasifier is also expected to be different than for ashes of high-rank coals.

Sulfur Content

Low-rank coals contain on the average a significantly lower sulfur content than higher-rank coals. Although few low-rank coal sources are low enough in sulfur to eliminate the need for sulfur removal in the acid gas recovery section, the lower sulfur content is expected to reduce the operating requirements of acid gas scrubbers. This is particularly true for selective H_2S scrubbing systems, and a lower capital cost for these systems may also be expected.

3.5.3.6 Current Projects

In this section, leading commercial and developmental gasifiers are described and their development status discussed relative to both foreign and domestic activities. The specific effects of low-rank coals on each gasifier type are compared and contrasted to the use of higher rank coals in the same gasifier. The unique environmental control considerations which may be associated with each specific gasifier type are also indicated. Unfortunately, in many individual cases specific data concerning the performance of low-rank coals is not available and therefore remains uncertain.

3.5.3.6.1 Fixed Bed Gasifiers

Single Stage Gasifiers

Lurgi Dry Bottom Gasifier

The first commercial demonstration of the Lurgi gasifier took place in Germany in 1936. Although there are no commercial scale installations in the United States, the 18 plants world-wide are summarized in Table 3.5.3.8. The dry bottom Lurgi process is shown schematically in Figure 3.5.3.9 and the gasifier is depicted in Figure 3.5.3.10. The gasifier is a vertical cylindrical steel pressure vessel varying from 8 to 12 feet in diameter and operating at 350 to 450 psig. Screened coal in the 1/4 to 1-1/2 inch size range is introduced into the system via a lockhopper mounted above the reactor. A motor driven distributor insures even allocation of the coal over the coal bed. The bed depth varies from seven to ten feet. Coal moves slowly downward under the influence of gravity and against a countercurrent flow of gas. Gasification of caking coals cannot be achieved without a rotating mechanical stirrer in the gasifier. Steam and either air or oxygen are injected at the bottom of the bed through the slowly rotating grate.

Raw gas is removed from the top or side of the unit at a temperature between 700 and 1100°F and flows into a scrubber cooler where it is washed by a circulating liquor stream. Typically, the gas passes through a waste heat steam generator and one or more additional cooling stages in which condensable materials such as oils, tars, and water vapor are knocked out of the process gas stream. The low- or medium-Btu gas leaving this section is usually now processed for acid gas removal and is then available as a fuel or synthesis gas.

Table 3.5.3.8

Lurgi Gasifier Installations

Location	Year	Fuel	Gasifier I.D.	Capacity (MM SCFD)	No. of Gasifiers
Hirschfelde, Central Germany	1936	Lignite	3'9"	1.1	2
Bohlen, Central Germany	1940	Lignite	8'6"	9.0	5
Bohlen, Central Germany	1943	Lignite	8'6"	10.0	5
Most., CSSR	1944	Lignite	8'6"	7.5	3
Zaluzi-Most., CSSR	1949	Lignite	8'6"	9.0	3
Sasolburg, South Africa	1954	Subbituminous	12'1"	150.0	9
Dorsten, West Germany	1955	Caking Subbituminous	8'9"	22.0	6
Morwell, Australia	1956	Lignite	8'9"	22.0	2
Daud Khel, Pakistan	1957	High Volatile Coal	8'9"	5.0	2
Sasolburg, South Africa	1958	Subbituminous	12'1"	19.0	1
Westfield, Great Britain	1960	Weakly Caking	8'9"	28.0	3
Jealgora, India	1961	Different Grades	N/A	0.9	1
Westfield, Great Britain	1962	Weakly Caking Subbituminous	8'9"	49.0	1
Coleshill, Great Britain	1963	Caking Subbituminous	8'9"	46.0	5
Sasolburg, South Africa	1966	Subbituminous	12'1"	75.0	3
Luenen, GFR	1970	Subbituminous	11'4"	1400 MM Btu/hr.	5
Sasolburg, South Africa	1973	Subbituminous	12'4"	190.0	3

Source: Reference 14,a

[a]References for sections 3.5.3.2-3.5.3.6 are listed following section 3.5.3.6.

Figure 3.5.3.9 Lurgi Gasifier

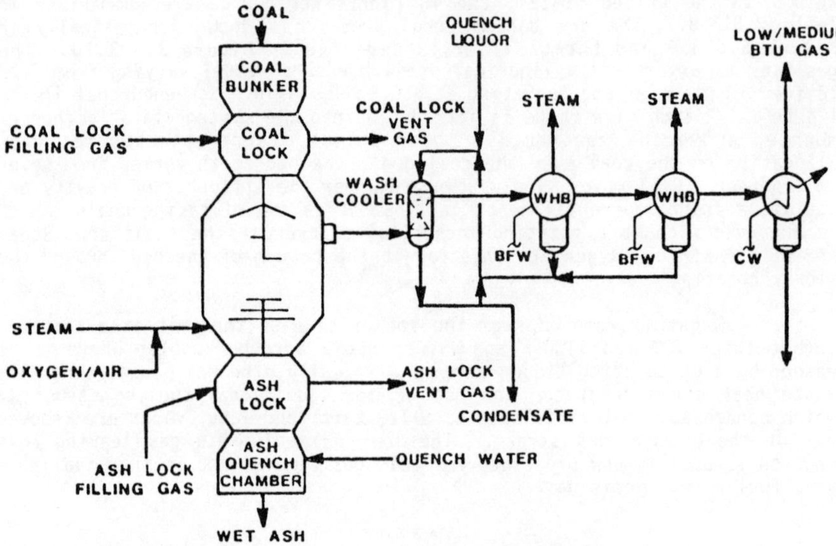

Source: Reference 5

Figure 3.5.3.10 The Dry Bottom Lurgi Gasifier

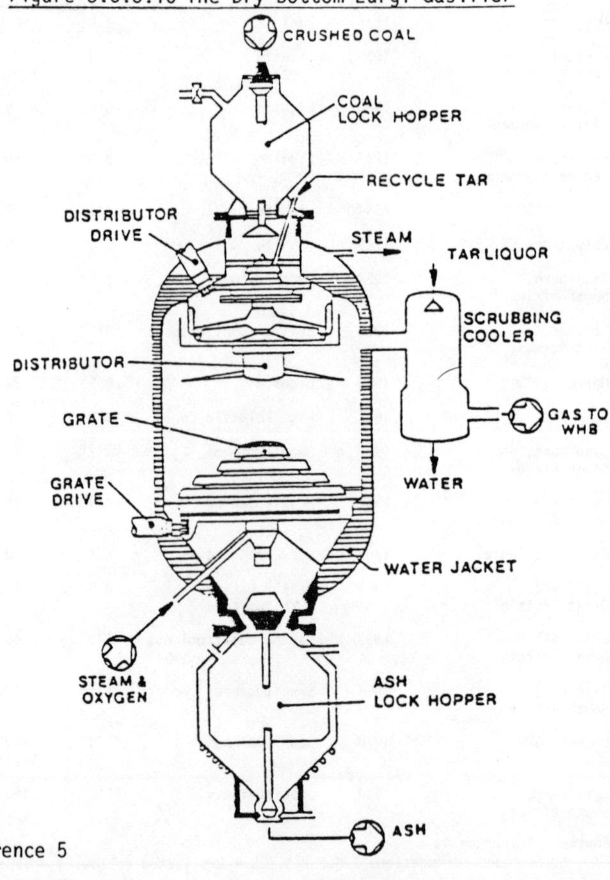

Source: Reference 5

The use of air or oxygen for gasification will depend upon the end use of the gas. Significant differences between these two operating modes can be noted in Table 3.5.3.9, where some selected operating data are provided. Comparison of the gas production rates reflects the nitrogen diluent present in gasification air. The dry raw gas composition indicates the higher nitrogen content for the airblown case. The higher hydrogen content of the oxygen blown product gas is due almost entirely to the absence of the nitrogen diluent in the gas. Methane content of the oxygen blown gas is considerably higher due to the higher partial pressure of hydrogen, which may form methane by reaction with carbon, carbon monoxide, or volatiles. Although comparisons are difficult due to differences in pressure, coal and steam rate, this may explain the fact that the carbon monoxide contents are equal in both cases, whereas we would expect CO to be higher in the oxygen blown case due to the lack of diluent. The higher gas heating value for the oxygen blown case reflects the absence of nitrogen diluent, but on an equivalent basis (if the nitrogen were removed from the air blown product gas) the air blown gas would have the higher heating value.

On a lb O_2/lb coal basis, the air blown case displays a lower consumption (a range of 0.3 to 1.0 lb O_2/lb coal for air versus 0.5 to 1.4 lb O_2/lb coal for oxygen). Temperature control is thus achieved at higher steam expense in the oxygen case, producing the same gas outlet temperatures.

The Lurgi dry ash gasifier has been selected for a number of proposed SNG plants in the U.S., including the Westco, Panhandle Eastern, NGPL, and Great Plains Gasification Associates (GPGA) projects. The following paragraphs provide preliminary design data on the GPGA plant proposed for Beulah, North Dakota.[14]

Phase 1 of the coal gas project will use a multiple train of Lurgi pressurized, fixed-bed gasifiers. These gasifiers, of the new Mark IV design, will operate at a pressure of about 430 psig, reacting steam and oxygen in countercurrent flow with coal. Present design for the Phase 1 plant contains 14 gasifiers. The material balance calls for operation of 11 or 12 gasifiers, each at an output of 52,000 normal cubic meters (about 1.84 million SCF) per hour of raw gas. This allows for a minimum of 2 spare gasifiers. Phase II will add 12 gasifiers, giving a total of 23 operating gasifiers and 3 spares. With an individual expected gasifier availability of 87 percent, GPGA anticipates an overall availability for the gasifier area of about 100 percent. The major maintenance item contributing to down time in this area is expected to be lock hopper valve assemblies, which will be replaceable on a unit basis.

The crude gas leaving the gasifiers has a heating value of 310 Btu (HHV). One-third of this gas is passed through shift converters, containing a sulfur-resistant cobalt molybdenum catalyst. Carbon monoxide reacts with steam, adjusting the H_2/CO ratio to slightly more than 3 to 1 in preparation for methanation. The two-thirds crude gas and the shifted gas are cooled, removing tar, tar oils, and other compounds such as ammonia and phenols.

Gas purification continues with a low-temperature methanol wash to remove carbon dioxide, sulfur compounds, and naphtha. Sulfur-containing gases from this Rectisol system will be converted to elemental sulfur in parallel Stretford units. Sulfur-free synthesis gas, on the other hand, passes to hot gas recycle methanators, which react carbon oxides and hydrogen over a reduced-nickel-base catalyst to form methane and water.

A clean-up methanator removes the remaining traces of carbon monoxide, leaving a gas with a higher heating value of 977 Btu per cubic foot (dry basis). The gas is cooled, dried, and compressed to pipeline pressure of 1,440 psig in two stages of compression.

Table 3.5.3.9

Lurgi Gasifier
Typical Operating Data

	Low BTU Gas	Medium BTU Gas
Gas Production, MM SCF/hr	2.3	1.3
Gas Composition, Volume %:		
Hydrogen	23	39
Carbon Monoxide	17	17
Methane	5	10
Nitrogen	39	1
Carbon Dioxide	15	31
Other	1	2
Total	100	100
Gas Higher Heating Value, BTU/SCF (Dry Basis)	198	298
Feed Requirements, lb/lb coal:		
Air	1.3-4.2	
Oxygen		0.5-1.4
Steam	0.6-1.3	2.0-7.0
Operating Conditions		
Gas Outlet Temperature, °F	700-1100	700-1100
Pressure	20 atm.	30 atm.
Residence time, minutes	60	60
Turndown Capability	4:1	4:1
Coal Feed:		
Type	Subbituminous	Bituminous
Size, Inches	0.25-1.25	0.25-1.25
Free Swelling Index	2	3
Ash Fusion Temperature, °F		
Feed Rate, tons/day		200-500
Thermal Efficiency, %:		93
Hot Raw Gas		68
Cold Clean Gas		
By-Product Rate, lb/lb coal:		
Tar		
Oil	0.07	0.08
	0.07	0.008

NOTE: Table may not be internally consistent due to averaging of values.

Source: Reference 14

BGC-Lurgi Slagging Gasifier

In 1955 the British Gas Council in collaboration with the Ministry of Power purchased an experimental three foot diameter slagging gasifier which the German Lurgi Company had been developing at Aberhauser-Holten.

The gasifier was erected at Solihull, England, and testing and development were conducted there until 1964 when further work was suspended indefinitely as a result of the North Sea gas discovery. In 1975, the British Gas Corporation resumed slagging gasification development at Westfield, Scotland under the sponsorship of Conoco Coal Development Company and other U.S. sponsors. The Westfield slagging gasifier is similar in design to the Lurgi dry ash gasifier (and in fact is a scaled down version of the dry Lurgi), but the Westfield test unit is somewhat smaller (6 feet) in diameter. Oxygen and steam injection are accomplished through tuyeres located in the gasifier wall toward the bottom of the bed. A water jacket encircles the vessel, and refractory lining is used inside the steel vessel wall. Slag is intermittently withdrawn from a slag tap hole at the bottom of the bed, below which it is quenched with water in an unlined vessel. Slag and quench water removal takes place through a slag lockhopper located below the quench vessel. Coal is introduced through a lockhopper and distributor arrangement, similar to the non-slagging version. Figure 3.5.3.11 shows a process flow diagram schematic for a proposed SNG demonstration plant based on high sulfur Ohio coal, sponsored jointly by Conoco and DOE.

Figure 3.5.3.11

British Gas/Lurgi Slagging Process for Proposed DOE/Conoco Ohio Demonstration Plant

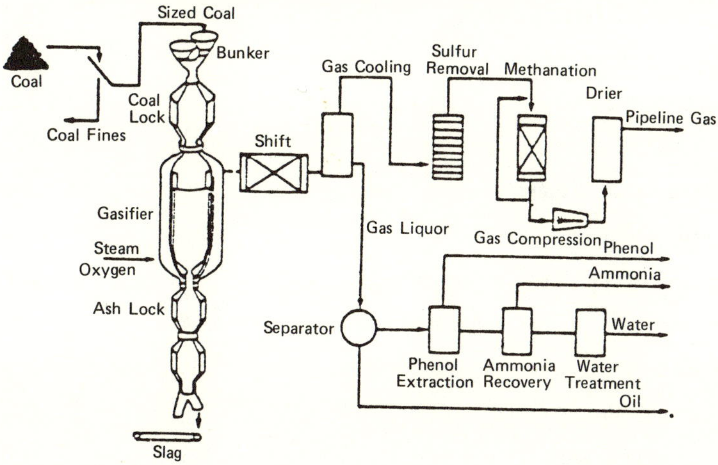

Source: Reference 5

In this design, the raw gasifier effluent passes through a shift conversion followed by cooling of the gas stream and condensation of the gas liquor. Wastewater treatment recovers oils, phenols, ammonia, and produces a treated water stream. The cooled gas stream is treated for acid gas removal and naphtha recovery, and sulfur is produced as a by-product of this processing sector. The clean, cooled gas is then available for methanation and drying before being sold as pipeline gas. Utility requirements are met by gasification of approximately 18 percent of the coal in a dry ash Lurgi gasifier, which is used to fire a boiler plant to produce steam and electricity.

Table 3.5.3.10 compares the slagging and non-slagging versions of the Lurgi gasifier. It is emphasized that data for the slagging unit have been provided from an experimental test model, and therefore may not be as representative of what can be expected of commercial scale operation as the data from the non-slagging version. A notable difference in crude gas compositions is in the content of carbon dioxide. The higher concentration of CO in the slagging case reduces hydrogen concentration by the water gas shift reaction, relative to the non-slagging case. The higher total concentration of CO and H_2 for the slagging gasifier reflects its crude gas higher heating value.

Table 3.5.3.10

A Comparison Between the Performances of a Commercial Lurgi Gasifier and the Experimental Slagging Gasifier

	Lurgi gasifier	Slagging gasifier Westfield
Operating pressure, psi	355	300
Cross-sectional area of shaft, ft^2	66	28
Fuel	Pittsburgh #8	Pittsburgh #8
Rank	902	902
Size range, in.	1-1/2 to 1	1-1/2 to 1
Ash (including flux), %	14.6	11.4
Moisture, %	15.6	12.7
Steam/oxygen ratio, vol/vol		1.10
Crude-gas composition, vol % (dry basis)		
CO_2	24.6	6.5*
CH_4	8.7	6.2*
CO	24.6	59.8*
H_2	39.8	26.3*
C_nH_m	1.1	0.5*
N_2	1.2	0.7*
	100.0	100.0
Calorific value, Btu/ft^3	309	345
Gasifier loading	Normal	Normal
Steam consumption, ton per ton of coal	1.46	0.31
Oxygen consumption, scf per ton of coal	12,474	13,580
Crude gas, except lock gas, MMBtu/ton	25.2	24.6
Fuel rating, lb/hr per ft^2	210	981
Thermal rating, MMBtu/hr per ft^2	2.36	11.3
Net tar yield, gal per ton of coal	10.8	15.4
Liquor yield, ton per ton of coal	1.36	0.25
Gasification efficiencies		
a. $\frac{\text{Crude gas}^a + \text{tar} + \text{oil} + \text{benzole}}{\text{coal gasified}} \times 100$	89.0	90.0
b. $\frac{\text{Crude gas}^a}{\text{Coal for gasification \& steam raising}} \times 100$	68.3	76.2
c. $\frac{\text{Crude gas}^a}{\text{Coal for gasification, steam raising and equivalent of power used for oxygen production}} \times 100$	62.6	68.3

[a]Crude gas excluding lock gas and benzole.

Source: Reference 5
*Source: Reference 28

One very significant factor in determining gasifier performance is the steam consumption of the non-slagging version. Steam is injected to limit the reaction temperature below conditions of slagging and clinker formation, and has an important effect on gasifier efficiency and waste liquor yield as is reflected by some of the other figures in the Table. Although gasification in the slagging gasifier occurs at temperatures above 2500°F, its oxygen consumption is only slightly higher than for the non-slagging version in which bed temperatures are of the order of 1300°F.

In choosing a coal feedstock for either the slagging or non-slagging Lurgi gasifier, several considerations apply:[5]

- Coal should range in size from 1/8 to 1-1/2 inch
- Moisture level should be below 35 percent
- Non-caking coals should be used unless a mechanical stirrer can be provided
- Up to 10 percent coal fines (<1/8") may be used if they are injected with the steam at the bottom of the bed

The characteristics of low-rank coals have the following implications with respect to their use in Lurgi gasifiers:

- The absence of any caking tendencies for low-rank coals allows their use in either version of the Lurgi gasifier without any mechanical stirring devices
- Low-rank coals having low ash fusion points may be susceptible to clinker formation in the non-slagging Lurgi if operating temperatures approach the initial deformation temperature
- Some low-rank coals have moisture contents in excess of 35 percent requiring drying before gasification. The tendency of low-rank coals to produce considerable quantities of fines during drying may limit the use of low-rank coals in the Lurgi process without some concurrent use for the fines generated
- The "low reactivity" of bituminous coal in a dry ash Lurgi is the reason given by Conoco for advocating the slagging design particularly for bituminous coal. By this criterion, low-rank coals are suitable for dry ash fixed bed gasifiers

Due to the significant production of tars, oils, phenols, naphthas, hydrocarbons, and ammonia, the Lurgi process requires extensive wastewater treating facilities. However, no other unique environmental problems are characteristic of the Lurgi gasifier.

GFETC Slagging Fixed Bed Gasifier

GFETC is operating the only slagging fixed bed gasifier in the United States. The pilot plant gasifier was first operated under the Bureau of Mines during the period 1958-1965 to demonstrate the feasibility of slagging operating and assess operational parameters. Operations were resumed in 1976 to investigate the environmental concerns associated with commercial scale fixed bed coal gasification facilities. In September 1978, gasifier operation was suspended for modifications and updating of the existing equipment and to expand the operational capabilities. The facilities as originally constructed were designed for gasification of non-caking coals. Experiments were generally of 8 to 12 hours duration because of equipment and personnel limitations. The present program provides modifications which will permit tests of up to 5 day duration and operational capability on Eastern bituminous coals. Operation of the pilot plant resumed during 1980. Figure 3.5.3.12 shows the details of the gasifier as modified.

Figure 3.5.3.12

Cross Section of Modified GFETC Slagging Gasifier

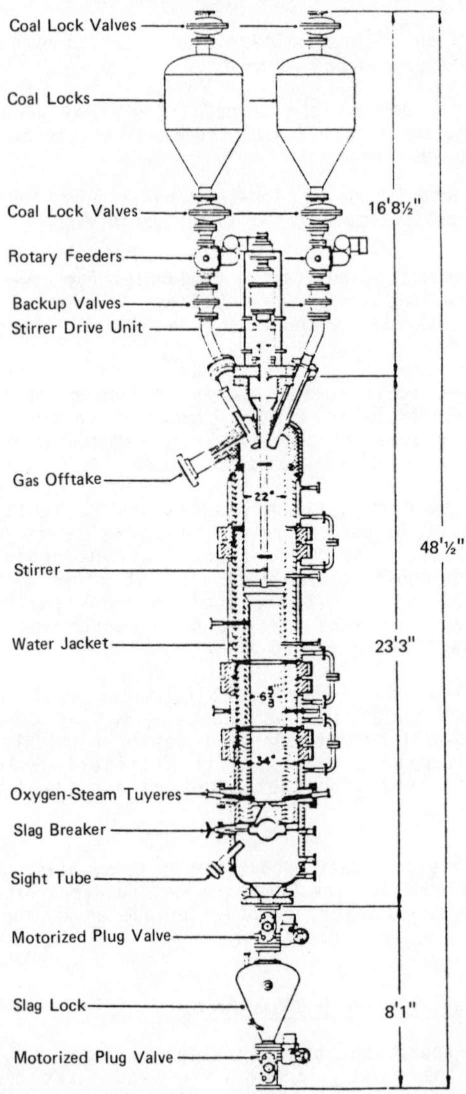

Source: Reference 15

Formerly, the unit had only one lockhopper and no stirrer. Figure 3.5.3.13 shows the process flow sheet for the modified pilot plant. Up to 1 ton/hr of lignite can be gasified in the unit at 400 psig operating pressure.

Coal moves by gravity flow from the lockhopper into the gasifier. The coal is continuously gasified by an oxygen-steam mixture introduced through four water-cooled tuyeres at the bottom of the gasifier. Molten slag is formed at the hearth and flows through a central one-inch diameter taphole into a water quench bath. Gas from the high-temperature reaction zone can be drawn through the taphole to aid slag flow. This gas is cooled and metered in a separate circuit.

The product gas leaving the gasifier contains water vapor and tar. It is scrubbed in a spray washer with condensed recycle liquor. The gas liquor is periodically discharged from the spray washer to a settling tank. The washed gas is then cooled to 60°F in an indirect cooler before being sampled, metered, and flared. The heating value of this gas is approximately 350 Btu/scf.

Figure 3.5.3.13

Process Flow Sheet of Modified Gasifier System

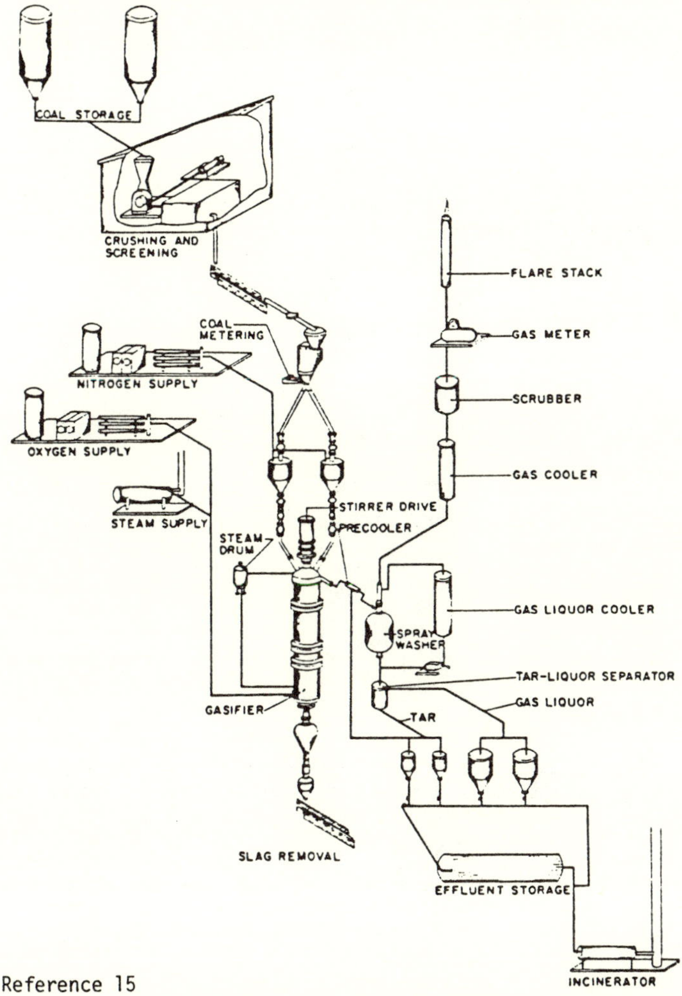

Source: Reference 15

Table 3.5.3.11 shows some of the results obtained in the GFETC unit (prior to the modifications) for four different low-rank coals at the same operating conditions.

The most critical operating problem with this gasifier has been the failure of hearth zone materials and interruption of slag flow. Various types of hearth plate materials in combination with different types of cooling coils and taphole inserts have been tested. Erosion occurred on all of the refractory hearth plates in areas that weren't sufficiently cooled.

Table 3.5.3.11

GFETC Slagging Gasification Results for Three Lignites and a Subbituminous Coal at the Same Operating Conditions

Run Number	RA-31	RA-40	RA-57 [1]	RA-43 [1]
Operating Pressure, lb/sq.in.	200	200	200	200
Coal	Baukol Noonan	Indian Head	Gascoyne	Rosebud
Moisture in Coal, pct	30.3	34.7	40.5	22.0
Test Conditions:				
Oxygen rate, std.cu.ft./hr.	4,000	4,000	4,000	4,000
Oxygen/steam molar ratio	1.0	1.0	1.0	1.0
Calculating period, hr.	11.7	8.9	3.6	1.3
Fuel Rate (as-charged), lb./hr	1,180	1,264	1,364	1,063
Fuel Rate (Maf), lb./hr.	751	747	709	709
Product Gas Rate, std.cu.ft./hr.	21,736	22,038	22,841	19,052
Slag Rate, lb./hr.	95	54	66	93
Oxygen Consumption:				
cu.ft./1000 cu.ft. gas	185.4	181.5	175.1	210.0
cu.ft./1000 cu.ft. CO + H_2	211.6	212.5	201.7	236.1
cu.ft./lb. maf fuel	5.33	5.35	5.54	5.65
lb./lb. maf fuel	0.45	0.45	0.48	0.48
Steam Consumption:				
lb./1000 cu. ft. gas	8.81	8.62	8.32	9.97
lb./1000 cu. ft. CO + H_2	10.05	10.09	9.58	11.22
lb./lb. maf fuel	0.25	0.25	0.14	0.27
Production, cu.ft./lb. maf fuel.				
Gas	28.72	29.49	32.22	26.89
CO + H_2	25.17	25.19	27.98	23.91
Product Gas Composition, pct:				
CO_2	6.4	8.4	7.7	4.1
H_2	30.5	32.5	31.3	26.9
CO	57.3	53.6	56.1	62.6
C_2 -C_4 hydrocarbons	0.7	0.6	0.6	0.9
CH_4	5.1	4.9	4.3	5.5
Gas Heating Value, Btu/SCF(HHV).	348	339	337	361

[1] Results determined on tests of limited duration. Gasifier may not have reached equilibrium.

Source: Reference 15

Results from testing refractory hearth plates demonstrated the need to develop an alternate approach to the problem. A water-cooled metal hearth plate was designed as shown in Figure 3.5.3.14. Water flow is maintained at a high rate to prevent film boiling. The water flow pattern through the hearth plate is shown in the figure. Fifteen runs using the water-cooled hearth plate were made and slag flow was achieved during each test. One test of 25 hours was terminated only due to exhaustion of the coal supply. This type of plate offers the best potential for extended operation. However, the design of this particular hearth plate is not scaleable to commercial size (i.e., 14 foot diameter) with its single slag tap hole. A different geometric configuration using multiple tapholes would be required for that scale.

Figure 3.5.3.14

GFETC Slagging Gasifier Hearth Design and Slag Flow Control System

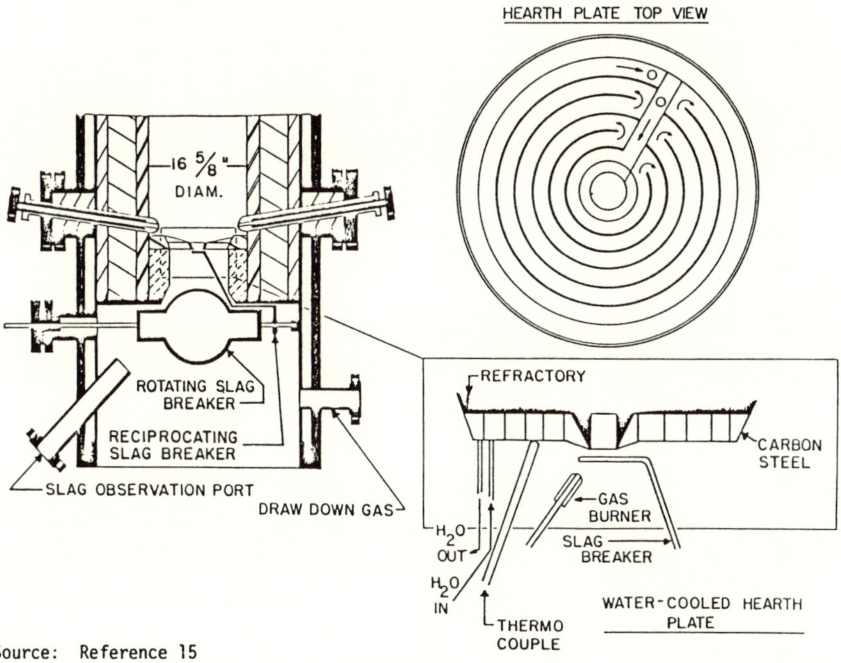

Source: Reference 15

The main thrust of the GFETC gasification project is to develop the support data base required for scaleup and demonstration of the slagging fixed-bed gasification process, with emphasis on environmental-related data on by-products and effluents. Tests will be run with a variety of coals to establish production rates of effluents, fate of trace elements, waste disposal constraints, and occupational health and safety factors. These studies will identify process limitations, and differences in products and effluents, as a function of coal source, type, chemical and physical properties, and moisture content.

Wellman-Galusha

The Wellman-Galusha coal gasification process was first used commercially in 1941 in Germany and is currently licensed by the McDowell-Wellman Engineering Company of Cleveland, Ohio. It is an atmospheric pressure reactor that can be operated with a steam/air or steam/oxygen gasifying medium. Eight systems are currently in operation in the United States. Some 150 gasifiers have been installed world-wide since its introduction in 1935. Table 3.5.3.12 summarizes the later installations of these gasifiers and their application. All of the earlier installations involved only one gasifier, whereas later plants have multiple units. Many gasifiers are still operating; the majority of these are air blown systems for a wide variety of applications using a wide range of coal types.

Table 3.5.3.12

Commercial Installations of McDowell-Wellman Gasifiers

Company/Location	No. of Units	Gasifier Diameter, ft.	Coal Type	Gas BTU/SCF	Application	Status
Gypsum Lime Ltd. Beechville, Ont.	1	10 with agitator	Bituminous	Low	Lime Kiln	Inoperative since 1965
Stelco Limited Beechville, Ont.	1	10 with agitator	Bituminous	Low	Heat Treating Furnace	Inoperative since 1963
Mississippi Lime, St. Genevieve, MO	1	10 with agitator	Bituminous	Low	Lime Kiln	Inoperative since 1964. In place.
Union Carbide, Nopco Chemical Division	1	3.5	Coke	Medium	Chemical Feedstock	Inoperative. In place.
NL Industries National Lead Division, South Amboy, NJ	1	2	Coke	Medium	Chemical Feedstock	Inoperative since 1966. In place.
Allied Chemical, Ltd. Corunna, Ont.	1	3	Bituminous	Low	Fuel Gas	Inoperative
New Jersey Zinc Ashtabula, Ohio	1	6.5 with agitator	Coke	Medium	Chemical Feedstock	Operating
U.S. Bureau of Mines Morgantown, West Virginia	1	3.5	Bit./Lignite	Low-Med.	Experimental with 300 psi pressure	Operating
U.S. Bureau of Mines Twin Cities, Minn.	1	10	Bit./Lignite	Low	Iron Ore Pellet Kilns	Operating
Olin-Mathieson Ashtabula, Ohio	1	5	Petr./Coke	Medium	Feedstock in Producing Phosgene	Operating
Riley-Stoker Worcester, Mass.	1	2	Bit. anthr.-coke, lignite	Low	Performing with with various coals	Operating
Glen-Gery Corp. Reading, PA	12	10 with agitator	Anthracite	Low	Brick Kilns	Operating
National Lime & Stone Co. Carey, Ohio	2	10 with agitator	Bituminous	Low	Lime Kiln	Operating
Pikeville Energy Center Pikeville, Ky.	2	6.5 with agitator	Bituminous	Low	Commercial Fuel Gas	1979 Operation
Can Do, Inc. Hazelton, PA	2	6.5	Anthracite	Low	Industrial Fuel Gas	1979 Operation
Taiwan Fertilizer Co. Formosa	7	10	Korean Anthr.	Medium	Ammonia Feedstock	Operating
Nickel Processing Corp.	14	10	Bituminous	Low	Industrial Fuel Gas	Operating

Source: Reference 14

The gasifier is a vertical cylindrical steel vessel with a diameter ranging from 1-1/2 to 10 feet. The vessel is either refractory lined or water jacketed for cooling. Crushed coal is fed to the gasifier through vertical feed pipes. The fuel bed can be unstirred or stirred. Agitation is a necessary requirement with caking coals but capacity increases of up to 35 percent can also be realized with the stirred bed. A steam and air/oxygen mixture is introduced from the bottom of the vessel and the product gas exits from the top of the unit. Gas leaving the vessel is at about 900-1200°F and can be used directly as a fuel gas after passing through a ceramic lined cyclone to remove particulate.

Table 3.5.3.13 shows typical operating conditions for air-blown and oxygen-blown gasification. The gas composition differs in hydrogen, carbon dioxide, and carbon monoxide content very nearly to the extent that nitrogen is present as a diluent in the air-blown gas product. Placed on an equivalent basis, the two designs consume similar amounts of oxygen per pound of coal. Operating conditions and other parameters are also similar.

Table 3.5.3.13
Wellman-Galusha Gasifier
Typical Operating Data

	Low BTU Gas	Medium BTU Gas
Gas Production, MM SCF/hr	0.3 - 0.4	
Gas Composition, Volume %:		
Hydrogen	19	36
Carbon Monoxide	25	47
Methane	1	1
Nitrogen	49	2
Carbon Dioxide	6	14
Other	-	-
Total	100	100
Gas Higher Heating Value, BTU/SCF (Dry Basis)	160 - 210	258 - 270
Feed Requirements, lb/lb coal:		
Air	3.5	
Oxygen		0.8
Steam	0.4 - 0.7	
Operating Conditions		
Gas Outlet Temperature, °F	800 - 1200	800 - 1200
Pressure	5-6 in. H_2O	5-6 in. H_2O
Residence time, minutes	120 - 500	120 - 500
Turndown Capability	4:1	4:1
Coal Feed:		
Type	Bituminous	Bituminous
Size, Inches	1.25 - 2.0	1.24 - 2.0
Free Swelling Index	no limit	no limit
Ash Fusion Temperature, °F	2100	
Feed Rate, tons/day	84	
Thermal Efficiency, %:		
Hot Raw Gas	91 - 95	91 - 95
Cold Clean Gas	75	75

NOTE: Table may not be internally consistent due to averaging of values.

Source: Reference 14

The DOE Morgantown Energy Technology Center has an experimental Wellman-Galusha gasifier modified for pressurized operation in the 100 to 300 psig range. The project's objective is to produce a low-BTU gas for testing, cleanup, and combustion systems. Figure 3.5.3.15 shows a process flow design diagram of this installation.

From the gasifier, the gas is passed through a cyclone and a tar condenser to remove particulates and tar. It is then scrubbed with water in a venturi scrubber, a disengaging chamber, and a direct cooler/scrubber for final cleanup and cooling. H_2S is removed and converted to sulfur in a high pressure Stretford plant. The gas is passed through a pressure letdown valve and vented.

Figure 3.5.3.15

Wellman-Galusha Gasifier at METC

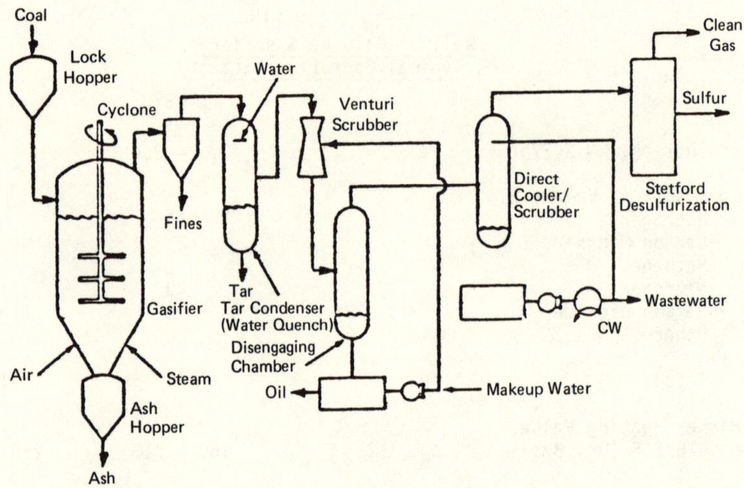

Source: Reference 14

The pilot plant test program will continue in FY 1980, and a tar combustor and novel waste water evaporator will be added. The products from the tar combustor will be reinjected into the producer; the waste water evaporator precludes the need for a conventional water treatment system.

Advantages claimed for the Wellman-Galusha design include the fact that it is a mature technology. Its high fuel inventory due to the fixed bed design provides for greater safety and a high turndown ratio. High efficiencies result from the countercurrent flow of gases and fuel and the long coal residence time (up to four hours) in the gasifier.

However, the advantages are moderated by several drawbacks. For example, the commercial sized unit has not been operated under pressurized conditions and therefore does not have the advantage of increased throughput. Coals with low ash softening points may require the addition of steam to maintain bed temperatures below slagging conditions, adversely affecting gasifier efficiency. In addition, high moisture content coals may cause a problem in handling, and can lead to large amounts of tar liquor condensation from the exit gas. Formation of such tars reduces the carbon available for conversion to combustible gas (a problem inherent in the fixed bed design).

Coal to be gasified must normally be crushed to 1 to 2 inches requiring a slight energy investment. As much as 20 percent of the coal may be 1 inch or smaller. Ash softening temperatures of 2200°F or higher are preferred; but may be as low as 1800°F if steam is added.

Low-rank coals also have the following characteristics which may affect their performance in the Wellman-Galusha gasifier:

- Moisture content - the high moisture level in some low-rank coals may cause problems in the handling of the crushed coal, and may reduce the temperature of the gasifier effluent. High levels of moisture in the feed coal will lead to increased production of tar liquor in the gasifier effluent.

- Ash characteristics - those low-rank coals having ashes with lower softening points will require close temperature control and possibly additional steam injection. This will reduce gasification efficiency, and increase tar liquor yield.

Special environmental considerations for the Wellman-Galusha gasifier should be directed to wastewater treating in cases where tar production is higher than average.

Riley-Morgan

The Riley-Morgan unit was developed by Riley-Stoker Corporation as an outgrowth of the Morgan Gas Producer, which was used in several commercial installations. A single demonstration unit started up in 1975 exists in the United States. The gasifier is a rotating vertical cylindrical refractory lined steel vessel 10-1/2 feet in diameter with a 6-1/2 foot bed depth. Ash is removed from the rotating ash pan via a helical plow. An air-steam mixture enters the bottom of the ash pan through a blast hood. Typical operating conditions are shown in Table 3.5.3.14.

Operating conditions, performance, and requirements are similar for both cases, with the exception of higher coal throughputs for oxygen blown operation.

One limitation many coals may face (including low-rank coals) is the requirement of a 2400°F minimum ash fusion temperature. Other coal feed requirements are as follows:

- Coal feed should range in size from 1-1/4 to 2 inches, or 3/4 to 1-1/2 inch for coke

- Fines may be accepted in quantities up to 10 percent of coal feed. This may impose a second limitation on the use of low-rank coals due to their tendency to form fines during coal preparation.

High rates of steam consumption in the oxygen blown case may limit efficiency, and will impose a requirement for extensive wastewater treatment facilities.

Table 3.5.3.14

Riley-Morgan Gasifier
Typical Operating Data

	Low BTU Gas	Medium BTU Gas
Gas Production, MM SCF/hr	0.4 - 0.5	0.36
Gas Composition, Volume %:		
Hydrogen	19	39
Carbon Monoxide	25	41
Methane	1	1
Nitrogen	49	1
Carbon Dioxide	6	18
Other	-	-
Total	100	100
Gas Higher Heating Value, BTU/SCF (Dry Basis)	185 - 201	262 - 305
Feed Requirements, lb/lb coal: (as fed to gasifier)		
Air	3 - 3.4	
Oxygen		0.6 - 0.7
Steam	0.6	1.3 - 1.5
Operating Conditions		
Gas Outlet Temperature, °F	1100	1100
Pressure	40 in. H_2O	40 in. H_2O
Residence time, minutes		
Turndown Capability	10:1	10:1
Coal Feed:		
Type	Bituminous	Bituminous
Size, Inches	1.25 - 2.0	1.25 - 2.0
Free Swelling Index	<8.5	<8.5
Ash Fusion Temperature, °F	2400	2400
Feed Rate, tons/day	90	156
Thermal Efficiency, %:		
Hot Raw Gas	88 - 90	
Cold Clean Gas	71 - 78	75
By-Product Rate, lb/lb coal:		
Tar	0.08	0.08
Oil	0.009	0.009

NOTE: Table may not be internally consistent due to averaging of values.

Source: Reference 14

Two Stage Gasifiers

Woodall-Duckham (Gas Integrale)

This two stage gasifier was developed by IL Gas Integrale of Milan, Italy, approximately 30 years ago for the production of industrial fuel gases from coal. Previous to its use in this capacity, it was used for approximately 20 years in a cyclic process. The gasifier is now licensed by Babcock Contractors of Pittsburgh, Pennsylvania.

Forty-seven Woodall-Duckham gasifiers have been operated to produce industrial fuel gas, 39 in public utility gas plants, and 24 in synthesis and water gas plants, all of which are outside the United States.

The gasifier is a vertical cylindrical steel structure with a rotating grate at the bottom of the vessel. A vertical cross section of the gasifier is shown in Figure 3.5.3.16. The grate is composed of concentric rings that distribute the air/steam mixture and also remove ash. The gasifier contains two zones: 1) a lower zone in which gasification takes place, and where a water jacket provides close temperature control, and 2) a refractory lined drying and distillation zone through which the coal descends before entering the gasification zone. Sized coal enters the vessel via a lockhopper and coal distributor. Coal is dried in the drying zone shown in the figure by a fraction of the gases generated in the gasification zone. The coal descends through the distillation zone where the coal is partially devolatilized. Negligible cracking of the tar occurs because of the slow heating rate. In the distillation zone, caking coals partially melt and resolidify as semi-coke and noncaking coals form a char. The semi-coke or char then enters the gasification zone where it is almost completely gasified with steam and oxidant in an unagitated countercurrent flow fixed bed. Incoming oxidizer and steam are preheated before contacting the coal by heat exchange with the descending ash. Ash is then removed from the vessel by the rotating grate, and falls into an ash lockhopper.

Figure 3.5.3.16
Woodall-Duckham/Gas Integrale Gasifier

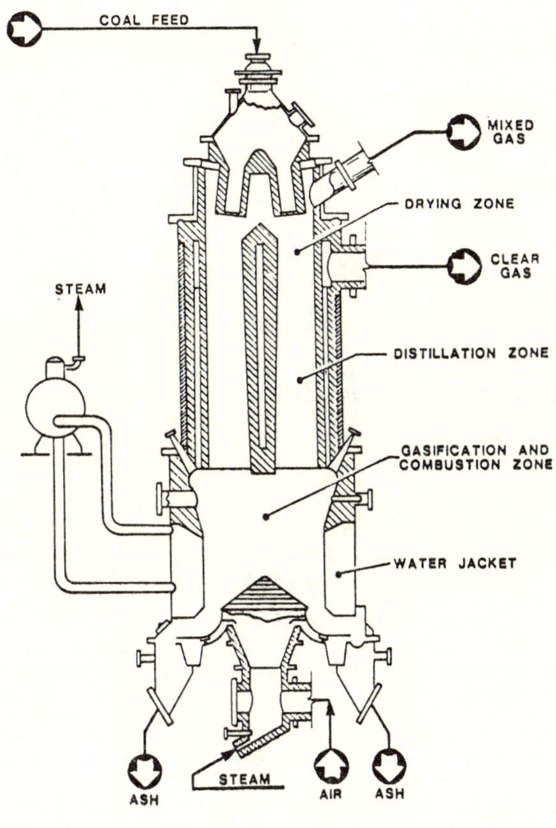

Source: Reference 14

The water jacket is a unique feature of the Woodall-Duckham gasifier, and functions primarily for temperature control, which reduces the steam requirement in the gasifier. Steam used in the gasifier is generated in the water jacket, and the ratio of steam to oxidizer is critical in determining the physical form of the ash when it is removed. A low steam injection rate results in clinker formation whereas high steam rates give a fluffy quality to the ash. Thus, the combined water jacket and steam injection format allows for higher efficiencies and lower amounts of waste liquor while still maintaining dry bottom operation.

Two gas streams results from this gasifier. A portion of the gas formed in the gasification zone is withdrawn from the lower portion of the gasifier, is termed "clean gas." "Top" gas is obtained from the top of the gasifier, and consists of the remainder of the gas from the gasification zone which has also passed through the distillation and drying sections. Distillation zone temperatures are controlled by varying the amount of clean gas withdrawn.

Typical operating data for this gasifier are given in Table 3.5.3.15. The gas composition data show a more than double hydrogen content for the oxygen blown case, indicating that this difference is not due entirely to the presence of the nitrogen diluent. High CO_2 levels are noted for the oxygen case, correspondingly reducing the carbon available for conversion to carbon monoxide. This is reflected in the gas heating values, the air blown case being lower than that for oxygen blown operation, but is higher than what would be obtained if the oxygen blown product gas were diluted similarly with nitrogen. Other operating and performance characteristics are similar. It should be noted that medium-Btu gas may be produced by operating the gasifier in a cyclic mode with air as the oxidant. As soon as the bed is heated to high temperatures, steam gasification of the bed is carried out until temperatures fall too low for reaction.

In 1977, ERDA chose two separate proposals involving the use or potential use of Woodall-Duckham gasifiers for small scale industrial applications. One project involves the Erie Mining Company for gasification of 500 TPD of high-sulfur, high-caking coals for heating iron-ore-pellet shaft furnaces in a taconite plant. The second project falls under the DOE's Gasifiers-in-Industry program, and proposes the use of a Woodall-Duckham gasifier for switching brick kilns from natural gas to low-Btu gas. The project is scheduled for start-up in 1980.

The Woodall-Duckham gasifier has the advantage of being commercially mature. In addition, its small reactor size makes it feasible for use on a small industrial scale. Predrying of the coal is not required, and its two stage operation yields a fairly high thermal efficiency. However, for large scale applications, the small reactor size would require many gasification trains. The process is currently limited to noncaking and weakly caking coals only.

The use of low-rank coals having low-ash fusion points may require a higher rate of steam addition than for other coals, due to the requirement of maintaining the ash in solid form. This will decrease thermal efficiency, and result in a higher rate of waste-liquor production. Fines production due to surface slacking of low-rank coals may be a problem, since coal should be double screened at 1/4 to 1 inch and 1/2 to 1-1/2 inch. Free swelling index of the coal feed must be below 2-1/2. Other low-rank coal characteristics of importance to their use in this gasifier are:

> Caking properties: Although a development program is underway to gasify caking coals, the absence of caking properties in low-rank coals is an advantage relative to currently existing commercial units.

Moisture content: Low-rank coals of high moisture content are not excluded from use in this gasifier.

Although tars and oils are produced by this process, no unique environmental problems are associated with the use of the Woodall-Duckham gasifier.

Table 3.5.3.15

Woodall-Duckham Gasifier
Typical Operating Data

	Low BTU Gas	Medium BTU Gas
Gas Production, MM SCF/hr	0.4	
Gas Composition, Volume %:		
Hydrogen	17	38
Carbon Monoxide	28	38
Methane	3	4
Nitrogen	47	2
Carbon Dioxide	5	18
Other	-	-
Total	100	100
Gas Higher Heating Value, BTU/SCF (Dry Basis)	175	280
Feed Requirements, lb/lb coal: (as fed to gasifier)		
Air	2.3	
Oxygen		
Steam	0.25	
Operating Conditions		
Gas Outlet Temperature, °F	250 - 1200	250 - 1200
Pressure	atm	atm
Residence time, minutes		
Turndown Capability	4:1	4:1
Coal Feed:		
Type	Bituminous	Bituminous
Size, Inches	0.25 - 1.0	0.25 - 1.0
Free Swelling Index	<2.5	<2.5
Ash Fusion Temperature, °F	>2200	>2200
Feed Rate, tons/day	72 - 84	
Thermal Efficiency, %:		
Hot Raw Gas	92	
Cold Clean Gas	77	
By-Product Rate, lb/lb coal:		
Tar	0.07	0.07
Oil	( )	( )

NOTE: Table may not be internally consistent due to averaging of values.

Source: Reference 14

Wellman-Incandescent

The two stage Wellman-Incandescent gasifier is mechanically identical to the International Furnace Equipment Company Ltd. gasifier that was commercially available in the 1950's. Currently, Applied Technology Corporation markets the system in the United States. Interesting features of this unit are its rotating drum, self-sealing feed system and its water seal between the ash pan and the reactor walls. Ash must be removed wet as a result of the water seal. Tar laden top gas is cleaned in a hot cyclone at 240°F and the bottom char gas is cleaned of dust in a separate cyclone at 1170°F. The combined stream has a heating value of about 200 Btu/scf at 690°F.

Coal used in the Wellman-Incandescent gasifier is subject to several limitations:

- Coal size should be from 2 to 3 inches
- Less than 10 percent of the feed can be between 5/16 inch and 2 inches, and less than 15 percent of the fines can be smaller than 5/16 inch
- Ash fusion temperature should be greater than 2200°F

The use of low-rank coals may face some limitations here because of the difficulty in gasifying fines (generated by surface slacking). In addition, low-rank coals with low ash fusion points may not be acceptable candidates for gasification.

Stoic

This South African designed two-stage gasifier has been in commercial operation for a number of years. The system is marketed in this country by Foster-Wheeler Energy Corporation.

The design is quite similar to other fixed bed two-stage designs and resembles the Wellman-Incandescent with its ash pan-gasifier vessel water seal. The unit has operated on bituminous coal and is described as suitable for operation with subbituminous coal and lignite.

The Stoic gasifier is part of the Department of Energy's Gasifiers-in-Industry program. A unit has been installed at the University of Minnesota to provide fuel gas for a campus heating and steam generation plant.

Operating conditions and fuel product characteristics are shown for a subbituminous feedstock in Table 3.5.3.16. The gasifier is air blown, and produces a fairly high heating value gas for this operational mode. Production of CO_2 is low, indicating efficient contacting of solid and gaseous reactants. Methane content is also fairly high at 2 to 3 percent for atmospheric operation. Consumption of oxygen in lb O_2/lb coal is reasonably low at 0.5, as compared to other air blown gasifiers, and steam consumption is not excessive.

Feed coal should be in the size range of 3/4 to 1-1/2 inches. The higher reactivity of low-rank coals is of little consequence in this gasifier due to the long solids residence time of approximately eight hours. Low-rank coals are suitable for use in this gasifier, and no special environmental problems are associated with its operation.

3.5.3.6.2 Fluidized Bed Gasifiers

Low/Medium Btu Gas Producers

Winkler

The Winkler gasifier was originally used in a commercial installation in 1926 in Leuna, Germany. Davy Powergas Inc. currently markets the system in this country. There are currently no installations in the U.S. but there have been 16 installations in foreign countries of which three are still in operation. Table 3.5.3.17 summarizes 16 installations. All of the later installations have been synthesis gas plants, indicating the declining trend of water gas (town gas) usage in favor of natural gas during the 1940's and 1950's. The only three presently operating gasifiers are located in India, Turkey, and Yugoslavia, and are all for synthesis gas production.

Table 3.5.3.16

Stoic Gasifier
Typical Operating Data

	Low BTU Gas
Gas Production, MM SCF/hr	0.4 - 0.5
Gas Composition, Volume %:	
Hydrogen	14 - 16
Carbon Monoxide	29 - 30
Methane	2 - 3
Nitrogen	47 - 51
Carbon Dioxide	2 - 4
Other	
Total	100
Gas Higher Heating Value, BTU/SCF (Dry Basis)	186 - 207
Feed Requirements, lb/lb coal:(as fed to gasifier)	
Air	2.1
Oxygen	
Steam	0.4
Operating Conditions	
Gas Outlet Temperature °F	750
Pressure	atm
Residence time, minutes	500
Turndown Capability	5:1
Coal Feed:	
Type	Subbituminous
Size, Inches	0.75 - 1.5
Free Swelling Index	<3
Ash Fusion Temperature, °F	
Feed Rate, tons/day	108
Thermal Efficiency, %:	
Hot Raw Gas	85 - 95
Cold Clean Gas	69 - 76
By-Product Rate, lb/lb coal:	
Tar	0.2
Oil	( )

NOTE: Table may not be internally consistent due to averaging of values.

Source: Reference 14

Table 3.5.3.17

Winkler Gasifier Installations

Location	Year	Product	Capacity Per Generator 1000 SCFH	No. of Gasifiers
Leuna-Werk Leuna, Germany	1926- 1930	Fuel Gas Water Gas	2240 1120	5
Braunkohle-Benzin AG Bohlen, Germany	1936	Water Gas	1030	3
Braunkohle-Benzin AG Magdeburg, Germany	1936	Water Gas	1030	3
Yahagi Japan	1937	Water Gas	330	1
Braunkohle-Benzin AG Zeitz, Germany	1938	Water Gas	840	3
Dai-Nihonyinzo-Hiryo Japan	1938	Synthesis Gas	520	2
Nippon Tar Japan	1938	Water Gas	520	2
Toyo-Koatsu Japan	1939	Synthesis Gas	560	2
Sudentenlandische Treibstoffwerke Brux, Czechoslovakia	1943	Water Gas	1030	5
*Fabrika Azotnih Jendinjenja Gorazde, Yugoslavia	1953	Synthesis Gas	190	1
Calvo Sotelo Puertollano, Spain	1954	Water Gas	350	1
Union Rheinische Braunkohlen Wesseling, Germany	1956	Synthesis Gas	450	1
Calvo Sotelo Puertollano, Spain	1957	Synthesis Gas	350	1
*Azot Sanyyii TAS Kutahya, Turkey	1959	Synthesis Gas	450	2
*Neyveli Lignite Corporation Madras, India	1959	Synthesis Gas	1550	3
Unión Rheinische Braunkohlen Wesseling, Germany	1960	Synthesis Gas	450	1

*Presently Operating

Source: Reference 14

Figure 3.5.3.17 shows the Winkler unit. The gasifier is of vertical cylindrical construction with a refractory lined steel shell. Coal is fed into the bed via variable speed screw feeders which are located in the lower portion of the gasifier. Steam and oxygen (or air) are charged through nozzles located at several levels in the fluid bed. The fluidized bed occupies only part of the gasifier volume, while the remainder serves as a disengaging zone. Secondary steam and oxygen injection is provided above the bed level to gasify unconverted fly carbon leaving the bed. Above this level, a radiant heat boiler is installed to recover heat from the gases before they leave the gasifier and to condense any molten or vaporized ash species which may have been carried out of the bed.

Since the reactants are intimately mixed, uniform temperature is quickly reached between the solids and the gases, and the coal gasification reactions approach equilibrium. The main reactions occurring in the bed are the combustion and water gas reactions. Because the bed is at a relatively high temperature, all of the tars and heavy hydrocarbons are gasified.

Figure 3.5.3.17

Winkler Gasifier

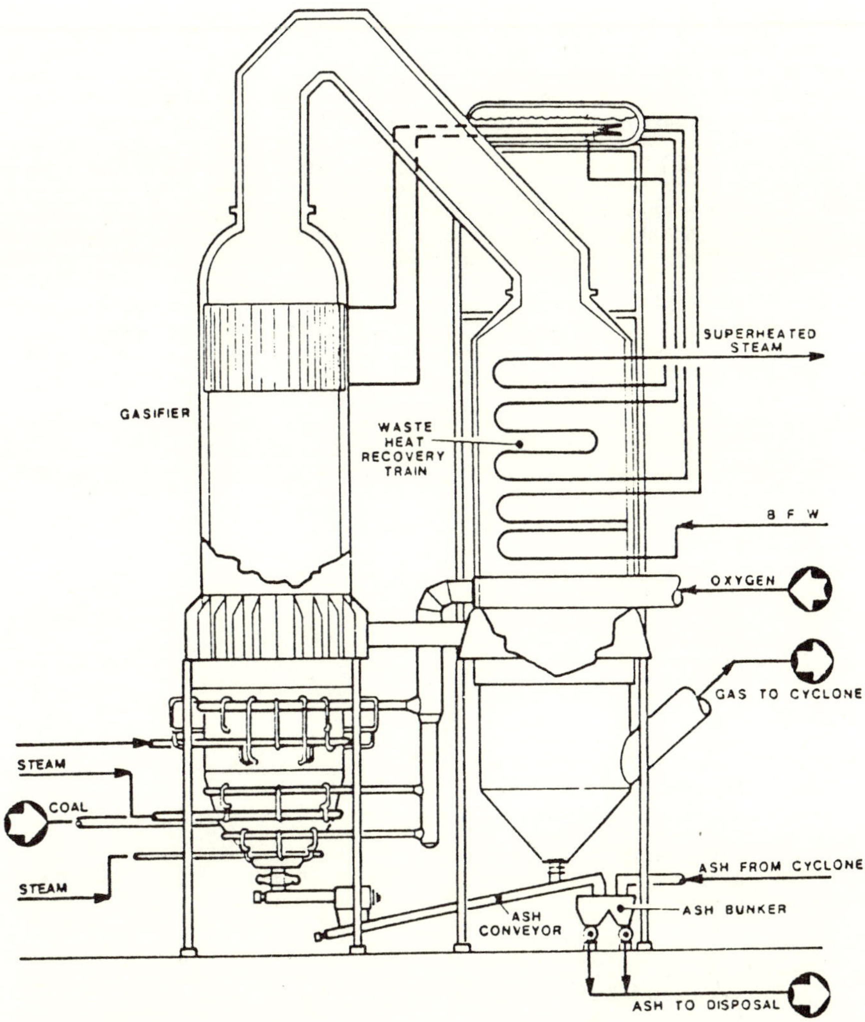

Source: Reference 5

Residual ash particles are segregated in the fluidized bed according to size and specific gravity. The heavier particles drop through the fluidized bed and pass into the ash discharge unit at the bottom of the gasifier, while the lighter particles are carried up through the bed with the product gas. This process results in approximately 30 percent of the ash being removed at the bottom and the remaining 70 percent being carried overhead in the product gas stream. Table 3.5.3.18 summarizes some typical operating data for both air and oxygen blown modes of the Winkler gasifier with lignite and subbituminous coals as feeds.

For the particular air-blown example presented, a relatively high (56%) concentration of nitrogen is noted. The concentrations of hydrogen, carbon monoxide, methane, and carbon dioxide in the oxygen blown case are proportionally higher to the extent that nitrogen is a diluent in the air blown case. The higher concentrations of the combustible components are reflected in the higher heating value of the gas product.

Consumption of oxygen is slightly higher in the air blown case in terms of lb O_2/lb coal, but it must be remembered that the comparison is between two different low-rank coals. Higher consumption of air is reflected in the nitrogen content of the low-Btu gas, which is higher than for most air-blown coal gasifiers. However, this results in a greater volume of gas passing through the reactor and therefore a higher gas outlet temperature due to the limited time available for heat transfer and more oxidant available for reaction. Thus, the air blown gasifier will have associated with it a considerably larger heat recovery system as a result of the greater gas volumes and their higher outlet temperature. Steam consumption for the oxygen blown case is considerably higher, due to the need for temperature control in the bed.

The original commercial gasifier operated at atmospheric pressure. Some testing has been done at 20 psig, with plans to go to pressures in excess of 200 psig. Present day commercial units would probably be designed for approximately 50 psig.

Table 3.5.3.18

Winkler Gasifier Typical Operating Data

	Low BTU Gas	Medium BTU Gas
Gas Composition, Volume %:		
Hydrogen	14	35
Carbon Monoxide	22	48
Methane	1	2
Nitrogen	56	1
Carbon Dioxide	7	14
Other	-	-
Total	100	100
Gas Higher Heating Value, BTU/SCF (Dry Basis)	126	288
Feed Requirements, lb/lb coal:		
Air	2.5	
Oxygen		0.5
Steam	0.2	0.4-0.7
Operating Conditions		
Gas Outlet Temperature, °F	1450	1150
Pressure	atm.	atm.
Coal Residence Time, Minutes	20-30 min.	20-30 min.
Turndown Capability	4:1	4:1
Coal Feed:		
Type	Subbitumious	Lignite
Sizes, Inches	$<$0.38	$<$0.3
Free Swelling Index	$<$ 2.5	$<$2.5
Ash Fusion Temperature, °F		
Feed Rate, tons/day	1100	
Thermal Efficiency, %:		
Hot Raw Gas		81.8
Cold Clean Gas	72	74.9

NOTE: Table may not be internally consistent due to averaging of values.

Source: Reference 1

Thermal efficiencies for oxygen-blown operation appear to be marginally higher than for air-blown operation. The Winkler gasifier is a commercially available design, with considerable operating experience at atmospheric pressure. In addition, the gasifier has a high throughput capacity with good operational stability and reliability. High turndown ratios can be imposed without affecting gas quality or reactor performance. Since the gas leaves the reactor at high temperatures relative to most fixed bed units, no significant production of tars, oils, and other such compounds takes place, greatly reducing the size and extent of wastewater treating required. A wider range of particle sizes can be tolerated, and all types of coal can be gasified without pretreatment except very strongly caking feedstocks. Most of the operational experience has been with lignite or subbituminous coals.

However, operation of the Winkler gasifier is faced with high carryover of ash and char. Ash recovery requires effective particulate removal equipment, and maintenance of good thermal efficiencies demands char recovery from the overhead solids and recycle of this carbonaceous matter to the gasifier. Finally, unproven performance of the gasifier at high pressures may limit the applications for which the Winkler gasifier may be considered.

Crushed coals of 3/8 inch size or smaller should be used in the gasifier, with a wide particle size distribution. Best operation of the gasifier is achieved with coals having a swelling index of 2.5 or less. The use of low-rank coals in the Winkler gasifier appears favorable, and the following considerations apply:

- Coals with high ash content or a fluctuating feedstock ash content may be used without upsetting gasifier operation. Ash content of feed coals to operating units has been as high as 50 percent
- Lignites of up to 18 percent moisture content have been handled and gasified without further drying. However, the limit on coal moisture content will vary for individual cases, and is determined by the economics of air or oxygen blown gasification
- Coals with high reactivity, such as low-rank coals, are preferred since their higher reaction rate reduces the amount of unconverted carbon carried overhead with the product gas
- Noncaking coals (low-rank coals) are preferred so that proper gas-solid contacting may be maintained, and proper particle size distribution be preserved
- Coal particles up to 3/8 inch are used, and fines can be handled in the bed under most circumstances. Friable coals may thus be used
- The maximum allowable temperature in the fluid bed depends on the ash softening point, and must be maintained 35 to 50°F below this level

Design of environmental equipment for the plant will be based on the plant's lower production rate of tars, oils, phenols, naphthas, and hydrocarbons. No other unique environmental considerations apply.

The High Temperature Winkler (HTW) Process[25]

Operation of the standard Winkler process was discussed in the previous section in light of its advantages and disadvantages. Several of the drawbacks discussed are not inherent to the fluidized bed configuration, and merit investigation as to design changes which may be feasible for improving gasifier performance. These limitations are:

- Operating pressure - The reactor is not operated under pressure. This means limited outputs per gasifier unit and high compression energy required for the raw gas.
- Operating temperature - Since the operating temperature in the reactor has to be below the ash softening point, only comparatively low operating temperatures can be used for different types of coal. Compared to the high temperature Winkler, low CO and H_2 contents result in the raw gas.
- Carbon conversion - The maximum carbon conversion rate obtained has been 90 percent.

A German firm, Rheinische Braunkohlenwerke AG (Rheinbraun) is developing the High Temperature Winkler process under the sponsorship of the Federal Minister for Research and Technology of the Federal Republic of Germany. The aim of this development is to operate a fluidized bed at elevated pressures and temperatures that are higher than have been known up to now for comparable types of coal in a Winkler-type gasifier.

The use of elevated temperatures raises the gasification rate. At the same time, the gas quality is improved (higher carbon monoxide and hydrogen contents).

The higher gasification temperatures involve the risk of ash fusion and this interferes with a smooth operation of the gasifier. Preliminary experiments on a bench-test scale of the "Institut für Eisenhüttenkunde" of Technische Universität, Aachen have shown, that by adding limestone, lime or dolomite, the fusion point of the ash, for example of Rhenish lignite, is raised and crumbled ash is formed in the fluidized bed of the gasifier. A desirable side-effect is that sulfur will be bound to the lime and thus discharged with the ash. Operation under pressure has the following advantages:

- The gasificiation rate increases. This permits units with higher capacities.
- The costs for providing the oxidizing agent under pressure are more than compensated by savings in product gas compression.

It is planned to raise the carbon conversion rate up to 95 percent by returning carbon containing flue dust to the gasifier.

A schematic diagram of the gasification system is shown in Figure 3.5.3.18.

Crushed coal, if necessary with the addition of limestone, is fed via a lockhopper system to the pressurized gasifier by means of a screw conveyor. The gasification agent, i.e. oxygen/steam, or air, can be injected into the gasifier at several different levels. It is possible to preheat the gasification agent. The coarse grains in the effluent gas are separated from the raw gas and fed back to the fluidized bed. The coarse ash particles which are collected on the bottom of the gasifier are withdrawn by means of a lockhopper system.

The use of low-rank coals in the HTW system may require special attention due to possibly lower ash fusion temperatures. Although this is also a consideration for other coals, low-rank coals may require especially high additions of refractory agents such as limestone. Additional problems with low-rank coals due to higher temperature and pressure operation are not anticipated.

Figure 3.5.3.18

High-Temperature Winkler Gasification

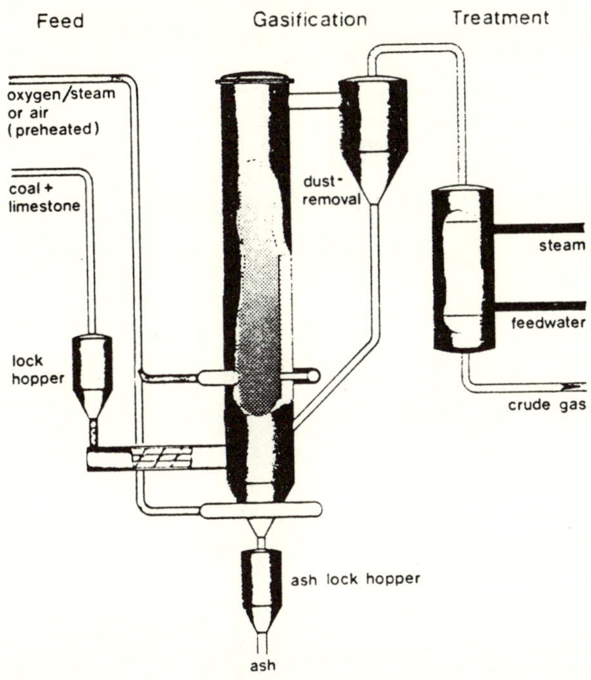

Source: Reference 25

U-Gas (Ash Agglomerating Gasifier)

The U-Gas process was developed by the Institute of Gas Technology under the sponsorship of the American Gas Association and the Office of Coal Research. Initial operation of a 4 foot diameter PDU was in 1974.

Figure 3.5.3.19 depicts the U-Gas gasifier. The pressurized fluidized bed reactor has a unique method for ash discharge. The bottom of the reactor is shaped like an inverted cone. Steam and oxygen enter and flow through a venturi pipe section at the very bottom tip of this cone. The hottest zone of the reactor bed is just above the oxygen inlet. Any particle of ash that has lost most of its carbon content and strays into this hot zone becomes sticky and adheres to the nearest ash pellet. Ash rich pellets grow by agglomeration until they reach a critical mass sufficient for them to fall out through the venturi throat. Ash pellets in the pilot plant are about 1/4 inch in diameter.

Figure 3.5.3.19

U-Gas Pilot Plant Reactor Cross Section

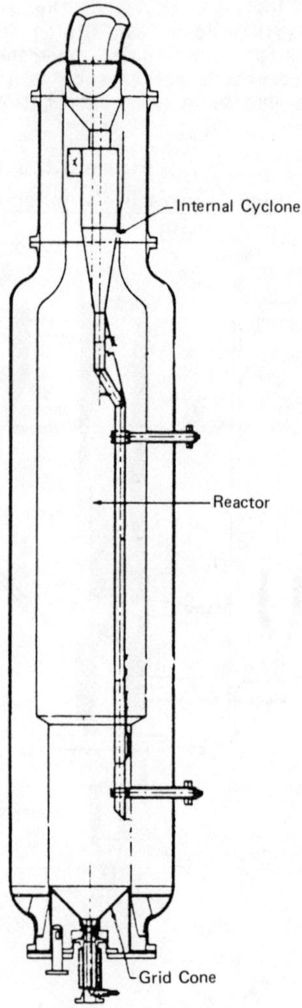

Source: Reference 26

Both air and pure oxygen have been used in test runs since 1974. Since pilot plant operating pressures are limited to 50 psig by mechanical design considerations, the potential pipeline gas applications of the U-Gas process have not been proven. However, the U-Gas reactor is being considered for the first reactor stage of the Hygas demonstration project discussed below. Table 3.5.3.19 provides general operating information on the U-Gas process.

The cold gas efficiency is somewhat higher for the oxygen blown operation compared to air blown gasification, although it is noted that the data were taken for different coals. Consumption of oxygen is considerably higher in the air blown case when put on a comparable lb O_2/lb coal basis (average 0.70 for air compared to 0.55 for oxygen). This results in hydrogen, carbon monoxide and methane contents which are lower than that which could only be explained by the presence of the nitrogens as a diluent. Heating value trends follow this gas composition difference. Other operating conditions in the gasifier are similar between air and oxygen blown cases.

Table 3.5.3.19

Operating Parameters for the U-Gas Process

	Air Blown Operation	Oxygen Blown Operation
Product Gas Compositon (dry basis) mol %		
CO	19.6	41.4
H_2	17.5	35.8
CO_2	9.9	16.1
CH_4	3.4	5.9
H_2S+COS	0.7	0.2
N_2+Ar	48.9	0.6
High Heating Value, Btu/SCF (dry basis)	154	320
Steam Consumption, lb/lb coal as fed to gasifier	0.2 - 0.6	0.2 - 0.6
Oxident Consumption lb/lb coal as fed to gasifier	2.8 - 3.3 (air)	0.55 (oxygen)
Gas Residence Time	1-2 sec.	1-2 sec.
Coal Residence Time	45-60 min.	45-60 min.
Turndown Ratio	3:1	3:1
Operating Pressure	50-350$^+$ psig	50-350$^+$ psig
Cold Gas Efficiency	75.7*	80.6**
Gas Outlet Temperature	1700-1900 °F	1700-1900°F

$^+$Higher pressure operation not yet demonstrated.

*Illinois bituminous coal.

**Montana subbituminous coal.

Source: Reference 26

The U-Gas gasifier is not commercially available. However, a demonstration-plant program is underway to establish the technical and economic feasibility of the process. After completion of this program, the gasifier will be available for commercial use. The unique method of ash agglomeration and removal for recovery of low carbon ash from a high carbon environment promises a significant advantage over established fluid bed technology. As a result of this technique, carbon conversions as high as

98 percent have been demonstrated in the test gasifiers, and the gasifier is flexible enough to handle a wide variety of coals and a wide range of particle sizes.

However, highly caking coal may not be suitable for use in the gasifier without pretreatment. Some uncertainty still exists concerning the viability of the full scale unit, since complete sets of pilot data are not yet available. In comparison to fixed bed gasifiers, efficiency may suffer somewhat due to unconverted carbon contents which can range from 5 to 6 percent (compared to <1% for many fixed bed designs).

In addition, the gasifier requires close control of the bed temperature to insure a proper fluidization and ash agglomeration. Thus, accurate control of the steam-to-oxidant ratio is critical, and disruptions in the coal feed rate or variations in coal properties could cause operational difficulties.

The use of low-rank coals in the U-Gas gasifier is subject to the following considerations:

- Reactivity - due to the tendency to produce unreacted carbon, highly reactive low-rank coals may be more suitable than others.
- Caking properties - low-rank coals are more desirable and will not require pretreatment since they do not display caking characteristics.
- High moisture - low-rank coals may affect the flow of crushed coal in freezing weather or if surface moisture is present, but will not be detrimental to gasifier operation.
- Ash softening point - some low-rank coals will require an adjustment to the ash softening temperature. The relationship which describes this change is entirely different from that for higher rank coals, and may require different additives.
- High concentrations of fines from crushed low-rank coals do not represent a problem, since they are returned to the bed by a series of cyclones.

No production of tars or oil type compounds takes place due to the high temperatures in the gasifier. Wastewater treatment is thus simplified. Further environmental considerations apply equally well to other gasification technologies.

Westinghouse

The design phase of a 15 TPD pilot plant began in 1972 and construction was started in January 1973 at Waltz Mill, Pennsylvania. Construction was mechanically completed in 1974, and hot flow operations have been conducted since 1975. The program for 1979 calls for integrated operation of a complete coal gasification system including the devolatilizer, the gasifier/agglomerator and associated support equipment. Figure 3.5.3.20 depicts the Westinghouse multi-stage gasification system. The gasifier and associated equipment consists of two vertical, cylindrical vessels, one a recirculating-bed devolatilizer/desulfurizer and an agglomerating, fluidized bed combustor/gasifier. Enthalpy required by the devolatilization vessel is provided in a hot gas stream arising from the partial oxidation of char in the gasifier/agglomerator.

Crushed, dried coal is fed at the bottom of the devolatilizer. The incoming coal then mixes with internally recycled solids consisting of char and dolomite, and is carried upward at 20 to 40 fps through a draft tube by the hot gas stream from the gasifier/agglomerator. This entrained flow of the fresh feed coal prevents agglomeration of caking coals as they devolatilize and pass through the plastic phase. As the solids approach the top of the draft tube, a dense dry char is formed and mixes with other char in the bed to descend in the annular space at weight rates approximately 20 to 50 times the coal feed rate.

Removal of sulfur bearing compounds may be accomplished by the addition of dolomite to the bed material. Sulfided dolomite is dense and settles to the lower portion of the bed to be withdrawn in the dolomite draw-off pot. Char is stripped from the dolomite and returned to the devolatilizer by a slip-stream of recycle product gas.

Figure 3.5.3.20

The Westinghouse Gasification System

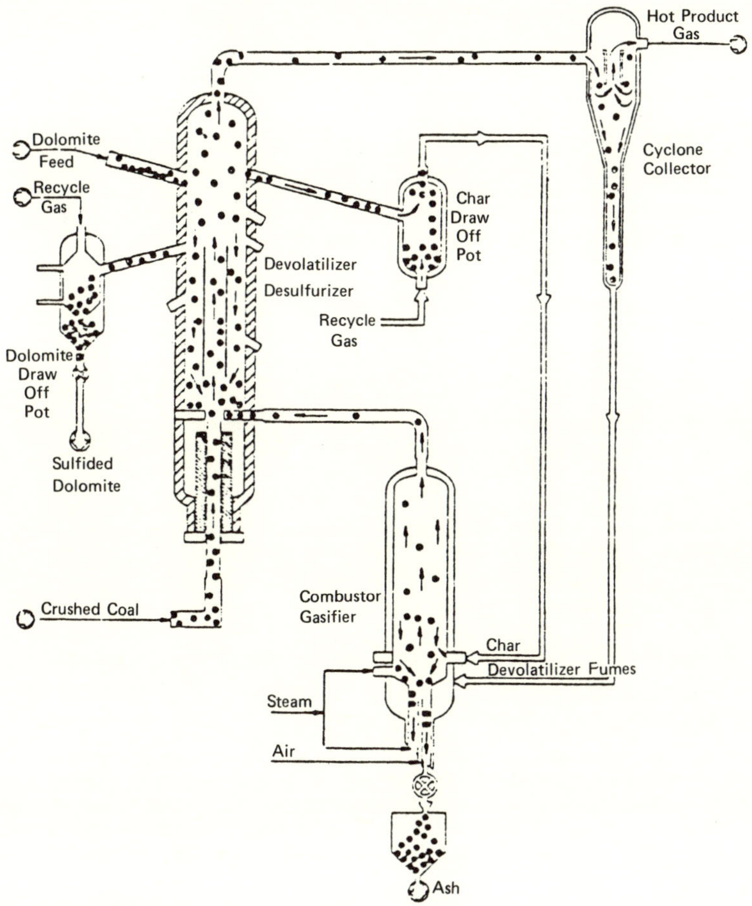

Source: Reference 5

Char gasification occurs in the agglomerating fluidized bed combustor/gasifier. Fines and char collected from the devolatilizer are oxidized to completion in the lower portion, providing the main heat source for the process. Ash agglomerates in this vessel at temperatures between 1800 and 2000°F for most coals, and settles to the lower leg for removal. Hot gases generated in the combustion process rise into the upper fluidized bed gasification section. High temperature steam gasifies char to form hydrogen and carbon monoxide, which is then charged to the devolatilizer to provide its endothermic heat requirements. Gas from the devolatilizer enriches the lean gas from the combustor/gasifier to produce a 130 Btu/scf product gas.

Table 3.5.3.20 gives operating data for the Westinghouse gasification process. Slightly higher than average contents of N_2 are present in the product gas, even though oxidant consumption is reasonably low.

Table 3.5.3.20

Operating Data for the Westinghouse Gasification System (Eastern Coal)

	Air Blown Operation
Gas Composition, mol % (dry basis)	
CO	19.2
CO_2	9.3
H_2	14.4
$H_2S + COS$	0.1
CH_4	2.8
$N_2 + Ar$	54.3
Gas Higher Heating Value, Btu/SCF (dry basis)	135
Feed Requirements	
Air consumption, lb/lb coal	2.2 - 2.8
Steam consumption lb/lb coal	0.2 - 0.4
Operating Conditions	
Gasifier/combustor, temperature, °F	1700 - 2000
Devolatilizer, temperature, °F	1500 - 1700
Pressure, psig	200 - 250
Gas Velocity, fps	2-3
Thermal Efficiency	
Hot raw gas	97%
Overall thermal efficiency*	94%

* cold clean gas

Source: Reference 5

Lower than average hydrogen contents combine with a fairly high CO_2 content to keep the heating value of the gas below average for air-blown gasifiers. Operation of the gasifier at 250 psig may allow sufficient pressure for use in a combined cycle plant, although it is likely that this major intended use for the Westinghouse system would dictate operating pressures in excess of 300 psig. Since efficiency is a critical parameter in the performance of combined cycle plants, the system's high efficiency is an important factor.

The Westinghouse system has achieved successful operation of the gasifier and devolatilizer. In addition, a wide variety of coals can be handled. Caking coals can be processed without pretreatment, coals of any sulfur content can be used, and ashes of any composition can be separated from the coal and effectively agglomerated.

However, complete integrated PDU demonstration is required before commercialization can be justified. As with the U-Gas agglomerating gasifier, the operating range may be somewhat limited, because of the need to maintain proper fluidization and ash agglomerating conditions. Solid entrainment may also be a problem. High-moisture coals may require drying, but not pretreatment before sizing. Erosion may be a problem in the devolatilization vessel (but has not been observed), due to the extremely high circulation rates of solids. Finally, separation of the unburned char from the dolomite must be achieved in the integrated system, or carbon conversion efficiency will be reduced.

The noncaking property of low-rank coals is not a particular advantage in this process since the Westinghouse gasification system tolerates caking coals. Since sulfur removal is accomplished by the addition of dolomite, lower contents of sulfur present in low-rank coals reduces the rate of consumption of dolomite. Fines content is not an extremely important factor, since fines are cycled to the combustor/gasifier. The presence of dolomite will tend to raise the ash softening temperature of lignite ash, while the opposite effect is observed for bituminous coal ash. This may be a significant consideration in choosing reaction parameters. The temperature viscosity dependence of the specific coal in question will also determine the allowable range of operating temperatures with respect to this characteristic.

Production and disposal of spent dolomite will be an environmental consideration not common to most other gasifiers. However, removal of sulfur in the gasification step obviates downstream recovery, thereby simplifying acid gas removal and the air, water and solid waste pollution problems associated with sulfur removal and recovery.

Fast Fluid Bed Process

The Fast Fluid Bed Process has been under development at the facilities of Hydrocarbon Research Inc. since 1976. The PDU was started up in 1978, and has the following design parameters:

- Coal Feed Rate: 600 lbs/hr
- Solids Density in Fast Bed: 7 lbs/cf
- Average Gas Velocity in Fast Bed: 10 ft/sec
- Average Temperature in Fast Bed: 1700°F
- Pressure: 150 psig

Gas/solid contact in the fast fluidization operational mode occurs at high gas velocities in the range of 7 to 20 feet per second and high solids loadings in the range of 10 to 20 pounds per cubic foot. This mode of operation can be considered intermediate between (1) the (standard) fluidized bed that operates in the range of 0.5 to 5.0 feet per second gas velocity and at solids loadings of 20 to 40 pounds per cubic foot; and (2) the entrained or dilute-phase transport mode that operates in the range of 30 to 60 feet per second gas velocity with a solid loading in the range of one to ten pounds per cubic foot.

Operation of the reactor in this flow regime provides the following advantages for the fast fluidized bed design:

- Higher gasification capacity by an order of magnitude over the standard fluid bed
- No formation of tars
- Higher turndown capability with a minimum efficiency loss
- Lower operating temperature than the entrained flow mode, allowing increased flexibility in the materials of construction
- Decreased potential for the formation of explosive mixtures
- Potential elimination of pretreatment for highly caking coals

A process flow diagram for the fast fluidized concept is provided in Figure 3.5.3.21.

Figure 3.5.3.21

The Fast Fluid-Bed Process

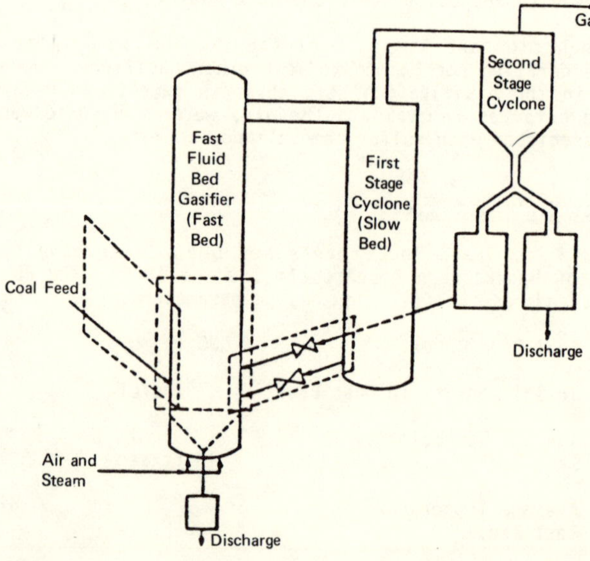

Source: Reference 20

In the fast fluid-bed process, -20 mesh coal is fed into the lower section of the gasifier. The incoming coal is mixed with char fed from a companion slow fluid-bed gasifier at a rate of ten parts char to one part feed coal. The coal and char react with air and steam fed into the bottom of the generator. The gasifier is operated at approximately 2000°F to 2400°F and 10 atm.

The gas and solids from the gasifier pass through a primary cylone to remove practically all solids, which are then discharged into the nitrogen-gas charged fluidized-bed reactor directly below. The char from the slow bed reactor is then fed into the fast fluid-bed reactor via a transfer leg. Ash from the fast fluid-bed reactor drops to the bottom and is discharged. Gas and particulate matter from the priamry cyclone are passed on to a secondary cyclone for further separation.

Initial testing of Eastern bituminous coal has taken place at 1700°F, and plans call for increasing reaction temperatures with a variety of coals.

The higher reactivity of low-rank coals is one factor to indicate their preferred use in this process. The low-ash fusion point of some low-rank coals, however, may limit the reaction temperature when these coals are gasified. Fines, as content and moisture content will not limit gasifier performance, but operation of the coal feeding system may be affected by the raw coal moisture content. The noncaking character of low-rank coals is not an advantage over caking coals in this gasifier, due to the fluid mechanics of the fast fluid-bed and the dilution of feed coal with recycle char.

No unique environmental considerations are associated with the fast fluid-bed process.

Integrated High Btu Gas Processes

HYGAS

The HYGAS process was developed by the Institute of Gas Technology in the late 1960's under American Gas Association and Office of Coal Research sponsorship. Initial pilot plant operation of a 75 TPD system began in 1974 and continues today. The HYGAS process is one of three selected for the design of a high Btu gas demonstration plant sponsored by DOE. A process flow diagram of the pilot plant is shown in Figure 3.5.3.22 and an enlarged view of the gasifier is presented in Figure 3.5.3.23.

Coal is crushed, screened for oversize, weighted, and fed to an agitated tank where it is slurried in light oil. A positive displacement pump feeds the slurry at high pressure into the gasifier. This reactor operates at 500 to 1200 psi. It has four internally connected, fluidized beds in which the reaction of gases with the coal is staged and in reverse flow. The upper bed dries the coal slurry. From there, the coal flows by gravity into a dilute phase riser stage which is the first step of hydro-gasification. Here, coal particles are heated to 1100°F by hot gases, which react with about 20 percent of the coal to produce methane. The partially reacted coal (now char) flows to the second gasification step, where it is heated in a fluidized bed to about 1700°F and is further gasified by the steam and hydrogen-rich gas rising from the steam-oxygen gasification stage below. The third gasification stage, at the bottom of the reactor (see U-Gas), receives the feed gases made up of steam and oxygen. The temperature here of 1800°F results from the balance of the heat in the feeds, the exotherm from the oxygen reaction and the endotherm from the steam-carbon reaction. The ascending stream of hot gas provides heat and hydrogen to the rest of the reactor. Fluidization is sufficiently vigorous in this state to prevent slagging of the ash.

Figure 3.5.3.22
HYGAS Pilot Plant Configuration

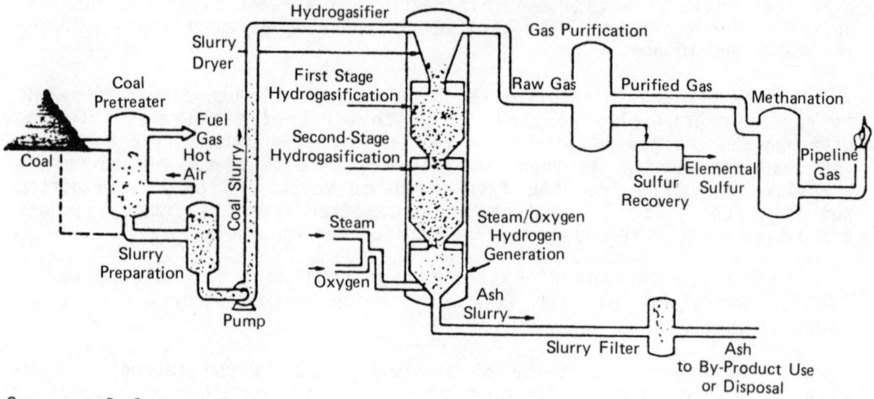

Source: Reference 5

Figure 3.5.3.23
HYGAS Gasifier

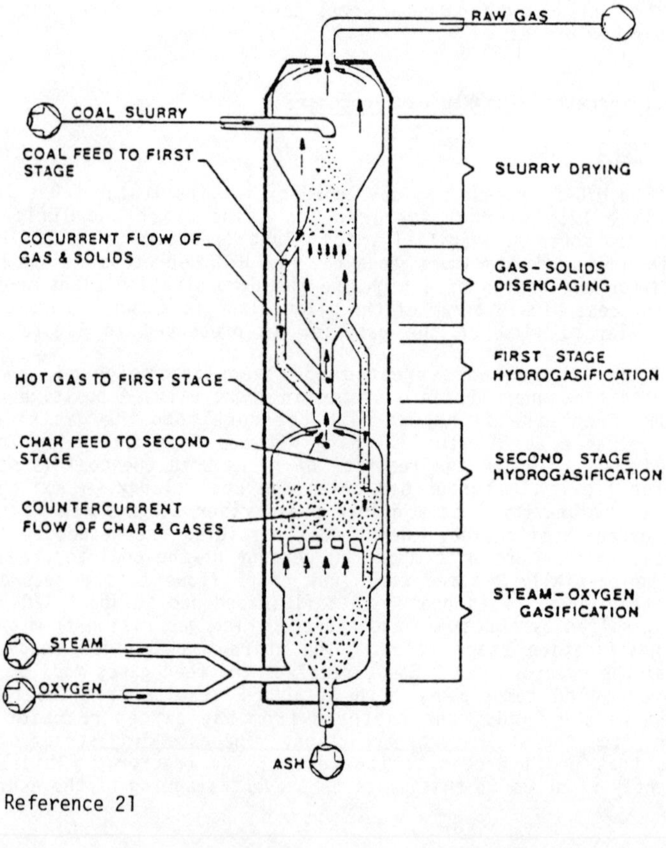

Source: Reference 21

The high-ash spent char from the lower bed of the reaction system vessel is discharged through a solids control valve and carried away by steam. A circulating pump mixes it with water and maintains a slurry of even consistency with up to 30 weight percent solids. The coal ash slurry is let down in pressure using a special tungsten-carbide coated valve. The slurry is filtered at low pressure, and the filtrate is recycled to the quench vessel.

A conceptual design for a commercial scale demonstration plant is now being completed by Procon Inc. for construction consideration. Additional details of the HYGAS process are shown in Table 3.5.3.21.

Table 3.5.3.21

Operating Parameters for the HYGAS Gasifier

(Montana Subbituminous Coal)

Gas Composition, mol % (dry basis)		
	CO	26.1
	CO_2	24.1
	H_2	30.7
	CH_4	16.6
	higher hydrocarbons	1.7
	N_2+Ar	0.2
	H_2S+COS	0.2
	NH_3	0.4
Heating Value, Btu/SCF (dry basis)		375
Pressure, psig		~1160
Feed Requirements		
Oxygen, lb/lb coal		0.22 - 0.25
Steam, lb/lb coal (as fed to gasifier)		1.0 - 1.2
Thermal Efficiency		
hot gas		74%

Source: Reference 5

The most notable feature about the gasifier effluent is the high methane content, due to the high pressure of operation. Carbon dioxide is also quite high, and CO is relatively low for an oxygen blown gasifier. The gas heating value is high, reflecting the high methane content. Oxygen consumption is low at approximately 0.23 lb/lb coal (as fed to the gasifier), but would be somewhat higher for a bituminous coal (0.24 to 0.26 lb/lb coal). Steam consumption is quite high, since this is used as a reactant in the process.

The HYGAS process has the advantage of having been tested with a variety of coals, from lignite subbituminous to bituminous coals. The slurry feed system used in the HYGAS process prevents the loss of gases experienced in lockhopper operation. The high content of methane in the raw gas reduces the size of the downstream methanation equipment. Approximately two-thirds of the total methane production occurs in the gasifier. Thermal efficiency of the HYGAS process is enhanced by lower temperature operation which limits sensible heat losses, by the slurry feed system which reduces compression requirements associated with lockhopper operation, and by high methane production in the gasifier. In general, the series arrangement of the fluidized beds provides sufficient inventory of fuel to ensure safe and steady operation.

The HYGAS process is not yet developmentally mature enough to proceed with the commercial scale design. The process is oriented to methane production, which is an advantage in SNG applications but not in some low- or medium-Btu gas uses. Caking coals will require pretreatment to prevent blockage of slurry discharge and transfer lines, as well as agglomeration in the reactors. Although the efficiency of conversion of coal to methane is higher in the HYGAS process than for systems based on fixed bed gasifiers, consistent operation above 80 percent coal to gas conversion for bituminous coal is being sought in the current test program. One factor limiting conversion is the carbon content of the ash withdrawn from the gasifier, which is projected to be approximately 10 to 25 percent, representing a 2 to 3 percent loss of feed carbon. The use of low-rank coals may prove to be advantageous in the HYGAS process. It is possible that their higher reactivity may reduce the amount of carbon lost in the ash withdrawal stream. Low-rank coals also maintain the advantage over agglomerating coals (which require pretreatment) of not plugging the slurry discharge and high temperature transfer lines. However, the tendency for dried low-rank coals to reabsorb moisture may limit solids content in slurry lines. Fines are not a problem in the HYGAS process, which uses pulverized coal as a feedstock.

Although ash from the HYGAS gasifier contains approximately 10 to 25 percent carbon, it may still be disposed of by landfill. Wastewater streams undergo moderate treatment and are recycled to the gasifier. Toxic element studies have been completed for the plant, and acid gases will undergo conventional removal and sulfur recovery to protect process catalysts and meet environmental standards.

Synthane

In 1961, work was started at the Pittsburgh Energy Technology Center to develop a high-Btu gas from coal process. A 75 TPD pilot plant was completed in 1975 by Rust Engineering, but further development on the concept has been stopped. A process flow diagram is shown in Figure 3.5.3.24.

Figure 3.5.3.24
Synthane Flow Schematic

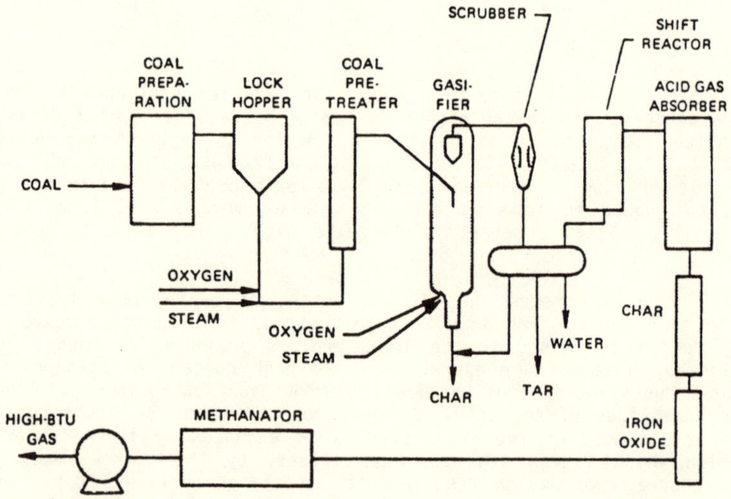

Source: Reference 22

Coal, crushed to -20 mesh, is dried, pressurized to approximately 70 atm, and is transferred into the fluidized-bed pretreater by means of high pressure steam and oxygen. Pretreatment provides a mild oxidation of the coal particle surface so that caking coals will not agglomerate in the gasifier. The coal overflows from the pretreater into the top of the fluidized-bed gasifier about 10 ft. above the bed level, falls through hot gases rising from the fluidized bed, and is devolatilized. This devolatilization contributes significantly to the methane yield. Steam and oxygen enter the gasifier just below the fluidizing gas distributor. The gasification reaction occurs within the fluidized bed. Unreacted char flows downward into a bed fluidized with steam and is removed through lockhoppers. The product gas is passed through a venturi scrubber and a water scrubber to remove carryover ash, char, and tars. The concentration of hydrogen and carbon monoxide in the gas is then adjusted to a three-to-one ratio in a shift converter. The acid gases are absorbed in a hot-potassium-carbonate scrubber. The purified gas is then sent to a methanator for final upgrading.

Table 3.5.3.22 provides additional operating information for the process.

Table 3.5.3.22

Synthane Process Data

OWNER DESIGNER	Bureau of Mines, Pittsburgh Energy Technology Center (PETC) 4800 Forbes Ave. Pittsburgh, Pa. 15213
LOCATION	Bruceton, Pennsylvania
DESCRIPTION	Synthane process combined gas flow from fluidized bed pretreater and fluidized bed gasifier Pressure 1000-1500 psig; temperature 1800°F
SIZE	75 TPD
STATUS	In pilot plant operational stages
YEARS OPERATION	Since 1976
COAL TYPE	All types
MAJOR PRODUCTS	SNG precursor
ENVIRONMENTAL STATUS	Effluent emission and solid characterization and monitoring are ongoing
OXYGEN CONSUMPTION	0.3 tons per ton of coal

Source: Reference 9

Two methanation systems have been proposed: a tube wall reactor, in which the catalyst is sprayed onto tube walls of an exchanger, and a hot gas recycle methanator, in which the catalyst is sprayed onto metal plates.

Tests on Rosebud coal and caking bituminous coals were successful; however, the plant has been deactivated on the basis that the technical and economic potential of the process did not warrant continued development.

CO_2 Acceptor

The CO_2 Acceptor process was developed for operation on lignite and subbituminous coals by the Consolidation Coal Company/Conoco Coal

Development Company starting in the early 1960's. The development of the CO_2 Acceptor process includes successful operation of a 40-TPD pilot plant to identify and resolve process and equipment problems relevant to commercialization. Most major operating problems in the pilot plant were solved, and the technology can be considered a candidate for a future demonstration plant.

The novel feature of the CO_2 Acceptor process is that it uses the reversible exothermic carbonation-endothermic calcination reactions of an acceptor (limestone or dolomite) circulated between two reactors (gasifier and regenerator) to provide heat for the gasification process. CO_2 reacts with MgO·CaO inside the fluidized bed gasifier. The MgO·$CaCO_3$ produced is regenerated in a char-fueled, air-blown fluidized bed.

The features of this process include:

- Production of an SNG suitable for conversion to pipeline gas without the use of oxygen
- Use of exothermic energy from the carbonization reaction
- Hydrocarbons other than CH_4 absent in the SNG
- Near-total carbon utilization (more than 99%)
- H_2 to CO ratio in excess of three to one; i.e., shift reaction not needed before methanation
- Minimal CO_2 and H_2S concentrations in the SNG

The main limitations of the process are the mechanical operating problems that attend the integrated and balanced operation of the multiple fluid beds.

A CO_2 Acceptor Process Flow diagram is shown in Figure 3.5.3.25.

In this process, lignite or subbituminous coal is reduced to 8 by 100 mesh, dried to 5 percent moisture, fed into a fluidized-bed preheater, and conveyed to a lockhopper system. The preheated coal is fed to the bottom of a fluidized-bed gasifier operating at 140 psia and 1500°F. The acceptor, calcined limestone or dolomite, is fed at 1870°F from a regenerator to the top of the gasifier's fluid bed. Steam is injected at the bottom. Devolatilization and gasification of coal take place in the presence of steam, CO, H_2, and the acceptor. Char from the gasifier is fed to the fluidized bed of the regenerator. The regenerator also receives streams of air, steam, and a partially carbonated acceptor from the gasifier and sends waste gas to an energy recovery system. The circulating solids, introduced as limestone or dolomite, evolve CO_2 and absorb energy in the regenerator, accept CO_2, and release sensible and chemical energy in the gasifier. The heat needed to regenerate the acceptor is obtained through the combustion of char.

The gasifier product gas flows through a steam-generating heat exchanger, a quench tower, a scrubber, an acid gas removal system, purification, and finally to a packed bed methanator. The technical feasibility of this process has been demonstrated by pilot plant operation with lignite and subbituminous coal.

From 1972 to 1977, more than 42 runs were made by Consolidation Coal Company in the pilot plant at Rapid City, South Dakota. Many successful runs, including one of 251 hours of operation at steady-state temperature conditions and a demonstrated process efficiency of 77 percent, were achieved.

Figure 3.5.3.25

CO_2 Acceptor Process

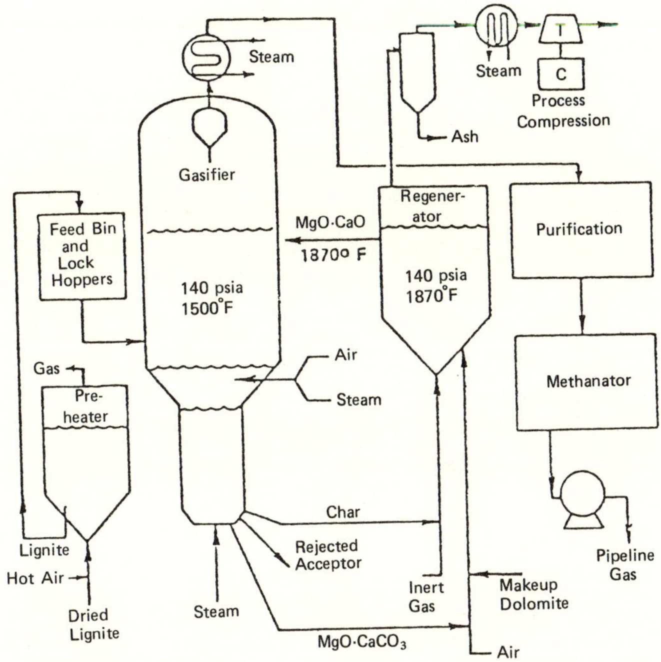

Source: Reference 22

Typical operating data for the CO_2 acceptor process are given in Table 3.5.3.23. Very high hydrogen contents and relatively high methane contents characterize the gasifier raw gas composition, and provide a fairly high heating value. Oxidant consumption in the regenerator is reasonably low on a lb O_2/lb coal basis, but steam consumption is relatively high because this is a reactant stream.

Technically, the process is attractive on several points. For example, coal fines may be used in the gasifier. In addition, oxygen is not required to produce a synthesis gas with quite a high heating value, and water requirements are greatly reduced compared to the Lurgi dry ash process. Due to high temperature operation, tars and oils are not produced, and minimal contamination from H_2S, CO_2 and higher hydrocarbons would be present in a SNG product from the CO_2 acceptor process. Very high carbon conversion efficiencies have been achieved.

The process is limited to the use of low-rank coals due to their high reactivity. The difference in reactivity for low-rank coals is probably nowhere as evident as it is in the CO_2 acceptor process. Higher rank coals cannot be economically gasified due to their decreased reactivity relative to low-rank coals. However, low-rank coals must be dried to a maximum of 5 percent moisture by weight, which is a difficult and expensive pretreatment step, at least for the higher-moisture lignites.

The waste products of this process include dry ash, acid gases, sour water, and spent dolomite. Dolomite disposal is the only environmental feature of the plant not common to most other coal gasification plants.

Table 3.5.3.23

Operating Data for the CO_2 Acceptor Process

(lignite feedstock)

Gas Composition, mol % (dry basis)	
CO	15.5
CO_2	9.1
H_2	58.8
CH_4	13.7
H_2S + COS	0.0
N_2 + Ar	2.9
Raw Gas Heating Value, Btu/SCF	385
Feed requirements	
air, lb/lb coal (regenerator)	2.3
steam, lb/lb coal (gasifier) (as fed to units)	1.1
Operating Conditions	
Pressure, psig	150
Temperature, ^oF	
regenerator,	1850
gasifier	1500
gas velocities, fps	1.8 - 2.4 (gasifier)
Thermal Efficiency	
cold gas	77%

Source: Reference 5

Exxon Catalytic Gasification[23]

Since 1968, the Exxon Research and Engineering Company has been developing (under government sponsorship) the Catalytic Coal Gasification process to produce substitute natural gas. The Process Development phase of work is currently being performed under joint DOE/GRI sponsorship. This phase of the project is slated to continue until March 1981, and a major task in this phase is the demonstration of continuous operation of a one TPD Process Development Unit.

The process produces high concentrations of methane directly in the gasifier, so that the methane can be separated as SNG without a methanation step. It is based on the principle that the thermodynamic equilibrium for the reaction of steam and carbon in the presence of H_2 and CO favors high yields of CH_4 at lower temperatures. To achieve acceptable rates of reaction at appropriate temperature levels, a catalyst must be added.

The catalysts being studied by Exxon for use in this process are basic and weakly acidic salts of potassium. The use of potassium salts has three major benefits with respect to coal gasification:

- The rate of steam gasification is increased:

$$2CH + 2H_2O \longrightarrow 2CO + 3H_2$$

- Swelling and agglomeration of caking coals is reduced

- Methanation equilibrium is promoted:

$$2CO + 2H_2 \rightleftharpoons CH_4 + CO_2$$

The three catalyst features are utilized in the process flow diagram of Figure 3.5.3.26. Coal is crushed and dried before being impregnated with catalyst solution (about 15% catalyst in coal). Water from the catalyst addition is removed in an entrained flow drying column, and the prepared coal is then fed to the gasifier by means of a lockhopper arrangement. Gasification takes place at about 1300°F and 500 psia in a simple fluid-bed reactor. Due to the presence of the catalyst, reaction rates are increased, but gas velocity and gas production per unit of reactor volume is low. Also, because catalytic gasification takes place at relatively low temperatures, high methane concentrations are achieved due to the favorable shift in equilibrium achieved.

Reactor effluent flows through a series of cyclones for fines recovery, which are then recycled to the gasifier. A final cyclone is used after high temperature heat recovery, and the last remaining particulate matter is removed in a venturi scrubber. Acid gases are removed from the clean low temperature gas stream with conventional recovery systems. Product methane is recovered from the CO and H_2 components of the gas by cryogenic distillation, and the methane so produced is available for pipeline distribution. The carbon monoxide and hydrogen are recycled to the gasification reactor. Thus, the only net products are CO_2 and CH_4, along with minor amounts of H_2S and NH_3. Therefore, the net reaction may be written as:

$$\text{Coal} + H_2O \longrightarrow CH_4 + CO_2 \qquad \Delta H \approx 0$$

The overall reaction is only slightly endothermic, and only small quantities of heat are required for coal preheating and to account for heat losses.

Due to the rather high cost of the catalyst, the economics of the Catalytic Gasification process are sensitive to the amount of catalyst which can be recovered and reused in the gasifier. The catalyst leaves the gasifier with the ash material, and approximately 70 percent of it is water soluble, (but depends on the initial loading and nature of the coal ash), suggesting a water leach recovery step. However, any recovery of the remaining catalyst, most of which is in the form of potassium aluminosilicate, will require added measures such as the addition of calcium hydroxide. In any event, make up catalyst in the form of potassium hydroxide will be required, and is added in the catalyst recovery step.

Several advantages may be associated with the Exxon Catalytic Gasification process:

- Since high yields of methane are produced directly, no shift or methanation steps are required

- Problems associated with slagging operation are eliminated

- Caking coals are more easily gasified due to the presence of the catalyst

- Tars and oils are not produced, simplifying wastewater treatment

Figure 3.5.3.26

Flowplan for Exxon Catalytic Coal Gasification Process

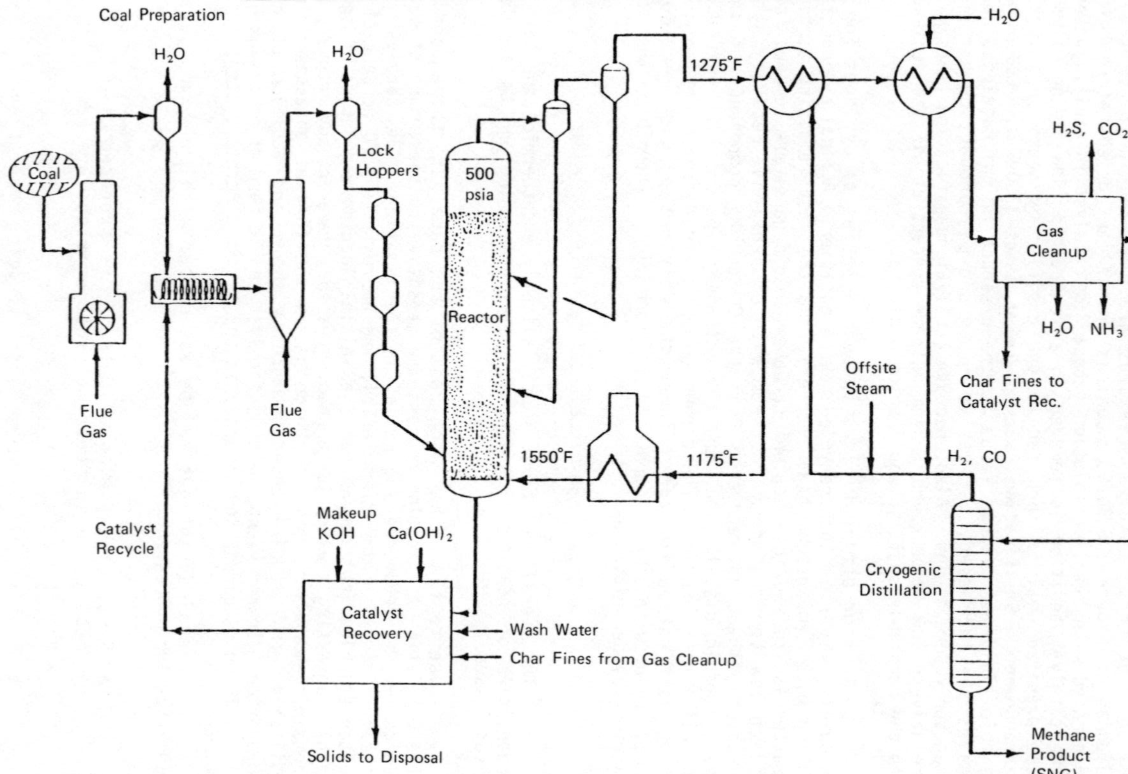

Source: Reference 23

- Low temperature gasification permits the application of existing technology to recover high level heat from the gasifier effluent
- Moderate reaction conditions mitigate materials and operating problems
- Oxygen is not required for gasification.

The process, however, is still in a developmental state. Technically and economically, adequate catalyst recovery must yet be achieved, and continuous reliable operation must be demonstrated. In addition, despite the use of a catalyst, gas production rates remain low; only about 10 percent of the capacity of the slagging Lurgi (per unit of cross sectional area) has been demonstrated in the Exxon gasifier.

The use of low-rank coals in this process is subject to several considerations:

- The different chemical composition of low-rank coal ash may have a different effect on catalyst activity or recovery than that associated with higher rank coals. The predominantly basic nature of most low-rank coal ash may leave more of the basic catalyst in a water soluble form.
- The leaching properties of low-rank coal ash can be expected to affect the catalyst recovery process, since soluble sodium or magnesium salts from the ash would be expected to build up in the catalyst recovery wash. Up to a point, this could be tolerated, since the coal-ash alkali would be expected to have some catalytic effect. However, at a point where the most potent catalytic effect of the potassium was unacceptably diluted, it would be necessary to increase blowdown from the recovery circuit and increase catalyst make-up rate, to control concentrations.
- Higher ash contents present in some low-rank coals may increase the difficulty of catalyst recovery due to the higher volume of inert material requiring processing.
- The high moisture content of some low-rank coals may complicate the drying and catalyst impregnation step.
- Higher reactivities of low-rank coals may increase the already high output of the catalytic gasifiers.

The effect of unrecoverable catalyst on the solid waste disposal problem will have to be addressed as part of the environmental impact assessment and control technology development for this process.

3.5.3.6.3 Entrained Flow Gasification

Texaco

The Texaco entrained flow reactor is a one-stage vertically mounted cylindrical pressure vessel. Coal, oxidizer, and steam are reacted under slagging conditions in vertical downflow. After passing out of the refractory lined partial oxidation chamber, the products are quenched with water and some of the entrained slag is removed in a slag quench bath, which is then discharged through the slag pot (see Figure 3.5.3.27). An alternate method for gas cooling is also under consideration. Here, raw gas leaves the gasifier at high temperature and is cooled in a high pressure steam generator. Choice of gas cooling options depends on the end use of the gas product.

Figure 3.5.3.27
Texaco Gasifier

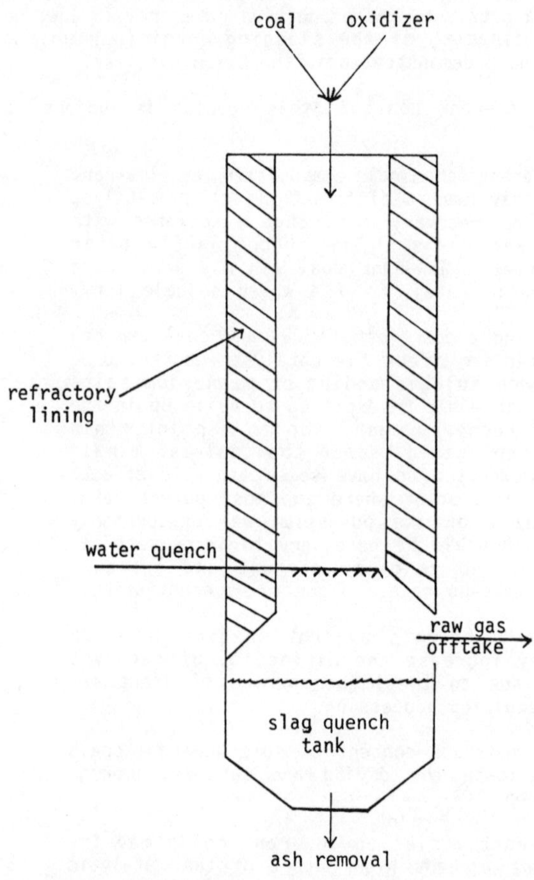

The Texaco gasifier has been developed as a result of many years of related experience in the partial oxidation of petroleum residual oils. Most of the work on the gasifier has been done with oxygen as the oxidizer, because this is the operational mode preferred by the majority of potential industrial clients Texaco has contacted.[24] However, air-blown gasification tests have been done in the past, and Texaco is currently engaged in air-blown testing for a client. The Texaco gasifier has been considered as a gas producer for combined cycle power generation, and part of the present test program is to provide data for this operating mode. Texaco claims that the gasifier operates equally well with air or oxygen.

A ten-fold scaleup from the Montebello unit has been successfully operated for over a year in Oberhausen, West Germany. Another important aspect of this unit has been the demonstration of a waste heat boiler at the gasifier outlet. The waste heat recovery system, the performance of which is of critical importance in determining plant efficiency, is of the radiant box type. Radiant box exchangers are relatively expensive due to their large size.

Dow Chemical is currently in the final stages of construction on a 150 TPD air-blown Texaco gasifier to provide fuel gas for an existing combustion turbine at their Plaquemine, Louisiana plant.

Detailed design of a 1000 TPD commercial sized unit is now underway for 100MW combined cycle power plant to be located at Southern California Edison's Cool Water station. Startup is slated for 1984-85. The $28 million design phase of the project is being funded jointly by SCE, EPRI, and Texaco.

Because of the very high mass throughputs and very short contact times between coal and oxidizer in the Texaco gasifier, the high reactivity of low-rank coals is expected to enhance the operational efficiency. Since the Texaco gasifier relies on a refractory lining, any benefit attributed to the alkaline nature of low-rank coal ash (such as reduced refractory degradation) on ceramics will apply here.

Since one operational configuration of the Texaco gasifier calls for a slurry feed system, the use of low-rank coals may be limited for the Texaco design. The solids content of the water slurry is limited by the high moisture content of the raw coal, or in the case of dried coal, by the reabsorption of moisture. It may be necessary to utilize a drying process that limits reabsorption of moisture. The use of low-rank coals which may have acceptable slag viscosities at lower temperatures than higher rank coals may be advantageous. Operation of the gasifier at lower temperatures can mean decreased consumption of oxygen and possibly lower production rates of CO_2, lowering the capital cost associated with the oxygen plant and the acid gas removal equipment. However, this is very coal specific, and will not apply to low rank coals having high ash fusion temperatures. Based on the limited data available, low-rank coals with high ash calcium levels experience a drastic lowering in viscosity once the melting point has been reached. Thus, acceptable slag viscosities may be reached without increasing temperatures far beyond the melting point for these coals.

The Texaco gasifier displays several advantages over other gasifiers currently available or under development. Since the device operates at relatively high temperatures, formation of tars, oils, phenols, naphthas and ammonia is expected to be minimal, limiting the extent of the wastewater and by-product handling facilities required.

Other advantages of the Texaco gasifier include its ability to accept any type of coal. The system is also capable of very high coal throughputs as a result of its short residence time. This feature is noted in comparison to most other entrained flow gasifiers, and is particularly significant when comparing fluidized bed and fixed bed reactors. The extensive commercial experience Texaco has had in residual oil gasification gives their coal gasifier an important head start in development status.

However, larger scale operations are needed to commercially demonstrate the process with coal. In addition, the coal feed slurry system is complex and a potential source of problems in large scale operations. Slurry pumping is by positive displacement, meaning erosion from coal particles and higher maintenance costs over periods of long continuous operation. Due to the presence of water in the slurry, heat is absorbed in gasification which effectively raises the amount of oxygen required to maintain operational temperatures. This results in a decrease in gasifier efficiency and production of more carbon dioxide in the gasifier, and hence a larger acid gas removal system downstream compared to dry feed injection. Quenching of the raw gasifier effluent can also result in a thermal loss. Oxygen-blown gasification implies close control of oxygen and coal feed rates to maintain reaction temperatures at slagging conditions without producing too much carbon dioxide. Close control of oxygen rate is also highly important from a safety standpoint.

Koppers-Totzek

The atmospheric pressure Koppers-Totzek (K-T) process was originally developed by Heinrich Koppers GmbH under the sponsorship of the Republic of West Germany. Subsequently, the commercial process was developed by Friedrich Totzek of Essen and Koppers Company, Inc. (USA) in 1949 following U.S. Bureau of Mines demonstration in 1948. No installations exist in the U.S. but some 22 foreign plants have been started up since 1949. Nineteen of these utilize coal as feedstock and are listed in Table 3.5.3.24.

Table 3.5.3.24
Koppers-Totzek Coal Gasifier Installations

Location	Fuel	No. Gasifiers	Capacity (MM SCFD)	Use of Synthesis Gas	Year of Order
Mazingarbe Works Paris, France	Coal	1	5.58	Methanol Ammonia	1949
Typpi Oy, Oulu, Finland	Coal Peat	3	5.21	Ammonia	1950
Nihon Suiso Kogyo Kaisha Tokyo, Japan	Coal	3	7.82	Ammonia	1954
Nitrogen Works Coruna, Spain	Lignite	3	9.0	Ammonia	1954
Typpi Oy, Oulu, Finland	Coal Peat	2	5.21	Ammonia	1955
Amoniaco Portugues Lisbon, Portugal	Gasoline, Lignite	2	6.3	Ammonia	1956
Nitrogenous Fertilizer Plant Ptolemais, Greece	Lignite Bunker-C	4	23.45	Ammonia	1959
Nitrogen Works Coruna, Spain	Lignite Naphtha	1	6.5	Ammonia	1961
Chemical Fertilizer Company Lampang, Thailand	Lignite	1	8.0	Ammonia	1963
Azot Sanayil, T.A.S. Kutahya, Turkey	Lignite	4	28.85	Ammonia	1966
Kobe Steel Ltd. Zambia, Africa	Coal	1	7.98	Ammonia	1967
Nitrogenous Fertilizers Ptolemais, Greece	Lignite	1	6.15	Ammonia	1969
The Fertilizer Corporation Ramagundam Plant, India	Coal	4	74.45	Ammonia	1969
The Fertilizer Corporation Talcher Plant, India	Coal	4	74.45	Ammonia	1970
Nitrogenous Fertilizers Ptolemais, Greece	Lignite	1	9.0	Ammonia	1970
The Fertilizer Corporation New Delhi, India	Coal	4	74.45	Ammonia	1972
AE & CI Ltd. Johannesburg, S.A.	Coal	6	80.0	Ammonia	1972
Indeco Chemicals Ltd. Lusaka, Zambia	Coal	1	8.22	Ammonia	1974
Indeco Chemicals Ltd. Lusaka, Zambia	Coal	2	16.44	Ammonia	1975

Source: Reference 14

All K-T foreign installations have been devoted to syngas production for ammonia or hydrogen manufacture. Unlike several other gasifiers in a mature state of development, installations of the K-T system have occurred as late as 1975, with the largest installations occurring in 1969, 1970, and 1972. Earlier units used a variety of fuels other than coal, including peat, gasoline, Bunker-C oil, petroleum gases, naphtha, and lignite. Due to changing economics for some of these fuels, later installations have coal and lignite exclusively as feedstocks.

Figure 3.5.3.28 shows a cross section of the gasifier vessel and Figure 3.5.3.29 shows a process flow diagram of the marketed system.

The K-T gasifier is horizontal and ellipsoidal in shape, with two heads shaped as truncated cones mounted on the ends. A waste heat boiler is mounted on the top to recover heat from the hot effluent gases. Two adjacent burners are installed in each of the heads. Coal, steam, and oxygen are injected through the four burners. The gasifier shell has a double-walled construction; the inner shell is protected from the high temperatures of gasification by a castable refractory lining. Heat escaping through the refractory is recovered by water circulating through the annulus between the inner and outer shells. The annulus is connected to a steam drum. K-T also makes four-headed gasifiers which employ burner heads 90° apart and a total of eight burners. These larger units resemble intersecting ellipsoids and also have a double-walled construction.

Figure 3.5.3.28

Koppers-Totzek Gasifier

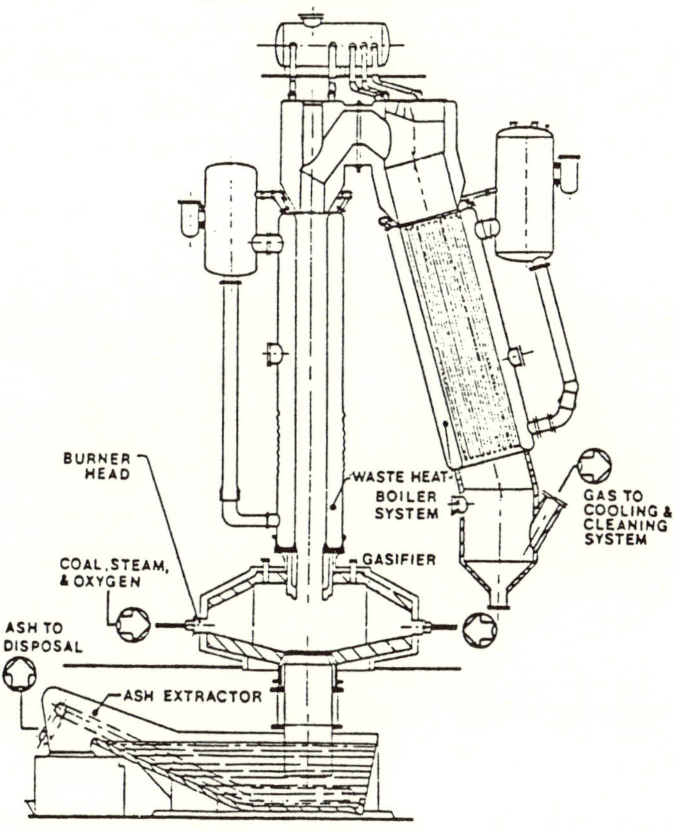

Source: Reference 21

Figure 3.5.3.29

Koppers-Totzek Coal Gasification System

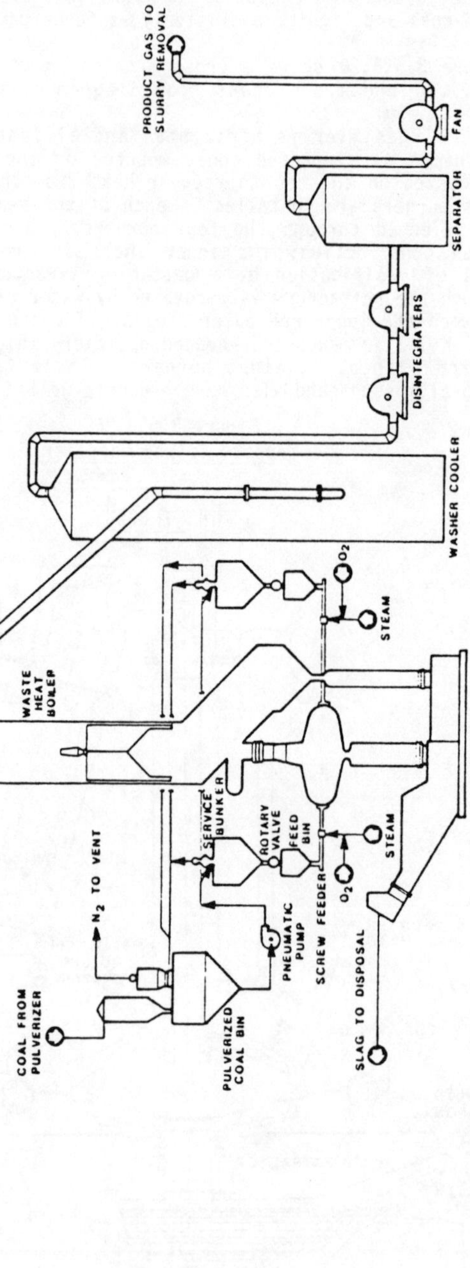

Source: Reference 5

Coal to be gasified is pulverized to minus 200 mesh and dried, depending on rank, to moisture contents of 2 to 8 percent. A variable speed screw feeder moves the pulverized coal to the injection nozzles (burners) where it contacts a premixed stream of oxygen and steam. This mixture of fuel and oxidizer is injected into the gasifier at velocities high enough to prevent flashback. The multiple headed design provides turbulence in the reaction zone, improving product gas composition by breaking up rich and lean pockets of fuel and oxidizer, and providing increased safety through continuous ignition.

Coal is oxidized in the gasifier, producing a temperature of approximately 3500°F with reaction completion in one second. Heat losses and endothermic reactions occurring within the gasifier reduce the gasifier effluent to approximately 2700°F. Approximately 50 percent of the molten slag flows down the gasifier walls and drains into a slag quench tank where it solidifies into a granular form. The remainder of the ash leaves the gasifier as fine slag particles entrained in the gasifier effluent.

The entrained slag is solidified at the gasifier exit and the gas cooled a few hundred degrees by water sprays. Solidification prevents particulate matter from adhering to the waste heat boiler tubes. Sensible heat in the gas is recovered in the waste heat boiler and high pressure steam is generated. Gas leaving the waste heat boiler is water-scrubbed and cooled in a washer cooler and Theisen disintegrator system. The particulate matter is reduced here to a negligible amount. Water from the gas cooling and cleaning system is sent to a clarifier. The separated sludge is sent to disposal along with slag from the gasifier; the water is sent to a cooling tower and then recirculated to the cooling and cleaning system. The scrubbed and cooled gas is processed in a sulfur removal system, resulting in a clean medium-Btu product gas.

Coals of any rank and a wide variety of ash fusion points may be handled in the gasifier. Coals having a large percentage of refractory ash may require the addition of fluxing agents to lower ash fusion temperatures and promote slag flow. The moisture content of the raw coal may affect its transfer in pneumatic conveying lines, but generally the pulverizing and simultaneous drying operation reduces the moisture content below 8 percent. Thus, the effect of the coal moisture content on gasifier performance is usually kept within these bounds. Coals with very high initial moisture contents may be difficult or uneconomic to reduce to low moisture contents but may nevertheless be processed by the gasifier if appropriate adjustments are made in the steam/oxygen ratio.

Typical operating parameters for the K-T gasifier are given in Table 3.5.3.25. The gas is high in hydrogen and carbon monoxide, but has no methane due to the high temperature and low pressure in the gasifier. As such, the gas heating value may be slightly lower than many fixed or fluid-bed gasifiers. Consumption of oxygen is also relatively high at 0.9 lb/lb coal, increasing the size of the oxygen plant relative to other gasifier types. Steam consumption is quite low, allowing the gasification temperature to remain high. The turndown capability of the K-T gasifier is somewhat limited relative to others.

The K-T gasifier has the advantage of being commercially available with over 25 years of successful commercial operating experience. Many types of fuel may be gasified, including a full range of coal types with wide ranges of ash and moisture contents. Gasifier steam consumption is low, and zero in the case of lignite feeds. Wastewater treating facilities are simplified due to the absence of any tars, oils, phenols, naphthas or hydrocarbons in the raw gasifier effluent. Carbon conversion is generally higher than in fluid-bed units, ranging from 92 to 95 percent for bituminous coals, and nearly 100 percent for lignites. Loss of efficiency due to the water quench may be partially offset by the use of a cold gas quench system patented by the Koppers Co., Inc.

Table 3.5.3.25
Koppers-Totzek Gasifier
Typical Operating Data

	Medium BTU Gas
Gas Production, MM SCF/hr	1 - 1.7
Gas Composition, Volume %:	
Hydrogen	36
Carbon Monoxide	52
Methane	-
Nitrogen	1
Carbon Dioxide	10
Other	1
Total	100
Gas Higher Heating Value, BTU/SCF (Dry Basis)	290
Feed Requirements, lb/lb coal:(as fed to gasifier)	
Air	
Oxygen	0.9
Steam	0 - 0.4
Operating Conditions	
Gas Outlet Temperature, °F	2700
Pressure	atm
Residence time, minutes	<1
Turndown Capability	2:1
Coal Feed:	
Type	any
Size, Inches	70% < 200 mesh
Free Swelling Index	no limit
Ash Fusion Temperature, °F	
Feed Rate, tons/day	600
Thermal Efficiency, %:	
Hot Raw Gas	
Cold Clean Gas	75
By-Product Rate, lb/lb coal:	
Tar	0
Oil	0

NOTE: Table may not be internally consistent due to averaging of values.

Source: Reference 1

However, the gasification system faces certain disadvantages. The atmospheric pressure operation limits the number of potential applications for this technology. Although coal moisture content does not limit operation of the gasifier, pneumatic conveying problems may result if surface moisture contents are too high. Oxygen consumption of this gasifier is higher than many others, especially nonslagging devices. Insufficient oxygen supply can result in clinker formation. Ash removal can be a problem even in the presence of sufficient quantities of oxygen. This may occur if the initial deformation, fusion and fluid temperatures of the ash are close together and in the region of 2700°F, causing solidification of ash on the gasifier walls. In these cases, the addition of fluxing agents may prevent the adhesion problem. A thermal loss which is not easily recoverable occurs in the water quench at the gasifier outlet since the latent heat of vaporization for the water can only be recovered at low temperatures. However, the product gas must be cooled below the ash deformation temperature to prevent slag buildup on heat recovery equipment. Finally, the direct contact of sour product gases on heat exchanger surfaces at high temperatures may cause corrosion problems.

The use of low-rank coals in the K-T gasification process is associated with some important considerations:

- The higher reactivity of low-rank coals means high carbon conversions and therefore higher process efficiencies
- High moisture content low-rank coals do not require steam injection as do higher-rank coals thereby improving process efficiencies
- According to the Koppers Co., high moisture coals may be difficult to dry to levels suitable for handling in the pneumatic conveying system. Therefore, each feedstock must be evaluated with respect to this problem
- Since caking coals may be gasified equally well as noncaking coals, low-rank feedstocks do not hold this advantage relative to the K-T gasifier
- Gasification of low-rank coals at lower temperatures may reduce the efficiency loss in the water quench step by decreasing quench water rates

No unique environmental considerations are associated with the K-T gasifier.

BI-GAS

The Bi-gas process was originally developed by Bituminous Coal Research, Inc. under the sponsorship of the Office of Coal Research and the American Gas Association. A PDU was completed in 1971 and a pilot plant employing a two stage entrained flow gasifier was completed in 1976. An air blown version of the Bi-gas gasifier is proposed by Foster-Wheeler Engineering Co., but is in a very immature state of development.

A schematic diagram of the gasifier is shown in Figure 3.5.3.30. The unit is of vertical cylindrical construction with an outer pressure shell. The gasifier consists of three separate zones: 1) a slag quench zone in the bottom; 2) stage 1, where char burners are located; and 3) stage 2, where coal feed is injected and gasified.

The walls of stage 1 are made of vertical closely spaced, water-cooled tubes covered with dense refractory. The tubes are bent inward to form a throat which aids in separation of the molten slag formed in stage 1. The walls of stage 2 consist of an inner layer of precast dense refractory and an outer layer of lightweight insulating refractory to minimize heat losses. Temperature control is achieved in the vessel by a row of vertical wall tubes located just inside the pressure shell. Water circulation in the tubes minimizes hot spots, H_2S corrosion, and hydrogen damage.

Recycle conveying gas transfers coal into the two gasifier injection nozzles. Steam is also injected in the nozzles by a separate annular region. The two streams combine at the tip of the nozzle and join the rising hot gases from stage 1 converting the coal into synthesis gas, methane, and char.

The char is then transported by steam to the three char burners in stage 1. Here recycled char is oxidized with oxygen and steam to produce a hot gas which enters stage 2 to gasify and entrain the incoming coal. Steam is introduced near the throat between the stages to solidify any entrained molten slag. Slag separation is enhanced by the swirling motion of the gas, imparted by the directional firing of the char burners. Molten slag separated in this manner then flows down the walls of stage land into the slag quench zone. The molten ash shatters into a granular form as it contacts the water, and is educted through slag lockhoppers.

Figure 3.5.3.30

BI-GAS Gasifier

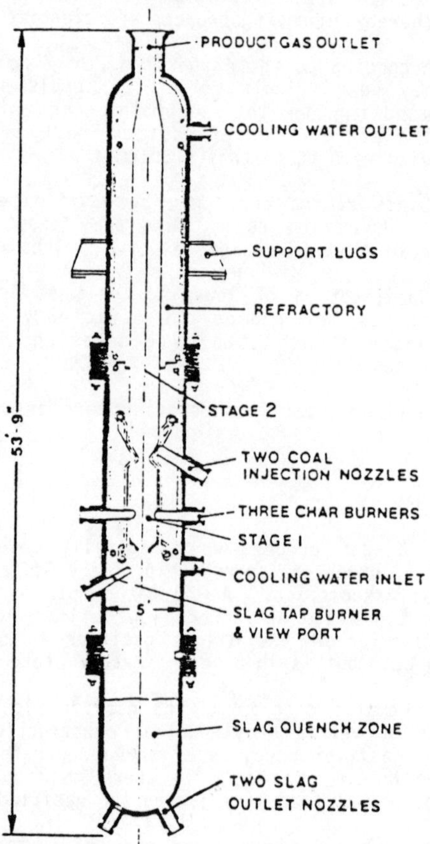

Source: Reference 21

Figure 3.5.3.31 shows the integration of the Bi-gas reactor in a high-Btu gas plant. Gas and char leaving the gasifier are quenched by atomized water and fed to a cyclone separator. Raw gas leaves the cyclone for further processing, while the solid material is recycled to the gasifier.

Uncollected char and ash are scrubbed from the raw gas in a water scrubber which also cools the gas and condenses water vapor. The char-free gas product is the medium-Btu gas product, which may then be further processed for methane production.

Figure 3.5.3.31
BI-GAS Process

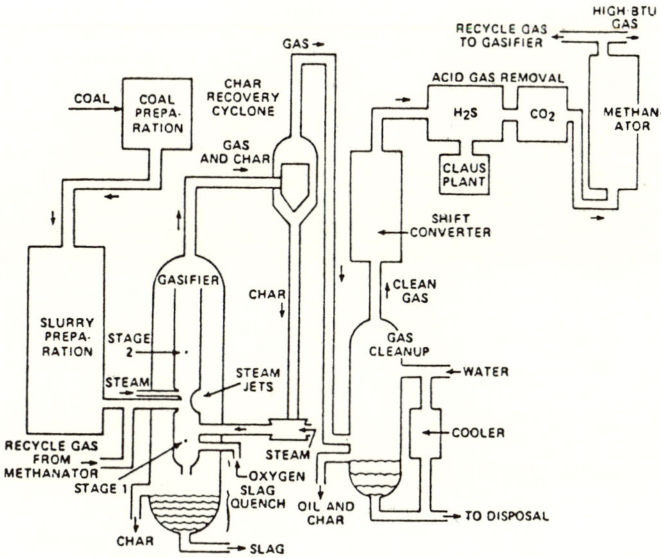

Source: Reference 22

Table 3.5.3.26 presents operating data for the Bi-gas reactor.

Due to the high pressure of operation, the cooled raw gas has a fairly high methane content, which accounts for its relatively high heating value. However, large amounts of CO_2 are present, which affect the size of the acid gas removal unit required before methanation. Steam and oxygen consumption rates are moderate in comparison to other gasifier types.

Several advantages may be cited for the Bi-gas system:

- A 120 TPD pilot plant has been operated
- The two stage gasifier is an integral unit and should be suitable for scale-up
- Pressurized operation yields a synthesis gas product that may be transmitted by pipeline
- Nearly all coals may be handled successfully
- No by-product tars, oils or other similar compounds are produced
- Char recycle allows high carbon conversions.

However, several limitations face the process:

- Steady operation has not been achieved in the gasifier, and other operational difficulties have plagued the pilot plant

- The high pressure operation used to promote methane formation is unnecesarry for synthesis gas production
- Temperature control in the slagging region must be reliable to maintain adequate flow and avoid blockages
- Control of the proper fuel/oxygen ratio may be difficult due to the low concentration of fuel in the gasifier at any given time
- Coals with high ash fusion points may require fluxes to maintain slag flow at gasification temperatures
- An energy penalty must be paid to prepare and process the coal-water slurry. High rates of erosion in the char and coal eductors is expected and erosion of refractory material in the vicinity of the coal and char injectors may occur.

Table 3.5.3.26
Bi-Gas Reactor Operating Data

Gas Composition, mol % (dry basis)	Montana Coal	Oxygen-Blown Operation Pittsburgh Coal
CO	12.74	29.3
CO_2	23.46	21.5
H_2	42.46	32.0
CH_4	6.74	15.7
H_2S + COS	0.08	0.8
N_2 + Ar	14.52	0.7
High Heating Value, Btu/SCF (dry basis)	246	356
Steam Consumption, lb/lb coal (as fed to gasifier)		0.4
Oxygen Consumption, lb/lb coal (as fed to gasifier)		0.5
Operating Temperature, °F		
Stage 1		2300 - 3000
Stage 2		1700
Operating Pressure, psig		500 - 1500
Coal Residence Time, sec		
Stage 1		2
Stage 2		10
Cold Gas Thermal Efficiency		69%
Coal to High-Btu Product Gas	62.9	

Source: Reference 5

The use of low-rank coals in the Bi-gas gasifier is subject to several considerations:

- Since caking coals may be gasified, low-rank coals hold no advantage with respect to this characteristic
- Char recycle is expected to be lower for low-rank coals due to their higher reactivity
- Reabsorption of moisture by dried low-rank coals, or coals with high initial moisture contents may prevent attaining high concentrations of coal in the water slurry used to feed the gasifier.

Waste streams from the Bi-gas process consist mainly of granular ash. Small amounts of char and oil are produced, but no unique environmental problems are associated with this gasifier.

Combustion Engineering

Combustion Engineering, Inc. proposes an entrained flow, atmospheric pressure gasifier to produce low-Btu gas for electric power generation. Conceptual design studies were initiated in 1972 to develop a gasifier, the design of which was funded in 1974 by the Office of Coal Research to be part of a 120 TPD pilot plant. The pilot plant was completed in December 1977, and is now in operation.

Figure 3.5.3.32 shows a schematic of the Combustion Engineering gasifier. The vessel is of vertical cylindrical construction and is designed for atmospheric-pressure operation. Coal is injected at two points, the lower area being a combustion section where a swirling motion is imparted to the hot gases. Here, complete combustion of coal and recyle char occurs in a slagging mode. These gases then rise to the second section where they contact another coal stream resulting in devolatilization, gasification and entrainment of this second coal feed. The walls of the reactor in the gasification section are refractory lined and water cooled, producing high pressure steam. In a commercial size plant, raw gas exiting the gasifier is cooled in a waste heat boiler for steam generation. The cooled gas is then sequentially processed in a spray dryer, a cyclone and a scrubber for removal of particulate matter. Char and ash collected in this manner are ultimately recycled to the combustion section of the gasifier. The cooled, particulate-free raw gas is then sent to acid gas removal, where sulfur is recovered and a clean low-Btu gas is produced.

Table 3.5.3.27 presents operating data for the Combustion Engineering gasifier.

Due to the high operating temperatures and low pressures in the gasifier, no methane or higher hydrocarbons are present in the gasifier effluent. The gas heating value of 127 Btu/scf is somewhat lower than other air blown gasifiers, and is due to the high consumption of air. However, the gasifier consumes no steam, representing an efficiency plus. Slagging conditions are achieved in the combustion zone, and the gasification products remain at a fairly high temperature before leaving the gasifier, thus cracking any large molecules into H_2, CO, and CO_2. Because of this, no tars, oils, or similar compounds are produced. The cold gas efficiency listed in the table represents an isolated gasifier, and hence credit is not given for sensible heat in the gas below the reactor outlet temperature. An integrated plant would recover this heat and hence would show a higher efficiency.

Figure 3.5.3.32

Combustion Engineering Gasifier

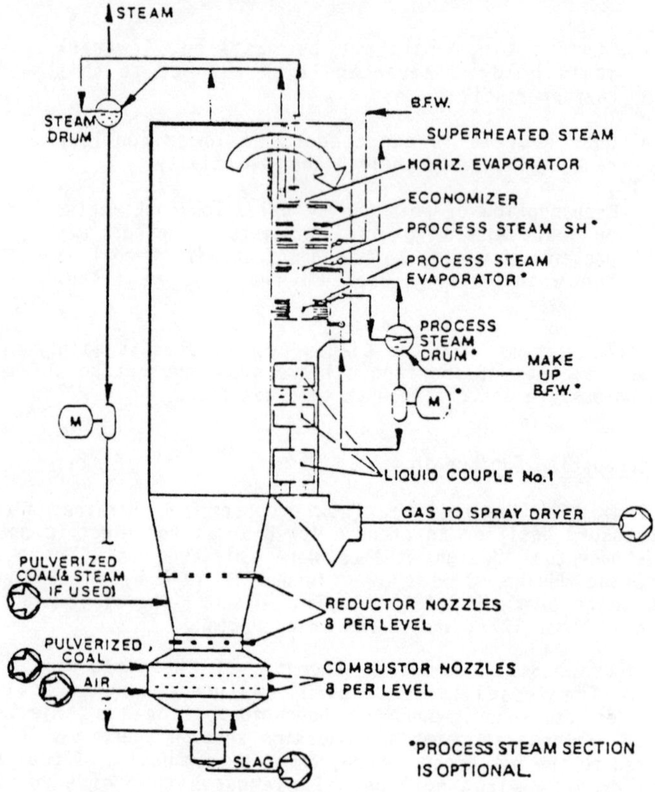

Source: Reference 5

The Combustion Engineering gasifier has several advantages:

- The viability of the concept has been proven in a 120 TPD pilot plant, and should be suitable for scale up to commercial size
- The atmospheric pressure operation of the gasifier is well suited to simple systems for coal feeding and ash removal
- The system can process virtually any kind of coal
- Tars, oils, phenols, naphthas and hydrocarbons will not be produced due to the high operating temperatures
- Ash removal is simplified due to the slagging condition
- Char recycle will provide for a high carbon conversion efficiency.

Table 3.5.3.27

Operating Data for the Combustion Engineering Gasifier

Product Gas Composition, mol % (dry basis) (Kentucky bituminous coal)	Air-Blown Operation
CO	17.0
H_2	7.0
CO_2	22.1
CH_4	0.0
N_2 + Ar	53.3
H_2S + COS	0.6
High Heating Value, Btu/SCF	127
Steam Consumption, lb/lb coal (as fed to gasifier)	0.0
Air Consumption, lb/lb coal (as fed to gasifier)	4.5
Operating Conditions	
Temperature, °F	
Combustion Zone	3000
Gasifier Outlet	1700 - 1800
Pressure	atmospheric
Cold Gas Efficiency*	60%

*Energy recovery will be higher in an integrated plant.

Source: Reference 5

The system also faces several limitations:

- Although pilot plant operation has been demonstrated, commercial operation has not
- Product gas compression will be required for most applications
- Close control over the fuel/air ratio must be maintained due to low gasifier inventories
- Flux addition may be required when gasifying coals with a high refractory ash content
- The recycle of char may represent additional energy requirements, thermal loss and a source of erosion in the system
- Heat transfer surfaces may require frequent cleaning to maintain adequate transfer coefficients.

The use of low-rank coals in this gasifier is associated with several considerations:

- The higher reactivity of low-rank coals may be an important factor in attaining high single pass carbon conversions, thus limiting the amount of char recycle required

- The tendency of low-rank coals to produce fines is not a drawback in this application, since pulverized coal is the required form of coal feed

- The noncaking character of low-rank coals is not advantageous in the Combustion Engineering gasifier, since caking coals can be processed equally well

- High moisture content low-rank coals may require more air to achieve slagging conditions, since most other coals will be dried to 1 or 2 percent moisture during the pulverizing process.

Waste products of the process include quenched slag, acid gases and sour water. No unique environmental considerations apply to the Combustion Engineering system.

References - Sections 3.5.3.2 - 3.5.3.6

1. Economic Studies of Coal Gasification Combined Cycle Systems for Electric Power Generation, prepared by Fluor, Inc., EPRI Report No. AF-642, Palo Alto, CA, 94304, January 1978.

2. Gluckman, M.J., N.A. Holt, S.B. Alpert, and D.F. Spencer. The Near Term Potential for Gasification - Combined Cycle Electric Power Generation, paper presented at the Sixth Energy Technology Conference, February 1979.

3. Economics of Current and Advanced Gasification Processes for Fuel Gas Production, prepared by Fluor, Inc., EPRI Report No. AF-244, July 1976.

4. Economics of Fuel Gas from Coal - An Update, prepared by Fluor, Inc., EPRI Report No. AF-782, May 1978.

5. Hartman, H.F., et al. Low-Btu Coal Gasification Processes, Volume 2, Selected Process Descriptions, ORNL/ENG/TM-13V2, November 1978.

6. Miller, C.L. The Current Technological and Economic Status of High-Btu Gas From Coal, prepared for Short Course on Coal Characteristics and Coal Utilization, Pennsylvania State University, University Park, Pa., October 22-26, 1979.

7. Economics of Texaco Gasification - Combined Cycle Systems, prepared by Fluor, Inc., EPRI Report No. AF-753, April 1978.

8. Schora, F.C., Jr. The Challenge of Solids Handling in Coal Gasification, paper presented at Second Symposium on Storage and Flow of Solids, Chicago, Illinois, September 19, 1972.

9. International Coal Technology Summary Document, Department of Energy, HCP/P-3885, December 1978, P. 41.

10. Jahnig, C.E. and R.B. Bertrand. Environmental Aspects of Coal Gasfication, CEP, August 1976.

11. Handbook of Gasifiers and Gas Treatment Systems, The Dravo Corporation, Pittsburgh, Pennsylvania, February 1976.

12. Comparative Evaluation of High and Low Temperature Gas Cleaning for Coal Gasification - Combined Cycle Power Systems, by Stone and Webster Engineering Corp., EPRI Report No. AF-416, April 1977.

13. Wastewater Treatment in Coal Conversion, Interagency Energy/Environment R&D Program Report, by Industrial Environmental Research Laboratory, Research Triangle Park, N.C., 27711, EPRI Report No. 600/7-79/133, August 1979.

14. Fodor, R.J., Overview of the Great Plains Gasification Associates Coal Gasification Project, Symposium on Technology and Use of Lignite, Grand Forks, N.D., May 1979, DOE GFETC/IC-79/1, 1979, pp. 298-216.

15. Ellman, R.C., L.E. Paulson, D.R. Hajicek, and T.G. Towers. Slagging Fixed-Bed Gasification Project Status at the Grand Forks Energy Technology Center; presented at the Tenth Biennial Lignite Symposium sponsored by the U.S. DOE Grand Forks Energy Technology Center and the University of North Dakota, Grand Forks, N.D., May 30-31, 1979, 41 pp.

16. Magee, E.M., C.E. Jahnig, and C.D. Kalfaldelis. The Environmental Influence of Coal Liquefaction, Exxon Research and Engineering Co., Linden, New Jersey.

17. Roe, D.E., A.W. Karr, K.R. Lemmerman, and J.E. Sinor. Solid Fuels For U.S. Industry, Volume II, Cameron Engineers, March 1979.

18. Attari, A. The Fate of Trace Constituents of Coal During Gasification, Report EPA 650/2-73-004, August 1973.

19. Lapalucci, T.L., et al. Chlorine in Coal Combustion, Bureau of Mines, RI7260, May 1969.

20. Fossil Energy Program Summary Document, U.S. Department of Energy, Document No. DOE/ET-0087, March 1979.

21. Coal Demonstration Plants, U.S. Department of Energy, No. DOE/ET-0069/2, January 1979.

22. Fossil Energy Research Program of the Energy Research and Development Administration, ERDA 77-33, FY 1978, April 1977.

23. Gallagher, J.E. and C.A. Euker, Jr. Catalytic Coal Gasification for SNG Manufacture presented at the Sixth Annual International Conference on Coal Gasification, Liquefaction and Conversion to Electricity, at the University of Pittsburgh, July 31-August 2, 1979.

24. Dillingham, E. Texaco Montebello Research Laboratory, personal communication.

25. Franke, F.H., E. Pattas, and W. Adlhoch. First Experimental Results on Operation of the High Temperature Winkler Process in a Semi-Technical Plant, Symposium on Technology and Use of Lignite, Grand Forks, N.D., May 1979, DOE GFETC/IC-79/1, 1979, pp. 278-287.

26. Patel, J.G. and D. Leppin. The U-Gas Process, presented at the Sixth Annual International Conference on Coal Gasification, Liquefaction and Conversion to Electricity at the University of Pittsburgh, July 31-August 2, 1979.

27. Willson, W., et al. Alkali Carbonate and Nickel Catalysis of Coal Steam Gasification Adv. in Chem. Series 154, "Coal Gasification: 1974."

28. Phase 1: The Pipeline Gas Demonstration Plant, Annual Technical Progress Report, July 1, 1977-June 30, 1978, Conoco Oil Company, DOE Contract No. EF-77-C-01-2543, Document No. FE-2542-12.

29. Parkinson, G. "How to Get Water Out of Lignite, Wood and Peat," Chemical Engineering, V. 85, No. 7, March 27, 1978, pp. 77-78.

30. Jones, F.L. The Great Plains Gasification Associates Project - SNG from Lignite, presented by the American Natural Service Co. at the Fifth Annual International Conference on Coal Gasification, Liquefaction and Conversion to Electricity, University of Pittsburgh, Pittsburgh, Pa., August 1-3, 1978.

3.5.3.7 SELECTED REFERENCES

In order to aid the reader in selecting useful general or comprehensive references from among the many listed, selected references are discussed below.

1. Economic Studies of Coal Gasification Combined Cycle Systems for Electric Power Generation, prepared by Fluor, Inc., EPRI Report No. AF-642, Palo Alto, CA, 94394, January 1978. Economics of Fuel Gas from Coal - An Update, prepared by Fluor, Inc., EPRI No. AF-782, May 1978.

Mass and energy balances for the conceptual flow diagrams of four coal gasifiers in fuel gas generation applications are presented. Estimates of the economic performance of each design are also presented, and are based on capital and operating cost estimates. The updated version (AF-782) presents the results of an economic screening study for British Gas Corporation oxygen blown slagging coal gasification process to produce intermediate-Btu fuel gas, and an update of the economic sections of AF-244, which included air and oxygen blown Lurgi moving bed, U-Gas fluidized bed and Combustion Engineering's entrained process for fuel gas production.

The conclusion reached in the supplementary report is that, within the accuracy of the study, fuel gas costs projected for the moving bed process (using the BGC slagger) are competitive with costs projected in earlier studies based on fluidized bed and entrained processes. The major assumption underlying this conclusion is that the BGC gasifier will operate successfully on a commercial scale, and in the manner assumed by the performance estimates used in the study.

These reports are excellent for establishing comparisons between gasifier types, taking into account the complete system required for fuel gas production.

2. Hartman, H.F., et al. Low-Btu Coal Gasification Processes, Volume 2, Selected Process Descriptions, ORNL/ENG/TM-13V2, November 1978.

A survey was made of 102 reported processes that produce low and intermediate Btu gas from coal. Volume 1 contains the summary, screening, and comparison material resulting from the survey. Concise summaries are provided for 47 processes and include status, operating conditions, and a description of the gasifier. Characteristics of different types of gasification processes are compared and specific comparisons are made for the processes that were investigated in depth. Other process considerations such as potential applications, problem areas, economics, and environmental considerations are discussed. Volume 2 contains the detailed descriptions and information on the 21 more promising processes that were selected. The report summarizes work completed from July 1976 through August 1978 and can be used as a source of information for additional comparisons and evaluations.

3. Handbook of Gasifiers and Gas Treatment Systems, The Dravo Corp. Pittsburgh, Pennsylvania, February 1976.

Similar to the above reference, Low-Btu Coal Gasification Processes, the Dravo handbook contains gasifier descriptions, and summaries of operating conditions and status. In addition, gas treatment systems for removing H_2S and CO_2 from gasifier raw gas streams are covered. Due to its earlier date of publication, it should be used as a source of information on gasifiers only if the desired data is not available in the above reference.

4. Comparative Evaluation of High and Low Temperature Gas Cleaning for Coal Gasification-Combined Cycle Power Systems, by Stone and Webster Engineering Corp., EPRI Report No. AF-416, April 1977.

This screening-type study examined the incentives for developing hot gas clean-up technology for use in coal gasification combined cycle power plants. The iron-oxide process developed at MERC was used as the study basis. The process is designed to remove H_2S from the raw gas at 1000°F. Five process and economic evaluations were performed; air and oxygen blown dry ash Lurgi gasification air and oxygen blown entrained flow gasification (Foster Wheeler), and moving bed slagging gasification (oxygen blown) by BGC.

The study identified a large incentive for hot gas clean-up applications for dry ash Lurgi systems, but none for the advanced gasifiers.

This study is a good introduction to the technical and economic tradeoffs and other considerations pertaining to hot gas clean-up technology.

5. Wastewater Treatment in Coal Conversion, Interagency Energy/Environmental Research Laboratory, Research Triangle Park, N.C. 27711, EPRI Report No. 600/7-79/133, August 1979.

A thorough discussion of leading systems applicable to coal conversion wastewater treatment is presented. The report includes detailed performance characteristics and process design data. The suitability of each system to various wastewaters is discussed, and economic data is presented to aid designers in selecting appropriate treatment systems.

6. Ellman, R.C., et al. Slagging Fixed-Bed Gasification Project Status at the Grand Forks Energy Technology Center, Symposium on Technology and Use of Lignite, Grand Forks, N.D., May 1979, DOE GFETC/IC-79/1, 1979, pp. 236-277.

The Grand Forks Energy Technology Center has been a leader in research on fixed bed coal gasification for years. The goal of the current GFETC gasification program is to develop data which will support commercial and demonstration applications of fixed bed gasification. The paper describes the gasifier and a series of 50 tests which were made on it using a variety of lignite and subbituminous coals. Typical reported data include oxygen consumption, steam consumption, gas production and composition and operating conditions. Experimental technique is also presented.

3.5.4 Coal Liquefaction

3.5.4.1 Introduction and Summary

Coal liquefaction processes break down the molecular structure of the coal and increase the hydrogen-to-carbon ratio of the products to form liquids. The technology (embodied in many different processes) has existed since the early 1900's, but has been applied commercially only in the context of severe national need (i.e., in the German war effort and in South Africa). In essence, the technology has been, and remains today, too costly to be justified on other than national security grounds.

It now appears that just such a consideration applies in the United States. Domestic oil production has begun to decline and will continue to do so. Demand for oil continues to grow (domestically and worldwide) and increasing supplies can only be provided by the Middle East. To the extent that political instability, supply interruptions, and Soviet influence continue to characterize that area, our increasing dependence is equivalent to increasing vulnerability. Therefore, coal liquefaction processes are of immediate national interest, whether they are cost-competitive or not. The questions to be considered in this study are: (1) how do these processes best apply to low-rank coals; (2) what are the unique problems encountered in converting lignite and subbituminous coals to liquids; and (3) what research, development, and demonstration projects should be funded by the Federal government to enhance the utilization of low-rank coals in these processes?

The only commercial liquefaction plant operating today is the SASOL installation in South Africa. The process utilized is indirect liquefaction, which consists of gasifying coal with oxygen, cleaning and shifting the gas, and synthesizing a large number of liquid products from CO and H_2 in catalytic reactors. Compared to other liquefaction processes, this method is expensive. However, it has the advantages of being commercially proven and reliable, and of producing relatively clean, simple chemicals from the basic building blocks CO and H_2. Variations on the SASOL process include producing methanol from the synthesis gas, and in turn producing high-octane gasoline from the methanol by a highly selective catalytic process being developed by the Mobil Corporation.

Indirect liquefaction, therefore, is well suited for producing high quality, light liquid fuels and chemicals suitable for a wide variety of uses (e.g., transporation fuel). It is not an efficient or cost-effective process for producing liquid boiler fuels. Because indirect liquefaction is a gasification process as far as its interaction with coal is concerned, low-rank coals are in many respects preferred feedstocks for this technology in comparison with bituminous coals. The reasons for this are explained in Section 3.5.3, Coal Gasification, as are the unique problems and key issues associated with gasification of low-rank coals. The problems and research needs associated with manufacture of chemicals and fuels from synthesis gas are not unique to low-rank coals in any way, and are not considered in this study.

Direct liquefaction processes are designed to react coal with a process derived solvent and a reducing gas (hydrogen or synthesis gas) at high temperature (750-880°F) and pressure (1500-4000 psig). Initially the coal's weakest bonds break thermally to form free radicals that must be stabilized by hydrogen to produce a first set of products (gas, light oil, asphaltenes, and preasphaltenes or asphaltols). Additional hydrogenation brings about defunctionalization of the initial products. Suitably designed catalysts can be used to influence rate and selectivity in both primary liquefaction and secondary treatment for upgrading products. The solvent used for slurrying the feed coal and donating hydrogen atoms to free radicals is a portion of the plant's product oil which is recycled. In some processes this recycle solvent is catalytically hydrogenated to improve its hydrogen donor quality.

These processes are not in commercial use today; only a few complete, integrated, large-scale pilot plants have been operated continuously in recent years. Although many difficult engineering and operational problems remain to be solved, the individual process steps are relatively well characterized. One of the most difficult of these is to consistently achieve "solvent balance" in a continuous plant with a variety of coals; in other words it has not always been possible to produce, separate, and recycle the desired fraction of product oil at the desired rate to keep the plant operating continuously.

Direct liquefaction processes typically produce a synthetic crude oil consisting of a huge quantity of very complex organic compounds boiling over a wide range. Cyclic compounds predominate; these can be hydrocracked and hydrogenated to form high-quality distillate fuels, naphtha, etc. Alternatively, the raw liquid (which may be a solid at room temperature) can be burned directly as a low-sulfur, low-ash boiler fuel.

Three direct liquefaction processes are currently receiving major attention and funding - Solvent Refined Coal (SRC-I and -II); H-Coal; and Exxon Donor Solvent (EDS). Two SRC pilot plants (50 TPD and 6 TPD) have been operating for several years. Large H-Coal and EDS pilot plants have just been constructed and will be operated during 1980. All three of these projects have selected bituminous coals as their primary design and test feeds; although in each case (particularly EDS) some significant work with low-rank coals has been done and is planned. However, it is fair to say that none of these projects is addressing the problem of optimizing a direct liquefaction process to best utilize the unique properties of low-rank coals.

U.S. low-rank coals are distinguished from higher rank coals by high moisture and oxygen contents, low sulfur contents, and a finely dispersed alkaline ash content. These properties affect the liquefaction behavior of low-rank coals in terms of reactivity, reductant requirement, product yields and quality, solids accumulation in reactors, and catalyst life. The effects of the unique properties of low-rank coals are not sufficiently well defined at present to permit direct application of the leading liquefaction processes developed for bituminous coals. Current understanding of the important effects of these properties is summarized below, and explored in more detail in the discussion of liquefaction key issues which follows.[1,4]

Oxygen and sulfur content - Higher contents of functional groups containing oxygen and the generally low sulfur in low-rank coals affect liquefaction chemistry, hydrogen consumption, catalyst selection, and to some degree, product distribution.

Reaction with carbon monoxide - Low-rank coals, particularly lignite, react very rapidly with carbon monoxide. The chemistry of the reactions includes both removal of part of the coal oxygen (probably the carbonyl) and interaction with coal moisture to provide "in situ" hydrogen. This provides the basis for conversion processes for low-rank coals based on synthesis gas rather than more expensive hydrogen.

Mineral matter - The alkaline and alkaline earth mineral constituents in low-rank coals may catalyze liquefaction reactions which is particularly important in the SRC and EDS processes where no catalyst is added to the liquefaction reactor. The dispersed character of the mineral matter in low-rank coals may cause catalyst poisoning in systems using synthetic catalysts. Also, ash agglomerates containing carbonates may be a problem in processes operated on low-rank coals.

High moisture content - Excess moisture content in low-rank coals adds to the pressure of the liquefaction process unless it is removed. Drying tends to deactivate the coal because of surface oxidation and collapse of the pore structure. These factors are unique to low-rank coals in considering research strategies to lower pressure in a liquefaction process. Also, evolution of large amounts of CO_2 in the process affects both process pressure and the related problem of cleaning recycle gas.

High product viscosity - It is not known which characteristic(s) of low-rank coals cause the liquefaction vacuum bottoms product to be very viscous. However, this is a consistently observed phenomenon which requires more severe operating conditions (higher temperature, longer residence time) to remedy.

In summary, the properties of low-rank coals can be expected to importantly affect both the process design and the overall cost of liquefaction. Positive characteristics of low-rank coals are their relatively low cost, high reactivity, natural catalytic activity, and rapid reaction with CO. Negative characteristics are the requirement for some additional reducing gas, lower liquid yields, and possibly a requirement for higher process pressure.

Key Issues

The key issues associated with application of direct liquefaction processes to low-rank coals are listed below and discussed in the sections which follow:

1. Optimization of Process Variables for Low-Rank Coals
2. Optimization of Recycle Solvents
3. Control Reactor Solids (Calcium Carbonate)
4. High Bottoms Viscosity
5. Techniques for Drying Coal Without Loss of Reactivity
6. Catalytic Effects of Mineral Matter
7. Regenerable or Disposable Catalysts
8. Two-Stage Liquefaction of Low-Rank Coals
9. Minimization of Water Requirement
10. Economics of Low-Rank Coal Liquefaction

1. Optimization of Process Variables for Low-Rank Coals

The significant differences in coal properties and liquefaction behavior as a function of rank imply that the optimal process configurations and conditions will also be significantly different for coals of different ranks. The available experimental evidence tends to support this hypothesis, as summarized below. Directionally, the effects of the major process variables on liquid yields and quality in the leading liquefaction processes have been characterized for at least a few low-rank coals. However, integrated process optimization studies (requiring extensive bench- and PDU-scale data gathering, pilot-scale process operability testing, and economic tradeoff analyses) have not been performed.

The tendency in the current major DOE liquefaction projects has been to test low-rank coals at conditions only marginally different from those used for bituminous coals. This may not be sufficient to identify optimal processing schemes for the low-rank coals, since these processes have been designed for bituminous coals. In contrast, the liquefaction projects at GFETC and the University of North Dakota have been exclusively concerned with low-rank coals. However, these efforts are orders of magnitude smaller than the SRC, EDS, and H-Coal projects, and do not incorporate all of the R&D components (mentioned above) required to resolve this key issue.

The process optimization objectives include: (1) production of high yields of high-quality liquids (with low viscosity being the primary concern for the heavy liquid product); (2) reasonably low hydrogen consumption; (3) low gas yields; (4) minimization of potential operating problems; and (5) maximization of overall process economics, as measured by rate of return on a project.

The primary process variables which affect these objectives are the liquefaction reactor temperature, residence time, and pressure; the hydrogen donor quality (and other properties) of the recycle solvent; the use of synthesis gas versus purified hydrogen; and the solvent-to-coal ratio in the reactor slurry. In addition, many other unit operations in the liquefaction plant impact, or are impacted by, the process optimization problem. Examples are the gas handling and cleanup circuit, the heavy liquids/solids processing train, and the hydrogen (or syngas) generation unit. The effects of many of these variables are summarized below:

Reactor Temperature

Several studies have shown that in the presence of a good donor solvent, higher reactor temperature combined with shorter residence time can improve product yield and viscosity. Tests at GFETC on lignite have shown that as temperature is increased from 440°C to 480°C:[1]

a. The heavy bottoms yield remains about the same while its molecular weight distribution shifts towards lighter compounds, and thus its viscosity decreases.

b. The distillate yield increases from about 30 wt. percent (on MAF coal) to about 60 wt. percent.

c. Total net liquid yield (MAF basis) increases from about 60 wt. percent to 95 wt. percent.

d. The yield of C_1-C_4 gases (MAF basis) increases from about 7 percent to 20 percent.

e. Synthesis gas consumption increases from about 1.2 percent to 2.5 percent.

Similarly, process variable studies on Wyodak Subbituminous coal by Exxon showed that the C_4-1000°F liquid yield at 880°F (471°C) and 25 minutes residence time was equal to that obtained at 840°F (449°C) and 60-80 minutes residence time.[2] An economic evaluation showed that the higher temperature operation was somehwat more attractive; however, it was eliminated from further consideration in the EDS program due to potential operability problems.

a. Use of a heavier solvent to suppress vaporization would involve modification of the atmospheric pipestill overhead system.

b. High film temperatures in the slurry furnace tubes would increase the risk of coking, plugging, and tube failure.

In summary, operating at higher temperature may be advantageous with low-rank coals if high solvent quality can be maintained and potential operating problems can be solved.

Reactor Residence Time

EDS process data indicate that at a constant temperature, the yield of C_4-1000°F liquids tends to increase, peak, and then decrease as liquefaction resident time increases.[3] The maximum occurs at different residence times for different coals (see Figure 3.5.4.1). At 840°F, the yield from Monterey (Illinois) bituminous coal peaks at about 40 minutes; Big Brown (Texas) lignite produces the highest yield at about 20 minutes; and the liquid yield from Wyodak subbituminous levels off at residence times above 60 minutes. In addition, for a given coal, the residence time at which the C_4-1000°F yield is maximized can shift dramatically as the temperature is changed. For Wyodak coal at 880°F, the maximum yield occurs at a residence time below 25 minutes.

Figure 3.5.4.1

Yield Response to Liquefaction Residence Time

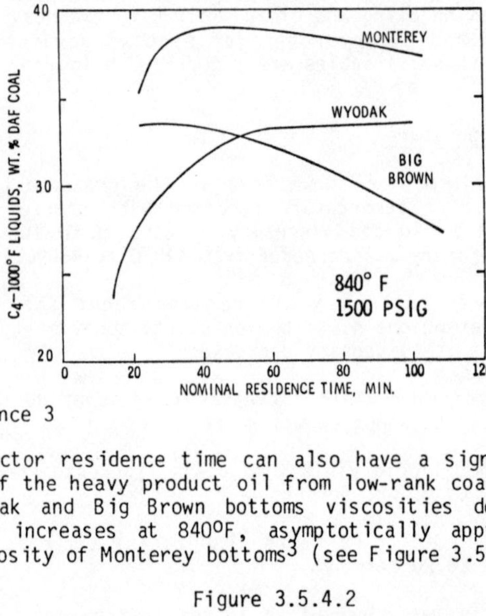

Source: Reference 3

The reactor residence time can also have a significant effect on the viscosity of the heavy product oil from low-rank coals. EDS data show that both Wyodak and Big Brown bottoms viscosities decline rapidly as residence time increases at 840°F, asymptotically approaching the uniformly low viscosity of Monterey bottoms[3] (see Figure 3.5.4.2).

Figure 3.5.4.2

Effects of Coal Type and Process Conditions on Viscosity of EDS Residual Liquefaction Bottoms

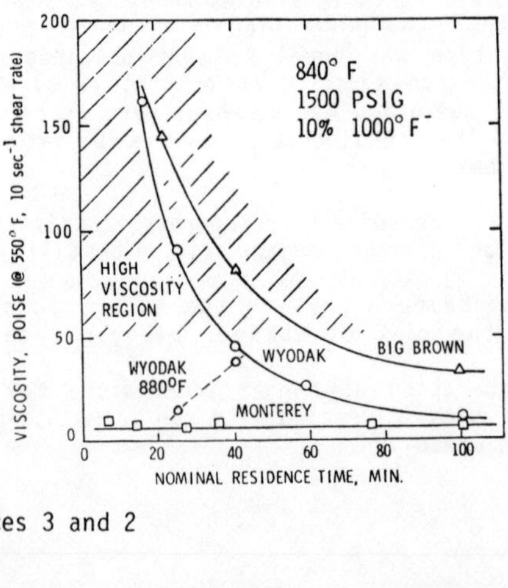

Source: References 3 and 2

An increase in temperature also reduces the bottoms viscosity. Data on lignites from batch autoclaves and a continuous stirred tank reactor at GFETC corroborate the effects of residence time and temperature on viscosity; these data also provide some interesting insights on the effects of residence time distributions in the different reactors.[1]

Solvent Hydrogen Donor Quality

Positive effects of operating at higher temperature have been cited; these are strongly dependent on the presence of a good hydrogen donor in the reactor. The GFETC experiments which showed dramatic increases in distillate and total liquid yields with temperature using tetralin in anthracene oil as the solvent, showed equally dramatic decreases in these yields with temperature using anthracene oil alone. Tetralin is a known hydrogen donor; anthracene has no donor quality.

Catalytic hydrogenation of the recycle solvent to give it good hydrogen donor capabilities is the key to the EDS process. The importance of solvent quality is also recognized in the SRC process development work. The production of a high-quality recycle solvent is sufficiently important and complex that it is discussed separately as a key issue below.

However, one "systems integration" problem that exists in low-rank coal liquefaction using synthesis gas does affect the solvent hydrogenation step. In a process based on synthesis gas, hydrogen remains relatively unreacted while most of the carbon monoxide is consumed. It may be important to find optimum means for using hydrogen-rich purge gas to catalytically hydrogenate recycle solvent.

Finding the best conditions for liquefaction of low-rank coals will require more exhaustive testing than has been accomplished thus far over the domain of the three principal variables: temperature, reactor residence time, and hydrogen donor quality.

Synthesis Gas versus Hydrogen

Low-rank coals react preferentially with carbon monoxide in the presence of hydrogen and water removing coal oxygen as CO_2 and forming in situ hydrogen by the water-gas shift reaction. This provides the opportunity to design a liquefaction process using synthesis gas rather than purified hydrogen as the reactant. This should be advantageous, since it eliminates the shift and hydrogen purification units from the recycle gas system, and reduces overall hydrogen consumption.

The use of synthesis gas for liquefaction has been the primary mode of operation in the GFETC and UND projects, and is believed by those researchers to be the preferred operating procedure when using lignite.[4,5] Experiments in UND's Solvent Refined Lignite PDU indicated that the total conversion of coal, and yields of light oil and SRL, increased as the fraction of hydrogen in the feed gas mixture was increased.[5] However, the yield of "effective solvent" for recycle was maximized at about 50/50 CO/H_2. Overall, the maximum yield of solid SRL and solvent was favored by temperatures of 820 to 830°F and 50/50 H_2/CO.

The use of synthesis gas has also been briefly examined during the development of the EDS and SRC processes. During Exxon's "predevelopment" phase, synthesis gas and water were found to be as effective as pure hydrogen as a liquefaction treat gas when a donor solvent was used. The percentage of hydrogen donated by the solvent was higher in synthesis gas liquefaction.[7] The final configuration selected for the EDS process utilizes hydrogen, however, even when low-rank coals are the feedstocks.

In tests at the SRC Wilsonville facility, a North Dakota lignite was processed with pure hydrogen and with two mixtures of CO and H_2.[8] In apparent contrast to results at UND, higher conversion was obtained as the CO concentration was increased. However, because a larger percentage of the CO reacted than H_2, an imbalance was felt to exist (presumably this would require the removal of a hydrogen-rich purge stream from the reactor system). Also, the yield structure obtained from the lignite under all conditions led to a recycle solvent deficiency. Similarly unfavorable results were obtained on a Belle Ayr subbituminous coal, using hydrogen as the treat gas.[9] SRC yields were relatively low, solvent balance was not achieved, and high yields of gas (particularly CO_2 and H_2O) were produced. These unfavorable results were at least partly the result of air-drying the feed coal to 7 percent moisture, which oxidized the coal and collapsed its pore structure.

Reactor Pressure

The driving force for liquefaction reactions is a function of the hydrogen partial pressure in the reactor. Because of the high oxygen and moisture contents of low-rank coals, large amounts of CO_2 and H_2 enter the vapor phase in the reactor, diluting the concentration of hydrogen. Thus a higher reactor pressure is required to obtain a specified hydrogen partial pressure, compared to a bituminous coal case in which the hydrogen is more concentrated.

Plant costs increase significantly at higher pressure. The liquefaction reactor requires a thicker shell, and possibly a more costly alloy of reduced strength to avoid susceptibility to high-pressure hydrogen attack.[2] Furnace, separation, and hydrogen makeup compression costs are also increased. In short, there is a strong incentive to decrease the reactor pressure.

Until recently, the liquefaction experiments at GFETC have been conducted at about 4000 psig. Recent work at higher temperature, shorter resident time, and with good donor solvents has determined that reducing the reactor pressure to 3000 psig causes only minimal differences. At 2000 psig, the overall conversion was lower, but yields of distillate oil were high.[10] The Solvent Refined Lignite tests at the University of North Dakota were largely conducted at a reactor pressure of 2500 psig. In both of these projects, the lignite was fed with its full moisture content (typically 30-40 percent); however, the recent GFETC work mentioned above has incorporated partial drying of the coal with promising results.

In comparison, the EDS and SRC processes, optimized for bituminous coals, operate at nominal reactor pressures of 1500 psig. Exxon's extensive work with Wyodak subbituminous coal, and limited work with Texas and North Dakota lignites, has largely been conducted at this relatively low pressure level. The use of a slurry drying step in the EDS process is very likely the major factor allowing successful operation with low-rank coals at 1500 psig. Economic studies by Exxon have determined that increasing the pressure is so costly that it is not justified, even under the optimistic assumption that the solvent hydrogenation system could be eliminated with no yield or operability debits.[2]

Solvent-to-Coal Ratio

The concentration of coal in the reactor feed slurry affects yields and operability. In the absence of these effects, higher concentrations would be desirable because less recycle solvent would have to be processed and pumped. However, as solids concentration increases, the ratio of hydrogen donor to coal decreases. Potential plugging problems also become more likely. GFETC data show that the yield of heavy bottom (the "soluble but non-distillable fraction") increases with coal concentration in the slurry. However, the effect on overall process yields is relatively small.[1] Typical ranges for coal concentration in the various processes are 35-45 wt. percent (or solvent-to-coal ratios of 1.2:1 to 2:1).

2. Optimization of Recycle Solvents

As described previously, the recycle solvent quality is a critical variable which impacts on coal liquefaction yields and product quality. The primary function of the donor solvent is to prevent repolymerization of coal derived free radicals, and coking. As illustrated in Figure 3.5.4.3, the donor solvent contains hydroaromatic compounds such as tetralin that donate their hydrogen to thermally decomposed coal fragments, forming stable product molecules. In addition to donating hydrogen, the solvent is the vehicle for transporting the coal into the liquefaction reactor and for promoting the dissolution of the coal particles.

Figure 3.5.4.3
Donor Solvent Reactions

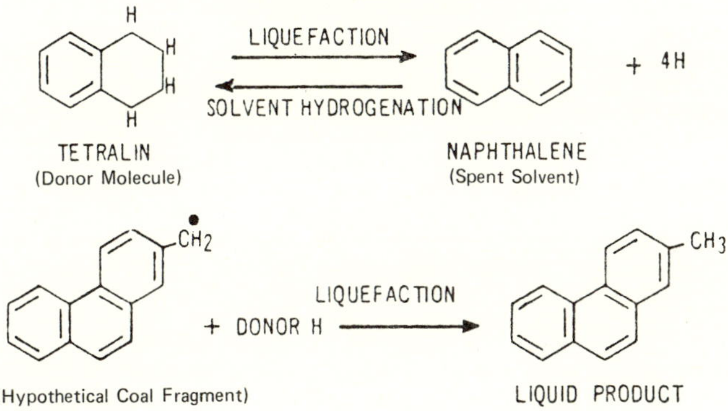

Source: Reference 7

The issues relative to optimization of recycle solvents for low-rank coal liquefaction include:

a. Establishing criteria for solvent effectiveness in a liquefaction process optimized for low-rank coals.

b. Developing an accurate index for solvent quality applicable to low-rank coal liquids.

c. Determining the effectiveness of low-rank coal recycle solvents as a function of reaction conditions, separation techniques, upgrading.

The criteria for solvent effectiveness (in addition to H-donor quality) would be determined by the process requirements. For example, high temperature operation with low-rank coals requires a solvent with a sufficiently low vapor pressure to remain substantially in the liquid phase. The choice of a coal drying technique might also impact on the preferred solvent properties. The choice of boiling range of the recycle fraction may be constrained by the process yields; i.e., the process must produce enough of the desired fraction for recycle in a continuous mode. Data from the SRC, SRL, and EDS continuous units indicate that solvent balance is tenuous with low-rank coals.

The development of an accurate solvent quality index is a prerequisite to reproducible process variable studies and valid comparisons between different investigations. Exxon's proprietary Solvent Quality

Index was developed from extensive studies of model compounds and process-derived solvents.[7] It varies with the degree of hydrogenation of the solvent and with the coal. Control of the index is achieved by adjusting the hydrotreater temperature and the solvent-to-coal ratio.[2] Exxon determines the Solvent Quality Index by mass spectrometry.[11] Other analytical techniques, such as NMR, might be preferable.

The determination of most effective recycle solvents for low-rank coals will require significant efforts. Recent batch autoclave work by GFETC and UND has shown that a heavy process-derived solvent gives better yields of distillable oil than hydrogenated anthracene, tetralin plus anthracene, etc.[10] However, the additional yield appears to come from conversion of the heavy solvent rather than the oil.

Similar evidence that certain high-boiling liquids have unique hydrogen-transferring abilities has been noted in the SRC development work with bituminous coals. In the Wilsonville pilot plants, there has been a strong correlation between the presence of nondistillable liquids in the recycle solvent, and higher coal conversion.[12] (This has been attributed at least in part to the increased presence in the reactor of catalytic mineral matter, e.g., pyrites.)

In the EDS process, the recycle solvent is a relatively light distillate fraction (440/850°F boiling range). The selection of this fraction was made primarily to minimize catalyst deactivation (by heavier liquids) in the solvent hydrogenation reactor.[7] Both bituminous and subbituminous coal liquefaction yields are strongly dependent on solvent quality, as shown for a generalized case in Figure 3.5.4.4 (with and without gaseous hydrogen present in the reactor). Bottoms viscosity is similarly affected, increasing rapidly as the solvent quality index decreases.[2] Low-rank coals tend to produce solvents with more saturates (poor hydrogen donors) when hydrotreated in the EDS process. However, results of EDS process variable studies on Wyodak coal are typically reported at an index value of 4; this implies that a high-quality recycle solvent can be produced from this coal.[2]

Figure 3.5.4.4

Effect of Solvent Donatable Hydrogen

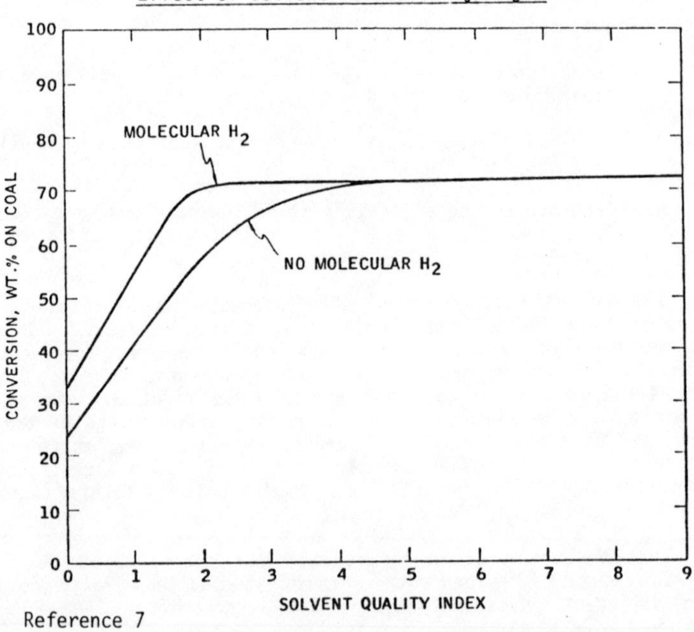

Source: Reference 7

3. Control of Reactor Solids (Calcium Carbonate)

A widely experienced operating problem with low-rank coals has been the accumulation of solids in the liquefaction reactor. These solids are observed both as wall scale and as free-flowing agglomerates. Problems caused by the accumulation of solids include:

a. Reduction of available reactor volume and slurry residence time.

b. Reduced conversion and more viscous bottoms, which can lead to solidification and plugging.

c. Plugging of the reactor or downstream process by the reactor solids.

In general, there are two categories of solids which accumulate in reactors:[13]

a. Coke-like carbonaceous materials, typically formed at high temperature when solvent donor quality is low or depleted.

b. Minerals and unreacted macerals, which in bituminous coals are typically reduced forms of pyrite, plus quartz, anhydrite, and calcite; in low-rank coals, these solids are predominantly calcium carbonate, which grow continuously in shell-like fashion on walls or around inert particles.

The growth rate of calcium carbonate deposits is a function of the ion-exchangeable calcium content of the coal. Low-rank coals typically contain 1-2 wt. percent calcium in the form of salts of humic acids. (Bituminous coals typically contain one-tenth this amount of calcium, in the form of calcite and gypsum, which are stable under liquefaction conditions). The calcium humates decompose in the reactor to form calcium carbonate.

Exxon has studied the formation of these deposits during low-rank coal liquefaction and has developed several control techniques.[14] Tests with Wyodak subbituminous coal, Big Brown lignite, and Indian Head lignite showed essentially the same behavior. Both wall scale and oolites (particles of clay, silica, etc., which have a uniform layer of calcium carbonate growing around them) accumulate in the reactor, relatively unaffected by temperature, space velocity, solvent quality, or hydrogen rate. A significant increase in the accumulation rate was noted, however, when pressure was increased from 1500 to 2500 psig. The wall scale is located mostly in the first half of the reactor, with the heaviest deposits in the zone corresponding to 15-25 minutes residence time.

Control techniques developed by Exxon include periodic withdrawal of free-flowing solids, acid washing of reactor walls during shutdowns, and coal pretreatment by ion exchange. The first two techniques have apparently have been very effective, and are used as the basis for reactor solids control in Exxon's Study Design Update for Wyoming coal.[15] Pretreatment with sulfuric acid, sulfur dioxide, and ferrous sulfate (to form stable calcium sulfate) has been effective in reducing calcium carbonate scale in Exxon's bench scale units. However, in the larger Recycle Coal Liquefaction Units, pretreated coal gave no significant reduction in the amount of solids accumulated (although the percentage of calcium carbonate in the solids did decrease).[2]

Techniques tried unsuccessfully by Exxon included the addition of surfactant materials to the reactor, and feeding the coal at its as-received moisture level instead of dried.[2,14]

Solids accumulations were routinely observed and studied in the UND Solvent Refined Lignite PDU.[15] Two apparent mechanisms were observed:

a. Settling of clay minerals such as feldspar and quartz along with large particles of pyridine-insoluble organic material.

b. Formation and settling of calcium/magnesium carbonate crystals, such as calcite and dolomite.

Similarly, reactor plugging and retention of agglomerated solids in the dissolver were observed at the SRC Wilsonville pilot plant during their tests of subbituminous coal and a lignite.[8,9] It should be noted that the three facilities mentioned above in which calcium carbonate deposition has occured utilize tubular reactors with no internals and cocurrent upflow of slurry and gas at low space velocities.

No buildup of calcium carbonate has been detected in the stirred batch and continuous reactors at GFETC. However, coke formation has been a problem under certain operating conditions (not related to calcium carbonate deposition). Reactor coke material from a recent CSTR run showed no enrichment of calcium, magnesium, iron, or sodium.[17] The deposit was apparently formed by polymerization, not agglomeration. After approximately three days of operation, the upper 1/3 of the reactor was solid coke. The bottom of the reactor is stirred by marine propellers. GFETC has postulated that the high CO_2 and H_2O concentrations in this reactor might account for the absence of calcium carbonate. It is also possible that turbulence in the stirred reactor prevents agglomeration.

4. High Bottoms Viscosity

Another widely observed problem in the liquefaction of low-rank coals has been the production of an extremely viscous bottoms fraction. This is a problem both in terms of product quality (e.g., fuel oil atomization specs), and in plant operability. For example, in the EDS process, high viscosity adversely affects the operation of the vacuum tower and the feed system to bottoms processing (pumping problems occur at viscosities above about 50 poise).[3] One remedy to this problem is to change the bottoms cut point to include approximately 15 percent of the 1000°F material. This lighter material is not consumed in the FLEXICOKING process, so no significant loss of distillate oil yield occurs.

Exxon data on the viscosities of three liquefaction bottoms products were shown on Figure 3.5.3.2 in the discussion of the effects of reactor residence time. At 840°F, the low-rank coals must be subjected to long residence times in order to bring their bottoms viscosities down to comparable level with the bituminous coal bottoms. An alternative is to increase the reactor temperature. At 25 minutes residence time, Wyodak bottoms viscosity declines from 160 poise at 840°F (449°C) to 12 poise at 880°F (471°C).[2]

Extensive work at GFETC molecular weight distributions of liquefaction bottoms using calibrated gel permeation chromatography has shown:[18,1,5]

a. The viscosity is a direct function of the average molecular weight and the concentration of pre-asphaltene material.

b. Temperature has the greatest effect on the molecular weight distribution, with the major peak shifting from 1500 MW to 250 MW as temperature is increased from 404°C to 500°C.

c. At temperatures of 435 to 450°C, significant reductions in molecular weight have been observed with increasing residence time, especially in the greater than 750 MW material.

d. The MW distribution is also shifted downward by increasing the tetralin (hydrogen donor) content in the feed; this effect is more pronounced at 404°C than at 460°C or 500°C.

In summary, although low-rank coals do have a problem with high bottoms viscosity compared to bituminous coals, solutions do exist. These involve either the selection of appropriately severe reactor conditions, or changing the distillation cut point to include some lighter material.

5. Techniques for Drying Coal Without Loss of Reactivity

The high moisture content of low-rank coals acts as an expensive diluent in the gas phase of the liquefaction reactor, increasing both the size and the required pressure of the reactor. Therefore, removal of the inherent moisture from the coal before feeding it to the reactor is desirable. However, conventional air-drying tends to deactivate the coal because of surface oxidation and collapse of the pore structure.

Penetration of the donor solvent into the pore structure of the coal is of crucial importance to the achievement of high conversion.[19] The physical interaction of the solvent with the coal matrix causes extensive fracture (chemical comminution) of the particles in the early stages of liquefaction. This is critical because to prevent repolymerization, a source of hydrogen atoms for capping off free radicals must have a high probability of being within a few Angstrom units of any molecular event inside a particle in which a thermal decomposition occurs.

Although conventional drying collapses the pore structure of the coal, it also is known to expose highly reactive sites to oxidation. Therefore is should be possible to remove the inherent moisture in an inert liquid or gaseous medium which would preserve or enhance the reactive surface area of the coal for liquefaction.

The EDS process employs a slurry drying step which appears to be of interest in this regard. Exxon developed the process to simultaneously dry the coal while preparing the coal slurry feed for liquefaction.[7] (In addition to the effects of conventional drying on liquefaction reactivity, Exxon cites the large energy consumption, particulate cleanup requirements, and potential fire and explosion hazards as other disadvantages of conventional air drying.) In this scheme, crushed coal and hot recycle solvent are fed separately to a well-stirred tank maintained at about 250-350°F and near atmospheric pressure. Moisture evolves from the coal particles and bubbles up through the hot solvent. No oxygen comes in contact with the hot coal. The evolving moisture generates a foam, so space is provided for this. The evolved moisture along with some stripped solvent vapor is withdrawn from the slurry drier and condensed. Condensed solvent is returned to the drier while the water is removed for subsequent water treating. Adequate residence time is provided to dry the coal slurry to less than 4 wt. percent on dry coal feed. Heat required to vaporize the water from the coal is provided by recycling a portion of the slurry through a heat exchanger.

This simple drying process integrates efficiently into the feed system of the liquefaction plant. It appears to provide the intimate contact desired between the solvent and the coal, prior to the removal of the coal moisture. The yields obtained in the EDS process from subbituminous coal and lignite are sufficiently high to indicate that the reactivity of the coal is not impaired by the drying.

In contrast, the limited testing that has been done on low-rank coals at the SRC Wilsonville pilot plant has involved pre-drying of the coal in hot air or nitrogen. The North Dakota lignite was nitrogen-dried from 35-38 percent to 21-23 percent moisture.[8] The conversions obtained in these tests were quite poor; solvent deficiency was experienced; and the effect of added CO to the treat gas was directionally inconsistent with results from GFETC and UND. The Belle Ayr subbituminous coal was pulverized and dried to 7 percent (method not reported) and "probably oxidized."[9] Again, results were poor compared to bituminous coals tested in the Wilsonville plants, as well as to subbituminous coals tested in other liquefaction facilities.

In summary, finding a method to dry low-rank coals without deactivating them for liquefaction is an important aspect of the overall process optimization problem. The slurry drying scheme employed in the EDS process appears to be an effective and efficient means of accomplishing this objective. Comparative data should be obtained on liquefaction process yields and operating parameters when using this technique to achieve various levels of moisture removal from different coals.

6. Catalytic Effects of Low-Rank Coal Mineral Matter

The finely dispersed alkaline and alkaline earth mineral constituents in low-rank coals may catalyze liquefaction reactions. This would be particularly important in the SRC and EDS processes where no catalyst is added in the liquefaction reactor. The effect has been observed qualitatively in the UND Solvent Refined Lignite PDU, when liquefaction yields of Rosebud (higher ash) coal were compared with Decker (lower ash) coal. The Rosebud coal produced lower net gas yields, and higher net liquid yields (28 percent versus 22 percent of the net liquids), with overall SRL yields remaining about the same.

More recently, UND has conducted preliminary batch experiments in a program designed to identify the catalytic effects of low-rank coal mineral matter.[17] So far, any such effects have proved elusive. In fact, lower ash coals gave higher conversions than higher ash Beulah lignite under the same conditions. More work needs to be done on the chemistry of this natural catalysis. The effects appear to be small enough to be masked by other differences in coal properties.

7. Regenerable or Disposable Catalysts

In liquefaction of high-sulfur Eastern bituminous coals, higher distillate yields obtained when heavy liquid is recycled have been attributed to the increased presence of mineral matter.[12] In particular, pyrite or its reduced forms (pyrrhotite, Troilite) appear to have high catalytic activity in SRC-type processes. This catalytic activity achieves extensive hydrocracking of the dissolved coal.[20] The mineral residue activity varies, but all high sulfur Eastern bituminous coals tested have shown good conversions at relatively mild reaction conditions.

This observation has prompted researchers to consider the use of added pyrites as regenerable or disposable catalysts in liquefaction of coals which have low concentrations of these materials. It has been shown that adding pyrite can reduce reaction temperatures without significant loss of liquefaction yields. Alternatively, adding about 5 percent pyrite increases the amount of liquid fuel that can be recovered from bituminous coals about 10 to 15 percent at constant reaction conditions.[21] The goal of the current DOE effort is to find more active catalyst species that will allow a reduction in catalyst concentration.[22] In particular, synthesis of a colloidal, highly active form of iron sulfide is being pursued.

The relevance of this work to improved liquefaction of low-rank coals needs to be determined. Because low-rank coals generally require more severe processing (temperature or residence time) to achieve acceptable product yields and quality, the potential usefulness of an effective inexpensive catalyst is high. Possible interactions of the added compounds with the alkaline mineral matter typical of low-rank coals would have to be determined.

8. Two-Stage Liquefaction of Low-Rank Coals

DOE and EPRI are investigating a two-stage liquefaction process that operates at relatively mild conditions to produce a high yield of high-quality distillate fuels.[12,13] Overall hydrogen consumption would be comparable to other direct liquefaction processes. Three key elements comprise the process:

a. A relatively mild noncatalytic solvent extraction operation (e.g., 750-850°F, low residence time).

b. Solvent deashing to prepare a selective recycle solvent stream with simultaneous rejection of solids.

c. A mild catalytic hydrotreating operation to produce high-quality distillate fuels.

While apparently more complex than existing single-stage processes, the two-stage process may prove to be more economical because operating conditions are less severe and hydrogen is used more selectively. Relatively little hydrogen is used in the extraction stage; more is used in the hydrotreating stage to crack the extract to distillate and remove sulfur and nitrogen. One of the key features of the process is that the solvent deashing rejects some heavy liquids with the ash. Hydrotreating the clean extract is much easier than treating filtered SRC that contains heavy liquids (i.e., hydrogen consumption and catalyst fouling rates are lower).

This interesting process should be explored from the special perspective of low-rank coals. For example, answers to the following questions may reveal unique opportunities for these feedstocks:

a. Given the high reactivity of low-rank coals with carbon monoxide, would the process be better with syngas or purified hydrogen in the solvent extraction stage? High-purity hydrogen would in any event be manufactured in the plant for hydrotreater use.

b. Would the high bottoms viscosity obtained with low-rank coals preclude the use of low-severity solvent extraction conditions because of difficulty in pumping the raw extract? Alternatively, would the two-stage process be effective in overcoming the high bottoms viscosity problem through the combination of solvent deashing and hydrocracking?

c. Given the possibility that low-rank coal mineral matter could poison or accelerate the fouling of supported catalysts (e.g., in the H-Coal process), would the use of the two-stage process be an effective alternative to H-Coal for low-rank coals? The solvent extraction and de-ashing steps are in effect a "super-cleaning" coal preparation process to provide acceptable feed for the hydrogenation step.

9. Minimization of Water Requirements

Coal liquefaction plants utilize comparable quantities of fresh water and produce similar amounts of dirty water to coal gasification plants.[11] Therefore the same general considerations apply, as discussed in Section 3.5.3.1. Net plant water consumption can be adjusted over a wide range by the designers in response to local requirements and costs.

As in coal gasification plants, the largest consumer of water is evaporative cooling. The liquefaction section of the plant does not account for the major heat load; the driving energy for turbines and compressors does. Specifically, cooling of the turbine condensors and compressor interstages accounts for almost 60 percent of the water evaporated (when high use of wet cooling is made). In arid regions, when a combined wet/dry cooling system is employed, the total evaporated water requirement can be reduced about one half. This point is illustrated in Table 3.5.4.1, which shows relative water consumption figures for various levels of wet and dry cooling in "typical" liquefaction, SRC, and oil shale plants. The ranges in each case are functions of the region (and corresponding coal type) assumed, as illustrated in Figure 3.5.4.5 for the "high wet cooing" case. Although the specific liquefaction design assumed for these estimates (Synthoil) is no longer of interest, specific comparisons were made with H-Coal and EDS processes and no major differences were noted.[11]

Of the direct heat losses in a liquefaction plant, the one most affected by the coal rank is the energy needed to dry the coal for liquefaction. This is a large heat load, and is directly proportional to the coal moisture. Note, however, that in a slurry drying processs such as the one employed by Exxon, the moisture removed from the coal is captured, condensed, and routed to the phenol extraction unit. As in gasification, this coal-derived water represents a very significant fraction of the plant's net water requirement when high-moisture low-rank coals are utilized. (Recovery of this water was not assumed in the calculations leading to Table 3.5.4.1 and Figure 3.5.4.5. However, a significant effect of coal oxygen content on the net process water requirements was noted, presumably because this oxygen was assumed to be removed by reaction with hydrogen and the resulting water recovered. This effect would not be as noticeable if syngas were utilized in place of hydrogen, because the coal oxygen would then be removed predominantly as CO_2.)

Table 3.5.4.1
Water Consumption for Standard Size Coal Liquefaction and SRC Plants[a]
(Gallons per 10^6 Btu of fuel product)

	Minimum Practical[b] Wet Cooling		Intermediate Wet Cooling		High Wet Cooling	
	net water consumed	cooling water consumed	net water consumed	cooling water consumed	net water consumed	cooling water consumed
Synthoil	11-16	7-8.2	12-16	7.9-9.1	17-20	12.9-15
SRC	5-10	2.7-2.8	5-11	3.2-3.3	10-15	7.3-7.5
Oil Shale (Paraho Direct)	-	-	19	11.6	-	-

Notes: [a]See Figure 3.5.4.5 for explanation of standard size plants and ranges. All plants are at the minemouth; water consumption for extraction and revegetation operations, etc., is included.
[b]In "minimum practical wet cooling," wet cooling handles 10% of the cooling load on turbine condensers and 50% on interstage coolers. In "intermediate wet cooling," wet cooling handles 10% of the turbine condenser load and all of the interstage cooler load. All loads are wet cooled in the "high wet cooling" case

Source: Reference 11

Figure 3.5.4.5

Net Water Consumption for Standard Size[a] Coal Liquefaction, Clean Coal, and Oil Shale Plants (FC, Four Corners Region; FU, Fort Union Region; PR, Powder River Region; A, Appalachian Basin; I, Illinois Basin; GR, Green River Formation[b]

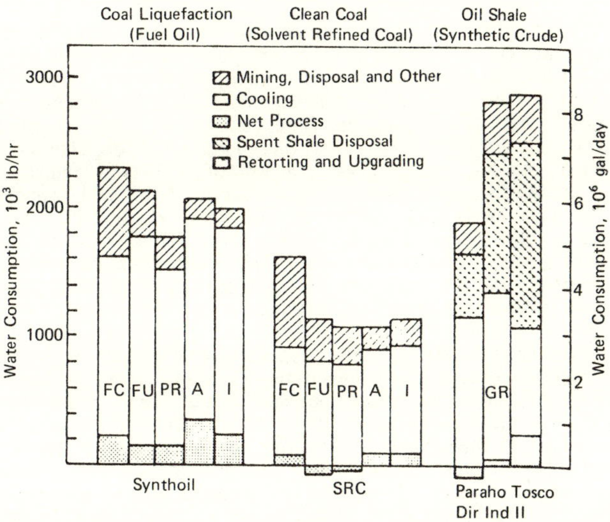

Notes: [a]Standard size plants are as follows: Synthoil - 50,000 B/D of fuel oil output (3.1×10^{11} Btu/day);SRC - 10,000 T/D of SRC (3.2×10^{11} Btu/day; Oil Shale - 50,000 B/D of synthetic crude (2.9×10^{11} Btu/day).

[b]Coals and their respective moisture and ash contents (weight basis) assumed for each region are as follows: Four Corners Subbituminous 12.4% moisture, 25.6% ash; Fort Union Lignite - 36.2% moisture, 8.6% ash; Powder River Subbituminous - 30.4% moisture, 7.8 ash; Appalachia Bituminous - 2.3% moisture, 9.7% ash; and Illinois C Bituminous - 16.1% moisture, 7.4% ash.

Source: Reference 11

For example, in the case of Fort Union Lignite illustrated in the Table and Figure, the net water consumption of the plant would be approximately 40 to 50 million lb/day (4.8 to 6 million gallons/day) depending on the relative use of wet and dry cooling (but not assuming extreme use of either). Coal moisture entering the plant would be approximately 23 million lb/day (31,600 t/d lignite at 36.2 percent moisture). If the coal slurry were dried to less than 4 percent moisture (dry coal basis) and the water recovered as in the EDS process, approximately 20 million lb/day of water would be supplied from this source. In comparison, the case of Appalachian bituminous coal shows essentially the same net water consumption of 40-50 million lb/day in the standard size liquefaction plant. However, coal moisture entering the plant amounts to only 0.74 million lb/day (16,000 t/d coal at 2.3 percent moisture).

In summary, it appears fortunate that high-moisture, low-rank coals are available as liquefaction feedstocks in arid regions of the country. This coal moisture can be utilized, along with increased air cooling and purification and recycle of wastewater, to substantially reduce the very large water demands of these plants. The costs of these conservation measures are significant but relatively small fractions of the product costs (approximately 1 percent for moderate designs, but never more than 5 percent in any case).[11] The issue is therefore one of engineering design and economic trade-offs which will differ for each individual case according to technical, economic, environmental and political constraints.

10. Economics of Low-Rank Coal Liquefaction

In virtually all coal conversion technologies which have been proposed to date, the purchase cost of the raw coal has been by far the most significant operating cost. As such, the dramatically lower prices which currently are available for low-rank coals compared to bituminous coals suggest that their use as feedstocks for coal liquefaction may be economically attractive. However, the issue is complicated by the fact that low-rank coals produce lower yields of saleable products per ton of material processed. In addition, the substantially different properties displayed by low-rank coals will most likely dictate that optimal liquefaction process conditions are different from those found for bituminous coals. These process modifications will affect both capital and operating costs, in certain cases representing increases and in other cases decreases relative to higher-rank coal feedstocks.

A completely satisfactory assessment of this issue would require detailed economic comparisons based on optimal plant designs for the two coals. However, the pilot plant operating and cost data required for such an analysis are not available at this time. In the absence of such data, some indication of the economic driving force may be obtained by considering the relative values of the feed and product streams in the two cases (i.e., the value added). This type of evaluation is extremely rough, and treats the liquefaction plant itself as a "black box" with no assumed differences in capital or operating costs. Implicit in the analysis are product yield structure which are significantly different for the two feedstocks, as suggested by experimental data.

To illustrate the complexity of the issue and the sensitivity of the calculations to assumptions made, two comparative examples of liquefaction economics are presented here. The first example is based on the SRC-II process and is strictly a "value added" comparison of the performance of a lignite and a bituminous coal feedstock. It incorporates the current best estimate by GFETC of an attainable yield structure from the North Dakota lignite in an SRC-II plant optimized for the feed. The second example is based on detailed engineering/economic screening studies by Exxon for three coals in the EDS process. In this case, process cost estimates are available as a result of many manyears of engineering analysis and linear programming model optimization; however, the EDS process configuration and yields applied to the low-rank coals probably do not represent optimal conditions.

The SCR-II example is derived from a report by Gulf Mineral Resources Company[20] examining the economic viability of a commercial-sized plant feeding 33,500 tons per day of West Virginia coal. The feed and product flow rates assumed for this plant (Case A) are shown in Figure 3.5.4.6. A similar input/output estimate of the yield pattern for a lignite-fed SRC-II process was made based on results of lignite liquefaction experiments conducted at the Grand Forks Energy Technology Center and University of North Dakota. Assuming that the SRC-II process could be modified to produce these yields, one lignite-fed comparison case (B) was sized to produce the same yield of naphtha/light distillate oil as in the bituminous coal-fed plant. This yield structure is also shown in Figure 3.5.4.6.

One difficulty in comparing the liquefaction economics of two distinctly different feedstocks is the selection of an "equivalent" plant size for the second coal. Several logical definitions of equivalency with the base case could be used. The cases considered in this analysis are the following:

- Case A - Bituminous Coal-base case - 33,500 T/D (as received).
- Case B - Lignite - naphtha/light distillate production equal to Case A - 42,633 T/D (as-received)
- Case C - Lignite - fuel oil production equal to Case A - 67,520 T/D (as received)
- Case D - Lignite - total product BTU output equal to Case A - 69,331 T/D (as received)
- Case E - Lignite - moisture-free tons/day of coal feed equal to Case A - 51,300 T/D (as received)

Figure 3.5.4.6

SRC-II Yield Structures Assumed for the Value Added Comparison

CASE A: Bituminous Coal Feed (base case)(1)

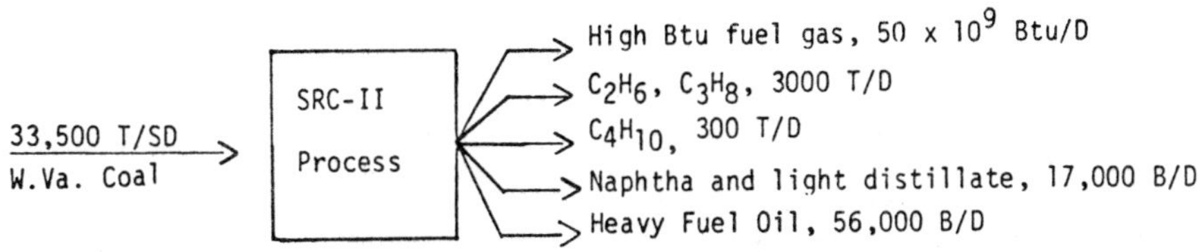

By-Product: Tar Acids, 240 B/D

CASE B: Lignite Coal Feed - Equal Naphtha/Light Distillate Production(2)

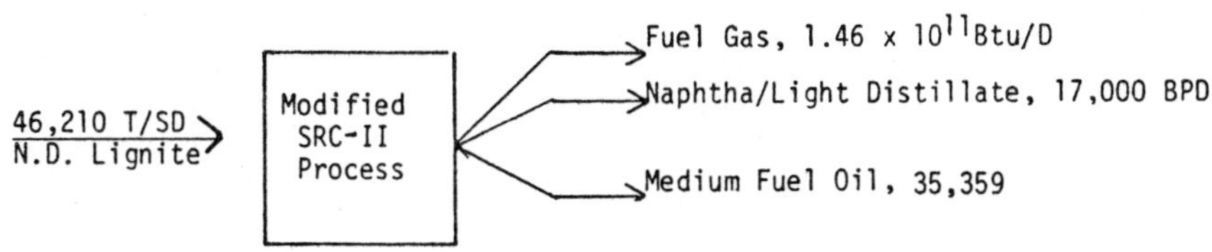

By-Products: Phenols, Cresols, 111 TPD
Gasifier Tar, 305 TPD

Notes: 1. Source - Reference 20
2. Source - Personal communcation from W.G. Willson and G.G. Baker (GFETC), February 1980

As indicated, cases B through E represent lignite-fed plants of widely varying size. This implies that the economic comparison is very sensitive to the definition of equivalency that is selected. Similarly, the comparison is very sensitive to the coal and product prices assumed. Table 3.5.4.2 shows the standard set of prices assumed for all cases, based on February 1980 market values.

Table 3.5.4.2
Coal and Product Prices Assumed for the SRC-II Economic Comparison

Coal Prices		(Approximate $/10^6 Btu)
1. West Virginia Bituminous, $/ton	35.00	(1.34)
2. North Dakota Lignite, $/ton	11.50	(0.86)
Product and By-Product Prices		
1. Fuel Gases and C_1-C_4 hydrocarbons, $/10^6 Btu	4.00	(4.00)
2. Naphtha and Light Distillates, $/gal	0.80	(5.93)
3. Fuel Oil, $/gal	0.70	(5.00)
4. Heavy Oil, $/gal	0.60	(3.95)
5. Phenols/cresols, $/lb	0.20	(13.00)
6. Gasifier tar, $/ton	80.00	(2.42)
7. Tar Acids, $/lb	1.50	(100)

As stated previously, this economic comparison is limited to a calculation of the "value added," or the revenues received from product sales minus the expenditures for coal feedstock, in each base. Purely from an economic viewpoint, it is hoped a rough indication of the desirability of lignite vs. bituminous coal as a liquefaction feedstock will be revealed by this analysis. Table 3.5.4.3 shows the results of applying the standard coal and product prices to cases A through E. Figure 3.5.4.7 plots the "bottom line" of this analysis, graphically showing the sensitivity of the value added comparison to the size of the lignite-fed plants.

The calculations indicate that liquefaction of lignite generates a "value added" somewhat less than that of bituminous coal when the two plants are fed at the same moisture-free coal rate. When the plants are sized to produce equal total product heating values or equal heavy fuel oil yields (i.e., when the lignite plant is made larger), lignite generates slightly more value added. Because the assumed lignite yield structure is skewed toward the lighter products, relatively less lignite is required to generate an equal yield of naphtha/light distillate (compared to the other lignite cases), creating less value added in that case.

The sensitivity of these results to changes in the assumed coal costs is illustrated in Figure 3.5.4.8. The horizontal dotted line indicates the breakeven value added at the base case bituminous coal cost of $35/ton. For example, when the plants are sized for equal total-BTU output (Case D), lignite produces more value added as long as its price is below $12.50/ton (with bituminous at $35/ton).

If capital and operating cost estimates are available for each of the cases, a discounted cash flow analysis would reveal the most attractive investment. In the absence of such estimates the "value added" cash flows can be used to indicate the conditions under which a lignite-fed plant would be a more attractive investment than the bituminous coal-fed plant. This is illustrated in Table 3.5.4.4. The total investment estimated by Gulf Mineral Resources Company for the bituminous coal-fed plant is $1.725 billion; annual operating costs (ex coal) are $96 million. Ignoring taxes, depreciation, etc., the net present value of this assumed 20-year project (discounting all cash flows at 15 percent is $1.39 billion.

Table 3.5.4.3

Value Added Calculations for Lignite and Bituminous Coal in the SRC-II Liquefaction Process

Case	A	B	C	D	E
Coal	W.Va.Bituminous	N.D.Lignite	N.D.Lignite	N.D.Lignite	N.D.Lignite
Description	Base Case	Equal Naphtha/ Light Distillate Prod.	Equal Fuel Oil Production	Equal Total Product Btu Output	Equal T/D of Moisture-Free Coal[1]
Coal Feed Rate, T/D:					
(as received)	33,500	42,633	67,520	69,331	51,300
10^{11} Btu/day	8.78	5.46	8.65	8.88	6.57
Production:					
Fuel Gas, 10^{11} Btu/D	0.50	0.33	0.53	0.54	0.40
Naphtha/Light Distillate, B/D	17,000	17,000	26,924	27,646	20,456
Fuel Oil, B/D	56,000	35,359	56,000	57,502	42,547
Total, 10^{11} Btu/day (including by-products)	6.49	3.99	6.32	6.49	4.80
Cash Flows, 10^6\$/Year:[2]					
Product Revenues	946	523	829	851	630
Coal Cost	352	147	233	239	177
Value Added	594	376	596	611	453
Incremental Value Added Compared to Case A	-	(218)	2	17	(141)

Notes:

[1]Coal moisture contents are assumed to be 2.3% for W.Va. Bituminous and 36.2% for N.D. Lignite.

[2]300 operating days per year.

Figure 3.5.4.7
Sensitivity of SRC-II Value Added Comparison to Selection of "Equivalent" Plant Size

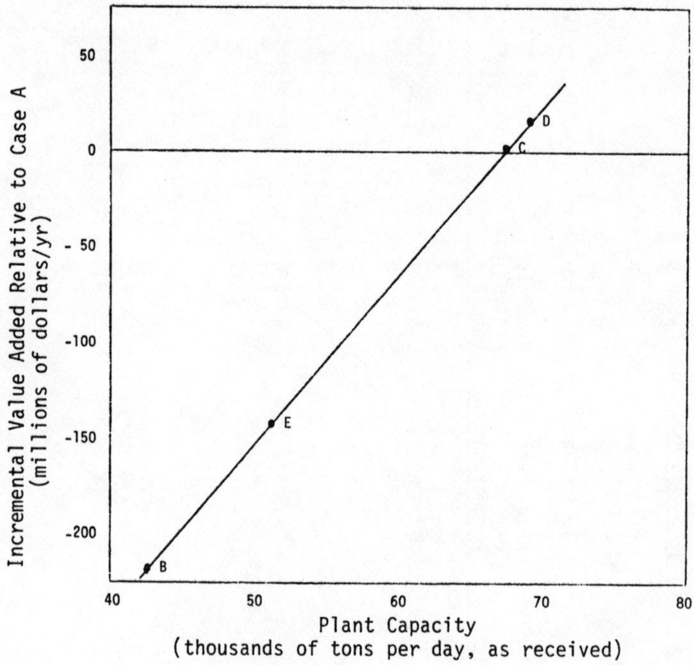

Figure 3.5.4.8
Sensitivity of SRC-II Value Added Comparison to Coal Prices

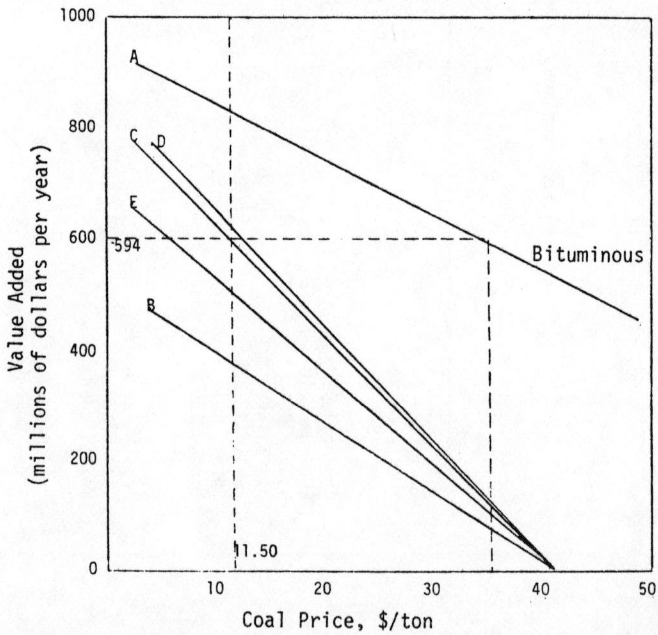

Table 3.5.4.4

Calculation of Plant Investments to Equalize Project Net Present Values

Case	A	B	C	D	E
Coal	W.Va.Bituminous	N.D.Lignite	N.D.Lignite	N.D.Lignite	N.D.Lignite
Description	Base Case	Equal Naphtha/ Light Distillate Production	Equal Fuel Oil Production	Equal Total Product Btu Output	Equal T/D of Moisture-Free Coal
Annual Cash Flows, 10^6/year:					
Value Added	594	376	596	611	453
Operating Costs	(96)	(80)	(126)	(130)	(96)
Total	498	296	470	481	357
Present Value of Annual Net Cash Flows at 15% (20 years), 10^6 $	3117	1853	2942	3011	2234
Plant Investment, 10^6 $	(1725)	(461)	(1550)	(1619)	(842)
Net Present Value, 10^6	1392	1392	1392	1392	1392
% Increase (Decrease) of Plant Investment vs Case A	-	(73)	(10)	(6)	(51)

Costs have not been estimated for the lignite-fed plants. However, in the Table we have assumed for illustrative purposes that the operating costs of lignite Plant E are $96 million/year, and that in the other plants they are proportional to the lignite feed rate. We have then calculated the plant investment in each case that gives the project a net present value of $1.39 billion, thus making it an "equally attractive investment" to Case A. For example, if an "equal product Btu"-sized lignite liquefaction plant (Case D) can be built for less than 6 percent under the cost of the bituminous plant, it will provide a larger return to the investor.

It is important to recognize that this analysis is useful at best to establish a trend. The effect of process design on coal consumption and yield pattern should not be overlooked, and has received only a brief consideration in preparing this exercise. Further work should be oriented at the effects of process optimization and yield selectivity on process economics. At a minimum, it can be concluded that liquefaction of lignite warrants further development work and evaluation along with higher ranks of coal.

The second example of comparative liquefaction economics is a summary of Exxon's process variable screening studies on the EDS process.[15] An economic comparison was made among selected "best cases" for three coals: Big Brown (Texas) lignite, Illinois #6 (Monterey) bituminous; and Wyodak (Wyoming) subbituminous coal. The comparison includes estimated differences in coal prices and location effects. All cases reflect Illinois product/byproduct values and utility costs. A summary of the three "best cases" for the once-through FLEXICOKING/partial oxidation EDS configuration is shown in Table 3.5.4.5.

Table 3.5.4.5

Exxon Donor Solvent Engineering Process Variables Screening Studies "Best Case" Economic Comparison (Once Through Flexicoking/Partial Oxidation)

Coal	Big Brown	Illinois	Wyodak
Liq'n Conditions	840/40	840/40	840/60
Case Number	854	563	851
Investments, $MM TEC(1)			
Liq'n/Sol. Hyd.(2)	1037	897	1110
H_2 Generation	458	416	429
Fuel Generation	372	327	382
Other	949	1025	958
Location (Credit)Debit(3	(563)	-	303
Total	2253	2665	3182
Operating Costs, $MM/yr			
Coal(4)	168	204	59
By-products			
C_2-	(125)	(139)	(140)
Sulfur, NH_3(5)	(12)	(25)	(5)
Catalyst/Chemicals	3	8	3
Purchased Utilities(6)	44	50	48
Investment Related(7)	507	601	716
Total	585	699	681

(continued)

Table 3.5.4.5 (continued)

Exxon Donor Solvent Engineering Process Variables
Screening Studies "Best Case" Economic Comparison
(Once Through Flexicoking/Partial Oxidation)

Coal	Big Brown	Illinois	Wyodak
Revenues, $MM/yr			
C_3	44	41	40
C_4	24	40	37
Naphtha	234	260	257
LSFO	131	233	143
RSFO (Scrubber Bottoms)	36	27	33
Total	469	601	510
Net Cash Flow, $MM/yr	(116)	(98)	(171)
Sensitivities on Net Cash Flow			
I. No Location Credit/Debit	(243)	(98)	(103)
II. Big Brown Lignite Cost			
$15.50/T	(93)	(98)	(171)
$20.50/T	(140)	(98)	(171)
III. $CaCO_3$ Cleanup			
Controlled Scaling	(129)	(98)	(185)
SO_2 Pretreatment	(154)	(98)	(203)

Notes:

[1]1985$ basis. Investments are based on the 1975/1976 EDS Study Design and its configuration (20 kT/SD dry coal feed). Same Process Development Allowances (PDA) for all coals.

[2]10% was added to the liquefaction/solvent hydro investments in the Big Brown and Wyodak cases to reflect an increase in solvent-to-coal ratio from 1.2 (Illinois coal) to 1.6.

[3]Big Brown cases include a 20% location credit over a Western Illinois location. A $300 MM location debit was added to the Wyodak cases to reflect a Wyoming location relative to Western Illinois.

[4]Illinois coal at $27.80/T; Wyodak coal at $6.50/T; Big Brown lignite at an average estimated value of $18.00/T.

[5]By-product prices are 1975/1976 EDS Study Design values of $80/LT for sulfur, $228/T for ammonia based on Illinois coal.

[6]Utility prices are 1975/1976 EDS Study Design values of 3.5¢/KWH for purchased power, 20¢/kgal for raw water based on Illinois coal.

[7]Reflects Illinois capital recovery factor of 17.2% of TEC. (15% current $ DCF return on investment)

Source: Reference 15

As shown, the Texas lignite-fed plant is estimated to be 15 percent less expensive to construct than the Illinois bituminous plant, which in turn has a 16 percent lower investment cost than the Wyoming subbituminous plant. When the location credits/debits are removed, the three plants are seen to be nearly equivalent in cost (lignite 6 percent higher than bituminous; subbituminous 8 percent greater than bituminous). When the annual revenues, operating costs, and coal costs are considered, and the investment costs are annualized by a capital recovery factor, the net cash flows for all three cases are negative. Illinois coal provides the best (least negative) cash flow of $98 million/year); the lignite plant's cash flow is 18 percent lower at $116 million/year); and the subbituminous plant is 74 percent lower at ($171 million/year). As shown at the bottom of the table, these estimates are very sensitive to the location credits and coal costs assumed. The costs of calcium carbonate reactor solids control are also seen to be substantiated in the sensitivity analysis.

Exxon's conclusion from this analysis is stated as follows:[15]

"The overall results of this screening study indicate that, in general, it is more economic to use Illinois coal as feed to an EDS plan rather than Big Brown or Wyodak coals. With estimated location effects included and excluding any operability problems, net cash flows of "best case" Big Brown and Wyodak cases are lower than Illinois by $20 million/year and $60 million/year, respectively. Only under the most optimistic circumstances can Big Brown become about a standoff with Illinois coal. Without the significant location factor credit, Big Brown becomes the least attractive coal."

As the previous analysis of SRC-II economics showed, this type of comparison is very sensitive to the definition of "equivalency" used in establishing the relative plant sizes. Table 3.5.4.6 shows the coal feed and product flow rates utilized by Exxon in their analysis. The plants are sized on an equal moisture-free coal feed basis of 20,000 T/D. (Subsequent Exxon studies of economics of scale have indicated a desirability to increase the Illinois coal feed rate from 24,000 to 30,000 T/D.[2] This is analogous to Case E in the SRC II study. The value added in each case is indicated in Table 3.5.4.6. as well. In Figure 3.5.4.9, the sensitivity of the value added calculations to the low-rank coal plant sizes are plotted for the same cases considered in the SRC-II study. The same trends are observed in the EDS cases (compare with Figure 3.5.4.7), with two noticeable differences:

a. The low value of Wyodak coal ($6.50/ton) gives it a substantial value added advantage over the Illinois coal in all cases; conversely, the realtively high value assigned to Big Brown lignite ($18.00/ton) makes it compare poorly.

b. The relative value added of Cases E and B are reversed in the EDS process compared to the SRC-II process. This is because at equal moisture-free coal feed rates, the EDS process gives lower naphtha yields for the low-rank coals while the SRC-II process gives a higher naphtha/light distillate yield from the lignite compared to bituminous coal. This difference may reflect 1) a real difference between the two processes; or 2) a failure to optimize yields in one of the cases; or 3) merely estimating error.

The availability of plant capital and operating cost estimates for the EDS cases allows a more rigorous examination of the effects of the "equivalent" plant size assumption. A well-accepted rule of thumb is that plant capital costs increase in proportion to a fractional exponent as plant capacity increases:

$$\frac{\text{Investment 2}}{\text{Investment 1}} = \left(\frac{\text{Capacity 2}}{\text{Capacity 1}}\right)^{0.6}$$

The exponent varies for different types of equipment, but 0.6 is a typical value and is selected here for illustrative purposes. Figure 3.5.4.10 shows the EDS investment costs and annual net cash flows as a function of the low-rank coal plant capacities, assuming:

a. Investment costs increase with capacity to the 0.6 power.

b. All operating costs and credits increase linearly with capacity (Note, however, that the annual net cash flow as calculated by Exxon includes the annualized investment cost, which follows assumption (a). This annual net cash flow is analogous to the net present value concept in the SRC-II analysis (Table 3.5.4.4), in that it incorporates all project cash flows, appropriately discounted, into a single "figure of merit.")

The figure indicates that as larger low-rank coal plant sizes are considered (to produce total product BTU or fuel oil yields equal to the bituminous plant), the increases in annual value added more than compensate for the increase in plant capital and operating costs. For example, increasing the Big Brown plant size from 30,000 to 40,000 T/D raises its investment cost to the level of the 24,000 T/D Illinois coal plant ($2.7 billion). The two plants then produce equal amounts of total product heating value. The net cash flow for the Big Brown plant would be ($83 million) per year, or $15 million higher than the Illinois case. This is in contrast to the Exxon base case of $18 million lower net cash flow for the Big Brown plant, when feeding equal amounts of moisture-free coal.

Table 3.5.4.6

Feed and Product Rates Used in Exxon Screening Studies

Coal	Big Brown	Illinois	Wyodak
Coal Feed Rate, T/SD[1]			
- Moisture-Free Basis	20,000	20,000	20,000
- As Received	30,000	24,000	29,000
Product Rates, B/SD[2]			
C_2/C_4 LPG	6,700	8,000	7,600
Naphtha	19,800	22,000	21,700
Fuel Oil	19,100	29,800	20,200
	45,600	59,800	49,500
Value Added, 10^6\$/year[3]	301	397	451

Notes:

[1]Calculated from coal costs given in Table 3.5.4.5.

[2]Illinois product rates as reported for the 1975/1976 EDS Study Design, in Reference 2 (page 147). Big Brown and Wyodak product rates ratioed from Illinois rates based on costs given in Table 3.5.4.5.

[3]Product revenues minus coal costs given in Table 3.5.4.5.

This sensitivity analysis does not justify a conclusion that Big Brown lignite is a preferable liquefaction feedstock compared to Illinois bituminous coal; nor does the Exxon analysis justify the conclusion that lignite is the "least attractive coal." In fact, the various net cash flow figures are generally within the "noise" - the ± 20 or 30 percent level of confidence that can be reasonably attributed to this type of analysis at best. Finally, it is doubtful that the operating conditions and yield structures assigned to the low-rank coals in Exxon's analysis reflect optimal values (e.g., use of syngas, higher temperature and pressure, etc.). In conclusion, the preceding analysis constitutes a strong argument in favor of more focused R&D work on low-rank coal liquefaction, including engineering/economic parameter studies and tests of selected low-rank coals in the large pilot plants now being constructed.

Figure 3.5.4.9

Sensitivity of EDS Value Added Comparison to Selection of "Equivalent" Plant Size

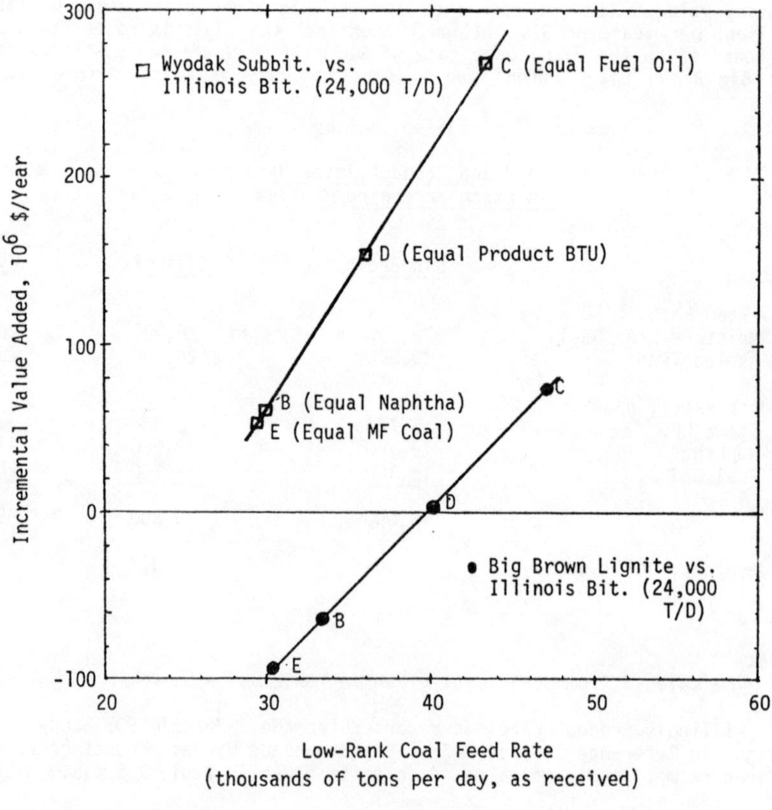

Figure 3.5.4.10
Sensitivity of EDS Investment and Net Cash Flow Comparison to Selection of "Equivalent" Plant Size

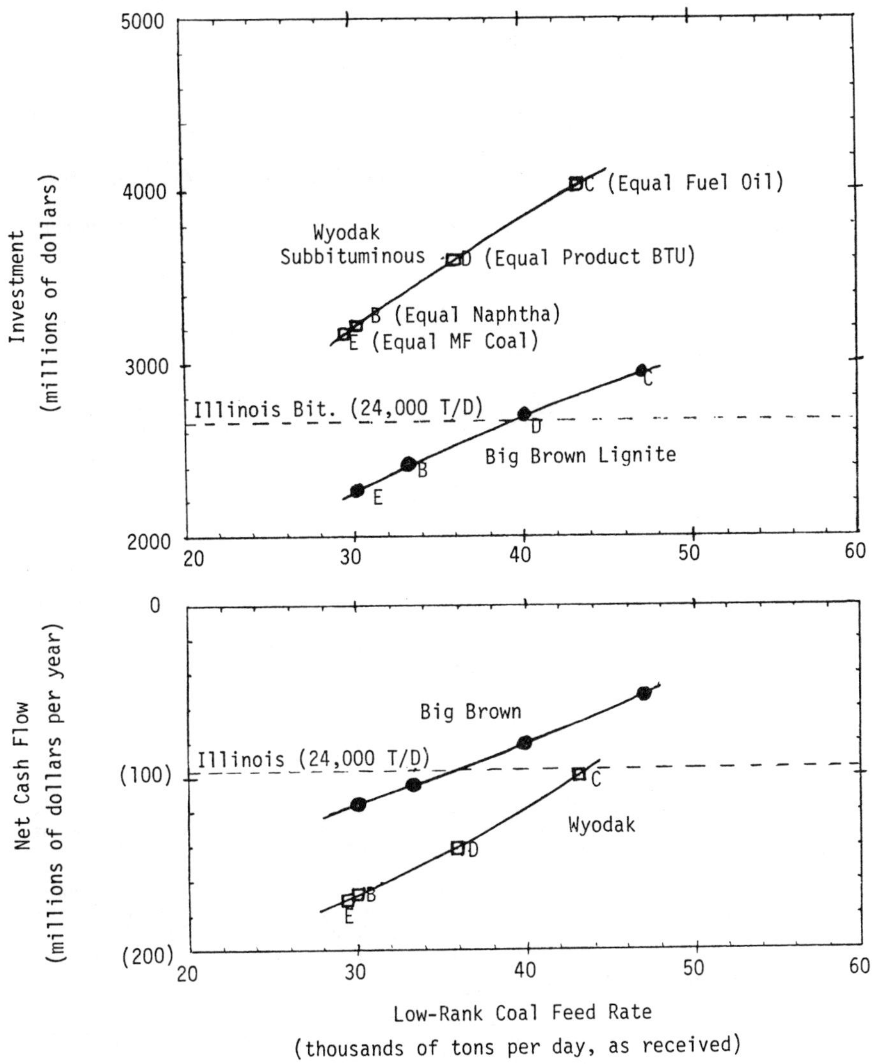

References - Section 3.5.4.1

1. Willson, W.G., C.L. Knudson, G.G. Baker, T.C. Owens, and D.E. Severson. Application of Liquefaction Processes to Low-Rank Coals, Symposium on Technology and Use of Lignite, Grand Forks, N.D., May 1979, DOE GFETC/IC-79/1, 1979, pp. 170-206.

2. Epperly, W.R. EDS Coal Liquefaction Process Development, Phases III B/IV, Annual Technical Report, July 1, 1977-June 30, 1978; FE-2893-17, September 1978, 456 pp.

3. Mitchell, W.N., K.L. Trachte, and S. Zaczepinski. Performance of Low-Rank Coals in the Exxon Donor Solvent Process, Symposium on Technology and Use of Lignite, Grand Forks, N.D., May 1979, DOE GFETC/IC-79/1, 1979, pp. 157-169.

4. Sondreal, E.A. Utilization Potential for Low-Rank U.S. Coals in Liquefaction and Gasification, Presented at the Short Course on Coal Characteristics and Coal Utilization, Penn State, October 22-26, 1979, 51 pp.

5. Sondreal, E.A., C.L. Knudson, J.E. Schiller, and T.H. May. Development of CO-Steam Process for Liquefaction of Lignite and Western Subbituminous Coals, Symposium on Technology and Use of Lignite, Grand Forks, N.D., May 1977, GFERC/IC-77/1, pp. 129-158.

6. Severson, D.E., A.M. Souby, and J.J. Harris. Process Development of Lignite Liquefaction, Symposium on Technology and Use of Lignite, Grand Forks, N.D., May 1977, GFERC/IC-77/1, pp. 159-192.

7. Epperly, W.R. EDS Coal Liquefaction Process Development: Phase IV, Summary of EDS Pre-Development, (1966-1975) Interim Report; FE-2893-16; July 1978, 50 pp.

8. Solvent Refining of Indiana V Coal and North Dakota Lignite, Prepared by Hydrocarbon Research, Inc. Lawrence Township, N.J., EPRI AF-666, Project 779-4, Final Report, January 1978, 64 pp.

9. Lewis, H.E., et.al. Solvent Refined Coal (SRC) Process: Operation of Solvent Refined Coal Pilot Plant at Wilsonville, Alabama, Annual Report, January-December 1977; FE-2270-31, May 1978, 205 pp.

10. Willson, W.G. Grand Forks Energy Technology Center Liquefaction Project, Quarterly Report, April to June 30, 1979; September 1979.

11. Probstein, R.F. and H. Gold. Water in Synthetic Fuel Production - The Technology and Alternatives, Water Purification Associates, The MIT Press, Cambridge, MA, 1978.

12. Wolk, R., N. Stewart, and H. Lebowitz, Coal Liquefaction, Fossil Fuel and Advanced Systems Division R&D Status Report, EPRI Journal, September 1979, pp. 38-39.

13. Jenkins, R.G. Unwanted Deposits in Liquefaction Reactors, Presented at the Short Course on Coal Characteristics and Coal Utilization at Penn State University, October 22-26, 1979.

14. Stone, J.B., K.L. Trachte, and S.K. Poddar. Calcium Carbonate Deposit Formation During the Liquefaction of Low-Rank Coal, Exxon Research and Engineering Company paper presented at the ACS Symposium on Coal Structure, Honolulu, Hawaii, April 2-6, 1978, ACS Div. of Fuel Chem. Preprints, v. 24, No. 2, pp. 225-262.

15. Epperly, W.R. EDS Coal Liquefaction Process Development, Phase IV, Quarterly Technical Progress Report, January 1-March 31, 1979; FE-2893-29, June 1979, 267 pp.

16. Severson, D.E. Solvent Refined Lignite Process Development, Final Technical Report, FE-4189-15.

17. Willson, W.G. Grand Forks Energy Technology Center Liquefaction Project, Quarterly Report, January 1 to March 31, 1979, September 1979.

18. Knudson, C.L., J.E. Schiller, and A.L. Ruud. Temperature Effects on Coal Liquefaction: Rates of Depolymerization and Product Quality as Determined by Gel Permeation Chromatography, Organic Chemistry of Coal, ACS Symposium Series 71, edited by John W. Larsen, Washington, D.C., 1975.

19. Given, P.H. The Dependence of Coal Liquefaction Behavior on Coal Characteristics, Paper in Short Course on Coal Characterists and Coal Conversion Processes, The Penn State University, October 22-26, 1979.

20. Schmid, B.K., J.C. Koenig, and D.M. Jackson. Economic and Market Potential for SRC II Products, Presented at the Sixth Annual International Conference on Coal Gasification, Liquefaction and Conversion to Electricty, University of Pittsburgh, Pittsburgh, PA, July 31-August 2, 1979.

21. Chemical and Engineering News, August 6, 1979, P. 19.

22. Fossil Energy Program Summary Document FY 1981, Draft Assistant Secretary for Fossil Energy, U.S. Department of Energy, January 1980.

23. Neuworth, M.B. and E.C. Moroni. Development of a Two-Stage Liquefaction Process, DOE.

3.5.4.2 Technology Description

Coal liquefaction is a term applied to a large variety of processes that yield liquid fuel products by breaking down the molecular structure of the coal and increasing the hydrogen-to-carbon ratio of the product to one typical of a liquid. Processes have been available for converting coal into liquids since about 1927 when a commercial coal hydrogenation plant was started up in Leuna, Germany. Subsequently, numerous coal conversion processes were developed that converted coal either directly or indirectly (via synthesis gas) to liquid products.

Commercial scale conversion plants were used in Germany during the second World War, and in this sense, coal liquefaction made a valuable contribution to Germany's war effort by supplying a source of synthetic petroleum. The German process was improved and demonstrated on a pilot plant scale in this country after the war by the U.S. Bureau of Mines.[2]

Today, only one commercial scale plant remains in operation. The SASOL installation near Johannisberg, South Africa has been producing coal-derived liquids since 1956. A second installation, SASOL II, is scheduled for initial shakedown runs in late 1979. Both SASOL plants are based on the indirect liquefaction method, which first produces a clean synthesis gas using oxygen blown Lurgi coal gasification, cleaning and treatment steps. The Fischer-Tropsch synthesis process is then used to convert the gas to a variety of liquid products.

To demonstrate the difference in hydrogen content of solid, and gaseous fuels, consider the following data:

FUEL	HYDROGEN TO CARBON MOLAR RATIO
Solids	
Coal	0.7-1.0:1
Liquids	
Crude Oil	1.7-1.8:1
Gasoline	2:1
Gases	
Methane	4:1

Among coal liquefaction process, there are various techniques for contacting hydrogen with the coal to raise the atomic hydrogen-to-carbon ratio. Liquefaction processes may be categorized according to the method of contacting employed:

- Direct Hydrogenation - In these processes, feed coal is slurried with a process-generated solvent and hydrogen mixed directly at reaction conditions with the slurry. The reaction may be catalyzed.

- Donor Hydrogenation - Here, coal is slurried with a process-generated solvent as in the direct hydrogenation case, but the solvent has been processed to match the hydrogen source.

- Indirect Liquefaction - Coal is gasified to produce synthesis gas (primarily CO and H_2) which is converted to liquids via various synthesis techniques.

Processes representing each of these different methods will be discussed in the subsection on current projects.

For those processes which employ hydrogenation directly or with a donor solvent, a generalized process flow diagram is presented in Figure 3.5.4.11. A general summary of important process features for these systems is given below.

- The choice of process conditions for the available feedstock must be such that an adequate amount of distillable oil for use as a solvent is generated by the process.

- Pulverization of the coal usually is not necessary, since extensive disintegration takes place during the solvation process.

- In cases where catalysts are used, the reactor contains two solid phases, (coal and catalyst) a liquid and a gas phase. For some high moisture content low-rank coals, enough free water may be present to form a second liquid phase.

- In order to limit reactor residence times, rapid mass transport of solvent to coal and to catalyst, and hydrogen to catalyst, as well as rapid reaction rates, are desirable.

- Deactivation or poisoning of the catalyst due to carbon deposition, or precipitation of certain inorganic agents, will occur over time.

- Regardless of which liquefaction process is chosen, a certain amount of gas will be produced in the system. This gas may be sold as a plant by-product, or it may be used for hydrogen generation. All uses of the gas will first require sulfur removal. Excess production of gas at the expense of liquid yield is undesirable.

- Separation of product from ash and unreacted carbonaceous matter must be one of the unit operations in the process. Suitable choices in many cases are filtration, centrifugation, or distillation.

- Abrasion of process equipment and piping will be a function of the process conditions and the amount and nature of the mineral content of the coal.

- Achieving the desired yield and composition of products is dependent upon choice of feedstock and process conditions.

Figure 3.5.4.11

Generalized Flow Diagram
for Direct Coal Liquefaction Processes

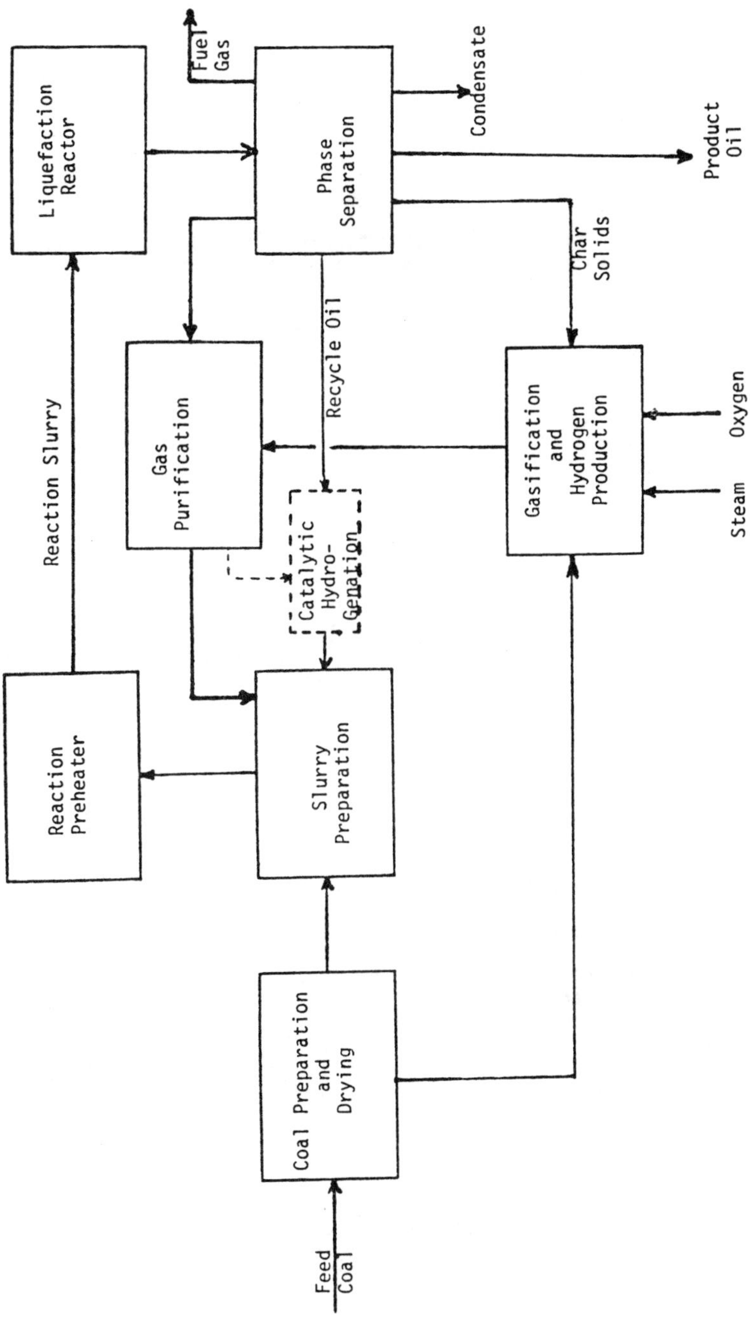

- Environmental considerations including the quantity and characteristics of effluents and the selection and sizing of cleanup facilities will be unique to the feedstock and process conditions.

Engineering design decisions on many of these factors must be made based on knowledge of the properties of the coal feedstock and their affect on the process.

3.5.4.3 Environmental Control Technology

The environmental impacts from liquefaction are very similar to those of gasification. The major areas of concern are:

- Effluents to air from ponds and cooling towers
- Leaching of solids
- Water clean-up
- Combustibles control in acid gas removal
- Trace elements
- Stream analyses for trace materials

Blowdown water from wet cooling towers contains a high concentration of dissolved solids, possibly including toxic compounds added to retard fouling of the cooling tower. If not suitable for direct discharge to ambient waters, use of an evaporative pond to accumulate the salts is an alternative. Use of direct air coolers and dry cooling towers would reduce the quantity of blowdown.

"Sour" process water contains ammonia, hydrogen sulfide, and dissolved hydrocarbons such as phenols and cresols. For the major identified pollutants, commercial control technology is available, including the following:[19]

- Physical separation of oil and water streams.
- Steam stripping to remove volatiles.
- Phenol recovery by non-proprietary oil extraction or by proprietary process is such as the Lurgi Phenosolvan or the U.S. Steel Kempro process.
- Sulfur recovery through a number of processes for removal, conversion to elemental sulfur, and cleanup of tail-gas streams.
- Biological oxidation to remove residual amounts of dissolved salts, phenol, ammonia, etc.
- "Polishing" operations, such as activated carbon adsorption, if required, to remove residual amounts of refractory organics which are not biodegradable.

Substantial quantities of sulfur and nitrogen compounds may remain in the liquefied product and require additional refining. This will increase the number of individual cleanup processes required and, hence, plant emissions; however, it will reduce environmental problems associated with the end use of the fuel.

The principal solid waste from coal liquefaction will be the residue formed from the mineral matter present in the coal feedstock. Direct liquefaction processes will produce a char or filter cake containing unreacted carbon that can be burned or gasified. All liquefaction processes will discharge a final char, ash or slag from the gasification and combustion units with properties which depend on the coal and the process specifics. Landfill disposal is possible, but leaching characteristics of the solid wastes must be considered in selecting landfill procedures.

Other solid wastes will include spent catalysts, organic sludge from biological wastewater facilities, and inorganic sludges from flue-gas desulfurization or evaporation ponds. These wastes may require specialized treatment to avoid transmission of soluble components, including trace elements (e.g., arsenic, chromium, mercury, molybdenum, and selenium) to local water supplies.

The available technologies for handling of airborne, liquid and solid waste effluents are treated in the sections on Direct Combustion and Gasification, and apply equally to coal liquefaction.

Coal liquids are known to have carcinogenic or toxic effects. A key issue to resolve as a coal liquefaction industry grows is the development of better techniques to identify and minimize health effects associated with worker or public exposure to coal liquids.

3.5.4.4 Effects of Low-Rank Coal Properties[a]

Low-Rank coals display certain properties which may affect their behavior during liquefaction, in both positive and negative directions. These properties are of prime importance in determining the technical and economic feasibility of their use in specific processes, and in some cases, the product distribution which can be obtained.

Oxygen Content

Low-rank coals contain higher contents of oxygen-bearing functional groups. This together with the generally low sulfur content of these coals will affect liquefaction chemistry, hydrogen consumption, catalyst selection and to some degree, product distribution in direct solvent based processes. Indirect liquefaction schemes are not sensitive to the molecular functional groups that are present, but rather to the total amount of each element and the resulting composition of the synthesis gas fed to the liquefaction process.

Reaction with CO

Low-rank coals, particularly lignite, have been observed to have high reaction rates with carbon monoxide. The reaction chemistry involves both removal of part of the coal oxygen (probably the carbonyl group) and interaction with the moisture in the coal to provide a direct in-situ source of hydrogen. This fact forms the basis for direct liquefaction processes based on injection of synthesis gas in the slurry rather than more expensive hydrogen. Although this fact has a much lesser signifacance for direct liquefaction processes using hydrogen as a reactant, processes which generate gas in the reaction step will be subject to these interactions between coal and CO if the gas generated in that step contains carbon monoxide.

---

[a]Much of this section is adapted from Reference 5.

Coal Mineral Matter Catalysis in Direct Liquefaction

It is suspected that inorganic components in coal such as the alkali metal and alkaline earth elements and pyrites may catalyze different liquefaction reactions. This is of particular importance in those processes which do not rely on catalyst addition to the system, such as the Solvent Refined Coal and the Exxon Donor Solvent processes. Alkali metals have a special role in some of the mechanisms proposed for the CO-coal reactions. The available results of past tests on the effects of coal minerals do not provide quantitative measurements or definite conclusions for low-rank coals. Additional research is required to define the chemistry of this natural catalysis, to optimize the process conditions for a given feedstock, and to determine the most suitable feedstock for a given process. Such research might also identify which additive catalysts would best compliment the natural catalytic activity.

Catalyst Poisoning

In liquefaction systems using synthetic catalysts, some catalyst poisoning may result from the dispersed mineral matter in the coal (reference 3). Indirect liquefaction processes will not be affected, nor will processes which do not use added catalysts.

Calcium Carbonate Deposits

Calcium carbonate deposits in the liquefaction reactor appearing as wall scale and as free flowing agglomerates have been identified specifically with low-rank coals.[4,10] The original form of the calcium is largely salts of humic acids within the organic matrix, which decompose during liquefaction to form calcium carbonates. The extent of the deposits appears to be related to the ion-exchangeable calcium content of the coals, and there is some evidence that it may be related to reactant gas composition and pressure.

Calcium carbonate deposition can create operational problems in any direct liquefaction process. The extent of the problem in relation to operating conditions is being assessed by several investigators. Control of the problem can be achieved by periodic withdrawal of carbonate agglomerates and acid washing of reactor internals during downtimes. Alternatively, the coal can be pretreated with sulfur dioxide to convert the calcium to calcium sulfate, which is stable under liquefaction reaction conditions.

High Bottoms Viscosity

The vacuum bottoms produced from low-rank coals typically display high bottoms viscosity, which can affect the operability of the vacuum tower and the bottoms recycle and processing units due to increased pumping difficulty. Viscosity has been shown to be reduced by longer residence times in the reactor when using low-rank coals; whereas for bituminous coals the viscosity shows only a slight dependence on residence time, and the vacuum bottoms are easily pumpable under conditions of normal reactor throughput rates.

Dilution has also been suggested as a means of controlling bottoms viscosity. Addition of moderate amounts of $1000^{o}F$ distillate in the bottoms will lower viscosity significantly, but some consumption of these lighter components may occur during recycle or bottoms processing.

High-Moisture Content

Moisture content of coal feedstocks is an important characteristic affecting liquefaction performance for direct liquefaction systems.

Excess moisture increases the total operating pressure required to maintain a given partial pressure of hydrogen or carbon monoxide.[3,5] In addition, carbon monoxide may react with excess moisture in the coal to form carbon dioxide and hydrogen. Hydrogen may be less reactive in converting coal than the original CO under some conditions therefore, eliminating moisture from the feed coal may reduce the total operating pressure required for liquefaction.

However, drying low-rank coals has been observed to reduce their activity with respect to liquefaction, possibly because of surface oxidation and collapse of the pore structure. This effect is particularly true when synthesis gas is used as a reactant. These factors are unique to low-rank coals, and are significant when process designers contemplate strategies to reduce pressures.

Hydrogen Donor Activity

For processes using a donor solvent, the hydrogen donor activity is a variable of prime importance. For systems based on synthesis gas as a reactant, carbon monoxide is consumed preferentially to hydrogen, creating a hydrogen rich purge gas. Optimum use of this purge gas for catalytic hydrogenation of the donor solvent may be important to the process economics.

Choice of Reactant Gas

The choice of reactant gas is important in determinig the yield and product distribution for lignite liquefaction. Yields of liquid products have been shown to be significantly higher when using synthesis gas and coal with its natural moisture content, when compared to pure hydrogen and pure carbon monoxide atmospheres.[3] Gaseous yields were proportional to the concentration of carbon monoxide in the feed gas, due to CO_2 production by the water gas shift reaction. However, CO_2 may also be produced directly from the coal. Production of CO_2 by this mechanism is desirable due to its effect of decreasing the oxygen content of the liquid products. Production of CO_2 by the water gas reaction may or may not be desirable, depending upon whether or not the hydrogen produced reacts to form additional liquid products. The highest gas yield was obtained from the pure carbon monoxide atmosphere and the as recieved coal, due to the catalytic activity of the mineral matter and the presence of natural moisture (as a comparison, samples of acid leached lignite were tested which had lower mineral matter contents due to the acid treatment). Production of methane in the liquefaction reactor is undesirable since it is done at the expense of the liquid yield, and is a larger consumer of hydrogen.

The properties of low-rank coals can thus be expected to strongly affect the operation of coal liquefaction processes and the overall process economics. Low-rank coals as feedstocks for liquefaction are favored by their low cost, high reactivity, natural catalytic activity, and reactivity with carbon monoxide. Negative attributes include their tendency to form calcium carbonate deposits, the higher viscosity of their liquefaction bottom product, the requirement for additonal reducing gas, possibly the operation of liquefaction reactors at higher pressures, and the higher gas yields due to high initial oxygen and moisture contents.

3.5.4.5 Optimizing Liquefaction Yield from Low-Rank Coals

This section presents information from reference 5. The discussion covers the reactions of CO versus H_2, the effect of temperature on liquefaction yields, the role of the hydrogen donor solvent, and the combined effect of temperature and reactor backmixing on the molecular weight and the viscosity of the residuum.

Other techniques exist for yield improvement which are not presented in this discussion. These include improvements in catalysts, (such as the use of one-pass throwaway catalysts or recoverable gaseous catalysts), optimum temperature staging, and multiple step liquefaction.

Reactions of CO and H_2 with Lignite

Carbon monoxide will react in preference to H_2 from a starting mixture of both with lignite and water.[3] The CO can, in general, react in three ways (net reactions):

1. $CO + H_2O \rightleftharpoons CO_2 + H_2$
2. $CO + \text{Lignite·O} \longrightarrow CO_2 + \text{Lignite}$
3. $CO + \text{Lignite} + H_2O \longrightarrow CO_2 + \text{Lignite-2H}$

These do not address mechanism. Figure 3.5.4.12 shows the amount of CO that reacted at 4000 psi and 440°C (final temperature) with water, dry lignite, and moist lignite, corresponding respectively to the three generalized equations above. Very little CO reacted to form gaseous hydrogen (curve 1). However, the reaction with dry lignite (curve 2) was about one third of that with moist lignite. Curve 2 is believed to represent primarily the reaction of CO with coal oxygen to produce CO_2. Curve 3 includes both reaction with coal oxygen and hydrogenation of the coal. The mechanism for reaction 3 has been proposed to involve the formation of alkali formates as intermediates.[20,21]

Figure 3.5.4.12

Consumption of CO Versus Time at 440°C; 4 Moles of CO Charged at 4000 psi

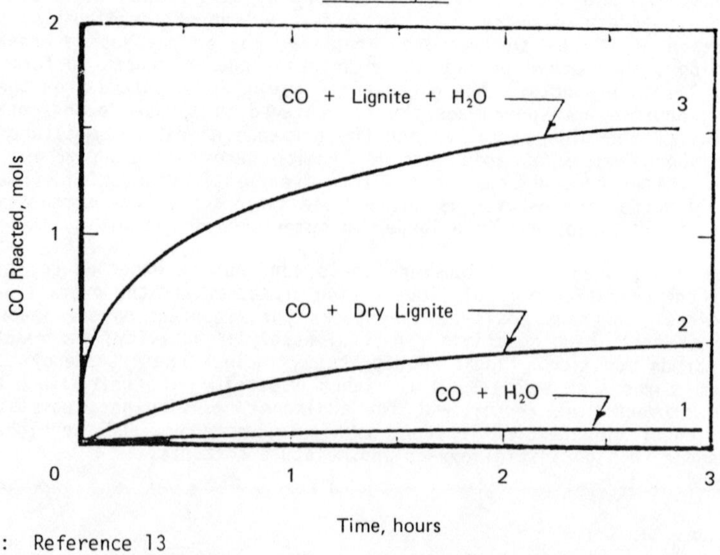

Source: Reference 13

The general question has been raised as to whether the extent of reaction of CO with lignite and moisture is typically controlled by rate, equilibrium, or stoichiometry. For example, an HRI study on lignite[12] proposed that conversion was controlled by the simultaneous equilibria of formates and the water gas shift reaction.

In experiments at GFETC, the extent of reaction by CO was shown to be roughly proportional to the amount of lignite charged, with essentially all of the CO reacting in the presence of excess lignite (Figure 3.5.4.13) The shaded regions in the figure represent the production of gaseous H_2; gaseous hydrogen production by the water gas shift reaction was almost negligible when excess lignite was present. It should be noted that heating from 350° to 470°C occurred during this experiment, which caused the final conversion to be controlled by the higher temperature. The conclusion for these conditions is that the final consumption of CO was controlled by stoichiometry; the CO consumption on a time scale of minutes was controlled by rate.

The consumption of hydrogen by lignite is increased as temperature rises between 400° and 470°C. However, even at the higher temperature, H_2 consumption is greatly suppressed by the competing reaction of CO.[13]

Figure 3.5.4.13

Consumption of CO Versus Time During Heating From 350 to 470°C and Reacting at 470°C, 4500 psi

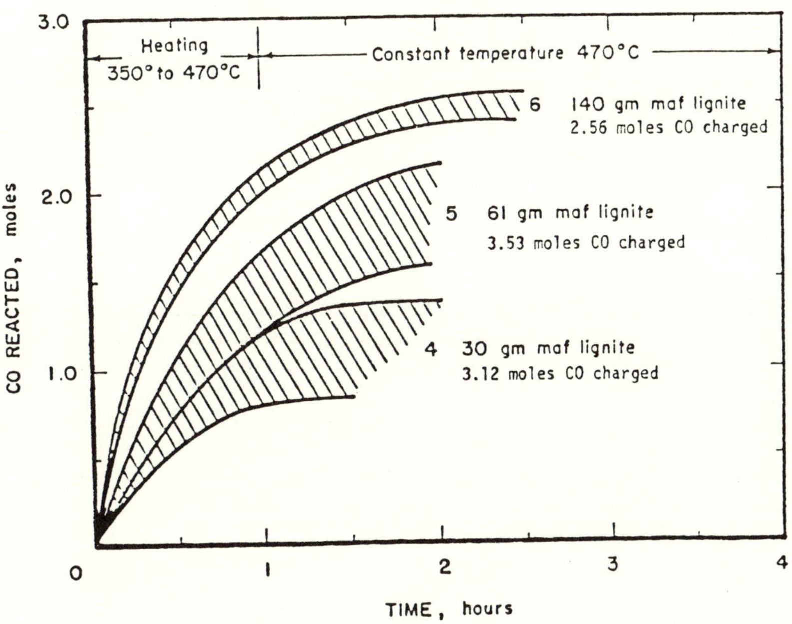

Source: Reference 13

The Effect of Temperature on Yields

Of all variables that have been studied at GFETC, temperature has the greatest effect on liquefaction yields from lignite and synthesis gas. The effect of increasing temperature can be a monotonically increasing yield or a maximum followed by a reduced yield, depending on the hydrogen donor quality of the solvent used.[13] Figures 3.5.4.14 and 3.5.4.15

illustrate this fact for the total liquid yield and net distillate yield, respectively, which either peak at about 460°C or continue increasing out to 480°C depending on the addition of tetralin (a hydrogen donor) to the anthracene oil solvent. These experiments were performed on 40 percent coal slurries in a continuous flow stirred tank reactor at a nominal pressure of 4000 psi (50/50 H_2/CO feed gas) and a reactor residence time of 30 minutes.

Figure 3.5.4.14

Net Liquid Yields Versus Reaction Temperature for 40 pct Coal Slurries

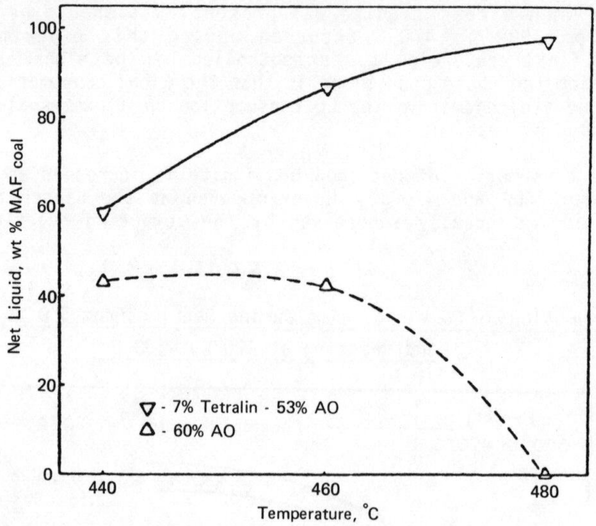

Source: Reference 14

Figure 3.5.4.15

Net Distillate Yields Versus Reaction Temperature for 40 pct Coal Slurries

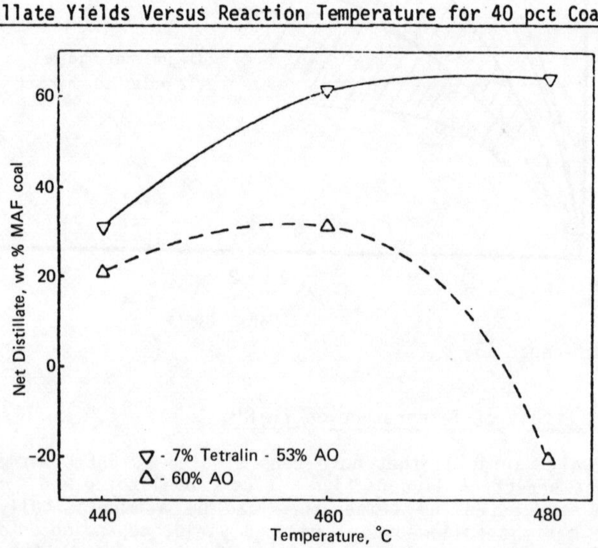

Source: Reference 14

Figure 3.5.4.16 shows that the consumption of reducing gas increased from 1+ to 2.5 percent of MAF coal (H_2 equivalent) between 440° and 480°C; this trend for anthracene oil with added tetralin is generally representative regardless of the hydrogen donor property.[14]

The yield of hydrocarbon gases (C_1 to C_4) as shown in Figure 3.5.4.17, increase from the range of 5 to 10 percent to the range of 20 to 25 percent of MAF coal between 440° and 480°C.[14] The two curves indicate that gas yield is greatest for the case of low hydrogen donor quality (where, in this case, the donor activity of the solvent for the lower curve was enhanced both by distilling off the light oil [400°F] and by adding tetralin). The conclusion from this set of data is that enhanced donor activity favorably increases liquid yield while reducing gas yield.

Figure 3.5.4.16

Reducing Gas Consumption Versus Reaction Temperature For 40 pct Coal Slurries

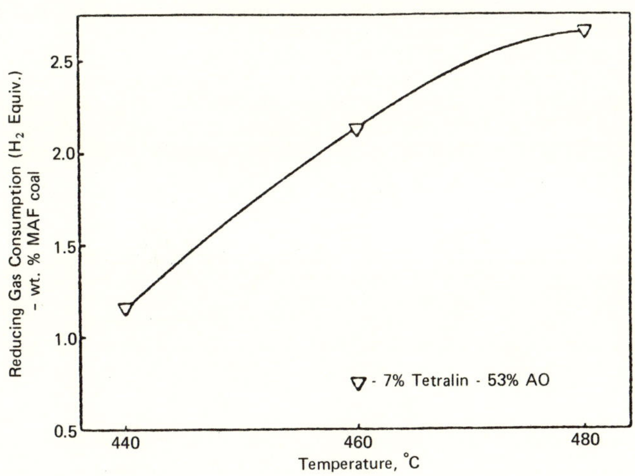

Source: Reference 14

Figure 3.5.4.17

Net Hydrocarbon Gas Yield Versus Reaction Temperature for 40 pct Coal Slurries

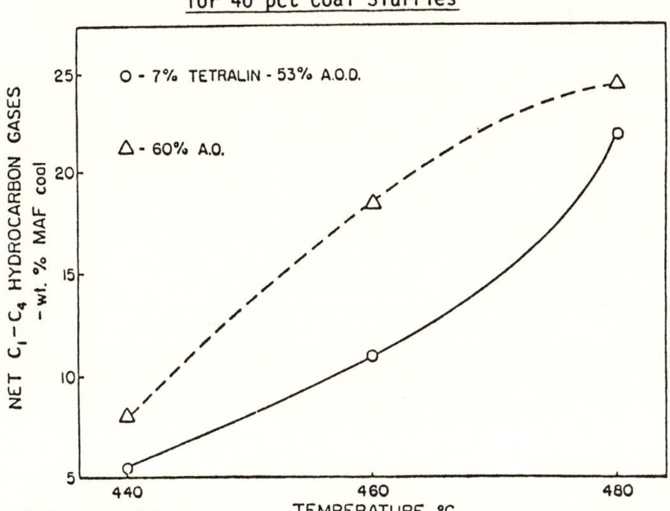

Source: Reference 14

The use of high liquefaction temperatures to optimize yield and increase reactor throughput is somewhat controversial because of the following factors: 1) the reduced yield which occurs without sufficient hydrogen donor activity; 2) the coking and plugging that attends item 1; 3) the higher gas yields; and 4) the alternative of using catalysts. To find the best conditions for low-rank coals will require more exhaustive testing than has been accomplished thus far over the domain of the three principal variables: A) temperature, B) reactor residence time, and C) hydrogen donor quality. The tests must be extended to include process-derived solvents that have well-characterized hydrogen donor quality. To support this work, analytical techniques must be developed to provide a common basis for ranking donor qualities between different investigations.

Molecular Weight Effects

The determination of molecular weight distribution (MWD) has been a principal analytical method in the GFETC liquefaction project. The procedure used is calibrated gel permeation chromatography.[15] Typical results for two reactor samples are shown in Figure 3.5.4.18. The UV absorbance is plotted against the molecular weight (the latter being equated with an elution time of known molecular weights). Since absorbance is proportional to molar concentration, and since the molar absorptivities were found to be similar, the graph is an estimate of the distribution of mole fraction according to molecular weight. Note the MW scale is logarithmic.

Figure 3.5.4.18

MWD of the THF Soluble, Non-Volatile Fraction of Reactor Slurry Samples Obtained at 404° and 500°C

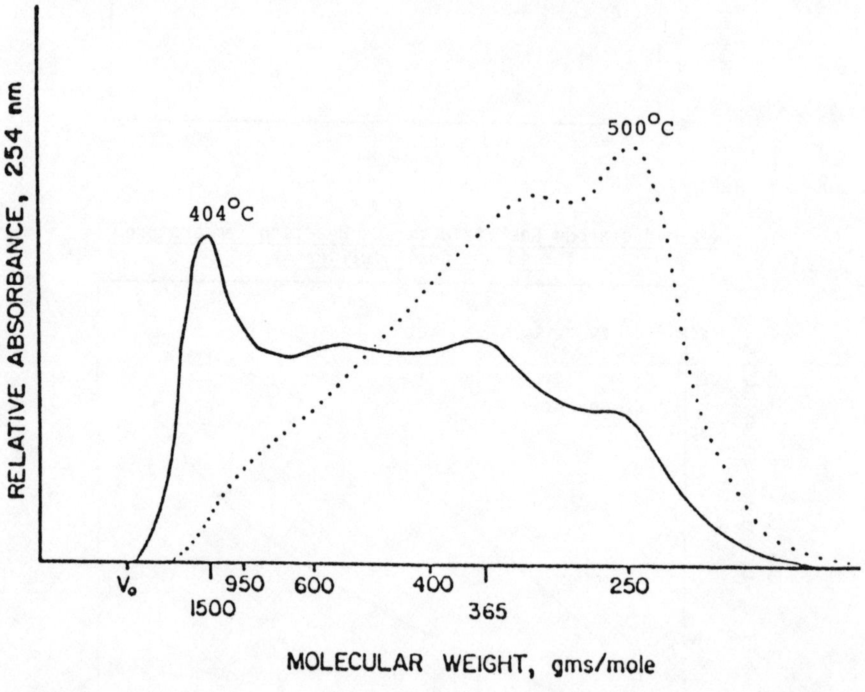

Source: Reference 14

The material on which MW determinations are made is the coal-derived liquid fraction that is THF soluble but non-distallable, corresponding roughly to vacuum bottoms in the EDS process. In previous batch experiments, the amount of this fraction has been found to remain quite constant under almost all reaction conditions,[13] indicating this fraction of the product is formed rapidly and is stable under the observed conditions. Weak bonds exist that rupture upon increased thermal agitation, typically yielding two lower molecular weight but still non-distillable fragments. Temperature has the greatest effect on the MWD, with the major peak shifting from 1500 MW to 250 MW as temperature is increased from 404°C to 500°C in the continuous stirred tank reactor (CSTR).

The means used to express the molecular weight data is to consider the ratio of UV absorbances at high versus low molecular weight. The ratio for 950 MW versus 280 MW (A_{950}/A_{280}) is correlated with reactor temperature in Figure 3.5.4.19; different curves are shown for batch and CSTR data. It may be observed here that the batch unit is more effective in reducing the ratio at temperatures above 440°C, but that the CSTR is more effective at lower temperatures. The MWD is also shifted downward by increasing the tetralin content in the feed; this effect is more pronounced at 404°C than at 460° or 500°C. The level of coal loading in the feed slurry thus far shows little effect on the MWD.

Figure 3.5.4.19

Ratio of HPLC Ultra Violet Absorbance at 254 mm of 950 MW to 280 MW Material Versus Reactor Temperature for Reaction Times Under 30 Minutes

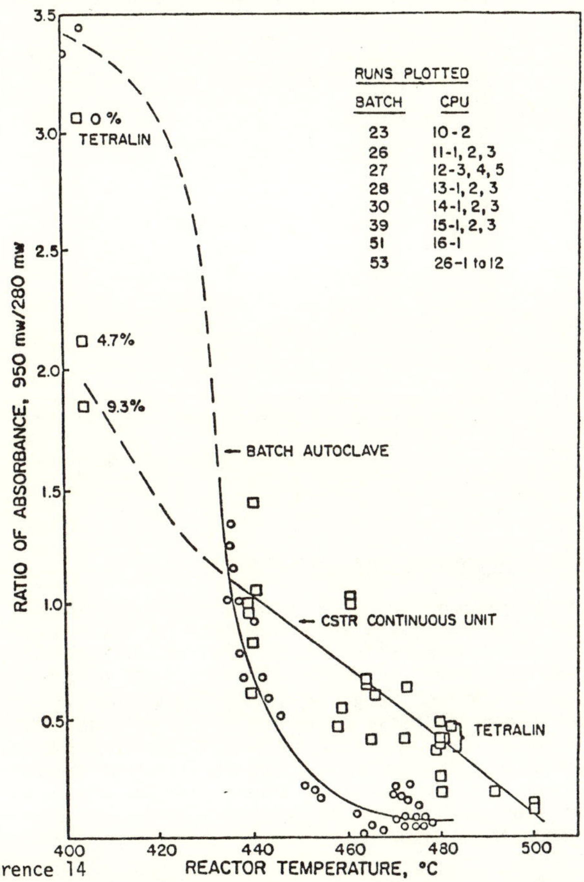

Source: Reference 14

The importance of a change in molecular weight is principally found in its correlation with residuum viscosity. As discussed previously, this is a potential problem in liquefying low-rank coals in some processes. Figure 3.5.4.20 illustrates the sensitivity of the viscosity of distillation residues dissolved in anthracene oil to molecular weight.[13] Based on this correlation, the previously noted change in molecular weight between 404°C and 500°C would cause a many-fold reduction in residuum viscosity.

Figure 3.5.4.20

Correlation of Viscosity with Molecular Weight for Various Distillation Residues (25 pct w/w) in Anthracene Oil

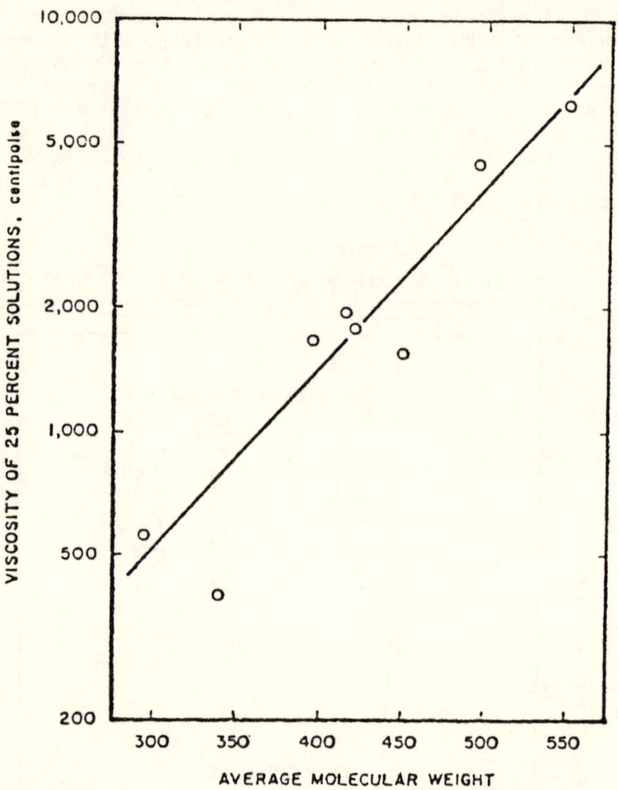

Source: Reference 13

Effect of Residence Time Distribution

The interpretation of differences in MWDs requires that we note the residence time distributions (TRDs) for the CSTR and the batch reactor. The RTD of the batch reactor is a delta function at the time representing the duration of the batch experiment - that is, all of the material stays in the reactor for the same length of time. The RTD for the CSTR is represented by an exponential decay curve - that is, for each unit of entering feed, the increments leaving during equal time segments thereafter are reduced exponentially with time so the largest increment leaves in the first time segment but some small amounts remain indefinitely. We should also note that the depolymerization does progress with time even though it is principally temperature dependent.[15]

The effect of the TRD on MWD is illustrated in Figure 3.5.4.19 showing UV absorbance ratios versus temperature. Note that progressive depolymerization causes this ratio to be lowered. At low temperatures where time dependence is greater, the internal backmix of the CSTR causes some of the high-MW material to be internally recycled over a sufficient time to generate the low-MW fractions required to reduce the ratio. At high temperatures, where the depolymerization reaction is very rapid and conversion (MWD change) is nearing completion, the predominant effect in the CSTR is the nearly immediate departure of a fraction of the feed material in an unconverted state; in this case, the CSTR results in a lesser reduction in the ratio. A further complication of this argument is introduced by the action of tetralin (or other hydrogen donor materials), which releases hydrogen only in the early stage of a batch reaction but is continuously replenished in a CSTR so that some "old reactant" is in contact with "fresh hydrogen donor."

In conclusion, a CSTR should be less effective than a batch or plug-flow reactor with adequate radial gas-slurry contact in achieving a once-through high conversion from high to low molecular weight. This has been generally assumed in liquefaction, and the current results support this assumption. However, the ideal reactor system may be far more complex; for example, we can envision a tubular reactor with a staged side input of a hydrogen donor and with separation and recycle of the fraction of the coal material that has not depolymerized. We believe that optimization of reactor systems for coal liquefaction has a long way to go.

In the immediate future, the continuous process liquefaction unit at GFETC will be modified by installing a tubular reactor in place of the CSTR. In addition, the system flow pattern will be changed to include recycle - possibly both around the reactor and back to the slurry preparation tank. These modifications will: 1) permit line-out on coal-derived solvent during continuous operation; 2) extend the analysis of reactor design factors; and 3) facilitate continuous experiments on the recycle of mineral residues and heavy organic fractions as an extension of part of the work at the Chemical Engineering Department at the University of North Dakota. Plans are being developed to study fractionation and selective hydrogenation of recycle streams and use of disposable catalysts.

3.5.4.6 Current Projects

The current array of coal liquefaction development programs in the United States consists of bench, pilot, and demonstration projects with a major emphasis on direct hydrogenation and donor hydrogenation processes. The primary thrust addresses two markets: low grade fuels (boiler-type fuels low in sulfur and ash) and high grade fuels (gasoline, heating oils and chemical feedstocks). A secondary emphasis, but still of major importance, is the environmental impact considerations associated with both the processes and their products.

The discussion of processes will be divided into three main categories: direct hydrogenation, donor hydrogenation, and indirect liquefaction. Each project discussion will address several pertinent topics:

- Process Description - a discussion of the liquefaction processing sequence encountered by the coal as it moves from storage to liquid products.

- Development Status - the current development or commercial state of the technology.

- Effects of Low-Rank Coal Properties - an assessment of the unique effects of using low-rank coals as feedstocks for the liquefaction process under consideration.

- Environmental Control Technology - unique environmental control needs for the liquefaction process under consideration.

3.5.4.6.1 Direct Hydrogenation

Processes which fall under the classification of Direct Hydrogenation are characterized by a processing sequence which involves slurrying the feed coal in a process derived oil or solvent, and reacting the resulting coal slurry with hydrogen gas at high temperatures and pressures. Reaction of hydrogen with the coal is usually enhanced catalytically, but this feature is not a requirement.

H-Coal

The H-Oil process was originally developed by Hydrocarbon Research Inc. (HRI) as a means of converting heavy oil residues to lighter fractions. H-Coal, an extension of the H-Oil technology, is a catalytic hydroliquefaction process that converts coal to either a boiler fuel or a refinery syncrude.

A process flow diagram is shown in Figure 3.5.4.21. Coal is first dried and then slurried with a process generated recycle oil. Addition of compressed hydrogen takes place after the slurry has been pumped up to a pressure of 200 atm. The mixture is then sent through a fired preheater and charged continuously to the bottom of the ebullated-bed catalytic reactor which operates at 850°F and 3000 psi (maximum). The upward flow of the reaction mixture maintains the cobalt-molybdenum catalyst pellets in a fluidized state (catalyst activity is maintained by the semi-continuous addition of fresh catalyst and the withdrawal of spent catalyst). Temperature control is achieved externally by manipulation of the degree of preheating of the reaction mixture. The reactor effluent, which contains some unconverted coal but negligable amounts of catalyst, is flashed. The flash bottoms are further processed in a hydroclone, a liquid-solid separator and a vacuum distillation column. The hydroclone overhead stream consists of recycle solvent which is pumped back to the slurry preparation unit. A solids laden residue consisting of unreacted coal and very heavy liquid is recovered from the solid-liquid separation step and is used in the hydrogen generation plant.

The vapor product leaving the top of the reactor is cooled to condense the heavier components. Gas treatment recovers light hydrocarbons, ammonia and hydrogen from the vapor stream, leaving a hydrogen-rich gas which is compressed and recycled to the coal slurry.

Final products of the process consist of the light hydrocarbon gas separated from the reactor overhead and flash gases, and an array of distillate materials ranging from light and heavy distillate fuels to heavy residuum.

Depending upon the product desired, the rate of hydrogen consumption may vary. For example, boiler fuel will be produced at relatively high reactor throughputs and low hydrogen consumption rates. Sythetic crude oil requires more hydrogen, resulting in a lower yield of bottoms product.

The DOE H-Coal pilot plant is scheduled for start-up in 1980. Table 3.5.4.7 provides some general information on the pilot plant project. The major objective of the pilot plant experimental program will be to establish process yield and material balance data as a function of operating conditions and feedstock characteristics.

Figure 3.5.4.21

H-Coal: Process Flow Diagram

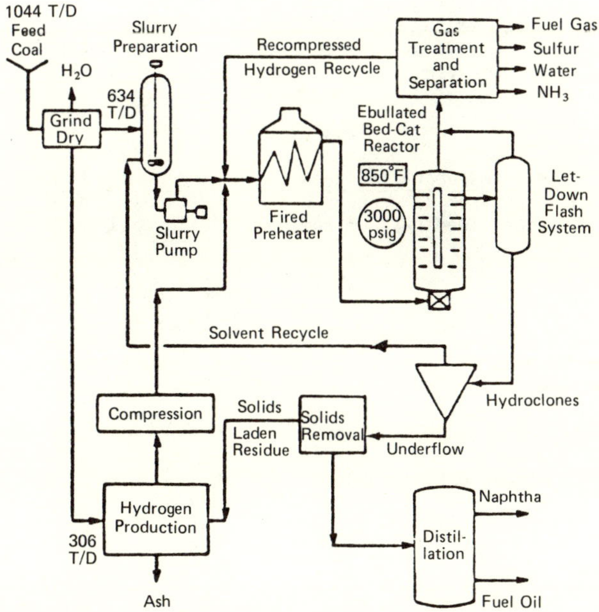

Source: Reference 6

Table 3.5.4.7

H-Coal Pilot Plant Project

OWNER/DESIGNER	Ashland Oil Company & Hydrocarbon Research Inc.
CONSTRUCTION	Badger Plants, Inc.
SPONSORS	DOE, EPRI, Ashland Oil Inc., Standard Oil of Indiana, Conoco Coal Development Co., Commonwealth of Kentucky
LOCATION	Catlettsburg, Kentucky
DESCRIPTION	• Oil slurry feeding • Reaction in a catalytic (Co/Mo) ebullated-bed reactor • Pressure 3000 psig • Temperature 850°F
SIZE	600 TPD
YEARS OPERATION	H-Coal Project began in 1965; Pilot plant construction began in 1977
COAL TYPE	All types
MAJOR PRODUCTS	Syncrude & Heavy Boiler Fuel
ENVIRONMENTAL STATUS	Plant will not significantly affect the environment in the site locality

Source: Reference 7

A large number of PDU test runs sponsored by both industry and government have been made on coals ranging from bituminous to lignite. Design data for the 600 TPD DOE/Multi-sponsor pilot plant being constructed at Catlettsburg, Kentucky are available for Wyodak subbituminous and Illinois bituminous test coals. Table 3.5.4.8 summarizes this test data, and illustrates several differences in process conditions and products arising from feedstock variations.

Table 3.5.4.8

Selected Operating Results for the H-Coal Process
(Data taken from References 8 for operations in the syncrude mode at a coal feed rate of nominally 31 lb coal/hr/ft^3 reactor.)

	Wyodak Subbituminous Coal	Illinois No. 6 Bituminous Coal
Reactor temperature, ^oF	816-825	816-849
Reactor pressure, psi	2900	2590-2755
Catalyst usage, lb coal/lb catalyst	120-740	60-1464
Product distribution, pct dry coal:		
CO_2	3-5	0.5-1
H_2O (net production)	13-17	6-13
C_1-C_3	8-9	9-13
C_4-400^oF naphtha	17-19	17-28
400^oF-975^oF oil	15-26	15-28
975^oF & soluble residual	11-12	10-24
Unconverted coal	12-20	4-7
Ash	7	10-13
NH_3	0.4-0.7	0.7-1.1
H_2S	0.1-0.3	2.3-2.8
Coal Analysis:		
As-charged basis:		
Moisture	1.6-20.4	0.6-2.9
Ash	7.1-8.7	10.4-11.5
Dry-ash-free basis:		
C	74.3[1]	79.2
H	5.3	5.2
O	18.7	10.3
S	0.5	3.7
N	1.2	1.5

	Wyodak Subbit. Coal		Illinois No. 6 Bit.Coal	
Product Analysis:	640^{o+}F Vac. Dist.	975^{o+}F Vac. Bottom	650^{o+}F Vac. Dist.	975^{o+}F Vac. Bottom
C	89	74	89	65
H	7.3-8.5	4.3-5.4	8.5-8.9	3.7-4.3
S	<0.1	0.2-0.9	.03-.12	1.5-2.8
N	0.3-0.5	1.0-1.1	0.5-0.6	0.8-1.2

Source: Reference 4

Although reaction temperature is similar, the reaction pressure is notably higher for the subbituminous coal. Yields of CO_2 and H_2O are also higher for the low-rank coal, as expected due to the higher initial moisture content of the feedstock. Gas and oil yields are similar, but the Illinois coal is expected to have a higher maximum naphtha yield. Vacuum residuum is anticipated to have a higher range in the Illinois coal case, while unconverted coal is significantly higher using the subbituminous coal. Due to the higher sulfur content of the higher rank coal, more H_2S is expected in the reactor and flash overhead streams, and in the distillate bottoms.

Based on general knowledge of low-rank coal properties and the results obtained in other liquefaction studies, the use of low-rank coals in the H-Coal process might be expected to have the following effects:

- Catalyst deactivation due to deposition of calcium carbonate, possibly requiring pretreatment of the feed coal.
- Mineral agglomeration in the reactor (primarily $CaCO_3$), requiring periodic discharge of flocs or coal pretreatment.
- Production of a high viscosity residuum causing difficulty in separation and pumping operations.

However, a more recent report on PDU tests of Wyodak coals indicates that calcium precipitation was not a problem, and that similar vacuum resid bottoms viscosities were obtained at similar levels of catalyst activity as for higher rank coals.[9]

Environmental pollutants expected from the H-Coal process are expected to be similar to those considered in Section 3.5.3.2. No unique effluents are attributed to this process design.

Synthoil

Work on the Synthoil concept was performed by the U.S. Bureau of Mines at the Pittsburgh Energy Research Center under sponsorship of the Bureau of Mines, The Office of Coal Research, and the Bethlehem Steel Corporation. The objective of the effort was to develop a method of producing a low-sulfur, low ash synthetic fuel.

Figure 3.5.4.22 is a simplified Synthoil flow schematic.

Figure 3.5.4.22

Flow Schematic for the Synthoil Concept

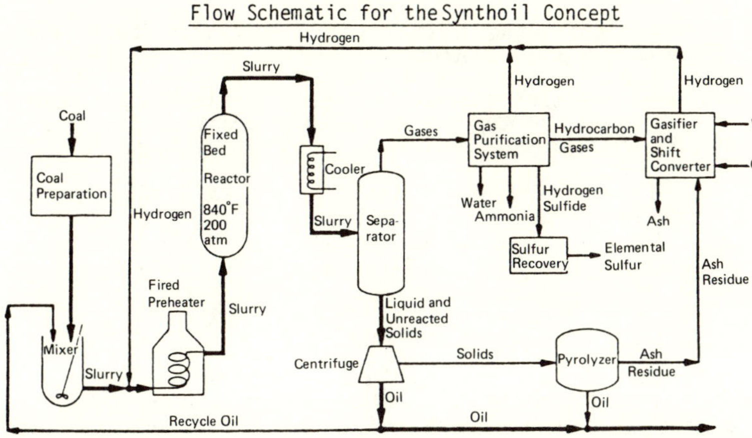

Source: Reference 11

The coal preparation section includes crushing, grinding and drying after which the coal is slurried with a portion of the product oil. The coal slurry is pumped up to reaction pressure, charged with hydrogen and preheated in a fired preheater. From the preheater, reaction mixture enters a fixed-bed catalytic reactor packed with catalyst pellets of cobalt molybdate on a silica-alumina substrate. The reactor effluent stream is then cooled and passed to a separator drum where the liquid and unreacted solids are separated from the volatile components.

The separator bottoms stream is centrifuged, and the recovered solids are fed to a pyrolysis unit. The liquid stream from the centrifuge is a non-polluting fuel oil product, part of which is recycled to the coal slurrying mixers; the remainder is blended with oil from the pyrolysis unit to comprise the final product stream. The solid residue from the pyrolysis unit contains ash and carbonaceous matter, which is sent to a gasifier for recovery of the remaining carbon. The gasifier effluent is shifted for hydrogen production.

The gases from the separation vessel are treated in a gas purification train for recovery of ammonia, hydrogen sulfide, hydrocarbon gases, and water, leaving the hydrogen rich stream available for recycle to the coal slurry. Gas purification takes place at process pressure to minimize the cost of hydrogen recycle. Hydrocarbon gases recovered in the gas purification step may be steam reformed and shifted for hydrogen generation, or sold as a plant by-product. Hydrogen sulfide is processed for sulfur recovery, and may be sold with ammonia as by-products.

Construction of a 10 TPD Process Development Unit at the Pittsburgh Energy Technology Center is essentially complete, but current evaluation of the process potential has indicated that further development work should not be continued. The PDU is being maintained in a stand-by condition.

Although previous experiments have concentrated on high-sulfur Eastern feedstocks, the use of low-rank coals in the Synthoil process is anticipated to have the following effects due to differences in coal properties:

- Increased hydrogen consumption may be associated with low-rank coal feedstocks, due to the higher oxygen.
- Some catalyst poisoning may result from the mineral matter present in the coal feed.
- Low-rank coals having high concentrations of ion-exchangeable calcium (such as Wyodak subbituminous coal) may cause reactor scale formation and catalyst deactivation due to precipitation of calcium carbonate.
- Higher operating pressures may result from feedstocks having higher moisture contents.
- High moisture and oxygen content coals may produce larger quantities of CO_2 and wastewater, increasing the size of the facilities required for handling these components.

Waste products from a Synthoil plant would be primarily ash residue, treated wastewater, CO_2 vent streams, and cooling tower effluents. There are no unique environmental problems associated with the Synthoil concept.

The SRC Process (Solvent Refined Coal)

The earliest work on the solvent refining of coal was carried out in Germany in the 1920's. Work in the United States on the SRC concept has been continuing since 1966 when the Office of Coal Research awarded a multiyear contract to the Pittsburgh and Midway Coal Company (a subsidiary of Gulf Oil Co.) which included funds for a 50 TPD pilot plant.

The concept of solvent refined coal has actually been executed in two different versions. The primary intent of the SRC I process is to produce a low-sulfur, low-ash solid fuel. The SRC II process is an extension of the original SRC technology to higher reactor temperatures with slurry recycle, promoting greater hydrogen addition to the feedstock and producing a substantial yield of liquid, instead of a solid product.

Figures 3.5.4.23 and 3.5.4.24 depict both versions of the SRC process. Pulverized coal is mixed with a process-derived solvent in a slurry mix tank. The slurry is then combined with hydrogen and pumped through a fired preheater. The heated slurry is then charged to a dissolver where at conditions of 850°F and 1500 psia, up to 93 percent of the carbon containing material in the feed is dissolved. In the dissolver, several changes occur simultaneously: coal is depolymerized and hydrogenated, resulting in an overall decrease in molecular weight; the solvent is hydrocracked to form lower molecular weight hydrocarbons ranging from methane to light oil; and much of the organic sulfur is hydrogenated to hydrogen sulfide.

From the dissolver, the reaction mixture is charged to a separation vessel where the vapor components are separated from the slurry of undissolved solids and coal solution. The raw gas is then processed for hydrogen recovery and gas desulfurization. Recovered hydrogen is recycled with the effluent from the slurry mix tank. Hydrocarbon gases are recovered and the hydrogen sulfide is processed for sulfur recovery.

In the SRC I process, the bottoms stream from the gas separator is filtered for removal of the ash and unreacted coal from the product. In a commercial plant, the solids would be sent to a gasifier and subsequent shift conversion for hydrogen generation. Process solvent is recovered from the product by distillation, leaving a molten residue which is solidified to the SRC product.

Figure 3.5.4.23

SRC-I (Solid) Process

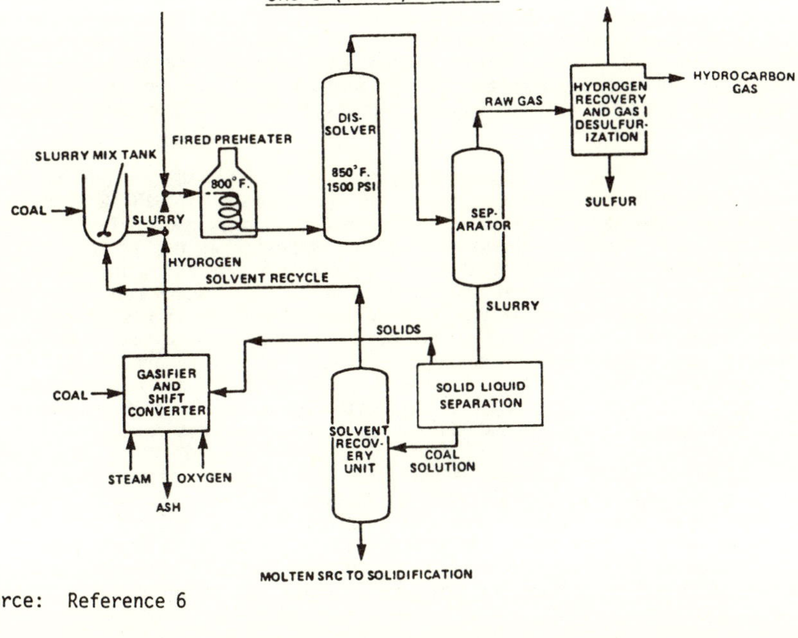

Source: Reference 6

Figure 3.5.4.24

SRC-II (Liquid) Process

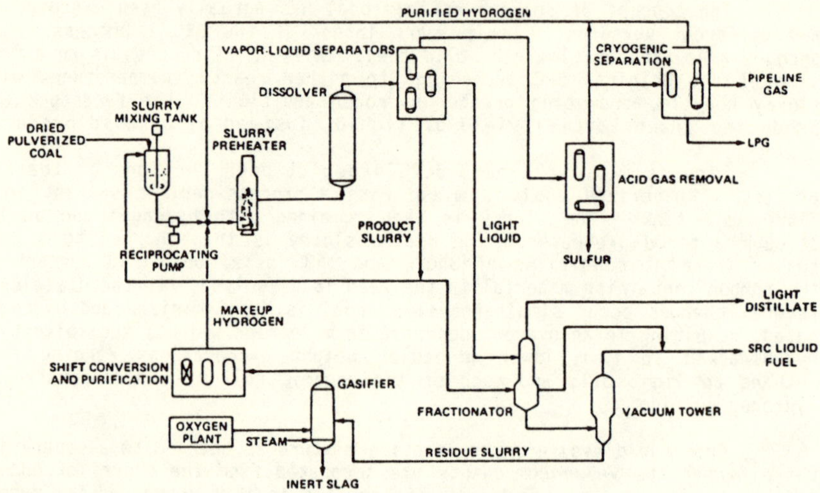

Source: Reference 6

In the SRC-II process, part of the dissolver product slurry is recycled as solvent. The remaining slurry is combined with a light liquid stream recovered from the vapor liquid separator and distilled to produce a range of products. As a result of the higher dissolver severity in this version of the process, the extent of hydrogenation is greater, giving a higher yield of liquids. However, the quantity of unconverted coal and vacuum residuum is controlled so that a balance is achieved between available gasifier feedstock and process hydrogen requirements.

Two pilot plants are currently in operation, one a 50 TPD unit at Ft. Lewis, Washington, and the other 6 TPD plant at Wilsonville, Alabama. The Ft. Lewis work has generated data which will provide the design basis for planned demonstration plants for solid and/or liquid products. The Wilsonville pilot plant has provided supplemental screening of various coals and produced improvements in solid-liquid separation techniques. Supporting research at various facilities rounds out the SRC development effort.

A comparison of the performance of lignite at the University of North Dakota Solvent Refined Lignite PDU (shut down in 1978) with Kentucky high sulfur coal at the Ft. Lewis pilot plant is provided in Table 3.5.4.9. Note that the lignite tests were based on sythesis gas rather than the pure hydrogen used in the case of the Ft. Lewis runs. Operating conditions were similar with the exception that the lignite tests were run in a higher range of pressure. The maf yield of solvent refined coal was in a slightly lower range in the case of lignite, primarily due to the lower carbon content and higher oxygen content of the feed. The properties of the products were similar with the exception of the O, S, and N contents. The net gas yield from the lignite was much higher due to the production of CO_2, from the coal oxygen and the CO in the reactant gas feed.

Pilot plant problems associated with lignite operation concerned solvent balance and accumulation of solids in the dissolver. The net liquid yield from lignite was quite variable, and solvent balance was achieved only at 2500 psi. The effectiveness of the solvent as measured by lignite conversion was reduced if light oil ($<400^o$) was recycled. Adequate yield of an effective solvent ($>400^oF$) was favored by high pressure, intermediate reactor temperatures of 820° to 830°F, and use of a 50/50 H_2/CO feed gas[3].

The use of a synthesis gas containing carbon monoxide is a preferred operating procedure when using lignite. In cases where CO and H_2 are both present, the carbon monoxide is usually consumed in larger amounts.[3] Work conducted at the University of North Dakota and at Grand Forks Energy Technology Center has indicated that a mechanism involving direct reaction of CO to rupture oxygen linkages in lignite is operative at temperatures above 800°F.[5]

Effluents from SRC plant operation are expected to be typical of liquefaction processes.

Table 3.5.4.9

Selected Operating Results for the SRC-I (SRL) Process
(Data taken from references 22 and 3 for operations at the Ft. Lewis pilot plant and U. of N. Dak. PDU.)

	Operation on Lignite at UND		Operation on KY No. 9 & 14 Coal at Ft. Lewis	
Coal Analysis:	As Fed	MAF	As Fed	MAF
Moisture	31.50		1.45	
Ash	5.81		9.89	
C	45.20	72.10	70.60	79.60
H	2.93	4.67	5.09	5.74
O	13.45	21.45	7.62	8.59
S	0.45	0.72	3.84	4.33
N	0.64	1.02	1.54	1.74
Heating value, Btu/lb	7500 (est)		12,700	
Product Analysis:				
C	87.40		87.20	
H	5.67		5.77	
O	4.86		3.93	
S	0.3-1.0		0.75	
N	1.07		2.15	
Ash	0.1-0.3		0.16	
Heating value, Btu/lb	15,990		16,000	
Fusion point, °F	327		324	
Specific gravity	1.24		1.24	
C/H	15.40		15.10	
Operating Conditions, Typical Range:				
Feed gas, pct H_2 in CO/H_2	25-75		100	
Feed slurry, pct coal	32-38		35-40	
Coal space rate, lb/hr/ft^3	14-70	(max range)	29-99	
Reactor temperature, °F	778-862		840-870	
Reactor pressure, psi	1500-2500		1027-1528	
Gas feed rate, SCF/ton	20.1-26.1		14.0-21.3	
H_2 consumption, % MAF coal	1.0	75/25 H_2/CO		
CO consumption as H_2 equivalent, pct MAF coal	1.1	75/25 H_2/CO	--	
Total Coal Conversion, pct MAF COAL	77.0-94.0		93.0-94.0	
SRC Yield, pct MAF coal	47.0-64.2		58.3-67.3	
Liquid Yield, pct MAF coal	6.2-13.3		6.6-17.7	
Net Gas Yield, pct MAF coal	16.7-44.1		4.9- 8.7	

Liquefaction Applications Using Synthesis Gas

The CO-Steam concept was initially studied at the Pittsburgh Energy Research Center under the U.S. Bureau of Mines and developed further by the Grand Forks Energy Technology Center specifically for coals of high moisture contents and high reactivities; i.e., low-rank coals. Due to the high reactivity of these coals, with carbon monoxide, synthesis gas is indicated to be the preferred reducing gas in place of hydrogen.

The process flow diagram of a CO-Steam type process can be similar to SRC I, SRC II or EDS depending on the desired products. Pulverized coal is typically slurried with recycle oil in a mixing tank and charged to the reactor after being mixed with the gaseous reactant (CO or synthesis gas). In early tests the reactor was operated at 800°F and 4000 psi, but more recent results indicate that 860°F and 3000 psi may be closer to optimum. Water for the reaction is provided by the high moisture content of the coal. Carbon monoxide reacts with this in-situ water to form hydrogen for liquefaction by the water gas shift reaction.

$$CO + H_2O \rightleftharpoons H_2 + CO_2.$$

CO-Steam was designed to liquefy low-rank coals without the use of a catalyst, although throw-away type catalysts or coal mineral matter may be used. In addition to the savings on catalyst costs, low purity carbon monoxide or synthesis gas may be used as a reactant, representing a potential cost savings relative to the use of more expensive pure hydrogen in other liquefaction processes.

The liquefaction research facilities established at GFETC since 1975 have included a system for studying reaction kinetics, a 5-lb coal/hour continuous process unit, and an array of analytical instrumentation for determining elemental and molecular compositions.

The early emphasis in the CO-Steam work at GFETC was to reduce product viscosity while maintaining high liquid yield at minimum reaction times. It was found that higher reaction temperatures and shorter residence times were together effective in reducing the molecular weight of the vacuum bottoms product without an unacceptable production of methane or other hydrocarbon gases. However, the higher temperatures required (460 to 500°C) could not be achieved unless the solvent introduced with the coal had a suitable hydrogen donor activity.[14]

Recent emphasis has shifted away from boiler fuel toward interest in distillate products as a means of avoiding the problem of ash-solids separation. The optimization of distillate yield will represent a balance between the depolymerization of heavy ends and the coincident formation of gaseous products. Determination of the optimum combination of time, temperature, and recycle which produce this result for specific coals will be an important goal of future research. Research will also continue on improving the breakdown of high molecular weight lignite structures with recycle and high temperature techniques.

No unique environmental problems are associated with the CO-Steam approach.

Advanced Processes (Zinc Chloride ($ZnCl_2$) Catalyst)

The Zinc Chloride catalyst development project is part of the DOE's Third Generation Processes Program to develop advanced processes that offer significant advantages over second generation processes currently in pilot or demonstration scale phases.

The zinc chloride catalyst process is designed to convert bituminous and subbituminous coal into distillates (in the gasoline range) by severe catalytic hydrocracking. The process may be applied either to coal as a one-step process or to coal extract as a two-step process. The process configuration will be set by economic considerations related primarily to the extent of catalyst recovery. Figure 3.5.4.25 shows a process flow diagram.

Figure 3.5.4.25

Zinc Chloride Catalyst: Process Flow Diagram

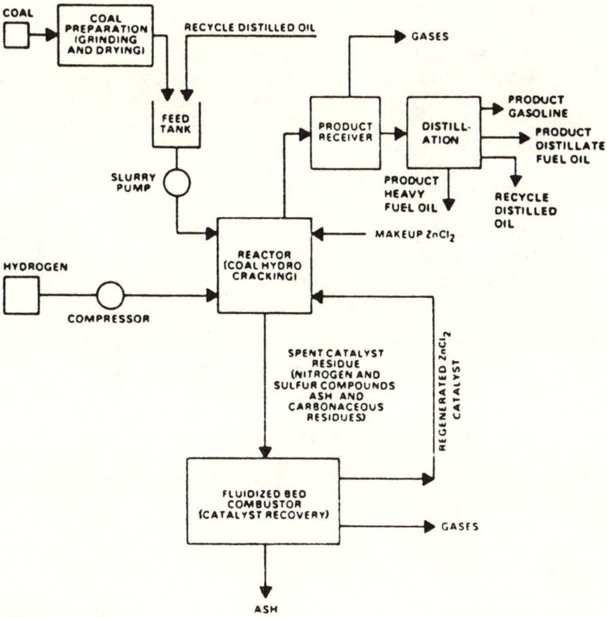

Source: Reference 17

In the process, coal is dried and pulverized before introduction to a feed tank where it is slurried with a process-derived recycle oil. The slurry feed proceeds to the hydrocracking reactor where it is mixed with hydrogen and the $ZnCl_2$. The reactor operated between 675°F and 825°F and between 1500 and 3500 psig. In the reactor, coal is cracked to distillates primarily in the gasoline range. All products go to a receiver where gas is separated from the liquid which is further processed to produce gasoline. The gasoline and light fuel oil distillates are essentially solid-free.

From the hydrocracking reactor, the solid discharge containing spent catalyst, nitrogen, sulfur compounds, ash, and carbonaceous residue is fed to a fluidized-bed combustor which operates at 1700°F and 2 psig. $ZnCl_2$ is separated from the rest of the residue as a vapor, condensed, and recycled back to the hydrocracking reactor together with the fresh zinc chloride.

The use of molten zinc halide as a catalyst for production of low sulfur gaseous and liquid fuels was discovered under office of Coal Research sponsorship by the Consolidation Coal Company (a subsidiary of Continental Oil Co.) during the period of 1963 to 1968. Work being performed by the Conoco Coal Development at Library, Pennsylvania, is aimed at developing a process which selectively produces high-octane product and achieves a high catalyst recovery by regeneration.

Major activities were conducted by Conoco on a 100-lb/hr bench-scale unit during the period 1975 to 1977 to study the conversion of subbituminous coal into distillate fuels, and to investigate zinc chloride catalyst regeneration. This work was successfully completed. Primary products from the bench-scale experiments have been characterized as 90 Research Octane Number (RON) gasoline and a low-sulfur, low-nitrogen content fuel oil. Primary and secondary zinc chloride catalyst regeneration achieved a minimum catalyst recovery of 99.5 percent. Other accomplishments included the development of a mathematical model describing coal reaction kinetics in the hydrocracking reactor.

In June 1978, construction of 1 TPD PDU at Library, Pennsylvania, was completed. Shakedown of the hydrocracking and catalyst regeneration sections is finished, and material balance operation feeding SRC-I product from the Ft. Lewis pilot plant begun in January 1979 is now complete. The PDU is developmentally ready to accept a coal feedstock, but funds have not been made available for the continuation of the project. However, the tests on a 2 lb/hr bench-scale unit included Montana Rosebud subbituminous and Kentucky bituminous coals, and provide some basis for comparison. Eastern coals have shown marginally higher liquid yields, but the difference is probably due to variations in the activity of the catalyst. No problems were encountered with calcium carbonate precipitation with the Montana feedstock. No difference in the viscosity of the liquid products was noted, possibly because almost all of the products produced were in the gasoline range.

One of the technical difficulties encountered in previous work is the corrosive products which can form as a result of the presence of zinc chloride. There is considerable uncertainty regarding the materials of construction, in particular the metals and ceramics for use in the high-temperature zone, the zinc chloride condenser in the catalyst recovery system, and any areas in contact with moist hydrogen chloride. One important limitation is that the process cannot tolerate coals with high alkali contents, such as U.S. lignites.

Unique environmental considerations may arise from the disposal of solid materials containing high concentrations of zinc compounds. Other environmental concerns associated with this process are common to other coal liquefaction processes.

Disposable Catalyst Hydrogenation

This process is also part of the DOE's Third Generation Processes Program and is being investigated by the Pittsburgh Energy Technology Center at their Bruceton, Pennsylvania, facility. The process is directed toward the use of inexpensive single pass catalysts which avoids the costly catalyst recovery and regeneration steps. A sulfur-free fuel oil is produced by the process.

Figure 3.5.4.26 is a simplified process flow diagram for the Disposable Catalyst Hydrogenation concept. Coal is dried and mixed with a process-derived oil and the disposable catalyst. The paste is compressed to 2000-4000 psi and preheated. Recycle gas and make-up hydrogen are added to the paste before hydrogenation reactions take place in a pressure vessel in liquid phase. Gaseous and sludge discharge from the reactor are sent to a gas separator and letdown system. Hydrocarbons are recovered from the gas stream, and a hydrogen rich gas stream is recycled. The gaseous fuel product is composed of light hydrocarbons. The oil is distilled to obtain liquid fuel and pasting oil, while the residue and heavy oil are processed to recover light fractions, paste oil, and heavy liquid products.

The major thrust in the process research is catalyst selection. During FY 1980 promising leads will be further pursued in the laboratory and larger scale testing will be intensified. A major effort will involve extended operations of a 1200 lb/day coal liquefaction facility at PETC. An independent engineering evaluation of this process will be conducted and the process compared with other processes already in the pilot plant stage.

Figure 3.5.4.26

Disposable Catalyst Hydrogenation: Process Flow Diagram

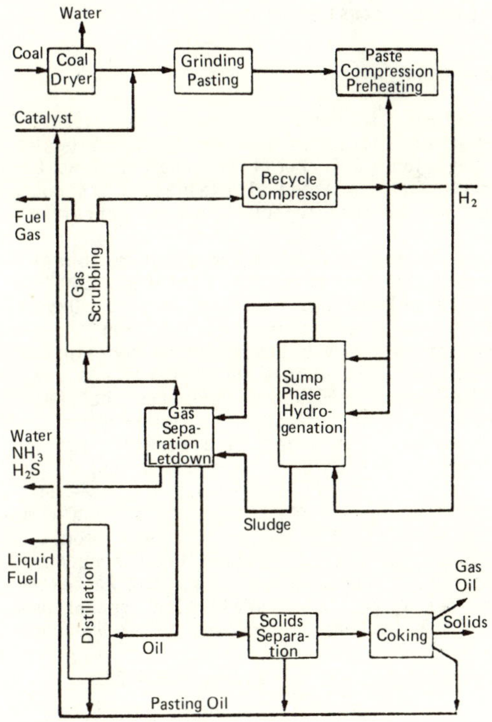

Source: Reference 17

The effects of using low-rank coals in the disposable catalyst hydrogenation process are expected to be similar to their behavior in other liquefaction systems:

- It may be economically advantageous to use synthesis gas in place of hydrogen in the process, due to the higher rate of consumption of reducing gas when low-rank coals are gasified, as well as the higher reaction rate with gases containing carbon monoxide.

- When the catalyst selection is made, compatibility with the mineral matter present in low-rank coals must be assured. One phase of the catalyst development program may be to identify specific catalysts which maximize the liquefaction rate of low-rank coals.

- Formation of calcium carbonate deposits in the reactor when using low-rank coals having high contents of ion-exchangeable calcium may require an oil filtration step before distillation.

- Unlike other catalytic systems where the catalyst activity must be maintained for long periods of time, catalyst deactivation due to precipitation of calcium carbonate is unlikely to be a problem.

- A higher bottoms viscosity resulting from some low-rank coal feedstocks may shift the split between recycle oil, liquid fuel products and gas output of the plant.

Disposal of the solid waste effluent from the plant may be complicated depending on the type of catalyst chosen for the process. Catalyst selection may also influence the yield and product spread, which may have unique environmental considerations.

Flash Liquefaction

The Flash Liquefaction process is another element of the DOE's Third Generation Processes Program, and is being investigated by the Rocketdyne Division of Rockwell International at Canoga Park, California. The development work is aimed at evaluating the concept of direct hydrogenation of pulverized coal injected into an entrained flow reactor.

A simplified process flow diagram appears in Figure 3.5.4.27. Pulverized coal is entrained in a heated (1500°F) hydrogen stream and injected into a cylindrical reaction vessel. Oxygen may be introduced to the reactor if required, since the reaction is endothermic and an additional source of heat may be required in certain cases. Rapid reaction of the coal with the hydrogen gas phase takes place at reactor conditions of 1800°F and 1000 psi. Residence times for the reacting mixture are of the order of 10 to 100 milliseconds in the reactor, after which the product mixture is quenched with water in a separate vessel. Unreacted coal, char and tar products are recovered in a third vessel, while the vapor phase is separated at approximately 500°F and 100 psi and sent to a condenser. Condenser overhead is scrubbed, and would in a commercial scale plant be processed for recovery of hydrogen sulfide, ammonia, hydrocarbon gases, and hydrogen for recycle as a reactor feed stream. Condenser bottoms is processed for oil product recovery, and the wastewater recovered is treated for reuse or discharge. Full scale plant operations would most likely use the char and tar products to generate the hydrogen used in the process.

Figure 3.5.4.27

Flash Liquefaction: Process Flow Chart

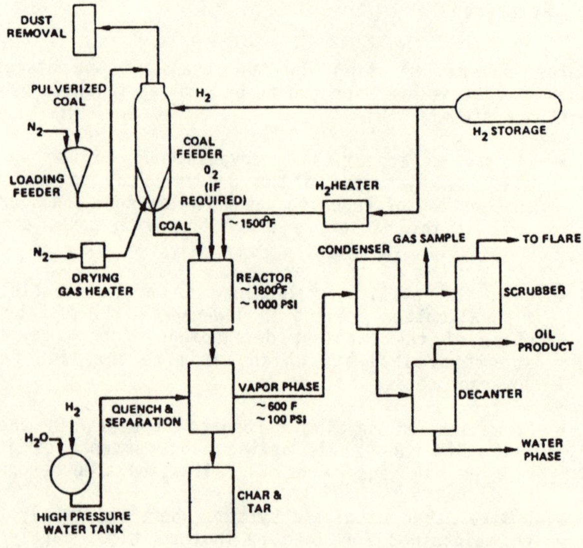

Source: Reference 17

Some of the key process variables which will affect the yield and quality of the liquids produced are reactor residence time, reactor temperature and pressure, ratio of hydrogen to coal feed (hydrogen recycle rate) and oxygen consumption (if any).

Current activities include expansion of the reactor system to a full process development unit which is now in the start-up phase. Operations will continue in FY 1980 with additional equipment to recover the benzene, toluene and xylene liquids which have relatively high commercial value. (Production of these compounds is probably due to short reaction times and incomplete hydrogenation of the coal matrix, with breakage of only primary bonds). Further design and economic studies will be conducted to permit evaluation of scaled-up operations. Data from extended operations will be provided for pilot plant design.

Although test data on the use of low-rank coals in the flash liquefaction process are not available, several considerations may apply to the utilization of these feedstocks in this process:

- The tendency of low-rank coals to produce heavy oils having a high viscosity may be somewhat exacerbated by the extremely short residence times in the entrained flow reactor. Higher temperatures may be required to offset this tendency, meaning increased consumption of oxygen and hydrogen, and larger gas yields.

- Any tendency of low-rank coals to form calcium carbonate deposits will not affect operation of the reactor due to the entrained flow condition.

- Increased hydrogen consumption may be a feature of using low-rank coals, regardless of the reaction temperature requirements. Due to the short residence times available in the reactor and the higher reaction rate of carbon monoxide with low-rank coals, the use of cheaper synthesis gas might be a fruitful direction for exploration in the research program. Enhanced yeilds may be especially significant with synthesis gas due to the short reaction times.

- Catalyst activity of the ash in low-rank coals may be especially significant in this process as compared to the reactivity of higher rank coals, due again to the short residence times.

The environmental impact of this process is expected to be similar to other coal liquefaction processes.

3.5.4.6.2 Donor Hydrogenation

The detailed chemistry and mechanisms of coal liquefaction are still largely unknown. In certain cases, however, hypotheses regarding the function of the slurrying oil are well substantiated by scientific evidence. The term "donor" applies to a solvent which chemically donates atomic hydrogen from within the molecular structure of the solvent itself. Although other solvents provide a mechanism by which molecular hydrogen in the gas phase can be brought into contact with reactive sites within the coal matrix, this process is largely only physical absorption and mass transfer. It is this characteristic which distinguishes physical solvent activity from the chemical activity of a true donor solvent.

Exxon Donor Solvent (EDS) Process

The Exxon Donor Solvent process has been under development since

1966 by the Exxon Research and Engineering Co. The process consists of non-catalytic coal liquefaction with hydrogen and a hydrogen donor solvent followed by catalytic solvent re-hydrogenation in a separate vessel.

A simplified process flow diagram is shown in Figure 3.5.4.28. Coal is ground and slurried with the recycle donor solvent. The slurry is heated in a fired heater, and preheated gaseous hydrogen is added. The reaction is carried out at approximately 800°+F and 2000 psi in a tubular reactor having no internals. Products from the liquefaction reactor are sent to several stages of separation for recovery of gas, naphtha, middle distillates, and bottoms comprised primarily of unreacted coal and mineral matter. Solid-liquid separation is achieved by distillation.

Figure 3.5.4.28
Exxon Donor Solvent: Process Flow Diagram

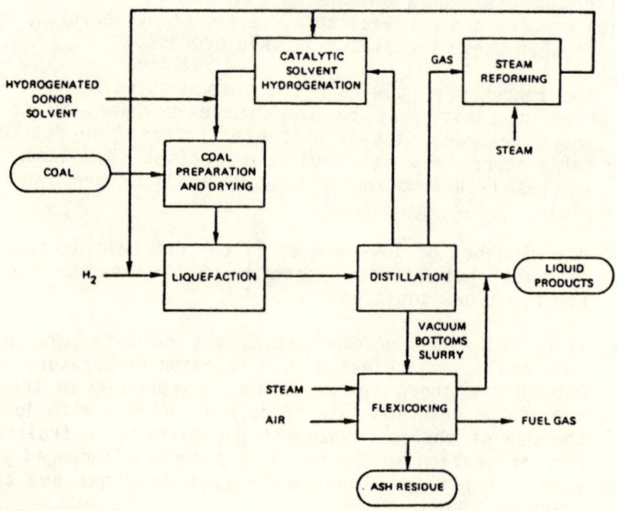

Source: Reference 17

The heavy bottoms from distillation are sent to a flexi-coker to produce additional liquids and low-BTU gas for in-plant fuel use. Hydrogen for in-plant use is provided by steam reforming of C_1-C_2 gases produced in the process. The hydrogen is recycled to the liquefaction and solvent hydrogenation sections.

A portion of the middle distillate product is sent to the solvent hydrogenation step, using a catalytic fixed-bed reactor to produce donor solvent to be recycled to the slurry preparation step. Depending on the ultimate product utilization, the primary liquid products may be further refined.

The plant is balanced in that it is self-sufficient in both process fuel and H_2 requirements. The process gives high yields of low-sulfur liquids. For example, the liquid yield from Illinois bituminous coal is 2.6 barrels of C_4/1000°F liquids per ton of dry coal feed.

A 250 tpd pilot sponsored by DOE and EPRI, among others, is currently under construction in Baytown, Texas; it is scheduled for completion in 1980.

The EDS process handles a full range of coals from bituminous through subbituminous to lignites. The product yields obtained with these three coals at preferred operating conditions are related to the coal composition. The low-rank coals, which contain higher organic oxygen, yield less liquid product and more water and carbon dioxide. Operations at 50 lb/day and 1 TPD pilot plants on various coals have shown that low-rank coals are more difficult to process than bituminous coals. Calcium carbonate deposits form in the liquefaction reactor as wall scale and free flowing agglomerates. In addition, the viscosity of the vacuum bottoms is higher for low-rank coals, which adversely affects the operation of the vacuum tower and feed system to bottoms processing.

Solid agglomerates form during liquefaction as a result of calcium humates decomposing to form calcium carbonate. The extent of deposits is related to the ion exchangeable calcium content of the coals. Two methods have been demonstrated for control: 1) periodic withdrawal of agglomerates from the reactor with acid washing of wall scale during shutdowns, and 2) coal pretreatment with sulfur dioxide to convert the calcium to calcium sulfate which is stable under liquefaction conditions. The high viscosity of the EDS liquefaction bottoms can be controlled by increasing the liquefaction residence time, which can be up to appoximately one hour for Texas lignite. Alternatively, viscosity can be decreased by changing the bottom distillation cut to increase the $1000^{\circ}F^{-}$ fraction from 10 to 15 percent. The additional liquids which remain in the bottoms go to the Flexicoking process and are recovered. Flexicoker operation produces a range of liquid products and solid coke from its heavy feed.

Product and by-product identification and characterization are ongoing; emission and effluent monitoring and solids characterization are planned.

3.5.4.6.3 Indirect Liquefaction

Indirect conversion of coal to liquids refers to initial gasification of the coal followed by a series of synthesis processes to yield liquid products. The Fischer-Tropsch process converts purified synthesis gas (H_2 + CO) to a variety of liquids. The Mobil process produces a high-octane gasoline product from methanol or synthesis gas.

The section will describe only the synthesis processes used to produce liquid products. Discussion of gasification systems which are capable of producing synthesis gas may be found in Section 3.5.3.

The Fischer-Tropsch Process

In 1923, Fischer and Tropsch produced a mixture of alcohols, aldehydes, fatty acids and hydrocarbons by the catalytic hydrogenation of carbon monoxide at high temperatures (800-1000°F) and high pressures (1500-2000 psi).[18]

A 1000 tpy plant was completed in Germany in 1933 followed three years later by a 30,000 tpy plant. In 1939, nine plants existed in Germany and one in France. South African interest in Fischer-Tropsch began in 1935 and in 1956 the first SASOL (South African Coal, Ore, and Gas Corp. Ltd) Plant was completed. SASOL I remains the only commercial scale coal liquefaction plant in existence. (The SASOL II plant was more than 80 percent complete in late 1979, and construction of Sasol III already begun.)[16]

Two synthesis processes are integrated in the first SASOL plant. The ARGE synthesis process, which was developed by Lurgi-Ruhrchemie is a medium pressure fixed bed catalytic system. The SYNTHOL process which was developed by Kellogg is a medium pressure fluidized-bed reaction system. Both reaction schemes are based on iron catalyst.

A process flow diagram illustrating the complexity of the Fischer-

Tropsch Synthesis Process as used at SASOL I is shown in Figure 3.5.4.29. The total production of liquid products is about 9000 B/D.

Figure 3.5.4.29

Process Flow Diagram, SASOL I

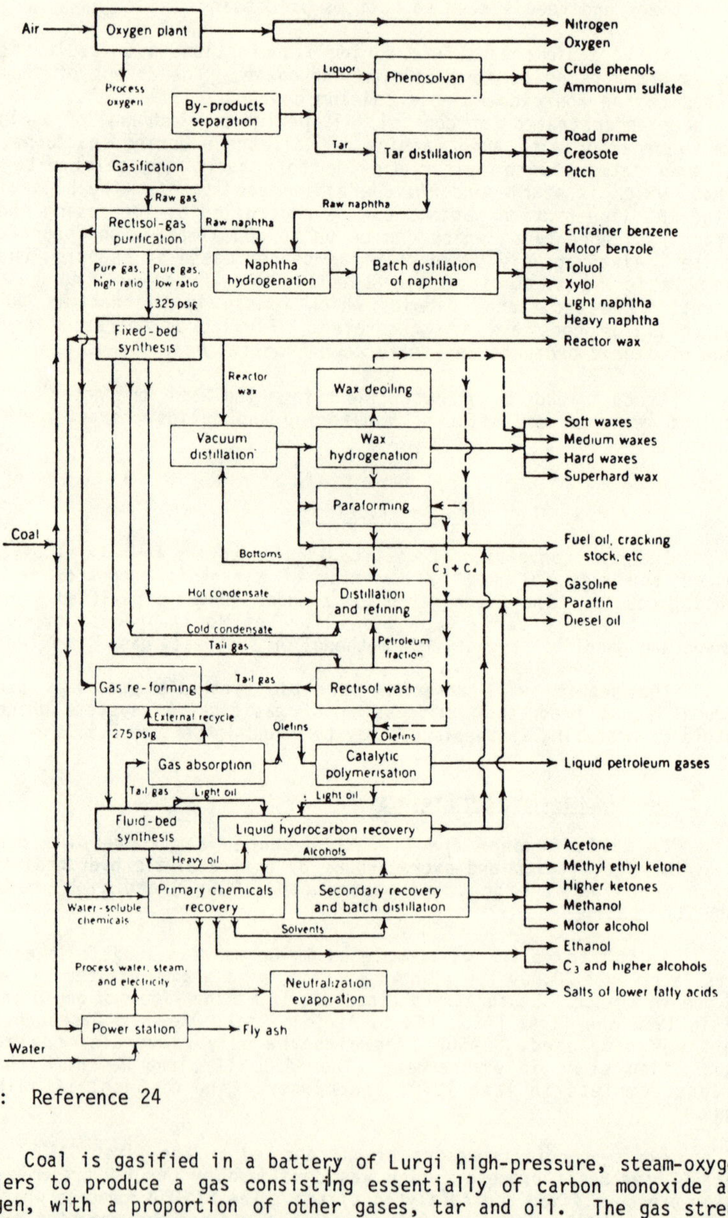

Source: Reference 24

Coal is gasified in a battery of Lurgi high-pressure, steam-oxygen gasifiers to produce a gas consisting essentially of carbon monoxide and hydrogen, with a proportion of other gases, tar and oil. The gas stream from the gasifiers is quenched to remove tar and oil and purified by the Rectisol (Lurgi) process at very low temperature which uses a single solvent (methanol) to remove the last traces of tar and oil, carbon dioxide, hydrogen sulfide, organic sulfur, ammonia and phenol. The purified synthesis gas stream is partitioned and a part of the gas is passed through a fixed-bed catalytic reactor (Arge synthesis). Feed gas has a H_2/CO ratio of about 2:1 and synthesis occurs under conditions of 430°F and 360 psi. The products of the Arge synthesis are straight-chain, high-boiling hydrocarbons, with some medium-boiling oils, diesel oil, L.P.G., and oxygenated compounds such as alcohols.

The other portion of the synthesis gas goes to the Synthol plant (Kellogg synthesis) which is a fluidized-bed catalytic (iron) reactor. In this reactor, catalyst is circulated along with the synthesis gas. Gas and catalyst leaving the reactor are separated in cyclones and the catalyst is recycled. Operating conditions are 600° - 625°F and 330 psi. A portion of the Synthol plant tail gas is reformed with steam to increase the H_2/CO ratio to about 3:1, and is mixed with the fresh synthesis gas.

Synthol reactor effluent gas is quenched in a scrubbing tower where the remaining catalyst dust is removed and returned to the reaction zone in the form of a heavy slurry oil.

The raw products from synthesis require certain treatment and then final purification to make the specification products. From the gas phase, valuable hydrocarbon and chemical products are scrubbed out and recovered. The oil phase is treated catalytically to remove dissolved oxygenates and then distilled into gasoline and fuel oil fractions. The remaining liquor is distilled and fractionated to produce chemical products. Heavy alcohols to pentanol are also recovered.

In 1974, SASOL II was conceived, followed by SASOL III in early 1979. Each plant, will be capable of producing approximately 58,000 B/D of liquid transportation fuels, and will employ only the Kellogg Synthol Process to optimize gasoline yields. SASOL II is expected to start up in 1981, SASOL III in 1983.

In the United States, research interest in the Fischer-Tropsch concept has continued since the mid 1940's, with work sponsored by the Bureau of Mines and the Office of Coal Research. However, despite the advanced developmental state of the process, economic justification cannot be given for its commercialization in the United States. In addition, the Fischer-Tropsch technology produces many chemical products which do not fit well in U.S. fuel markets.

Subbituminous coals are feedstocks for the SASOL plants. The effects of low-rank coal usage on the liquefaction process are primarily observed in the gasification train, which were discussed in section 3.5.3.5. Beyond the gasification step, the effect of coal rank is soley the determination of the gas composition, which may be varied according to the gas clean-up conditions and any shift reaction imposed on the gas. Feed composition to the ARGE and SYNTHOL units is fixed by process requirements.

Environmental considerations of the Fischer-Tropsch process involve effluents from both the coal gasification and the various liquid synthesis and recovery processes.

Mobil

The Mobil Research and Development Corporation, sponsored by the Department of Energy, is developing a synthesis process for the production of gasoline-range liquids from coal. Two alternative schemes have been proposed. In one, methanol would be synthesized and then converted to synthetic gasoline; the second proposal involves the direct formation of gasoline from synthesis gas. In the presence of a highly selective catalyst, aromatic hydrocarbons can be formed from methanol, possibly through the intermediate formation of ethylene.

A simplified process flow schematic depicting the fluid-bed process is presented in Figure 3.5.4.30.

A mixture of methanol liquid and vapor reacts in a fluid bed reactor in the presence of a catalyst to form gasoline and water. The product mix is condensed, and the gasoline and aqueous phases are separated. Provision is made in the reactor system for catalyst regeneration and for the addition of fresh catalyst.

The process is attractive because the gasoline produced has a high octane rating and is free from heavy ends, obviating further product upgrading. Under the best conditions achieved to date, over 75 percent of the hydrocarbon product is in the C_5^+ range. Mobil estimates that the process would have an overall thermal efficiency of approximately 66 percent when based on Lurgi coal gasification and including energy credits for LPG, methane and other useful process by-products.[18]

Figure 3.5.4.30
Mobil Indirect Liquefaction

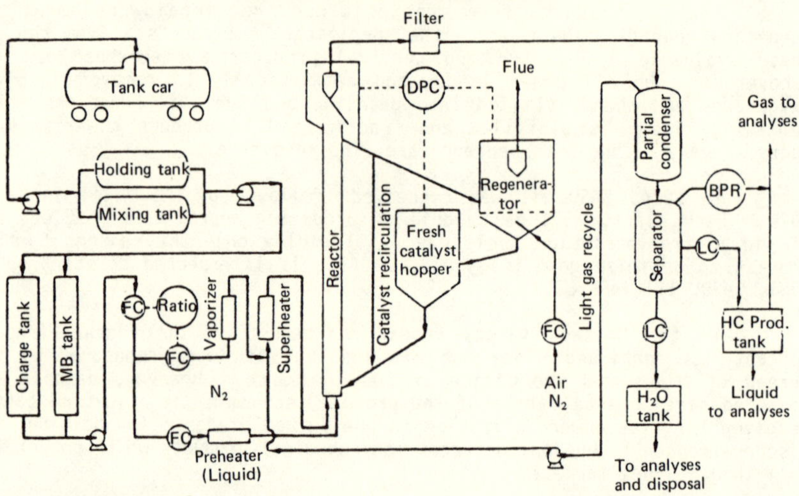

Source: Reference 6

Mobil is also working on other catalysts which may promote the gasoline formation directly from synthesis gas. This would simplify the process and reduce overall costs. Even at the present state of development, the overall economics are claimed to be better than for the SASOL process because a gasoline type product is produced without the range of by-products for which markets may not be available. Construction of a 15 ton products/day pilot plant is under consideration. The project would be based on the design of a smaller unit which was operating successfully at the end of 1977 in Paulsboro, N.J.

The use of low-rank coals in the Mobil liquefaction process must be considered from the standpoint of their effects on the gasification system These effects of low-rank coals were addressed in section 3.5.2.5. Environmental problems arising from the Mobil process involve effluents from both the coal gasification and liquid synthesis.

3.5.4.6.4 Foreign Liquefaction R&D Products

A list of foreign Research and Development projects for liquefaction is presented in Table 3.5.4.10. This work covers a large variety of processes, some similar to those discussed in the preceding sections, and others utilizing novel techniques (e.g., supercritical gas extraction).

Table 3.5.4.10

Foreign Coal Liquefaction Research and Development Projects

Country	Company	Development Stage	Process	Product
UK	National Coal Board Cheltenham	2-liter stirred autoclave work. A 0.75 t/d extraction plant and a 0.025 t/d hydrogenation plant are operational.	Solvent extraction (anthracene Oil) followed by catalytic hydrogenation.	Light oil, middle oil, heavy oil (for recycling)
UK	National Coal Board, Cheltenham	Batch extraction followed by semi-continuous work. A 0.1 t/d plant is operational. Pilot plant of 25 t/d envisaged.	Supercritical gas extraction. Coal constituents pass into gas phase leaving involatile residue of char and mineral matter. Coal extract is precipitated and separated from the extracting gas.	Extract yields up to 50% coal, to be converted to chemical feedstocks and liquid fuels.
Germany	Bergbau-Forschung GmbH, Essen	0.25 to 0.5 t/d unit to test feasibility of producing heavy and medium oil. Also examining possible catalytic effect of mineral matter in coal liquefaction.		Product oil will contain less than 0.5% sulfur.
Germany	Ruhrkohle AG, Essen (Sub-contractor STEAG)	Evaluating Stearns-Roger design study for construction of 6000 t/d demonstration plant based on SRC-2 process. Completion by end of 1979.		
Germany	Schering AG, Bergkamen (Sub-contractor Lurgi-Mineraloltechnik GmbH Frankfurt)	Study concerning synthesis of raw materials for chemical industry by means of Fischer-Tropsch process. Pilot plant with Kolbel liquid-phase slurry reactor is operating.		
Germany	Union Rheinische Braunkohlen Kraftstoff, Wesseling (Sub-contractor SASOL and German engineering companies)	To make mainly petrochemical raw materials via Fischer-Tropsch synthesis or related processes, using brown coal as feedstock.		
Germany	Lurgi Ruhrgas	The technique is being applied to lignite on an industrial scale of 1600 t/d at Lukavac, Yugoslavia, and in other large-scale plants around the world. Extent to which tar products are used to make synthetic oils by hydrogenation is unknown.		

(continued)

Table 3.5.4.10 (Continued)

Foreign Coal Liquefaction Research and Development Projects

Country	Company	Development Stage	Process	Product
USSR	Institute of Fossil Fuels	Liquid-phase hydrogenation of brown coal/oil paste is being carried out in 6-liter reactors. (40% coal plus 60% paste-forming agent.)	Use of petroleum oils as hydrogen-donor solvents, organic additives as polymerization inhibitors and employing active catalyst enables process to take place at low pressure and with low H_2 consumption.	High-boiling oil fuel; low-boiling oil, refined to gasoline.
Japan	Osaka Gas Company, Torishima, Osaka	0.25 t/d pilot unit was due for expansion into 24 t/d facility to go on stream summer 1975. 40 t/d asphalt treating demonstration plant now being built. Stanford Research Institute carried out feasibility study of process.	Some coal powder is added to crude oil or asphalt and the mixture reacted at high temperature and pressure. Next the slurry is cracked, then the lighter fractions are flashed off and the bottoms are separated into liquid and solids by centrifuge. Pitch is separated as distillation residue.	LPG/Naphtha, Kerosene, Heavy fuel oil, Binder, Solids.
Japan	Fuel Research Institute of Japan	Pilot Plant work on solvent extraction of coal using a coal-tar oil. Research work on use of petroleum heavy oils. 250 t/d plant designed to make a SRC and convert it to electrode carbon in delayed coker, but construction costs were too high.		SRC constituted by weight of coal.
Japan	Electric Power Development Corporation	1 t/d unit in Nagasaki. A 40 t/d pilot plant expected by 1981, a 40 t/d solvent treatment plant by 1984, and a 40 t/d direct liquefaction plant by 1985.	Coal is treated with asphalt as solvent in absence of catalyst.	SRC as black pitch brittle material.
Japan	Mitsui Group of Companies	5 t/d SRC pilot plant operating at Ohmuta works of Mitsui producing 3 t/d SRC. Mitsui to join Gulf Oil to commercialize process, probably in Australia. A semi-commercial scale plant was built in Milke, Kyushu with a capacity of 1500-2000 t/d.	SRC	
Japan	Sumitomo Coal Mining Company	Operating 2 t/d plant in Akabira, Hokkaido, using solvent extraction process. Hopes to sponsor one of the U.S. extraction processes.	SRC	

(continued)

Table 3.5.4.10 (Continued)

Foreign Coal Liquefaction Research and Development Projects

Country	Company	Development Stage	Process	Product
Japan	Taikeiyo Coal Mining Company	Plan to build 6000 t/d coal extraction-hydrogenation plant in Kushiro, Hokkaido.	SRC	
Japan	Consortium of 10 Japanese companies	Plan to build commercial liquefaction plant yielding 7000 t/d by 1985.	Exxon Donor Solvent	
Poland	Central Mining Institute	Fluidized-bed, low-temperature carbonization plant to treat 24 t/d non-caking hard coal is being built, and plans made for 360 t/d plant. Research work on hydrogenation of primary fluidized-bed tar to diesel fuels.	Hydrogenation of Fluidized Bed primary tar.	Light oils, 28.2% of product by volume.
Poland	Central Mining Institute	Continuous solvent extraction/ hydrogenation of coal in 0.1 t/d plant. 1 t/d plant has been built to enable extraction/hydrogenation and also direct hydrogenation of coal to be examined in ebullated-or fixed-catalyst beds of high throughput. Plant could also be used to treat fluidized-bed tar. Next step is 500 t/d plant prior to erection of commercial plant.	Solvent used is hydrogenated anthracene oil, recycle oil and LT oil. Extracts hydrogenated in two stages.	1st stage gave 21.4% light oil. 2nd stage, 15.4% light oil. Aim is Syncrude; 0.5%S.
Australia	Mitsui Group of Japan and Gulf Oil Corp. of USA	10,000 t/d commercial plant planned and expected to operate by 1981, with operating conditions similar to PAMCO plant.	SRC	
Australia	New South Wales Government	Considering building plant to make 4000 t/d gasoline. Feasibility study being made by Gas Developments Corporation (Institute of Gas Technology, USA, subsidiary).	Coal Liquefaction	
Australia	Australian Coal Industry Research Laboratories Ltd.	Production of synthetic oil and chemicals from coal by batch hydrogenation process. Autoclave work on solvent extraction with anthracene oil. A 0.01 to 0.02 t/d continuous reactor is planned.	Coal Liquefaction	
India	Central Fuel Research Institute, Dhanbad	1 t/d pilot plant to be run on Assam coals. 25 t/d plant to follow. Project supported by NPRC of India and Indian Oil Corporation.	Hydrogenation	

Source: Reference 18

References - Sections 3.5.4.2 - 3.5.4.6

1. Lowry, H.H., (editor). Chemistry of Coal Utilization - Supplementary Volume, John Wiley & Sons, Inc., New York, 1963.

2. Given, P.H. The Dependence of Coal Liquefaction Behavior on Coal Characteristics, Presented at the Short Course on Coal Characteristics at Penn. State University, October 22-26, 1979.

3. Severson, D.E., A.M. Souby, and J.J. Harris. Process Development of Lignite Liquefaction, Symposium on Technology and Use of Lignite, Grand Forks, N.D., May 1979, DOE GFETC/IC-79./1, pp. 157-169.

4. Mitchell, W.N., K.L. Trachte, S. Zaczepinski. Performance of Low-Rank Coals in the Exxon Donor Solvent Process; Symposium on Technology and Use of Lignite, Grand Forks, N.D., May 1979, DOE GFETC/IC-79/1, 1979, pp. 157-169.

5. Sondreal, E.A. Utilization Potential for Low-Rank U.S. Coals in Liquefaction and Gasification, Presented at the Short Course on Coal Characteristics and Coal Utilization. Penn. State, October 22-26, 1979.

6. Fossil Energy Research and Development Program of the United States Department of Energy, FY 1979, DOE/ET-0013(78), March, 1978, 476 pp.

7. International Coal Technology Summary Document, prepared by the U.S. Department of Energy with the assistance of TRW Energy Systems Planning Division, December 1978.

8. H-Coal Integrated Pilot Plant. Prepared by Hydrocarbon Research, Inc., Trenton, N.J. EPRI report AF-681, Project 238-1, Final Report vols. 1 and 2, March 1978, 46 and 438 pp.

9. H-Coal PDU Tests on Illinois #6 and Wyoming Coals. Prepared by Hydrocarbon Research, Inc., Lawrenceville, N.J., EPRI AF-1143-SR, Special Report, July 1979, 206 pp.

10. Jenkins, R.G. Unwanted Deposits in Liquefaction Reactors; Presented at the Short Course on coal characteristics and coal utilization at Penn. State University, October 22-26, 1979.

11. Coal Conversion. 1977 Technical Report., U.S. Department of Energy Report # DOE/ET-0061/1, June 1978.

12. Solvent Refining of Indiana V Coal and North Dakota Lignite. Prepared by Hydrocarbon Research, Inc., Lawrence Township, N.J., EPRI AF-666, Project 779-4, Final report, January 1978, 69 pp.

13. Sondreal, E.A., C.L. Knudsen, J.E. Schiller, and T.H. May. Development of the CO-Steam Process for Liquefaction of Lignite and Western Subbituminous Coals, Symposium on Technology and Use of Lignite, Grand Forks, N.D., May 1977, GFETC/IC-77/1, 1978, pp. 129-158.

14. Willson, W.G., C.L. Knudson, and G.G. Baker, T.C. Owens, and D.E. Severson. Application of Liquefaction Process to Low-Rank Coals, Symposium on Technology and Use of Lignite, Grand Forks, N.D., May 1979, DOE GFETC/IC-79/1, 1979, pp. 170-206.

15. Knudson, C.L., J.E. Schiller, and A.L. Ruud. Temperature Effects on Coal Liquefaction: Rates of Depolymerization and Product Quality as Determined by Gel Permeation Chromatography; Organic Chemistry of Coal, ACS Symposium Series 71, edited by John W. Larsen, Washington, D.C., 1975.

16. Heylin, M. "South Africa Commits to Oil-from-Coal Process," C&E News, September 17, 1979.

17. Fossil Energy Program Summary Document, U.S. Department of Energy, Document # DOE/ET-0087, March 1979.

18. Liquid Fuels from Coal, a National Coal Board Report, Prepared by Planning, Assessment and Development Branch of the Coal Research Establishment with Operational Research Executive, August 1978.

19. International Coal Technology Summary Document, Department of Energy, HCP/P-3885, December 1978.

20. Appell, H.R., R.D. Miller, and I. Wender. On the Mechanism of Lignite Liquefaction with Carbon Monoxide and Water, Presented at Div. of Fuel Chem., 163rd National Meeting, ACS, Boston, Mass., April 10-14, 1972.

21. Appell, H.R., E.C. Moroni, and R.D. Miller. CO-Steam Liquefaction of Lignite, Presented at the Div. of Fuel Chem. 169th National Meeting, ACS, Philadelphia, Penn., April 6-11, 1975, v. 20, No. 1, pp. 58-65.

22. Compilation and Assessment of SRC Experience: Data Book. Prepared by Bechtel National, Inc., San Francisco, Calif., EPRI AF-1019, Project 987-1, Final Report, March 1979, 301 pp.

23. Fossil Energy Research Program of the Energy Research and Development Administration, FY 1978, ERDA 77-33 April 1977.

24. Braunstein, H.M., E.D. Copenhauer, and H.A. Pfuderer (Editors), Environmental, Health, and Control Aspects of Coal Conversion: An Information Overview, Volume I, ORNL,/EIS-94, April 1977.

3.5.4.7 SELECTED REFERENCES

In addition to the preceding complete reference list, a few selected reference abstracts are provided below which may be of general interest for readings in coal liquefaction:

1. Willson, W.G., C.L. Knudson, G.G. Baker, T.C. Owens, and D.E.Severson. Application of Liquefaction Process to Low-Rank Coals, Symposium on Technology and Use of Lignite, May 1979, Grand Forks, N.D., DOE GFETC/IC-79/1, pp. 170-206.

   The Grand Forks Energy Technology Center and the University of North Dakota researchers are engaged in liquefaction research to develop a scientific and engineering data base on the liquefaction behavior of low-rank coals necessary to apply major developing processes to these distinctly different coals. The process initially reacted CO with a raw lignite to produce an "ash-free" solid or heavy liquid. Emphasis has been changed toward the production of a distillable liquid; and synthesis gas--equimolar CO/H_2--has been substituted for CO.

   In the newly constructed 5-lb coal/hr continuous process unit, synthesis gas, raw lignite and anthracene oil solvent were reacted at elevated temperatures in single pass tests in a continuous-stirred tank reactor (CSTR). Product yield fractions were correlated with percent coal in the feed slurry, hydrogen donor (tetralin) concentration and temperature. The molecular weight of the soluble but non-distillable yield fraction was markedly reduced by increasing temperature, but this effect was less pronounced in a CSTR reactor with internal backmix than in a batch reactor.

In batch autoclave systems, work has been conducted to establish rates and product distributions from several liquefaction solvents. Two solvents were chosen for subsequent tests aimed at determining the catalytic effects of diverse mineral matter in 8 different low-rank coals. Preliminary tests to determine product yield and quality in a single pass for some of the diverse coals have been performed.

2. Sondreal, E. A., C. L. Knudson, R. S. Majkrzak, and G. G. Baker. Liquefaction of Lignite by the CO-Steam Process, Preprints, AIChE Meeting, Miami, November 1978, 36 pp.
The rapid reaction of synthesis gas with lignite plus combined water was investigated in a 5-lb. coal/hr. continuous process unit at high temperatures (up to 500°C) and in the presence of a hydrogen donor additive (tetralin). Product yield fractions were correlated with temperature, tetralin level, and percent coal in the feed slurry. The molecular weight of the soluble but non-distillable yield fraction was markedly reduced by increasing temperature, but this effect was less pronounced in a CSTR reactor with internal backmix than in a batch reactor. Future work will involve fractionation and recycle of the high molecular weight product and hydrogenation of recycle to permit high-temperature operation without coking. This research will support the application of the SRC-II and Donor Solvent processes to lignite.

3. "Process Engineering Evalutions of Alternative Coal Liquefaction Concepts", prepared by the Ralph M. Parsons Co., Pasadena, CA. EPRI document AF-741, Volume 1, April 1978.
Process engineering evaluations were prepared for two coal liquefaction concepts. These evaluations were made within the framework of complete self-sustained plants which require only coal and water as feed materials. The two liquefaction concepts considered were the non-catalytic Solvent Refined Coal process and a generic catalytic hydroliquefaction process referred to as the CHL process in this report. The engineering designs and capital cost estimates were developed for a southern Illinois location using the same general process configuration and supporting units for both liquefaction systems. The designs are based on the use of commercially demonstrated technology wherever possible. This emphasis on current technology provides a basis for assessing the cost of a first generation plant that could be constructed today without further improvements from the 1975 technology base. These initial designs are intended to be base cases for assessing the cost impact of process improvements that will be incorporated into subsequent designs.

Two cases representing different processing severities were developed for each liquefaction process. The more severe operations produce products with a lower sulfur content and higher liquid yield. The designs were developed for plants with a nominal product capacity of 50,000 fuel oil equivalent (FOE) bbls/day (FOE bbl = 6.3×10^6 BTU). The total plant investments range from \$1,097 MM to \$1,173 MM for the two SRC cases, and from \$1,228 MM to \$1,445 MM for the two CHL cases. The higher capital costs were associated with the more severe operations.

The cost of on-site power generation represents a significant proportion of the overall plant capital cost. To assess the implication of this, two additional cases were developed using purchased power. These cases are based on the low severity SRC process, and are presented in a separate supplement to this report. The use of purchased power can reduce the total plant investment in the range of 12 to 17 percent.

4. Mitchell, W. N., K. L. Trachte, S. Zaczepinski. Performance of Low-Rank Coals in the Exxon Donor Solvent Process, Symposium on Technology and Use of Lignite, Grand Forks, N.D., May 1979, DOE GFETC/IC-79/1, 1979, pp. 157-169.

The Exxon Donor Solvent Coal Liquefaction Process (EDS) handles a full range of coals ranging from bituminous through subbituminous to lignites. The overall process performance based on the 50 lb/day recycle coal liquefaction unit (RCLU) and the 1 ton/day coal liquefaction pilot plant (CLPP) is summarized as a function of process conditions and coal rank.

Special emphasis is placed on the conversion and yield response of the range of coals demonstrated in the EDS process to date. In addition to the liquefaction potential, the operability issues associated with operating a low rank coals are addressed. More specifically, the relationship between the operating severity and the liquefaction bottoms viscosity is explored in detail. Also, the calcium carbonate scale deposition and agglomerates formation in the process reactors is covered. As part of this discussion, process and mechanical solutions to this problem are summarized.

5. Severson, D.E. Solvent Refined Lignite Process Development Final Report, June 16, 1977-June 15, 1978, DOE FE-4189-15, January 5, 1979, 96 pp.

Project Lignite was initiated in 1972 at the University of North Dakota to determine the appropriate technological approach to the conversion of Northern Great Plains Lignite to premium solid, liquid and gaseous fuels. Laboratory and bench-scale work were carried out and a continuous process development unit (PDU) was constructed and operated to demonstrate the feasibility of continuous liquefaction of lignite. Economic studies were made, information was developed on catalytic hydrogenation of solvent refined lignite.

The continuous process originally envisioned was the two stage conversion of lignite to fuel liquids, with solvent refined lignite as an intermediate and useful fuel product. The first stage was a unit converting 50 pounds per hour of raw lignite into solvent refined lignite by reaction with synthesis gas (carbon monoxide and hydrogen) in the presence of a hydrogen donor solvent that was continually regenerated from the lignite during processing. The mineral separations system was a countercurrent decantation process carried out at elevated temperature and pressure. The second stage was to be the catalytic hydrogenation to premium liquid fuels of the solvent refined lignite produced in the first stage. However, this latter step was not implemented, and the program was directed at developing the liquefaction and solid-liquid separation stages.

During its final year of operation, the project included in the liquefaction section of the Process Development Unit, together with some operation of the gas clean-up and recycle sections, the evaluation of the solid/liquid separation section, corrosion coupon testing, the determination of solids accumulation on the dissolvers, the production of clean solvent refined lignite, and data collection and analysis.

3.5.5 Pyrolysis

3.5.5.1 Introduction and Summary

Pyrolysis is defined as the transformation of a substance into one or more substances by the application of heat alone. This technique has been historically applied to many different materials, including but not limited to wood, coal and petroleum feeds. The technique is still important today in the petroleum refining industry, and has been used successfully here and in other industries both for waste disposal and as a method of producing commercially important products.

Although the term pyrolysis implies a reduction in the size or complexity of the feedstock chemical composition, pyrolytic changes may also involve chemical isomerization and the formation of higher molecular weight fractions, generally through free radical recombination of cracked molecules. This applies to coal pyrolysis, and hence control of the product distribution produced by pyrolysis is achieved by manipulation of the rate, degree and length of time the sample is exposed to the pyrolysis conditions.

A distinction is made between low-temperature carbonization (generally 840 to 1300°F) and high-temperature carbonization (generally greater than 1650°F) of coal. Low-temperature carbonization results in the production of a solid, smokeless fuel (char) coupled with a high yield of coal tar. The char is a soft, black porous material, which still contains considerable volatile material. High-temperature carbonization is used for the production of coke by the metallurgical industry. Generally, the discussion of pyrolysis here is directed at low-temperature carbonization.

The relative proportions and quality of gas, liquid and char produced from a pyrolysis process depend on the feedstock composition. For example, carbonization tests at 500°C have shown that for lignites more than three-fourths of the original heat content is concentrated into char which represents less than half of the original sample weight. By comparison, higher rank coals do not show this same upgrading in char quality. The result is that low-rank coal pyrolysis chars have nearly the same heating value as do their higher-rank counterparts. Although the gross heating values are similar, low-rank coal chars are reactive to the point of pyrophoric behavior, and also display enhanced reactivity to steam and hydrogen relative to higher rank chars. These char properties may in certain cases require special handling considerations, but also provide opportunities for use in processes where high reactivity is advantageous. The lower sulfur content of low-rank coals is an advantage retained in the char, which may therefore make it more widely applicable or require a reduced degree of environmental control than bituminous coal chars.

Operation of the pyrolysis reactor will also be affected by the rank of coal being fed to the unit. The noncaking characteristic of low-rank coals is an advantage in fixed bed or fluidized bed pyrolysis reactors. On the other hand, initial decomposition temperatures increase with rank, suggesting that optimal temperature levels and profiles will be different when low-rank instead of high-rank coals are used as feedstocks.

Three domestic coal pyrolysis projects, COED, Toscoal, and Occidental Research Corporation Flash Pyrolysis, are described in addition to foreign work currently underway. Different reactor schemes are used in each process, representing different heat transfer mechanisms. The COED concept calls for a series of fluidized beds in which heat transfer is effected by a hot gas stream. Ceramic balls are used in the Toscal design to provide the pyrolysis heat in a rotating retort. The Occidental Research Corporation process uses hot char as a heat carrier; the pyrolysis reaction time is very short, which minimizes the production of gas and also prevents the formation of very high molecular weight fractions due to repolymerization.

Key Issues for Low-Rank Coal Pyrolysis

The unique properties of low-rank coals raise the following key issues with respect to pyrolysis process development:

1. Economic feasibility of pyrolysis processes for low-rank coals.

2. Improved analytical methods for low-rank coal pyrolysis studies.

3. Improved pyrolysis product properties and yields.

In addition, coal pyrolysis technology shares with gasification and liquefaction technologies the need to minimize potential health effects of coal liquids.

Economic Feasibility

Because of their high inherent moisture and oxygen contents, low-rank coals produce larger yields of gas and water and relatively low yields of liquid and char products when pyrolyzed, as compared to bituminous coals. On the other hand, pyrolysis upgrades the heating value of the solid product from low-rank coal to a much greater extent than the upgrading achieved with bituminous coal. These differences, as well as differences in the properties and behavior (such as reactivity) of char and tar produced from low-rank coal as compared to bituminous coal, raise questions as to the comparative economic feasibility of using low-rank coals as feedstocks for pyrolysis.

One concept would be to use a pyrolysis plant primarily as a means of low-rank coal beneficiation: produce a highly reactive, high-heating-value char for power plant fuel; utilize the tar fraction either as chemical feedstock or as binder for briquetting the char; and utilize the offgas locally as fuel gas. Both the transportation economics and the powerplant economics would be improved by using the upgraded fuel. The question is whether these savings would be great enough to pay for the profitable construction and operation of the pyrolysis plant. Process-, coal-, and site-specific studies are needed to define these tradeoffs.

Improved Analytical Methods

Pyrolysis products from low-rank coals are highly complex, and their properties can be influenced for better or worse by selection of process conditions. Such properties as reactivity of char and stability, corrosivity and viscosity of pyrolytic oil are difficult to measure on a consistent basis and relate to the chemical compositions of these substances. Basic studies to improve analytical methods for the characterization of pyrolysis products from low-rank coals would aid process developers in their efforts to produce improved products.

Product Properties and Yields

One of the factors that has kept coal pyrolysis from commercial use despite many years of research is the difficulty of marketing its products. Upgrading of product quality can be approached through modification of the pyrolysis process conditions, or through operations such as hydrotreating of oil and briquetting of char. However, additional process steps such as these add substantially to the product costs and tend to decrease their yields. Concepts such as flash pyrolysis and hydropyrolysis are being developed with some success, as described below.

3.5.5.2 Technology Description

Pyrolysis, carbonization and retorting are terms used in a general way to describe the basic operation of heating coal to decompose and separate it into various products. Pyrolysis can be conducted over a wide range of processing conditions and can produce products with a wide range of characteristics. Although the physical and chemical changes occurring during pyrolysis are complex, the technique itself is relatively simple in terms of processing complexity and control for separating coal into several components of potentially improved value.

The products of pyrolysis (low-temperature carbonization) are char, tar, and gas, listed in order of decreasing yield. An aqueous fraction containing some organic material, such as carboxylic acids, is also produced; but, generally, this fraction does not have a positive economic value. Particularly with low-rank coals, a considerable increase in heating value of the char and tar is accomplished by removing the moisture of the coal.[1,a]

The char (also commonly designated as "semicoke") is usually a soft, black, porous material, either finely divided or in the form of lumps, which still contains some volatile material. In general, this product is suitable for use in combustion or gasification processes, and in some cases for blending into metallurgical coke.

The tar is a highly complex, oxygenated material that varies widely in composition and properties depending on the process used. It represents the products of repolymerization of free radicals formed during thermal decomposition of the coal. Generally the tar is highly viscous (semisolid or solid) at room temperature. It can be utilized as fuel oil, after varying degrees of further processing. It contains many chemical species which may be valuable as a source of chemicals.

The gas produced from pyrolysis contains varying amounts of hydrogen, carbon monoxide, carbon dioxide, hydrocarbons (primarily methane), and water vapor.

The relative yields of the liquid and gaseous pyrolysis products depend on three major factors:

- Proportions of volatile matter present in the coal
- Reaction temperature
- Residence time of the coal and pyrolysis products within the reactor

The technology of pyrolysis is characterized by a great number and variety of processes that have been utilized or proposed during the past several hundred years. The operating conditions vary greatly, depending on the coal properties, products desired, and many other factors. The products from these processes necessarily vary in chemical and physical characteristics, their properties depending largely on feed composition and the treatment to which they have been exposed. For char especially, the size may vary from a fine powder to large lumps, including, in some processes, the preparation of a shaped product or briquette.[2]

Vertical retorts include both batch and continuous types, with some processes resembling current high-temperature coke ovens. They are typically of the fixed bed or moving bed type, and can be heated either indirectly (through the walls) or directly (by hot gases). The continuous flow reactors are not well suited to highly caking coals (see section 3.5.3.2 for a discussion of the same problem in gasification reactors).[2]

---

[a]References for Sections 3.5.5.2-3.5.5.5 are listed in Section 3.5.5.6.

Fluidized or entrained retorts are well adapted to continuous processing of large tonnages of coal, and are applicable to practically all types of coals. Carbonizing fine coal while fluidized by or entrained in a stream of heated gas produces a finely divided char, which may be utilized for boiler firing. In addition, such chars have found application in the preparation of metallurgical coke blends.[2] A notable example of fluidized bed carbonization is the Parry process developed by the U.S. Bureau of Mines in cooperation with the Texas Power and Light Company.[3]

Horizontal retorts consist essentially of long steel vessels, often cylindrical, through which the coal is moved as carbonization proceeds. The cylinders may be either rotated or fixed, and either direct or indirect heating may be used. These types of retorts have problems with highly caking coals unless pretreatment or other steps are taken to prevent adhesion. The Disco process employs rotating horizontal retorts. A commercial plant using this process near Pittsburgh, Pa., converted cleaned bituminous coal fines into lump char suitable for domestic use.

A more recent development in the field of pyrolysis is the technique of "flash" pyrolysis, which uses an entrained-flow type of reactor. A significant increase in yield of volatiles is obtained by this technique, which entails very rapid heating of the coal particles (on the order of thousands of degrees per second) coupled with very short residence times (on the order of seconds or fractions of a second). The short residence time prevents the thermal decomposition of the tars to gases. The rapid heating minimizes the repolymerization of the pyrolysis products and usually results in the production of ethylene-rich gases. (Maintaining high temperatures for extended periods results in high yields of gas.)

The coal decomposition rate is very temperature-sensitive, with a maximum occurring at a distinct temperature for each coal (see Figure 3.5.5.2). The heating rate also has a strong influence on the rate of decomposition, as illustrated in Figure 3.5.5.3.

Figure 3.5.5.2

Decomposition Rates of Various Coals

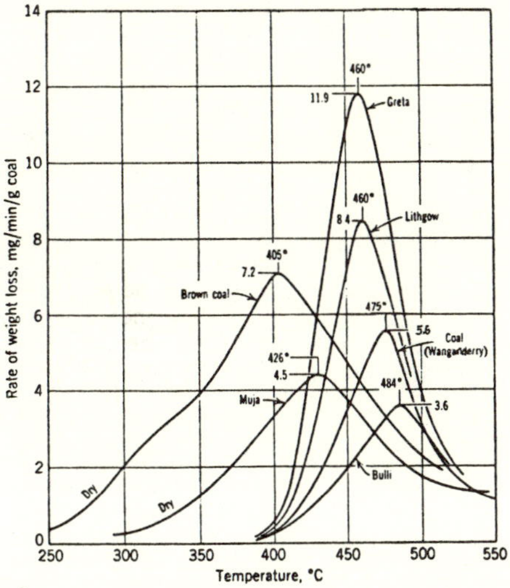

Source: Reference 2

Figure 3.5.5.3

Effect of Heating Rate on Rate of Decomposition

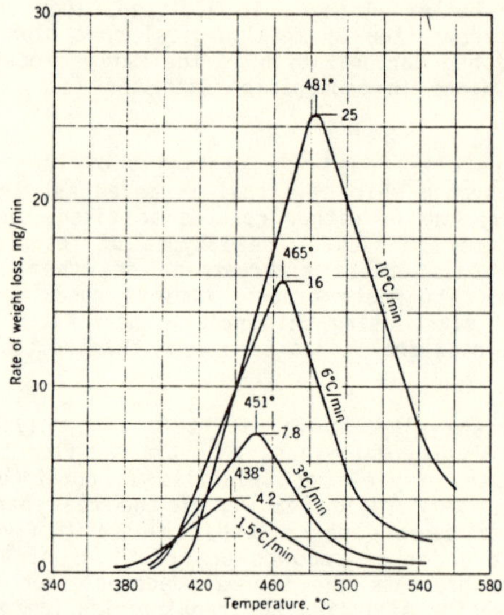

Source: Reference 2

Extensive work on carbonization (pyrolysis) of U.S. low-rank coals was conducted by V.F. Parry and co-workers at the Bureau of Mines during the 1940's and 1950's.[3,4] Five lignites, nine subbituminous coals and six bituminous coals were tested in a fluidized-bed pyrolysis unit. Low-rank, nonagglomerating western coals were found to be particularly amenable to high-capacity fluidized carbonization. Figures 3.5.5.4 and 3.5.5.5 graphically show the relationship between product yields and temperature of carbonization. As temperature is increased, the degree of decomposition increases correspondingly; this effect diminishes above about 600°C.

From Figure 3.5.5.4 it is seen that the gases released at the low temperatures of carbonization (150°C) appear to be mainly adsorbed carbon dioxide and methane. Decomposition products resulting from the destruction of the coal particle itself do not appear in the gas until carbonization has progressed beyond 200°C. The total gas produced per unit of coal, at 250°C carbonization temperatures, is of the order of 0.01 cubic foot per pound of moisture-and ash-free coal carbonized.[4]

Light oil appears in significant quantities at carbonizing temperatures as low as 200°C. Figure 3.5.5.5 shows that the maximum tar plus light oil is obtained at about 500°C and indicates that very little condensable oil product, if any, is produced beyond 500°C. The gaseous products from the carbonization reaction beyond 500°C are augmented by an increase in the hydrogen fraction. At the highest temperatures of carbonization (1,000°C and above), the principal gaseous product becomes hydrogen.[4]

Summarized data on the properties of the tars produced from the various coals tested are shown in Figure 3.5.5.6. An interesting result is that, when averaged, the tar fraction boiling to 270°C is approximately the same for all ranks of coal studied; yet there is wide variation in the composition produced from coals of the same rank. For ex-

Figure 3.5.5.4

Relationship Between Gas Composition (Oxygen-Nitrogen-Free Basis) and Temperature of Carbonization from Assay Pyrolysis of Elkol Subbituminous Coal

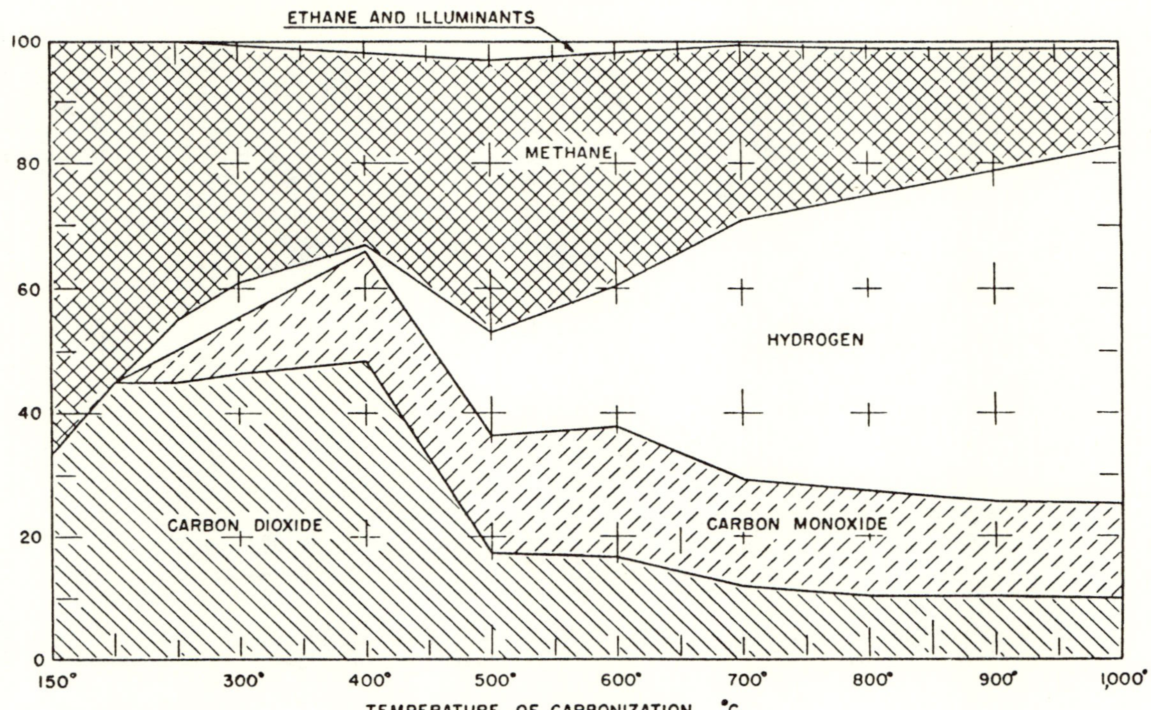

Source: Reference 4

Figure 3.5.5.5

Relationship Between Yield of Products and Temperature of Carbonization from Assay Pyrolysis of Elkol Subbituminous Coal

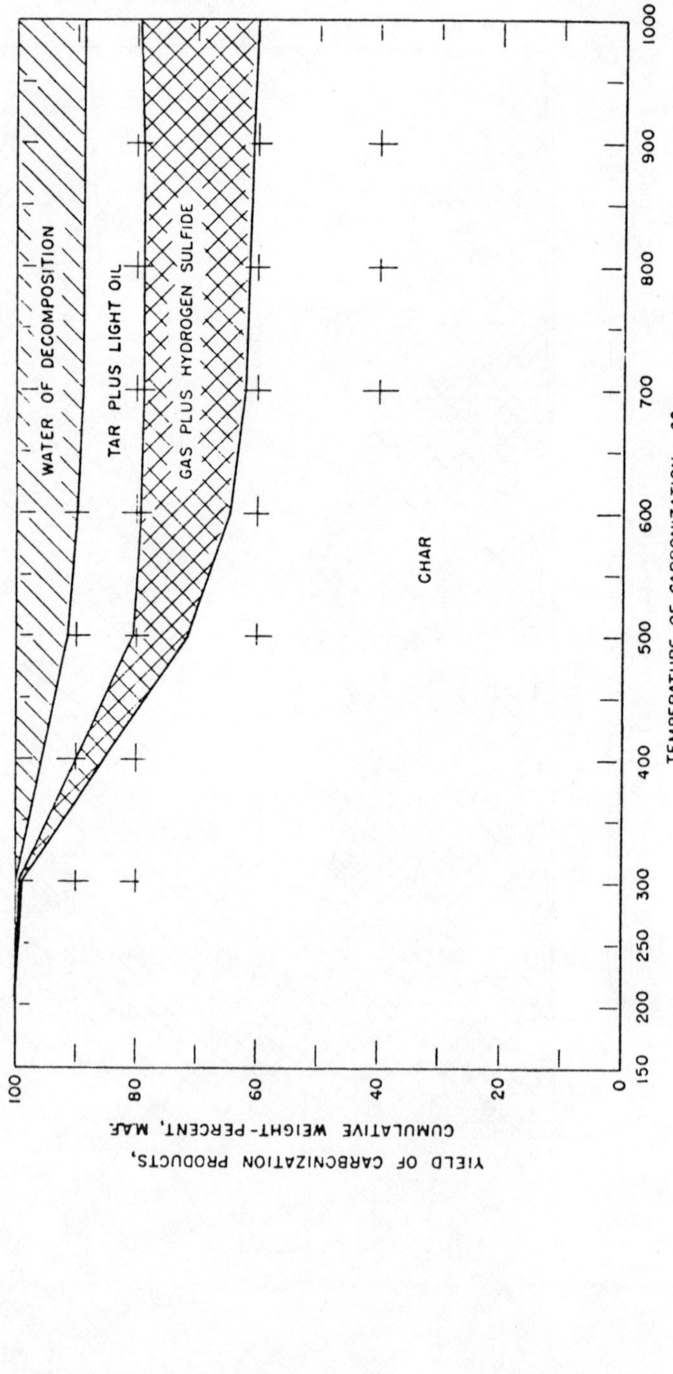

Source: Reference 4

ample, the Dakota Star lignite produced a distillate rich in tar acids boiling above 270°C while the Sandow lignite tar distillate was rich in neutral oils above 270°C, as was the Delcarbon high-volatile B bituminous coal. Some of these differences were attributed by the researchers to differences in carbonization conditions, which can "overrule" the influence of coal rank.[4]

Figure 3.5.5.6

Effect of Coal Rank on Low-Temperature Tar Composition (atmospheric distillation)

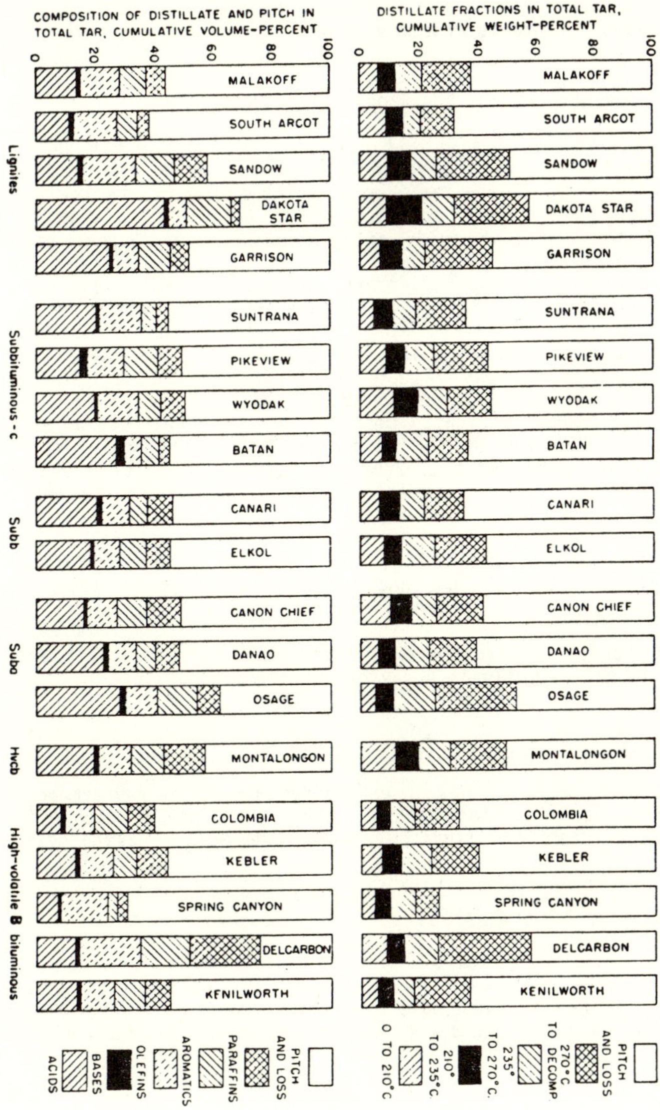

Source: Reference 4

3.5.5.3 Environmental Control Technology

The environmental problems associated with pyrolysis plants are very similar to those faced in gasification and liquefaction processes, which were discussed in sections 3.5.3.3 and 3.5.4.3. In particular, the waste water cleanup technology described in the gasification section will be of major importance to pyrolysis plants. Foul odors arising from cooling tower operation or other unit operations may also be a significant consideration.

As is the case in coal gasification and liquefaction, pyrolysis processes produce liquids that are known to have carcinogenic or toxic effects. Better techniques to identify and minimize health effects associated with worker or public exposure to coal liquids will be needed as the coal-based synfuels industry grows.

3.5.5.4 Effects of Low-Rank Coal Properties

Coal rank has a significant influence on the behavior of coal during pyrolysis, and on the yields and properties of the products. The dominant properties of low-rank coals in this regard are their high inherent moisture and oxygen contents, which translate to relatively high yields of water and low yields of char compared to higher rank coals.

Generally speaking, all coals having an ASTM classification up to and including high-volatile C bituminous can be considered suitable fuels for rapid carbonization at low temperature in a fluidized or entrained state. When such coals are carbonized, the heating value of the char is approximately 14,000 Btu per pound from any of the fuels considered, because of all the water and much of the oxygen have been evolved during carbonization. Despite the great variation in physical and chemical properties of the different lower rank coals in their raw state, the chars produced therefrom are substantially of the same basic quality. They are characterized by their high reactivity and ease of ignition.[4]

On the operational side, low-rank coals have the significant advantage of being non-caking, and therefore amenable to almost any kind of reactor design without the need for pretreatment.

The various factors which distinguish low-rank coals from high-rank coals in their pyrolysis behavior are itemized below.

Initial Decomposition Temperature

The rank of coal has a pronounced impact on the temperature at which thermal decomposition begins. For brown coals decomposition begins at about 270°C; for lignites, 300°C; and weakly caking coals, about 400°C.

Yields of Primary Products

The yield of primary products of distillation from different coals varies widely. Table 3.5.5.2 shows the analysis of and products from seven United States coals varying in rank from lignite to high-volatile bituminous A. The carbonization data on these coals were obtained from Fischer assay tests at 500°C[5]. The table shows that the yield of tar and light oil varies from 9.3 to 44 gallons per ton, depending upon the rank of the coal. It is interesting to note the distribution of heat in the products from the low-rank coals compared with those from the high rank coals. In the lignites, 76 to 80 percent of the heat remains in the char, which represents about 45 percent of the weight of the raw coal, but in the bitumimous coals the char represents about 70 percent of the heat and about 70 percent of the yield on a weight basis. Carbonization upgrades the lignites considerably by driving off inert gases and water, so the heating value of the char is just slightly lower than the heating value of chars from the higher rank coals.

Table 3.5.5.2

Yields of Primary Products From Low-Temperature Carbonization Assay of Typical Coals at 500°C

	Bituminous			Subbituminous		Lignite	
Source	Penna.	Utah	Illinois	Wyoming	Wyoming	Texas	N. Dakota
Classification:	Hvab2/	Hvbb2/	Hvbb2/	Suba2/	Subb2/	Lig.2/	Lig.2/
B.t.u. per lb. MAF1/	14,810	14,430	14,350	13,360	12,640	12,900	12,000
Oxygen, percent MAF	7.5	11.4	10.0	15.6	18.1	17.0	22.0
Analysis, percent:							
Moisture	2.0	4.6	7.9	20.6	32.6	35.2	37.0
Volatile matter	41.3	44.4	32.1	35.5	28.7	28.2	27.0
Fixed carbon	48.7	46.4	47.7	40.9	32.9	28.2	30.2
Ash	8.0	4.6	12.3	3.0	5.8	8.4	5.8
	100.0	100.0	100.0	100.0	100.0	100.0	100.0
B.t.u. per pound ...	13,320	13,100	11,540	10,200	7,790	7,260	6,800
Yields, percent:							
Char	69.7	65.0	72.0	58.2	48.0	45.7	45.0
Tar + L.O.	17.8	19.6	10.0	10.4	6.7	7.3	4.0
Gas	5.8	5.8	5.0	5.4	7.4	7.0	8.2
Water	6.7	9.6	13.0	26.0	37.9	40.0	42.8
	100.0	100.0	100.0	100.0	100.0	100.0	100.0
Tar + L.O. gal./ton.	41.0	44.0	24.3	24.0	15.4	17.0	9.3
Potential heat distribution, percent:							
Char	69.5	68.0	77.0	76.0	77.0	76.0	80.7
Tar + L.O.	22.0	23.0	14.3	17.3	14.6	17.0	10.7
Gas	7.0	7.0	7.2	6.3	6.0	5.0	6.6
	98.5	98.0	98.5	99.6	98.0	98.0	98.0
Char......B.t.u./lb.	13,270	13,700	12,350	13,300	12,500	11,430	12,100

1/ Moisture- and ash-free.
2/ Hvab = high-volatile A bituminous; Hvbb = high-volatile B bituminous; Suba = subbituminous A; Subb = subbituminous B; Lig. = lignite.

Source: Reference 5

Table 3.5.5.3 presents some additional yield data to illustrate the same points, in this case also indicating wide ranges of variation observed for coals of the same rank.

Char Properties

As indicated above, the chars from low-rank coals are typically only slightly lower in heating value than the chars from high-rank coals. The most noticeable properties of low-rank coal chars are their high reactivity and ease of ignition. Chars of low-rank coals tend to be pyrophoric; that is, room-temperature char can be so reactive that it begins to burn when exposed to air. Low-rank coal char is also typically a dusty material, as compared to the agglomerated material produced from caking coals. These differences represent both problems (storage and handling) and opportunities (use in systems where very high reactivity is important). Low-rank coal char is also significantly more reactive to steam and hydrogen (e.g., in gasification) than char from high-rank coals.

Table 3.5.5.3

Average Yields and Range of Yields of Fischer Assay of Various Coals

(As-received basis; maximum temperature, 500°C)

Rank of Coal*	Coke (%) No. of Samples	Coke (%) Average	Coke (%) Range	Tar (gal/ton) No. of Samples	Tar (gal/ton) Average	Tar (gal/ton) Range	Light Oil (gal/ton) No. of Samples	Light Oil (gal/ton) Average	Light Oil (gal/ton) Range
Sa	–	–	–	2	0.7	–	2	0.03	–
Lvb	17	89.7	85.8-93.3	17	8.6	6.3-12.7	17	1.02	0.73-1.61
Mvb	30	83.3	77.4-90.4	30	18.9	9.7-25.6	30	1.67	1.07-2.30
Hvab	134	75.5	68.8-81.4	134	30.9	22.9-40.7	134	2.29	1.48-3.28
Hvbb	11	70.4	66.0-73.2	12	30.3	24.3-43.1	11	2.18	1.55-3.44
Hvcb	7	67.1	65.4-68.6	12	27.0	18.5-38.8	7	1.88	1.29-2.71
Hvcb or Suba	1	59.1	–	5	20.5	18.4-24.4	5	1.65	1.36-1.93
Suba	–	–	–	2	17.8	17.6-18.0	2	1.35	1.33-1.36
Subb	6	57.6	54.8-59.9	7	15.4	13.2-16.7	7	1.33	1.14-1.55
Lignite	1	36.5	–	7	15.2	6.7-27.0	7	1.19	0.63-1.89
Cannel	7	58.8	44.1-69.0	8	73.5	53.7-108.3	8	5.06	3.66-7.42

Rank of Coal*	Gas (ft^3/ton) No. of Samples	Gas (ft^3/ton) Average	Gas (ft^3/ton) Range	Water (%) No. of Samples	Water (%) Average	Water (%) Range
Sa	–	–	–	–	–	–
Lvb	17	1,760	1,600-1,960	17	3.2	1.1-6.6
Mvb	30	1,940	1,390-2,240	30	4.1	2.8-7.0
Hvab	134	1,970	1,690-2,360	134	6.0	3.0-9.2
Hvbb	10	2,010	1,660-2,420	11	11.1	10.2-13.1
Hvcb	7	1,800	1,560-2,070	7	15.9	12.0-19.1
Hvcb or Suba	1	2,660	–	1	23.4	–
Suba	–	–	–	–	–	–
Subb	6	2,260	1,830-2,760	6	27.8	23.3-30.4
Lignite	1	2,100	–	1	44.0	–
Cannel	2	1,810	1,500-2,120	7	3.7	2.0-4.8

*Sa is semianthracite; Lvb is low-volatile bituminous; Mvb is medium-volatile bituminous; Hvab is high-volatile A bituminous; Hvbb is high-volatile B bituminous; Hvcb is high-volatile C bituminous; Suba is subbituminous A; and Subb is subbituminous B.

Source: Reference 2

Tar Properties

The chemical composition and physical properties of tars produced in devolatilization of lignites and subbituminous coals are significantly different from tars produced from high-rank coals. Because of the great number and complexity of chemical compounds present, it is difficult to present these differences accurately without going into a great deal of detail. In addition, the significance of different tar properties is a strong function of the intended utilization of the product. A complete characterization of a low-temperature tar from a North Dakota lignite is presented in Reference 6.

Caking Characteristics

As higher-rank coals are heated, they pass through a temperature range in which they become plastic and agglomerate to form coke. Low-rank coals do not have this property. Both lignites and subbituminous coals may, therefore, be utilized directly without pretreatment in most pyrolysis processes. If caking coals are to be pyrolyzed, this property must be destroyed by a separate process treatment; or the particles so diluted in the reactor that agglomeration is prevented; or mechanical stirring must be used to break agglomerates.

Mineral Matter Effects

The inorganic components of the coal remain in the char fraction, and the effects associated with the highly dispersed, alkaline nature of low-rank coal mineral matter will therefore be observed primarily in the processes in which the char is utilized. These effects are discussed in the sections on combustion, gasification, and liquefaction presented earlier in this report.

3.5.5.5 Current Projects

Coal pyrolysis processes in the United States and Canada have operated largely on an experimental basis, with two exceptions. The two processes that have been utilized commercially are the Disco process of the Consolidation Coal Company, located near Pittsburgh, Pennsylvania, and the stoker carbonization process, operated by Shawinigan Chemicals Ltd. in Canada.

In addition to these plants, the Parry process was operated experimentally on lignite by the Texas Power and Light Company at the power installation for the Aluminum Company of America at Rockdale, Texas. The Lurgi-Spulgas process is used by Husky Corporation at Dickinson, North Dakota to produce lignite char for barbeque briquet manufacture. A process for producing a metallurgical formed coke from Wyoming subbituminous coal is operated by the FMC Company and the United States Steel Corporation at Kemmerer, Wyoming.[2] Production of a low-temperature char is one of the first steps in this process.

There are three major pyrolysis R&D projects currently underway in the U.S.:

- COED
- TOSCOAL
- Occidental Research Corporation Flash Pyrolysis

Other research and development work is being conducted on advanced heating techniques employing electric arcs and laser radiation but these are in a very early stage of development.[7]

COED

Project COED (Char-Oil-Energy-Development) was initiated by FMC Corporation at Princeton, New Jersey in 1962 under OCR sponsorship. It is a process for converting coal to char, oil, and gas by reacting the coal in a multistage, fluidized bed system as illustrated in Figure 3.5.5.7. The pyrolysis-derived oil is hydrotreated to produce a synthetic crude oil. The product gas can be upgraded to a pipeline gas or utilized as syngas or fuel. The char product can be utilized as a fuel for power generation if the sulfur content is low enough, or gasification with sulfur removal. (Project COGAS addressed the gasification of char from this process.)

Dried crushed coal is treated in four fluidized bed stages at successively higher temperatures until a major fraction of the volatile matter of the coal is evolved. Heat for this pyrolysis is obtained by burning a portion of char with oxygen in the last stage. Hot gases from the last stage flow countercurrently to the coal, acting as the fluidizing gas and heat supply for the third and second stages in order. Hot char from the fourth and third stages is recycled to supplement the heat from the gases. The first stage fluidizing medium is supplied by burning a portion of the char or gas with air. Gas and oil are recovered by cooling and condensing the volatiles from the pyrolysis. A 36 TPD pilot plant has been operated since August 1979.[8]

Figure 3.5.5.7

Schematic Flow Diagram of the COED Process

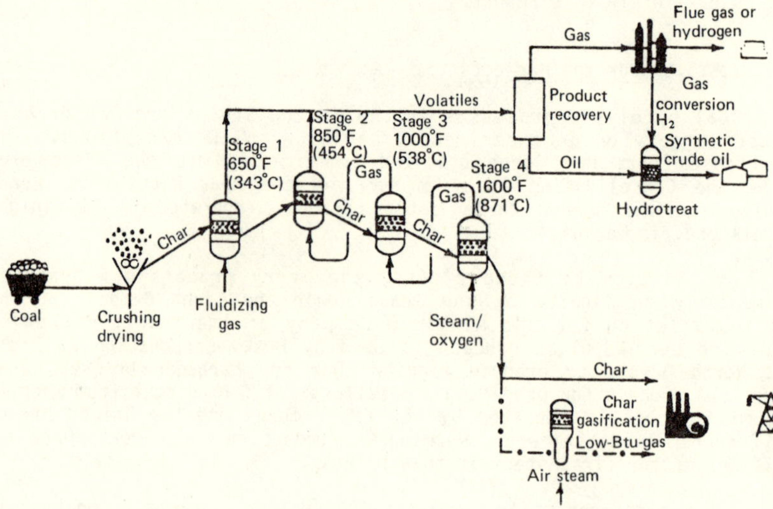

Source: Reference 1

Synthetic crude-oil yields from the COED process vary from 1 to 1.5 barrels per ton of bituminous coal, depending on the coal feed type. The off-gas yield is 8000 to 10,000 standard cubic feet per ton of coal with a heating value of 500 Btu/SCF. Char yield is about 0.5 tons per ton of coal. Gasification of this char yields about 56,000 standard cubic feet per ton of coal with a heating value of 220 Btu/SCF.[1]

Over 20,000 tons of coal have been processed in the pilot plant: high-volatile bituminous coals from Colorado, Utah, Illinois, and Kentucky; subbituminous coals from Wyoming; and lignite from North Dakota. Reliable operation of the four fluidized beds with the transfer of solids and gases between them was demonstrated over a variety of processing conditions.

Gaseous products from the process could be sold as fuel gas or reacted to produce hydrogen. The hydrogen would be used in hydrotreating the liquid products to remove nitrogen, sulfur, and oxygen. Hydrotreating upgrades the gravity, pour point, and viscosity of the liquid product to levels comparable to petroleum crude oil. The char could be utilized directly for power generation or reacted with air and steam in the COGAS process to produce low-Btu gas.[1]

TOSCOAL[6]

The Toscoal process is an offspring of the Tosco II oil-shale retorting process developed by The Oil Shale Corporation. The Toscoal process objective is to upgrade the low heating value of coal, especially low-sulfur western coal, through pyrolysis.

The Toscoal process scheme utilized in a 25 ton-per-day pilot plant is shown in Figure 3.5.5.8. In the process, coal is heated and pyrolyzed

Figure 3.5.5.8

Oil Shale Corporation TOSCOAL Pyrolysis-Type Fuel Gas From Coal Process

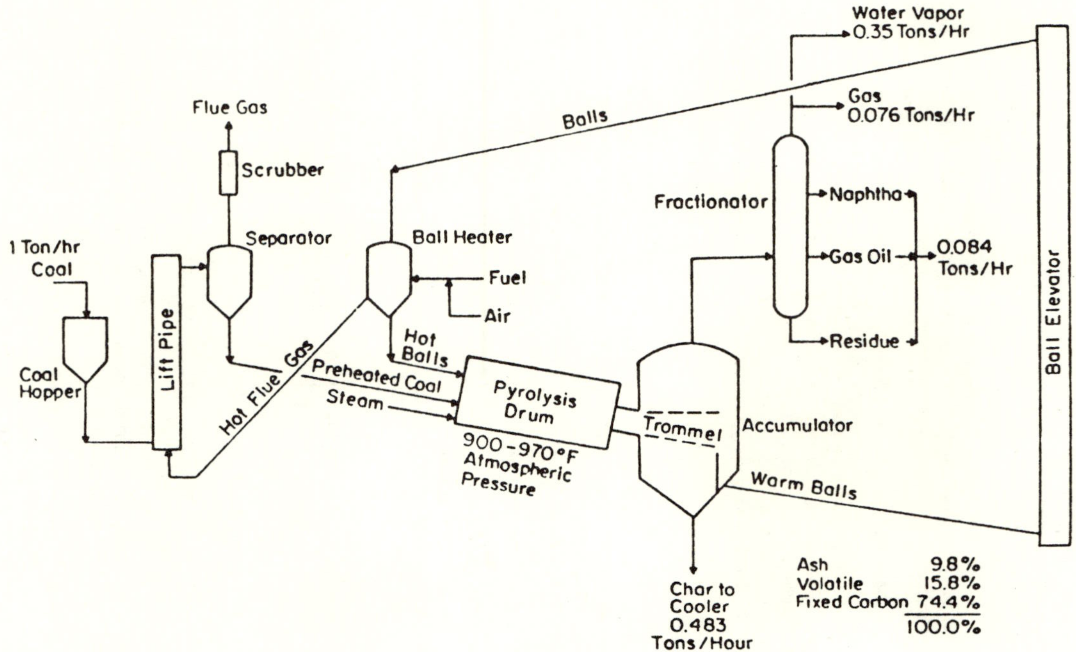

Source: Reference 1

by contact with heated ceramic balls in a rotating pyrolysis drum or retort. The ceramic balls are separated from the char in a trommel screen at the drum exit. The ceramic balls are conveyed to the ball heater and reheated for another cycle through the retort. The hot char from the retort is cooled in a rotary-tube cooler.

Feed coal is dried and preheated in the lift pipe by entrainment in a stream of hot flue gas from the ballheater. The preheating step reduces the thermal load on the retort and increases retort throughput for a given circulation of ceramic balls. The preheat step also provides an efficient way to recover waste heat from the ball-heater flue gas. Supplemental fuel is required in the preheater to dry coals with high moisture content. The optimum coal preheat temperature is generally the highest temperature attainable without excessive hydrocarbon vapor emission from the preheat scrubber.

Coal feed size of 3/4-inch X 0 has been used in the pilot plant, although coal sized to pass 1/2-inch is preferred. Upper feed size was limited in the pilot plant by the lifting capacity of the entrained preheater. Coal fines do not adversely affect retort operation, but crushing with minimum fines production is desirable to avoid heavy dust-collector loading. The coal decrepitates and is attrited in both the preheater and retort so that the average particle size of the char is less than 1/4-inch.

Tar vapor is cooled and condensed in the fractionator in which top temperature is maintained so that all pyrolysis water and steam condensate are collected in the overhead accumulator. The aqueous phase from the accumulator is treated for chemical recovery and/or for disposal. Gas from the accumulator can be treated for sulfur removal and sold or used as fuel for the ball heater.

The Toscoal process has considerable flexibility in pyrolysis temperature and in the distribution and characteristics of the products produced. Pyrolysis temperature of 800 to 1000°F (427 to 538°C) is the range of general interest. At retorting temperatures above 1000°F, the char (from subbituminous coal) contains less than 16 percent volatile matter and would require modifications in most combustion systems to be used satisfactorily. Retort throughput declines and operating costs increase as retort temperature is increased. Tar yield increases with retort temperature, but the rate of increase declines considerably at temperatures above 1000°F. Gas yield also increases with temperature, and the rate increase above about 1000°F corresponds with probable cracking of the tar.

At temperatures below 800°F, tar and gas yields diminish sharply, and the principal product is essentially dried coal, containing substantially all the initial volatile matter.

The Toscoal process has been applied only to noncaking coals. Caking coals would require some oxidative pretreatment before processing in the Toscoal process.

Results of tests conducted on a Wyoming subbituminous coal in the Toscoal 25-ton-per-day pilot plant are shown in Tables 3.5.5.4 through 3.5.5.7. The chars produced from subbituminous coals in the temperature range of 800 to 1000°F are relatively reactive, and require care in storage and transportation to avoid spontaneous ignition.

Since air or flue gas is excluded from the Toscoal retort, the gas produced is not diluted and has a high heating value. The exclusion of air also simplifies tar condensation.

Table 3.5.5.4

Toscoal Pilot Plant Test Results

Feedstock: Wyoming Subbituminous Coal

Proximate Analysis (wt %)		Ultimate Analysis (wt %)	
Moisture	30.0	Carbon	46.4
Ash	5.3	Hydrogen	2.8
Volatile matter	30.7	Oxygen	14.7
Fixed carbon	34.0	Nitrogen	0.7
Gross heating value, Btu/lb	8,140	Sulfur	0.3
		Chlorine	0.0
		Moisture	30.0
		Ash	5.3

Retort Product Yield (lb/ton of as-mined coal)

Retort Temperature	800°F (427°C)	900°F (482°C)	970°F (521°C)
Char	1,049	1,012	969
Gas (C_4 and lighter)	119	157	126
scf/ton	1,250	1,777	1,625
Oil (C_4 and heavier)	114	143	186
gal/ton	13.2	17.4	21.7
Water*	702	702	702
Unaccounted	16	14	17

*Yield of retort water estimated from Fischer assay of feed coal. Total water yield in pilot plant includes both retort water and steam condensate from retort seals.

Table 3.5.5.5

Toscoal Char Properties

Char Properties	Retort Temperature: 800°F (427°C)	900°F (483°C)	970°F (521°C)
Proximate, wt %			
Moisture	0.0	0.0	0.0
Ash	12.4	10.0	9.8
Volatile matter	25.3	19.7	15.9
Fixed carbon	62.3	70.3	74.3
Total	100.0	100.0	100.0
Ultimate, wt %			
Carbon	68.8	74.7	77.5
Hydrogen	3.4	3.0	2.9
Oxygen	13.3	11.8	8.3
Nitrogen	1.0	1.2	1.3
Sulfur	0.5	0.2	0.3
Chlorine	0.0	0.0	0.0
Moisture	0.0	0.0	0.0
Ash	12.4	10.0	9.8
Total	99.4	100.9	100.1
Other data			
Equilibrium moisture, wt %	10.0	10.8	9.9
Hardgrove grindability	78.2	49.1	45.6
Heating values			
Gross, Btu/lb	11,826	12,560	12,963
Net, Btu/lb	11,516	12,280	12,693
Bulk density (¼" x 0) packed, lb/ft^3	51.2	48.8	47.8

Source: Reference 1

Table 3.5.5.6

Toscoal Gas Analysis

	Retort Temperature 800°F (427°C)	900°F (483°C)	970°F (521°C)
Components, mol %			
H_2	0.8	1.0	7.8
CO	18.0	17.3	18.4
CO_2	51.1	42.3	36.4
H_2S	1.7	1.3	0.3
C_1	16.9	22.0	24.9
C_2	3.6	4.7	4.4
C_2-	1.9	1.9	2.4
C_3	1.3	2.2	1.2
C_3-	1.6	3.7	1.6
iC_4	0.1	0.1	0.0
C_4	0.3	2.0	1.1
C_5	1.0	1.8	0.7
C_6	0.7	0.6	0.4
C_7	0.5	0.1	0.3
C_{8+}	0.2	0.0	0.1
Total	99.7	101.0	100.0
Average molecular weight	35.9	35.0	30.6
Weight percent carbon	40.5	45.9	44.7
Heating values, calculated:			
Gross, Btu/scf	534	717	630
Net, Btu/scf	494	663	580
Heating values, calculated with CO_2 and H_2S removed:			
Gross, Btu/scf	1,113	1,234	995
Net, Btu/scf	1,029	1,138	920

Table 3.5.5.7

Toscoal Oil Properties

Oil Properties	Retort Temperature 800°F (427°C)	900°F (482°C)	970°F (521°C)
Ultimate wt %			
Carbon	81.4	80.7	80.9
Hydrogen	9.3	9.1	8.7
Oxygen	8.3	9.4	9.3
Nitrogen	0.5	0.7	0.7
Sulfur	0.4	0.2	0.2
Chlorine	0.0	0.0	0.0
Ash	0.0	0.2	0.1
Total	99.9	100.3	99.9
Heating values			
Gross, Btu/lb	16,590	16,217	15,964
Net, Btu/lb	15,740	15,372	15,160
API gravity			
Primary oil	7.9	4.5	1.9
Primary oil calculated with C_4 and heavier components of gas added	13.2	12.1	6.2
Pour point	90°F (32.2°C)	100°F (37.8°C)	95°F (35.0°C)
Conradson carbon, wt %	7.6	9.9	11.4
Distillation, vol %			
2.5	413°F (211.8°C)	420°F (216°C)	390°F (199°C)
10	490°F (254°C)	475°F (246°C)	405°F (207°C)
20	575°F (301.5°C)	550°F (288°C)	455°F (235°C)
30	645°F (340.5°C)	625°F (329.5°C)	545°F (285°C)
40	710°F (371°C)	700°F (371°C)	640°F (388°C)
50	765°F (407°C)	775°F (413°C)	725°F (385°C)
Viscosity (Saybolt Universal Seconds)			
180°F (82°C)	122	123	128
210°F (99°C)	63	66	69

Source: Reference 1

Occidental Research Corporation Flash Pyrolysis

Occidental Research Corporation has been developing a flash pyrolysis process for producing liquids and gases since 1969. The process is based on the concept of partial gasification in which direct yields of methane and other hydrocarbons are obtained by rapid pyrolysis of coal.[8]

In the process, crushed coal is introduced into a pyrolysis vessel in a stream of recycled gas where it is pyrolyzed at 1100°F through contact with hot char from the char heater vessel. Effluent from the pyrolysis reactor is sent to a series of cyclones for separation of gas and entrained char. A portion of the char is cooled as product char. The remaining char is sent to the char heater where a portion is burned with air to heat the contents to 1400°F. Hot char is cycled to the pyrolysis reactor.

The gas from the cyclones is cooled and scrubbed. Tar is removed in a gas-liquid separator and after acid-gas removal, a portion of the gas is separated as product gas (700 BTU/SCF) to be utilized as fuel gas or upgraded to pipeline quality by methanation. The remaining portion of the gas is sent to a hydrogen plant for generation of hydrogen which is used in the tar hydrotreater for the production of liquid products.[7] A flow diagram of the process is shown in Figure 3.5.5.9.

Figure 3.5.5.9

Occidental Research Corporation Flash Pyrolysis Process Schematic

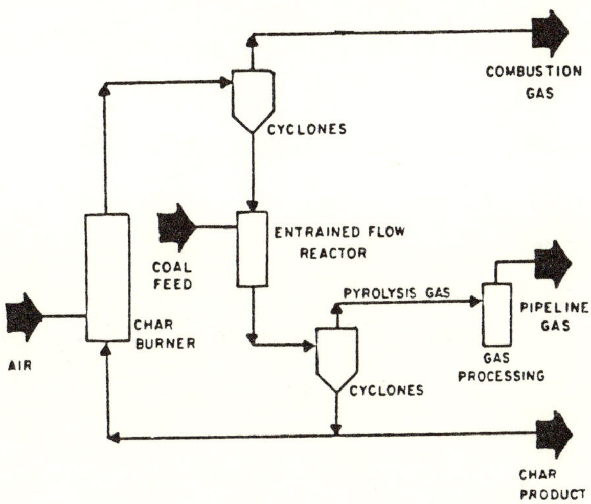

Source: Reference 9

The objective of rapid heating is to minimize the production of gas. The product yields and compositions from PDU runs with a western subbituminous coal (Table 3.5.5.8) are given in Tables 3.5.5.9 through 3.5.5.12. The gas yield increases with temperature but a limit exists at 1750°F due to the low ash softening temperature of this coal.

Table 3.5.5.8
Analysis of Coal Used in Occidental Research Corporation Flash Pyrolysis PDU Baseline Runs
(Subbituminous Coal, Monarch Seam, Big Horn Coal Company, Pit No. 1, Sheridan County, Wyoming)

Proximate, wt.%	As Fed, lb.	Moist. Free, lb.	M.A.F.[a] lb.
Moisture	18.45	--	--
Ash	4.99	6.1	--
Volatile Matter	33.77	41.4	44.1
Fixed Carbon	42.79	52.5	55.9
	100.00	100.0	100.0
Ultimate, wt.%			
C	57.02	69.02	74.5
H	5.90	4.72	5.0
O[b]	30.70	17.52	18.7
N	1.01	1.24	1.3
S	.37	.45	.5
Ash	5.0	6.13	--
	100.00	100.00	100.00
Fisher Assay, wt.%			
Char	56.8	69.6	67.1
Tar	8.9	10.9	11.8
Water	26.3	9.7	10.5
Gas	8.0	9.8	10.6
	100.0	100.0	100.0

Notes: [a]Moisture, ash free
[b]By difference
Source: Reference 10

Table 3.5.5.9
Overall Plant Material Balance,
Occidental Research Corporation PDU Run 122-77
Basis: 100 lbs. of coal as fed, inert gas-free

Input, lbs.	As Fed, lb.	Dry Basis, lb.	M.A.F., lb.
Coal	100.0	100.0	100.0
Air	45.5	52.8	56.6
Total	145.5	152.8	156.6
Output, lbs.			
Tar	6.1	7.1	7.6
Gas	16.3	18.9	20.3
Char	34.2	36.6	35.7
Water	18.8	5.7	6.1
Flue Gas	63.1	73.3	78.6
Total	138.5	141.6	148.3

Source: Reference 10

Table 3.5.5.10

Pyrolysis Gas Stream Composition
Occidental Research Corporation PDU
Run 122-77

Composition, Vol. %	Pyrolysis Gas[a]
CO	19.3
CO_2	21.7
H_2	33.5
CH_4	16.7
C_2H_4	3.2
C_2H_6	1.6
C'3s	3.0
C'4s	0.1
C'5s	--
H_2S	1.0
H_2O	--
O_2	--
N_2	--
Ar	--
Volume, SCF[c]	289
Weight, lbs[b]	16.3
Avg. Molecular Weight	21.4
Heating Value (HHV), Btu/scf	522

Notes: [a]Inert gas-free, dry basis
[b]No diluted air, dry basis
[c]Per 100 lbs. coal as fed

Table 3.5.5.11

Char Compositions[a]
Occidental Research Corporation PDU
Run 122-77

Pounds: Per 100 pounds as fed coal	34.23
Ultimate, (As Recovered), wt.%	
C	78.70
H	1.46
O[b]	2.10
N	0.99
S	0.61
Ash	16.10

Notes: [a]Initial start-up inventory
[b]By difference

Source: Reference 10

Table 3.5.5.12

Tar Composition,[a]
Occidental Research Corporation PDU
Run 122-77

Source:	Tar Collection System				
	1st Stage			2nd Stage	Vent Losses
	In	Out	Net Product Tar		
Pounds Per 100 feed coal	10.01	12.80	2.79	2.32	0.98
Ultimate, (As Recovered), wt.%					
C	91.11	88.30	78.26	84.61	87.44
H	5.87	6.37	8.14	7.75	8.07
O[b]	1.36	3.52	11.25	3.46	3.14
N	1.11	1.26	1.80	3.10	0.45
S	0.55	0.55	0.55	1.08	0.18
Ash	--	--	--	--	0.72

Notes: [a]Water and Char free
[b]By difference

Foreign Projects

A 50 ton per day pilot plant is presently in operation in the United Kingdom. The plant began operation in April of 1974 and is operated by the British Coal Utilization Research Association (BCURA). This system utilizes circulating char as the heat-carrier combined with slagging combustion to facilitate ash separation with minimum carbon loss. Char from the pyrolysis section of the COGAS process is fed to the gasifier where it is gasified with steam utilizing heat supplied by recycled hot char from the air-blown combustor. Char fines separated from the flue-gas stream are fed to the combustor. Coarse char is separated from the product-gas stream and is recycled to the lift tube where the hot combustion gases reheat it for recycling to the gasifier.

A pilot and full scale plant are scheduled for 1980 and 1985 respectively by the Australian Company CSIRO. They will use a flash pyrolysis process using no hydrogen or catalyst. Laboratory test units are presently operating using 0.5 tons per day fluidized and entrained bed reactors. The primary product from the plants will be char.

In West Germany there is considerable research and development aimed at low temperature carbonization of brown coal. Anticipated end use of the char produced would be as a sintering fuel, an adsorption material, as a source of carbon in the production of carbide and phosphorus or in the manufacture of ferro alloys. A 100,000 ton/year demonstration plant for the production of fine coke has been built recently and tests to produce form coke are being carried out on a 5 ton/day capacity pilot plant.[11]

References - Section 3.5.5

1. Considine, D.M., Ed. Energy Technology Handbook, McGraw-Hill Co., New York, 1977.

2. Lowry, H.H., Ed. Chemistry of Coal Utilization - Supplementary Volume, John Wiley & Sons, Inc., New York, 1963.

3. Parry, V.F. Low-Temperature Carbonization of Coal and Lignite for Industrial Uses, Bureau of Mines RI 5123, April 1955.

4. Gomez, M., J.B. Goodman, V.F. Parry. Low-Temperature Tar From Fluidized Carbonizing Reactors, Bureau of Mines RI 5302, February 1957.

5. Goodman, J.B., M. Gomez, V.F. Parry, and W.S. Landers. Low-Temperature Carbonization Assay of Coal in a Precision Laboratory Apparatus, Bureau of Mines Bulletin 530, 1954, 24 pp.

6. Fowkes, W.W., C.M. Frost, J.J. Hoeppner, W. Beckering, P.G. Freeman, and R.W. Youngs. An Examination of Low-Temperature Tar from a North Dakota Lignite, U.S. Bureau of Mines RI 5813, 1961, 27 pp.

7. Smith, Howard I., and G.J. Werner. Coal Conversion Technology, Noyes Data Corporation, Park Ridge, N.J., 1976

8. Evaluation of Coal Conversion Processes to Provide Clean Fuels, Part II, University of Michigan, PB-234-203, Electric Power Research Institute, February 1974.

9. Longanbach, J.R., and F. Bauer, "Fuels and Chemicals by Pyrolysis," Fuels and Chemicals, Presented at ACS Symposium Series 32, as Industrial and Laboratory Pyrolyses, eds., L.F. Albright and Billy L. Crynes, Washington, D.C., 1976.

10. Krell, E.W., et al. Flash Pyrolysis Coal Liquefaction Process Development, Annual Report for January-December 1977, Occidental Research Corporation, DOE FE-2244-20, 1978.

11. __________ Brown Coal Utilization, Research and Development, 204 pages, Verlag Die Braunkohle, Dusseldorf, Germany.

Appendix–Publications on the Technology of Low-Rank Coal 1939-1979

Compiled by Charles C. Boley, formerly Staff Engineer, Grand Forks Energy Technology Center, Grand Forks, ND

INTRODUCTION

The Grand Forks Energy Technology Center (GFETC) is one of five Energy Technology Centers in the U.S. Department of Energy (DOE). Designated as the lead agency in DOE on utilization technology for low-rank coal (lignite and subbituminous coal), GFETC has cooperated closely with industry in defining problems and performing research on low-rank coal for more than 40 years.

Early work was conducted in cooperation with the University of North Dakota, in Grand Forks. The present Center was dedicated in 1951 and was originally designated as the Charles R. Robertson Lignite Research Laboratory, reporting to the U.S. Bureau of Mines (Department of the Interior). Under several other titles, including Lignite Experiment Station, Grand Forks Lignite Laboratory, Grand Forks Coal Laboratory, and Grand Forks Energy Laboratory, the Center continued as a part of the Bureau of Mines until 1975, when it was administratively absorbed by the newly formed U.S. Energy Research and Development Administration (ERDA). In 1977 ERDA was, in turn, absorbed by DOE, and the Center's designation was adjusted to Grand Forks Energy Research Center, and subsequently to the present Grand Forks Energy Technology Center.

This appendix lists all technical publications that have been prepared and presented by the research staff of GFETC, under its present or an earlier title, from 1939 through 1979. The appendix also includes all publications written by the staff of the former Denver Coal Research Laboratory (Bureau of Mines), the functions of which were absorbed in 1965 by the then Grand Forks Coal Research Laboratory. Included in the listing are patents that have been granted to staff members on the basis of their research at the Center, as well as theses for advanced degrees granted to graduate students by the University of North Dakota, Montana State University, and Montana College of Mineral Science and Technology on the basis of research sponsored by the Center.

In accordance with the Center's major area of past and continuing effort, this booklet includes primarily publications related to low-rank coals. These publications have been grouped into six major interest areas, where they are listed by author for convenience of the reader. In cases where a single publication is of significant interest in more than one major area, it is listed in each. Finally, to complete the record of the Center's technical contributions, publications not related to low-rank coals are listed in a separate group.

GFETC PUBLICATIONS ON LOW-RANK COAL

Physical and Chemical Properties of Low-Rank Coal and Coal Products, Including Petrography and Analytical Techniques

1. Adams, R.L. Rheological Characteristics of Leonardite Gels and Lignite Slurries. M.S. Thesis, Univ. of N.D., Grand Forks, 1963, 38 pp.

4. Beckering, Willis. Intramolecular Hydrogen Bonding to π-Electrons in Ortho-Substituted Phenols. J. Phys. Chem., v. 65, 1961, pp. 206-208.
5. Beckering, Willis, and W.W. Fowkes. Analysis of Polyhydric Phenol Mixtures. Anal. Chem., v. 30, No. 8, 1958, pp. 1336-1338.
6. Beckering, Willis, and W.W. Fowkes. Infrared Spectra of Hydroxy-Aromatic Organic Compounds. BuMines RI 5505, 1959, 100 pp.
7. Beckering, Willis, and W.W. Fowkes. Infrared Spectra of Hydroxy-Aromatic Organic Compounds (Supplement to RI 5505). BuMines RI 5806, 1961, 34 pp.
8. Beckering, Willis, C.M. Frost, and W.W. Fowkes. Infrared Examination of Carbon-Hydrogen Stretching Frequency in Pyrocatechols, Guaiacols, and Phenols. Anal. Chem., v. 36, 1964, pp. 2412-2414.
9. Beckering, Willis, C.M. Frost, and W.W. Fowkes. Cell with Four Variable Compartments for Rapid Infrared Analysis. BuMines RI 6787, 1966, 8 pp.
10. Beckering, Willis, H.L. Haight, and W.W. Fowkes. Examination of Coal and Coal Ash by X-Ray Techniques. Sympos. on Techn. and Use of Lignite, Grand Forks, N.D., May 1969. BuMines IC 8471, 1970, pp. 89-102.
11. Beckering, Willis, and A.L. Severson. Matrix Corrections for X-Ray Fluorescent Analysis of Ash Elements in Coal. Pittsburgh Conf. on Anal. Chem. and Applied Spectroscopy, Cleveland, March 1975, 14 pp.
13. Benson, S.A. Examination of Low-Rank Coals and Ash Deposits Using Thermal Analytical Procedures. Proc., Dakota Acad. of Sci., v. 33, Apr. 1979, p. 14.
14. Benson, S.A., and Willis Beckering. Characteristics of Coals by Differential Scanning Calorimetry. Canadian Chem. Conf., Vancouver, June 1979, 18 pp.
15. Boteler, D.C. Assay Carbonization of Lignite from the Northern Great Plains Province. M.S. Thesis, Univ. of N.D., Grand Forks, 1966, 65 pp.
16. Boteler, D.C., W.R. Kube, and C.C. Boley. Bench-Scale Carbonization of Lignite from the Northern Great Plains Province: A Statistical Approach. BuMines RI 7386, 1970, 32 pp.
17. Cavanaugh, J.E. A Mineral and Elemental Analysis of the Moorhead Coal Field, Montana. M.S. Thesis, Mont. Coll. of Mineral Sci. and Tech., Butte, 1970, 70 pp.
18. Cooley, S.A., and R.C. Ellman. Analyses of Coal and Ash of Lignites and Subbituminous Coals of Eastern Montana. In "Energy Resources of Montana," Mont. Geol. Soc. Bull., 1975, pp. 143-158.
19. Crystal, J.T. The Removal of Sodium from Lignite by Ion Exchange. M.S. Thesis, Univ. of N.D., Grand Forks, 1970, 24 pp.
22. Fowkes, W.W. Low-Temperature Tar as a Raw Material. Sympos. on Techn. and Use of Lignite, Grand Forks, N.D., Apr. 1961. BuMines IC 8164, 1963, pp. 69-79.
23. Fowkes, W.W. A Note on the Mineralization of Lignite. Fuel, v. 51, No. 2, Apr. 1972, pp. 165-166.
24. Fowkes, W.W. Leonardite. In "Mineral and Water Resources of North Dakota," U.S. Geol. Surv. and U.S. BuReclam., GPO, 1973, 252 pp., pp. 72-75; also in N.D. Geol. Surv. Bull. 63, 1973.
25. Fowkes, W.W. Some Characteristics of Natural and Coal-Derived Humates: With Bibliography. DOE GFERC/IC-75/1, 1975, 67 pp.
27. Fowkes, W.W., and C.M. Frost. Leonardite: A Lignite Byproduct. BuMines RI 5611, 1960, 12 pp.
28. Fowkes, W.W., C.M. Frost, J.J. Hoeppner, Willis Beckering, P.G. Freeman, and R. W. Youngs. An Examination of Low-Temperature Tar from a North Dakota Lignite. BuMines RI 5813, 1961, 27 pp.

29. Fowkes, W.W., and J.J. Hoeppner. Sulfur in Lignite: Form and Transformations on Thermal Treatment. BuMines RI 5626, 1960, 15 pp.
30. Fowkes, W.W., J.J. Hoeppner, and Robert McMurtrie. Some Notes on the Determination of Moisture in Low-Rank Fuel (Lignite). Fuel, v. 36, No. 4, 1957, pp. 469-474. From BuMines RI 5215, 1956.
31. Fowkes, W.W., T.D. Wilhelm, and W.R. Kube. Sample Preparation and Electron Microprobe Examination of Lignite Ash Deposits. Metallography, v. 2, 1969, pp. 209-225.
32. Freeman, P.G. FIA Analysis of Coal Tar. ACS, Div. of Gas and Fuel Chem. Preprints, v. 2, No. 1, Sept. 1960, pp. 87-90.
33. Freeman, P.G. Catechols from the Carbonization of North Dakota Lignite. ACS, Div. of Fuel Chem. Preprints, v. 7, No. 1, 1963, pp. 139-143. Also in Proc., N.D. Acad. of Sci., v. 18, 1964, pp. 104-108.
35. Gomez, Manuel, J.B. Gayle, and A.R. Taylor. Heat Content and Specific Heat of Coals and Related Products. BuMines RI 6607, 1965, 45 pp.
36. Gomez, Manuel, and J.B. Goodman. Distillation Assays of Missouri River Basin Coals. BuMines RI 5009, 1953, 9 pp.
37. Gomez, Manuel, J.B. Goodman, and V.F. Parry. Low-Temperature Tar from Fluidized Carbonizing Reactors. BuMines RI 5302, 1957, 44 pp.
38. Gomez, Manuel, J.B. Goodman, and V.F. Parry. General Properties of Low-Temperature Tar. BuMines Bull. 569, 1958, 31 pp.
39. Gomez, Manuel, Kathleen Hazen, and E.A. Sondreal. Prediction of Sodium Concentration in Lignite. BuMines RI 7420, 1970, 17 pp.
40. Gomez, Manuel, W.S. Landers, J.L. Shultz, and A.G. Sharkey. Distribution of Phenols in Low-Temperature Tar from Low-Rank Coals. BuMines RI 6586, 1965, 29 pp.
41. Goodman, J.B., Manuel Gomez, and V.F. Parry. Determination of Moisture in Low-Rank Coals. BuMines RI 4969, 1953, 20 pp.
42. Goodman, J.B., Manuel Gomez, and V.F. Parry. Laboratory Carbonization Assay of Low-Rank Coals at Low, Medium, and High Temperatures. BuMines RI 5383, 1958, 49 pp.
43. Goodman, J.B., Manuel Gomez, V.F. Parry, and W.S. Landers. Low-Temperature Carbonization Assay of Coal in a Precision Laboratory Apparatus. BuMines Bull. 530, 1953, 24 pp.
44. Hoeppner, J.J., M.L. Opland, and W.W. Fowkes. Carbonization Characteristics of Some North-Central United States Lignites. BuMines RI 5260, 1956, 21 pp.
45. Honea, F.I., and E.A. Sondreal. Ash Properties and Combustion Characteristics of Gulf Province Lignites. SME/AIME Fall Meeting, Tucson, Ariz., Oct. 1979, (Minisymposium Preprint, Coal--Series No. 79-021), 10 pp.
46. Karner, F.R., and F.S. O'Toole. Secondary Pyrite, Barite, and Gypsum on Fracture Surfaces in Lignite at the Baukol-Noonan Mine, North Dakota. Proc., N.D. Acad. of Sci., v. 33, Apr. 1979, p. 71.
47. Karner, F.R., Gary Winbourn, S.F. White, and A.K. Gatheridge. Sodium and Calcium in Overburden, Lignite, and Underclay at the Beulah and Baukol-Noonan Mines, Mercer and Oliver Counties, North Dakota. Proc., N.D. Acad. of Sci., v. 33, Apr. 1979, p. 70.
48. Kauphusman, J.R. A Study of the Relative Reactivity of Carbon from Coal Using the Bouduard Reaction. M.S. Thesis, Univ. of N.D., Grand Forks, 1972, 62 pp.
49. Kube, W.R. Sorption of Water Vapor by Thermally Treated North Dakota Lignite. Proc., N.D. Acad. of Sci., v. 6, 1952, pp. 83-85.

50. Kube, W.R. Sorption of Water Vapor by Thermally Treated Lignite at Different Relative Humidities. Chem. and Eng. Data Ser., v. 2, No. 1, 1957, pp. 46-51.
51. Landers, W.S., J.B. Goodman, and D.J. Donaven. Low-Temperature Carbonization Assays of Coals and Relation of Yields to Analyses. BuMines RI 5904, 1961, 41 pp.
52. McDonald, J.L. Binding Properties of Leonardite in Lignite and Lignite Char Agglomerates. M.S. Thesis, Univ. of N.D., Grand Forks, 1963, 51 pp.
53. McNally, J.R. Activated Carbon from North Dakota Lignite. M.S. Thesis, Univ. of N.D., Grand Forks, 1970, 46 pp.
54. Medaris, J.A. Feasibility Characteristics of Synthetic and Natural Coal Ashes. M.S. Thesis, Univ. of N.D., Grand Forks, 1960, 24 pp.
55. Miller, D.J. Quantitative Analysis of Coal-Derived Liquids by Low Voltage Mass Spectroscopy. Proc., Dakota Acad. of Sci., v. 33, Apr. 1979, p. 55.
56. Miller, D.J., J.K. Olson, and H.H. Schobert. Mass Spectroscopic Characterization of Tars from the Gasification of Low-Rank Coals. ACS National Meeting, Washington, D.C., Sept. 1979, 10 pp.
57. Nallaperumal, U. Studies on Thermal Solution of Low-Rank Fuels. M.S. Thesis, Univ. of N.D., Grand Forks, 1959, 36 pp.
58. Ness, H.M., A.F. Volesky, and S.Y. Johnson. Physical and Chemical Characteristics of Fly Ash and Scrubber Sludge from Some Low-Rank Western Coals. Ash Management Conf., Texas A&M Univ., College Station, Sept.1978, 34 pp.
59. Parry, V.F. Correlation of Analysis of Coal with the Products of Carbonization in Externally Heated Retorts. BuMines RI 3482, 1939, 37 pp.
60. Parry, V.F., and J.B. Goodman. Friability, Slacking Characteristics, and Low-Temperature Carbonization Assays of Subbituminous Coals of the Denver (Colo.) Region. BuMines RI 3457, 1939, 12 pp.
61. Parry, V.F., and W.S. Landers. Size Consist, Chemical Analysis, and Physical Properties of 2-1/2-inch Subbituminous Slack from the Denver (Colo.) Region. BuMines RI 3655, 1942, 30 pp. Also in Proc., Rocky Mt. Coal Mining Inst., 1942.
62. Paulson, L.E., Willis Beckering, and W.W. Fowkes. Separation and Identification of Minerals from Northern Great Plains Province Lignite. Fuel, v. 51, July 1972, pp. 224-227.
63. Paulson, L.E., and W.W. Fowkes. Changes in Ash Composition of North Dakota Lignite Treated by Ion Exchange. BuMines RI 7176, 1968, 18 pp.
64. Porter, R.B., and R.C. Ellman. Laboratory Scale Carbonization of North Dakota Lignite: Influence of Temperature and Atmosphere on Product Distribution and Activity. ACS, Div. of Fuel Chem. Preprints, v. 8, No. 3, 1964, pp. 15-24.
65. Sanderson, D.M. Reaction of Lignite and Oxygen at Low Temperature: Heat of Reaction and Rate of Oxygen Consumption. M.S. Thesis, Univ. of N.D., Grand Forks, 1966, 28 pp.
66. Schiller, J.E. Composition of Coal Liquefaction Products. Hydrocarbon Processing, Jan. 1977, 6 pp.
67. Schiller, J.E. Analysis of Solvent-Refined Coal, Recycle Solvents, and Coal Liquefaction Products. ACS Sympos. on Anal. Chem. of Tar Sands and Oil Shale, Div. of Anal. Chem. and Petroleum Chem., New Orleans, March 1977; in Advances in Chem. Series, No. 170, 4 pp.
68. Schiller, J.E. Nitrogen Compounds in Coal-Derived Liquids. Anal. Chem., v. 49, Dec. 1977, pp. 2292-2294.
69. Schiller, J.E., and C.L. Knudson. Effect of Solvent on Molecular Composition in Coal Liquefaction. Fuel, v. 57, Jan. 1978, pp. 36-40.

71. Schiller, J.E., and D.R. Mathiason. Separation Method for Coal-Derived Solids and Heavy Liquids. Anal. Chem., v. 49, July 1977, p. 1225-1228.
72. Schiller, J.E., and D.H. Neal. Preparation of Hydroaromatic Compounds for Mass Spectrometry Standards in Coal Conversion Analysis. DOE GFERC/RI-75/2, 1975, 3 pp.
74. Sondreal, E.A., W.R. Kube, and J.L. Elder. Characteristics and Variability of Lignite Ash from the Northern Great Plains Province. Sympos. on Techn. and Use of Lignite, Bismarck, N.D., Apr. 1965. BuMines IC 8304, 1966, pp. 39-50.
75. Sondreal, E.A., W.R. Kube, and J.L. Elder. Analysis of the Northern Great Plains Province Lignites and Their Ash: A Study of Variability. BuMines RI 7158, 1968, 94 pp.
76. Thelen, R.C. Production of Activated Carbon from North Dakota Lignite. M.S. Thesis, Univ. of N.D., Grand Forks, 1963, 53 pp.
77. Traverse, Alfred. Petrographic Analysis of Column Sample and Run-of-Mine Lignite from Zap, Mercer County, North Dakota. Bull. Geol. Soc. Amer., v. 63, No. 12, part 2, Dec. 1952, pp. 1306-1307.
78. Traverse, Alfred. Coal Microscopy, An Important Research Technique. Consulting Eng., v. 4, No. 1, 1954, pp. 38-41, 78-79.
79. Traverse, Alfred. Applications of Coal Microscopy. Consulting Eng., v. 4, No. 2, 1954, pp. 32-35, 74.
82. Traverse, Alfred. A Method for the Petrographic Analysis of Commercially Delivered Lignite. Econ. Geol., v. 49, No. 1, 1954, pp. 92-100.
83. Traverse, Alfred. Behavior of Petrographic Components of North Dakota Lignite in Preparation, Low-Temperature Carbonization, and Steam-Drying. Econ. Geol., v. 50, No. 1, 1955, p. 102 (abstract).
84. Traverse, Alfred. Pollen Analysis of the Brandon Lignite of Vermont. BuMines RI 5151, 1955, 107 pp.
85. Traverse, Alfred, and E.S. Barghoorn. Micropaleontology of the Brandon Lignite, An Early Tertiary Coal in Central Vermont. J. Paleontol., v. 27, No. 2, March 1953, pp. 289-293.
86. Wolfe, D.F. Noncombustible Mineral Matter in the Pawnee Coal Bed, Powder River County, Montana. M.S. Thesis, Mont. Coll. of Mineral Sci. and Tech., Butte, 1969, 64 pp.
87. Youngs, R.W. Humic Acids from Lignite. Sympos. on Techn. and Use of Lignite, Grand Forks, N.D., Apr. 1961. BuMines IC 8164, 1963, pp. 87-93.
88. Youngs, R.W., and W.W. Fowkes. Synthesis of Ethylmethylphenols. Proc., N.D. Acad. of Sci., v. 19, 1965, pp. 100-103.

Preparation, Pulverization, Storage, Handling, and Transportation

101. Allen, R.R., and V.F. Parry. Storage of Low-Rank Coals. BuMines RI 5034, 1954, 31 pp.
102. Belter, J.W. A Review of the Use of Pulsating Combustors for Drying and Conveying. Proc., 1st Internat. Sympos. on Pulsating Combustion, Sept. 1971, Univ. of Sheffield, Sheffield, England; ed. by D. J. Brown; 21 pp.
103. Belter, J.W., Leroy Dockter, and R.C. Ellman. Operating Experience with Lignite Fueled Pulse-Jet Engines. ASME Winter Meeting, Los Angeles, Nov. 1969, 5 pp.
104. Belter, J.W., and R.C. Ellman. Grindability Testing of Lignite and Correlation to Pilot Plant Pulverizer Operation. AIChE Sympos. on Size Comminution, Houston, Dec. 1963, 7 pp.
105. Dockter, Leroy, J.W. Belter, and R.C. Ellman. Pulverizing Lignite in a Ball Mill. BuMines RI 7010, 1967, 13 pp.

106. Ellman, R.C. Current Problems in Preparing Lignitic Coals. Proc., Coal Conf., Univ. of Mo. Sch. of Mines and Met., Bull. 85, Tech. Ser., Dec. 1954, pp. 76-86.
107. Ellman, R.C. Lignite Preparation. Sympos. on Techn. and Use of Lignite, Grand Forks, N.D., Apr. 1961. BuMines IC 8164, 1963, pp. 35-44.
108. Ellman, R.C., and J.W. Belter. Grindability Testing of Lignites. BuMines RI 5167, 1955, 38 pp.
109. Ellman, R.C., and J.W. Belter. Observations on Large Stockpiles of Lignite at Garrison Dam, Riverdale, N.D.; Progress for Fiscal 1955. BuMines Prelim. Report 109, 1955, 17 pp.
111. Ellman, R.C., and J.W. Belter. Problems in Freezeproofing North Dakota Lignite. Joint Solid Fuels Conf., Quebec, Oct. 1957.
112. Ellman, R.C., and J.W. Belter. Observations on Large Stockpiles of Lignite at Garrison Dam, Riverdale, N.D.; Progress Report for June 1, 1955, to January 1, 1958. BuMines Prelim. Report 117, 1958, 26 pp.
113. Ellman, R.C., and J.W. Belter. Pulverizing Lignite in a North Dakota Powerplant. BuMines IC 8032, 1961, 31 pp.
114. Ellman, R.C., and J.W. Belter. Elimination of Agglomeration by Freezing of Lignite or Other Moisture-Containing or Wetted Aggregates During Shipment. U.S. Pat. 3,243,889, Apr. 5, 1966, 2 pp.
115. Ellman, R.C., and J.W. Belter. Bigger Markets Mean More Preparation for Lignite and Subbituminous Coal. Mining Eng., v. 22, No. 3, March 1970, pp. 47-50. Originally presented as: Preparation Practices in the Lignite and Subbituminous Coal Industry, at SME Fall Meeting, Salt Lake City, Sept. 1969, Preprint No. 69-F-362, 15 pp.
117. Ellman, R.C., J.W. Belter, and Leroy Dockter. Observations on Large Stockpiles of Lignite at Garrison Dam, Riverdale, ND. Progress Report - January 1, 1958 to July 1, 1960. BuMines Prelim. Rpt. 134, 1960, 24 pp.
118. Ellman, R.C., J.W. Belter, and Leroy Dockter. Comparative Pulverization Tests on Five Lignites Using In-the-Mill Drying. BuMines Prelim. Rpt. 144, Apr. 1962, 18 pp.
119. Ellman, R.C., J.W. Belter, and Leroy Dockter. Effects of In-the-Mill Drying on Pulverizing Characteristics of Lignite. BuMines RI 6074, 1962, 18 pp.
121. Ellman, R.C., J.W. Belter, and Leroy Dockter. Pulverizing Lignite in a Pilot-Plant Test Unit. Proc., Amer. Power Conf., v. 25, 1963, pp. 513-524.
122. Ellman, R.C., J.W. Belter, and Leroy Dockter. Freezeproofing Lignite. BuMines RI 6677, 1965, 28 pp.
123. Ellman, R.C., J.W. Belter, and Leroy Dockter. Long-Term Storage of Lignite at Garrison Dam, Riverdale, N.D. BuMines RI 7037, 1967, 56 pp.
124. Ellman, R.C., J.W. Belter, and Leroy Dockter. Lignite Pulverization: A Status Report. Sympos. on Techn. and Use of Lignite, Grand Forks, N.D., Apr. 1967. BuMines IC 8376, 1968, pp. 29-39.
125. Ellman, R.C., Leroy Dockter, and J.W. Belter. Pulverizing Lignite in a Ring-Roller Mill. BuMines RI 7631, 1972, 25 pp.
126. Ellman, R.C., L.E. Paulson, and S.A. Cooley. Commercial-Scale Drying of Low-Rank Western Coals: Part I. Rail Shipment Test Observations. Sympos. on Techn. and Use of Lignite, Grand Forks, N.D., May 1975. DOE GFERC/IC-75/2, 1975, pp. 312-340.
127. Gomez, Manuel, J.B. Goodman, and V.F. Parry. Changes Induced in Low-Temperature Tar by Oxidation and Storage. BuMines RI 5227, 1956, 12 pp.

128. Goodman, J.B., V.F. Parry, and W.S. Landers. Storage of Subbituminous Slack Coal in Open Pits. BuMines RI 3915, 1946, 37 pp.
129. Hanson, A.M. Methods of Determining the Size Distribution of Pulverized Lignite. M.S. Thesis, Univ. of N.D., Grand Forks, 1960, 51 pp.
131. Lammers, G.C., R.R. Allen, D.J. Donaven, E.O. Wagner, and V.F. Parry. A Study of the Feasibility of Hydraulic Transport of a Texas Lignite. BuMines RI 5404, 1958, 39 pp.
132. Landers, W.S., and V.F. Parry. Domestic Storage of Subbituminous Lump Coal and its Performance in a Hand-Fired Furnace. BuMines RI 3759, 1944, 17 pp.
133. Parry, V.F. Questions and Answers on Storage of Coal in the Rocky Mountain Area. BuMines IC 7214, 1942, 9 pp.
134. Parry, V.F., and J.B. Goodman. Storage of Subbituminous Coal in Bins. BuMines RI 3587, 1941, 10 pp.
135. Parry, V.F., L.C. Harrington, and A.W. Koth. Preparation of Stable Nonslacking Fuel by Steam-Drying Subbituminous Coal and Lignite. Trans. ASME, v. 64, 1942, pp. 177-183.
136. Paulson, L.E. Lignite Transportation by Pipeline: A Study Simulating Flow Characteristics. M.S. Thesis, Univ. of N.D., Grand Forks, 1964, 48 pp.
137. Paulson, L.E., S.A. Cooley, and R.C. Ellman. Shipment, Storage, and Handling Characteristics of Dried Low-Rank Coals. Sympos. on Techn. and Use of Lignite, Grand Forks, N.D., May 1973. BuMines IC 8650, 1974, pp. 49-75.
138. Paulson, L.E., S.A. Cooley, C.J. Wegert, and R.C. Ellman. Experiences in Transportation of Dried Low-Rank Coals. SME Fall Meeting, Salt Lake City, Sept. 1975, Preprint 75-B-337, 20 pp. SME/AIME Transactions, Dec. 1976.
139. Paulson, L.E., and R.C. Ellman. Reduction of Sodium in Lignite by Ion Exchange: A Pilot Plant Study. DOE GFETC/RI 79/1, 1979, 50 pp.
140. Sondreal, E.A. Studies on Low-Temperature Oxidation of Lignite as Related to Storage. Sympos. on Techn. and Use of Lignite, Grand Forks, N.D., Apr.-May 1963. BuMines IC 8234, 1964, pp. 59-73.
141. Sondreal, E.A., and R.C. Ellman. Laboratory Determination of Factors Affecting Storage of North Dakota Lignite: Computer Simulation of Spontaneous Heating. BuMines RI 7887, 1974, 83 pp.
83. Traverse, Alfred. Behavior of Petrographic Components of North Dakota Lignite in Preparation, Low-Temperature Carbonization, and Steam-Drying. Econ. Geol., v. 50, No. 1, 1955, p. 102 (abstract).

Combustion, Including Emission Control and Ash Fouling

13. Benson, S.A. Examination of Low-Rank Coals and Ash Deposits Using Thermal Analytical Procedures. Proc., Dakota Acad. of Sci., v. 33, Apr. 1979, p. 14.
201. Bradley, W.J., S. Panico, D.L. Keairns, R.A. Newby, N.H. Ulerich, G.M. Goblirsch, and W.M. Heher. Effect of Subbituminous Western Coal Ash on AFBC Power Generating Costs. AIChE Nat'l Meeting, April 1979, 44 pp.
202. Bredahl, R.G. Differential Thermal Analysis of Lignite Ash. M.S. Thesis, Univ. of N.D., Grand Forks, 1965, 41 pp.
203. Bucher, W.E. A Study of the Bulk Electrical Resistivity Characteristics of Fly Ash from Lignite and Other Western Coals. M.S. Thesis, Univ. of N.D., Grand Forks, 1970, 52 pp.
204. Evans, R.N. Dielectric Properties of Fly Ash from Western Coals. M.S. Thesis, Univ. of N.D., Grand Forks, 1978, 86 pp.

31. Fowkes, W.W., T.D. Wilhelm, and W.R. Kube. Sample Preparation and Electron Microprobe Examination of Lignite Ash Deposits. Metallography, v. 2, 1969, pp. 209-225.

205. Freeman, P.G., and W.W. Fowkes. Low-Temperature Ashing of Lignite in an Oxygen Plasma. Proc., N.D. Acad. of Sci., v. 24, 1971, pp. 119-124.

206. Goblirsch, G.M., R.W. Fehr, and E.A. Sondreal. Effects of Coal Composition and Ash Reinjection on Sulfur Retention Burning Lignite and Western Subbituminous Coals. Proc., Fifth Internat. Conf. on Fluidized-Bed Combustion, v. 2, Washington, D.C., Dec. 1977.

208. Goblirsch, G.M., and H.M. Ness. Fluidized-Bed Combustion Tests on Lignites and Subbituminous Coals at the Grand Forks Energy Technology Center. DOE Sympos. on Environmental Control Activities, Washington, D.C., Nov. 1978, 18 pp.

209. Goblirsch, G.M., and E.A. Sondreal. Effects of Operating Parameters on the Sulfur Retention on Alkaline Ash During Fluidized-Bed Combustion of a North Dakota Lignite. Proc., Fluidized-Bed Combustion Technology Exchange Workshop, April 1977, pp. 111-121.

211. Goblirsch, G.M., and E.A. Sondreal. Fluidized Combustion of North Dakota Lignite. Sympos. on Techn. and Use of Lignite, Grand Forks, N.D., May 1977. DOE GFERC/IC-77/1, 1978, pp. 82-99.

212. Goblirsch, G.M., and E.A. Sondreal. Low-Rank Coal Atmospheric Fluidized-Bed Combustion Technology. Sympos. on Techn. and Use of Lignite, Grand Forks, ND, May 1979. DOE GFETC/IC-79/1, 1979, pp. 75-107.

213. Gronhovd, G.H. Progress and Problems in Combustion of Lignite from the Northern Great Plains Area. SME/AIME Meeting, Minneapolis, Sept. 1968, Preprint 68-F-351, 15 pp.

214. Gronhovd, G.H., Willis Beckering, and P.H. Tufte. Study of Factors Affecting Ash Deposition from Lignite and Other Coals. ASME Winter Meeting, Los Angeles, Nov. 1969, Preprint 69-WA/CD-1, 9 pp.

215. Gronhovd, G.H., A.E. Harak, and L.E. Paulson. Ash Fouling Studies of North Dakota Lignite. Sympos. on Techn. and Use of Lignite, Grand Forks, N.D., Apr. 1967. BuMines IC 8376, 1968, pp. 76-94.

216. Gronhovd, G.H., A.E. Harak, and P.H. Tufte. Ash Fouling and Air Pollution Studies Using a Pilot-Plant Test Furnace. Sympos. on Techn. and Use of Lignite, Grand Forks, N.D., May 1969. BuMines IC 8471, 1970, pp. 69-88.

217. Gronhovd, G.H., P.H. Tufte, and S.J. Selle. Some Studies on Stack Emissions from Lignite-Fired Powerplants. Sympos. on Techn. and Use of Lignite, Grand Forks, N.D., May 1973. BuMines IC 8650, 1974, pp. 83-102.

218. Gronhovd, G.H., R.J. Wagner, and A.J. Wittmaier. Comparison of Ash Fouling Tendencies of High- and Low-Sodium Lignite from a North Dakota Mine. Amer. Power Conf., Chicago, Apr. 1966, pp. 632-645.

219. Gronhovd, G.H., R.J. Wagner, and A.J. Wittmaier. A Study of the Ash Fouling Tendencies of a North Dakota Lignite as Related to Its Sodium Content. Trans., Soc. of Mining Engrs. Quarterly, Sept. 1967, pp. 313-322.

221. Honea, F.I., S.J. Selle, and E.A. Sondreal. Factors Affecting Boiler Tube Fouling from Western and Gulf Coast Lignites and Subbituminous Coals. Paper 73, NACE Corrosion/79 Sympos., Atlanta, Ga., March 1979, 42 pp.

222. Honea, F.I., S.J. Selle, E.A. Sondreal, and V.H. Menze. Effects of Overfire Air and Low Excess Air on NO_x Emissions and Ash Fouling Potential for a Lignite-Fired Boiler. Amer. Power Conf., Chicago, Apr. 1978, 20 pp.

45. Honea, F.I., and E.A. Sondreal. Ash Properties and Combustion Characteristics of Gulf Province Lignites. SME/AIME Fall Meeting, Tucson, Ariz., Oct. 1979 (Minisymposium Preprint, Coal--Series No. 79-021), 10 pp.

223. Kube, W.R., G.H. Gronhovd, and J.L. Elder. Large Powerplants Fired With Dakota Lignite. North Dakota Quarterly, Summer, 1967, pp. 87-93.

132. Landers, W.S., and V.F. Parry. Domestic Storage of Subbituminous Lump Coal and Its Performance in a Hand-Fired Furnace. BuMines RI 3759, 1944, 17 pp.

224. Lukes, E.B. Removal of Sulfur Oxides from Synthetic Flue Gas by Lignite Fly Ash. M.S. Thesis, Univ. of N.D., Grand Forks, 1969, 47 pp.

225. Majkrzak, R.S. Fluidized Bed Combustion of High-Alkali-Ash, Low-Sulfur Western Coals. M.S. Thesis, Univ. of N.D., Grand Forks, 1976, 80 pp.

226. Ness, H.M. Scrubber Tests Show Good Results--Energy Research Center Heads Program. Minnkota Messenger (Minnkota Power Coop., Grand Forks, N.D.), March 1977, 3 pp.

227. Ness, H.M., F.I. Honea, E.A. Sondreal, and Philip Richmond. Pilot Plant Scrubbing of SO_2 with Fly Ash Alkali from North Dakota Lignite. Sympos. on Techn. and Use of Lignite, Grand Forks, N.D., May 1977. DOE GFERC/IC-77/1, 1978, pp. 100-122.

228. Ness, H.M., Philip Richmond, Glenn Eurick, and R. Kruger. Power Plant Flue Gas Desulfurization Using Alkaline Fly Ash from Western Coals. EPA Fifth Annual Sympos. on Flue Gas Desulfurization, Las Vegas, Nev., March 1979.

229. Ness, H.M., and S.J. Selle. Control of Western Power Plant Sulfur Dioxide Emissions: Development of the Ash-Alkali FGD Process and Dry Adsorption Techniques at the Grand Forks Energy Technology Center. DOE Sympos. on Environmental Control Activities, Washington, D.C., Nov. 1978, 20 pp.

231. Ness, H.M., S.J. Selle, and O.E. Manz. Power Plant Flue Gas Desulfurization for Low-Rank Western Coals. Sympos. on Techn. and Use of Lignite, Grand Forks, N.D., May 1979. DOE GFETC/IC-79/1, 1979, pp. 117-146.

232. Ness, H.M., E.A. Sondreal, F.Y. Murad, and K.S. Vig. Flue Gas Desulfurization Using Fly Ash Alkali Derived From Western Coals. ERDA-EPA Interagency Energy-Environment R&D Program Report, No. EPA-600/7-77-075, July 1977.

233. Ness, H.M., E.A. Sondreal, and P.H. Tufte. Status of Flue Gas Desulfurization Using Alkaline Fly Ash From Western Coals. EPA Sympos. on Flue Gas Desulfurization, New Orleans, March 1976, 56 pp.

58. Ness, H.M., A.F. Volesky, and S.Y. Johnson. Physical and Chemical Characteristics of Fly Ash and Scrubber Sludge from Some Low-Rank Western Coals. Ash Management Conf., Texas A&M Univ., College Station, Sept. 1978, 34 pp.

234. Odenbaugh, M.L. Sintering Characteristics of Lignite Ash. M.S. Thesis, Univ. of N.D., Grand Forks, 1966, 41 pp.

235. Oppelt, W.H., W.R. Kube, R.B. Porter, T.W. Kamps, O.C. Ongstad, and E.F. Golob. Combustion of North Dakota Lignite in Domestic Heaters. BuMines RI 5681, 1960, 52 pp.

236. Parry, V.F., and R.D. Segur. Performance of Subbituminous Coal in a Typical Underfeed Domestic Stoker. BuMines RI 3557, 1941, 18 pp.

139. Paulson, L.E., and R.C. Ellman. Reduction of Sodium in Lignite by Ion Exchange: A Pilot Plant Study. DOE GFETC/RI-79/1, 1979, 50 pp.

63. Paulson, L.E., and W.W. Fowkes. Changes in Ash Composition of North Dakota Lignite Treated by Ion Exchange. BuMines RI 7176, 1968, 18 pp.

237. Quilici, F.J. Characteristics and Uses of Montana Fly Ash. M.S. Thesis, Mont. Coll. of Mineral Sci. and Techn., Butte, 1972; also, Mont. Bureau of Mines and Geol., Bull. 90, 1973, 29 pp.
238. Rindt, D.K., and G.G. Montgomery. Microscopic Studies of Ash Deposit and Coal Related to the Ash Fouling Problem. Proc., N.D. Acad. of Sci., v. 33, Apr. 1979, p. 58.
239. Rindt, D.K., S.J. Selle, and Willis Beckering. Investigations of Ash Fouling Mechanisms for Western Coals Using Microscopic and X-ray Diffraction Techniques. ASME Winter Meeting, New York, Dec. 1979, Preprint No. 79-WA/CD-5, 9 pp.
240. Selle, S.J. A Preliminary Study of the Effects of Temperature and Coal Source on Efficiency of Electrostatic Precipitation of Fly Ash from Western Coals. M.S. Thesis, Univ. of N.D., Grand Forks, 1970, 41 pp.
241. Selle, S.J. Electrostatic Precipitator Test Program at the Grand Forks Energy Research Center, Grand Forks, N.D., ERDA. EPRI Select Sympos., July 1975.
242. Selle, S.J., and G.H. Gronhovd. Some Comparisons of Simultaneous Stack Gas Particulate Determinations Using the ASME and EPA Methods. ASME Winter Meeting, New York, Nov. 1972, ASME Preprint No. 72-WA/APC-4; in J. of Eng. for Power, July 1973, pp. 145-149.
243. Selle, S.J., and L.L. Hess. Factors Affecting ESP Performance on Western Coals and Experience with North Dakota Lignites. EPA/EPRI Meeting on Particulate Control in Energy Processes, May 1976, San Francisco, 20 pp. In "Symposium on Particulate Control in Energy Processes," D. E. Blake, ed., EPA-600/7-76-010, Sept. 1976.
244. Selle, S.J., L.L. Hess, and E.A. Sondreal. Western Fly Ash Composition as an Indicator of Resistivity and Pilot ESP Removal Efficiency. APCA Annual Meeting, Boston, June 1975, 16 pp.
245. Selle, S.J., F.I. Honea, and E.A. Sondreal. Direct Use of Lignite for Power Generation. IGT Sympos. on New Fuels and Advances in Combustion Tech., Chicago, March 1979.
246. Selle, S.J., P.H. Tufte, and G.H. Gronhovd. A Study of the Electrical Resistivity of Fly Ashes from Low-Sulfur Western Coals Using Various Methods. APCA Annual Meeting, Miami Beach, June 1972, Preprint No. 72-107, 32 pp.
247. Sondreal, E.A., and R.C. Ellman. Fusibility of Ash from Lignite and its Correlation with Ash Composition. DOE GFERC/RI-75/1, 1975, 121 pp.
248. Sondreal, E.A., G.H. Gronhovd, and W.R. Kube. Research and Development Relating to Lignite Use in Power Production. Gulf Coast Lignite Conf: Geology, Utilization, and Environmental Aspects, Austin, Tex., June 1976. Conf. Proc., 1978, pp. 100-124.
249. Sondreal, E.A., S.J. Selle, P.H. Tufte, V.H. Menze, and V.R. Laning. Correlation of Fireside Boiler Fouling with North Dakota Lignite Ash Characteristics and Power Plant Operating Conditions. Amer. Power Conf., Chicago, Apr. 1977, 18 pp.
252. Sondreal, E.A., and P.H. Tufte. Comparison of Flue Gas Desulfurization for Eastern Versus Western U.S. Coals. SME/AIME Fall Meeting, Acapulco, Mexico, Sept. 1974, Preprint No. 74-F-322, 33 pp; abstracted in SME Trans., v. 258, June 1975, pp. 122-123.
253. Sondreal, E.A., and P.H. Tufte. Scrubber Developments in the West. Sympos. on Techn. and Use of Lignite, Grand Forks, N.D., May 1975. DOE GFERC/IC-75/2, 1975, pp. 65-147.
254. Sondreal, E.A., P.H. Tufte, and Willis Beckering. Ash Fouling in the Combustion of Low Rank Western U.S. Coals. Combustion Sci. and Techn., 1977, v. 16, pp. 95-110.

255. Sondreal, E.A., P.H. Tufte, and S.J. Selle. Wet Scrubbing of SO_2 with Alkali in Western Coal Ash. APCA Annual Meeting, Denver, June 1974, Preprint No. 74-272, 32 pp.
256. Stern, F.R. Bench-Scale Study of Sulfur and Nitrogen Oxides Adsorption by Nahcolite and Trona. M.S. Thesis, Univ. of N.D., Grand Forks, 1978, 81 pp.
257. Thunem, C.B. Laboratory Studies on the Applicability of Western United States Fly Ashes to Wet Scrubbing for SO_2 Removal. M.S. Thesis, Univ. of N.D., Grand Forks, 1975, 51 pp.
258. Tufte, P.H., and Willis Beckering. A Proposed Mechanism for Ash Fouling Burning Northern Great Plains Lignite. ASME Winter Meeting, New York, Nov. 1974, Preprint No. 74-WA/CD-3; also in J. of Eng. for Power, July 1975, pp. 407-412.
259. Tufte, P.H., G.H. Gronhovd, E.A. Sondreal, and S.J. Selle. Ash Fouling Potentials of Western Subbituminous Coals as Determined in a Pilot Plant Test Furnace. Amer. Power Conf., Chicago, Apr. 1976, 15 pp.
260. Tufte, P.H., E.A. Sondreal, K.W. Korpi, and G.H. Gronhovd. Pilot-Plant Scrubber Tests to Remove SO_2 Using Soluble Alkali in Western Coal Fly Ash. Sympos. on Techn. and Use of Lignite, Grand Forks, N.D., May 1973. BuMines IC 8650, 1974, pp. 103-133.
261. Wilhelm, T.D. Sample Preparation and Electron Microprobe Analysis of Lignite Ash Deposits. M.S. Thesis, Univ. of N.D., Grand Forks, 1968, 37 pp.

Gasification and Liquefaction, Including Environmental Aspects

300. Berg, Lloyd, and J.S. Malsam. Catalytic Hydrogenation Using KCl-$ZnCl_2$ Molten Salt Mixture as a Catalyst. U.S. Pat. 3,736,250, May 29, 1973, 2 pp.
301. Bodner, S.S. Effect of Catalysts on the Reduction of Subbituminous Coal by the CO-Steam Process. M.S. Thesis, Univ. of N.D., Grand Forks, 1977, 36 pp.
302. Bureau of Mines Staff. Lignite Gasification Plant Shown at Grand Forks, N.D. Chem. and Eng. News, v. 23, No. 14, July 25, 1945, pp. 1242-1244.
303. Burr, A.C., R.C. Ellman, J.J. Hoeppner, J.C. Holtz, T.W. Kamps, W.R. Kube, O.C. Ongstad, and W.H. Oppelt. Gasification of Lignite in a Commercial-Scale Pilot Plant: Progress Report for January 1, 1949, to June 30, 1950. BuMines RI 4997, 1953, 48 pp.
304. Burr, A.C., J.C. Holtz, A.W. Koth, and W.H. Oppelt. Gasification of Lignite in a Commercial-Scale Pilot Plant: Progress Report for 1947-48. BuMines RI 4940, 1953, 28 pp.
305. Chetrick, M.H. Undecomposed Steam in Lignite Gasification. BuMines RI 4738, 1950, 11 pp.
306. Chetrick, M.H. Thermal Requirements for the Gasification of Lignite in an Externally Heated Retort. BuMines RI 4957, 1953, 7 pp.
307. Elder, J.L. The Underground Gasification of Coal. Chap. 21, Supp. Vol., Chem. of Coal Utilization. Ed. by H. H. Lowry, Nat. Acad. Sci., Nat. Res. Council. John Wiley & Sons, 1962, pp. 1023-1040.
308. Ellman, R.C., M.M. Fegley, B.C. Johnson, L.E. Paulson, and H.H. Schobert. Slagging Fixed-Bed Gasification. Fourth Internat. Conf. on Coal Gasification, Liquefaction, and Conversion, Pittsburgh, Aug 1977, 30 pp.
309. Ellman, R.C., and B.C. Johnson. Slagging Fixed-Bed Gasification at the Grand Forks Energy Research Center. 8th Synthetic Pipeline Gas Sympos., Chicago, Oct. 1976, 11 pp.

311. Ellman, R. C., B. C. Johnson, and M. M. Fegley. Studies in Slagging Fixed-Bed Gasification at the Grand Forks Energy Technology Center. Tenth Synthetic Pipeline Gas Sympos., Chicago, Oct. 1978, 19 pp.

312. Ellman, R.C., B.C. Johnson, H.H. Schobert, L.E. Paulson, and M.M. Fegley. Current Status of Studies in Slagging Fixed-Bed Gasification at the Grand Forks Energy Research Center. Sympos. on Techn. and Use of Lignite, Grand Forks, N.D., May 1977. DOE GFERC/IC-77/1, 1978, pp. 207-247.

313. Ellman, R.C., L.E. Paulson, D.R. Hajicek, and T.G. Towers. Slagging Fixed-Bed Gasification: Project Status at the Grand Forks Energy Technology Center. Sympos. on Techn. and Use of Lignite, Grand Forks, N.D., May 1979. DOE GFETC/IC-79/1, 1979, pp. 236-277.

314. Ellman, R.C., and H.H. Schobert. Pilot Plant Operation of a Fixed-Bed Slagging Gasifier. ACS Sympos. on Anal. Chem. of Tar Sands and Oil Shale, New Orleans, March 1977, 13 pp.

315. Ellman, R.C., and E.A. Sondreal. Environmental Control Activities on Slagging Fixed-Bed Gasification at the Grand Forks Energy Technology Center. DOE Environmental Control Sympos., Washington, D.C., 1978, 18 pp.

316. Farnum, B.W., C.L. Knudson, and D.A. Koch. Products of Liquefaction of Lignite with Synthesis Gas Byproduct Slurry Recycle. ACS, Div. of Fuel Chem. Preprints, v. 24, No. 3, Sept. 1979, pp. 195-203.

318. Gronhovd, G.H. Design and Preliminary Operation of a Slagging, Fixed-Bed, Pressure Gasification Pilot Plant. Sympos. on Techn. and Use of Lignite, Grand Forks, N.D., Apr. 1961. BuMines IC 8164, 1963, pp. 16-29.

319. Gronhovd, G.H. Status of Lignite Gasification Research. Mo. River Basin Interagency Committee Meeting, Minot, N.D., Oct. 1961, 10 pp.

322. Gronhovd, G.H. Experiments in Slagging Gasification at Pressures to 400 Psig. Sympos. on Techn. and Use of Lignite, Grand Forks, N.D., Apr.-May 1963. BuMines IC 8234, 1964, pp. 11-27.

323. Gronhovd, G.H. Pilot Plant Experiments in Slagging Gasification. Chem. Eng. Progress, Sympos. Series 54, v. 61, 1965, pp. 104-113.

324. Gronhovd, G.H. Status of Coal Gasification in the United States. Proc., Sympos. on Energy and New Energy Sources, Univ. of Saskatchewan, Regina, Canada, 1973, 17 pp.

325. Gronhovd, G.H., A.E. Harak, and M.M. Fegley. Development of an Experimental Slagging, Fixed-Bed Gasifier for Production of Synthesis Gas at Pressures to 400 Psig. ACS Sympos. on Production of Hydrogen, New York, Sept. 1963, pp. 89-97.

326. Gronhovd, G.H., A.E. Harak, and M.M. Fegley. Some Results from Slagging, Fixed-Bed Gasification of Lignite at Pressures to 400 Psig. ACS Sympos. on Gas Generation, Philadelphia, Apr. 1964, pp. 193-206.

327. Gronhovd, G.H., A.E. Harak, M.M. Fegley, and D.E. Severson. Slagging Fixed-Bed Gasification of North Dakota Lignite at Pressures to 400 Psig. BuMines RI 7408, 1970, 40 pp.

328. Gronhovd, G.H., A.E. Harak, W.R. Kube, and W.H. Oppelt. Design and Initial Operation of a Slagging, Fixed-Bed, Pressure Gasification Pilot Plant. BuMines RI 6085, 1962, 50 pp.

329. Haab, G.R. Destructive Hydrogenation of Savage, Montana, Lignite. M.S. Thesis, Mont. State Univ., Bozeman, 1968, 56 pp.

331. Hajicek, D.R., and L.E. Paulson. Development and Operating Characteristics of a Gas Sampling System for the Grand Forks Energy Technology Center's Slagging Fixed-Bed Gasifier. Argonne National Lab Sympos. on Instrumentation and Control of Fossil Energy Processes, Denver, Aug. 1979, 13 pp.

332. Hansel, W.J. Size Reduction, Reliquefaction, and Combustion Characteristics of Solvent Refined Coal. M.S. Thesis, Univ. of N.D., Grand Forks, 1977, 88 pp.

333. Holtz, J.C. Gasification of Lignite in the Commercial-Scale Pilot Plant at Grand Forks. Proc., N.D. Acad. of Sci., v. 4, 1950, pp. 39-42.

334. Johnson, B.C., M.M. Fegley, R.C. Ellman, and L.E. Paulson. Gasification of North Dakota Lignite in a Slagging Fixed-Bed Gasifier. AIChE Annual Meeting, Miami Beach, 1978, 23 pp.

335. Johnson, B.C., H.H. Schobert, and M.M. Fegley. The Grand Forks Slagging Gasifier: Gas-Solid Reactions in the Real World. AIChE Sympos., Theory and Practice of Gas-Solids Contactors in Fossil Fuel Processing, New York, 1977. In Coal Processing Tech., v. 4, 1978, pp. 94-98.

336. Knudson, C.L., B.W. Farnum, and E.A. Kline. Polymerization of Phenanthrene Induced by 1-Methylnaphthalene at High Temperatures and Pressures. ACS National Meeting, Washington, D.C., Sept. 1979; Div. of Fuel Chem. Preprints, v. 24, No. 3, 1979, p. 278.

337. Knudson, C.L., J.E. Schiller, and A.L. Ruud. Temperature Effects on Coal Liquefaction: Rates of Depolymerization and Product Quality as Determined by Gel Permeation Chromatography. In "Organic Chemistry of Coal," ACS Sympos. Series 71, ed. by John W. Larsen, Washington, D.C., 1975, 15 pp.

338. Mayer, G.G., D.R. Hajicek, and P.G. Freeman. Correlating Process Parameters and Tar/Water Production for a Slagging Fixed-Bed Gasifier. Argonne National Lab Sympos. on Instrumentation and Control of Fossil Energy Processes, Denver, Aug. 1979, 14 pp.

339. Mayer, G.G., J.K. Olson, and H.H. Schobert. Standard Methods for the Analysis of Coal Gasification Effluents--Caveat Emptor. CIC National Meeting, Winnipeg, 1978, 9 pp.

55. Miller, D.J. Quantitative Analysis of Coal-Derived Liquids by Low Voltage Mass Spectroscopy. Proc., Dakota Acad. of Sci., v. 33, Apr. 1979, p. 55.

56. Miller, D.J., J.K. Olson, and H.H. Schobert. Mass Spectroscopic Characterization of Tars from the Gasification of Low-Rank Coals. ACS National Meeting, Washington, D.C., Sept. 1979, 10 pp.

342. Nguyen, Dat, Warren Scarrah, and Lloyd Berg. Liquefaction of Subbituminous Coal. Sympos. on Techn. and Use of Lignite, Grand Forks, N.D., May 1973. BuMines IC 8650, 1974, pp. 247-262.

343. Olson, J.D. Production of Liquids and Gases from Savage, Montana, Lignite by Hydrogenation with a Nickel-Tungsten Catalyst. Ph.D. Thesis, Mont. State Univ., Bozeman, 1969, 201 pp.

344. Olson, J.K., and H.H. Schobert. Evaluation of Methods for Determination of Sulfide in Coal Gasification Wastewater. ACS National Meeting, Miami Beach, 1978.

345. Olson, J.K., and H.H. Schobert. Production of C_1-C_4 Hydrocarbons in the Gasification of Some North Dakota Lignites. N.D. Acad. of Sci. Meeting, Grand Forks, N.D., 1978 (abstract).

346. Olson, J.K., and H.H. Schobert. An Overview of Effluents from the Gasification of Some Western Coals. Proc., N.D. Acad. of Sci., v. 33, 1979, p. 60.

347. Olson, J.K., and H.H. Schobert. Effects of Coal Structure and Processing Conditions on Organic Effluents from Slagging Fixed-Bed Gasification. ACS, Div. of Fuel Chem. Preprints, v. 24, No. 3, 1979, pp. 2-9.

348. Ongstad, O.C., M.H. Chetrick, and W.H. Oppelt. Cost Data for Gasification of Lignite in an Externally Heated Retort. BuMines RI 5272, 1956, 17 pp.

349. Oppelt, W.H. Drying and Gasification of Lignite at the Bureau of Mines Station, Grand Forks, N.D. School of Education Record, Univ. of N.D., v. 36, No. 4, Jan. 1951, pp. 106-111.

351. Oppelt, W.H. Slagging Slices Gasification Costs. Chem. Week, v. 83, No. 22, Nov. 29, 1958, pp. 53-54 (information release).

352. Oppelt, W.H. Slagging Fixed-Bed Pressure Gasification Pilot Plant. Gas Age, v. 127, No. 3, Feb. 2, 1961, pp. 34-35, 60.

354. Oppelt, W.H., and G.H. Gronhovd. Design and Preliminary Operation of a Slagging Fixed-Bed Pressure Gasification Pilot Plant. Trans. AIME (Mining), v. 220, 1961, pp. 154-160.

355. Oppelt, W.H., T.W. Kamps, G.H. Gronhovd, W.R. Kube, and Robert McMurtrie. Production of Crude Ammonia-Synthesis Gas from North Dakota Lignite in an Annular-Retort Gasifier. BuMines RI 5297, 1957, 29 pp.

356. Oppelt, W.H., W.R. Kube, M.H. Chetrick, T.W. Kamps, and E.F. Golob. Gasification of Lignite in a Commercial-Scale Pilot Plant: Progress Report for July 1, 1950, to December 31, 1951, and Summary of Work Previous to July 1, 1950. BuMines RI 5164, 1955, 80 pp.

357. Parry, V.F. Bureau of Mines Investigations on Gasification of Lignite and Subbituminous Coal. Skillings' Mining Rev., v. 37, No. 19, Aug. 21, 1948, pp. 1, 4, 15.

358. Parry, V.F., D.C. Gernes, J.B. Goodman, E.O. Wagner, A.W. Koth, W. L. Patty, and E. C. Yeager. Gasification of Lignite and Subbituminous Coal, Progress Report for 1944. BuMines RI 3901, 1946, 59 pp.

359. Parry, V.F., E.C. Gernes, E.O. Wagner, J.B. Goodman, and A.W. Koth. Gasification of Lignite and Subbituminous Coal, Progress Report for 1945-46. BuMines RI 4128, 1947, 69 pp.

361. Parry, V.F., E.O. Wagner, A.W. Koth, and J.B. Goodman. Gasification of Subbituminous Coal and Lignite in Externally Heated Retorts. Ind. and Eng. Chem., v. 40, No. 4, Apr. 1948, pp. 627-641.

362. Paulson, L.E., H.H. Schobert, and R.C. Ellman. Sampling, Analysis, and Characterization of Effluents from the Grand Forks Energy Research Center Slagging Fixed-Bed Gasifier. ACS National Meeting, Anaheim, Calif., 1978; in Div. of Fuel Chem. Preprints, v. 23, No. 2, 1978, pp. 107-118.

66. Schiller, J.E. Composition of Coal Liquefaction Products. Hydrocarbon Processing, Jan. 1977, 6 pp.

67. Schiller, J.E. Analysis of Solvent-Refined Coal, Recycle Solvents, and Coal Liquefaction Products. ACS Sympos. on Anal. Chem. of Tar Sands and Oil Shale, Div. of Anal. Chem. and Petroleum Chem., New Orleans, March 1977; in Advances in Chem. Series, No. 170, 4 pp.

363. Schiller, J.E. Preparation and Low-Voltage Mass Spectrometry Sensitivities of Methylated Polynuclear Aromatic Hydrocarbons. Anal. Chem., v. 49, July 1977, pp. 1260-1262.

68. Schiller, J.E. Nitrogen Compounds in Coal-Derived Liquids. Anal. Chem., v. 49, Dec. 1977, pp. 2292-2294.

364. Schiller, J.E., and C.L. Knudson. Sampling System for Mass Spectrometry of High Boiling Mixtures Using Gas Chromatography--Mass Spectrometry. Anal. Chem., v. 48, Feb. 1976, pp. 453-454.

69. Schiller, J.E., and C.L. Knudson. Effect of Solvent on Molecular Composition in Coal Liquefaction. Fuel, v. 57, Jan. 1978, pp. 36-40.

365. Schobert, H.H. Petrochemistry of Coal Ash Slags. 2. Correlation of Viscosity with Composition and Petrographic Class. CIC-ACS Joint Conf., Montreal, 1977; in ACS Div. of Fuel Chem. Preprints, v. 22, No. 4, 1977, pp. 143-148.

366. Schobert, H.H., D.L. Barbie, O.D. Christensen, and F.R. Karner. Petrochemistry of Coal Ash Slags. 1. Formation of Melilite and a High-Temperature Glass from a Calcium-Rich, Silica-Deficient Slag. CIC-ACS Joint Conf., Montreal, 1977; in ACS Div. of Fuel Chem. Preprints, v. 22, No. 4, 1977, pp. 133-142.

367. Schobert, H.H., B.C. Johnson, and M.M. Fegley. Carbonization Reactions in the Grand Forks Slagging Gasifier. ACS National Meeting, Miami Beach, 1978; in Div. of Fuel Chem. Preprints, v. 23, No. 3, 1978, pp. 136-141.

368. Sherick, L.S.C. Effect of Residence Time on the Reduction of Subbituminous Coal and Lignite by the CO-Steam Process. M.S. Thesis, Mont. State Univ., Bozeman, 1977, 96 pp.

369. Sondreal, E.A. Utilization Potential for Low-Rank U.S. Coals in Liquefaction and Gasification. Pa. State Univ., Short Course on Coal Characteristics and Utiliz., Univ. Park, Oct. 1979, 51 pp.

371. Sondreal, E.A., C.L. Knudson, R.S. Majkrzak, and G.G. Baker. Liquefaction of Lignite by the CO-Steam Process. Preprints, AIChE Meeting, Miami, Nov. 1978, 36 pp.; condensed in AIChE CEP Tech. Manual, pp. 191-192.

372. Sondreal, E.A., C.L. Knudson, J.E. Schiller, and T.H. May. Development of CO-Steam Process for Liquefaction of Lignite and Western Subbituminous Coals. Sympos. on Techn. and Use of Lignite, Grand Forks, N.D., May 1977. DOE GFERC/IC-77/1, 1978, pp. 129-158.

373. Speakman, R.F. Analysis of Liquids from the Catalytic Hydrogenation of Savage, Montana, Lignite. M.S. Thesis, Mont. State Univ., Bozeman, 1968, 43 pp.

374. Tang, I.N. Studies on Thermal Solution of Low-Rank Fuel: 2. Leonardite in Hydrogen Atmosphere with Tetralin. M.S. Thesis, Univ. of N.D., Grand Forks, 1960, 25 pp.

375. Willson, W.G., C.L. Knudson, G.G. Baker, T.C. Owens, and D.E. Severson. Application of Liquefaction Processes to Low-Rank Coals. Sympos. on Techn. and Use of Lignite, Grand Forks, N.D., May 1979, 38 pp. DOE GFETC/IC-79/1, 1979, pp. 170-206.; also in Ind. Eng. Chem., Product R&D, No. 18, Dec. 1979, 14 pp.

Drying and Carbonization

102. Belter, J.W. A Review of the Use of Pulsating Combustors for Drying and Conveying. Proc., 1st Internat. Sympos. on Pulsating Combustion, Sept. 1971, Univ. of Sheffield, Sheffield, England; ed. by D. J. Brown; 21 pp.

103. Belter, J.W., Leroy Dockter, and R.C. Ellman. Operating Experience with Lignite Fueled Pulse-Jet Engines. ASME Winter Meeting, Los Angeles, Nov. 1969, 5 pp.

401. Boley, C.C., and M.M. Fegley. The Use of Char to Improve the Physical Quality of Coke. Sympos. on Techn. and Use of Lignite, Grand Forks, N.D., May 1969. BuMines IC 8471, 1970, pp. 165-174.

402. Boley, C.C., and M.M. Fegley. Design and Operation of Two Refractory-Lined, Internally Heated, Entrained-Bed Carbonizers. DOE GFERC/RI-77/1, 1977, 38 pp.

15. Boteler, D.C. Assay Carbonization of Lignite from the Northern Great Plains Province. M.S. Thesis, Univ. of N.D., Grand Forks, 1966, 65 pp.

403. Boteler, D.C., and C.C. Boley. Partial Devolatilization of Coal in Two Entrained-Bed Carbonizers. BuMines RI 7843, 1973, 20 pp.

16. Boteler, D.C., W.R. Kube, and C.C. Boley. Bench-Scale Carbonization of Lignite from the Northern Great Plains Province: A Statistical Approach. BuMines RI 7386, 1970, 32 pp.

404. Burr, A.C., W.H. Oppelt, and T.W. Kamps. The Rate of Drying Lignite in Hot Oil at 315 F. Proc., N.D. Acad. of Sci., v. 2, 1949, pp. 32-36.

405. Carpenter, H.C., P.L. Cottingham, C.M. Frost, and W.W. Fowkes. Hydrocracking Low-Temperature Tar from a North Dakota Lignite. BuMines RI 6237, 1963, 13 pp.

406. Cooley, A.M., L.E. Paulson, and S.A. Cooley. Drying of Lignite in a Roto-Louvre Type Dryer. Conf. Canadian Inst. of Engrs, Univ. of Saskatchewan, Saskatoon, Oct. 1972, 13 pp.

407. Davis, J.D., and V.F. Parry. Carbonizing Properties of a Subbituminous Coal from Puritan Mine, Dacono, Weld County, Colorado. BuMines RI 3428, 1939, 32 pp.

118. Ellman, R.C., J.W. Belter, and Leroy Dockter. Comparative Pulverization Tests on Five Lignites Using In-the-Mill Drying. BuMines Prelim. Rpt. 144, Apr. 1962, 18 pp.

119. Ellman, R.C., J.W. Belter, and Leroy Dockter. Effects of In-the-Mill Drying on Pulverizing Characteristics of Lignite. BuMines RI 6074, 1962, 18 pp.

408. Ellman, R.C., J.W. Belter, and Leroy Dockter. Adapting a Pulse-Jet Combustion System to Entrained Drying of Lignite. Proc., 5th Internat. Coal Prep. Congress, Pittsburgh, Oct. 1966, pp. 463-474.

126. Ellman, R.C., L.E. Paulson, and S.A. Cooley. Commercial-Scale Drying of Low-Rank Western Coals: Part I. Rail Shipment Test Observations. Sympos. on Techn. and Use of Lignite, Grand Forks, N.D., May 1975. DOE GFERC/IC-75/2, 1975, pp. 312-340.

409. Fieldner, A.C., J.D. Davis, V.F. Parry, L.D. Schmidt, J.L. Elder, J.B. Goodman, W.S. Landers, and E.W. Goodwin. Tests on the Heliopore Coal-Carbonization Power-Plant Process. BuMines RI 3733, 1943, 36 pp.

22. Fowkes, W.W. Low-Temperature Tar as a Raw Material. Sympos. on Techn. and Use of Lignite, Grand Forks, N.D., Apr. 1961. BuMines IC 8164, 1963, pp. 69-79.

28. Fowkes, W.W., C.M. Frost, J.J. Hoeppner, Willis Beckering, P.G. Freeman, and R. W. Youngs. An Examination of Low-Temperature Tar from a North Dakota Lignite. BuMines RI 5813, 1961, 27 pp.

32. Freeman, P.G. FIA Analysis of Coal Tar. ACS, Div. of Gas and Fuel Chem. Preprints, v. 2, No. 1, Sept. 1960, pp. 87-90.

33. Freeman, P.G. Catechols from the Carbonization of North Dakota Lignite. ACS, Div. of Fuel Chem. Preprints, v. 7, No. 1, 1963, pp. 139-143. Also in Proc., N.D. Acad. of Sci., v. 18, 1964, pp. 104-108.

36. Gomez, Manuel, and J.B. Goodman. Distillation Assays of Missouri River Basin Coals. BuMines RI 5009, 1953, 9 pp.

127. Gomez, Manuel, J.B. Goodman, and V.F. Parry. Changes Induced in Low-Temperature Tar by Oxidation and Storage. BuMines RI 5227, 1956, 12 pp.

411. Gomez, Manuel, J.B. Goodman, and V.F. Parry. Effect of Oxidation on Carbonization Yields of a Texas Lignite. BuMines RI 5277, 1956, 17 pp.

37. Gomez, Manuel, J.B. Goodman, and V.F. Parry. Low-Temperature Tar from Fluidized Carbonizing Reactors. BuMines RI 5302, 1957, 44 pp.

38. Gomez, Manuel, J.B. Goodman, and V.F. Parry. General Properties of Low-Temperature Tar. BuMines Bull. 569, 1958, 31 pp.

412. Gomez, Manuel, W.S. Landers, and C.K. Boyd. Estimation of Low-Temperature Carbonization Product Yields: Utah Coal Fields. BuMines RI 6990, 1967, 32 pp.

40. Gomez, Manuel, W.S. Landers, J.L. Shultz, and A.G. Sharkey. Distribution of Phenols in Low-Temperature Tar from Low-Rank Coals. BuMines RI 6586, 1965, 29 pp.

413. Gomez, Manuel, and C.H. Prien. Pyrolysis of Coal and Shale (section, "Pyrolysis of Coal," by Gomez). Ind. Eng. Chem., v. 52, No. 8, Aug. 1960, pp. 717-723.

414. Gomez, Manuel, and C.H. Prien. Pyrolysis of Coal and Shale (section, "Pyrolysis of Coal," by Gomez). Ind. Eng. Chem., v. 53, No. 8, Aug. 1961, pp. 674-679.

415. Gomez, Manuel, and C.H. Prien. Pyrolysis of Coal and Shale (section, "Pyrolysis of Coal," by Gomez). Ind. Eng. Chem., Ann. Review Suppl. 1962, Sept. 1962, pp. 110-114.

416. Gomez, Manuel, and C.H. Prien. Pyrolysis of Coal and Shale (section, "Pyrolysis of Coal," by Gomez). Ind. Eng. Chem., Ann. Review Suppl. 1963, Dec. 1963, pp. 82-85.

417. Goodman, J.B., and R.S. Detrick. Thermal Refining of Low-Temperature Tar. BuMines RI 5534, 1959, 43 pp.

42. Goodman, J.B., Manuel Gomez, and V.F. Parry. Laboratory Carbonization Assay of Low-Rank Coals at Low, Medium, and High Temperatures. BuMines RI 5383, 1958, 49 pp.

43. Goodman, J.B., Manuel Gomez, V.F. Parry, and W.S. Landers. Low-Temperature Carbonization Assay of Coal in a Precision Laboratory Apparatus. BuMines Bull. 530, 1953, 24 pp.

421. Harrington, L.C., V.F. Parry, and A.W. Koth. Technical and Economic Study of Drying Lignite and Subbituminous Coal by the Fleissner Process. BuMines TP 633, 1942, 84 pp.

422. Hoeppner, J.J., W.W. Fowkes, and Robert McMurtrie. Removal of Moisture from Lignite in Inert-Gas Atmospheres. BuMines RI 5215, 1956, 25 pp.

44. Hoeppner, J.J., M.L. Opland, and W.W. Fowkes. Carbonization Characteristics of Some North-Central United States Lignites. BuMines RI 5260, 1956, 21 pp.

423. Johnson, B.C. A Study of the Carbonization and Activation Conditions for the Production of Water Treatment Carbon from North Dakota Lignite. M.S. Thesis, Univ. of N.D., Grand Forks, 1971, 64 pp.

49. Kube, W.R. Sorption of Water Vapor by Thermally Treated North Dakota Lignite. Proc., N.D. Acad. of Sci., v. 6, 1952, pp. 83-85.

50. Kube, W.R. Sorption of Water Vapor by Thermally Treated Lignite at Different Relative Humidities. Chem. and Eng. Data Ser., v. 2, No. 1, 1957, pp. 46-51.

424. Landers, W.S. Low-Temperature Carbonization of Coal and Lignite for Thermal Power Generation. Proc., Amer. Power Conf., v. 19, 1957, pp. 344-351.

425. Landers, W.S. Binderless Briquetting of Indian Lignite. Inst. for Briq. and Agglom., 6th Conf., Glacier Park, Mont., 1959; IBA Proc., v. 6, 1960, pp. 30-37.

427. Landers, W.S., Manuel Gomez, and E.O. Wagner. Entrainment Carbonization of Texas Lignite. BuMines Bull. 639, 1968, 44 pp.

428. Landers, W.S., and J.B. Goodman. Effect of Moderate Pressure on Low-Temperature Carbonization of Lignite. Fourth Internat. Conf. on Science of Coal, Le Touquet, France, May-June 1961.

51. Landers, W.S., J.B. Goodman, and D.J. Donaven. Low-Temperature Carbonization Assays of Coals and Relation of Yields to Analyses. BuMines RI 5904, 1961, 41 pp.

429. Landers, W.S., and V.F. Parry. Briquetting Properties of Fluidized Chars. Inst. for Briq. and Agglom., 3rd Conf., Banff, Alberta, 1953; IBA Proc., v. 3, 1954, pp. 37-47. Also in Univ. of Wyo. NRRI IC 6, 1953.

431. Landers, W.S., V.F. Parry, Manuel Gomez, E.O. Wagner, J.B. Goodman, and C. R. Nelson. Carbonizing Properties of Wyoming Coals. BuMines RI 5731, 1961, 74 pp.
432. Malsam, J.S. Coal Hydrogenation Studies Using a $KCl-ZnCl_2$ Molten Salt Mixture as a Catalyst. M.S. Thesis, Mont. State Univ., Bozeman, 1970, 46 pp.
52. McDonald, J.L. Binding Properties of Leonardite in Lignite and Lignite Char Agglomerates. M.S. Thesis, Univ. of N.D., Grand Forks, 1963, 51 pp.
434. Oppelt, W.H. The Mechanism of Drying Lignite. Proc., N.D. Acad. of Sci., v. 4, 1950, pp. 33-39.
349. Oppelt, W.H. Drying and Gasification of Lignite at the Bureau of Mines Station, Grand Forks, N.D. School of Education Record, Univ. of N.D., v. 36, No. 4, Jan. 1951, pp. 106-111.
435. Oppelt, W.H. Thermal Treatment of North Dakota Lignite at Controlled Temperatures. Proc., N.D. Acad. of Sci., v. 6, 1952, pp. 36-38.
436. Oppelt, W.H., J.P. Cooney, E.F. Golob, and W.R. Kube. Thermal Pretreatment and Pelletizing of North Dakota Lignite. BuMines RI 5382, 1958, 55 pp.
437. Oppelt, W.H., R.C. Ellman, O.C. Ongstad, and W.R. Kube. Experiments in Fleissner Drying of North Dakota Lignite. BuMines RI 5122, 1955, 24 pp.
438. Oppelt, W.H., E.F. Golob, J.P. Cooney, and W.R. Kube. Thermal Processing of North Dakota Lignite in a Rotary-Drum Retort. BuMines Prelim. Rpt. 99, Dec. 1954, 49 pp.
439. Oppelt, W.H., and W.R. Kube. Bench-Scale Experiments on Low-Temperature Carbonization of Lignite and Subbituminous Coals at Elevated Pressure. Trans. AIME (Mining), v. 220, 1961, pp. 126-133.
440. Oppelt, W.H., W.R. Kube, and E.F. Golob. Moisture Reduction, Decarboxylation, Pelletizing, and Water-Vapor Sorption of North Dakota Lignite. BuMines Prelim. Rpt. 76, Nov. 1952, 51 pp.
441. Oppelt, W.H., W.R. Kube, and T.W. Kamps. Drying North Dakota Lignite to 1,500 Pounds Pressure by the Fleissner Process. BuMines RI 5527, 1959, 30 pp.
59. Parry, V.F. Correlation of Analysis of Coal with the Products of Carbonization in Externally Heated Retorts. BuMines RI 3482, 1939, 37 pp.
442. Parry, V.F. Drying and Preheating Coal for Briquetting. Inst. for Briq. and Agglom., 1st Conf., Laramie, Wyo., 1949; IBA Proc., v. 1, 1950, pp. 103-108.
443. Parry, V.F. Low-Temperature Carbonization of Coal and Lignite for Industrial Uses. BuMines RI 5123, 1955, 27 pp.
444. Parry, V.F. Continuous Process for Drying, Preheating, and Devolatilization of Carbonaceous Materials. U.S. Pat. 2,773,018, Dec. 4, 1956, 8 pp.
60. Parry, V.F., and J.B. Goodman. Friability, Slacking Characteristics, and Low-Temperature Carbonization Assays of Subbituminous Coals of the Denver (Colo.) Region. BuMines RI 3457, 1939, 12 pp.
445. Parry, V.F., J.B. Goodman, and Manuel Gomez. Low-Temperature Distillation Assays of Representative Western United States Coals. Colo. Sch. of Mines Quart., v. 45, No. 2-A, Apr. 1950, pp. 133-162.
446. Parry, V.F., J.B. Goodman, and E.O. Wagner. Drying Low-Rank Coals in the Entrained and Fluidized State. Mining Eng., v. 1, No. 4, Apr. 1949, pp. 89-98 of Transactions.
135. Parry, V.F., L.C. Harrington, and A.W. Koth. Preparation of Stable Nonslacking Fuel by Steam-Drying Subbituminous Coal and Lignite. Trans. ASME, v. 64, 1942, pp. 177-183.

447. Parry, V.F., and W.S. Landers. Low-Temperature Carbonization of Lignite for a 240,000-kw Steam Electric Power Plant. World Power Conf., Rio de Janeiro, 1954.

449. Parry, V.F., W.S. Landers, and E.O. Wagner. Low-Temperature Carbonization of Lignite and Noncoking Coals in the Entrained State. Trans. AIME, v. 205, 1956, pp. 54-64.

450. Parry, V.F., W.S. Landers, E.O. Wagner, J.B. Goodman, and G.C. Lammers. Drying and Carbonizing Fine Coal in Entrained and Fluidized State. BuMines RI 4954, 1953, 43 pp.

451. Parry, V.F., and E.O. Wagner. Up-Grading Missouri Basin Coals by Flash Drying. 75th Anniversary Conf. of Colo. School of Mines; in CSM Quarterly, v. 45, No. 2B, Apr. 1950, pp. 57-66. Also in BuMines Prelim. Rpt. 30, June 1949.

452. Parry, V.F., and E.O. Wagner. Drying Fine Coal in the Entrained and Fluidized State. Trans. AIME, v. 187, Sept. 1950, pp. 974-982.

64. Porter, R.B., and R.C. Ellman. Laboratory Scale Carbonization of North Dakota Lignite: Influence of Temperature and Atmosphere on Product Distribution and Activity. ACS, Div. of Fuel Chem. Preprints, v. 8, No. 3, 1964, pp. 15-24.

453. Porter, R.B., W.H. Oppelt, and W.R. Kube. Low-Temperature Carbonization of Lignite and Subbituminous Coal: Effect of Hydrogen Atmosphere to 1000 Pounds Pressure. BuMines RI 5804, 1961, 25 pp.

367. Schobert, H.H., B.C. Johnson, and M.M. Fegley. Carbonization Reactions in the Grand Forks Slagging Gasifier. ACS National Meeting, Miami Beach, 1978; in Div. of Fuel Chem. Preprints, v. 23, No. 3, 1978, pp. 136-141.

454. Sondreal, E.A. Carbon Disulfide Production by Reaction of Elemental Sulfur With Carbonized Lignite and Wood Charcoal. Process Design and Development, Ind. and Eng. Chem., v. 4, No. 1, Jan. 1965, pp. 111-117.

83. Traverse, Alfred. Behavior of Petrographic Components of North Dakota Lignite in Preparation, Low-Temperature Carbonization, and Steam Drying. Econ. Geol., v. 50, No. 1, 1955, p. 102 (abstract).

455. Warfield, R.S., and C.C. Boley. Sampling and Coking Studies of Several Coalbeds in the Kokolik River, Kukpowruk River, and Cape Beaufort Areas of Arctic Northwestern Alaska. BuMines RI 7321, 1969, 58 pp.

General, Including Reserves, Mining, Plant Growth Studies, Agglomeration, and Land Reclamation

1. Adams, R.L. Rheological Characteristics of Leonardite Gels and Lignite Slurries. M.S. Thesis, Univ. of N.D., Grand Forks, 1963, 38 pp.

500. Belter, J.W., Leroy Dockter, and R.C. Ellman. Production of Activated Carbon in Dual Pulse Jet Engine System. U. S. Patent 3,668,145, June 6, 1972, 4 pp.

501. Boley, C.C. Progress Report on Development of ASTM Test Methods for Barbecue Briquets. Inst. for Briq. and Agglom., 11th Conf., Sun Valley, 1969; IBA Proc., v. 11, 1970, pp. 39-41.

502. Boley, C.C., and W.R. Kube. Research on Western Coals: Present Program of the Federal Bureau of Mines. Proc., Rocky Mt. Coal Mining Inst., 1966, pp. 53-62.

503. Boley, C.C., and W.R. Kube. Western Coals: Some Trends in Utilization and Research. Western Resources Papers, 1966, Univ. of Colo. Press, Boulder, pp. 195-221.

504. Brown, C.J. Effect of Coal-Derived Humic Acid on Growth and Chlorophyll Content of Chlorella Vulgaris. M.S. Thesis, Univ. of N.D., Grand Forks, 1969, 24 pp.

505. Bureau of Mines Staff. Technology of Lignitic Coals--Part 1: Summary of Industrial Development Possibilities; European Technology on Lignitic Coals; Occurrence and Properties; Lignite Mining in North Dakota; Preparation; Storage and Transportation. BuMines IC 7691, 1954, 142 pp.
506. Bureau of Mines Staff. Technology of Lignitic Coals--Part 2: Combustion--Power Generation; Carbonization; Gasification; Hydrogenation; Other Chemical Processes. BuMines IC 7692, 1954, 120 pp.
508. Burr, A.C., R.C. Ellman, and A.W. Koth. The North Dakota Lignite Industry. BuMines Prelim. Rpt. 67, 1952, 37 pp.
509. Elder, J.L. Lignite in the North Central States. Walker Announcer of Sci. Eqmt., No. 59/4/71, 1959, pp. 4-8, publ. by G.T. Walker & Co., Inc.
511. Elder, J.L., and W.R. Kube, compilers. Technology and Use of Lignite: Proc. of Bureau of Mines-Univ. of North Dakota Sympos., Grand Forks, N.D., Apr. 1961. BuMines IC 8164, 1963, 113 pp.
512. Elder, J.L., and W.R. Kube, compilers. Technology and Use of Lignite: Proc. of Bureau of Mines-Univ. of North Dakota Sympos., Bismarck, N.D., Apr. 1965. BuMines IC 8304, 1966, 124 pp.
513. Elder, J.L., and W.R. Kube, compilers. Technology and Use of Lignite: Proc. of Bureau of Mines-Univ. of North Dakota Sympos., Grand Forks, N.D., May 1969. BuMines IC 8471, 1970, 174 pp.
514. Elder, J.L., and W.R. Kube. Solid Fuel Resources and Utilization in North Dakota. Mining Record, Western Coal Edition, v. 83, No. 13, 1972, p. 7.
515. Ellman, R.C. Potential of Lignite in North Dakota. Morning Pioneer, Mandan, N.D., 2nd Ann. Industrial Edition, "North Dakota, The Awakening Giant," Feb. 1968.
27. Fowkes, W.W., and C.M. Frost. Leonardite: A Lignite Byproduct. BuMines RI 5611, 1960, 12 pp.
516. Freeman, P.G. The Effect of Some Coal-Derived Humates on the Growth of Tomato Plants in Hydroponic Culture. Sympos. on Techn. and Use of Lignite, Grand Forks, N.D., Apr. 1967. BuMines IC 8376, 1968, pp. 191-201.
517. Freeman, P.G. The Use of Lignite Products as Plant Growth Stimulants. Sympos. on Techn. and Use of Lignite, Grand Forks, N.D., May 1969. BuMines IC 8471, 1970, pp. 150-157.
518. Freeman, P.G., and W.W. Fowkes. Coal-Derived Humus: Plant Growth Effects. BuMines RI 7203, 1968, 16 pp.
519. Freeman, P.G., and W.W. Fowkes. Lignite and Agriculture: A Marriage of Industries? North Dakotan, v. 44, No. 9, Oct. 1969, pp. 18-19.
521. Frost, C.M., J.J. Hoeppner, and W.W. Fowkes. Source and Some General Properties of Humic Acids from Lignitic Materials. J. Chem. Eng. Data, v. 4, No. 2, 1959, pp. 173-176.
522. Gronhovd, G.H. Utilization Research. In "Mineral and Water Resources of North Dakota," U.S. Geol. Surv. and U.S. BuReclam., GPO, 1973, 252 pp., pp. 75-80; also in N.D. Geol. Surv. Bull. 63, 1973, pp. 75-80.
523. Gronhovd, G.H. ERDA's Research Program Relating to Lignite and Other Low-Rank Coals. Proc., Rocky Mt. Coal Mining Inst., 1977, pp. 38-60.
527. Gronhovd, G.H., and W.R. Kube, compilers. Technology and Use of Lignite: Proc. of Bureau of Mines-Univ. of North Dakota Sympos., Grand Forks, N.D., May 1973. BuMines IC 8650, 1974, 262 pp.
528. Gronhovd, G.H., and W.R. Kube. Coal Development in North Dakota, Montana, and Wyoming. Conf. on Fuels, Regina, Sask., Nov. 1974, 46 pp.

529. Gronhovd, G.H., and W.R. Kube, compilers. Technology and Use of Lignite: Proc. of Sympos. sponsored by U.S. Energy Res. and Devel. Admin. and Univ. of North Dakota, Grand Forks, N.D., May 1977. DOE GFERC/IC-77/1, 1977, 390 pp.

530. Gronhovd, G.H., and E.A. Sondreal. Technology and Use of Low-Rank Coals in the U.S.A. Seminar on Technologies for the Utilization of Low Calorie Fuels, Varna, Bulgaria, Apr. 1976, 40 pp.

531. Koch, G.S., and Manuel Gomez. Delineation of Texas Lignite Beds by Statistical Techniques. BuMines RI 6833, 1966, 38 pp.

532. Kube, W.R., and J.L. Elder, compilers. Technology and Use of Lignite: Proc. of Bureau of Mines-Univ. of North Dakota Sympos., Grand Forks, N.D., Apr.-May 1963. BuMines IC 8234, 1964, 128 pp.

533. Kube, W.R., and J.L. Elder, compilers. Technology and Use of Lignite: Proc. of Bureau of Mines-Univ. of North Dakota Sympos., Grand Forks, N.D., Apr. 1967. BuMines IC 8376, 1968, 201 pp.

534. Kube, W.R., and J.L. Elder, compilers. Technology and Use of Lignite: Proc. of Bureau of Mines-Univ. of North Dakota Sympos., Bismarck, N.D., May 1971. BuMines IC 8543, 1972, 145 pp.

535. Kube, W.R., and G.H. Gronhovd, compilers. Technology and Use of Lignite: Proc. of Sympos. sponsored by U.S. Energy Res. and Devel. Admin. and Univ. of North Dakota, Grand Forks, N.D., May 1975. DOE GFERC/IC-75/2, 1975, 371 pp.

536. Kube, W.R., and G.H. Gronhovd, compilers. Technology and Use of Lignite: Proc. of Sympos. sponsored by U.S. Dept. of Energy and Univ. of North Dakota, Grand Forks, N.D., May 1979. DOE GFETC/IC-79/1, 1979, 537 pp.

425. Landers, W.S. Binderless Briquetting of Indian Lignite. Inst. for Briq. and Agglom., 6th Conf., Glacier Park, Mont., 1959; IBA Proc., v. 6, 1960, pp. 30-37.

538. Landers, W.S. Metallurgical, Domestic, and Industrial Utilization of Low-Rank Coals: In "Natural Resources--Energy, Water, and River Basin Development," volume 1 of a compilation of United States papers prepared for the United Nations Conference on the Application of Science and Technology for the Benefit of the Less Developed Areas, 1963, GPO.

539. Landers, W.S. Utilization of North Dakota Lignite. Sympos. on Techn. and Use of Lignite, Bismarck, N.D., Apr. 1965. BuMines IC 8304, 1966, pp. 50-57.

429. Landers, W.S., and V.F. Parry. Briquetting Properties of Fluidized Chars. Inst. for Briq. and Agglom., 3rd Conf., Banff, Alberta, 1953; IBA Proc., v. 3, 1954, pp. 37-47. Also in Univ. of Wyo. NRRI IC 6, 1953.

52. McDonald, J.L. Binding Properties of Leonardite in Lignite and Lignite Char Agglomerates. M.S. Thesis, Univ. of N.D., Grand Forks, 1963, 51 pp.

541. McMurtrie, Robert. The Fuels of North Dakota. Fuels and Power Conf., Coll. of Eng'g, Univ. of N.D., Dec. 1953, 10 pp.

542. McMurtrie, Robert, and W.H. Oppelt. Developments in Lignite Technology and Utilization. Ind. and Eng. Chem., v. 48, No. 11, 1956, pp. 1996-2000.

543. Odenbaugh, M.L., and R.C. Ellman. Leonardite and Other Materials as Drilling-Fluid Dispersants and Viscosity Control Agents. BuMines RI 7043, 1967, 22 pp.

436. Oppelt, W.H., J.P. Cooney, E.F. Golob, and W.R. Kube. Thermal Pretreatment and Pelletizing of North Dakota Lignite. BuMines RI 5382, 1958, 55 pp.

438. Oppelt, W.H., E.F. Golob, J.P. Cooney, and W.R. Kube. Thermal Processing of North Dakota Lignite in a Rotary-Drum Retort. BuMines Prelim. Rpt. 99, Dec. 1954, 49 pp.

440. Oppelt, W.H., W.R. Kube, and E.F. Golob. Moisture Reduction, Decarboxylation, Pelletizing, and Water-Vapor Sorption of North Dakota Lignite. BuMines Prelim. Rpt. 76, Nov. 1952, 51 pp.
544. Parry, V.F. Trends in the Use of Energy in the Western States with Particular Reference to Coal. BuMines RI 3680, 1943, 44 pp. Also in Proc., Rocky Mt. Coal Mining Inst., 1942.
545. Parry, V.F. Trends in the Production of Coal and Energy in the Western States as a Result of the War. Proc., Rocky Mt. Coal Mining Inst., 1947.
546. Parry, V.F. Future Prospects for Western Fuel Supply. Western Industry, Jan. 1948.
547. Parry, V.F. Resume of Bureau of Mines Research and Development Work on Western Coals, 1942-47. BuMines RI 4171, 1948, 9 pp.
442. Parry, V.F. Drying and Preheating Coal for Briquetting. Inst. for Briq. and Agglom., 1st Conf., Laramie, Wyo., 1949; IBA Proc., v. 1, 1950, pp. 103-108.
548. Parry, V.F. Production, Classification, and Utilization of Western United States Coals. Nov. 1949 Meeting, Soc. of Economic Geologists. Econ. Geol., v. 45, No. 6, Sept.-Oct. 1950.
550. Parry, V.F., and J.B. Goodman. Briquetting Subbituminous Coal. BuMines RI 3707, 1943, 37 pp.
551. Parry, V.F., and W.S. Landers. Trends in Demand for Energy in the Western States. Proc., Rocky Mt. Coal Mining Inst., 49th Annual Meeting, Glenwood Springs, Colo., 1953, pp. 47-56.
552. Parry, V.F., W.S. Landers, and J.B. Goodman. Automatic Water Heating, Utilizing Subbituminous Coal. BuMines RI 3890, 1946, 11 pp.
553. Parry, V.F., W.S. Landers, E.O. Wagner, J.B. Goodman, and R.R. Allen. Utilization of South Arcot, India, Lignite. BuMines RI 5351, 1957, 34 pp.
554. Sondreal, E.A. Carbon Disulfide Production by Reaction of Elemental Sulfur with Wood Charcoal and Lignite Char. M.S. Thesis, Univ. of N.D., Grand Forks, 1962, 62 pp.
555. Sondreal, E.A., A.M. Cooley, and R.C. Ellman. Bench-Scale Production of Carbon Disulfide from Lignite Char and Sulfur. BuMines RI 6891, 1967, 20 pp.
556. Van Sant, J.N., and R.C. Ellman. Methods and Costs of Mining Lignite in North Dakota. BuMines IC 7891, 1959, 82 pp.
557. Wali, M.K., and P.G. Freeman. Ecology of Some Mined Areas in North Dakota. Chapter in N.D. Geol. Survey Educational Series 5, "Some Environmental Aspects of Strip Mining in North Dakota," ed. by M. K. Wali, 1973, 121 pp; pp. 25-47.
558. Wali, M.K., P.G. Freeman, A.L. Kollman, and Wilton Johnson. Overview of Reclamation in the West. Sympos. on Techn. and Use of Lignite, Grand Forks, N.D., May 1975. DOE GFERC/IC-75/2, 1975, pp. 294-311.
455. Warfield, R.S., and C.C. Boley. Sampling and Coking Studies of Several Coalbeds in the Kokolik River, Kukpowruk River, and Cape Beaufort Areas of Arctic Northwestern Alaska. BuMines RI 7321, 1969, 58 pp.
87. Youngs, R.W. Humic Acids from Lignite. Sympos. on Techn. and Use of Lignite, Grand Forks, N.D., Apr. 1961. BuMines IC 8164, 1963, pp. 87-93.
559. Youngs, R.W., and C.M. Frost. Humic Acids from Leonardite; A Soil Conditioner and Organic Fertilizer. Proc., N.D. Acad. of Sci., v. 17, 1963, pp. 76-82.

Publications on Other Than Low-Rank Coal

Allen, R.R. Effects of Retort-Immersion Depth on Gieseler Plasticity Determinations on Western Coals. BuMines RI 6559, 1964, 26 pp.

Boley, C.C., M.M. Fegley, and R.B. Porter. Use of Char to Improve Physical Quality of Coke from Sunnyside (Utah) Coal. BuMines RI 7817, 1973, 30 pp.

Boley, C.C., M.M. Fegley, and R.B. Porter. Use of Char to Improve Physical Quality of Coke from York Canyon (New Mexico) Coal. DOE GFERC/RI-77/2, 1977, 28 pp.

Boley, C.C., M.M. Fegley, and R.B. Porter. Effect of Charge Preheat on the Physical Quality of Coke from Three Western Coals. DOE GFERC/RI-77/3, 1978, 19 pp.

Boley, C.C., M.M. Fegley, R.B. Porter, and D.C. Boteler. Use of Char to Improve Physical Quality of Coke from Allen (Colorado) Coal. BuMines RI 7599, 1972, 21 pp.

Boley, C.C., and W.S. Landers. Entrainment Drying and Carbonization of Wood Waste. BuMines RI 7282, 1969, 15 pp.

Capp, J.P., J.L. Elder, C.D. Pears, R.W. Lowe, K.D. Plants, and M.H. Fies. Underground Gasification of Coal: Hydraulic Fracturing as Method of Preparing a Coalbed. BuMines RI 5666, 1960, 50 pp.

Elder, J.L., M.H. Fies, H.G. Graham, J.P. Capp, and E. Sarapuu. Field-Scale Experiments in Underground Gasification of Coal at Gorgas, Ala. BuMines RI 5367, 1957, 101 pp.

Frost, C.M., H.C. Carpenter, C.B. Hopkins, Jr., S.S. Tihen, and P.L. Cottingham. Hydrogenating Shale Oil and Catalytic Cracking of Hydrogenated Stocks. BuMines RI 5574, 1960, 17 pp.

Goblirsch, G.M., and E.A. Fletcher. Nitric Oxide Emissions from the Fluidized Combustion of Coke. Combustion Inst., Spring 1976 Meeting of Central States Section, 12 pp.

Gomez, Manuel, W.S. Landers, and E.O. Wagner. Entrained-Bed Carbonization of Highly Fluid Bituminous Coals. BuMines RI 7141, 1968, 34 pp.

Gomez, Manuel, J.G. Walters, and J.B. Gayle. The Effect of Carbonization Conditions and Coal Properties on Coke Size and Strength Parameters. BuMines RI 7024, 1967, 34 pp.

Gomez, Manuel, J.B. Walters, and J.B. Gayle. Dependence of Coking Time on Coal Properties and Carbonization Parameters. BuMines RI 7093, 1968, 28 pp.

Landers, W.S. Process for Continuous Carbonization of Coal. U.S. Pat. 3,455,789, July 15,1969, 8 pp.

Landers, W.S., Manuel Gomez, and C.C. Boley. Coking Properties of Selected Utah Coals and Blends. BuMines RI 6948, 1967, 55 pp.

Landers, W.S., Manuel Gomez, and D.J. Donaven. Bulk Density Studies on a Commercial Blend of Western Coking Coals. BuMines RI 6502, 1964, 33 pp.

Landers, W.S., Manuel Gomez, and E.O. Wagner. Coking Rate Study on a Commercial Blend of Western Coals. BuMines RI 6289, 1963, 26 pp.

Landers, W.S., E.O. Wagner, Manuel Gomez, C.C. Boley, and J.B. Goodman. Entrained-Bed Carbonization of Bituminous Coal: Tests on an Australian Coal. BuMines RI 6608, 1965, 51 pp.

Parry, V.F. Technical and Economic Study of Packaged Fuel. BuMines RI 3757, 1944, 45 pp.

Warfield, R.S., W.S. Landers, and C.C. Boley. Sampling and Coking Studies of Coal from Castle Mountain Mine, Matanuska Coalfield, Alaska. BuMines Open File Series, No. 7-66, 1966, 14 pp.

Wilson, M.W., L.L. Hirst, J.L. Elder, J.P. Capp, and J.J. Gentile. Simulated Underground Gasification of Coal and Electrolinking-Carbonization Method of Preparing Path in a Coalbed. BuMines RI 5605, 1960, 61 pp.

Acknowledgements

This study was directed by the Grand Forks Energy Technology Center (GFETC), which has the lead mission within the Department of Energy for technology "applications for low-rank coals." G.H. Gronhovd (Director) and E.A. Sondreal (Deputy Director) of GFETC provided technical direction and review of all aspects of the study. The work was performed by Energy Resources Company, Inc. (ERCO).

Because of the scope and complexity of the effort, GFETC enlisted a task force of recognized experts on the technical and regional issues germane to the study. These individuals are listed below; their contributions to the quality and direction of the study were highly significant. The task force met with the study team at four critical points to review interim results and to lead working groups which established the emphasis, priorities, and methodologies for the analysis. Primarily through the efforts of the task force members, useful data inputs and critiques of working draft materials were received from a number of organizations as the study progressed.

Task Force Participants

Participant	Affiliation
Dr. Martin A. Elliott Houston, Texas	Consultant Texas Eastern Gas Transmission Co.
Professor George R. Hill Salt Lake City, Utah	University of Utah Department of Chemical Engineering
Mr. James Jonakin Birmingham, Alabama	Consulting Engineer (Retired from Combustion Engineering, Inc.)
Mr. Paul W. Crutchfield Mr. David J. Beecy Washington, D.C.	U.S. Department of Energy Office of Policy and Planning
Professor Donald E. Severson Grand Forks, North Dakota	University of North Dakota Department of Chemical Engineering
Mr. David M. White Austin, Texas	Texas Energy and Natural Resources Advisory Council
Mr. Kurt Yeager Dr. Charles R. McGowin Palo Alto, California	Electric Power Research Institute

Individual contacts and contributions made during the course of the study are too numerous to list. The following (in addition to the task force members) contributed significantly to the review of part or all of the document: G.H. Gronhovd, E.A. Sondreal, W.G. Willson, and H.H. Schobert of GFETC; W.R. Kube of the University of North Dakota and GFETC; S. Alpert, K. Clifford, S. Ehrlich, T. Lund, C. Aulisio, D. Giovanni, and R. Wolk of the Electric Power Research Institute; W. McCurdy, S. Freedman, L. Miller, M. Kopstein, L. Ludwig, E. Burwell, W. Schmidt, M.N. Rosenthal, J. Nardella, and J. Turner of DOE; W.R. Kaiser of the University of Texas at Austin; and P. Averitt (retired) of the U.S. Geological Survey.

The ERCO Program Manager on this effort was Dr. John Kotowski. Mr. George Wiltsee was the Assistant Program Manager and Technical Director. Other ERCO personnel who provided major contributions to the effort include Paul Goodson, Randall Smith, Wayne Simmons, Barbara Acker, Jeffrey Feerer, Timothy Buscheck, and Myron Burr. In addition, special thanks should be extended to Lydia Felix and Jennifer Spinello of the administrative staff for their support and assistance in the preparation of this report.